Frontiers of Fundamental Physics

Frontiers of Fundamental Physics

Edited by

Michele Barone

Nuclear Research Centre
Demokritos, Greece

and

Franco Selleri

Università di Bari
Bari, Italy

Springer Science+Business Media, LLC

Library of Congress Cataloging-in-Publication Data

Frontiers of fundamental physics / edited by Michele Barone and Franco
  Selleri.
      p.   cm.
    "Proceedings of an international conference on Frontiers of
  fundamentl physics, held September 27-30, 1993, in Olympia, Greece"-
  -T.p. verso.
    Includes bibliographical references and index.
    ISBN 978-1-4613-6093-3   ISBN 978-1-4615-2560-8 (eBook)
    DOI 10.1007/978-1-4615-2560-8
    1. Physics. 2. Astrophysics. 3. Geophysics. I. Barone,
  Michele. II. Selleri, Franco.
  QC21.2.F76 1994
  500.2--dc20   94-38843
                                                        CIP

Proceedings of an International Conference on Frontiers of Fundamental Physics,
held September 27–30, 1993, in Olympia, Greece

ISBN 978-1-4613-6093-3

Preface

The Olympia conference **Frontiers of Fundamental Physics** was a gathering of about hundred scientists who carry on their research in conceptually important areas of physical science (they do "fundamental physics"). Most of them were physicists, but also historians and philosophers of science were well represented. An important fraction of the participants could be considered "heretical" because they disagreed with the validity of one or several fundamental assumptions of modern physics. Common to all participants was an excellent scientific level coupled with a remarkable intellectual honesty: we are proud to present to the readers this certainly unique book.

Alternative ways of considering fundamental matters should of course be vitally important for the progress of science, unless one wanted to admit that physics at the end of the XXth century has already obtained the final truth, a very unlikely possibility even if one accepted the doubtful idea of the existence of a "final" truth. The merits of the Olympia conference should therefore not be judged a priori in a positive or in a negative way depending on one's refusal or acceptance, respectively, of basic principles of contemporary science, but considered after reading the actual new proposals and evidences there presented. They seem very important to us.

The confrontation between different lines of research has accompanied science from its birth. Galileo's scientific ideas were heretical, not only with respect to the dominant religious and political powers of his times, but with respect to the academic establishments of the universities as well: Well known is the example of the astronomy professor who refused to look in the telescope, but many were the centers where the heliocentric ideas were rejected. The great results obtained by Kepler, Newton, and many others, slowly transformed Galileo's heresy into the orthodoxy of modern physical science.

Atomism had existed as an idea cultivated by few isolated people for about 2300 years when, at the end of the XIXth century, Ludwig Boltzmann presented his conception of an objectively existing atomic structure of matter. Almost all scientists surrounding him seemed to reject atomism, and the bitter struggle that went around this question probably contributed to the dramatic ending of his life (1906). In the same years, however, Albert Einstein and Jean Perrin obtained an atomistic description of Brownian motion and shortly afterwards atomism was fully accepted in physics, owing also to the discoveries made by Ernst Rutherford and Niels Bohr. In this way the isolated ideas of Boltzmann became the new orthodoxy.

The geophysicist Alfred Wegener was much laughed at for his 1912 proposal that the continents had shifted relatively thousands of km, that the Atlantic Ocean had opened as the Americas split from Africa and Europe, and that all the continents had once been united as a single supercontinent, Pangaea. Only after the confirmations found by Warren Carey in 1954 Wegener's discovery started to be accepted in the scientific community. Today there are so many independent proofs that continents have been united in the past that it seems impossible to doubt it. Here the new frontier has become the conjecture that the Earth radius has considerably increased in the past.

In spite of these well known examples science is of course not reducible to an endless confrontation between opposite ideas, since it deals with the material reality surrounding us and uses powerful methods that allow sometimes the scientists to understand true properties of the real world. Therefore part of the orthodoxy can also be considered as <u>valid knowledge</u>. Such are for example the following statements: the Sun is just a star; the Milky Way is our galaxy seen from inside; in outer space there are hundreds of millions of other galaxies; there is a molecular and atomic structure of matter, and a nuclear structure of atoms; there exist subatomic entities called electron, proton, pion, etc. Many other examples of valid knowledge could easily be given. It is a fact however that today's physics seems to contain more than just valid knowledge.

In books dealing with astrophysics and cosmology one often finds statements like: "the astronomer Edwin Hubble established beyond all reasonable doubt that the Universe is expanding", but Hubble himself wrote in several different occasions statements like the following one of 1939: "... the results do not establish the expansion as the only possible interpretation of redshifts". Moreover quasars are the objects with the largest observed redshifts, and should therefore be considered at the margins of the visible universe, but many independent pieces of observational evidence indicate that some of them are actually associated with nearby galaxies and that their redshifts cannot therefore be due to recessional motion. A recent amazing discovery is the so called "redshift quantization" phenomenon for spiral galaxies, and this is so difficult to explain within the standard cosmology that most people prefer to forget about it - a predictable reaction of modern scientific thinking confronted with radically new evidence. Important astrophysicists and cosmologists (Hannes Alfven, Halton Arp, Geoffrey Burbidge, Fred Hoyle, Jayant Narlikar, ...) have repeatedly argued that the observed redshifts of quasars and galaxies could well have an explanation radically different from the standard one based on *big bang*. In spite of all this the dominant view remains the idea that the only possible explanation of galaxy and quasar redshifts is based on the universal expansion

In relativity most people believe that the "luminiferous ether" of the XIXth century has been ruled out by Michelson-type experiments and by the development of the theory of special relativity. The situation is very different however, since

Poincaré and Lorentz were both defenders of the existence of ether, and Einstein himself after 1916 radically modified his previously negative attitude. For example in 1924 he wrote: "According to special relativity, the ether remains still absolute because its influence on the inertia of bodies ... is independent of every kind of physical influence." The minority group of people working today in the foundations of special relativity seems to be almost completely ether-oriented, and there are many proposing a reformulation of the theory along the lines dear to Lorentz: Simon Prokhovnik and the late John Bell are two examples. It has also become clear how such a reformulation should be carried out, after the 1977 realization that the conventional nature of the clock synchronization procedures opens the door to theories which are different from, but physically equivalent to special relativity.

Also general relativity has problems of fundamental nature, in particular those connected with the right-hand side of Einstein's field equations, where only the matter stress-energy tensor, but not the field stress-energy tensor, contributes to space-time curvature. This goes against the very fundamental conclusion of special relativity that all forms of energy are completely equivalent, and gives rise to a curious conservation law of rest mass, but not of energy-momentum. A very large number of theoretical physicists seem to be happy with calculations performed strictly within the standard formulation, in spite of the fact that it has been shown that Einstein's field equations do not lead to interactive N-body solutions, if N > 1. General relativity can be considered as a test-particle theory, and as such it explains the three classical tests, but in other respects seems sometimes not to be quite satisfactory. More about this can be found in these proceedings.

In our century the interplay between science and ideology has become more important than ever and the hystorians of physics have produced detailed reconstructions of the true scientific/cultural processes leading to the development of what we call "modern physics". From this work evidence has emerged for the existence of common cultural roots with philosophers such as S. Kierkegaard, M. Heidegger, A. Schopenhauer, and W. James. It is therefore not surprising that these philosophers developed ideas similar to some now prevailing in modern physics, in particular concerning the negative attitude toward the possibility of a correct understanding of the objective reality. In fact in quantum physics the standard teaching (after 1927) is that one cannot understand the atomic world in "classical" terms, that is by employing causal space-time descriptions. People active in the foundations of quantum physics believe instead that no good reason for such a pessimistic conclusion has ever been presented, and recall that Einstein, Planck, Schrödinger and de Broglie could not accept it. A group of participants in Olympia try accordingly to find new space-time models of elementary particles and/or to develop new mathematical tools useful for this task.

Bell's theorem states that any theory of the physical world based on the rather natural point of view of local realism must disagree at the empirical level with the predictions of quantum mechanics by as much as 42%. Experiments performed in the seventies and early eighties have produced results compatible with the existing quantum theory, but Bell's theorem has actually not been checked due to the introduction in the reasoning of arbitrary (but unavoidable, given the efficiency of the used apparata) additional assumptions. In this way a confusion has been produced between Bell's original inequality and the much stronger inequality violated in those experiments, forgetting that the latter owes the very possibility of being violated by the quantum theoretical predictions to the mentioned additional assumptions. In spite of the fact that Bell's theorem could allow in principle to decide who was right in the Einstein/Bohr debate, we still do not know the answer thirty years after the formulation of the theorem.

The confrontation between different points of view goes on, but a strange mutation seems to reduce its effects, since new ideas in fundamental physics find invariably difficulties in being accepted by the majority, no matter how well formulated and important they could be. While the ruling of the majorities is a fundamental feature of every democracy, it certainly does not apply to science where the great steps forward have always been made by isolated individuals. This dogmatic hardening risks today to make the scientific majorities impenetrable to a critical understanding of the foundations of contemporary scientific theories.

The existence of such attitudes within modern science has been observed by many physicists and also by the best epistemologists of our century. Thomas Kuhn, for example, wrote about the education of young physicists: "Of course, it is a narrow and rigid education, probably more so than any other except perhaps in orthodox theology." Karl Popper was worried about the poor standards of scientific confrontations and stated: "A very serious situation has arisen. The general anti-rationalist atmosphere which has become a major menace of our time, and which to combat is the duty of every thinker who cares for the traditions of our civilization, has led to a most serious deterioration of the standards of scientific discussion. ... But the greatest among contemporary physicists never adopted any such attitude. This holds for Einstein and Schrödinger, and also for Bohr. They never gloried in their formalism, but always remained seekers, only too conscious of the vastness of their ignorance." The understanding of the vastness of our ignorance was generally present in Olympia, but in all fairness we must add that one could also get glimpses of what our science could become in the future: in all cases these were very exciting moments.

The choice of Olympia for helding the conference was not casual: this is the place where the Olympic games of ancient times were held for something like 1,200 years. Wars were stopped when the games started and activities included reading of poems, and discussions about science and philosophy. Olympia is not only one of the most beautiful and interesting spots of the world, but also a positive symbol of the modern civilization.

The generous efforts of many people have made our conference possible. First of all we wish to thank Attanassios Kanellopoulos for his encouragment and for many useful suggestions. The elected member of the Parliament Crigno Kanellopoulos-Barone has generously helped us in establishing fundamental contacts in Olympia and elsewhere. The constant help of Georges Kanellopoulos has been of tremendous importance for the success of the meeting: we thank him warmly. We are also very grateful to the physics students Rossella Colmayer, Francesco Minerva and Gabriella Pugliese, who formed an efficient and charming secretariat.

Our thanks go also to the International Olympic Committee, and to its president Prof. X. Yzezezez, for allowing us to use, free of charge, the wonderful structures of the Olympic Academy where the conference was held. Mr. A. Bababab, representative of the Greek government, brought us welcome greetings and encouragment, and Prof. R. Rapetti, president of the Istituto Italiano di Cultura in Athens, stressed the European nature of the conference. The words of Mr. X. Kosmopoulos, major of Olympia, made the participants feel at home in his marvellous town.

Last but not least our gratefulness goes to the generous sponsors: the Greek Ministry of Culture, the General Secretary of Research and Technology of the Greek Ministry of Industry, the Università di Bari and, independently, the Physics Department of the Università di Bari, the National Tourist Organization of Greece, the Commercial Bank of Greece, the Ionian Bank of Greece, the Ellenic Industrial Development Bank S.A., and Glaxo A.E.B.E. Without their concrete help the Olympia conference would not have taken place.

M. Barone and F. Selleri

Contents

RELATIVITY : ENERGY AND ETHER

GEOPHYSICS : EXPANDING EARTH

FIELDS, PARTICLES : SPACE-TIME STRUCTURES

QUANTUM PHYSICS : DUALITY AND LOCALITY

EMPIRICAL EVIDENCE ON THE CREATION OF GALAXIES AND QUASARS

Halton Arp

Max-Planck-Institut für Astrophysik
Garching bei München, Germany

Simply the arrangement on the sky of extragalactic objects has long shown that the youngest, smallest quasars and compact galaxies have been created recently in the vicinity of older progenitor galaxies. Now high energy observations in X-rays and γ-rays confirm these connections and require the creation of matter as an ongoing process marked by an initially high intrinsic redshift.

The nearest superclusters of galaxies show creation along lines in space originating from the central, ejecting galaxy. String theory may be pertinent. The existence of preferred values of redshift (periodicity) rule out, again, an expanding universe. They also imply quantum mechanical effects at the m = 0 creation points of particulate matter. No theory has been advanced, however, which numerically predicts the quantization values.

Introduction

The Big Bang theory of the universe precludes any scientific observation of creation because the event is so remote in time and space. But even if we could observe this singular event at a distance of 15 bilion light years this age zero universe would supposedly surround us in every direction. That leads to the rather bizarre conclusion that we are, at this moment, "inside" a point that is so small it is dimensionless (the point from which the universe is supposed to have suddenly expanded).

Perhaps the conclusion is illogical enough to send us back to what we should have been doing all along – looking at the actual observations. If we do, we find that they all point to the incorrectness of one key assumption in the current theory. That assumption is that extragalactic redshifts measure velocities of expansion. If redshifts are not due to recessional velocities the expansion of the universe and Big

Frontiers of Fundamental Physics, Edited by M. Barone
and F. Selleri, Plenum Press, New York, 1994

Bang is wrong and consequently creation must take place throughout the universe in events which can be observationally (hence scientifically) studied.

Alignment of Quasars and High Redshift Galaxies Across Low Redshift Galaxies

The clear observational pattern that emerges from systematic study of the actual sky is that galaxies occur in groups. Large, dominant galaxies tend to be surrounded by smaller younger galaxies of somewhat higher redshift ($c\Delta z \lesssim 100 kms^{-1}$). Even younger, more active companions tend to have higher excess redshifts. The youngest, most compact galaxies and quasars tend to be associated with active galaxies in these groups and have the largest excess redshifts (from $c\Delta z \lesssim 100 kms^{-1}$ to $c\Delta z \lesssim 2$).

Statistically these associations are overwhelmingly significant (see for review Arp 1987). *In addition* there are numerous instances of interactions or connections between individual low redshift galaxies and high redshift compact galaxies and quasars (see for update Arp 1993). The obvious validity of these observations has not been accepted by influential astronomers because the evidence falsifies the assumption that redshift equals velocity and hence the expanding universe to which most scientists are committed.

As would be expected of any valid conclusion, more evidence is continually being discovered which confirms these empirical relationships between objects of widely varying redshifts. Judging from past behavior, the latest evidence will not sway the opinion of those whose interest lies with the status quo. But since the latest evidence deals with high energy X-rays and very high energy γ-rays it is of prime usefulness to those interested in real processes of matter creation in the universe.

NGC 4258

One of the most striking new observations is shown in Fig.1. The galaxy is an unusually active one, known to be ejecting hydrogen emission, proto spiral arms and radio material from an excited (Seyfert) nucleus. (van der Kruit, Oort and Mathewson 1972; Arp 1986a; Courtés et al. 1993). A spectacular result emerges from recent observations in X-ray wavelengths. (Pietsch et al. 1994). As Fig.1 shows, the two most conspicuous, point X-ray sources in the field pair exactly across the nucleus of this galaxy which is so well known for ejecting excited material. Any two X-ray sources in this field would only have about one chance in a thousand of accidentally pairing this exactly across the galaxy. But we have to multiply this by the small chance that two such strong X-ray sources would fall so close to the galaxy plus the extraordinary coincidence that the pairing would occur across one of the most striking examples known of an ejecting spiral galaxy. Altogether there is clearly negligible chance that the pair of X-ray sources is not associated with the galaxy. The authors of the new X-ray paper suggest they may be bipolar ejecta from the nucleus of NGC4258. The crowning result, however, is that both components of the X-ray pair are identified with blue stellar objects. One of these has been confirmed as a quasar of $z \sim .4$ (W. Pietsch, private communication) and the other is almost certainly a quasar, probably of comparable redshift.

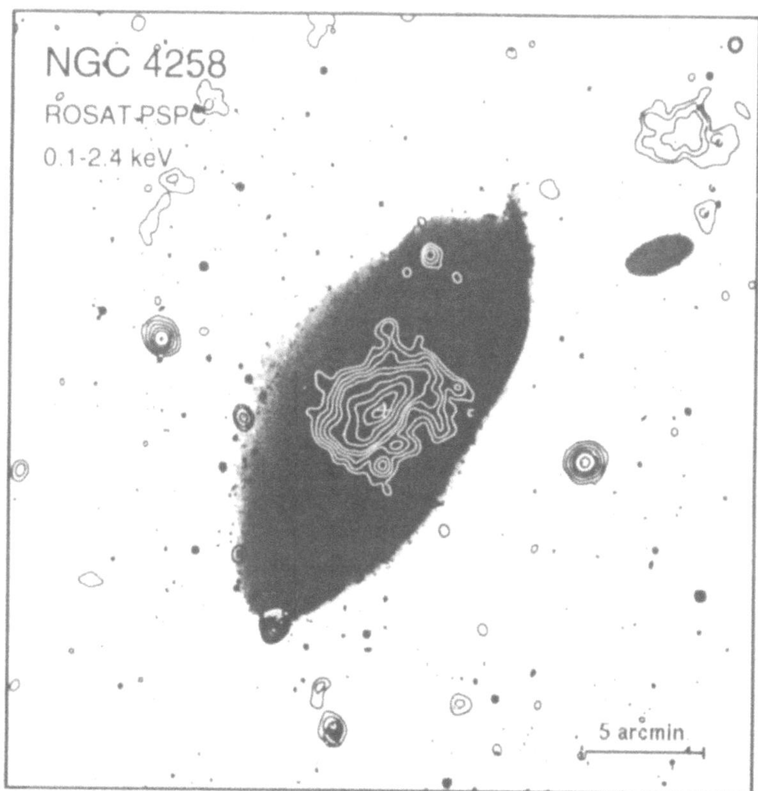

Figure 1. X-ray observations by W. Pietsch et al. (1974) of the active, ejecting, galaxy NGC4258. Conspicuous X-ray sources paired across the minor axis are identified with blue stellar objects, one of which has been confirmed as a quasar with the other being investigated.

The upshot of this one observation, by itself, is to confirm unequivocally that high redshift quasars are physically associated with and presumably ejected, from active, low redshift galaxies. This is far from the first example of this kind of association. The first one was discovered among the initial surveys of the brightest radio quasars.

3C273 and M87

Fig.2 shows that the brightest apparent magnitude quasars in the sky, 3C273, and the most active, bright radio galaxy (M87 = 3C274) – these two are aligned almost perfectly across the brightest galaxy in the Virgo Cluster (Arp 1967). The chance of such a configuration being accidental was calculated to be about one in a million. Many observational arguments point to the ejection origin of these two famous active objects from the central galaxy in the Virgo Cluster and, in fact, the origin of the whole cluster from this central point (Arp 1978). The Virgo Cluster is central to the Local Supercluster which is the largest aggregation of galaxies known in our sector of the universe.

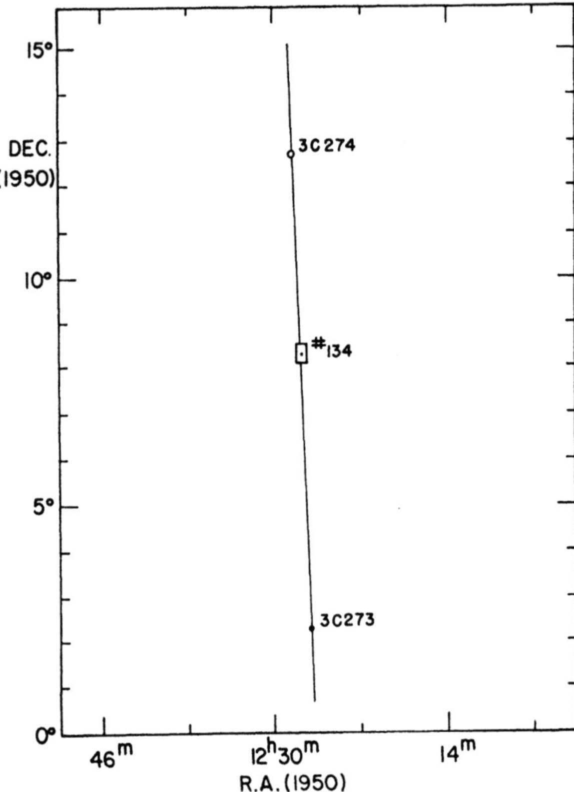

Figure 2. The brightest apparent magnitude quasar in the sky, 3C273 and the brightest jet radio galaxy 3C274 (M87) are aligned exactly across the brightest galaxy in the center of the Virgo Cluster, M49 (from Arp 1967; 1990). (# 134 from Atlas of Peculiar Galaxies = M49)

So just the original geometrical configuration on the sky showed 27 years ago that the quasar was a member of the relatively nearby Virgo Cluster despite its much higher redshift ($cz = c \times 0.16 = 48,000 kms^{-1}$ versus $cz \simeq 1000 kms^{-1}$ for the Virgo Cluster). Of course, during the following years all sorts of evidence accumulated to confirm that the quasars actually inhabited the Virgo Cluster. A brief summary of this evidence is as follows:

1) A class of relatively radio bright quasars was shown in 1970 to be strongly associated with bright galaxies in the Local Supercluster of which Virgo is the center (Arp 1970).

2) The brightest quasars in an objective prism survey by X. T. He et al. in 1984 were shown to be associated with the M87 region of the Virgo Cluster (Arp 1986b).

3) In the Palomar Survey of ultraviolet selected quasars brighter than V~ 16.2 mag., J. Sulentic showed in 1988 that these bright quasars were concentrated in the region of the Local Supercluster (Sulentic 1988).

4) Quasars with measured Faraday rotation show effects in the direction of the Virgo Cluster which require some to be in front of cluster (Arp 1988).

4

5) An extremely unusual, low density hydrogen cloud was discovered in the Virgo cluster by R. Giovanelli and M. Haynes in 1989. It lay only 45' distant from 3C273 and was elongated accurately back toward the position of 3C273. As a clinching property the famous nonthermal jet in 3C273 pointed down the length of this extended feature (Arp and Burbidge 1990). Since the cloud had redshift of $z = 1248kms^{-1}$ it was clearly a member of the Virgo cluster and its association with 3C273 therefore marked the latter as also a member.

6) When Hubble Space Telescope obtained spectra in the far ultraviolet of 3C273 it was found that lower redshift absorption lines were about an order of magnitude more numerous than expected from high redshift quasars in other directions (Weymann 1991). Although the conclusion was avoided, it was obvious that the extra absorption systems were most simply explained as objects in the Virgo cluster with a range of redshifts between that of the large galaxies in the cluster and the redshift of 3C273.

7) Most recent, high resolution images with Hubble Space Telescope (*Nature* 9 Sept. 1993) lead to an interpretation that the famous jet of 3C273 "... must be viewed in the plane of the sky nearly perpendicular to the line of sight" (Thomson et al. 1993). It is well known that, when placed at its redshift distance the quasar exhibits superluminal motion. The customary model invoked to avoid this difficulty is to have the jet aimed almost exactly at the observer. If this geometry is no longer possible then the only way to escape faster than light motion is to significantly decrease the conventional distance of 3C273.

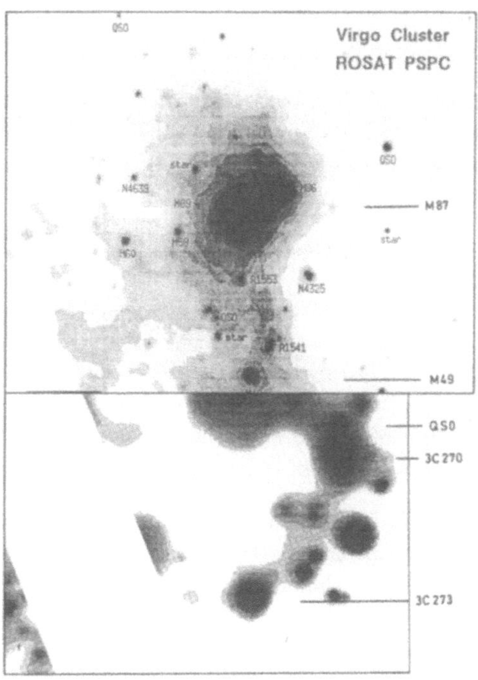

Figure 3. Low surface brightness X-rays connect M49 to M87 in the north and 3C273 to the south. Upper integration from ROSAT Sky Survey by Böhringer et al. 1994, lower integration by Arp from same survey.

But now the German X-ray satellite, ROSAT, has been observing famous objects in
very high energy bands and startling results have appeared. One result is the pair
of quasars across NGC4258 as just described. Another result, partially in press, is
shown below in Fig.3 (Böhringer, et al. 1994). A glance at the figure shows that
the previously known pairing of active objects across the central galaxy in the Virgo
Cluster is now confirmed by the new observation of high energy X-rays. An actual
continuous path of X-rays now connects M49 northward to M87 and southward to
3C273. Southward, in the direction of 3C273 the trail of X-rays leads to another
quasar of $z = .334$ and then to an active galaxy of $cz = 2075 kms^{-1}$ (3C270) and
finally, in a special analysis of an area extended further south by H. Arp (paper to be
submitted), the X-ray trail leads into 3C273 where it appears to end. The appears to
be the "smoking gun" where the smoke leads all the way from the active gun to the
bullet which it has ejected. It is difficult to imagine what further proof one should
hold out for.

The Bright Apparent Magnitude, Active Quasar 3C279.

Further south from 3C273 is a quasar which, although moderately faint in appearance
now, was much brighter only about 40 years ago. At that time it was comparable
with the brightest quasar in the sky, 3C273. Since for a long time it has been clear
that 3C273 was a member of the Virgo Cluster, it was highly probable that the
violently variable 3C279, falling very close in the sky to 3C273, was also a member.

Now confirmation of this has recently been obtained from observations at even higher
energy wavelengths, namely γ-rays. (The X-rays we have discussed are in the range
of photon energy from 0.1 to 2.0 keV whereas the γ-ray observations shown below in
Figs. 4 and 5 are in the range 0.7 to 20,000 Mev!)

Fig.4 shows observations published by a team of observers in the 0.7-30 Mev range
of γ-rays (Hermsen et al. 1993). These COMTEL observations in Fig.4 were then
later confirmed by the entirely independent EGRET observations in the higher Mev
range shown in Fig.5.

The startling aspect of the publication of these results was that despite the huge
team of scientists reporting the results none ventured to mention the extraordinarily
important fact that the two quasars, 3C273 and 3C279 were linked together by a
connection of γ-rays. The highest energy EGRET results were published in the form
of a color picture in *Sky and Telescope* (Dec. 1992 p. 634). The strong emission from
3C279 was clearly extended to the northwest and it must have been known that it
terminated on the position of 3C273. Yet the position of 3C273 was not plotted on
the picture nor any mention made of it in the text.

The intensity isophotes Fig.5 shown here were simply estimated and traced from
that color *Sky and Telescope* picture by the present author and the position of 3C273
and 3C279 indicated here by + signs. Although this situation has been discussed in
meetings and privately in 1993, to date the further γ-ray observations of this crucial
pair have not been released.

In Fig.6 the X-ray observations of the Virgo Cluster have been plotted to the same
scale as the γ-ray observations of 3C273 and 3C279. The extraordinary result is
that the major active galaxies in the Virgo Cluster lie along an X-ray delineated
extension which passes through the largest galaxy, M49 and extends southward to

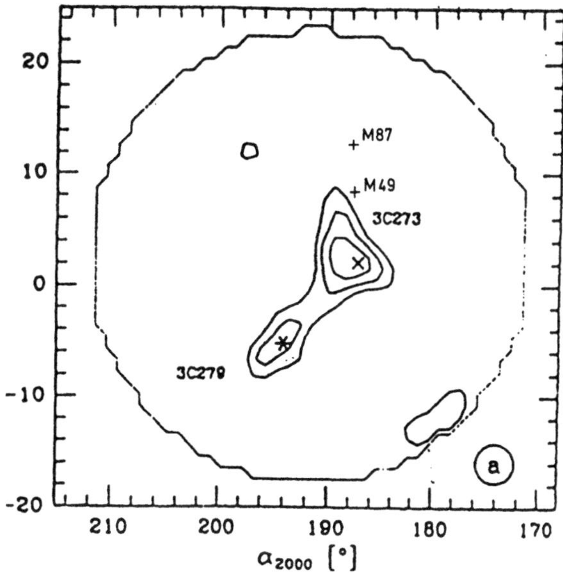

Figure 4 Observations of 3C273 and 3C279 in low energy γ-rays (Hermsen et al. 1992) from COMPTEL instrument aboard the Gamma Ray Observatory (GRO)

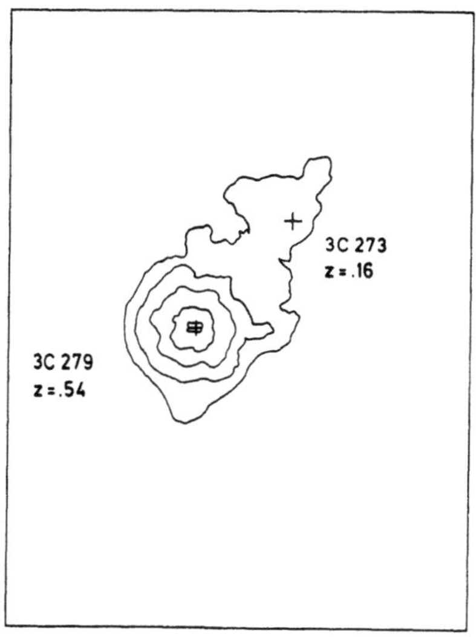

Figure 5 Observations of 3C279 by high energy, EGRET instrument aboard GRO (γ-rays 10 to 10^4 Mev). Isophotes of picture in Sky & Telescope (Dec. 1992) have been copied by present author who has added positions of 3C273 and 3C279 as crosses to show that these independent observations confirm the connection in γ-rays found in the observations by COMPTEL shown in Fig.4.

3C273. The X-rays are high energy ($\sim 1 - 2 keV$) and as they approach 3C273 the photons become harder until 3C273 is conspicuous in lower energy γ-rays. The final part of the connection to 3C279 is only in high energy γ-rays and 3C279 itself is most conspicuous in the highest of all observed energy γ-rays.

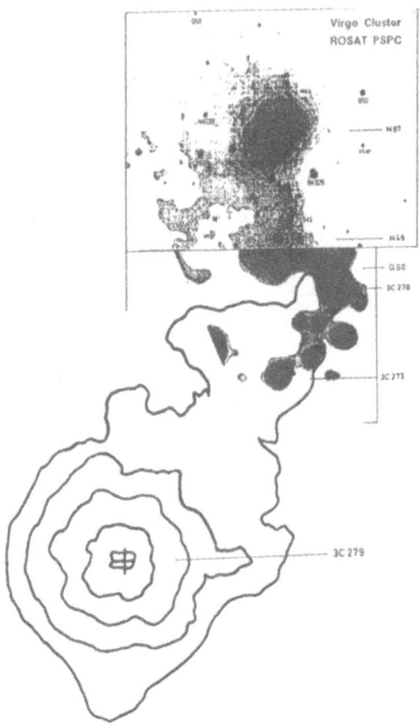

Figure 6. Plot of X-ray emitting material in Virgo Cluster which ends on the quasar 3C273. The higher energy γ-rays continue on to 3C279 and are shown by approximate isophotes. The connecting material appears to rise in energy toward the highest energy quasar, 3C279 ($z = .538$).

Trying to Understand the Observations

The first order result is to confirm decisively all the previous evidence that objects of widely disparate redshifts are physically grouped together in the same assocations. Further it is confirmed that the most compact, and hence youngest, objects such as quasars have the highest intrinsic (non-velocity) redshifts. Empirically this requires the younger age to be related to the cause of the intrinsic redshift.

Fortunately now there is known a solution to the field equations of general relativity which is *more general* than the traditional, Friedmann, expanding universe solutions. The more general solution allows creation of matter at any epoch in the universe and since the matter is created with initially low particle masses, the newly born matter has initially high intrinsic redshift which declines as it ages (Narlikar and Arp 1993). This theoretical interpretation accounts for the numerous discrepant redshift observations that have accumulated over the past 28 years (83 years if one wishes to count the unexplained systematic redshifts of young stars, the so called K effect).

In particular, the discoveries reported here of intense emission of very high energy x-rays and γ-rays from quasars linked to nearby galaxies shows especially clearly that the higher redshift of these objects is connected with their extreme youth. The point is that the emission is supposed to result from acceleration processes arising from travel of charged particles through magnetic fields (synchrotron radiation). But even for the X-ray wavelengths the decay time is of the order of only 50 yrs. (Arp 1994) and for γ-rays, correspondingly shorter. This marks the higher redshift objects as characteristically in a young, active stage where they are intermittently injecting high energy particles.

But the shocking surprise is that *the low density connections* between these young objects are emitting such short lived radiation. Until now the working hypothesis has been that the creation process takes place in the active nuclei of older galaxies. The new matter in compact form is then ejected in opposite directions in the form of high redshift quasars which evolve, as they age, into only moderate excess redshift companion galaxies.

The optical connections that are occasionally observed between the older galaxies and the higher redshift companions have naturally been supposed to consist of gas, dust and or stars from the older galaxy that have been entrained during the ejection process. *But now we see many more connections consisting of very high energy, short lived radiation.* The only possible suggestion would seem to be that very small "retarded cores" were also thrown out with the quasar and that the quasar has left a sparkling trail of rapidly decaying high energy radiation.

There are some difficulties with this model, however, which suggest the consideration of some fascinating alternative possibilities. The difficulties are:

1) The lifetime of the high energy radiation is so short that it would seem difficult to sustain the emission of the connection even for the relatively short lifetime of the ejected quasar. This radiation would have to live for at least $10^6 - 10^7$ years in what appears to be a low density environment.

2) Even with low ejection speeds some of the lower intrinsic redshift ejects should show observable blue shift and red shift differences as they are ejected toward and away from us. This situation has not been ruled out by the observations but for a long time it has been estimated to be an uncomfortable restraint.

3) Although there is abundant evidence for secondary and even tertiary ejection coming off at arbitrary angles to the original ejection lines, the development of great clusters like Virgo and Fornax seems to be in an appreciably broad filament stretching great distances and drifting somewhat irregularly from a straight line.

All this suggests a modification of the ejection hypothesis based on reconsideration of the assumption that creation of matter takes place only in point locations in space. Dislocations in spacetime along *lines* in space which enable the emergence of new matter would not seem to be forbidden and could possibly explain better the newer observations. (This immediately suggests string theory although that theory has not been developed to the extent that it could make predictions of actual events in the extragalactic realm.) The possible amelioration of the aforementioned difficulties which such a "white line" theory could offer are enumerated below:

1a) If matter wells up at one point in such a "fault line" in space this could represent the original galaxy. Later emergences, perhaps due to a creation signal from the original galaxy, could produce secondary creation along this same line. The useful point would be that smaller upwelling over an extended period all along the line could possibly account for the currently observed high energy connections between high and low redshift objects.

2a) Since the creations take place from preexisting locations in space there need be no velocities of transport from the original galaxy nuclei and the blue and redshifts from ejection velocities could be avoided (This latter is particularly important in the matter of quantization of redshifts which would place limits of $lsim 20 kms^{-1}$ on true translational velocities of galaxies in space).

3a) If secondary creation lines, in analogy with strings, move through space – where ever they intersect the original creation line may promote creation nodes. If later nodes produced younger quasars and compact galaxies, the ejection lines from these secondary objects would be situated at arbitrary angles to the original creation line as observed. This interpretation suggests that jets represent material under pressure guided out of active nuclei by creation lines.

Quantization of Redshifts.

The one problem that seems to present unresolvable contradictions at this time are the observed quantization of redshifts. Evidence has been available for a long time which establishes that the whole redshift plane is quantized – the quasars in large steps, the galaxies in smaller (Arp 1993). Recently the smallest quantization steps of 37.5 kms^{-1} seen by William Napier in the most accurate HI redshifts have become overwhelmingly statistically significant (Napier 1993).

It is tempting to connect this quantization with periodicity in the creation process. Since the matter is created with zero mass it transitions from a a quantum mechanical realm where discretization is expected. But if they are not all at the same distance, any intrinsic galaxy redshift would be smeared out by continuously changing look-back times if the distribution of galaxies were continuously spread throughout space.

This is presently what I would consider the most difficult unsolved problem in the subject of galaxies and galaxy creation.

Summary Comments

The observations push us irresistably toward a certain empirical picture of the creation of galaxies and quasars. This in turn opens exciting opportunities for theory. The creation processes of matter are no longer some kind of obscure miracle but we can actually measure the state of the matter from its radiation property at various stages in its evolution. In order to make progress, however, researchers must give up the arbitrary assumption that particle masses are constant in time. When the general case, $m = m(x, t)$ is taken as a starting point the general solution of the Einstein field equation corresponds very well to the observed phenomena. The general connection between age and redshift becomes natural and we can hope to trace the materialization of matter from the quantum mechanical field (or material vacuum) to its better known state in the form of large galaxies. The problem of ultimate destiny of matter in these old galaxies lies untouched. The prediction of observed quantization of redshifts as a function of fundamental cosmic parameters forms a formidable challenge.

But as a first step, before the vast majority of observers and researchers can undertake anything meaningful, they must admit the zero order result that extragalactic redshifts are not due to velocities. The empirical evidence on this point was already overwhelming and the new observation in high energy x-rays and γ-rays now render the evidence completely inescapable. The vast observational facilities, exponentiating publication and well funded theoretical schools will continue to produce misinformation until the crucial issue of the empirical disproof of the redshift assumption is faced.

References

Arp, H. 1967, Ap.J. 148: 321.

Arp, H. 1970, A.J. 75: 1.

Arp, H. 1978, Problems of Physics & Evolution of the Universe, Acad. Sci., Armenian SSSR, Yerevan 1978. p.65.

Arp, H. 1986a, IEEE Transactions on Plasma Science 14: 748.

Arp, H. 1986b, Astrophys. Astr. (India) 7: 77.

Arp, H. 1988, Astrophys. Lett. A 129: 135.

Arp, H. 1994, "ROSAT Survey of an Area 10 Degrees Square around the Active Radio Galaxy Cen A", A&A, in press.

Arp, H. and Burbidge, G. 1990, Ap.J. 353: L1

Arp, H. 1987, "Quasars, Redshifts and Controversies" Interstellar Media, Berkeley.

Arp, H. 1990, Astronomy Now 4: 43.

Arp, H. 1993, "Progress in New Cosmologies: Beyond the Big Bang" ed. H. Arp, C.R. Keys, K. Rudnicki, Plenum Press, New York p. 1-28.

Courtés, G. Petit H., Hua, C.T. et al., 1993 A&A 268: 419.

Hermsen, W. and 25 collaborators, 1992, A&A Suppl. in press.

Napier, W. 1993, Progress in New Cosmologies: Beyond the Big Bang, ed. H. Arp, C.R. Keys, K. Rudnicki, Plenum Press, New York.

Pietsch, W., Vogler, A., Khabaka, P., Jain, A. and Klein, K. 1994, A&A , in press.

Thomson, R.C., Mackay, C.D. and Wright, A.E. 1993, Nature 365: 133.

van der Kruit, P.C., Oort, J.H., Mathewson, D.S. 1972, Astron. Astrophys. 21, 169.

Weymann, R.J. 1991 "The First Year of HST Observations STCI", Baltimore May 1991, p.58.

PERIODICITY IN EXTRAGALACTIC REDSHIFTS

W.M. Napier

Royal Observatory
Blackford Hill
Edinburgh EH9 3HJ
Scotland, U.K.

ABSTRACT. Claims that the redshifts of galaxies are quantized at intervals of $\sim$24, $\sim$36 or $\sim$72 $\mathrm{km\,s^{-1}}$ are being subjected to rigorous statistical scrutiny using new, accurate redshift data. The results of this enquiry to date are reviewed. The presence of a global galactocentric periodicity $\sim 37.5 \pm 0.2$ $\mathrm{km\,s^{-1}}$ is confirmed at a high confidence level. A strong redshift periodicity of $\sim 71.1 \pm 1.3$ $\mathrm{km\,s^{-1}}$ also exists amongst the galaxies of the outer regions of the Virgo cluster.

INTRODUCTION

The expression 'fool's experiment' appears to have been coined by Charles Darwin to describe the investigation of a hypothesis which no sensible individual would regard as worth testing; Darwin himself often undertook such enquiries. One imagines that, for most astronomers, the testing of 'redshift quantization' belongs firmly to this category. In essence, the hypothesis developed originally by Dr. Tifft and colleagues (*e.g.* Tifft 1977, 1980, 1993; Tifft & Cocke 1984) is that the redshifts of galaxies tend to occur in multiples of $\sim$24, $\sim$36 or $\sim$72 $\mathrm{km\,s^{-1}}$, the latter periodicity being local (applying to the redshifts of binaries or clusters of galaxies), the former two being global (depending on morphological type, and applying to the galactic redshifts after subtraction of the component due to the solar motion around the centre of the Galaxy).

A clear verification of redshift quantization would have far–reaching consequences. In cosmology, the derivation of virial masses, and even the existence of dark matter, would be thrown in doubt; and in astrophysics, there would be a distinct shift of balance in the debate over the discordant redshifts claimed by Arp (this volume). In fact, is not clear that current cosmological and astrophysical paradigms are capable of accommodating the phenomenon. Evidently, only clearly derived, unambiguous and strong results will suffice if the phenomenon is to be taken seriously.

However, the Tifft quantization studies, which have relied heavily on histogram binning techniques, have raised questions about the *a posteriori* selection of binning intervals (*cf.* Cocke & Tifft 1991, Schneider & Salpeter 1992) and the criteria for selection of binary galaxies; while in the case of global periodicity, a signal is in essence maximized through varying three parameters (the three components of solar motion), and before statements can be made about the statistical significance of the claimed periodicities, the effects of this freedom have to be assessed. Further, it is not always clear to the reader why one sample of galaxies rather than another has been chosen, and the sceptical reader may be left wondering whether negative results have gone unreported. Finally, while it is reasonable to expect an initial hypothesis to be modified with the accumulation of new or better data, several such modifications have occurred, raising the question of which version one is trying to test. For example, in its most recent manifestation it is claimed that periodicities occur over a wide range, from 2.66 km s^{-1} upwards (but with peak power at $\sim$36.5 km s^{-1}: Tifft, *pers. comm.*).

These issues suggest that there is scope for a fresh approach to the quantization issue. In recent years there has been a great increase in the number of accurately measured HI profiles of galaxies; as a result, there is now a sufficient body of new data for the existence of the claimed periodicities to be settled one way or another. A colleague (Dr B.N.G. Guthrie) and I therefore embarked on a study of the quantization issue a few years ago. The philosophy adopted was to apply an objective, rigorous scrutiny to new and unbiased data, with the intention of publishing the results whether they turned out to be positive or negative. The current state of this project is summarised in the present paper: it is already clear that extragalactic redshifts are indeed strongly quantized along the lines claimed by Tifft and others.

Methodology

A hypothesis, once set up, may be tested against new, independent and unbiased data by asking whether they confirm a prediction unique to it. The need for lack of bias requires that any selection of the new data from a larger dataset should be carried out with prescribed, simple rules which will not affect the outcome of the enquiry. If it turns out that some modification of the original hypothesis gives a better fit to the new data, then the 'improved' hypothesis should be put to the test against further data, and so on: the protocol thus requires a clear alternation between 'playing hunches' and verifying them.

Technique

The statistic we have generally used in the study is $I=2R^2/N$, N the number of galaxies and R the length of the resultant vector in the Argand diagram when the data are wrapped round a drum of circumference P and each assigned a unit vector. A power spectrum is a plot of I against frequency $1/P$. For a uniform, random distribution of independent redshifts, and neglecting edge effects, the I–distribution has a mean value $\bar{I} = 2$, and the probability of exceeding some value I_o by chance is $p(I \geq I_o) = \exp(-I_o/2)$. This formula becomes inaccurate for extreme values of I, and the statistic is also biased and inconsistent. These problems may be circumvented by comparing the signal strength obtained for the real data with those obtained from large numbers of trials in which suitably constructed synthetic data are analyzed in identical fashion. A full discussion of the technicalities is given elsewhere (Guthrie & Napier, in preparation).

Hypothesis

Tifft & Cocke (1984) – TC hereinafter –claimed to observe periodicities of $\sim$24.2 and $\sim$36.3 km s^{-1} in the redshift distributions of spiral galaxies with narrow and wide HI profiles respectively. These periodicities were global, applying to galaxies distributed over the celestial sphere, and galactocentric, emerging only when the redshift component due to the Sun's motion around the centre of the Galaxy was subtracted from each heliocentric redshift. The differential redshifts in binary galaxies (Tifft 1980) and the Coma Cluster (Tifft 1977) were said to be quantized at intervals of $\sim$72 km s^{-1}.

The subsequent modifications of the above basic hypothesis have added to, rather than replaced, the above claims. A new dataset should therefore still show the above periodicities and we do not require to discuss the refinements in testing the above.

Samples

In our study so far, two spiral galaxy samples have been examined and are discussed here. The first comprises the nearby galaxies with the most accurately measured redshifts out to roughly the edge of the Local Supercluster; the second comprises the galaxies in the Virgo Cluster, which happens to be the nearest rich cluster of galaxies. We can see no bias in these choices of sample. Evidence for redshift periodicity is found in both of them.

THE LOCAL SUPERCLUSTER

Nearby Galaxies

Guthrie & Napier (1991) first tested the global periodicity hypothesis by examining galaxies within 1000 $km\,s^{-1}$ of the Galactic centre. The database employed was a recent catalogue of 6439 extragalactic redshifts compiled by Bottinelli $et\ al.$ (1990). In the spirit of keeping the selection criteria simple and unbiased, spiral galaxies were culled from the dataset according to the following rules: (i) the quoted accuracies were $\sigma \leq 4$ $km\,s^{-1}$; (ii) galaxies used by TC in formulating the hypothesis under test were excluded; and (iii) galaxies within 12° of M87 were excluded. The latter restriction applied because such galaxies might belong to the Virgo Cluster, which was the subject of a separate enquiry.

By hypothesis, the global periodicity is galactocentric; thus from each observed (heliocentric) redshift one must subtract $V_\odot \cos\chi$ corresponding to the motion $V_\odot$ of the Sun around the Galactic centre, χ the elongation of the galaxy from the solar apex. According to TC, the solar vector yielding the periodicities was ($V_\odot = 233.6$ $km\,s^{-1}$ $l_\odot = 98.6°$, $b_\odot = 0.2°$), and we first carried out a power spectrum analysis on the redshifts corrected for this vector. A prominent peak was found at a period P=37.1 $km\,s^{-1}$, within one of the ranges 35–37.5 $km\,s^{-1}$ then under test. No evidence was found for the 24.2 $km\,s^{-1}$ peak claimed by TC for narrow–line galaxies, but then only two galaxies in the list had narrow line profiles.

The peak had a value I=18.1 which, for a white noise distribution, has a single–trial probability $\sim 10^{-4}$ according to the exponential formula. About two independent trials were involved in searching within 35–37.5 $km\,s^{-1}$, while a signal in the range $\sim$24 $km\,s^{-1}$ (and perhaps $\sim$72 $km\,s^{-1}$, although not strictly part of the hypothesis), would no doubt have been regarded as significant. The signal was therefore real at a confidence level C$\sim$0.999 according to the formula.

An independent assessment of C was made by generating sets of 89 synthetic redshift data and determining the I_{max}-distribution in the range 35–37.5 $km\,s^{-1}$. For the synthetic redshifts, the positions of the galaxies over the sky were preserved. The redshift data were generated by adding, to each measured redshift, the correction for the TC solar vector and then a random displacement in the range 0–δV $km\,s^{-1}$, where δV was small compared with the range of the redshifts and the likely dispersion within any groups and associations within the dataset. Thus the synthetic data were created by applying a 'haze' of width $\leq \delta V$ to the real data, sufficient to obliterate any periodicity in the range under test but too small to have any other effect. Any significant difference between the real and the synthetic data could thus only be due to periodicity in the former.

For each of δV=80, 60, 40 and 20 km s^{-1}, 3000 sets of 89 data were constructed and their power spectra obtained. The distribution did not change appreciably until δV=20 km s^{-1}, corresponding to synthetic redshifts within $\pm$10 km s^{-1} of the real ones. Typically, one in a thousand trials yielded I–values of 18.1 or higher. Allowing for the two or three ranges under test, the periodicity hypothesis was thus again preferred over the null one (random distribution) at a confidence level C~0.997 or 0.998, a result of high statistical significance.

However, the motion of the Sun around the centre of the Galaxy is known with limited accuracy, and the $(\mathbf{V}_\odot, P)$ found by TC was based on only 40 broad–lined galaxies. It therefore remained possible that the periodicity would emerge more strongly for some other solar vector in the neighbourhood of the TC solution. Guthrie & Napier (1991) therefore used published estimates of the motion of the solar neighbourhood around the Galactic centre, taking account of the solar motion relative to the neighbourhood, a probable expansion, and the uncertainty introduced by warping of the Galactic disc, to obtain a solar vector $(V_\odot = 233 \pm 7$ km s^{-1} $l_\odot = 93 \pm 10°$, $b_\odot = 2 \pm 10°)$ – the errors cannot be taken too literally. Power spectra were obtained by varying the solar vector over a wide volume of $\mathbf{V}_\odot$-space, $60°$ by $60°$ in longitude and latitude, and 130 km s^{-1} in $V_\odot$, which adequately encompassed the error box of the solar motion. For each $\mathbf{V}_\odot$ a set of corrected redshifts was obtained and analyzed. Several high peaks were found, the two highest being I_{max}=29.2 for a periodicity P=37.2 km s^{-1} and I_{max}=28.0 for a periodicity P=37.5 km s^{-1}. The corresponding vectors were (228 km s^{-1}, $99°$, -3$°$) and (212 km s^{-1}, $94°$, -13$°$). Within the errors, these are reasonably close to the solar motion around the Galactic centre, but the speeds are significantly lower than estimates of the solar speed with respect to the Local Group, which lie in the range $250 \lesssim V_\odot \lesssim 310$ km s^{-1}.

The significance of the peaks was again assessed by comparison with closely similar synthetic data treated identically to the real data, and for each peak the periodicity hypothesis was preferred over the null one at a confidence level ~0.999.

One further test was applied to this dataset: if the apparent periodicity was a statistical artefact, it would not in general vary with the accuracy of the data. Trials on synthetic data, on the other hand, revealed that the measured strength of the signal is highly sensitive to the redshift dispersion (Fig. 7). The resulting 89 galaxies conveniently divided up into 40 with σ=2 or 3 km s^{-1}, and 49 with σ =4 km s^{-1}. For each of the two solar vectors above, trials were carried out in which 40 redshifts were randomly extracted from the 89 and I–values computed. Twenty thousand such trials were conducted, and the periodic signal was found to be significantly concentrated in the more accurate data, this conclusion having a confidence level 0.93 for the 228 km s^{-1} peak and 0.984 for the 212 km s^{-1} peak.

Combining these factors, we concluded (Guthrie & Napier 1991) that the field galaxies within 1000 km s^{-1} of the Sun have a redshift periodicity of $\sim$ 37.5$\pm$0.3 km s^{-1}. The probability that the periodicity occurred by chance was found to be $3 \times 10^{-6} \lesssim p \lesssim 3 \times 10^{-4}$.

Extension of the Sample

An unexpected consequence of the study out to 1000 km s^{-1} was that the periodicity emerged with respect to the Sun's local galactocentric motion: the Galaxy's motion within the Local Group, or its infall towards the Virgo Cluster, did not appear to be relevant. The phenomenon was thus nucleus to nucleus between galaxies, irrespective of large scale motions in the field. If this continued to hold out to greater distances, then the signal should appear with increasing strength as the search volume around

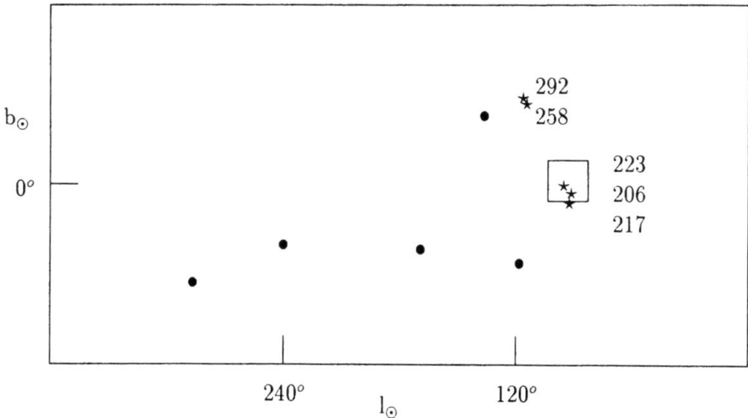

Figure 1. The ten highest peaks out of $\sim 10^6$ in a whole–sky search ($140\leq V_\odot \leq 360$ km s^{-1}), over $20\leq P \leq 200$ km s^{-1}. $\bullet = 24 \pm 3$ km s^{-1}, $\star = 37.5 \pm 0.2$ km s^{-1}. The formal error box of the solar galactocentric motion is shown.

the Galaxy is increased. We have therefore extended our analysis out to 2600 km s^{-1}, the edge of the Local Supercluster. There is no immediate reason to suppose that the periodicity should be confined to the LSC but the cut–off was convenient as one is running out of sufficiently accurate data beyond there. The criteria for selection of galaxies from the Bottinelli *et al.* dataset were as before (essentially, all accurate redshifts excluding those which TC had used). Two Virgo–like clusters (UMA and Fornax) are now incorporated in this enhanced volume, but only one eligible galaxy was found in them: for the sake of consistency, the Virgo Cluster having been excluded from the study, it too was excluded. The sample was now extended from 89 to 247 galaxies, and those with measured redshifts of extreme accuracy ($\sigma=2$ or 3 km s^{-1}) increased from 40 to 97. For practical reasons we concentrated on analyzing the 97 highly accurate redshifts.

In this extended analysis we varied the solar vector over the whole celestial sphere and over $140\leq V_\odot \leq 360$ km s^{-1} in speed, stepping in 2 or 3° intervals and in units of 5 km s^{-1}. For each pixel in this box, a power spectrum was constructed over the range 20–200 km s^{-1} and the highest peak recorded, irrespective of its period. About a million power spectra were generated in this search. The ten highest peaks are shown in Fig. 1. Five of the ten have P=37.5 km s^{-1} (Fig. 2), and three of them lie within or very close to the error box of the solar motion. Spearman ranking of the departures of individual redshifts from the periodicity show that all ten peaks are correlated, the three highest strongly so: thus a single underlying phenomenon is being detected. This whole–sky search confirmed that $\mathbf{V}_\odot$–space is not filled with all sorts of 'periodicities' in all sorts of directions, and that indeed the only outstanding phenomenon is the TC one.

The significance of these extremely high peaks (I$\sim$39) was assessed by constructing synthetic data as before. Thus each set of 97 artificial data was analyzed by varying the solar motion within a box of side 30° by 30° by 60 km s^{-1} centred on the galactocentric solar vector, and searching over 30–39 km s^{-1}. The highest peak out of the $\sim 10^4$ power spectra so constructed was recorded, and the procedure repeated for ten thousand sets of artificial data. The distribution of high peaks in the 10^4 search volumes was then

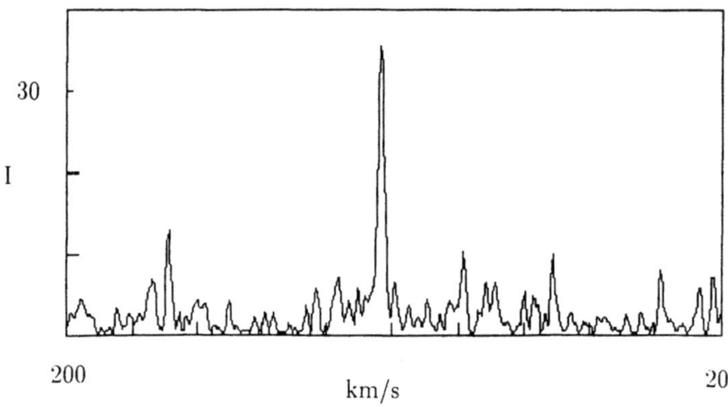

Figure 2. Power spectrum (I *vs* frequency) associated with the solar vector $(V_\odot, l_\odot, b_\odot) = 217 \text{ km/s}, 95^\circ, -12^\circ)$. The high peak occurs at 37.5 km s^{-1}.

compared with that obtained for the true data over the same search volume. Note that this procedure, involving as it does the overall power distribution, avoids extreme value statistics of any sort.

None of the 10^4 artificial datasets had statistical behaviour at all close to that of the real data. The real data, for example, threw up 25 peaks with $I > 20$ ($n_{20}=25$) and 12 with $I > 25$ within the search volume, whereas none of the artificial data did. Extrapolation shows that, roughly, the real redshifts differ from the random ones at about the million to one level (Fig. 3). Since periodicity is the only phenomenon which may be obliterated by the randomization procedure, it follows that the 97 galaxies possess a redshift periodicity of 37.5 km s^{-1} at about this confidence level.

A close examination of individual HI profiles revealed that a few of them had slightly asymmetric profiles, as might arise from foreground contamination. An objective criterion was used to reject 16 possibly contaminated galaxies from the list, and 24 TC galaxies which satisfied the other criteria were added. The resulting list of 103 galaxy redshifts constitutes the most accurate sample we currently have for the Local Supercluster. Its optimized power spectrum is shown in Fig. 4: $I\sim52$ for a periodicity 37.5 km s^{-1}. A histogram of the redshift differences for this sample is shown in Fig. 5: the periodicity is strong and coherent, with no sign of drift, from centre to edge of the LSC. We have not used this extraordinarily high peak to attempt a probability assessment; rather, it serves to confirm that the phenomenon in question is indeed a redshift periodicity.

Statistical Behaviour

The robustness of the result was tested in various ways.

(a) If the periodicity arose from some obscure statistical artefact, then it might be expected to behave erratically with respect to sample size, accuracy of data and magnitude of the optimum solar vector. Fig. 6 reveals that, for a fixed solar vector (217 km s^{-1}, 95°, -12°), the signal strength increases linearly with N, as expected theoretically for a real signal. The observed slope is consistent with a true dispersion $\sigma \sim 8 \text{ km s}^{-1}$. As can be seen from Fig. 7, this dispersion is rather critical for the detection of periodicity when the sample size is $N\sim100$, and this may account for the

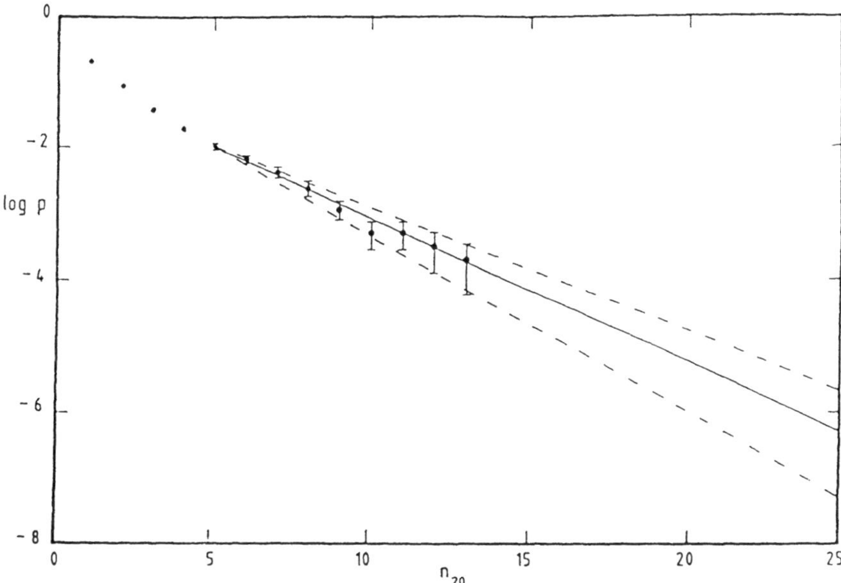

Figure 3. Probability that a set of randomized redshifts, constructed and analyzed as described in the text, would yield more than n_{20} spectral peaks.

difference in behaviour between the galaxies with a formal $\sigma \leq 3$ km s^{-1} and those with $\sigma = 4$ km s^{-1}, since unknown systemic errors of order several km s^{-1} probably exist in these measurements. In this larger sample too, the signal strength was found to concentrate strongly in the best data, at a significance level ~ 0.998.

(b) If, instead of holding the solar vector fixed, the optimum solar vector is derived as a function of sample size, the result shown in Table 1 is obtained. The period holds steady to within ± 0.15 km s^{-1} for $\mathbf{V}_\odot$ varying by only ± 2 km s^{-1} in speed and $\pm 1^\circ$ in direction, as the sample doubles in size from ~ 50 to ~ 100 redshifts. This is a remarkable degree of stability, difficult to reconcile with a statistical fluke or artefact. Trials on sets of random data with in-built periodicity revealed that, for $\sigma = 8$ km s^{-1}, the r.m.s. dispersions expected are 0.2 km s^{-1} in derived period, 3 km s^{-1} in $V_\odot$ and 1.2° in direction.

(c) The inclusion of the Virgo galaxies, or the arbitrary exclusion of 15 redshift calibrators from the list (Baiesi–Pillastrini & Palumbo 1986) made little difference to the result. Thus the signal strength is robust to modest changes in the dataset and cannot be attributed to a particularly favorable choice of sample.

Table 1. Optimized parameters as a function of sample size. The solution holds steady to $\Delta V_\odot = \pm 2$ km s^{-1}, $\Delta \theta = \pm 1^\circ$ and $\Delta P = \pm 0.15$ km s^{-1}, throughout the LSC.

V_{max}	N	$V_\odot$	$l_\odot$	$b_\odot$	P	I_{max}
1000	51	215	93	-13	37.8	30
1400	72	213	94	-13	37.7	31
1800	86	215	94	-13	37.7	36
2600	97	217	95	-12	37.5	38

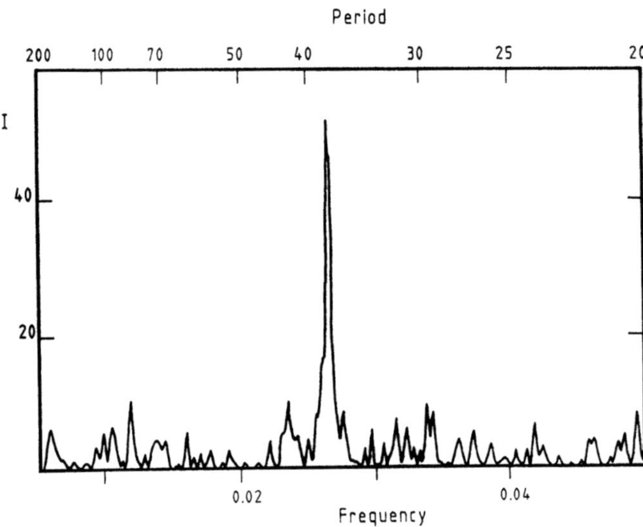

Figure 4. Power spectrum of the 103 most accurate, uncontaminated redshifts corrected for the optimum solar vector shown.

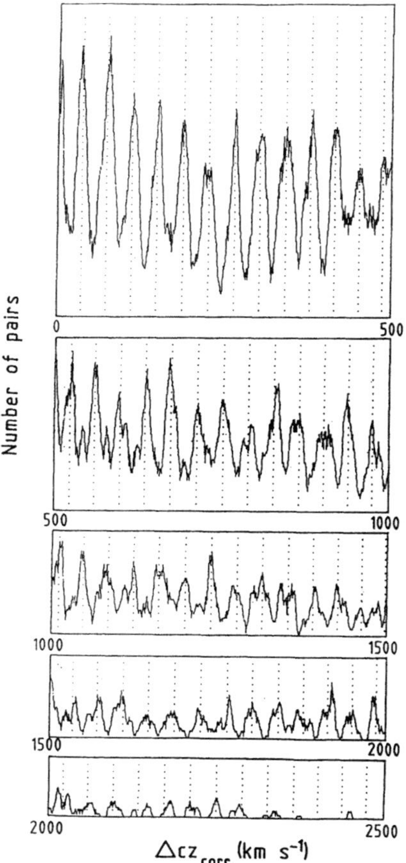

Figure 5. Two–point correlation function corresponding to the redshifts and optimum solar vector employed in the previous figure. Vertical dashed lines represent the best–fit periodicity, which seems to hold over the whole of the Local Supercluster.

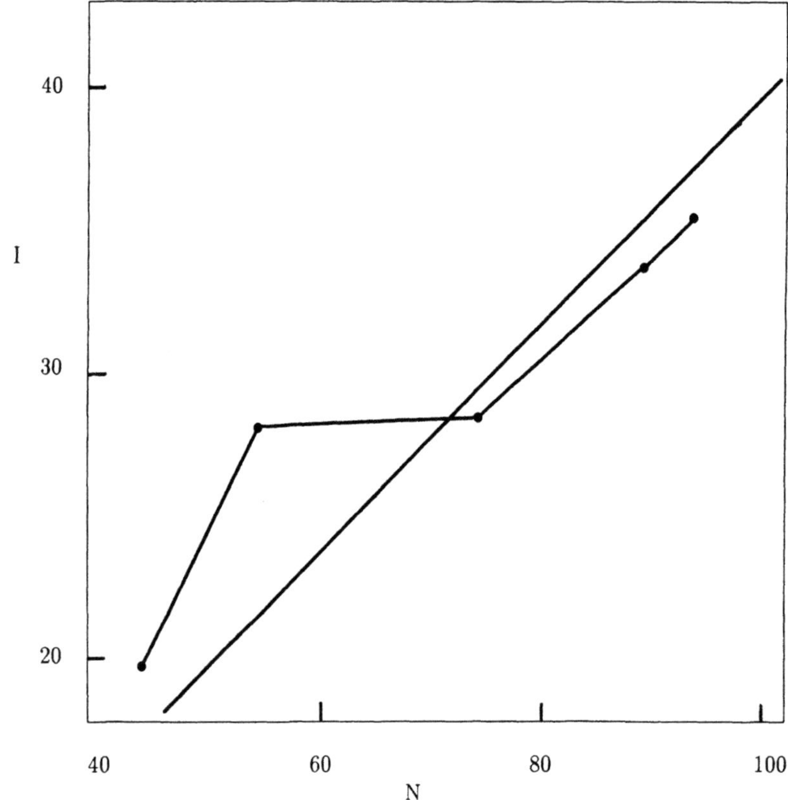

Figure 6. Signal strength I as a function of numbers N of galaxies. The dots are for galaxies out to 500, 1000, 1500, 2000 and 2500 $\mathrm{km\,s^{-1}}$. The straight line represents the mean behaviour of I(N) for an assumed true dispersion of $\sigma=8$ $\mathrm{km\,s^{-1}}$ about P=37.5 $\mathrm{km\,s^{-1}}$.

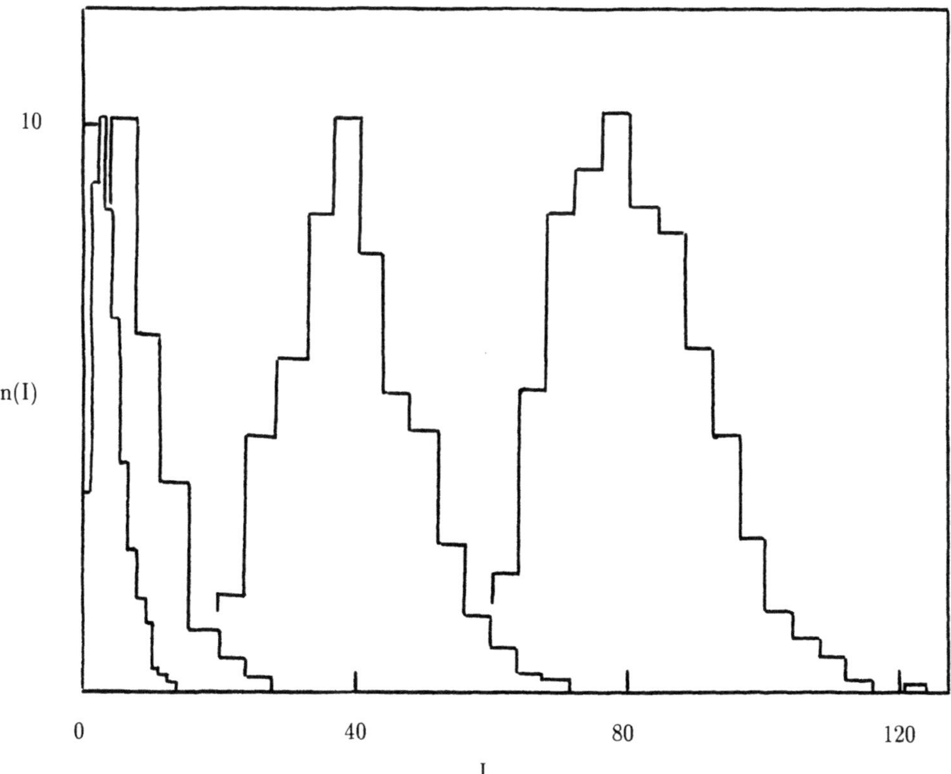

Figure 7. Power distributions n(I) obtained for synthetic datasets each containing 97 redshifts distributed with periodicity 37.5 $km\,s^{-1}$ and dispersions (left to right) $\sigma = 32,\ 12,\ 8$ and 6 $km\,s^{-1}$ respectively.

(d) The data were also divided by morphological type, radio telescope employed and celestial position; no correlation was found with any of these.

(e) Finally, a whole–sky search using synthetic data with P=37. 5 km s^{-1} and σ=8 km s^{-1} was conducted. The behaviour was found to have the same general characteristics as shown in Fig. 1: a few peaks with the inbuilt periodicity clustered around the galactocentric solar vector, and a scattering of peaks with fractional periods over the celestial sphere and with various $V_\odot$.

Groups and Associations

About half the galaxies in the sample of 97 belonged to loose groups or associations (Fouqué *et al.* 1992) containing a few bright galaxies (these groupings are preserved in the Monte Carlo simulations). The data were divided into two appropriate sets to explore whether the periodicity concentrated in either the field or group galaxies. The full sample of 247 spirals was used, enhanced to 261 by the addition of a few galaxies previously used by TC. Correlation analysis revealed a strong tendency for those galaxies which belonged to groups or clusters to possess the most accurately determined redshifts. The question arises whether the periodicity truly exists in clusters, or is simply detected preferentially there because cluster galaxies have been more accurately measured. A correlation analysis supports the latter at a confidence level $\sim$0.96.

There are 9 doubles, 6 triplets, 3 quadruplets and a quintuplet in the sample of 97, yielding 55 local differential redshifts within these small groups, of which 34 are independent. The differential heliocentric redshifts are plotted in Fig. 8a, and the galactocentric ones in Fig. 8b. Because of the small angular extent of the groups the galactocentric correction is now second order. In effect the large number of trials involved in varying $\mathbf{V}_\odot$ are replaced by a single trial, and so in effect the differential redshifts yield a parameter–free test of periodicity. It is clearly present, power spectrum analysis and comparison with Monte Carlo trials yielding a confidence level C $\gtrsim$ 0.9999.

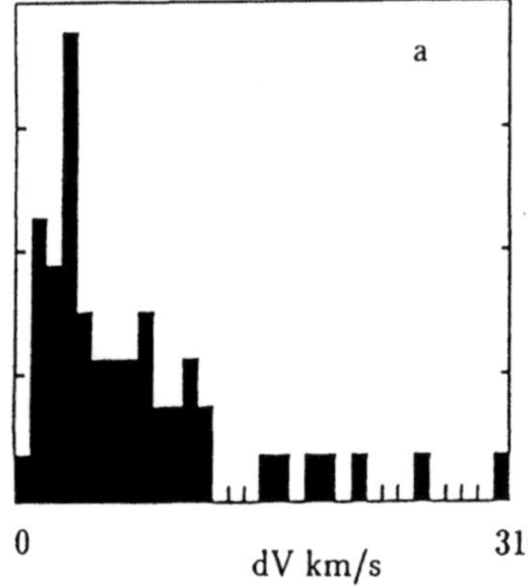
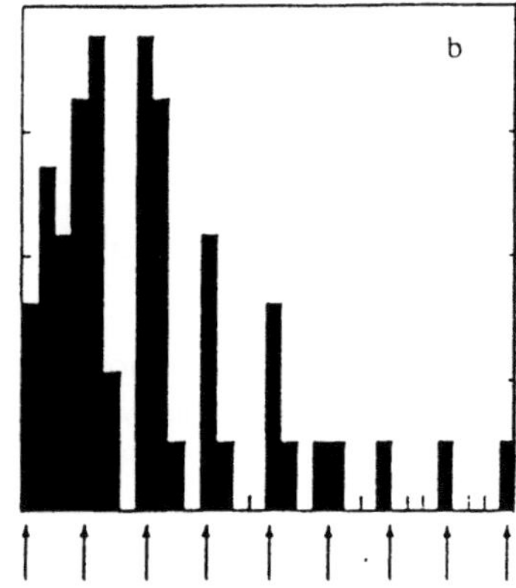

Figure 8. Histograms of differential redshifts dV for the 53 galaxies linked by group membership. (a) heliocentric redshifts; (b) redshifts corrected for $\mathbf{V}_\odot$ = (216 km s^{-1}, 93°, -13°). Binwidth is 10 km s^{-1}. Vertical arrows mark a periodicity of 37.6 km s^{-1} and zero phase.

Further trials involved scattering the groups around the LSC and confirmed that the coherence in phase of the periodicity, from one group to another, is real, and not an artefact induced by the optimization procedure. Thus the $\sim$37.5 km s^{-1} periodicity is a truly global phenomenon, the galaxies being in effect test particles whose group membership is incidental.

THE VIRGO CLUSTER REVISITED

The Virgo cluster is the nearest rich cluster of galaxies and, at the outset of the enquiry by Guthrie & Napier (1990), it had not been used in the formulation of the quantization hypothesis. It therefore constituted an unbiased and independent sample, suitable for the purposes of testing.

In their study, Guthrie & Napier (1990) first tested for quantization in a sample of 112 Virgo spirals with relatively well–determined redshifts, initially applying a correction using a solar apex (252 $\mathrm{km\,s^{-1}}$, $100°$, $0°$) to obtain the galactocentric redshifts. A signal was found in the range 70–75 $\mathrm{km\,s^{-1}}$ then under test, but at a confidence level only $0.96 \leq C \leq 0.99$. However it was found that this signal (P=71.1 $\mathrm{km\,s^{-1}}$) was strongly concentrated in 48 galaxies situated within the less dense parts of the Virgo cluster, galaxies in the core itself showing little sign of redshift periodicity. Allowing for the a *posteriori* nature of the finding, and the arbitrariness involved in defining 'core' and 'less dense' regions, the periodicity was confirmed at a confidence level $0.996 \leq C \leq 0.999$. A similar result was obtained when the solar vector was allowed to vary over the whole sky, the speed being maintained at $V_\odot = 252$ $\mathrm{km\,s^{-1}}$.

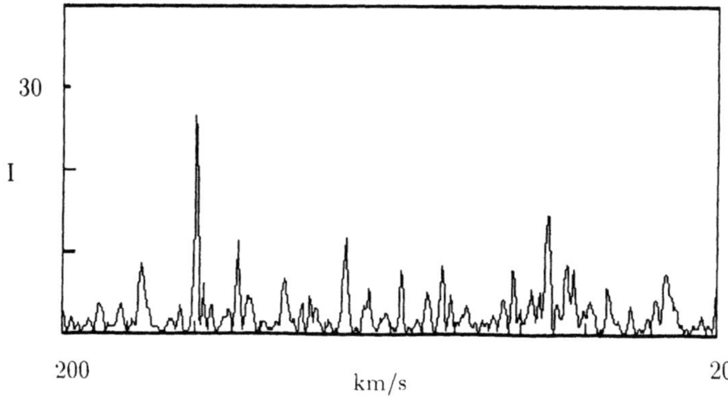

Figure 9. Power spectrum (I *vs* frequency) of 48 Virgo Cluster galaxies avoiding the core of the cluster. The high peak ($I \sim 26.5$) is at 71.1 $\mathrm{km\,s^{-1}}$.

The solar apex adopted in the above study was taken with respect to the Local Group. However the subsequent studies, described above, reveal that the global periodicity emerges strictly with respect to the Galactic nucleus: the motion of the Galaxy with respect to Local Group, Virgo Cluster or whatever seems not to be relevant. Thus in testing for periodicity within the Virgo Cluster, the Sun's galactocentric vector should have been subtracted. I have therefore repeated the analysis by conducting a box search within $\pm 30°$ and ± 20 $\mathrm{km\,s^{-1}}$ of (220 $\mathrm{km\,s^{-1}}$, $96°$, $0°$). A signal of strength $I \sim 26.5$ appears at P=71.1 $\mathrm{km\,s^{-1}}$ (Fig. 9), and varies little within the error box of the solar apex: thus there are no degrees of freedom within the error box of the latter and a single–trial probability $\exp^{-13.8} \sim 1.7 \times 10^{-6}$ is obtained (the exponential formula is valid to this height, against a white noise background: *loc. cit.*). Assuming $\sigma \sim 10$ $\mathrm{km\,s^{-1}}$, the periodicity which emerges is 71.1 ± 1.3 $\mathrm{km\,s^{-1}}$ which is, within the errors, the periodicity under test for a galaxy cluster.

The above significance level should be reduced by a factor of order five to allow for the *a posteriori* selection of low–density regions, and a further factor of perhaps two or three to allow for the possibility of redshift interdependence through the presence of binaries (*loc. cit.*). Thus the periodicity hypothesis is confirmed for the Virgo cluster at a confidence level $C \sim 1 - 2 \times 10^{-5}$. The earlier result yielding the same periodicity for a different vector arose because of the small angular extent of the Virgo Cluster; thus the differential solar apex correction is small and the signal is seen over a wide range of $\mathbf{V}_\odot$.

CONCLUSION

We have tested the prediction that redshifts show global periodicities $\sim$24.2 or $\sim$36.3 $\mathrm{km\,s^{-1}}$ after correction for the solar galactocentric vector. We find a strong periodicity of 37.5 ± 0.2 $\mathrm{km\,s^{-1}}$ to be present in accurate, independent redshift data. We have also used the Virgo Cluster to test the further prediction that a periodicity of $\sim$72 $\mathrm{km\,s^{-1}}$ occurs in clusters of galaxies; we find a periodicity 71.1 ± 1.3 $\mathrm{km\,s^{-1}}$ in the outer regions. The confidence levels of both these results are extremely high, and we conclude that extragalactic redshifts are quantized. Clearly there must be a transition regime between field galaxies, loose groups and rich clusters, but this matter has still to be explored.

The astronomer who wishes to build a cosmology based on quantized redshifts cannot be faulted on observational grounds. Thus the periodicities shown in Figs. 5, 8(b) and 9 are not 'statistical results': rather, they are the observed outcome of a single correction applied to the best heliocentric redshifts. Within the uncertainties, this solar correction is one which transforms our redshift catalogues to those which would be obtained by a civilization at the nucleus of our Galaxy. The phenomenon appears at about the expected strength for a given sample, it behaves as expected in respect of such matters as sample size, and it is robust to the choice of redshift data. If it is due to a gremlin in radio telescopes, then the gremlin concerned knows the galactocentric solar velocity.

Statistical analysis enters the issue when ascertaining whether there are sufficient degrees of freedom within the error box of the solar motion for a similar result to be derived from a random redshift distribution. This question can be settled by trials, and the answer is strongly in the negative. Thus the astronomer who wishes to maintain existing cosmological paradigms must first face the challenge set by the periodicities.

ACKNOWLEDGEMENTS

I am indebted to Bruce Guthrie for allowing him to describe some results in advance of publication, to Franco Selleri for the invitation to address the conference, and to numerous colleagues for stimulating discussions.

REFERENCES

Bottinelli, L., Gougenheim, L., Fouqué, P. & Paturel, G., 1990. Astr. Astrophys. Suppl. **74**, 391.
Cocke, W.J. & Tifft, W.G., 1991. Astrophys. J. **368**, 383.
Fouqué, P., Gourgoulhon, E., Charmaraux, P. & Paturel, G., 1992. Astr. Astrophys. Suppl. **93**, 211.
Guthrie, B.N.G. & Napier, W.M., 1990. MNRAS **243**, 431.
Guthrie, B.N.G. & Napier, W.M., 1991. MNRAS **253**, 533.
Schneider, S.E. & Salpeter, E.E., 1992. Astrophys. J. **385**, 32.
Tifft, W.G., 1977. Astrophys. J. **211**, 31.
Tifft, W.G., 1980. Astrophys. J. **236**, 70.
Tifft, W.G., & Cocke, W.J., 1984. Astrophys. J. **287**, 492.

QUASAR SPECTRA:
BLACK HOLES OR NONSTANDARD MODELS?

Jack W. Sulentic

Dept. of Physics & Astronomy
University of Alabama
Tuscaloosa, USA 35487

INTRODUCTION

The "Big Bang" model has ascended to a powerful position in modern cosmology over the past few decades. This position has become so strong that investigation of alternate ideas has almost ceased. Observational counter-evidence certainly exists (for reviews see e.g. Arp 1987; Sulentic 1987; Tifft 1987). The general belief is that this counter-evidence consists of misinterpreted data and false clues. One could easily get the impression that all of the observations fit easily into the accepted model. In fact, at least three new concepts have assumed great importance in preserving the Big Bang against observational and theoretical challenges. In temporal order of acclamation they are: 1) black holes; 2) merger phenomena and 3) dark matter. Gravitational accretion onto supermassive black holes was required as soon as it became generally accepted that the quasar redshifts were cosmological (i.e. proportional to their distance). It provides a mechanism for producing the enormous energies implied by the assumption that quasars are at their redshift distances. Mergers came upon the scene in order to account for nearby peculiar galaxies and for the increasing luminosity and size of many objects at higher redshift. Dark matter helps to bind the groups and clusters of galaxies as well as to explain the flat rotation curves in spirals. In reality, it can help to fit almost any observation into the conventional picture. What is the observational evidence for the above three phenomena?

Dark matter poses a very real threat to the observational astronomy profession. The fraction of matter in the universe thought to be invisible exceeds 90% in some recent estimates (e.g. Mulchaey et al. 1993). As this number converges toward 100%, astronomy may well cease to be an observational science. At the same time alternative explanations for the dynamical peculiarities are few with the most discussed being MOND (Milgrom 1983; see also Sanders 1987). Dark matter is sufficiently unconstrained at this time to handle most challenges to Big Bang theory that might arise in the forseeable future. An example of its tremendous versatility can be illustrated by the recent "rediscovery" of an association between higher redshift quasars and lower redshift galaxies (Rodrigues-Williams and Hogan 1993; see Sky and Telescope, November 1993, p.12). The reviews of counter-evidence cited above give extensive discussion to the past evidence for these associations. This evidence was long disputed as much because of its implications as for doubts about the observations. The excess reported in the latest study was so great that it was apparently necessary to invoke a closure density of dark matter in the galaxy clusters in order to forestall a crisis for conventional ideas.

In recent years, almost anything that looks peculiar or that shows a luminosity excess at some wavelength has been attributed to **merger activity**. In fact the observational definition for a merger is quite vague. We recently studied the observational properties of compact galaxy groups (Sulentic and Rabaça 1994ab). We found that these systems show very few of the accepted observational properties of mergers. At the same time very few potential merger remnants of past compact groups are observed. This is in spite of the fact that dynamical theory predicts that compact groups should be very unstable to collapse and coalescence (Barnes 1989). If the systems most susceptible to merging show such a low level of merger activity, how can the phenomenon be common in less dense environments? The observational evidence suggests that while mergers occur, they are relatively uncommon and cannot be used to explain most peculiar extragalactic objects.

There is no direct evidence for **black holes**. Indirect support comes from observation of rapid X and γ ray variations in active galactic nuclei (AGN$\equiv$ quasars, QSO's, Seyfert galaxies, broad line radio galaxies and BLLAC objects). Evidence for massive cores, believed to be inactive black holes, in the nuclei of nearby normal galaxies is also regarded as a form of indirect evidence. We report here on a study of the emission line properties of quasars. The motivations for this study were a) to study the frequency of occurence and magnitude of the internal redshift discrepancies observed in quasars and b) to use this data to critically test predictions of physical models for the central structure of AGN. Recently this work has become relevant to the first of the above three phenomena. The observation of double peaked emission lines in the quasar Arp 102b was interpreted as line emission arising from a radiating accretion disk (Chen, Halpern and Filippenko 1989). Direct observation of emission from an accretion disk would be tantamount to proof for the existence of black holes. Our unpopular conclusion was that the bulk of the data do *not* support this interpretation. We consider first the basic properties of quasar line spectra followed by the results of our study. This is followed by the results of our comparison with the predictions of line emitting accretion disk models. Finally we discuss evidence that the line shifts observed in AGN might arise from a non-Doppler cause.

EMISSION LINES IN AGN

Normal galaxies show spectra dominated by absorption lines that arise from the composite spectra of stars that represent their principal visual constituent. The advent of sensitive detectors in the past two decades has revealed the signature of emission from hot gas in most galaxies as well. There are two kinds of emission lines that are observed: 1) permitted lines arising from transitions following photoionization (the recombination spectrum characterized in the visual by the Balmer lines and 2) the "so-called" forbidden lines arising from transitions following collisional excitation. These lines are somewhat unique to astronomy because they arise in such low density emitting regions. The principal optical features are due to [OII], [OIII], [NII] and [SII]. The critical density for [OIII] $\lambda5007$A is about 10^6. AGN provided an introduction to UV spectroscopy long before the first satellite telescopes opened this domain to our direct study. The higher redshift objects opened up the spectral region with rest wavelenths between 1000 and 3000 A to our view. Higher ionization broad lines such as [CIV] and [CIII as well as Lyman α are redshifted into the visible region of the spectrum (e.g. we find Lyman α at about 4800A in a quasar with redshift z=3.0). The UV lines are often referred to as high ionization broad lines (HIL's) to distinguish them from the lower ionization lines observed at lower redshift

(LIL's). In summary, optical spectra of low redshift quasars show primarily NLR and LIL-BLR lines while high redshift quasars show primarily HIL-BLR lines.

Emission line widths in velocity units are typically less that 200 km s^{-1} for normal galaxies. It is the singular and unifying feature of AGN that the permitted lines are very broad (full width half maxima from 10^3 to 2×10^4 km s^{-1}). The forbidden lines show FWHM similar to or a little broader than the corresponding lines in normal galaxies. The Balmer lines also often show a narrow component. Figure 1 shows an example of a low redshift quasar spectrum in the region near 5000Å where we find both broad line (BLR) Balmer features (Hβ and γ) and narrow line (NLR) [OIII] features.

Broad lines in AGN show peculiarities that for a long time were only discussed privately in hushed tones. They were not even mentioned in the most recent textbook written on the subject (Weedman 1986). *Different* lines often show *different* redshifts in the *same* object. The range of redshifts in a single object can exceed 2000 km s^{-1}. At the same time the lines can show striking deviations from symmetry. The

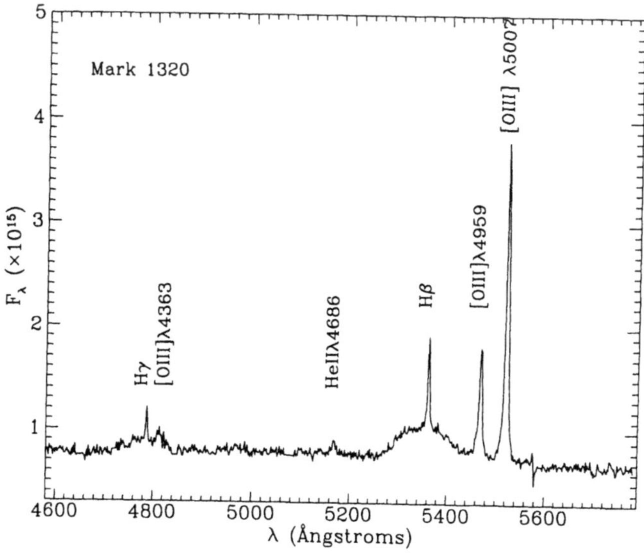

Figure 1. Spectrum of Markarian 1320 in the region between Hγ and [OIII]λ5007Å. Principal lines are identified. Note that Hβ shows both NLR and BLR components. 1Å$= 10^{-8}$ cm.

keen-eyed reader will have noted a redshift discrepancy in the spectrum illustrated in Figure 1. The NLR component of Hβ is centered at a higher wavelength than the BLR component in that object (Markarian 1320). Figure 2 shows an even more striking example (OQ208) where the Hβ component redshifts differ by 2700 km s^{-1}. Two kinds of internal line shifts are recognized: 1) red and blue shifting of the LIL BLR with respect to the NLR in low redshift AGN and 2) an apparent systematic blueshift (700-1000 km/s) of the HIL with respect to the LIL. The latter shift has to be inferred from observations of two different sets of AGN since both sets of lines (in the same object) have not, until recently, become accessible to study and measurement. Our study of the LIL vs. NLR shifts was the first of a recent flurry of activity in this area. The line shifts were originally noted by Gaskell (1982) and Wilkes (1984). It is clear that the shift and asymmetry properties of the AGN emission lines are important.

Are they giving us an invaluable clue into the internal geometry and kinematic of the line emitting region or is it possible that the shifts are evidence for a non-Doppler phenomenon? If the former is true then they pose an important challenge for current theories while the latter possibility is regarded as unthinkable. We consider both possibilities here.

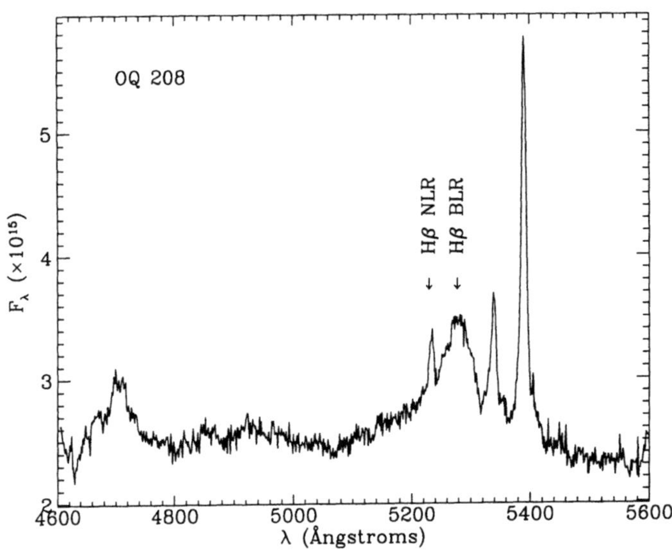

Figure 2. Spectrum of OQ208 in same wavelength region as Figure 1. Note the large NLR vs. BLR velocity shift ($\sim$2500 km s^{-1}).

SYSTEMATIC STUDY OF LINE SHIFTS AND ASYMMETRIES

Our first study (Sulentic 1989) focused on a comparison of the redshift of Hβ and [OIII]. Line asymmetry properties of Hβ were also measured. We chose these lines for two reasons. 1) Most published data involve optical spectra of low redshift AGN ($z\leq0.5$) where Hβ is one of the most prominent features. 2) We wanted to compare the (LIL) BLR vs. NLR redshift between lines that were close together but not too close to be severely blended with one another. Hα as a LIL line was ruled out because it is redshifted out of the visible at much lower redshift and is severely blended with NLR lines of [NII]. [OIII]λ4959, 5007Å are close to Hβ without excessive overlap in most cases.

There are two other important considerations for a study of this kind: 1) the standard of rest and 2) contamination by multiplets of FeII emission. There has been considerable confusion over which, if any, lines provide a measure of the "rest frame" for the quasar. We studied all available data that might be relevant which includes 21cm emission from neutral hydrogen (HI) surrounding the AGN and absorption line redshifts from "fuzz" surrounding some low redhift quasars. The latter is thought to be starlight from a galaxy believed to be "hosting" the AGN. This data showed agreement in redshift $\Delta V \leq 200$ km s^{-1} of the NLR redshift (Wilson and Heckman 1985). Usually the agreement was even better, leading us to adopt the NLR redshift from the lines [OIII]λ 5007 and NLR Hβ (recombination emission is often observed from the forbidden line region) as our zero point in the line shift study. Some recent studies have continued to use Hβ as a reference despite the fact that it shows large velocity excursions. Care should be exercized in using published data and in comparing their results with our work. The region of Hβ is infested with emission from myriads of lines arising from various multiplets of FeII. These lines inhibit accurate measurements of Hβ and [OIII] as well as preclude reliable determination of the continuum level in this region. They are also a problem in other regions of AGN spectra (notably the region near rest wavelength 2800Å). There are ways to model and correct for this contamination but we will not discuss them here. We avoid this discussion because our study focused on AGN where the presence of FeII emission was weak or absent. Figure 3 shows an example of an AGN with serious FeII contamination in this region.

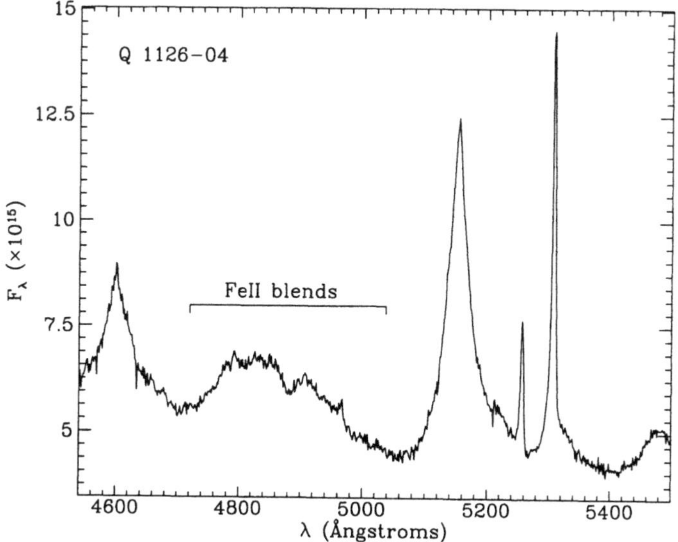

Figure 3. Spectrum of Q1126-04 in same wavelength region as Figure 1. Note the strong FeII emission blueward of Hβ. Data for Figures 1,2 and 3 were obtained at Kitt Peak.

In the end we quantified the Hβ emission line properties with the following measures: a) centroid redshift at 1/4, 1/3, 1/2 and 3/4 height; b) full width half maximum (FWHM). The centroid and width at half maximum were also measured for [OIII]. The LIL line shift was then measured with respect to [OIII] at each of the four heights in the line profile. In addition, an asymmetry index was defined as R= [C(3/4)-C(1/4)]/FWHM. Our first study involving 61 AGN was the largest for a sample of high resolution and S/N spectra. It revealed that LIL Hβ shows almost equal numbers of red and blue shifts (and asymmetries). This was a surprise because both were thought to be predominately red. There may still be an excess of red or blue shifts and/or asymmetries, but a larger sample will be needed to establish it. Figure 4 summarizes the kinds of line profiles that we were able to identify.

Most other recent studies (e.g. Corbin 1989; Francis et al. 1992) have focused on the HIL line properties derived from samples of high redshift quasars observed on the ground. They suggest that HIL are generally more symmetric than LIL and that HIL are blueshifted with respect to LIL. The magnitude of the latter shift (700-1000 km s^{-1}) could only be inferred indirectly since LIL (and NLR for that matter) features in the same objects were unobtainable. The most recent survey of the bright low redshift PG quasar sample (Boroson and Green 1992) confirm the basic results of our study. Despite a more sophisticated PCA analysis of their sample, these authors could not uncover any significant correlations between line properties that were not known (or suspected) previously. The next step is obvious; comparison of UV and optical spectra for HIL, LIL and NLR in the same objects. This has become possible because the UV sensitivity of the Hubble Space Telescope permits us to observe the HIL lines in the same sample that we have studied on the ground. Several groups, including our own, are engaged in such comparisons at this time.

IMPLICATIONS OF LINE STUDIES

What do we really know or think we know about quasars? We believe that they are powered by gravitational accretion onto a black hole. The only conventional alternative (the starburst model; Terlevich et al. 1992) argues that chain reactions of supernovae are responsible for the energy output. Anyone who has seen the spectrum of a supernova has been struck by the similarities with AGN emission spectra. The BLR emission line spectra in AGN are consistent with a gas at $T \sim 1\text{-}2 \times 10^4$K and the size of the BLR line emitting region is constrained by variablity studies to be in the range of 10-100 light days (1 light day$= 2.6 \times 10^{10}$ km.). The absence of broad forbidden emission lines like [OIII] indicates that $n \geq 10^9$ in the LIL-BLR, while the presence of weak [CIII λ 1909 indicates that $n \leq 10^{11}$ in the HIL-BLR. There is growing evidence for stratification in the BLR zone with HIL emitting region closer to the ionizing continuum source. Estimates of the covering factor yield small numbers, suggesting that the BLR is made up of a collection of clouds rather than diffuse gas. The predominance of forbidden lines in the LIL argues that it lies outside the BLR. Finally there is growing evidence, already implied by radio structures, that anisotropic emission may arise from jets or cones of radiation.

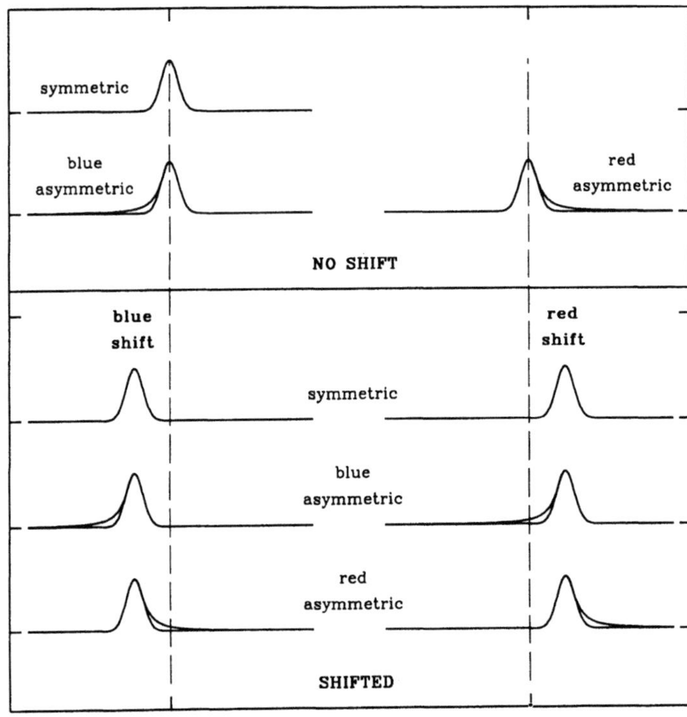

Figure 4. Combinations of emission line shift and asymmetry observed in our survey. Double peaked profiles are not included and were not found in our study.

Figure 4 shows that almost every possible combination of line shift and asymmetry is observed in AGN spectra. Both symmetric and asymmetric line profiles show (red and blue) line shifts. The only possible region of avoidance involves ones that are blue shifted and blue asymmetric at the same time (only one example was found).This *stochastic* property provides a very strong constraint on possible models for the kinematics and geometry of the emitting region. Favored models have invoked either a dominance of gravitational forces (infall or rotation) or radiative pressure driven outflow (ballistic models have also been considered). The stochastic nature of the line profile properties rules out such "single-force" models. It requires some hybrid explanation that can account for both red and blue shifts. We have considered the possibility that the BLR line radiation originates anisotropically (Zheng, Binette and Sulentic 1990) in a double stream model. Gaskell (1983) proposed a binary black hole model and Mathews (1993) recently proposed a model involving bouncing clouds. Any conventional model is a long way from adequately explaining the observations. It is usually possible to adequately model any single AGN but an attempt at generalization always leads to difficulties.

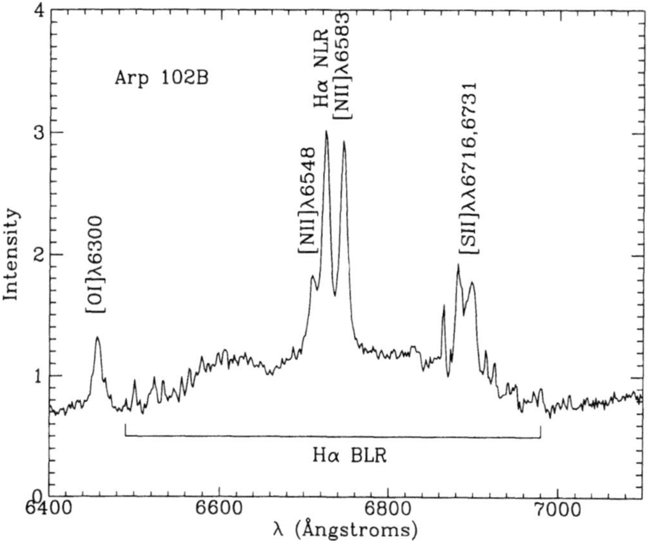

Figure 5. Spectrum of Arp 102B in the region of Hα. Data obtained at San Pedro Martir Observatory, Mexico.

Our results appear to deal a blow to models that invoke BLR line emission from accretion disks. Figure 5 shows the spectrum of Arp102b that generated considerable excitement a few years ago. It represents a kind of profile not found in our study. Its major characteristics are: 1) double peaked shape; 2) very large FWHM; 3) redshift of the line base and 4) stronger blue peak. It was quickly realized that these characteristics agree with predictions for emission originating from a rotating accretion disk. Arp102b immediately became a celebrity because it represented the nearest thing to direct proof for black holes in the centers of AGN. Detailed models for radiating disks were developed with full relativistic treatment (including gravitational redshift and Doppler boosting) (Chen and Halpern 1989). The fit was quite good and the standard model appeared to have a boost. This idea was attractive for other reasons 1) it provided an explanation for the strong FeII that was frequently seen and 2) helped solve an energy budget problem involving the ratio of LIL to HIL emission. Unfortunately double peaked line profiles are extremely rare among AGN. A few additional cases were cited but the fits to these profiles were much poorer. In addition, some of these objects showed blue shifted bases and amplified red peaks, both of which are forbidden by the model. Epicycles were added to save the model (spotty disks are popular). It was also argued that the disk emission was only one of several components to the observed emission lines.

We were troubled by the rarity of Arp102b type profiles in a class of objects where an accretion disk was believed to be a standard appliance. It seemed too miraculous that Arp 102b had come along to reveal nature's secret. We decided to look at the parameter space for line shifts and asymmetries predicted by relativistic Keplerian disks. We compared these predictions with the results of our line profile analysis. Figure 6 shows a comparison of the domain of shift and asymmetry parameters that are predicted and observed. We tried to consider a large range of viewing angles and Schwarzschild radii for the emitting portion of the disk. It is clear that the agreement is poor. One can argue that most of the emission in most objects arises from other non-disk sources. One can argue that our model for the radiating disk is not correct or adequate. This just brings us back to the points that Arp 102b suggests most of the emission comes from a disk and that our model fits Arp 102b very well. If all of the other AGN have a hidden disk and/or different geometry and radiative properties, Arp102b again becomes a miracle.

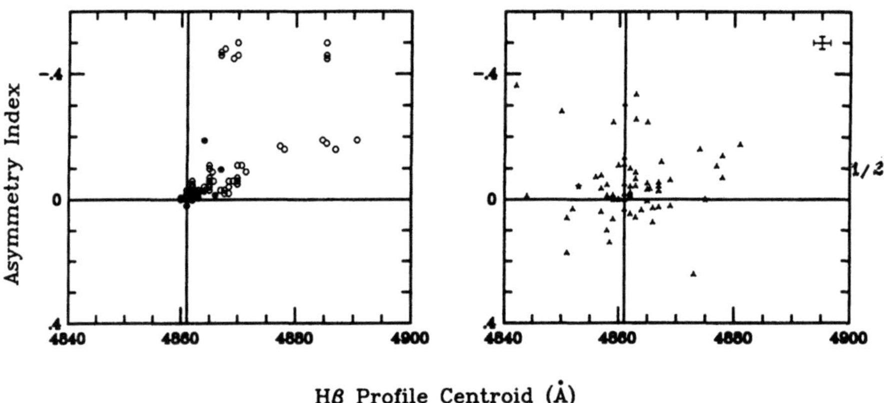

Hβ Profile Centroid (Å)

Figure 6. A comparison of the shift-asymmetry domain for line profiles at FWHM. Model predictions are on the left with observational results on the right.

NON-DOPPLER MECHANISMS

The distribution of shift and asymmetry values actually observed and displayed in Figure 6 deserves further comment. The distribution is smooth and peaked near zero or a little redward of zero. There is no evidence for bi-modality as might be expected if there were two classes of AGN: those with and without a significant disk contribution to the line emission. If we interpret the smoothness of this distribution to a single physical model, that model would not be dominated by accretion disk line radiation. That smoothness and stochasticity must be verified with larger samples in the future.

It is also worth noting that line shifts are observed for *all* classes of line profiles including those that show double peaks. If both shift and asymmetry were providing clues into the geometry and kinematics of the BLR region one might, albeit naively, expect shifts to be observed with only certain profile types. One might expect redshifted profiles to be associated with one or more classes and blue shifted with others. In making this observation, it is important to emphasize that the shifts are real and not simply an artifact of line asymmetry. One reason we know this is that symmetric profiles also show line shifts. This ubiquity of line shifts raise the question of whether the shift property is imposed from outside. In this case it might have nothing to do with structure or motion in the quasar. This raises the subject of a non-Doppler mechanism for the production of the line shifts. Another reason to look for new explanations comes from the difficulty of any model to deal with both red and blue LIL shifts, especially if both are common. Coupling this with a systematic HIL blueshift further complicates the problem. The mere suggestion of a non-Doppler mechanism is considered offensive by the establishment. This reaction is quite independent of whether the proposed mechanism comes from conventional physical ideas or not. Perhaps the reason for this is the fear that any non-Doppler mechanism is regarded as a threat to the entire redshift-distance relation.

Our work was originally motivated by the model for non-Doppler redshifts proposed by Wolf (1986, 1987). We discuss it here not so much as an advocate but more to demonstrate that new ideas and solutions might come from conventional physics. At this point we are only looking for a mechanism to explain internal redshift discrepancies. This solution may or may not have a bearing on the cosmological redshift in these objects. This model and subsequent laboratory verification demonstrated that the spectrum of light might not always be invariant on propagation from the source. The idea was that partial organization or coherence in a source could produce frequency shifts in the emitted radiation, that mimic the Doppler effect. The original idea involved a static phenomenon where fluctuations in a source (or field) distribution at different points in the source region were partially correlated. Such a process could only redistribute the flux within the original frequency envelope of a line. Therefore only red or blue shifts within the original envelope of frequencies could be produced. This mechanism would produce a frequency-dependent shift. Our observational study revealed that all but two of the 61 AGN showed shifts that satisfy the former requirement. A frequency dependence of the BLR redshift might well be present in AGN considering the large scatter of line shifts that are observed and the systematic difference between UV and optical shifts.

More recently Wolf and colleagues have investigated dynamical scattering processes (Wolf 1989; James, Savedoff and Wolf 1990; James and Wolf 1990). They conclude that "dynamic scattering in a random medium whose dielectric response function is suitably correlated in space and time" could give rise to red and blue shifts of almost any size. They originally considered only Gaussian correlation functions, but more recently (James and Wolf 1994) found that Doppler-like shifts could be generated from many random media with very different correlation functions. The relative stability of the NLR redshift suggests that, if a "Wolf" effect operates in AGN, it arises within the BLR (source) or in an envelope surrounding it (field). The dynamical model has two attractive features: 1) it produces line shifts that may be frequency independent and 2) the lines may be broadened. The best results have been produced in scattering media with a high degree of anisotropy. It is interesting that conventional models for the BLR and NLR regions have been moving in this direction for some time. BLR emission involved with jet or biconical structures is one of the few conventional ways to deal with the relatively common occurrence of both red and blue BLR line shifts. Thermal plasmas of the kind that produce the BLR radiation obey Lambert's law and should not possess any correlation properties. Therefore it is easiest to consider a component of synchrotron emission, which we know to be beamed in AGN, as the field responsible for producing the correlation induced line shifts. There is as yet no evidence for any partially coherent fluctuations of the kind required. Another problem lies in explaining why the mechanism produces almost equal numbers of red and blue shifts rather than some more restricted range of values. This may be less of a problem for a scattering model than for the conventional explanations involving source geometry and kinematics. We look forward to more detailed comparisons between observations and model predictions of the Wolf effect. In particular, the timescale of the expected fluctuations capable of producing the observed shifts might lead to a test. While spatial fluctuations will probably be unresolvable, temporal fluctuations might be accessible at some wavelength domain.

REFERENCES

Arp, H., 1987, "Quasars, Redshifts and Controversies", Interstellar Media, Berkeley.
Barnes, J.E., 1989, *Nature* 338:123.
Boroson, T., and Green, R., 1992, *Astrophys. J. Suppl. Ser.* 80:109.
Chen, K., and Halpern, J., 1989, *Astrophys. J.* 344:115.
Chen, K., Halpern, J., and Filippenko, A., 1989, *Astrophys. J.* 339:74.
Corbin, M.R., 1989, *Astrophys. J.* 357:346.
Francis, P., Hewett, P., Foltz, C., and Chaffee, F., 1992, *Astrophys. J.* 398:476.
Gaskell, M., 1982, *Astrophys. J.* 263:79.
Gaskell, M., 1983, in: "Quasars and Gravitational Lenses", Liège, p. 473.
James, D.F.V., Savedoff, M.P. and Wolf, E., 1990, *Astrophys. J.* 359:67.
James, D.F.V. and Wolf, E., 1990, *Phys. Lett. A* 146:167.
James, D.F.V. and Wolf, E., 1994, submitted to *J. Mod. Optics*.
Milgrom, M., 1983, *Astrophys. J.* 270:365.
Mulchaey, J.S., Davis, D.S., Mushotzky, R.F., and Burstein, D., 1993, *Astrophys. J.* 404:L9.
Rodrigues-Williams, L.L., and Hogan, C.J., 1993, *Bulletin Am. Astron. Soc.* 25:794.
Sanders, R., 1987, in: "New Ideas in Astronomy", F. Bertola, J.W. Sulentic and B. Madore, eds., Cambridge University Press, Cambridge, p. 279.
Sulentic, J.W., 1987, in: "New Ideas in Astronomy", F. Bertola, J.W. Sulentic and B. Madore, eds., Cambridge University Press, Cambridge, p. 123.
Sulentic, J.W., 1989, *Astrophys. J.* 343:54.
Sulentic, J.W., and Rabaça, C.R., 1994a, in: "Groups of Galaxies", O. Richter, ed., PASP Conference Series.
Sulentic, J.W., and Rabaça, C.R., 1994b, *Astrophys. J.*, July 10, in press.
Terlevich, R., Tenorio-Tagle, G., Franco, J. et al., 1992, *M.N.R.A.S.* 255:713.
Tifft, W.G., 1987, in: "New Ideas in Astronomy", F. Bertola, J.W. Sulentic and B. Madore, eds., Cambridge University Press, Cambridge, p. 173.
Weedman, D.W., 1986, "Quasar Astronomy", Cambridge University Press, Cambridge.
Wilkes, B., 1984, *M.N.R.A.S.* 207:73.
Wilson, A., and Heckman, T., 1985, in: "Astrophysics of Active Galaxies and Quasi-Stellar Objects", J.S. Miller, ed., University Science Books, Mill Valley, p. 39.
Wolf, E., 1986, *Phys. Rev. Lett.* 56:1370.
Wolf, E., 1987, *Nature* 326:363.
Wolf, E., 1989, *Phys. Rev. Lett.* 63:2220.
Zheng, W., Binette, L., and Sulentic, J.W., 1990, *Astrophys. J.* 365:115.

CONFIGURATIONS AND REDSHIFTS OF GALAXIES

M. Zabierowski

Technical University
50-370 Wrocław, Poland

I1. Arp's chains of peculiar objects and Tifft's bands of velocities form an important non-classical part of the subclustering branch in galaxy science, which (this branch) is theoretically and empirically progressive, is not degenerating.

2. Arp recognized chains of peculiar galaxies which are different from so-called 100 Mpc filaments - they are different also in a sense of procedures which preserve different sets of hypothesis which are not subject to falsification. The branch of Arp's investigations is less connected with the readjusting the long known, old-fashioned (conservative) and "obvious" claiming that galaxies are concentrated into grains as in the case of globulars, film cermets and generally into "things" known from a cermet-like heuristic which comes from the whole tradition of Earth-bound laboratory and technical experience and from the intuitive and Newtonian-Kantian mode of growth of star clumps (Einasto et al.,1989; Grabińska 1983, 1985, 1986, 1988, 1989, 1992, 1993; Grabińska and Zabierowski 1987; Rudnicki 1978; Lachiéze-Rey 1986).

3. The Tifft's bands hypothesis resembles the principle of quantum distribution of states, characterized by Zwicky as "an another statement which deals with numbers only". Is redshift proportional to k, where k=0, ± 1, ± 2,...?

4. Iwanowska (1989) was able to select five bipolar jets of galaxies in LG. She claimed explicitly that her plan of considerations favours the Arp's process but not the conventional isotropic or anisotropic collapse of the primordial diffusive matter.

5. It would be rather strange to defend that there are no Iwanowska's lines. It is very hard to think that the geometry of the five subgroups in LG is not highly special, that the geometrical considerations of Iwanowska are doubtful.

6. The residual velocity dispersion δ_{n-1} kms^{-1} for all galaxies classified by Iwanowska as members of the Arp's lines in LG is too great to warrant the stability of the lines and of the parts of these lines and is equal to 84. The dispersion ought to be several times smaller (Table I).

7. It is important to understand that - contrary to the tradition of searching the Virgo and other clusters - subclustering of the LG galaxies does not lead to any diminishing of the mean velocity dispersion δ_{n-1} which is as great for the "a", "b", "A", "B" and "C" jets as for the "all galaxies" from the scheme of Iwanowska (1989). It seems that subdivision of the LG galaxies increases the paradox "Arp's lines - great velocity dispersion" - Iwanowska did not recognized this serious discrepancy.

Frontiers of Fundamental Physics, Edited by M. Barone
and F. Selleri, Plenum Press, New York, 1994

Table I. The mean residual velocities and dispersion for five lines in LG solated by Iwanowska (1989)

name of the jet	mean residual velocity	dispersion	number of galaxies
	v (km/s)	δ_{n-1} (km/s)	n
a (Milky Way)	44	102	8
b (Milky Way)	51	78	15
A (Andromeda)	5	71	6
B (Andromeda)	72	120	4
C (Andromeda)	88	83	7
all galaxies	60	84	37

8. Table II shows that there is a certain possibility of substantial reduction of velocity dispersion δ_{n-1}. It is of great importance for understanding the progresiveness of such searches as given by Iwanowska (1989), of the new scheme in astronomy.

Table II. Five redshift states among Iwanowska's galaxies

number of state	"theoretical" per analogy with Tifft's formula	empirical	velocity dispersion	number of all galaxies in the kth state	
k	v_{Tifft} (km/s)	v (km/s)	δ_{n-1} (km/s)	n	$|v_{Tifft}-v|$
-1	-72	-58	8	5	14
0	0	1	17	11	1
1	72	65	16	9	7
2	144	141	20	9	3
3	216	219	24	3	3
instead of mixing of all states		instead of the mean			
-1, 0, ...,3		60	84		

9. Thus all galaxies (indicated by Iwanowska) are characterized by redshift regularity resembling the Tifft's formula v=k∗72 kms^{-1}, v stands for velocity, k is redshift band. Our hypothesis is easily for falsification because there are many galaxies without redshift value, thus additional galaxies could destroy the narrow value of the velocity dispersion $\delta_{n-1} \approx 10$ km s^{-1} and could destroy the idea of fragmentation of redshfift (k-regularity). All states of redshift are well separated in statistical and also individual meaning, we cannot get the other redshift state from the dispersion δ_{n-1}; δ_{n-1} cannot serve as a bridge between k-neighbours.

10. Many redshifts of LG galaxies have never been predicted by astronomers, they were observed but they contradicted the smallest common part of astronomical procedures - such galaxies had been considered as "uncertain". Now, the whole "uncertain" observational material is expected.

11. It is interesting that the Arp's branch evolved into the Tifft's science. This adaptability of the two branches each other is not a priori expected, it is a new problem per se and requires an explanation. A new objective situation appeared and it contributes substantially to the network of notions in extragalactic astronomy. It is a success.

II1. Arp did not expect short galaxy chains among normal galaxies, his famous works had been devoted to chains of peculiar galaxies. Recently Garncarek (1980, 1986, 1987, 1989) searched circular clustering, scattered distribution, pairs and chains of galaxies.

2. Circular clustering was defined by strongcompactness index of Garncarek z_4, chain shaped galaxies - by compactness index of Garncarek z_3. The indices correspond to Garncarek's G_3 and G_4 types of distribution. Garncarek had obtained the following values for z_3 and z_4 :

Table III. Garncarek's values for circular and oblong clustering of galaxies

R (Mpc)	0.2	0.4	0.6
z_3 (R)	6.5 ± 0.13	1.7 ± 0.14	1.4 ± 0.15
z_4 (R)	6.7 ± 0.13	2.2 ± 0.08	1.9 ± 0.08

It was widely accepted that 1^0 Garncarek's values of z_3 and z_4 mean that the clustering is an universal value at least in the Jagellonian Field, 2^0 the clustering drops with scale R, 3^0 circular-shaped clusters dominate over the chains. No anomaly reveals Garncarek's search.

3. However the truth is a bit different. z_4 is indeed greater than z_3 but the relative variability index $z_r = (z_4 - z_3) / z_{mean}$ is about ten! times greater for R=0.4 - 0.6 Mpc than for R=0.2 Mpc.

Table IV. The values of the relative variability index $z_r = (z_4 - z_3) / z_{mean}$

R (Mpc)	0.2		0.4		0.6
z_r (R)	0.03	$\ll$	0.26	$\approx$	0.25
z (R) = z_4 - z_3	0.2		0.5		0.4

Because z_4 and z_3 possess different sensitivities we have two parameters which cannot be directly compared. z_4 and z_3 form classes for itself. z_4 is intrinsic and z_3 is, too. Thus the relative variability index is necessary for proper understanding the phenomenon of clusters.

4. Between 0.2 Mpc and 0.4 Mpc a jump in the clustering properties is observed, an excess of chains is evident. Chains dominate the smaller scales of multiclustering. This or that individual chain can of course be an effect of projection (a standard argument against Arp's scheme in galaxy science), however the idea of accidentally projected background and foreground galaxies is absolutely useless to explain Garncarek's result.

5. It is strange that such small chains still exist, they are stable on the time scale of several percent of the standard (only normal galaxies are considered) age of galaxies. The other types of Garncarek distribution described by his non-isomorphic graphs G_1, G_2, and G_4 do not imply troubles.

6. Iwanowska advocated lines but irregular strings, rings, loops, etc. are possible too. There is no conventional explanation of chain anomaly. Why do the chains dominate small scale? Do the velocities of galaxies are less - tens and sometimes a hundred times than the conventional theory predicts?

III. The propositions given in this work contradict the standard understanding of observational cosmology. Such a direction requires a new metatheoretical plan of evaluation of astronomical results. In my opinion the most objective synthesis has been created and developed by Grabińska (1992, 1993). This new synthetic view on extragalactic astronomy is essential because the old picture on astronomy is absolutely insufficient. Astronomy calls for a new truth. Note that for advocates of the anthropic principle the lines of Iwanowska and similar configurations are rather an illusion than an astronomical fact (Paál 1980).

REFERENCES

Arp, H., 1969, *Sky and Telescope* 38, 2.

Arp, H., 1967, *Astrophysics Journal* 148, 321.

Garncarek, Z., 1980, Ph.D. Thesis, Jagellonian University.

Garncarek, Z., 1986, *Acta Cosmologica* 14, 101.

Garncarek, Z., 1987, in: "The Algorithms of Selected Problems in Cluster Analysis", Technical University at Kielce (K. Bereś and H. Bereś, Eds.), PL ISSN 0860-2077.

Garncarek, Z., 1989, *Acta Cosmologica* 16, 37.

Grabińska, T., 1983, Celebration of the Centenary (1882-1982) of the Birth of Thaddeus Banachiewicz, "Development of Techniques of Astronomical and Geodetic Calculations", Kraków, 20-21 May, Scientific Bulletins of Stanisław Staszic Academy at Kraków, *Geodezja* 1000, 84 (Jan Mietelski et al.,eds.).

Grabińska, T., 1985, *Astrophysis. Space Sci.* 115, 369.

Grabińska, T., 1986, The Hierarchical Structure of the Universe, in:"COSMOS-an Educational Challenge", ESA-SP 253, European Space Agency, Paris, p.303.

Grabińska, T., 1988, *Astrophysis. Space Sci.* 150, 75.

Grabińska, T., 1989, *Astrophysis. Space Sci.* 161, 347.

Grabińska, T., 1989, The Further Search of Voids in the Jagellonian Field , in: "From Stars to Quasars":Devoted to Prof. W. Iwanowska, S. Grudzińska and B. Krygier (eds.). Astronomical Observatory at Toruń, Toruń University of M. Kopernik, 125.

Grabińska, T., 1992, "Realism and Instrumentalism in Contemporary Physics", Wrocław Technological Institute Press.

Grabińska, T., 1993, "Theory, Model, Reality", Wrocław Technological Institute Press.

Grabińska, T., Zabierowski, M., 1987, in: "Evolution of Galaxies", Proc. 10th European Regional Astronomy Meeting of the IAU, Praha, 24-29 August, Jan Palouś (ed.), 441.

Einasto, J. et al., 1989, *Monthly Notices Roy. Astron. Soc.* 238, 155.

Iwanowska, W., 1989, in: "From Stars to Quasars", ibid., p. 159.

Lachiéze-Rey, M., 1986, Fractals and Galaxy Distribution, Prep. IRF-SAp CEN, Saclay,1

Paál, G., 1980, *Acta Geolog. Scientar. Hungar.* 23, 121 and 129.

Rudnicki, K., 1978, *Irish Astr. Journal* 13, No. 7-8, 241.

Tifft, W.G., 1973, Fine Structure within the Redshift- Magnitude Correlation, PRE 64.

Zwicky, F., 1933, *Phys. Rev.* 43, 1031.

ISOMINKOWSKIAN REPRESENTATION OF COSMOLOGICAL REDSHIFTS AND THE INTERNAL RED–BLUE–SHIFTS OF QUASARS

Ruggero Maria Santilli

The Institute for Basic Research, Box 1577, Palm Harbor, FL 34682, USA
E-mail: ibrrms@pinet.aip.org

1. STATEMENT OF THE PROBLEM

1.1: The main hypothesis. In this paper we study the cosmological quasar redshift and their internal redshifts and blueshifts via a new geometry, called *isominkowskian geometrY,* which is constructed as a covering of the Minkowskian geometry for the representation of electromagnetic waves and extended particles propagating within inhomogeneous and anisotropic physical media. The complementary *isoeuclidean* and *isoriemannian geometries* are also indicated.

Recall that: 1) homogeneity and isotropy of empty space are the geometric pillars of the conventional Doppler law; 2) quasars chromospheres are *inhomogeneous* (because of the local variation of the density) and *anisotropic* (because of the intrinsic angular momentum which creates a preferred direction in the physical medium, the underlying vacuum remaining homogeneous and isotropic); and 3) light is emitted in the interior of the quasars and propagates in their large chromospheres (of the order of millions of radial km) before reaching empty space.

The isominkowskian geometry implies a generalization of the Doppler law, called *isodoppler law,* which predicts: 1) a *frequency–dependent redshift* for inhomogeneous and anisotropic media of low density such as atmospheres and chromospheres (in which case light *loses* energy to the medium); 2) a *frequency–dependent blueshift* for inhomogeneous and anisotropic media of very high densities, such as those in the core of the quasars (in which case light *acquires* energy from the medium); and 3) *lack* of any shift for light propagating in homogeneous and isotropic media such as water.

Our main hypothesis is that the difference between the cosmological redshift of quasars over that of the associated galaxies is entirely reducible to the redshift of light while traveling in the quasar chromospheres *before* reaching empty space, thus permitting the quasars to be at rest with respect to the associated galaxies (or being expelled at small, thus ignorable speeds), while the internal quasar red/blue/shifts is due to the particular frequency dependence of the redshift itself. According to this hypothesis, the quasar cosmological redshifts and their internal red/blue/shifts are due to interior physical characteristics of the quasars and, more specifically, to the inhomogeneity and anisotropy of their chromospheres, i.e., to the *departures from the geometry of empty space.*

1.2: Experimental verifications. In this paper we show that the isominkowskian geometry provides a numerical representation of: I) the data by Arp [1] on the cosmological redshift of quasars, thus reducing them at rest with respect to the associated galaxy, as confirmed by a number of gamma spectroscopic data establishing the physical connection of the quasars with the associated galaxy; II) the data by Sulentic and others (see [2] and quoted literature) on the quasar internal red/blue/shifts; and III) the redshift of Fraunhofer lines of light from the inhomogeneous and anisotropic chromosphere of our Sun (see Marmet's studies [3] and vast literature therein).

Moreover, the isominkowskian geometry identifies intriguing interconnections between the seemingly different data I, II, III, and permits the prediction of novel, experimentally verifiable effects, such as the prediction that the dominance of red of Sun light at sunset is partially (but not entirely) an isoredshift due to the inhomogeneity and anisotropy of our atmosphere. This prediction is supported by the fact that the sky at the zenith is not red, in which case the increase in redness at the horizon would be completely explainable with conventional means (scattering, absorption, etc.). Instead, the dominance of *blue* at the zenith and of *red* at the horizon supports the isominkowskian geometry.

In this paper we also present of a number of experimental verifications of the isominkowskian geometry in particle physics which are indirectly, yet significantly related to the quasar red/blue/shifts, such as the anomalous behaviour of the meanlife of unstable hadrons with speed whose structure is fully equivalent to the isodoppler law, the data on the Bose–Einstein correlation for the UA1 experiments at CERN, the anomalous total magnetic moment of few–body nuclei, and others.

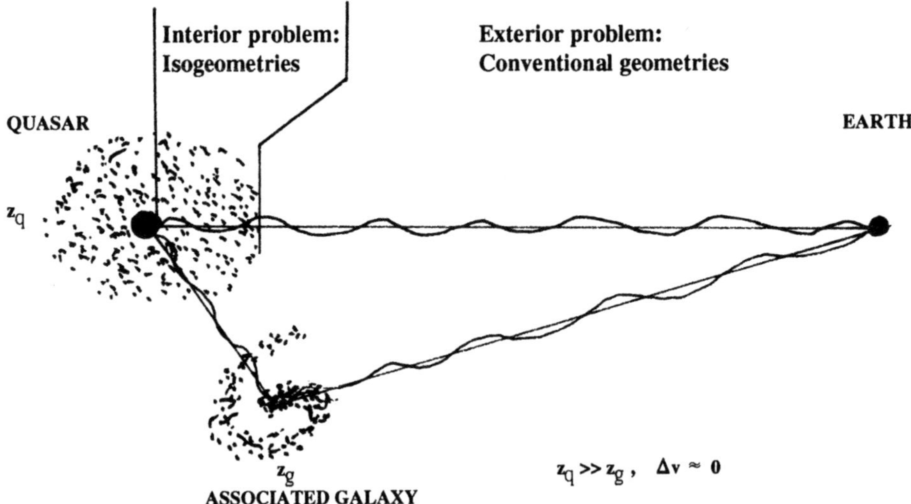

Fig. 1. A schematic view on the main hypothesis of this paper (Sect. 1.1) according to the original proposal [10].

Above all, this paper is intended to stimulate the experimental resolution of the now vexing problem of the quasar shifts via novel direct experiments, such as measure the isoredshift predicted for light from distant stars passing through the Sun's chromosphere, or a planetary atmosphere, or measure the predicted isoredshift component of the Sun's light at sunset by following a sufficient number of Fraunhofer lines from the zenith to the horizon.

All these measures, if confirmed, would provide final evidence that a portion (but not necessarily all) of the cosmological redshift of quasars is of interior geometric character due to the departures from the homogeneity and isotropy of space caused by the inhomogeneity and anisotropy in their environment. The separate problem of the cosmological redshift of galaxies is only briefly considered.

1.3: Connection with alternative theories. Numerous *alternative theories* (i.e., of non–Doppler character) have been submitted (see [1–3] and review [4]) such as Arp's theory of the *creation of matter* in the quasars, Marmet theory based on photon scattering, and others. These theories are capable of representing the cosmological quasar redshift, although their capability to represent the internal red/blue/shift and other recent evidence is under study.

The continuation of the study of these alternative interpretations is encouraged here because each one adds valuable information to the other and, in the final analysis, all quantitative interpretations may well result to be deeply interconnected.

For instance, Arp's theory emerges from our studies in a new light because the creation of matter may ultimately result to be an interplay between matter and antimatter which is prohibited in conventional geometries, but permitted in our isogeometries because of an inner conjugation indicated later on. Similarly, Marmet's representation of the data on the Sun's chromosphere [3] may essentially result to be an *operator* counterpart of our *classical* studies. We regret the inability to study these interconnections in detail at this time for lack of space.

1.3: A historical distinction. An aspect of fundamental relevance for the studies of this paper is the historical distinction between the *exterior dynamical problem* (i.e., electromagnetic waves and point-like test bodies moving in the homogeneous and isotropic vacuum), and the *interior dynamical problem* (i.e., electromagnetic waves and extended test bodies moving within inhomogeneous and anisotropic physical media). This distinction was introduced by the founders of analytic dynamics, and kept up to the first part of this century (see, e.g., Schwartzschild's *two* papers [4], the first famous one on the *exterior* problem and the second little known paper on the *interior* problem, or early treatises in gravitation, e.g., ref.s [6], the first with a preface by Einstein).

Regrettably, the above distinction was progressively relaxed, up to the current condition of virtual complete silence in the specialized literature. This is due to the belief that the interior problem can be reduced to to the exterior form, which is certainly admissible as an *approximation* (see Schwartzchild's [5] insistence on the approximate character of his solution for the *interior* problem). The point is that such a reduction cannot be exact, as established by the so-called *No-Reduction Theorems* [7] which prove that an interior system (such as a satellite during re-entry in Earth's atmosphere with a monotonically decaying angular momentum) simply cannot be consistently reduced to a finite collection of ideal elementary particles each in a stable orbit with conserved angular momentum.

With the clear understanding that the Minkowskian and Riemannian geometries are exactly valid in empty space, the above theorems establish their *inapplicability* (rather than "violation") for interior conditions on numerous, independent, topological, analytic, geometric and other grounds. For instance, interior systems are *nonlinear in the velocities* (a missile in atmosphere has a drag force nowadays proportional to the tenth power of the speed and more), *nonlocal-integral* (because the shape of the test body directly affects its trajectory, thus calling for integral terms), and *nonpotential* (because the notion of potential has no mathematical or physical meaning for contact interior forces and the systems are *variationally nonselfadjoint* [7]). The inapplicability of the Minkowskian and Riemannian geometries for these interior conditions is so evident to require no additional comment. The only scientifically meaningful issue is the construction of appropriate covering geometries specifically conceived for interior conditions.

1.4: Insufficiencies of the conventional interpretation. The conventional interpretation of quasars redshifts is based on the celebrated *Doppler law*

$$\omega = \omega_0 (1 - v \cos \alpha / c_0) \gamma, \quad \gamma = (1 - v^2 / c_0^2)^{-\frac{1}{2}}, \qquad (1.1)$$

where α is the angle between the direction of light and of motion of the source and c_0 is the speed of light in vacuum. The *redshift* $\Delta\omega = \omega - \omega_0 < 0$, is therefore reduced to the computation of the speed v of the quasars with respect to Earth (see, e.g., ref.s [6]). Note that such interpretation is: 1) purely classical, 2) relativistic without gravitational corrections, and 3) based on the assumption that light is emitted by the quasars and propagates immediately in vacuum without any effect when passing through the chromospheres.

The theoretical insufficiencies of law (1.1) for interior conditions are beyond credible doubts. The homogeneity and isotropy of empty space are known to be the geometric pillars for the derivation of the law. Its inapplicability for light propagating within inhomogeneous and anisotropic atmospheres is then unquestionable.

Astrophysical insufficiencies of law (1.1) for the interpretation of the data on quasars redshift began to emerge with the discovery of the quasars themselves, and then progressively increased in time [1,2,3]. Among the most visible inconsistencies we recall [loc. cit.] galaxies younger than their stars, galaxies older than the life of the universe, discrete variations of redshift, quasars evolving into galaxies, speeds in excess of those permitted by Einsteinian theories, etc.

These and other inconsistencies have now reached such dimension and diversification to call for a revision of the fundamental *geometries* used in the description of the universe.

1.5: Bibliographical notes. The isogeometries were constructed by this author to satisfy the following conditions 1) have a structure which is nonlinear (in coordinates, velocities and any needed additional quantity), nonlocal-integral (in all needed variables), nonpotential, inhomogeneous and anisotropic; 2) preserve the axioms of the original geometry at the abstract level so as to permit a geometric unification of exterior and interior problems; and 3) be coverings of conventional geometries, thus admitting the latter as particular cases when motion returns to be in vacuum.

The methods for the construction of the isogeometries were proposed by the author back in 1978 [8] when at the Department of Mathematics of Harvard University under DOE support. They are called *isotopies* from Greek terms meaning "preserving the topology", and interpreted as axiom-preserving (the broader *genotopies* [8] are reviewed for brevity, see in this respect the contribution by Jannussis [24] in these proceedings). These methods essentially permit nonlinear–nonlocal–nonhamiltonian, but axiom-preserving generalizations (called *liftings*) of any given mathematical or physical structure, as outlined in Sect. 2.

Isotopies were first applied to the lifting of classical Hamiltonian mechanics and Lie's theory into covering theories [7,8]. The first isotopic lifting of the Minkowskian geometry was proposed is ref. [9] of 1982. The isotopic lifting of the Riemannian geometry was first proposed in ref. [10] of 1988, jointly with the proposal to elaborate Arp's data [1] (Fig. 1.1). Such elaboration was subsequently conducted by Mignani in ref. [11]. A detailed study of the isogeometries first appeared in ref.s [12]. Ref.s [13] provide a classical presentation of the isogeometries with ref.s [14] giving the operator counterpart. Mathematical reviews are available in ref.s [15–17], an independent physical review is available in ref. [18]. A review of the isogeometries is available in ref. [19]. Preliminary, yet significant verifications are provided in ref.s [25–36]. A comprehensive presentation of the content of this paper is provided in ref. [37] for flat and in ref. [38] for curved isogeometries.

2: BASIC NOTIONS ON ISOTOPIES

2.1: Isotopies of the unit. The fundamental isotopies are the liftings of the n-dimensional unit I = diag. (1, 1,...., 1) of contemporary geometries into an n×n-dimensional matrix $\hat{1}$ whose elements have the most general possible, nonlinear and nonlocal dependence on time t, coordinates x, their derivatives of arbitrary order $\dot{x}$, $\ddot{x}$, ..., and any needed additional interior quantity, such as the frequency ω of the wave, the local density μ, the local temperature τ, the local index of refraction n, etc. [7,8]

$$I = \text{diag.} (1, 1, ..., 1) \quad \rightarrow \quad \hat{1} = \hat{1}(t, x, \dot{x}, \ddot{x}, \omega, \mu, \tau, n, ...) . \qquad (2.1)$$

under the condition (necessary for an isotopy) of preserving the original axioms of I. The above liftings have been classified into five topologically significant classes called *Kadeisvili's Classes* I–V [19,20]. In this paper we shall only consider liftings of Kadeisvili's Class I (with generalized units $\hat{1}$ that are smooth, bounded, nowhere degenerate, Hermitean and positive-definite), which characterize isotopies properly speaking), and of Class II (the same as Class I but with negative-definite isounits). For brevity we shall limit ourselves to brief comments on the remaining Class III (the union of Class I and II), IV (holding for singular isounits representing gravitational collapse) and V (with arbitrary, e.g., discrete, isounits).

The isotopies of the unit demand, for consistency, a corresponding, compatible lifting of all associative products AB among generic quantities A, B, into the *isoproduct* [8]

$$A B \quad \rightarrow \quad A * B = A T B, \quad T = \text{fixed} \; , I A = A I \equiv A \rightarrow I * A = A * I \equiv A, \quad \hat{1} = T^{-1}, \quad (2.2)$$

whose isotopic character is ensured by the preservation of associativity, A(BC) = (AB)C → A*(B*C) = (A*B)*C. Under the above conditions, $\hat{1} = T^{-1}$ is called the *isounit* and T is called the *isotopic element*. Note the necessity, e.g., in number theory, of lifting the product whenever the (multiplicative) unit is lifted and viceversa.

2.2: Isotopies of fields. The isotopies of the unit $I \rightarrow \hat{1}$ and of the product $AB \rightarrow A*B$ demand the lifting of conventional fields F(a,+,×) of real numbers R, complex number C and quaternions Q with generic elements a, conventional sum + and product a×b : = ab, into the so-called *isofields* [12]

$$F(a,+,*) \quad \rightarrow \quad \hat{F}(\hat{a},+,*), \quad \hat{a} = a\hat{1} , \quad \hat{a} * \hat{b} = \hat{a} T \hat{b} = (a\,b)\hat{1}, \; \hat{1} = T^{-1} . \qquad (2.3)$$

with elements $\hat{a} = a\hat{1}$ called *isonumbers*, conventional sum + and isoproduct (2.2), under the condition (again necessary for an isotopy) of preserving the original axioms of F. All operations in F must be generalized for $\hat{F}$. We have *isosquares* $\hat{a}^2 = \hat{a}*\hat{a} = \hat{a}T\hat{a} = a^2\hat{1}$, *isoquotient* $\hat{a}/\hat{b} = (a/b)\hat{1}$, *isosquare roots* $\hat{a}^{\frac{1}{2}} = a^{\frac{1}{2}}\hat{1}$, etc. (see [12,14] for detailed studies).

The above liftings are nontrivial inasmuch as they imply the inapplicability under isotopies of the entire mathematical formulations of conventional geometries. As an illustration, statements such as "two multiplied by two equals four" are generally incorrect for isogeometries. In fact, for $\hat{1}$

= 3, "two multiplied by two equals twelve", with the understanding that the very notion of integer number is generally lost in favor of an integro-differential notion, e.g., $\hat{2}$ = 2exp{N∫dxψ†(x)φ(x)} as for the Cooper pair of electrons in superconductivity with wavefunctions ψ and φ (see Sect. 5.5).

2.3: Isotopy of metric spaces. Liftings I → $\hat{1}$, AB ⇒ A*B and F → $\hat{F}$ then require the isotopies of vector, metric and pseudo-metric spaces, evidently because they depend on the field in which they are defined. In fact, real metric or pseudo-metric spaces S(x,g,R) with Hermitean metric g over R must be subjected to the liftings into the so-called *isospaces* (first introduced in [9])

$$S(x,g,R) \Rightarrow \hat{S}(x,\hat{g},\hat{R}), \quad \hat{g} = Tg, \quad \hat{1} = T^{-1}, \quad x^2 = (x^t \hat{g} x)\hat{1} \in \hat{R}. \tag{2.4}$$

under the condition, again, of preserving the original axioms of S(x,g,R). In particular, *the basis of a metric (or, more generally, vector) space is preserved under isotopies* [12], thus including the preservation of the basis of a Lie algebra. This results in nonlinear and nonlocal (in x, $\dot{x}$, $\ddot{x}$, ...) generalization of the original space, yet such that $\hat{S}(x,\hat{g},\hat{R})$ ≈ S(x,g,R).

We have indicated earlier the loss of conventional numbers under isotopies. When passing to isospaces, one should keep in mind the loss of conventional functional analysis into a covering formulation called *functional isoanalysis* [20]. In fact, the very notion of angle is lost under isotopies (see next section), thus implying the consequential loss of trigonometry, Legendre polynomials, etc. in favor of suitable, unique (and intriguing) covering notions [14].

2.3: Lie-isotopic theory. The preceding liftings demand a corresponding compatible lifting of all branches of Lie's theory into the so-called *Lie-isotopic theory* first submitted in [8] and then studied in ref.s [12–20]). We can here mention only the lifting of the envelope ξ(g) of a Lie algebra g and related exponentiation in terms of the original (ordered) basis {X_i} of g

$$\xi: \quad \hat{1}, \quad X_i * X_j \ (i \leq j), \quad X_i * X_j * X_k, \quad (i \leq j \leq k), \dots\dots, \quad i, j, k = 1, 2, ..., n, \tag{2.5a}$$

$$e_\xi^{i\,\hat{w}*X} = \hat{1} + (i\,\hat{w} * X)/1! + (i\,\hat{w}*X)*(i\,\hat{w}*X)/2! + .. = \{e^{iXTw}\}\hat{1} = \hat{1}\{e^{iwTX}\}, \tag{2.5b}$$

the lifting of Lie algebra g ≈ [ξ(g)]⁻ with familiar Lie theorem, such as the 2-nd theorem [X_i , X_j]$_\xi$ = $X_i X_j - X_j X$ = $C_{ij}{}^k X_k$, into the *Lie-isotopic algebras* $\hat{g}$ ≈ [$\hat{\xi}$(g)]⁻ with *Lie-isotopic theorems* [8], e.g.,

$$\hat{g}: [X_i, X_j]_{\hat{\xi}} = X_i * X_j - X_j * X_i = X_i T X_j - X_j T X_i = \hat{C}_{ij}{}^k(t, x, \dot{x}, \ddot{x}, \omega, \mu, \tau, n,...)* X_k, \tag{2.6}$$

where the $\hat{C}$'s, called *structure isofunctions*, are restricted by the *Third Isotopic Theorem* [8,16]; the lifting of transformations and related (connected) Lie groups G into the *Lie-isotopic transformation groups* [8]

$$\hat{G}: x' = \hat{U}(\hat{w}) * x, \quad \hat{U}(w) = \prod_k e_\xi^{i\,X_k * \hat{w}_k} = \hat{1}\{\prod_k e^{iw_k TX_k}\} = \{\prod_k e^{iX_k Tw_k}\}\hat{1}, \tag{2.7b}$$

$$\hat{U}(0) = \hat{1}, \quad \hat{U}(\hat{w}) * \hat{U}(\hat{w}') = \hat{U}(\hat{w}') * \hat{U}(\hat{w}) = \hat{U}(\hat{w} + \hat{w}'), \quad \hat{U}(\hat{w}) * \hat{U}(-\hat{w}) = \hat{1}, \tag{2.7a}$$

$$\{e_\xi^{X_1}\}*\{e_\xi^{X_2}\} = e_\xi^{X_3}, X_3 = X_1 + X_2 + [X_1, X_2]_{\hat{\xi}}/2 + [(X_1 - X_2), [X_1, X_2]_{\hat{\xi}}]_{\hat{\xi}}/12 + .. \tag{2.7c}$$

the lifting of the conventional representation theory into the *isorepresentation theory of Lie-isotopic algebras and groups* (which is structurally nonlinear, nonlocal and noncanonical); and other liftings [13,14].

Note the preservation of the Lie algebra axioms by the isotopic product [A, B]$_\xi$ = ATB − BTA. Note also the nontriviality of the isotopic theory from the appearance of the nonlinear-integral quantity T in its exponentiation (2.7a). We should also note that, even though structurally nonlinear, nonlocal and noncanonical, the Lie-isotopic theory verifies the axioms of linearity, locality and canonicity at the isotopic level and, for this reason, it is called *isolinear, isolocal and isocanonical*. Note finally that all nonlinear-nonlocal-noncanonical theories *always* admit an *identical* isolinear-isolocal-isocanonical reformulation with evident advantages.

2.5: Isosymmetries. The Lie-isotopic transformation groups are turned into symmetries of isospaces, called *isosymmetries*, via the following:

Theorem 2.1 [21]: *Let G be an N-dimensional Lie group of isometries of an m-dimensional, metric or pseudo-metric, and real or complex space* S(x,g,F) , F = R *or* C,

$$G: x' = A(w) x, \quad (x'-y')^t A^t g A (x'-y') \equiv (x-y)^t g (x-y), \quad A^t g A = A g A^t = g, \tag{2.8}$$

and T is derived from the deformed metric $\hat{g} = Tg$ (see the example in the next section). Note also that there is no need to verify isoinvariance (2.9) because ensured by the original invariance (2.8).

It is also easy to prove that $\hat{G} \approx G$ for all Class I isotopies (but not so for other Classes for which in general $\hat{g} \approx [\xi(g)]^- \not\approx g$). This property identifies one of the primary applications of isosymmetries, the reconstruction of exact symmetries when believed to be conventionally broken. In fact, in ref.s [21] one can see the reconstruction of the exact rotational *symmetry* at the isotopic level $\hat{O}(3) \approx O(3)$ for all ellipsoidical deformations of the sphere. In ref. [9] one can see the reconstruction of the exact Lorentz *symmetry* at the isotopic level $\hat{O}(3.1) \approx O(3.1)$ for all signature preserving $(T > 0)$ deformations of the Minkowski metric $\hat{\eta} = T\eta$. See ref.s [13,14] for the reconstruction of additional exact symmetries.

2.6: Inequivalence of the Lie and Lie–isotopic theories. Despite the isomorphism $\hat{G} \approx G$, Lie and Lie–isotopic symmetries are inequivalent on numerous counts, such as: 1) G is customarily linear–local–canonical, while $\hat{G}$ is nonlinear–nonlocal–noncanonical; 2) the mathematical structures underlying $\hat{G}$ and G (fields, spaces, etc.) are structurally different; 3) $\hat{G}$ can be derived from G via *nonunitary* transformations under which

$$UU^\dagger = \hat{1} \neq I, \quad U(AB - BA)U^\dagger = A'TB' - B'TA', \quad T = (UU^\dagger)^{-1} = T^\dagger, \quad A' = UAU^\dagger, \quad B' = UBU^\dagger. \qquad (2.10)$$

The above inequivalence also emerges in the *isorepresentation theory* [14] , e.g., because the spectra of eigenvalues of the same operator are different in the two theories (due to the necessary isotopy of eigenvalue equations $H|b> = E^\circ|b> \rightarrow H*|\hat{b}> = HT|\hat{b}> = \hat{E}*|\hat{b}> \equiv E|\hat{b}>, \ E \neq E^\circ$). Also, weights, Cartan tensors, etc. acquire a nonlinear–nonlocal–noncanonical dependence on the base manifold, etc.

2.7: Isodual conjugations and antimatter. The generalization of the unit permits the identification of a new antiautomorphic conjugation $\hat{1} \rightarrow \hat{1}^d = -\hat{1}$ introduced in [21] under the name of *isoduality*. This map implies the existence of isodual images of all quantities of Class I (fields, spaces, algebras, groups, etc.) into corresponding forms of Class II.

In particular, any positive number m or isonumber $\hat{m} = m\hat{1}$ is mapped into the *isodual number* $m^d = m1^d = -m$ or *isodual isonumber* $\hat{m}^d = m\hat{1}^d = -\hat{m}$, while the *isodual isonorm* is given by $\lceil \hat{m} \rceil^d = (m \, T \, m)^{\frac{1}{2}} \hat{1}^d = -\lceil \hat{m} \rceil$ and it is *negative-definite*. The most intriguing properties of isodual spaces and isodual symmetries is that they describe particles with *negative-definite energy moving backward in time*.

Recall that antiparticles originated from the negative-energy solutions of conventional relativistic equations, although such solutions were abandoned because the behaviour of the systems was unphysical in our space–time. Isodual spaces and isodual symmetries provide a fundamentally novel approach because the interpretation of the same negative–energy solution in isodual spaces is now fully physical [13,14].

The isogeometries therefore permit a novel cosmological conception of the structure of the universe in which, for the limit case of an equal distribution of matter and antimatter, all total quantities, such as total energy, total time, etc., are identically null (see ref. [38] for brevity).

3: ISOMINKOWSKIAN GEOMETRY

3.1: Isominkowskian spaces. Consider an electromagnetic wave propagating first in empty space (exterior relativistic problem), then throughout our atmosphere (interior relativistic problem). As well known, *the Minkowski space*

Then, the infinitely possible isotopes $\hat{G}$ of G characterized by the same generators and parameters of G and new isounits $\hat{1}$ (isotopic elements T), leave invariant the isocomposition on the isospaces $S(x,\hat{g},\hat{F})$, $\hat{g} = Tg$, $\hat{1} = T^{-1}$,

$$\hat{G}: \quad x' = \hat{A}(w) * x, \ (x'-y')^\dagger * \hat{A}^\dagger \hat{g} \hat{A} * (x'-y') = (x-y)^\dagger \hat{g} (x-y), \ \hat{A}^\dagger \hat{g} \hat{A} = \hat{A} \hat{g} \hat{A}^\dagger = \hat{1} \hat{g} \hat{1}, \qquad (2.9)$$

The above results yield the "direct universality" of the Lie–isotopic symmetries, i.e., their capability of providing the invariance of all infinitely possible deformations $\hat{g} = Tg$ of the original metric g (universality), directly in the x–frame of the experimenter (direct universality). Note also the simplicity of the explicit construction of the desired isotransformations via rule (2.7) where w are the conventional parameters, X are the conventional generators in their adjoint representation

$$M(x,\eta,R): \quad x = (\mathbf{x}, x^4), \quad x^4 = c_0 t, \quad x^2 = x^\mu \eta_{\mu\nu} x^\nu = x^1 x^1 + x^2 x^2 + x^3 x^3 - x^4 x^4, \quad \eta = \text{diag.}\ (1, 1, 1, -1) \quad (3.1)$$

geometrizes the homogeneity and isotropy of empty space and, as such, it is exactly valid for exterior conditions.

The *isominkowski space* (first submitted in [9]) is intended to geometrize the inhomogeneity and anisotropy of interior conditions. It is constructed via two simultaneous liftings, that of the Minkowski metric η into the *isometric* $\hat\eta = T\eta$ of Class I and the joint lifting of the unit of $M(x,\eta,R)$, $I = \text{diag.}\ (1, 1, 1, 1)$, into the $4{\times}4$–dimensional isounit $\hat1 = T^{-1}$, and we shall write

$$\hat M(x,\hat\eta,\hat R): \quad \hat\eta = T(x, \dot x, \ddot x, \mu, \tau, n, \ldots)\,\eta, \quad \hat1 = T^{-1} > 0, \quad x^{\hat2} = (x^\mu \hat\eta_{\mu\nu} x^\nu)\hat1 \in \hat R. \quad (3.2)$$

Note that isospaces $\hat M(x,\hat\eta,\hat R)$ have the most general possible nonlinear–nonlocal–noncanonical structure because the functional dependence of $\hat\eta$ remains unrestricted. The isometric can always be (although not necessarily) diagonalized for Class I, resulting in isoseparation of the type

$$x^{\hat2} = x^1 b_1^2(x, \dot x, \ldots) x^1 + x^2 b_2^2(x, \dot x, \ldots) x^2 + x^3 b_3^2(x, \dot x, \ldots) x^3 - x^4 b_4^2(x, \dot x, \ldots) x^4, \quad b_\mu > 0, \quad (3.3)$$

Despite evident structural differences, the joint liftings $\eta \to \hat\eta = T\eta$ and $I \to \hat1 = T^{-1}$ imply that the *isominkowskian space is locally isomorphic to Minkowskian space*, $\hat M(x,\hat\eta,\hat R) \approx M(x,\eta,R)$ [9,12,14]. Owing to the positive–definiteness of the isotopic element T, it is easy to see that $\hat M(x,\hat\eta,\hat R)$ and $M(x,\eta,R)$ coincide at the abstract level. Exterior and interior descriptions are therefore *different realizations* of the same abstract geometric axioms. This is the central geometric property which is assumed for the description of both, exterior and interior relativistic problems, and which carries intriguing consequences, as we shall see.

3.2: Characteristic quantities of physical media. The b–quantities (at times also expressed in the form $b_\mu = 1/n_\mu$) are called the *characteristic quantities* of the medium considered. The inhomogeneity of the medium can be represented via an explicit dependence of the b's on the local density, and the anisotropy can be represented via different values among the b's, the factorization of a preferred direction of the medium, and other means.

When the local behaviour is needed at one given interior point, one needs the full nonlinear–nonlocal dependence of the b's. This is illustrated, e.g., by the local speed of light at one given point when passing though our atmosphere which is given by $c = c_0 b_4 = c_0/n_4$, where $n_4 = b_4^{-1}$ (the *local index of refraction*) has a rather complex functional dependence on local quantities.

When the global behaviour throughout a given physical medium is requested, the characteristic quantities can be averaged into constants, $b^\circ_\mu = \text{Aver.}\ (b_\mu)$, or $n^\circ_\mu = \text{Aver.}\ (n_\mu)$, $\mu = 1, 2, 3, 4$. This is evidently the case for the *average speed of light* throughout our atmosphere $c = c_0/n^\circ_4$, in which case n°_4 is the *average index of refraction*. Note that $b_\mu \equiv b^\circ_\mu \equiv 1$ in vacuum.

A first intuitive understanding of the isominkowski spaces can be reached by noting that the characteristic functions $b_\mu = 1/n_\mu$ essentially extend the local index of refraction $1/n_4$ to all space–time components. Equivalently, by recalling that physical media are generally opaque to light, *the isotopies $M(x,\eta,R) \to \hat M(x,\hat\eta,\hat R)$ essentially extend to all physical media the geometric structure of light in vacuum*. In this sense, the characteristic constant b°_4 geometrizes the density of a given medium, while the constants b°_k geometrize the internal nonlinear–nonlocal effects.

It is evident that different physical media necessarily require different isounits $\hat1$. This occurrence is similar to the need of infinitely possible Riemannian spaces in general relativity in order to represent the infinitely possible astrophysical masses. The point here is that each mass admits infinitely possible isounits, trivially, because each mass can be realized in infinitely possible different densities, sizes, chemical compositions, etc.

3.3: Isominkowskian geometry. It is the geometry of isospaces $\hat M(x,\hat\eta,\hat R)$ and possesses novel characteristics as compared to the conventional geometry. Their understanding requires the knowledge of the inapplicability mentioned in Sect. 2 of the notion of angles, trigonometry and functional analysis at large in favor of covering isotopic notions.

To study the main characteristics, let us consider first the *isoeuclidean geometry* which is evidently the space–component of the isominkowskian geometry. Consider the *isoeuclidean subspace* $\hat E(x,\hat\delta,\hat R)$ in the 1–2 plane with diagonal isometric and separation

$$\hat{E}(x,\delta,\hat{R}): \quad x^2 = x^1 b_1^2(x, \dot{x}, \ddot{x}, \ldots)x^1 + x^2 b_2^2(x, \dot{x}, \ddot{x}, \ldots)x^2 = \text{inv.} \tag{3.4}$$

As one can see, this space is curved in the most general possible form, that is, with curvature dependent on local coordinates x, velocities $\dot{x}$, accelerations $\ddot{x}$, etc. (see next section). The loss of the conventional angles then follows from the evident loss of intersecting straight lines.

At this point, the isotopies play a central constructive role. Recall that the original space is flat. Its image under isotopy is then *isoflat*. Similarly, the images of straight lines are *isostraight* i.e., verify the axioms of straight lines in isospace. This implies the possibility of reconstructing angles under isotopies which is not possible for Riemann. The use of the isotopies of the group of rotation [21] permits the identification of the unique isotopic image $\hat{a}$ of a conventional angle a in $\hat{E}(x,\delta,\hat{R})$ given by $\hat{a} = ab_1b_2$. This permits the construction of the isotopies of conventional trigonometry, here called *isotrigonometry,* which is based on the following isofunctions and related properties

$$\text{isosin } \hat{a} = b_2^{-1} \sin(a\, b_1\, b_2), \quad \text{isocos } \hat{a} = b_1^{-1} \cos(a\, b_1\, b_2), \tag{3.5a}$$

$$b_1^2 \text{ isocos}^2 \hat{a} + b_2^2 \text{ isosin}^2 \hat{a} = \cos^2 \hat{a} + \sin^2 \hat{a} = 1. \tag{3.5b}$$

Note the deformation of the argument $a \to \hat{a}$ as well as of the magnitude of trigonometric functions $1 \to b_k^{-1}$ which are intriguing for certain (e.g., nuclear) deformations of potential wells and wavefunctions [14]. The rest of the isotrigonometry can then be constructed accordingly. The extension to the three–dimensional isoeuclidean case is consequential and it is omitted here for brevity [14].

We consider now the hyperbolic isoplane 3–4 with isoinvariant

$$\hat{M}(x,\hat{\eta},\hat{R}): \quad x^2 = x^3 b_3^2(x, \dot{x}, \ddot{x}, \ldots)x^3 - x^4 b_4^2(x, \dot{x}, \ddot{x}, \ldots)x^4 = \text{inv}. \tag{3.6}$$

The isotopic image $\hat{v}$ of a hyperbolic angle (speed) v is then given by $\hat{v} = vb_3b_4$, as provable via the use of the isorepresentations of $\hat{O}(1.1)$ [13,14], with corresponding *isohyperbolic functions* and related properties

$$\text{isosinh } \hat{v} = b_4^{-1} \sinh(v\, b_3\, b_4), \quad \text{isocosh } \hat{v} = b_3^{-1} \cosh(v\, b_3\, b_4), \tag{3.7a}$$

$$b_3^2 \text{ isocosh}^2 \hat{v} - b_4^2 \text{ isosinh}^2 \hat{v} = \cosh^2 \hat{v} - \sinh^2 \hat{v} = 1. \tag{3.7b}$$

We are now equipped to indicate a most important feature of the isominkowskian geometry, the reconstruction at the isotopic level of exact straight lines, perfect circles and conventional light cones. The loss of the notion of straight line and its reconstruction under isotopy has been indicated earlier. The preservation of perfect circles can be seen as follows. Recall that, by conception, isotopies of Class I map the circle into the infinite families of ellipses (3.4) with semiaxes b_k^2. But the unit is jointly lifted from $I = \text{diag. }(1, 1)$ to $\hat{1} = \text{diag. }(b_1^{-2}, b_2^{-2})$. We then have the deformation of each semiaxis $1 \to b_k^2$ with the joint deformation of the unit $1 \to b_k^{-2}$. The original circle therefore remains a perfect circle in isospace, while the ellipses emerge only when the figure are projected in our space (see ref.s [13,14] for details).

We now outline the preservation of the light cone under isotopy. Let us first recall that, in the physical reality, the speed of light is not a "universal constant", but a locally varying quantity with a rather complex functional dependence on density, index of refraction, etc. As a result, the "light cone" in interior problems is not a "cone", but a rather complex hypersurfaces. The understanding of the isominkowskian geometry requires the knowledge that the "deformed cone" of the physical reality is mapped into a perfect cone in isospace, called *light isocone,* and the locally variable speed is mapped precisely into the original, constant speed of light in vacuum c_0. Consider the isolight cone $x^2 = 0$ in the 3–4 plane, Eq. (3.7). Then, the isotrigonometry yields $\Delta x = D\, b_4\, \sin\hat{a}$, $\Delta t = D\, b_3 \sin \hat{a}$, and

$$\Delta x / \Delta t = D\, b_4 \sin \hat{a} / D\, b_3 \sin \hat{a} = (b_4 / b_3)\, c_0, \quad \hat{a} = a\, b_3\, b_4, \tag{3.8}$$

from which we recover the conventional expression in empty space $\text{tang } \hat{a} = c_0 = \text{const.}$ This occurrence is an expression of the overall unity of physical and mathematical thought achieved by isotopic technique because they allow the use of the *same* light cone for motion in vacuum with constant speed c_0 and motion in interior conditions with variable speed $c = c_0 b_4$.

3.4: Isolorentz and isopoincare' symmetries. Necessary complements of the isominkowskian geometry are given by the isotopies $\hat{O}(3.1)$ and $\hat{P}(3.1)$ of the Lorentz $O(3.1)$ and Poincaré $P(3.1)$ symmetries, respectively. They were constructed for the first time in ref. [9] via the Lie-isotopic theory and then studied in details in monographs [13,14] to which we must refer for brevity. We can only recall the *isolorentz transformations* here presented for $\hat{\eta} = $ diag. $(g_{11}, g_{22}, g_{33}, -g_{444})$ with conventional functions for simplicity (rather than isofunctions)

$$x'^1 = x^1, \qquad x'^2 = x^2, \tag{3.9a}$$

$$x'^3 = x^3 \cosh[\, v\,(\hat{g}_{33}\hat{g}_{44})^{\frac{1}{2}}\,] - x^4 g_{44}(g_{33}\,g_{44})^{-\frac{1}{2}} \sinh[v\,(g_{33}\,g_{44})^{\frac{1}{2}}] = \hat{\gamma}(x^3 - \beta x^4), \tag{3.9b}$$

$$x'^4 = - x^3 g_{33}(g_{33}\,g_{44})^{-\frac{1}{2}} \sinh[v\,(g_{33}\,g_{44})^{\frac{1}{2}}] + x^4 \cosh[\, v\,(\hat{g}_{33}\hat{g}_{44})^{\frac{1}{2}}\,] = \hat{\gamma}(x^4 - \beta x^3), \tag{3.9c}$$

$$\beta^{2\wedge} = v^k g_{kk} v^k / c_0 g_{44} c_0, \qquad \hat{\gamma} = |\,1 - \beta^2\,|^{-\frac{1}{2}}, \tag{3.9d}$$

which are easily constructed via rule (2.7) with $w = v$, X given by the conventional Lorentz generators in adjoint representation, and $T = $ diag. $(g_{11}, g_{22}, g_{33}, g_{444}) > 0$. Note the unity and mutual consistency of the algebraic and geometric isotopies. In fact, the latter predict the hyperbolic angle $\hat{v} = v\,(g_{33}g_{44})^{1/2}$ which turns out to be exactly that provided by the Lie-isotopic theory. The addition of the isorotations and isotranslations is done via similar rules and with similar algebraic-geometric consistencies (see [13,14,21] for brevity).

Note that the absolute value is necessary in the definition of $\hat{\gamma}$, Eq.s (3.9d) because $v^2 = v_k b_k^2 v_k \,{}^>_< c_0{}^2$. This is the first contact we have in this paper with the joint representation of redshift *and* blue shift (see below).

As expected, isotransformations (3.9) have the most general possible nonlinear-nonlocal-noncanonical structure (in which case they are called *general isolorentz transforms*) because of the arbitrariness in the functional dependence of the $g_{\mu\mu}$ terms, as needed for the form-invariance of isoseparation (3.3). Yet the isolorentz symmetry is locally isomorphic to the original symmetry, as expressed by their formal similarities with conventional Lorentz transformations, and confirmed by the isotopic commutation rules [13,14].

Note finally that general isotransforms (3.9) are nonlinear and therefore *noninertial,* as expected for interior conditions. Nevertheless, when passing to the outside and studying the global behaviour via the average of the b's to constants b°_μ, they reacquire the conventional linear and therefore inertial character (in which case they are called *restricted isolorentz transforms*).

3.5: Isominkowskian classification of physical media. Recall that there is an infinite variety of interior physical conditions for each given astrophysical mass. This variety is classified by the isominkowskian geometry into nine different types which play a fundamental role in practical applications (Sect. 5). Consider for simplicity the global interior cases with space isotropy $b^\circ_1 = b^\circ_2 = b^\circ_3$. We then have the *isominkowskian classification* into: Type I for $b^\circ_3 = b^\circ_4$ ($\hat{\beta} = \beta$, $\hat{\gamma} = \gamma$), II for $b^\circ_3 > b^\circ_4$ ($\hat{\beta} > \beta$, $\hat{\gamma} < \gamma$) and III for $b^\circ_3 < b^\circ_4$ ($\hat{\beta} < \beta$, $\hat{\gamma} > \gamma$). Each of these types is then divided into three subcases depending on whether $b_4 = 1, < 1, > 1$.

The following identifications are known at this writing. Type I.1 ($b_3 = b_4 = 1$) is therefore empty space. Type I.2 ($b_3 = b_4 < 1$) represents the homogeneous and isotropic water with index of refraction $n^\circ = b^\circ_4{}^{-1}$ and speed of light $c = c_0/n^\circ < c_0$. Type II.2 ($b_3 > b_4 < 1$) represents our inhomogeneous and anisotropic atmospheres with low density. Type II.3 ($b_3 < b_4 > 1$) represents the media of the highest possible density, such as those in the interior of a star (or, equivalently, in the interior of a hadron). Additional identifications are under study, e.g., for conductors (Type II.1), superconductors (Type I.3), intermediately heavy astrophysical atmospheres (Types III.1 and 2), etc. [13,14].

3.6: Isospecial relativity. The abstract identity between spaces and isospaces $M(x,\eta,R) \approx \hat{M}(x,\hat{\eta},\hat{R})$ and between symmetries and isosymmetries $O(3.1) \approx \hat{O}(3.1)$, implies the isotopies of all basic postulated of the special relativity, called *isopostulates,* originally proposed in [9] and studied in detail at the classical level in [13] and at the operator level. in [14].

A new relativity for interior conditions therefore emerges from the isominkowskian geometry, the isopoincaré symmetry, and the isopostulates, called *isospecial relativity* [9,13,14]. It is a covering of the special relativity in the sense that: A) it describes structurally more general systems (nonlinear–nonlocal–noncanonical systems of the interior problem), B) via structurally more general methods (isotopic methods); and C) admits the conventional special relativity as a particular case whenever motion returns to be in vacuum for which $b_\mu = 1$. Moreover, the special and isospecial relativities coincides, by construction, at the abstract level. Readers not familiar with isotopic techniques should therefore be warned that possible criticisms on the isospecial relativity for interior conditions essentially are criticisms on the conventional relativity in vacuum.

A significant property of the isospecial relativity in its most general possible formulation of Kadeisvili Class V is its *direct universality* in the sense of applying for all possible deformations $\hat\eta = T\eta$ of the Minkowski metric (universality), directly in the frame of the observer (direct universality). This property has significant experimental relevance. As we shall see in Sect. 5, numerous noneinsteinian time evolutions exist in the literature which, being different, create evident problems in their experimental test. Such problems are eliminated by the geometric unification of all seemingly different laws into a unique isotopic law.

Another general property of the isospecial relativity of Class I is its abstract identity with the general relativity for the isotopic element dependent on the local coordinates only, $T = T(x)$, $\hat\eta = T\eta = \eta(x)$. This property can be better seen from the fact that *the isopoincaré symmetry for the isotopic element* $T(x)$ *characterizes the symmetry of all possible Riemannian metrics* $\hat\eta(x)$. As an illustration, the nonlinear symmetry of the Schwartzchild line element is given by merely plotting its $g_{\mu\mu}$ elements in isosymmetry (3.9). The same holds fro all possible Riemannian line elements. The geometric unification of the special and general relativities then follows. The point important for this paper is that such unification is a mere basis for broader interior treatments because isotopic methods naturally hold for arbitrary dependence $T(x, \dot{x}, \ddot{x}, \omega, \mu, \tau, n, \dots)$.

3.7: Isodoppler red/blue/shifts. The prediction of the isospecial relativity most important for this paper is that light propagating within inhomogeneous and anisotropic media experiences an alteration of its conventional Doppler's effect according to the *isodoppler law* (for $\alpha = 0$)

$$\hat\omega = \hat\gamma\,\omega_0 = \frac{\omega_0}{|\,1 - v_k\,b_k^2\,v_k\,/\,c_0\,b_4^2\,c_0\,|^{\frac{1}{2}}}. \tag{3.10}$$

As one can see, the isospecial relativity has the following predictions: **Types I.1, I.2, I.3** (empty space or water) have *no deviation from the Doppler shift*, as verified in the physical reality in which light does not lose energy to the medium; **Types II.1, II.2, II.3** (such as our atmosphere) have an *isoredshift*, that is, a shift toward the red *in addition* to the Doppler shift due to the *loss* of energy to the medium; and **Types III.1, III.2, III.3** (such as hyperdense quasars atmospheres or the interior of hadronic matter) have an *isoblueshift* due to the *acquisition* of energy from the medium.

In order to reach a form of the isodoppler law applicable to astrophysics, we assume for simplicity the space–isotropy $b_1 = b_2 = b_3 = b$, we recall the dependence of the index of refraction b_4 from the frequency and assume the factorizability of such a dependence in the β term. We can therefore write $\beta^2 = \beta(b^2/b_4^2) = \beta[b^{\circ 2}/b^\circ_4{}^2]\,f(\omega_0)$, where b° and b°_4 are constants and $f(\omega_0) \gtrsim 1$ is the factorized frequency dependence. Law (3.10) for the global behaviour of light through quasar chromosphere can be written in one of the forms

$$\hat\omega = \hat\gamma\,\omega_0 = \frac{\omega_0}{|\,1 - \beta\,[\,b^{\circ 2}\,/\,b^\circ_4{}^2\,]\,f(\omega_0)\,|^{\frac{1}{2}}}. \tag{3.11a}$$

$$\hat\omega^2 - \omega_0^2 \equiv \beta\,[\,b^{\circ 2}\,/\,b^\circ_4{}^2\,]\,f(\omega_0)\,\hat\omega^2, \qquad \hat\omega - \omega_0 \approx -\tfrac{1}{2}\,\beta\,[\,b^{\circ 2}\,/\,b_4^2\,]\,\omega_0\,f(\omega_0). \tag{3.11b}$$

The *astronomical redshift of quasars* is then due to the property for a basic frequency, usually 4680 A$^\circ$ [2],

$$B^\circ = [\,b^{\circ 2}\,/\,b^\circ_4{}^2\,]\,f(\omega_0)\,\big|_{\omega_0 = 4680\ A^\circ} = \text{const.} > 1, \tag{3.12}$$

The *internal red/blue/shift of quasars* is then due to the full use of law (3.11a) which shows that frequencies smaller or bigger than the basic frequency 4680 A$^\circ$ have proportionately different shifts which are expected to have an approximate Gaussian behaviour owing to the condition $f(\omega_0) \gtrsim 1$.

For the sun's chromosphere we recall the experimental information (Sect. 5) that the velocity dependence is restricted to the space components b_k. In this latter case, the global averaging must be done on the expression βB resulting in the form $K°f(\omega_0) = < v^2 b(v)^2/c_0 b_4{}^2 >f(\omega_0)$, $f(\omega_0) \gtrsim 1$, with isodoppler law

$$\hat{\omega} = \hat{\gamma} \omega_0 = \frac{\omega_0}{|1 - K° f(\omega_0)|^{\frac{1}{2}}}, \quad \hat{\omega}^2 - \omega_0{}^2 \equiv K° \hat{\omega}^2 f(\omega_0), \quad \hat{\omega} - \omega_0 \approx -\tfrac{1}{2} K° \omega_0 f(\omega_0). \tag{3.13}$$

The comparison of the above laws with astrophysical data is done in Sect. 5.

3.8: Other predictions. The isospecial relativity has a number of other *novel* predictions for interior conditions (i.e., predictions not possible for the special relativity) which can be experimentally tested with contemporary technology, such as the *isodilation law* [9]

$$\tau = \hat{\gamma} \tau_0 = \tau_0/|1 - v_k b°_k{}^2 v_k / c_0 b°_4{}^2 c_0|^{\frac{1}{2}}, \tag{3.14}$$

which is confirmed by available experimental data on the behaviour of the meanlife of unstable hadrons with speed (Sect. 5), or the *isoequivalence law*

$$\hat{E} = m c^2 = m c_0{}^2 b°_4{}^2, \tag{3.15}$$

verified by preliminary experiments on the chemical synthesis of hadrons [36] and other data [36,14].

3.9: Isodual relativities. By recalling the antiautomorphic maps $1 \to 1^d = -1$, and $\hat{1} \to \hat{1}^d = -\hat{1}$ and their characterization of antiparticles (Sect. 2.7), isotopic methods identify *four* different relativities: the **conventional special relativity** on $M(x,\eta,R)$ with invariant $P(3.1)$ for the description of particles in vacuum; the **isodual special relativity** on the isodual Minkowski space $M^d(x,\eta,R^d)$ with isodual Poincaré symmetry $P^d(3.1)$ for the description of antiparticles in vacuum; the **isospecial relativity** on isominkowski spaces $\hat{M}(x,\hat{\eta},\hat{R})$ with isopoincaré symmetry $\hat{P}(3.1)$ for the description of particles within physical media; and the **isodual isospecial relativity** on the dual isominkowski spaces $\hat{M}^d(x,\hat{\eta}^d,\hat{R}^d)$ with *isodual isopoincaré symmetry* $\hat{P}^d(3.1)$ for the description of antiparticles in interior conditions.

The working hypothesis in which the total matter is equal to the total antimatter then leads to a structurally novel view of the universe in which *the total energy, the total time and other total characteristics of the universe (as the sum of those for matter and antimatter) are identically null,* a view confirmed by the isotopies of Riemann [23].

3.10. Connections with the studies by Arp, Sulentic, Marmet, and others. We indicated in Sect. 1 that Marmet theory [3] can ultimately result to be an operator version of the isodoppler formulation. A similar interconnection exists with Sulentic studies [2], and with other approaches.

A most intriguing interconnection appears to exist between the isodoppler representation and Arp's theory [1] achieving a non–Doppler redshift via the creation of matter. This latter view is faced with known problematic aspects and understandable resiliency in the physics community when considered within the context of conventional relativities *alone.* This scenario is altered by the isodual relativities. In fact, conventional relativities represent both matter and antimatter in the same space–time, with ensuing difficulties for the creation of matter from nothing. In our covering isorelativities antimatter is represented in a *separate isodual universe,* which is known not to be isolated from our universe because of the finite transition probabilities between positive– and negative–energy solutions of conventional relativistic equations. Rather than the creation of something from nothing, Arp's theory on the creation of matter acquires a different light in a cosmology with null total energy, time and other quantities [38] because it may result to be an interchange of energies between the two universes. We regret the inability to study these interconnections in more details at this time.

3.11: Applications. Numerous applications of the isospecial relativity are now available at the classical, operator, statistical and levels [13,14]. In Sect. 5 we shall outline only those experimental applications which are directly or indirectly related to the quasars red/blue/shifts. It may be important for an overall view to outline below other applications.

The simplest possible application is a *static* one, the representation of a straight rod when penetrating in water [14]. As well known, the rod *appears* to bend when entering in water, but in the reality it remains straight. Thus, the angle α of rod bending in water measured from the outside does not coincide with the physical angle $\hat{\alpha}$ in the interior of water. This occurrence is directly represented by the simplest possible case of isoeuclidean geometry with line element (3.5) in which $b_1 = b_2 = b°$ owing to the homogeneity and isotropy of water. The value $b°$ is then determined by the relation $\hat{\alpha} = ab°^2$. In short, the isoeuclidean geometry corrects the error in our perception that the rod is bent by keeping it straight.

The simplest possible *dynamical* application is the classical relativistic particle in a resistive medium without potential interactions [13]. Consider a free, classical, extended, relativistic particle with Lagrangian $L = mc$ and Minkowskian geodesic $d^2x^\mu/ds^2 = 0$. The penetration of the particle within a resistive medium is described by the *same* Lagrangian although now written in isominkowski space $L = mc = mc_0b_4$. The infinitely possible resistive forces due to shape, density, temperature, speed, etc. cannot be represented by central conception with the Lagrangian because they are nonpotential. They are then represented by the infinitely possible isotopies of the unit $I \rightarrow \hat{I}$. The understanding of the isospecial relativity requires the additional knowledge that *the motion of the extended particle in interior conditions remains fully geodesic*, i.e., in isospace we still have $d^2x^\mu/ds^2 = 0$. In summary, the two structurally different trajectories (one free and the other with contact interactions, one linear–local–potential and the other nonlinear–nonlocal–nonpotential) are completely unified, and solely differentiated by the selection of the unit. The point is that all geometric, algebraic and analytic axioms are the same.

A deeper inspection soon reveals possibilities of physical applications for the isospecial relativity which are simply beyond any descriptive capacity of Einsteinian theories [13,14]. In fact, the isounit of the preceding example admits the factorization $\hat{I} = \hat{I}_0 \text{diag.} (b°_1^{-2}, b°_2^{-2}, b°_3^{-2}, b°_4^{-2})$. Thus, *the Lagrangian* $L = mc$ *in isospace can directly represent the actual "nonspherical" shape of the test body considered*, such as a spheroidal ellipsoid with semiaxes $b°_1^2, b°_2^2, b°_3^2$ (or arbitrary shapes with a nondiagonal isounit). The term $b°_4$ geometrizes the density of the test body and the factor $\hat{I}_0$ represents the drag force. Such a representation is manifestly impossible with the conventional relativity even after quantization. But these are the beginning of the capabilities of the isospecial relativity. A still deeper inspection shows that *the same Lagrangian* $L = mc$ *in isospace can represent all infinitely possible "deformations" of its original "nonspherical" shape*, e.g., via a dependence of the $b°_k$–quantities on pressure, speed, etc., which is manifestly impossible for conventional relativities even after first, second or third quantization.

These and other features we cannot report here for brevity (see ref.s [13,14]) have permitted the isospecial relativity to resolve some of vexing problems in contemporary physics, such as the first achievement of an exact numerical representation of the total magnetic moments of few–body nuclei [14] which have still remained unexplained in their entirety despite studies over three quarter of a century. The isotopic treatment is simply given by representing protons and neutrons as extended and, therefore, deformable. This implies the deformability of their charge distributions depending on the physical conditions at hand and, thus, of their intrinsic magnetic moments. The anomalies in total magnetic moments then merely represent the (generally small) deformations of the constituents in a nuclear structure. The point is that these deformations are simply beyond any possibility of the special relativity.

We should also mention the resolution of another vexing problem of contemporary physics permitted by the isospecial relativity, that of quark confinement [14]. Current trends assume the same Minkowski and Hilbert spaces for the interior and exterior problems of hadrons. A finite probability of quarks tunneling free is then inescapable from the uncertainty principle irrespective of the infinite character of the potential barrier, which is contrary to experimental evidence. Now, the isotopic $S\hat{U}(3)$ symmetry is isomorphic to the conventional $SU(3)$, and the quantum numbers of the two theories are identical, thus rendering the isotopic theory fully compatible with existing experimental data. Moreover, the use of the conventional relativity for the exterior and the isospecial one in the interior easily permits the two Hilbert spaces to be *incoherent*, in which case the transition probability for free quarks is rigorously proved to be identically null even for collisions with infinite energy and no potential barrier at all (as hinted by asymptotic freedom).

It is important also to understand that the isospecial relativity is applicable in fields *beyond* physics, e.g., in theoretical biology. An unexpected and suggestive application along the latter lines is in conchology [14]. Consider the growth of sea shells with minimal complexity, e.g., with one bifurcation [22]. Such a growth can indeed be inspected with our Euclidean perception of physical reality. Nevertheless, computer simulations show that sea shells should crack during their growth if strictly represented in our Euclidean or Minkowskian spaces [22]. On the contrary, their growth is normal if represented in isoeuclidean or isominkowskian space, that is, with a conventional Lagrangian over a generalized unit. The representation of the bifurcations themselves is controversial in Euclidean or Minkowskian spaces because requiring *discontinuous transformations into negative times* [22], while the same can be continuously represented via our isorelativities of Kadeisvili Class III. Note that the dimension of the the space is not altered. The generalization is in the structure of the geometry, as advocated in this paper.

The latter example clearly identified the limitation of our perception of Nature, and suggests caution before claiming final knowledge based on our manifestly limited three Eustachian tubes, not only in biophysics, but also in physics and astrophysics.

4: ISORIEMANNIAN GEOMETRY

4.1: Isoriemannian geometry and its isodual. The cosmological implications of this paper are studied in the separate paper [38]. We here merely mention that the isotopies and isodualities apply also to the Riemannian geometry resulting in covering structures admitting in the tangent space the isominkowskian geometry and its isodual.

4.2: Gravitational isodoppler shifts. The aspect important for this paper is that the isodoppler shift is also additive to the gravitational redshift as in the relativistic case. Our study of the quasar red/blue/shifts can therefore be restricted to the isodoppler law (3.11) because the gravitational treatment would yield conventional gravitational corrections (when appropriate).

4.3: Isogeneral relativity and its isodual. The above studies imply a step-by-step generalization of Einstein exterior gravitation for test particles in vacuum into a dual form, one called *isogeneral relativity* or *isogravitation* for short, for interior gravitational problems of matter, and the other called *isodual isogravitation* for the interior gravitational problem of antimatter. The interested reader may consult ref.s [13,38]. The aspect important for this paper is that conventional gravitational theories possess no universal symmetry, as well known. On the contrary, isogravitation is based on the same symmetry at the foundation of the isodoppler law, the isopoincaré symmetry. Experimental confirmations of the isodoppler law within physical media would therefore have direct gravitational and cosmological implications.

5: REPRESENTATION OF QUASARS COSMOLOGICAL AND INTERNAL SHIFTS

5.1: Representation of Arp's data [1]. Isodoppler law (4.10) was originally submitted by this author in memoir [10] of 1988 to avoid the violation of Einstein's relativities under Einsteinian exterior conditions in vacuum, e.g., to avoid speeds of matter in vacuum higher then the speed of light. The main hypothesis of Sect. 1.1 can now be more technically expressed via the characterization of quasars chromospheres with isominkowskian media of Type II.2 with $b_1 = b_2 = b_3 > b_4$, $b_4 < 1$, $\hat{\beta} > \beta$, $\hat{\gamma} < \gamma$ and average speed of light $c = c_0 b_4 = c_0/n^\circ < c_0$, with consequential natural redshift $\hat{\omega}' = \hat{\gamma}\omega < \omega' = \gamma\omega$. The elaboration of Arp's data was then suggested in [10].

Numerical calculations along this proposal were done by Mignani in ref. [11] of 1992 by confirming that isodoppler's law (4.11) can indeed reduce the speed of the quasars all the way to that of the associated galaxies. This was submitted as a limiting case in which the difference between the quasars redshift and that of the associated galaxy is entirely of isotopic nature. It is understood that quasars can indeed be expelled from their associated galaxies, but at Einsteinian speeds $v \ll c_0$. This latter case implies a small correction of the b-quantities and can therefore be ignored.

The isotopic elaboration of Arp's data was conducted in ref. [11] via the relation

$$B^\circ = \frac{b^\circ_3}{b^\circ_4} = \frac{(\Delta\omega' + 1)^2 - 1}{(\Delta\omega' + 1)^2 + 1} \times \frac{(\Delta\hat{\omega}' + 1)^2 - 1}{(\Delta\hat{\omega}' + 1)^2 + 1}, \tag{5.1}$$

where $\Delta\omega'$ represents the measured Einsteinian redshift for galaxies, and $\Delta\hat{\omega}'$ represents the isotopic redshift for quasars according to law (4.11a), with resulting numerical values [1]

GAL.	$\Delta\omega'$	QUASAR	B	$\Delta\hat{\omega}'$
NGC	0.018	UB1	31.91	0.91
		BSOI	20.25	1.46
NGC 470	0.009	68	87.98	1.88
		68D	67.21	1.53
NGC 1073	0.004	BSO1	198.94	1.94
		BSO2	109.98	0.60
		RSO	176.73	1.40
NGC 3842	0.020	QSO1	14.51	0.34
		QSO2	29.75	0.95
		QSO3	41.85	2.20
NGC 4319	0.0056	MARK205	12.14	0.07
NGC 3067	0.0049	3C232	82.17	0.53

$$(5.2)$$

The above results provide a clear confirmation of the isospecial relativity and underlying isominkowskian geometrization. In fact, the data show that all B values are positive and bigger than one, exactly as predicted by the geometrization of Type II.2.

The identification of the individual values $b°_3 = <b_k>$ and $b°_4$ requires at least one additional experimental value, such as the average speed of light in the quasar chromospheres which would evidently fix $b°_4$. Then $b°_3$ could be computed from the B−ratios. As an indication, the assumption for quasar UB1 of the *average* speed of light in its chromosphere $c = 0.80 \ c_0$ would yield the value $b_3 \approx 40$.

The problem of the apparent speed of the galaxies is not considered in the above analysis because it is a separate issue. The reader should be aware that isogeometries imply three independent corrections to the current estimates of the distance of galaxies from us: 1) A correction due to a possible isoredshift of light in the interior of the galaxies; 2) Another correction due to the fact that space can be considered as empty only at the local (say, planetary) level because at intergalactic distances space itself becomes an ordinary medium (since it is filled up with dust, electromagnetic waves, particles, etc.), thus requiring a second, relatively smaller isotopic correction in the redshift; and 3) The very notion of distance is altered by the isogeometries [37,38]. Intriguingly, *each* of the above corrections implies a *decrease* of the current estimates on the distance of galaxies from us.

Under *limiting* conditions, these corrections are indeed capable of interpreting the cosmological redshift itself as being of entirely isotopic origin, thus yielding a new cosmological conception of the Universe as being unlimited, composed of essentially stationary galaxies of matter and antimatter and with a number of novel features, such as without any need for 'the 'missing mass" (from the isoequivalence law (3.15), see ref.s [37,38]).

It should be stressed that current data are insufficient to rule out the "big bang" theory, in which framework the isotopies merely yield corrections to the current estimates on the explosion of the Universe.

5.2: Representation of Sulentic data [2]. The *cosmological* redshift represented in ref. [11] is essentially that of isodoppler law (3.11a) under values (3.12) for a basic frequency such as 4680A°. The representation of Sulentic [2] *internal* red/blue/shift requires the full use of law (3.11a) with the explicit frequency dependence. The assumption of a Gaussian realization of $f(\omega)$ then leads to the isotopic behaviour

$$\hat{\omega}^2 - \omega^2 = k_1 \omega^2 e^{-k_2 (\omega - \omega_o)^2}, \qquad (5.2)$$

where k_1 and k_2 are positive constants. Numerous fits of the experimental data are then possible. As an indication, the values $k_1 = 10$ and $k_2 = 1$ yield a preliminary, yet meaningful representation of Sulentic data of Table 4, p. 61, ref. [2]. Note the shift of the center of the Gaussian as indicated by current data. Needless to say, a more accurate representation can be derived when additional measures are available such to permit the identification of the function $f(\omega)$.

5.3: Representation of Marmet's data [3]. The data on the redshift of spectral lines from the sun's chromosphere as studied by Marmet [3] and others are some of the most direct experimental confirmations of the isotopic character of the quasars redshift.

The latter data can be interpreted via essentially the same isodoppler law, only referred to form (3.13) because of the need of the different average since the sun is moving at low speed with respect to our laboratory. In fact, in first approximation, law (3.13) reproduces Marmet's expression (6), p. 240, ref. [3] identically

$$\omega \, / \, \Delta\omega \; = \; \Delta\lambda \, / \, \lambda \; \approx - \, 2 \, / \, K°f(\omega)\big|_{\omega=const.} \; \approx - \, 2.73 \times 10^{-21} \, T^2 N_c \, , \qquad (5.3)$$

where T is the temperature of the sun's chromosphere, and N_c is the average number of collisions of photons in a given column density. Note the emergence of a dependence on the frequency which is expected to be experimentally verifiable and which, if confirmed, would establish the possibility of resolving the problem of quasar red/blue/shifts via spectroscopic measures on the Sun.

5.4: Representation of timelife behaviour. The isospecial relativity has additional experimental verifications indirectly related to the quasar red/blue/shifts which, as such, are significant for this paper. The first one is the isotopic behaviour (3.14) of the meanlife of unstable hadrons with speed which, if confirmed, would provide a clear verification of the *structure* of the isodoppler law (3.10).

Blochintsev and his school [25] pioneered the hypothesis that the nonlocal internal effects expected in the hadronic structure from mutual penetrations of the wavepackets of the constituents can manifest themselves via departures from the Minkowskian behaviour of the meanlife of unstable particle with speed, and computed a generalized law. The problem was subsequently studied by several authors [26], resulting in additional different laws.

This author submitted in [9] the isominkowskian geometrization of the physical medium in the interior of hadrons with isotopic law (3.14) which was proved by Aringazin [27] to be "directly universal", i.e., including all possible generalizations [25,26] via different expansions in terms of different parameters and with different truncations.

The first phenomenological verification was provided in calculations [28] on deviations from the Minkowskian geometry inside pions and kaons conducted via standard gauge models in the Higgs sector. These phenomenological studies resulted in the deformed Minkowski metric inside hadrons $\hat{\eta}$ = diag. $\{(1 - \alpha/3), (1 - \alpha/3), (1 - \alpha/3), - (1 - \alpha)\}$, which is precisely of the isominkowskian type with numerical values

$$\text{PIONS } \pi^{\pm}: \quad b°_1{}^2 = b°_2{}^2 = b°_3{}^2 \cong 1 + 1.2 \times 10^{-3} \, , \; b°_4{}^2 \cong 1 - 3.79 \times 10^{-3} \, , \qquad (5.4a)$$

$$\text{KAONS } K^{\pm}: \quad b°_1{}^2 = b°_2{}^2 = b°_3{}^2 \cong 1 - 2 \times 10^{-4}, \; b°_4{}^2 \cong 1 + 6.1 \times 10^{-4}, \qquad (5.4b)$$

Note the *change in numerical value* of the isotopic element in the transition from pions to kaons, which is necessary because of the change of the density (recall that all hadrons have approximately the same size, but different rest energies, thus having different densities and different isounits).

The first direct experimental verification was reached by Aroonson et al. [29] who measured a clear nonminkowskian behaviour of the meanlife of the $K°$ in the energy range 30–100 GeV. Subsequent direct experiments conducted by Grossman et al. [30] confirmed the Minkowskian behaviour of the meanlife of the same particle in the *different* energy range 100–350 GeV (see review [18]).

These seemingly discordant experimental measures were proved to be unified by the isominkowskian geometrization of the $K°$-particle by Cardone et al. [31] via phenomenological plots of both measures [29,30] in the range 30–350 GeV resulting in the following characteristic $b°$-values

$$b°_1{}^2 = b°_2{}^2 = b°_3{}^2 \approx 0.909080 \pm 0.0004, \qquad b°_4{}^2 \approx 1.002 \pm 0.007 \, , \qquad (5.5a)$$

$$\Delta b°_k{}^2 \cong 0.007, \qquad \Delta b°_4{}^2 \cong 0.001 \, , \qquad (5.5b)$$

which are of the same order of magnitude of values (5.3b). Measures (5.4b) also confirm the prediction of the isominkowskian geometry in the range 30–400 GeV that the $b°_4$ quantity, being a geometrization of the density, is constant for the particle considered (although varying from hadron to hadron with the density), while the dependence in the velocities rests with the b_k-quantities.

the latter analysis is important inasmuch as it establishes the possible existence of an isodoppler shift even for a medium at rest in which $< v^2/c_o{}^2 > = 0$, but $<v^2 b(v)^2/c_o{}^2 b_4{}^2> \neq 0$.

5.4: Representation of Bose-Einstein correlation. Another important verification has been recently achieved via theoretical [32] and experimental [33] studies on the Bose-Einstein's correlation. These studies provide a direct verification of the basic isominkowskian geometrization of physical media and, as such, are significant for the quasars red/blue/shifts.

Evidence establishes that no correlation exists for particles interactions when admitting effective point-like approximations. The Bose-Einstein correlation therefore appears to be due precisely to the *extended* character of the wavepacket of particles, which results in an evident *nonlocal* structure of the interactions at very small distances. The use of the isominkowskian geometrization for the interior of the p-$\bar{\text{p}}$ fireball results in the two-point Boson isocorrelation function on $\hat{M}(x,\hat{\eta},\hat{R})$, ref. [32], Eq. (10.8), p. 122,

$$\hat{C}_{(2)} = 1 + \frac{K^2}{3} \sum_\mu \hat{\eta}_{\mu\mu} \, (e^{-q_t^2 / b^\circ{}_\mu{}^2}, \qquad \hat{\eta} = \text{Diag.} \, (b^\circ{}_1{}^2, b^\circ{}_2{}^2, b^\circ{}_3{}^2, - b^\circ{}_4{}^2), \qquad (5.6)$$

where q_t is the momentum transfer and the term $K = b^\circ{}_1{}^2 + b^\circ{}_2{}^2 + b^\circ{}_3{}^2$ is normalized to 3, under the sole approximation, also assumed in conventional treatments, that the longitudinal and fourth components of the momentum transfer are very small. Phenomenological studies conducted in [33] via the UA1 data at CERN confirm model (5.5) in its entirety, and identify the numerical values

$$b^\circ{}_1 = 0.267 \pm 0.054 , \quad b^\circ{}_2 = 0.437 \pm 0.035 , \quad b^\circ{}_3 = 1.661 , \quad b^\circ{}_4 = 1.653 \pm 0.015 . \qquad (5.7)$$

These measures have the following important implications: A) They confirm the nonlocal-nonhamiltonian origin of the correlation, which is at the foundation of these studies; B) They confirm the isominkowskian geometrization for the p-$\bar{\text{p}}$ fireball; C) they provide a numerical value of $b^\circ{}_4$ for particles of the density of the p-$\bar{\text{p}}$-fireball for use in isoequivalence principle (3.12) (see below); D) They confirm the capability of the isotopies of directly representing the nonspherical shape of the fireball and all its deformations; and E) They prove the reconstruction of the exact Poincare' symmetry under nonlocal-nonhamiltonian interactions.

5.5: Cooper pair in superconductivity. This is a clear physical systems beyond any realistic capability of Einsteinian theories because it consists of two electrons of the *same* charge experiencing an *attractive* interaction. Animalu [34] has shown that the use of the isominkowskian geometry representing the mutual wave-overlapping of the two electron (with isounit given in Sect. 2.2) permits a quantitative interpretation of the attractive interactions in the Cooper pair which is in excellent agreements with numerous experiments (see [34] for brevity).

5.6: Chemical synthesis of hadrons. The isominkowskian geometry also permits a speculative, yet intriguing prediction, the cold fusion/chemical synthesis of protons and electrons into neutrons (plus neutrinos). It is essentially allowed by the rest energy of the electron when inside the hyperdense medium in the interior of the proton and computed via isoequivalence principle (3.15) with numerical value $b^\circ{}_4 = 1.653$ from data (5.6). This permits a representation of all characteristics of the neutron [35]. This prediction has received a preliminary, yet direct experimental verification by don Borghi et al [36]. If confirmed, the event would permit the chemical synthesis of all unstable hadrons from lither (massive) hadrons. Moreover, it would permit the artificial disintegration of unstable hadrons, such as the artificial disintegration of peripheral neutrons in a nuclear structure, with realistic possibilities of a new technology, called *hadronic technology*, because based on mechanisms in the interior of individual hadrons. See Vol. III of ref.s [14] for other experimental verifications.

5.7: Proposed experiments. A number of experiments have been proposed in classical mechanics, astrophysics and particle physics to test the isominkowskian geometry and related isospecial relativity such as:

Experiment 1 [13]: measure the redshift of light from a quasar just before and then after passing through a planetary atmosphere or the sun's chromosphere. The isominkowskian geometry predicts in this case an additional redshift. The average data (5.2) yield $<B^\circ> = 72.78$, $<\Delta\hat{\omega}'> = 1.15$, $<\Delta\omega'> = 0.01$, thus characterizing the average isoshift $<\Delta\hat{\omega}'> - <\Delta\omega> = 1.14$. The assumptions that the quasar atmospheres are 10^5 denser than the atmosphere of Jupiter (or of Earth), and that the isotopic effect is proportional to the density in first approximation, lead to the estimate of the isoredshift in Jupiter's atmosphere of the order of $<\Delta\hat{\omega}'_{Jupiter}> \approx 1.14{\times}10^{-5}$ which is fully measurable. For smaller ratios of the densities of the quasars and planetary atmospheres, the effect evidently becomes bigger.

Experiment 2 [13]: Follow a sufficient number of Fraunhofer lines of sun light from the zenith to the horizon to see whether or not the tendency toward the red is in part an isoredshift. The numerical estimates of the preceding experiment also apply to Earth's atmosphere, yielding a measurable effect.

Experiment 3 [14]: Finalize the behaviour of the meanlife of unstable particles with speed [29,30]. As indicated earlier, any deviation from Minkowskian time dilation is a confirmation of the corresponding isodoppler behaviour for frequencies owing to its direct universality [27].

Acknowledgments: The author would like to thank Jack Sulentic for invaluable comments. Thanks are also due to Asterios Jannussis, Michele Barone and Horst Wilhelm for a critical reading of an earlier version of the manuscript.

REFERENCES

1. H. Harp, *Quasars redshifts and controversies*, Interstellar media, Berkeley (1987) (see also contributed paper to this volume)
2. J. W. Sulentic, Astrophy. J. 343, 54 (1989) (see also contributed paper in this volume)
3. P. Marmet, IEEE Trans. Plasma Science **17**, 238 (1988) (see also Phys. Essays, 1, 24 (1988), IEEE Trans. Plasma Science **18**, 56 (1990) and **20**, 958 (1992) and references quoted therein)
4. H.C.Arp, G. Burbidge, F.Hoyle, V.J.Nardikar and N.C.Wicramasinghe, Nature **346**, 807 (1990).
5. K. Schwartzchild, Sitzber. deut. Acad. Wiss., Berlin. Kl. Math.-Phys. Tech. 424-434 (1916)
6. P. G. Bergmann, *Introduction to the Theory of Relativity*, Dover Publ., New York (1942); C. Møller, *The Theory of Relativity*, Oxford Univ. Press (1952)
7. R. M. Santilli, *Foundations of Theoretical Mechanics*, Vol. I (1978), Vol. II (1983), Springer-Verlag (1978); and Hadronic J. Suppl. **1**, 662 (1985)
8. R. M. Santilli, Hadronic J. **1**, 223 and 574 (1978; Phys. Rev. **D20**, 555 (1979)
9. R. M. Santilli, Lett. Nuovo Cimento **37**, 545 (1983) (for a recent presentation see J. Moscow Phys. Soc., in press)
10. R. M. Santilli, Hadronic J. Suppl. **4A**, issue no. 3 (1988)
11. R. Mignani, Phys. Essays **5**, 531 (1992)
12. R. M. Santilli, Algebras, Groups and geometries **8**, 169 and 275 (1991), and **10**, 273 (1993)
13. R. M. Santilli, *Isotopic Generalization of Galilei's and Einstein's Relativities*, 2-nd edition, Ukraine Academy of Sciences, Kiev, in press
14. R. M. Santilli, *Elements of Hadronic Mechanics*, Vols. I, II, III, Hadronic Press, in press
15. J. V. Kadeisvili, Elements of Lie-Santilli Theory, Acta Appl. Math., in press
16. D. S. Sourlas and G. T. Tsagas, *Mathematical Foundations of the Lie-Santilli Theory*, Ukraine Academy   of Sciences, Kiev, (1993)
17. J. Lôhmus, E. Paal and L. Sorgsepp, *Nonassociative Algebras in Physics*, Estonian Academy of Sciences, Tartu, in press
18. A. K. Aringazin, A. Jannussis, D. F. Lopez, M. Nishioka and B. Velianoski, *Santilli's Lie-Isotopic Generalization of Galilei's and Einstein's Relativities*, Kostarakis Publ., Athens (1990)
19. J. V. Kadeisvili, *Santilli's Isotopies of Contemporary Algebras, Geometries and Relativities*, 2-nd edition, Ukraine Academy of Sciences, Kiev, in press
20. J. V. Kadeisvili, Algebras, Groups and Geometries **9**, 283 and 319 (1992)
21. R. M. Santilli, Hadronic J. **8**, 25 and 36 (1935)
22. C. Illert, *Foundations of Theoretical Conchology*, Hadronic Press (1992); and Hadronic J. in press
23. R. M. Santilli, Integral isotopies of the Riemannian geometry, in *Analysis, Geometry and Groups: A Riemann legacy Volume*, H. M. Srivastava and T. M. Rassias, Editors, Hadronic Press, in press
24. A. Jannussis, contributed paper in this volume
25. D. I. Blochintsev, Phys. Lett. **12**, 272 (1964)
26. L. B. Redei, Phys. Rev. **145**, 999 (1966); D. Y. Kim Hadronic J. **1**, 1343 (1978);
27. A. K. Aringazin, Hadronic J. **12**, 71 (1989)
28. H. B. Nielsen and I. Picek, Nucl. Phys. **B11**, 269 (1983)
29. B. H. Aroonson et al. Phys. Rev. **D28**, 476 and 495 (1983)
30. N. Grossman et al., Phys. Rev. Lett. **59**, 18 (1987)
31. F. Cardone, R. Mignani and R. M. Santilli, J. Phys. **G18**, L61 and L141 (1992)
32. R. M. Santilli, Hadronic J. **15**, 1 (1992)
33. F. Cardone and R. Mignani, Univ. of Rome preprint no. 894 (1992), subm. for publ.
34. A. O. E. Animalu, Hadronic J. **14**, 459 (1990); and Hadronic J. in press (1994)
35. R. M. Santilli, in *Proceedings of the International 1993 Workshop on Symmetry Methods in Physics*, G. Pogosyan et al, Editors, JINR, Dubna, in press, and JINR Communication E4-93-352 (1993)

36. C. Borghi, C. Giori, and A. dell'Olio, Hadronic J. **15**, 239 (1992)
37. R. M. Santilli, in *The Mathematical Legacy of Hanno Rund*, J. V. Kadeisvili, Editor, Hadronic Press (1993)
38. R. M. Santilli, in *Analysis, Geometry and Groups: A Riemann Legacy Volume*, H. M. Srivastava and Th. M. Rassias, Editors, Hadronic Press (1993)

THE RELATIVISTIC ELECTRON PAIR THEORY OF MATTER
AND ITS IMPLICATIONS FOR COSMOLOGY

Ernest J. Sternglass

Radiation Physics Laboratory
Scaife Hall RC- 406
University of Pittsburgh
Pittsburgh, PA 15261

INTRODUCTION

Although atomic and nuclear physics have made enormous advances during the last century, there has been an increasing crisis in fundamental physical theory with regard to the nature of the ultimate constituents of matter which appear to have both particle and wave-like properties and occur in a bewildering variety of types and masses. Since there is now overwhelming evidence that the universe is expanding from a highly compact state that appears to have had the dimensions of a single proton or less, further progress in understanding the origin of the universe and its structure cannot be made until the problem of the nature of the fundamental particles is resolved.

If one examines the spontaneous decay process of all the newly discovered unstable mesons and baryons, one finds that ultimately they all lead to the production of electrons, positrons, protons and radiation quanta such as photons and neutrinos. Even the proton is known to annihilate with an anti-proton into mesons that in turn decay to electrons and their oppositely charged anti-particles, positrons. It is therefore the purpose of the present paper to outline an approach to the theory of nuclear particles which reduces the number of truly elementary entities to only a single type, and to summarize the resulting implications for cosmology.

Of all the hundreds of matter particles on the sub-atomic scale discovered during the last century that have a clearly defined rest-mass and charge measured in the laboratory, only the electron and its anti-particle possess the properties required for a truly elementary entity [1]. These properties are (1) a very high degree of stability when isolated, (2) identity of all physically measurable properties under the same conditions and (3) absence of any experimental evidence for internal structure, or divisibility into fragments of smaller rest-mass, size or charge, so that the particle interacts with all other particles as if it were a mathematical point entity, requirements that have in fact been met in collisions as high as millions of times the energy needed to create an electron - positron pair.

The only condition under which an electron disappears is when it annihilates with a positron, resulting in purely electromagnetic radiation quanta. Likewise, electrons can be created only together with their anti-particles so that charge is conserved under all known physical conditions. Moreover, electrons and positrons can be created from electromagnetic photons of sufficient energy in the well-known pair-production process. Therefore, it is possible to assume that the mass of the electron is of purely electromagnetic origin as postulated by a number of investigators shortly after its discovery, or that its mass resides entirely in its surrounding field. This allows one to regard these particles as nothing more than centers of force or stable concentrations of electromagnetic field energy without any "hard cores" of "ponderable matter" of unknown nature along the lines arrived at by Motz [2].

This view immediately removes the apparent contradiction between the particle and wave aspects of material particles since electrons are in effect taken to be stable, extended wave pulses. Thus, they can produce interference patterns characteristic of a two-slit system even when they are fired at the slits one at a time without the need to give up our usual space-time mode of describing natural phenomena on the atomic scale [3].

It also follows that all forces, from chemical to nuclear, must utimately turn out to be understandable in terms of the known characteristics of the electron and positron as the centers of an extended field throughout which their mass-energy is distributed. Since the gravitational interaction can be understood in terms of the distribution of mass-energy that determines the local deviations from a flat Euclidean space according to Einstein's General Theory of Relativity regardless of the nature of the mass-energy, even gravitational forces must ultimately turn out to be explainable in terms of the properties of the electron and positron alone. Furthermore, since one must not introduce qualitatively new kinds of matter or forces in order to describe the more massive nuclear particles, one can only use purely geometric or dynamic considerations to explain their larger masses.

Frontiers of Fundamental Physics, Edited by M. Barone
and F. Selleri, Plenum Press, New York, 1994

THE RELATIVISTIC PAIR MODEL FOR NUCLEAR PARTICLES

Of all the new nuclear particles discovered in the 1940s the neutral pion or π_0 that is known to decay into two gamma rays bears the most obvious resemblance to positronium composed of an e^+ and e^- in a Bohr-type orbit. This led Fermi and Yang to consider a model for the π_0 consisting of a proton and an anti-proton in 1949 [4]. However, within another decade, experiments had revealed that the proton had internal structure, ruling out this early model. There was however a way that a high mass could be explained with only two electrons, namely if there were a previously unknown e^+e^- state with very high relativistic momentum and energy. This did in fact turn out to be the case, provided one makes the assumption that the force between the two particles is that calcuated by an observer at rest with respect to either one of the two particles [5] first suggested by Einstein in his paper on the theory of special relativity as the only way to arrive at a symmetrical result for the force between two moving particles. One then arrives at the existence of a minimum approach distance which except for a velocity dependent correction factor that ranges from 1 to 1/4 between low and high velocities equals the classical electron radius r_{cl} $= e^2 /2m_0c^2 = 1.4 \times 10^{-13}$ cm, where e is the charge of the electron, m_0 is its mass and c is the speed of light.

Quantization of the orbital angular momentum as in the Bohr model leads to a mass of $(2/\alpha) - 2$ or 272.072 times the electron mass m_0 for the ground state. Here α is the fine-structure constant $e^2/\hbar c = 1/137.036$ first used by Sommerfeld in the relativistic treatment of the Bohr model. When corrected for the effect of the magnetic moment spin-spin and spin orbit interaction of $8m_0$ this leads to a mass of $264.072\ m_0$ close to the observed π_0 mass of $264.116\ m_0$ and life of 2×10^{-16} seconds against annihilation into two gamma rays based on the analogy to positronium, again in surprising agreement with the most recent observed value of 0.87×10^{-16}.

Besides the π^0 which has spin 0 as a result of the spins of the electron and positron being parallel, there is also a spin 1 or $\pi^0{}_1$ meson in which the spins are anti-parallel whose mass is $12\ m_0$ larger due to the difference in the magnetic spin-spin and spin-orbit interactions, giving a mass of about $276.072\ m_0$ [5], close to the observed mass of the charged pions of $273.126\ m_0$, even without a correction for the addition of a charge [6]. As discussed below, the $\pi^0{}_1$ has a much longer life than the π^0 so that it forms the basic building block of all other particles.

Thus, using only Bohr's hypothesis that a system with minimum angular momentum $\hbar$ is stable against radiation, one arrives at a model of a meson of the right order of mass and size of nuclear particles. Moreover, the strength of the force was found to be $\gamma_{12} e^2 / r_{12}^2$ where γ_{12} is given by $(1-\beta_{12}^2)^{-1/2}, \beta_{i2} = v_{12} / c$ and r_{12} is the distance between the particles. For the ground state, γ_{12} is equal to $(2/\alpha)$ or 274.072 so that the relativistic interaction is of the correct order to explain the great strength of the nuclear force.

This surprising result is due to the high relative velocity, leading to a large magnetic interaction between two parallel currents that is much stronger than the ordinary electrostatic interaction. In this way, the strong nuclear force is seen to be a form of the electromagnetic interaction, so that this model unifies these two forces, a goal that supersymmetry theory has been trying to achieve by postulating a whole new family of particles.

Using the basic relativistic electron pair model of the neutral pion as the analog of the Bohr positronium atom, it becomes possible to describe the heavier mesons as quasi-molecular systems composed of pions with a binding strength $2(2/\alpha)\ e^2$, a Yukawa type of short-range potential of 75.7 Mev with a range of $\lambda_{\pi\pi} = \hbar/2\ M_\pi\ c = r_{cl}$ that follows from the solution of the Klein-Gordon equation, and a zero-point energy $2\ M_\pi\ c^2$ at a fixed distance of r_{cl} between them as determined by the minimum approach distance between the e^+ and e^- [6,7]. Thus, the model gives a mass of 484 Mev for the simplest molecular system consisting of two pions compared with the observed values of 494 to 498 Mev for the K-mesons. Due to a numerical coincidence, the same mass is also obtained for a 3π, close packed structure. Likewise, a 4π planar structure is found to have a closely similar mass of 490 Mev, thereby explaining why one observes both 2 and 3 pion decays from what appears to be a single particle.

The model also gives a series of still higher short-lived systems as rotationally excited or resonant states of these molecular structures [6,7]. Thus, the first excited spin 1 state of the the two- pion system gives a predicted mass of 764 Mev vs. an observed mass of 770 Mev for the ρ meson of spin 1, and the three pion close-packed system has one of its three spin 1 states at 753 Mev compared with an observed mass of 782.6 Mev. In all the cases worked out at the time [6,7], the calculated masses agreed to within $\pm 6\%$ with the measured values, even though in these simple models, the effects of magnetic moment interactions had not been taken into account.

Moreover, the excited states of baryons that decay to protons with the emission of mesons have masses that fit some of the same basic molecular rotator levels as the meson states [8], something that is to be expected if the nucleons contain mesons. In particular, the simplest of the "mesonic molecules" , the two pion system whose ground state mass is equal to that of the K-mesons, seems to be basic to the structure of baryons since the mass of the proton of 938.28 Mev is just slightly smaller than the mass of two K^0 -mesons, 995.41Mev, which is also consistent with the fact that the proton and anti-proton annihilate into K-mesons as well as pions. Thus, the proton appears to be composed of two K^0 mesons held together by the exchange of a highly relativistic positron [9] in a state with a binding energy of 57.13 Mev in such a way that it allows internal rotational excitations as well as an unusually high degree of stability, now known to result in a half-life of more than 10^{32} years [10]. This results in a three quark structure, in agreement with early group-theoretial conclusions by Gell-Mann and others based on the then known groupings of baryons into multiplets which suggested that the baryons belong to an SU(3) dynamic group. But unlike the standard model that postulates fractional charges and cannot explain the exact equality of the proton and positron charge, the electron-pair model requires that the two charges are the same.

Thus, the electron-pair model for the proton explains the excited states produced in the course of pion-proton collisions as short-lived states in which a π is temporarily bound to one of the 2π components of the proton, forming a 3π close packed system whose excited states fit the observed states to better than 3.7% [8]. Moreover, this model of the proton is further supported by the fact that the excited states predicted for the case where a π is temporarily bound at both ends of the proton fit another set of predicted excited states to better than 1.3%.

Still further, it is found that the hyperons that have the same spin 1/2 as stable nucleons fit excited states in which two pions or a $K_{2\pi}$ is temporarily bound at the two ends of the basic proton structure so as to form K-mesons of the 4π planar type rotating in opposite directions. As a result, the additional rotational momenta cancel each other, leading to total angular momenta equal to that of the proton. And again, the observed masses agree to within a few percent with the predicted 4π rotational states. Finally, two different sets of excited hyperon states that are observed fit single and double- ended rotational states to the same high degree of accuracy [8].

It remains to discuss the charged mesons, especially the relation between the muon and the pion that decay with the emission of neutrinos. As mentioned above, it is the longer lived spin 1 π°_1 with its anti-parallel spins and parallel magnetic moments of the two charges that forms the charged mesons. It is the net magnetic moment of the π°_1 that allows a charge to be bound to the pair- system[7]. In the case of the muon that has a spin 1/2, the spin of the added charge must oppose the orbital momentum of the spin 1 π°_1. At the same time, the magnetic moment of the added charge must be anti-parallel to that of the π° so that the sign of the charge that can be attached is fixed by the sense of the orbital angular momentum relative to the spins. But there are two different cases for this orientation : one in which the orbital angular momentum is parallel to the spin of the e^+, and one in which it is parallel to the spin of the e^-. In the former case, only a positive charge can be bound; in the latter case only a negative charge. Thus, while the π° is its own anti-particle, the π°_1 can exist in two possible states: matter and anti-matter states of opposite charge, and this property appears to be related to the origin of isotopic or isobaric spin , as well as to the existence of only one kind of stable matter in the universe, as will be dicussed later on.

The muon mass of 206.768 m_0 is very close to the calculated value of 206.7 m_0, and so is the life-time of 2.197 x 10^{-6} sec close to the theoretical life-time of 2.210 x 10^{-6} sec [7]. However, instead of emitting 3 gamma rays of spin 1 together with an e^+ or e^-, the $\mu^{\pm}$ emits two neutrinos of spin 1/2. The reason is that due to the relativistic orbital velocity of the pair-system, the usually small Sommerfeld precession becomes equal to the orbital angular velocity so that the angular momentum of the precessing reference frame in which the Kepler orbits are closed takes on the value $\hbar/2$, leaving $\hbar/2$ for the orbital angular momentum relative to the precessing frame [5]. Thus, in these highly relativistic pair systems, orbital angular momenta are quantized in units of $\hbar/2$ so that the central π°_1 annihihates by emitting two neutrinos of opposite helicity that together carry away the angular momentum h. Moreover, due to the extremely relativistic motion of the two charges in the π°_1, their mass is increased and the source size and magnetic moments are reduced by γ_{12}, so that the rate of radiation is much less than for low velocity states [7], thereby explaining the origin of the weak interaction and its relationship to the electromagnetic interaction as originally described by Fermi [11] and since then observed in high energy experiments.

As for the decay of the $\pi^{\pm}$ to a $\mu^{\pm}$, this is explained in terms of the present model as a transition from an excited state of the charge bound magnetically to the central pair with the emission of a single neutrino of spin $\hbar/2$. The theoretically calculated mass of this excited state is 67.8 m_0 larger than the muon mass, giving a pion mass of 274.5 m_0, while the theoretical mean-life is 2.49 x 10^{-8} sec, in good agrement with the latest measured value of 2.60 x 10^{-8} sec. It is of interest that the single neutrino emitted in the decay of the $\pi^{\pm}$ to a $\mu^{\pm}$ differs from the two neutrinos emitted when the muon decays to an electron or positron in that the neutrino produced in the pion decay carries away an orbital angular momentum 1/2 $\hbar$ relative to the precessing reference frame of the π°_1. Since the charge in the pion has to move in a direction opposite to that of the precessing frame, it carries only a small zero-point angular momentum relative to the laboratory rather than a spin $\hbar/2$. This appears to explain why the muon neutrino and the electron neutrino differ in their interaction with matter [7].

Considering that these models require only the fundamental constants of electromagnetic and quantum theory e, m_0, c and $\hbar$ and no arbitrary or adjustable parameters, the agreement with the masses, spins and life-times of the muon, the pion and with the masses and spins of the heavier mesons and baryons must be regarded as strong support of the pair model, as well as for the assumption that the electronic charge is the smallest in nature, in agreement with the fact that all searches for fractional charges have failed [12]. Thus, these results favor the suggestion of Han and Nambu that the quarks may be integrally charged, and it supports the hypothesis arrived at by a number of theorists that the many different types of quarks are not truly elementary particles but instead complex structural systems composed of more fundamental entities, namely relativistic electrons and positrons.

IMPLICATIONS FOR COSMOLOGY

It was the unexpected discovery of yet another class of mesons in 1974, namely the J/ψ of mass 3097 Mev, shortly followed by the discovery of the Y with a mass of 9460 Mev, that provided experimental evidence for a possible link to the initial high density state of an expanding universe first proposed by Lemaitre in 1932 [13]. The most surprising characteristics of this new family of mesons created in high energy collisions such as those of electrons and positrons were their high density, far above that of ordinary nucleons, and their long life compared with all previously known baryon and meson resonance states that decayed in a matter of 10^{-23} to 10^{-22} sec. Thus they produced extremely sharp, narrow resonances of less than 0.1 Mev in width, compared with 10s to 100s of Mev for the previously known massive hadrons. Moreover, they did not just decay into hadrons but also directly into electron and muon pairs without the emission of neutrinos. More unusual still, the heavier Y, instead of having a shorter life-time as in the case of ordinary hadronic resonances of increasing mass, actually proved to be longer lived than the lower mass J/ψ. Thus, the possibility arose that this type of particle might provide a physical model for the "primeval atom" postulated by Lemaitre as the immensely dense, massive "seed" of the universe from which all matter particles and the various cosmological structures evolved in a series of division processes.

Since these new particles could be created in collisions between an e^+ and an e^- and since they sometimes decayed into an e^+ and e^- as well as into spin 1/2 muon pairs of opposite charge, it was natural to assume that they might be highly excited states of the π^0_1, or massive positronium-like states of two spin 1/2 particles [14]. A closely related suggestion was that the J/ψ might be composed of a pair of quarks of a new type called charmed or c quarks, forming a massive "charmonium" system [15]. Likewise, when the Y was discovered, it was postulated that it in turn was composed of yet another pair of quarks, called beauty, bottom or b quarks.

However, it seemed simpler to assume that these particles were excited states of a single type of elementary entity that had already been able to explain the masses and life-times of the other hadrons instead of postulating a new type of particle every time a new resonance in electron positron collisions was observed. This possibility was supported by the fact that the J/ψ turned out to have a mass very close to the 22nd excited state of the π^0, which has a predicted mass of 3104 Mev, within only 7 Mev or 0.2% of the observed J/ψ mass. At very nearly the same mass lies a system composed of two $K_{2\pi}$ mesons excited to the ρ or $J = 1$ rotational state [7] whose theoretical mass is 764 Mev using $M_\pi = 140$ Mev, giving a total mass of 3056 Mev. Thus, with a central π^0_1 of mass 141 Mev bound with an energy of 100 Mev, such a system would provide the necessary spin 1 state into which the J/ψ could decay, in agreement with the most frequent decay modes involving ρ, π, ω and K mesons [16].

Likewise, the Y mass is close to the $n = 67$ level of the π^0_1, or at 9452 Mev, only 8 Mev from the observed mass of 9460 Mev. Again there is a nearby spin 1 system into which the state can decay to form hadrons, composed of two excited $K_{2\pi}$ in a $J = 5$ state for which the mass is 4684 Mev [7]. Two such systems plus 141 Mev for a central π^0_1 give 9509 Mev so that a binding energy of 49 Mev yields the observed mass of 9460 Mev.

Thus, one can make the hypothesis that the Lemaitre atom is a highly excited, long lived state of the basic relativistic π^0_1 pair system [17], with a mass equal to that of the universe M_u at a density equal to the maximum possible one [18], namely the Planck density $\rho_{Pl} = c^5/\hbar G^2 = 5.177 \times 10^{93}$ gm. Given the effective volume V_{eff} in which the field energy is confined, which in the present case must be of the order of nuclear particle dimensions, one can calculate a theoretical mass of the Lemaitre atom or of the universe from the relation $M_u = \rho_{Pl} V_{eff}$.

There are two ways in which one can arrive at a value for V_{eff} [17]. One is to calculate the mean value of the radius in which the field energy of the relativistic pair is distributed by analogy to the case of a single electron in quantum theory following Motz [2], namely the geometric mean of the inner radius of the electron r_{cl} and the outer radius of the field given by $(2/\alpha) r_{cl}$, the deBroglie wavelength. In the case of the π^0_1, this effective radius $<R_{ee}>$ is the mean of $r_{cl}/4$ and $(2/\alpha) r_{cl}/4$, or $(2/\alpha)^{1/2} r_{cl}/4 = 4.14 r_{cl}$. One can then calculate the volume $<V_{ee}>$ using the non-Euclidean expression for the extremely massive relativistic states as $2\pi^2 <R_{ee}>^3 = 1400\ r_{cl}^3$.

The second way is to use the Dirac large number defined for the present case of electrons, $N_D = e^2/m_o^2 G = 1.667 \times 10^{42}$. By eliminating G, the Planck density $\rho_{Pl} = M_u/V_{eff}$ becomes $N_D^2 (c^5 m_o^4/\hbar e^4)$. Next, defining the Eddington number $N_E = M_u/m_o$ one obtains $N_E = V_{eff} (c^5 m_o^3/\hbar e^4) N_D^2$ and since both N_E and N_D are dimensionless, the term in the bracket must represent the inverse of a volume V_o, given by $(\hbar e^4/m_o^5 c^5)$. This can be rewritten as $(1/\alpha)(2 r_{cl})^3 = 1096\ r_{cl}^3$ and is seen to be close to the value $<V_{ee}>$. For the volume V_o one gets the radius $3.8\ r_{cl}$. Taking the average of the two effective radii, one obtains the value $3.98\ r_{cl}$ or close to $4\ r_{cl}$.

Since the ratio $<V_{ee}>/V_o$ is 1.27 and therefore close to unity, it follows that the relation between N_E and N_D^2 derived above takes on the simple form $N_E = N_D^2$ which is the "Large Number Relation" that Dirac believed would eventually require a physical explanation. It means that the origin of this relation can be explained if all matter is composed of electrons, and that the universe began with all its mass concentrated in a volume V_o.

Thus, the relationship $(e^2/m_o^2 G)^2 = M_u/m_o$ holds exactly, so that the mass of the physical universe is finite and can be calculated when e, m_o and G are known, giving $M_u = e^2/m_o^3 G^2 = 1.7632 \times 10^{85}\ m_o$ or $9.4551 \times 10^{81}\ M_p$, where M_p is the proton mass. This serves as a first test of the Lemaitre model since the luminous mass of the universe is of the order of $10^{80}\ M_p$, so that the theoretical value is consistent with the fact that visible matter represents only about 1% of the total mass [19]. Another test is whether the size of the universe is consistent with the present Hubble radius of $10\text{-}15 \times 10^9$ light years. That this is the case may be seen from the fact that the Schwarzschild radius $R_s = 2GM_u/c^2$ of the universe is 2480×10^9 light years, large enough so as not to lead to a significant deceleration of the expansion at the present epoch, in agreement with observation.

The model explains why N_D is also a measure of the size of the universe in units of nuclear particle dimensions. Taking the expression for the Schwarzschild radius R_s and substituting $N_D^2 m_o$ for M_u and $e^2/m_o N_D$ for G, one obtains $R_s = 4 r_{cl} N_D = <R_{ee}>(M_u/m_o)^{1/2}$. According to Lemaitre's hypothesis, galaxies, stars and all other cosmological structures arise from similar-sized "seeds" which in the present model are relativistic pair-systems of lower mass to which the same laws apply. Thus, they all have the same value for $<R_{ee}>$ and V_o, so that leads to still another testable prediction, namely that the sizes of cosmological systems or their spacing should be proportional to $M_{ee}^{1/2}$. Such a relation has indeed been found to hold empirically for the large, diffuse systems [21]. However, inspection of a logarithmic plot of size vs. mass indicates that they are not uniformly distributed but tend to cluster strongly around certain average values that tend to be about a factor of 1000 apart. Thus superclusters typically have a mass of a few hundred to a few thousand galaxies, galaxies tend to have masses some thousand times that of dwarf galaxies, and so on down to stellar associations and stars. In terms of the present model, this suggests that the decay process of the original Lemaitre atom took place in 27 major stages of some 10 divisions by 2 since $2^{10} = 1024$. To go from the mass of the original pair down to the mass of hadrons therefore takes some 270 divisions, which in the present model is a form of internal pair production known to lead to the formation of pairs in nuclei when enough energy is available and other decay mechanisms are forbidden

As to the physical reason why such a halt or slow-down in the divison process might occur whenever a thousand pairs have been formed, this appears to be connected with the number of pairs that can fit into the volume V_0 in which the fields are strong enough to allow pair-creation to take place, just as in the case of so-called geons or self-stabilized vortex rings of pure electromagnetic field energy investigated by Wheeler [22]. From the fact that the spacing of pairs as confirmed by the rotational levels of the excited hadron states is equal to r_{cl} and the fact that the size of V_0 and V_{eff} lies between 1096 and 1400 r_{cl}^3 it appears that on average some 10 divisions by 2 fill this critical volume, and the next division process forces ejection from the rotating cluster that is held together by strong gravitational forces when each pair has the mass of a star, a galaxy or even a supercluster.

If one calculates the masses of successive clusters of 1024 pairs according to the simple relation $M_{ee} = M_u / (1024)^n$, one arrives at a mass of 7.40×10^{15} solar masses M_0 typical of superclusters for n =3; at 7.23×10^{12} M_0 typical of large galaxies for n=4; at 7.06×10^9 M_0 typical of globular clusters for n=5; at 6.73×10^3 M_0 seen for large stellar associations for n = 7; and at 6.5 M_0 characteristic of the more massive stars for n = 8. These masses are about ten times greater than the average luminous masses for these objects, again suggesting that the present model is consistent with the existence of significant amounts of dark matter. This may either be in the form of baryonic matter such as brown dwarfs, planets and smaller objects in the halos of galaxies as have been recently reported using gravitational lensing [23,24,25], or it may be in the form of as yet undecayed massive pairs still trapped in the nuclei of the larger systems such as superclusters and galaxies [20]. But according to the present theory this dark matter cannot be composed of any new exotic particles or neutrinos of finite rest-mass since, according to the electron pair model of matter, the neutrinos are pure electromagnetic radiation quanta.of spin 1/2.

Because of the relationship beween the masses and sizes of cosmological objects, there is a high regularity in the sizes or the distances between them [20]. Thus galaxies are predicted to have an average distance of 4.73×10^6 light years between their centers when the supercluster to which they belong is relatively old so that it has collapsed to a flattened form and stopped expanding at its high initial rate of expansion. Because according to the present model all systems including the largest must rotate just as planetary systems and galaxies do [17,20], the galaxies in superclusters such as our own are in quasi-stable equilibrium while participating in the overall expansion rate of the universe. This would explain the recent finding by Napier [26] that the red-shifts seem to be quantized as if the galaxies in the local supercluster were arranged in a lattice of about the predicted spacing that maintained its regularity while expanding . Moreover, because the masses of all systems differ by factors of two from the average values due to fluctations in the decay, the model also implies that the velocities of the stars and gas clouds in the arms of galaxies should show evidence of quantization as has in fact been reported by Tifft [27].

Other interesting aspects of the Lemaitre model that appear to resolve certain theoretical and observational problems faced by the standard model and discussed elsewhere in more detail include the following :
1) The problem of explaining the uniformity of the universe in different regions too far apart to have ever communicated with each other after the Big Bang does not exist in the Lemaitre model since the evolution takes place from a single pair over a long period of some 10^{13} years before the Big Bang when baryons formed [20,30].
2) The need for extremely "finely tuned" initial conditions at the Big Bang to achieve a precisely "flat" universe balanced between indefinite expansion and recollapse does not exist since all cosmological structures rotate so that any mass value can lead to stable states just as is observed for planetary systems and galaxies [17,20,21].
3) There exists no initial singularity since the initial pair has a finite size, nor do black holes contain singularities. There are no infinite physical quantities since all charges, masses, angular momenta and velocities are finite [31].
4) The problem of the early production of too much helium in the presence of a high density of ordinary baryonic matter does not exist because during the first moments after neutrons have been formed when the temperature is just high enough to fuse nucleons into helium , some 99% of all matter is still trapped in the massive central nuclei of the expanding systems, unavailable for the production of helium and other low mass elements [29,30].
5) There is no difficulty in explaining the very large coherent motions of galaxies since superclusters as well as all larger systems rotate like galaxies, leading to average rotational velocities of superclusters of 1171 km/s [30].
6) The problem of explaining how protons or ordinary matter survived the Big Bang without annihilating with anti-matter does not arise since the initial pair had either one sense of orbital motion relative to the spin of the electron or the other as explained above for the π^0_1, thus determining whether baryons or anti-baryons form.
7) The problem of understanding the presently observed ratio of photons to nucleons is solved by the possibility of calculating the kinetic energy with which baryons and pions are created in the final, 27th step of the decay process. This motional energy heats the universe to an initial temperature of about 10^{13} K and produces some 2.8 $\times 10^9$ thermal background radiation quanta per nucleon at the time when radiation decouples from matter [29,30].

8) Quasars and their tendency to occur more frequently near galaxies than expected [32,33] find a natural explanation as delayed ejections [34,35] of newly evolving cosmological structures from the extremely dense central nuclei of the more massive systems, as first suggested on observational grounds by Ambartsumian [36] but widely disbelieved because there was no theory for such enormously dense forms of matter. Therefore they seem to be "mini-bangs" in which galaxies and stars continue to be born, producing energetic baryons and relativistic jets of electrons and positrons accompanied by radiation, including powerful gamma rays [29] as first envisioned by Lemaitre [13].

There are other interesting aspects of the Lemaitre model dealt with elsewhere[17,29,30] such as the fact that the model requires the local value of G to be inversely proportional to the square-root of the mass of the relativistic pair. Thus, the local gravitational constant within $R_s(M_{ee})$ increases by N_D or 10^{42} times from its initial to its final value of $(1/\alpha)^{1/2}$ (e^2/m_o^2) close to that arrived at by Motz [2] for a gravitationally stabilized electron while the overall valueof G for the universe remains constant. This causes the value of the Planck mass $M_{Pl} = (\hbar c/G)^{1/2}$ to decrease 21 orders of magnitude from its large initial value to the geometric mean of m_o and $(1/\alpha)m_o$ while the Planck length $(\hbar / M_{pl} c)$ increases to $(2/\alpha) r_{cl} /4$ or to the outer size of the π^o_1 field. Therefore, the minute scale of superstrings becomes equal to that of hadrons so they that can now be identified with the lines of force between the relativistic charges confined within $R_s(M_{ee}$, which for the π^o_1 becomes $(2/\alpha) r_{cl} /4$. As a result, superstrings are brought from a hypothetical miniature world back to that of actual nuclear particles where they originated.

CONCLUSION

It appears that we may now be on the verge of reviving the hope that all physical processes can be understood in terms of an electrodynamic theory of matter, modified by the discovery of the quantization of orbital angular momentum and spin. Since all matter and forces are aspects of the properties of electrons in the present model, one can return to the relativity theory of Poincaré and Lorentz in which motions take place relative to an absolute reference frame or an ideal, fluid-like space-time continuum. This fruitful concept of an ether was taken from Anaximander in ancient Greece and revived by Descartes and Newton. Unfortunately, it was briefly abandoned by Einstein in 1905 because of the particle-like action of photons only to be found necessary for his General Theory ten years later [37,38]. However, such an absolute reference frame is required for the concept of rotation of the first electron pair and its stabilization by a centrifugal force in the absence of any other matter. But once such a plenum is assumed to exist, it is possible to consider the ultimate entities to be "quantized superstrings", or stable vortices in an ideal fluid as originally envisioned by Helmholtz, Faraday and Maxwell.

The Lemaitre model will soon be subjected to an important test by the Space Telescope as it looks back in time at the most distant galaxies. These newly forming "proto-galaxies" must turn out to be small and extremely bright objects similar to quasars that have since then expanded to their present spiral shape from a massive central source rather than by collapse from a diffuse gas. If this model continues to pass its tests, then there is hope that the universe can continue to exist forever since rotation will keep the systems of galaxies in equilibrium.

REFERENCES

1. E. J. Sternglass, in: " Proc. 10th Int Congr. Hist. of Science", Hermann, Paris (1962),p.355
2. L. Motz, Nuovo Cimento, 26 : 672 (1962)
3. E. J. Sternglass, in: "Horizons of a Philosopher", J.Frank et al., eds., E.J.Brill, Leiden (1963), p. 422
4. E. Fermi, and C. N. Yang, Phys. Rev. 76:1739 (1949)
5. E. J. Sternglass, Phys.Rev.123:391 (1961)
6. E. J. Sternglass, in " Nucleon Structure", R.Hofstadter and L. I. Schiff,eds., Stanford U., Stanford (1964), p. 340
7. E. J. Sternglass, Nuovo Cimento 35:227 (1965)
8. E. J. Sternglass, in: "Resonant Particles", B. A.Munir,e d., Ohio University, Athens (1965), p.33.
9. E. J. Sternglass, Int. J. Theor. Phys. 17:347 (1978)
10. M. Aguilar-Benitez et al, "Particle Data Booklet", North Holland, Amsterdam (1986), p.18
11. E. Fermi, Zeits. Phys. 88:161 (1934)
12. L.W. Jones and O.W. Greenberg, in "Nuclear Physics and Particle Source Book", S.B. Parker, ed., McGraw-Hill, New York (1987)
13. G. Lemaitre, Nature 128: 701 (1931), also "The Primeval Atom", Van Nostrand, New York (1950)
14. E. J. Sternglass, Bull. Am. Phys. Soc. April 1975 Meeting.
15. T. Applequist and H.D. Politzer, Phys.Rev.Lett. 34:43 (1975)
16. M. Aguilar-Benitez et al, "Particle Data Booklet", North Holland, Amsterdam (1986), p.37
17. E.J. Sternglass, Lett. Nuovo Cimento 41:208 (1984)
18. V.L.Ginzburg, Comments Astron. Astrophys. 3:7 (1971)
19. R.P. Kirshner, A. Oemler and P. L. Schecter1, Astrophys. J., 84, 95 (1979)
20. E. J. Sternglass, in:"Testing the AGN Paradigm", Am. Inst. Physics, New York (1992)
21. B. J. Carr and M.J. Rees, Nature: 278, 605 (1979)
22. J..A. Wheeler, Phys. Rev. 97: 511 (1955)
23. C. Alcock et al., Nature (1993)(In press)
24. M. Spiro, et al., Nature (1993)(In press)
25. B. Barczynski, et.al., (1993) Acta Astronomica (In press)
26. W. Napier (These Proceedings)
27. W.G. Tifft, Astrophys. J. 206:38 (1976)

28. V. F. Litvin et al., Astrophys. Space Sc. 202:33 (1993)
29. E. J. Sternglass, in "Proc. 2nd Compton Symposium", Am. Inst. Physics, New York (1994)
30. E. J. Sternglass, (Unpublished manuscript)
31. E. J. Sternglass, Annals N. Y. Acad. Science 480:614 (1986)
32. G. Burbridge, A, Hewitt, J.V. Narlikar, P. Das Guptas, Astrophys. Suppl. 74:675 (1990)
33. H. C.Arp, G.,Burbridge, F., Hoyle, J. V.,Narlikar and N. C. Wickramasinghe, Nature 346:807(1990)
34. I. D. Novikov, , Astron. Zh. 41:1075 (1964)
35. Y. Ne'eman, Astrophys. J., 141:1303 (1965)
36. V. A.1.Ambartsumian, in "La structure et l'evolution de l'univers", Institut International de Physique Solvay, ed. R. Stoops, Couder
37. A. Grunbaum, Philos. Rev. 66:525 (1957)
38. L. Kostro (These Proceedings)

ARE QUASARS MANIFESTING A DE SITTER REDSHIFT?

John B. Miller[*] and Thomas E. Miller[†]

[*] 6 Benham Street
   Danbury CT 06811, USA

[†] 2 Riverside Drive
   Cooperstown NY 13326, USA

ABSTRACT

In 1929, Edwin Hubble wrote in his classic paper[1] demonstrating a correlation between redshift and distance, "The outstanding feature, however, is the possibility that the velocity-distance relation may represent the de Sitter effect...." Since the discovery of quasars more than thirty years ago, many more-or-less plausible explanations for the quasar redshift have been proposed. Although the de Sitter redshift was the first known cosmological redshift, it has not yet been considered as a possible etiology for the redshift of quasars. We address the question, "Is it possible that the quasar redshift is a de Sitter redshift?" Perhaps the asymptotic character of a gravitational de Sitter redshift[2] could help explain the quasar phenomenon: objects with high redshifts that appear to be almost as bright as objects with intermediate redshifts. Reconsidering the possibility of a nonlinear de Sitter redshift-distance relation, we find quasar intrinsic brightness to be rather ordinary. Given a de Sitter redshift-distance law, intrinsic brightness is found to be independent of redshift over five orders of magnitude.

INTRODUCTION

The de Sitter redshift was most popular in the twenties[3] and has now become obscure. (The de Sitter solution is distinct from the Einstein-de Sitter solution.) There are many different formulations of the de Sitter solution, depending on the choice of coordinate transformations.[4] In order to use the inverse-square law so that coordinate distance equals luminosity distance, we choose coordinates (r, θ, ϕ, t) that preserve the Euclidean formula for surface area: $A = 4\pi r^2$. Any other coordinate transformation entails a surface area formula that does not obey strict inverse-square-law dimming.

THEORY

Given the relativistic spacetime metric

$$ds^2 = -\gamma^{-1} dr^2 - r^2 d\theta^2 - r^2 \sin^2\theta \, d\phi^2 + \gamma \, dt^2, \qquad (1)$$

the de Sitter solution to the Einstein field equations can be written as

Frontiers of Fundamental Physics, Edited by M. Barone
and F. Selleri, Plenum Press, New York, 1994

$$\gamma = 1 \ - \ (r/R)^2, \tag{2}$$

with $R^2 = 3/(8\pi\rho)$, where the constant R is the radius of spacetime curvature and ρ is the mean mass density of the universe (assuming simplified units so that $G = c = 1$). Redshift z is defined as

$$z \ = \ \lambda_r/\lambda_0 \ - \ 1 \ = \ \gamma^{-1/2} \ - \ 1, \tag{3}$$

where λ_0 is the wavelength of the unshifted photon and λ_r is the wavelength of the photon observed at a distance r. Distance can then be given as a function of redshift,

$$r \ = \ R \ [1 - (z + 1)^{-2}]^{1/2}. \tag{4}$$

Equation (4) is the de Sitter redshift-distance law—the de Sitter equivalent of the Hubble law (Fig. 1).

Assuming inverse-square-law dimming, absolute magnitude M and apparent magnitude m are related to distance r by

$$M + C \ = \ m - 5 \ \log_{10} (r) \ , \tag{5}$$

where C is a constant determined by observation. According to the Hubble law, redshift is directly proportional to distance, thus

$$M_H + C_H \ = \ m - 5 \ \log_{10} (z) \ , \tag{6}$$

where M_H is Hubble absolute magnitude and C_H is an observational constant that incorporates the Hubble constant. Combining equation (4) and equation (5) gives the de Sitter version of the magnitude-redshift relation

$$M_D + C_D = \ m - 2.5 \ \log_{10} [1 - (z + 1)^{-2}] \ , \tag{7}$$

where M_D is de Sitter absolute magnitude and C_D is an observational constant that incorporates the de Sitter radius R.

METHODS

To compare the Hubble and de Sitter laws in the context of observation, we examined large astronomical catalogs of objects (galaxies and quasars) with published measurements of both redshift and apparent magnitude[5,6,7]. We transformed apparent magnitude to absolute magnitude according to the Hubble and de Sitter laws using equation (6) and equation (7) respectively. Figure 2 shows magnitude plotted versus redshift: first the raw data or apparent magnitude, second the absolute magnitude assuming a Hubble law, and third the absolute magnitude assuming a de Sitter law. As a statistical assessment, a least-squares-fit line is drawn for each set of data in the plots and the slope and the square of the correlation coefficient are given.

RESULTS

The line in the Hubble plot (Fig. 2*b*) is down-sloping, while the line in the de Sitter plot (Fig. 2*c*) is nearly horizontal. Almost all objects fall within a band spanning about ten magnitudes vertically (for a given redshift). This entire band is clearly down-sloping assuming a Hubble law, but is roughly horizontal assuming a de Sitter law. De Sitter absolute magnitude is practically uncorrelated with redshift, while Hubble absolute magnitude is fairly strongly correlated with redshift. This is found to be the case not only for the combined catalogs, but also for galaxies and quasars considered separately (Table 1).

DISCUSSION

Some increase in intrinsic brightness is to be expected as a result of Malmquist bias: more distant objects will tend to be brighter because we will only be able to see distant objects if they are intrinsically brighter than closer objects. However, Malmquist bias is generally thought to be inadequate to explain quasar intrinsic brightness. Of

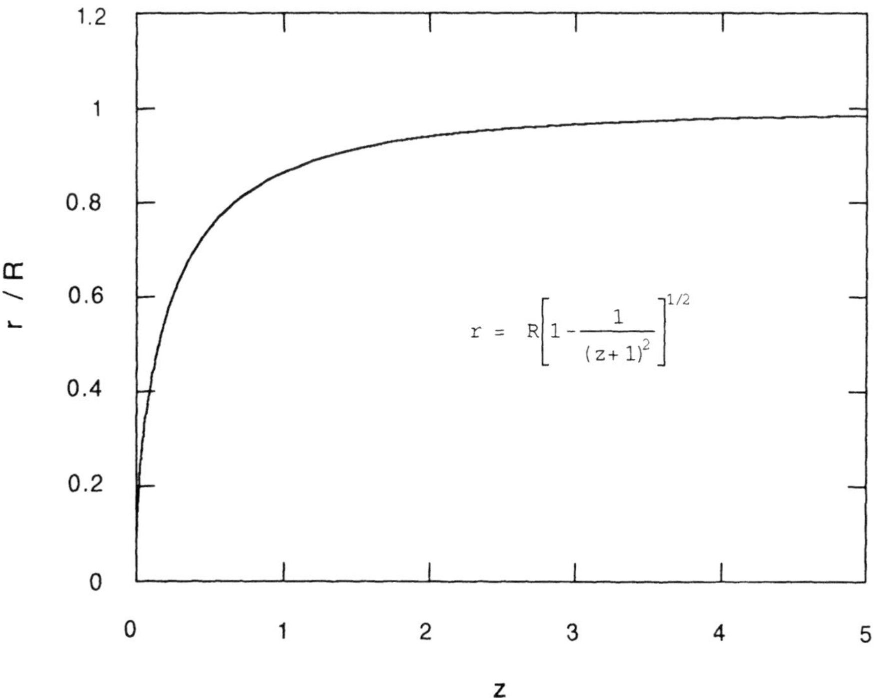

Figure 1. Distance (r/R) plotted versus redshift (z) in a de Sitter universe (equation 4).

course, the downward slope of the least-squares-fit line in the Hubble plot (Fig. 2*b*) can be explained by luminosity evolution. That is, as we look out into space, we look back to a time when objects existed that were intrinsically much brighter than anything currently (or locally) extant.

An alternative is the original de Sitter redshift. Given the asymptotic character of the de Sitter redshift at large distances, quasars with high redshifts may not be much more distant than are intermediate-redshift objects. Since absolute magnitudes are roughly the same at all redshifts assuming a de Sitter law, quasars may be considered in the context of a nonlinear de Sitter redshift-distance law without invoking luminosity evolution.

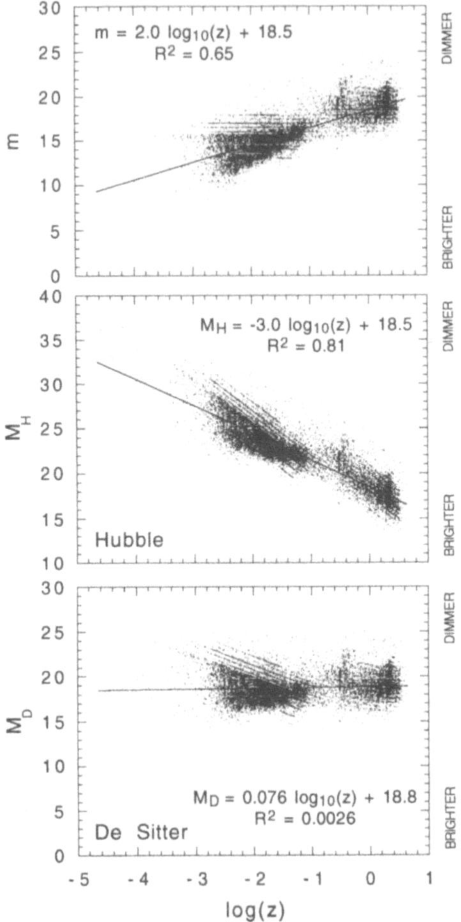

Figure 2. *a*, Apparent magnitudes of galaxies[5,6] and quasars[7] versus $\log_{10}(z)$. The corduroy texture represents lines of iso-apparent magnitude. *b*, Absolute magnitudes obtained by assuming a linear Hubble redshift-distance relation and applying equation (6) to the data set in Fig. 2*a*. The actual values are arbitrary, since the constant C_H was set equal to zero. This affects the intercept, but not the slope nor the correlation coefficient. *c*, Absolute magnitudes obtained by assuming a nonlinear de Sitter redshift-distance relation and applying equation (7) to the data set in Fig. 2*a*. The actual values are also arbitrary, since C_D was set equal to zero.

Table 1. Comparison of slope, intercept, and correlation.[†]

	Number	Hubble			De Sitter		
		Slope	Intercept	R^2	Slope	Intercept	R^2
Galaxies[5,6]	16118	-3.0	18.4	0.60	-0.27	18.2	0.011
Quasars[7]	4197	-3.6	18.6	0.52	0.33	19.0	0.0089
Total	20315	-3.0	18.5	0.81	0.076	18.8	0.0026

[†]Galaxies are taken from the *CfA Redshift Catalogue* (n = 15597, z < 0.33) and the *ZBIG Catalog* (n = 521, z > 0.33). Quasars are taken from *A New Optical Catalog of Quasi-Stellar Objects*. Slope refers to the slope of a least-squares linear regression curve for a scatter plot of absolute magnitude versus $\log_{10}(z)$. Intercept gives the $\log_{10}(z) = 0$ intercept of this line, assuming $C_H = C_D = 0$. R^2 is the squared correlation coefficient (which is distinct from the de Sitter radius R). Only objects with published measurements of both redshift and apparent magnitude are included. A few blue-shifted galaxies have been excluded.

REFERENCES

1. E. Hubble, A relation between distance and radial velocity among extra-galactic nebulae, Proc. natn. *Acad. Sci. U.S.A.* 15:168 (1929).
2. W. De Sitter, On Einstein's theory of gravitation and its astronomical consequences (third paper), *Mon. Not R. astr. Soc.* 78:3 (1917).
3. P. Kerszberg. "The Invented Universe: the Einstein-De Sitter Controversy (1916-17) and the Rise of Relativistic Cosmology," Clarendon, Oxford (1989).
4. A.S. Eddington. "The Mathematical Theory of Relativity, Second Edition," University, Cambridge (1924).
5. J. Huchra. "CfA Redshift Catalogue, Selected Astronomical Catalogs, Volume 1, CD ROM," Astronomical Data Center, NASA Goddard Space Flight Center, Greenbelt, Maryland, USA (1990).
6. J. Huchra and C. Clemens. "ZBIG Catalog," Personal Communication (1991).
7. A. Hewitt and G. Burbidge. "A New Optical Catalog of Quasi-Stellar Objects, Selected Astronomical Catalogs, Volume 1, CD ROM," Astronomical Data Center, NASA Goddard Space Flight Center, Greenbelt, Maryland, USA (1989).

WHAT, IF ANYTHING, IS THE ANTHROPIC COSMOLOGICAL PRINCIPLE TELLING US?

Silvio Bergia

Dipartimento di Fisica, Università di Bologna
I.N.F.N., Sezione di Bologna
Via Irnerio, 46 - 40126 Bologna - Italy

INTRODUCTION

There can be no doubt that the evolution undergone by our ideas concerning the universe from ancient until very recent times has been characterized, on the one hand, by the removal of the earth, and of mankind, from the privileged position attributed to both of them in old cosmologies, and, on the other hand, by the rejection of any teleological point of view.

Authors presenting or discussing the Anthropic Cosmological Principle, however, have often claimed that those ideas ought to be reconsidered; in other instances, the explicit claim has been replaced by a more or less explicit message. That we should seriously consider the possibility of a teleology of nature is the impression one receives from reading the most extensive book on the Anthropic Cosmological Principle appeared in the literature, the treatise by J.D.Barrow and F.J.Tipler, published in 1986 (BARROW and TIPLER 1986). The book's review written by W.H.Press for Nature was significantly titled "A place for teleology?" After recalling that "it is an understatement to say that teleology has been out of fashion in science in the past century, and probably not much of an exaggeration to say that the present scientific paradigm rejects the teleological hypothesis vehemently, categorically and usually with contempt", Press went on: "Now there comes a book - no crackpot tract, but a scholarly, philosophically sophisticated and mathematically high-brow monograph - that says we've all made a big mistake: there really is a place for teleology and related concepts in today's science. At least (the authors ask), give us a chance to present arguments drawn in the main from modern theoretical cosmology, which may convince the reader of an astounding claim: there is a grand design in the Universe that favours the development of an intelligent life." (PRESS 1986, p.315).

Are we really obliged, by the arguments presented under the heading "Anthropic Cosmological Principle" [1] to accept this conclusion? I will argue that we are not.

We may reject the conclusion *a priori* advocating the postulate that science is not concerned with final causes, and that adopting a teleological viewpoint is a one man/woman decision, which depends on his/her general philosophical attitudes

Frontiers of Fundamental Physics, Edited by M. Barone
and F. Selleri, Plenum Press, New York, 1994

and may in turn contribute to shape them, but which should not creep into his/her scientific work. However, this attitude may be dubbed dogmatic and we may be invited instead to face the evidence revealed by the authors who have developed the field.

I will therefore take a different attitude and in fact examine some of this evidence. This examination leads, in my opinion, to the conclusion that the universe is indeed, in a way to be specified, "critical with respect to biology". The conclusion could be seen as a manifestation of the presence of final causes; I will however argue that the situation is similar to that concerning the evolution of life on earth. Therefore, an extended Darwinian attitude can be adopted (and in fact *ought* to be adopted by scientists) in analysing it. The Darwinian outlook and teleology, as I will stress once more further on, are rigorously antithetic. The confusion produced by the studies on the anthropic principle has mainly arisen, in my opinion, from the circumstance that some presentations have unduly mixed the two points of view. This is, it seems to me, only part of the confusion that has very often surrounded the subject. There is in fact, as I will argue, more to it. To begin with, in what sense, if any, are we here dealing with a principle? My second purpose is to analyse the formulation of WAP with the preliminary aim of clarifying its nature as a proposition, that is to say from a *logical-syntactic* point of view. This will allow me to analyse its status also from the *methodological* point of view, with particular attention to its predictive power, which will be compared with that of the ordinary propositions of physics. Its interest from this point of view is however limited due to its reduced *explicative power*. This difficulty is usually overcome, as I will argue, by operating a gradual shift in the meaning of the principle. This is brought about in two steps: the first one consists in viewing the WAP as *an ordering principle*. This step is to be considered as a positive one, as it ultimately is the one that permits us to identify, as I will argue, the objective message conveyed by the considerations which are labeled under the term Anthropic Principle: as already mentioned, this step consists in the discovery that the universe is critical with respect to biology. The second shift of meaning is brought about when one subreptitiously makes the reader develop the feeling that the principle gives a causal explanation. Since, in this case, it would be the matter of a "final cause", one would then ineluctably open the doors to teleology.

In the exposition I will find expedient to reverse the order in which the two main points are dealt with. Hence Section 1 is devoted to analysing of WAP as a proposition; Section 2 deals with the difficulties, both objective and subjective, which are encountered when trying to ascribe an explicative power to WAP; Section 3 presents some of the evidence for the anthropic arguments and examines WAP as an ordering principle; Section 4 analyses the usual presentations, pointing out where they become misleading for lack of precision in separating Darwinian from teleological arguments; it will also be argued there that the so-called Strong Anthropic Principle represents the natural outcome of the above-mentioned process of contamination and that it amounts to stating the teleology of the universe.

1. THE WEAK ANTHROPIC PRINCIPLE AS A PROPOSITION

The term "Anthropic Principle" is due to Brandon Carter, of Cambridge, who coined it in 1974, in the two versions of "Weak Anthropic Principle" (WAP) and "Strong Anthropic Principle" (SAP). I will refer here to the formulations given of both by Barrow and Tipler in their book.

I will initially concentrate my attention on WAP, and come back later to the strong version. According to Barrow and Tipler, WAP states that:

"The observed values of all physical and cosmological quantities are not equally probable but they take on values restricted by the requirement that there exist sites where carbon-based life can evolve and by the requirement that the universe be old enough for it to have already done so" (BARROW and TIPLER 1986, p. 16).
Referring to formulations such as this, Wheeler, in his Foreword to the book, observes:

"What is the status of the anthropic principle? Is it a theorem? No. Is it a mere tautology, equivalent to the trivial statement, 'The universe has to be such as to admit life, somewhere, at some point in its history, because we are here'? No. Is it a proposition testable by its predictions? Perhaps. Then what is the status of the anthropic principle? This is the issue on which every reader of this fascinating book will want to make his own judgement." [2]

We have here a testimony of the fact that, even to the experts, it is not altogether clear what the actual meaning of the principle is.

Before entering into any detail as regards the points I wish to analyse, let me briefly mention two questions, implied by the statement of the principle, on which I will not be able to dwell here. Note, first of all, the appearance of the notion of probability in an unusual context. Indeed, this notion makes sense only if a set of alternatives is given from which the extraction of a particular choice is possible. As P.C.Davies phrases it, one should "envisage a huge collection of possible universes - a world ensemble - each varying slightly from the others so that somewhere among the ensemble would be a universe in which every conceivable value for each fundamental constant, and every conceivable initial arrangement of matter and motion, where realized to within a certain accuracy." (DAVIES 1982, p.123). The idea of the "world ensemble" can be, and has been criticized. G.Toraldo di Francia, for instance, observes: "What precisely are the possible worlds?...I think those scholars, like W.V.O.Quine, to be right, who doubt that it makes sense to speak of unrealized possibilities. The strong suspicion arises that, if they have not be realized, it is because they were not possible" (TORALDO 1990, p.27). The discussion of this point is certainly central for an analysis of the anthropic principle, but I will here sacrifice it in order to be able to deal at some length with the aspects mentioned above. I will thus acritically assume throughout, for the sake of the discussion, that the notion of world ensemble is a legitimate one. Note also that the statement makes an implicit reference to evolutionary cosmologies, such as the standard hot big bang cosmology, insofar as it alludes to the age of the universe. The anthropic principle allegedly provides arguments in favour of such cosmologies. The essential point seems to be the rough coincidence between the Hubble time and the age of a typical star. I have not yet analysed here the real content of the anthropic principle, but let us assume, for the sake of argument, that it implies that the value of the Hubble time is restricted by the conditions necessary for the existence of man. An essential condition is then that the universe be old enough for there to exist elements heavier than hydrogen. Now, heavier elements are synthesized in the interior of stars; as a consequence, the Hubble time of an inhabited universe cannot be shorter than the age of a typical star. On the other hand, if the Hubble time were much greater than the age of a typical star, most of the stars whose planets may sustain life would already be dead. [3]

Putting aside, for the moment, the question as to whether the anthropic principle *requires* such a restriction, it seems to me that this is not an argument in favour of an evolutionary cosmology, or it is so only to the extent that it finds it (qualitatively) self-consistent. An evolutionary cosmology is characterized by a typical time, the Hubble constant, or the age of the universe: it is found that its value is consistent with characteristic astrophysical times; no typical time exists in a steady-state cosmology and thus no problem of self-consistency arises there. What seems to me to be true, is that, since anthropic arguments deal with the origin and development of life, they are bound to be to some extent evolutionary. A steady-state cosmology cannot, and does not, of course, exclude evolution, at least on a local scale, and anthropic arguments can be used within its framework. However, an evolutionary view at the scale of the universe is only provided by a big bang cosmology, which is therefore more homogeneous with the anthropic viewpoint. For this reason I will consider a hot big bang cosmology as the natural background for the discussion that follows and will no longer come back on the issue of the comparison between the two cosmologies.

My purpose here is, first of all, to analyse the formulation of WAP with the preliminary aim of clarifying its nature as a proposition, that is to say from a *logical-syntactic* point of view. In J.Rosen's comment, WAP, in the version of Barrow and Tipler, "seems very incontroversial. It is simply recognition of the fact that the situation we observe in the universe must be consistent with our existence" (ROSEN 1988, p.416). My impression is that one could replace Rosen's "incontroversial" with "empty", at least until a point which I consider essential is not clarified.

Actually, a lot depends on the value one attributes to the phrase "take on values restricted by the requirement". It seems to me that presentations of WAP are often biased by the ambiguities concerning this point. It is possible to read the sentence as the simple statement of a matter of fact. In this case it does not in fact predicate anything on the subject "the observed values of all physical and cosmological quantities", and justifies the verdict of logical emptiness. On the same grounds, it is on the other hand possible to read it as a proposition having a prescriptive value: in other words, "take on" is read as *"must take on"*. It should of course be stressed that this value is in fact attributed to the sentence merely from a logical-syntactic point of view for the purpose of attributing significance to the proposition. Obviously, no direct operative character is, or can be, attributed to the prescription.

This is a first, and evidently not irrelevant, discrimination. It therefore seems to me inacceptable that much of the current literature leaves it unsolved, thus justifying Wheeler's uncertainties. It we adopt the second attitude, it becomes possible to answer the question, essential for a correct collocation of WAP from the methodological point of view, as to whether a testable proposition, or a set of testable propositions, can be associated with WAP.

To tackle this question, I will refer to a particular phenomenon: carbon nucleosynthesis in stars.[4] The synthesis (of three helium nuclei) takes place at a particular stage of the stellar evolution, both through both a direct and an "autocatalytic" reaction, proceeding via an intermediate beryllium 8 step. These reactions *must* proceed resonantly to produce a non negligible yield of carbon, such as to justify the actual presence of carbon (and, incidentally, of life) in the universe. That *must*, which follows from the assumption of the prescriptive character of the principle's statement, is the clue the example provides us with to identify the kind of proposition we are looking for. On the basis of the datum "existence in the universe of carbon-based forms of life", which in turn requires the presence of relevant quantities of this element, and on the basis of our knowledge of the processes of stellar nucleosynthesis, we can state that certain nuclear reactions, which can be studied in the laboratory, must take place in a specific way, and namely resonantly. From a logical point of view, we are therefore in the presence of a *prediction* following from two premises. Up to this point, the term "prediction" has been used in a purely logical and a-temporal sense. From the point of view of the temporal sequence of events, one must distinguish between the *post-diction* of what must have taken place in the interiors of stars (time here is that of natural history) and the *pre-diction*, in a strict temporal sense, of the features of a process we can study in the laboratory (for which time is that of the laboratory life). The first one acts in an anti-temporal way, hence in a different sense from that which is customary in physics, where"the future is deduced from the past" (GALE 1981,

p. 117), while the second one acts in a sense that is perfectly homogeneous to what normally occurs in physics; and it is interesting to note that the history, not of the universe, but of scientific activity, has followed this logical path. In 1953, Fred Hoyle, on the basis of considerations of the type reported above, formulated the prediction that, in order for the mentioned rections to be effective at the cosmological level, the carbon nucleus must have a resonant level lying near 7.7 MeV, whose existence was soon verified experimentally (BARROW and TIPLER 1986, p. 252).

Therefore, it is not at the level of the temporal sequence of events "formulation of a hypothesis" and "experimental verification of the hypothesis" that a proposition, like the one in which Hoyle's prediction could be condensed, differentiates itself from those of the type usually employed in physics. The difference can be seen, as will be discussed in the next section, when one tries to attribute an explicative power to the proposition.

2. EXPLICATIVE POWER

The most marked difference between WAP, interpreted in this way and the ordinary propositions of physics is best appreciated when one wonders about the explicative power of such a proposition. Can we say that the existence of life on earth *explains* the carbon's resonant state? Pretending to consider this as an explanation, observes Rosen, runs into two difficulties, one of a subjective and one of an objective nature.

The objective difficulty is what Rosen calls "the invariant context problem". "For an explanation to be an explanation, at the very least what is being explained must follow logically from what is explaining; in our case, the existence of Homo sapiens must be *sufficient* (emphasis added) for the coupling constants to have the value they do, or the actual values of the coupling constants must be *necessary* (emphasis added) for our existence , or if the values were different we could not exist". Now, can we state with certainty that, if those constants had not their actual values, man could not exist? Yes, but only if it is "tacitly assumed that no other aspect of the universe, no other law of physics, not even the form of the interaction equation, is varied. (ROSEN 1988, p. 417)" Without the assumption of the invariance of the context, the *sufficiency* of the existence of Homo sapiens, or, which is the same, the *necessity* of the constants having their actual values, does not arise.

The subjective difficulty "is that we physicists just do not feel that any explaining is being done". We want indeed "that which is explaining to be more fundamental, simpler, more general, and more unifying than that which is being explained, and we would also like the former to be the cause of the latter. In the present case, none of these seems to hold" (ROSEN 1988, p. 417). As I have anticipated, I will argue that the difficulty experimented in accepting that the principle has an explicative power is usually overcome through a gradual shifting of its meaning.

3. THE ANTHROPIC PRINCIPLE AS AN ORDERING PRINCIPLE

What has been said up to this point may leave one unsatisfied, as if the subject matter had only been touched on. One tends to ask: is it all there? In fact, it is not. When reading the current literature on the anthropic principle, one gradually discovers that people are saying something more, and different, than what is contained in a proposition like that stated the WAP. What, explicitly or implicitly is being said is rather: *we have discovered* that "the observed values of all physical and cosmological quantities are not equally probable but they take on values restricted by the requirement that there exist sites where carbon-based life can evolve and by the requirement that the universe be old enough for it to have already done so." In order to clarify this point, let us enunciate a logically equivalent proposition, which however highlights the aspect of interest:

We have discovered that, if the cosmological quantities had not all the observed values, there could not exist a carbon-based life.

I must say that a large part of the considerations appearing in the literature under the generic heading "Anthropic Principle" mostly stress the essential role that specific aspects of the cosmic evolution and of the laws and constants of nature have had and have in determining the conditions for the appearance of life on earth.

The first instance I would like to briefly discuss is provided by water, whose chemical and physical structure, as Henderson had already noted in 1913, make it "a uniquely useful liquid and the basis for living things" (quoted in BARROW and TIPLER 1986, p. 524). In particular, Henderson pointed out that "the expansion of water in freezing is essential for life if it is to evolve in a constant environment. If ice were not less dense than water, it would sink on freezing. The coldest water in a lake or ocean would congregate near the bottom and there freeze. Ice would accumulate at the bottom; the amount would become greater each year as more ice formed during the winter and did not melt during the summer. Finally, all the lakes and oceans would be entirely frozen" (cited in BARROW and TIPLER 1986, p.533). Now, all properties of water can be understood in terms of the atomic structure of the water molecule and of the chemical bonds that keep it together (*ibidem*, p. 526). In particular, the shape of the molecule as determined by these bonds is that of an isosceles triangle in which the H-O-H angle is 104.5 degrees (*ibidem*, p. 526). Moreover, water molecules tend to form highly directional hydrogen bonds with each other. The above value of the bond angle is only slightly less than that of the ideal tetrahedral angle (109.5 degrees); and the anomalous density of ice is ultimately due to this fact (BARROW and TIPLER 1986, p. 532-533).

As a second instance, we may recall that the possibility of carbon-based life rests upon a coincidence regarding the relative strength of the strong and electromagnetic forces. As discussed by Barrow and Tipler, "a 50% decrease in the strength of the nuclear force...would adversely affect the stability of all the elements essential to living organisms and biological systems. Similarly, holding the strong force constant...the stability of carbon requires the fine structure constant α to be less than ~ 0.1" (BARROW and TIPLER 1986, p.326-327).

Finally, let me consider another instance concerning once again nucleosynthesis. To what has already been said the stellar nucleosynthesis of carbon, a lot more may be added as regards primordial nucleosynthesis. Just to give an example, we may recall that, in adition to deuteron, a proton-neutron bound state, the nuclear strong interaction could give rise to a "diproton", a two-proton bound state: for its existence, the strong interaction would have to be only slightly stronger, in order to overcome the electric repulsion. However, as stressed for instance by Barrow and Tipler, the existence of this bound state would have catastrophic consequences, since all the hydrogen of the universe would have been burnt into helium during the initial phases of the big bang so that stars would not have been formed and the synthesis of the heavy elements would not have taken place (BARROW and TIPLER 1986, p.322).

On the other hand, the information upon which considerations of this kind are based is not new, as it has been available, often for quite a long time, in chapters of atomic and nuclear physics or astrophysics. It is clear then that, if this is the case, the term "principle" is being used improperly: from a strictly logical point of view, in fact, one is here stating no principle at all. What is being done, is that available information is being taken and ordered according to a thread which illustrates the above mentioned link between the physical world, its evolution and laws, and conscious life, instead of leaving it where it belongs naturally, that is in chapters of the various pertinent disciplines. In the best of circumstance, therefore, one is dealing with an *ordering principle*, with a *standpoint* from which to look at known facts according to a new perspective: the perspective dictated by the wish to understand what is necessary for life.

This operation, on the other hand, adds significance to the known facts, not by increasing their information content, but rather by making one see their relevance in a new context. It leads, and I have already used the term on purpose, to a *discovery*, the discovery that:

The universe is critical with respect to biology. [5]

Note, by the way, that this version provides an adequate framework for the self-consistency argument concerning the Hubble time discussed in Section 1.

The third instance reported above shows, however, that this statement is incomplete, and should in fact be extended. What the example shows is that the criticality of the universe with respect to biology cannot be disjoined from its criticality in a wider sense. Even minor changes in the values of the fundamental constants would not only lead to a universe incapable of sustaining life: in fact they wouldn't lead to any universe at all! This aspect has been discussed at some length in particular by I.L.Rozental (ROZENTAL 1980). For these reasons Rozental prefers to speak of a "principle of effectiveness" (or "appropriateness"). While agreeing on the necessity of the extension, I will stick for convenience to the terminology in use.

This seems to be the actual message (extension to a wider criticality included) conveyed by the complex of elements and arguments which fall under the label of "Anthropic Principle". And, from the point of view of its objective content, it is almost immaterial whether we are dealing with a principle or with a discovery. There is in fact no doubt that this way of looking at things adds significance to aspects, at times apparently marginal, of the laws of nature.

4. DARWINISM VERSUS TELEOLOGY

But the question is: do we want to infer some general conclusion from the mere observation that the universe is critical with respect to biology (or in a wider sense)? If we start speculating about this possibility, the message conveyed by the anthropic principle becomes more suggestive and intriguing, but also more ambiguous.

The starting point is the feeling that a kind of conspiracy must have been operating in order for life to have emerged in some place in the universe. This feeling is enhanced by a first evaluation of the probability of the evolutionary processes, required for the development of human kind having taken place on a cosmic and astrophysical scale and in the biosphere. The possibility of reasoning in terms of probability at the cosmic level depends, of course, on accepting the idea of the "world ensemble" briefly discussed above. Even if "it is hard to quantify the improbability of the choice of our perceived world, because...we do not know how to measure probabilities between possible universes", this idea authorizes us to state that "our world is indeed extremely unlikely on *a priori* grounds..." (DAVIES 1982, p. 123).

It is difficult not to be influenced by these considerations, to the extent that they could lead us to adopt a teleological view, which could be expressed in the following terms: "The existence of life in the universe is made possible by such a sequence of highly improbable events that it appears to be the result of a project".

However, this feeling is by no means new. For centuries throughout the course of history, human kind has never ceased wondering about the fitness of the terrestrial environment (on the whole, that is neglecting places like cold and hot deserts, the polar regions, etc.) for the purposes of human life.

The advent of Darwinism implied a deep change in this attitude. No wonder if the biosphere appears suitable to host man's life: had it not reached, in the course of its evolution, a condition favourable to the development of what is necessary to life, the evolutionary process which eventually led to man could not have taken place. The change of attitude implied by the Darwinian viewpoint has been lucidly expressed, for instance, by L.Gustavsson, in the following terms: the conception about the origin of the species "to whom Charles Darwin gave a conclusive form...states that certain systems like, for instance, the mammals' internal organs, are not made as they are because [final cause] something obliged them to take the aptest form, but rather because [efficient cause] modifications of the external conditions acted as a filter with respect to the manifold of theoretically possible evolutions, leaving go through only those suited to survive and to gradually transform. The teleological principle had been led from the status of universal principle to that of a simple relation between a biological system and its transitory natural environment. Teleology could be transformed in a normal causality relation" (GUSTAVSSON 1990).

There is therefore no need to have recourse to teleology to understand the fitness of the environment: it was not the environment to conspire to produce human kind, but it is man himself who is the result of an adaptation to the environment. Or, man is not the final cause of the earth in the universe, but it is the universe and the earth which are the efficient cause of man.

Now, the essential point seems to me to be that these considerations can be transferred from the ambit of the biosphere to that of cosmology, as defined by the standard model of a hot big bang. The only difference of principle that arises seems to derive from the fact that most of the *a priori* conceivable sets of laws and data, and the corresponding evolutionary paths, would not lead to any universe at all. However, this seems to me to produce only a non essential extension of the Darwinian viewpoint.

I am thus claiming that one should adopt in cosmology a Darwinian viewpoint, which essentially means replacing final causes with efficient ones, and thus, in this case, is satisfied with the conclusion that not only life could not have possibly arisen other than in a suitable universe, but also that it could not have arisen *without* a universe.

We can summarize this discussion in the following terms: the statement that "the observed values of all physical and cosmological quantities are not equally probable but they take on values restricted by the requirement that there exist sites where carbon-based life can evolve" is traced back to the consideration that, would it not have been so, the evolutionary process that we can retrospectively reconstruct could not have taken place. "Discovery" and assumption of the viewpoint do not add anything to WAP, either as a verifiable proposition, or set of propositions, or as an explicative proposition. The only novelties are in the discovery and the adoption of the viewpoint. This seems to me to be what, if anything, the anthropic cosmological principle is telling us. And I would like to stress that neither the discovery nor the viewpoint necessarily outline a crime of apostasy with respect to the current theoretical and epistemological framework, nor do they necessarily imply abandoning the secular attitude which, until one has proof to the contrary, rejects teleology as redundant and extraneous to natural sciences.

On the other hand, adopting a teleological viewpoint seems to be the only logically possible alternative to adopting a Darwinian viewpoint. It should be stressed that the issue is that of a *global alternative*, of an antithetic attitude, and not of a *variant*, or of a kind of integration of that viewpoint. This conclusion should be clear enough. And it must be said that there does not exist *any* version of the anthropic principle of which I am aware that immediately and explicitly reduces to a statement of teleology. However, some presentations have unduly mixed the two points of views. I believe this to be the essential cause of the confusion produced by the studies on the anthropic principle.

The contamination of the Darwinian viewpoint begins when one leaves almost inadvertently creep into the discussion the idea that WAP gives an explanation in causal terms. As I have said, we encounter a subjective and objective difficulty to accept that WAP provides an explanation. "To explain" probably means to us, first of all, to find out a causal link between *explanans* and *explanandum*. Can we say that the existence of man "explains" in this sense the values of the physical constants? It does not seem necessary to associate this idea with WAP. Nevertheless, a good deal of the anthropic arguments are pervaded by the question that looks for a cause: "Why?" And, what is worse, they let the reader develop the feeling that the question has found an answer. As stressed by Press, Barrow and Tipler's book is full of implicit question marks: "< Why > do the physical constants have the values that they do? < Why> is the Earth the size and temperature that it is?" And so on. In this way, he goes on, "the seductive trap that the authors are setting is already clear: as soon as as we accept the < why > formulation of questions that the WAP allows us to address, we have entered a receptive state of mind for the
Strong Anthropic Principle (SAP): The Universe must have those properties which allow life to develop within it at some stage in its history" (PRESS 1986, p.315).
Or, I would rather say, we have already taken a teleological viewpoint. Indeed, I must confess that I do not understand the meaning one wants to attribute to the SAP in the above statement: and, if we try to attribute one to it, we soon discover that it amounts necessarily to claiming that the universe has a purpose. For, once more, what does that "must" mean? At the logical-syntactic level it cannot but express a necessity; but, at this level, it expresses nothing more than what is already contained in WAP, once it has been decided that as an empty proposition it does not interest us. It is then clear that the authors want the implication to be read in another way. And it does not seem to be anything else but that of a cause-effect implication, as in fact suggested by Press when he reads the statement as an answer to the various < why's >. I therefore think that the SAP either is equivalent to the WAP, or answers the question: "Why does the universe have the properties that it has?" in the terms: "The universe has the properties that it has *in order that* [final cause] life might develop in it at some stages of its history."

Authors stating SAP seem to maintain that it is a proposition still possessing scientific relevance, situated somewhere in between WAP and the direct statement of teleology. On purely logical grounds, it appears *a priori* doubtful that an intermediate position may exist between so diametrically opposed points of view such as the Darwinian and the (neo)teleological ones. My conclusion is that, even if SAP does not *immediately* appear as a statement of teleology, it cannot have any other meaning.

Rosen also considers the as "meaningless within the context of science" (ROSEN 1988, p.416). The possibility of formulating such a proposition appears illusory. That a good deal of the arguments one finds under the generic heading of anthropic principle seem to make reference to it is the result of a methodological error.

FOOTNOTES

[1] Unless otherwise stated, I am here referring to the so-called Weak Anthropic Principle, or WAP for short; I will come back to the Strong Anthropic Principle (SAP) in due time.

[2] J.A.WHEELER, Foreword to BARROW and TIPLER 1986, p. vii.

[3] This argument was put forward by R.Dicke some thirty years ago: see Dicke 1961.

[4] This example is analysed in BARROW and TIPLER 1986, p.252.

[5] My attention was drawn towards this statement of the Anthropic Principle by A.Masani; see, for instance, MASANI 1987.

REFERENCES

J.D.Barrow, F.J.Tipler, 1986, "The Anthropic Cosmological Principle," Clarendon, Oxford.

P.C.Davies, 1982, "The Accidental Universe," Cambridge University Press, Cambridge.

R.Dicke, 1961, Dirac's cosmology and Mach's principle, *Nature* 192:440.

G.Gale, 1981, The anthropic principle, *Sci. Am.* 245:114.

L.Gustavsson, 1990, L'enigmatico principio antropico, *Lettera Internazionale* 24:44.

A.Masani, 1987, Il principio antropico *Il Nuovo Saggiatore*, nuova serie, 2:63.

W.H.Press, 1986, A place for teleology?, *Nature* 320:315.

J.Rosen, 1988, The anthropic principle II, *Am. J. Phys.* 56:415.

I.L.Rozental, 1980, Physical laws and the numerical values of fundamental constants, *Sov. Phys. Usp.* 23(6):296.

G. Toraldo di Francia, 1990, "Un Universo Troppo Semplice," Feltrinelli, Milano.

LARGE ANOMALOUS REDSHIFTS AND ZERO-POINT RADIATION

P. F. Browne

Department of Physics
UMIST
Manchester M60 1QD

INTRODUCTION

In de Sitter space-time using Robertson coordinates the Hubble
redshift can be interpreted as gradual decrease of photon frequency.
How this decrease of frequency might occur was suggested previously
(Browne, 1962). The radiation field for each Planck oscillator is
quantized gravitationally as a field of gravitons of minute constant
energy. Scattering of a graviton from one field to another toward the
equilibrium blackbody spectrum results in a redshift, $d\omega = - A\int \omega U dl$,
where distance dl is propagated in a medium with radiant energy density
U and A is a constant. A cross section, previously suggested (Browne,
1976), provides a theoretical value for A. The law becomes the Hubble
redshift $d\omega/\omega = - dl/R$ if $U = K\rho_0 c^2$, where ρ_0 is the gravitational mass
density required to close the universe at radius R and K is the ratio
of inertial to gravitational mass (a dimensional constant with value
unity). It is argued that zero-point radiation (vacuum fluctuations)
have renormalized energy density $K\rho_0 c^2$.
In quasars U is large enough to yield anomalous redshifts
comparable with the Hubble redshift of the sources, but the sources
must be at large distances in order to have the large values of U.

THE PHOTON AS A GRAVITON FIELD

Planck's radiation oscillators are quantized gravitationally as
different eigenstates of a fundamental oscillator of frequency equal to
the Hubble constant H ($\omega_0 = c/R = H$). The wavelength, $\lambda_0 = \pi c/\omega_0 = 2\pi R$,
is the circumference of the universe. Then

$$\hbar\omega = (n + \tfrac{1}{2})\hbar\omega_0 \tag{1}$$

where n is a very large integer. A photon is treated as a field n of
gravitons of energy $\hbar\omega_0$. Writing $\hbar\omega_0 = 2\epsilon$, we have

$$\epsilon (R/c) = \hbar/2 \tag{2}$$

Thus ϵ may be interpreted as a basic uncertainty in all energies
arising because measurement times cannot exceed the Hubble age R/c. The
spectrum of Planck oscillators is quasi-continuous because ϵ is
unmeasurable. Selection rules, $\Delta n = \pm 1$, constrain the photon to lose

Frontiers of Fundamental Physics, Edited by M. Barone
and F. Selleri, Plenum Press, New York, 1994

gravitons one at a time, which is quasi-continuously.

The process of interest is scattering of a graviton from photon field n to photon field n'. The probability for such a scattering is proportional to nn'. Introducing cross section σ_o, the rate of scatterings from field n to isotropic radiation of the medium with energy density U $(= N\hbar\omega_o)$ is

$$dn/dt \;=\; -\,n\sigma_o cN \;=\; -\,n\sigma_o cU/\hbar\omega_o \tag{3}$$

Noting that $dt = dl/c$, $dn/n = d\omega/\omega$, and $\omega_o = c/R$, we obtain

$$d\omega/\omega \;=\; -\,(\sigma_o R/\hbar c)U dl \tag{4}$$

For σ_O we modify the Thomson cross section for scattering of radiation by a free electron. We replace the electrostatic radius of the electron $d = e^2/mc^2 = 2.82 \times 10^{-13}$ cm by the gravitational radius of the electron a_o (Browne, 1976a), obtaining

$$\sigma_o \;=\; 8\pi a_o^2/3 \tag{5}$$

$$a_o \;=\; (2K)^{-1}(G\hbar/c^3)^{1/2} \;=\; 8.1 \times 10^{-34} \text{ cm} \tag{6}$$

G being the gravitational constant, and K the ratio of gravitational to inertial mass (Browne, 1977). At radius a_o the electron's gravitational and electrostatic fields become equal (Browne, 1977; see also Arnowitt et al., 1960). By Gauss's theorem, the sources of these equal fields are the real and imaginary charges $(\hbar c)^{1/2}$. From (5) and (4),

$$d\omega/\omega \;=\; -\,AU dl \tag{7}$$

where

$$A \;=\; 2\pi GR/3K^2 c^4 \;=\; 3.19 \times 10^{-21} \text{ erg}^{-1} \text{ cm}^2 \tag{8}$$

the A value being for $H = 50$ km s^{-1}Mpc^{-1} $(R = 1.85 \times 10^{28}$ cm).

Graviton scattering permits radiation trapped in a cavity with perfectly reflecting walls to attain the equilibrium blackbody spectrum in the absence of matter, which is not possible by recognized interactions.

ZERO-POINT RADIATION FIELD

In order to obtain the Hubble redshift as a special case of (7) it is necessary to consider zero-point radiation (vacuum fluctuations). It will be assumed that the vacuum fluctuations have the character of classical electromagnetic radiation, which is the view of an increasing number of authors seeking to interpret quantum mechanics (Welton, 1948; Marshall, 1963; Power, 1966; de la Pena-Auerbach and Garcia-Colin, 1968; Boyer, 1975; 1978). The fluctuations are standing wave modes with random relative phases. Their energy density, $\hbar\omega/2$ per mode, is

$$U_O \;=\; \int U_{o\omega}\,d\omega \;=\; \int (\hbar\omega/2)(\omega^2 d\omega/\pi^2 c^3) \;=\; \hbar\omega^4/8\pi^2 c^3 \tag{9}$$

The divergence of U_O as ω tends to infinity can be eliminated by including the negative gravitational potential energy of the zero-point radiation which also diverges. The total energy density is

$$K\rho_o c^2 \;=\; U_O \;-\; \rho_o \Phi(0)/2 \tag{10}$$

where $\Phi(r)$ is the Newtonian potential inside a uniformly dense sphere of radius R_N. Tolman (1934) has shown that the gravitational mass of disordered radiation is twice that of matter with the same energy density due to potential energy associated with pressure. Thus

$$\Phi(r) \;=\; 2\pi GR_N^2(2U_O/Kc^2)(1 - r^2/3R_N^2) \tag{11}$$

The radius of the universe in Newtonian cosmology R_N is related to that in de Sitter cosmology by $3R_N^2 = R^2$. Converting to R,

$$K\rho_o c^2 = U_o (1 - 2\pi G\rho_o R^2/3Kc^2) \tag{12}$$

As U_o tends to infinity, ρ_o remains finite only if

$$2\pi G\rho_o R^2 = 3Kc^2 \tag{13}$$

which expresses that ρ_o closes the universe at radius R. If zero-point radiation closes the universe, the current search for "missing dark matter" can be abandoned.

Zero-point radiation is not measurable because the most sensitive detector is in equilibrium with it. Only superimposed radiation due to finite temperature is measurable. Adding the zero-point and Planck spectra gives,

$$U_{o\omega}(T) = U_{o\omega}(0)(1 + x)/(1 - x) \tag{14}$$

where $x = \exp(-\hbar\omega/kT)$, and where $U_{o\omega}(0)$ is given by (8).

We postulate that radiation from a distant astronomical source loses gravitons to zero-point radiation as it propagates to the Earth. Then we substitute into (7) $U = K\rho_o c^2$, where ρ_o is given by (13). Then (7) and (8) yield $AU = 1/R$, so that (7) becomes

$$d\omega/\omega = - dl/R, \qquad \omega_2 = \omega_1 \exp(-l/R) \tag{15}$$

which is Hubble's law in differential and integrated forms.

Quantitatively, $K\rho_o c^2 = 6.9 \times 10^{-7}$ erg cm^{-3}, taking $H = 50$ km s^{-1} Mpc^{-1} which implies $R = 1.85 \times 10^{28}$ cm. For comparison, the 2.735 °K cosmic blackbody radiation yields $U = 4.23 \times 10^{-13}$ erg cm^{-3}.

DEFLECTION OF LIGHT

If a photon loses a graviton to radiation propagating transversely to its path, then conservation of momentum requires that the photon gain transverse momentum $\hbar\omega_o/c$, which implies that it has suffered angular deflection ω_o/ω. Interaction with isotropic radiation yields zero mean deflection, but after loss of N gravitons, where $N = \delta\omega/\omega$ for redshift $\delta\omega$, the angular spread is

$$\langle\delta\theta\rangle = N^{1/2}\omega_o/\omega = (\omega_o\delta\omega)^{1/2}/\omega \tag{16}$$

$\langle\delta\theta\rangle$ is undetectable because ω_o is so small.

If radiation of the medium is unidirectional and transverse to the ω-beam, beam deflection becomes measurable. Let the ω-beam pass within distance y of a star of radius R at temperature T. Then the transverse component of radiation flux has energy density $U_\perp = \sigma T^4(R/r)^2(y/r)$ at distance r from the star's center, where σ is the Stefan-Boltzmann constant. Noting $r^2 = l^2 + y^2$,

$$\delta\theta = \delta\omega/\omega = A\int U_\perp dl = 2AR^2\sigma T^4/cy \tag{17}$$

In the extreme case of a supergiant with $T = 40,000$ °K and $R = 100 R_\odot$ we find $\delta\theta = 44.5(R/y)$ arc sec, which is detectable.

ANOMALOUS REDSHIFTS

The redshift formula (7) integrates to $\omega_2/\omega_1 = \exp(- A\int Udl)$, where ω_1 is the frequency at source and ω_2 that at reception. In the case of a quasar at distance d, U can be large enough for the local redshift to exceed the Hubble redshift $\omega_2/\omega_1 = \exp(-d/R)$. The two redshifts are equal when $AUl \simeq d/R$. If F is the flux density at Earth from a quasar at distance d, then radiant energy density at source is $U \simeq Fd^2/l^2c$. The condition for equality of the redshifts therefore becomes

$$Fd/l = c/AR = 500 \quad \text{erg s}^{-1}\text{cm}^{-2} \tag{18}$$

Typically $F \simeq 10^{-11}$ erg s^{-1}cm^{-2} and $l \simeq 10^{13}$ cm (from variability time scales – for example, see Kunieda et al., 1990). Then (17) yields equality of the redshifts for $d \simeq 5 \times 10^{27}$ cm. If the source is more distant the anomalous redshift is dominant, and if closer the Hubble redshift is dominant. The Hubble component of the redshift must be known in order to determine d.

Evidence for anomalous redshifts of quasars has been accummulating for some years (Arp, 1987). Often a quasar is associated with a galaxy with considerably smaller redshift. The association may be a very faint filament joining the bright quasar to the galaxy. Previously (Browne, 1985; 1993), it was proposed that quasar emission is beamed along the axis of a magnetic vortex tube (MVT), which trails after the galaxy. Then the very bright quasar is seen where the line of sight is tangent to the MVT, and the faint filament is seen where the bent MVT becomes transverse to the line of sight.

DE SITTER COSMOLOGY

Einstein field equations, because of their general covariance, are applicable only to a perfectly isolated system, and the smallest such system for $r > a_o$ is a universe defined by $r < R$ (Browne, 1977; 1979). Clearly the source energy-momentum tensor $T_{\alpha\beta}$ in the Einstein equations is not open to free choice in the case of a universe. Rather should one seek a source involving only geometrical quantities. The obvious choice is to equate $\kappa T_{\alpha\beta}$ to the cosmological term $\Lambda g_{\alpha\beta}$. The constant Λ alone imposes a scale length on the description. Then the quations become

$$R_{\alpha\beta} = \Lambda g_{\alpha\beta} \tag{19}$$

The static spherically symmetric interior solution yields the metric of de Sitter space-time,

$$ds^2 = (1 - r^2/R^2)c^2dt^2 - (1 - r^2/R^2)^{-1}dr^2 - r^2d\sigma^2 \tag{20}$$

where $d\sigma^2 = d\theta^2 + \sin^2\theta\, d\phi^2$ and $\Lambda = 1/R_N^2 = 3/R^2$. Geodesic equations of radial motion for a particle released from rest at the origin are

$$dr/cd\tau = \pm r/R, \qquad dt/d\tau = (1 - r^2/R^2)^{-1} = c/c(r) \tag{21}$$

where $ds = cd\tau$ and where $c(r) = (1 - r^2/R^2)c$ (from $ds = 0$). Thus the universe expands with respect to reference system (r, t). However, in the case of a perfectly isolated system it is impossible to distinguish between expansion of the system and contraction of the reference system (Browne, 1979).

The time t_1 at which a wave crest is emitted by a source at radius r and the time t_2 at which it is received at the origin are related by $t_2 = t_1 + \int dr/c(r)$. By differentiation and use of (21) it follows that $\delta t_2/\delta t_1 = 1 \pm r/R$. Converting from δt to $\delta\tau$ by (21),

$$\omega_2/\omega_1 = \delta\tau_1/\delta\tau_2 = (1 \pm r/R)/(1 - r^2/R^2)$$

$$= D(1 - r^2/R^2)^{-1/2} \qquad D = (1 \pm \beta)/(1 - \beta^2)^{1/2} \tag{22}$$

where $\beta = (dr/dt)c(r)^{-1} = r/R$ from (21). The factor D is the Doppler effect in special relativity. The additional factor is a gravitational redshift. Thus Hubble's effect is partly Doppler and partly gravitational. However, its interpretation is different for a different reference system, as we now show.

The transformation of coordinates

$$\overline{r} = r(1 - r^2/R^2)^{-1/2}\exp(-ct/R)$$

$$\overline{t} = t + (R/2c)\ln(1 - r^2/R^2) \tag{23}$$

converts the de Sitter metric to the form

$$ds^2 = c^2 d\overline{t}^2 - \exp(2c\overline{t}/R)(d\overline{r}^2 + \overline{r}^2 d\sigma^2) \tag{24}$$

for which the geodesic equations are

$$d\overline{r}/d\overline{t} = 0, \qquad d\overline{t}/d\tau = 1 \tag{25}$$

Thus matter is at rest relative to $(\overline{r}, \overline{t})$. This is apparent also from the integral of (21) which is the first equation (23) with $\overline{r}$ = const.

The radial velocity of light (from ds = 0) is $c(\overline{t}) = c \exp(-c\overline{t}/R)$. The times of emission and reception of light signals now are related by $\overline{r} = \int d\overline{t}/c(\overline{t})$ between limits $\overline{t}_2$ and $\overline{t}_1$. Now $\overline{r}$ is a constant so that differentiation yields $\delta\overline{t}_2/c(\overline{t}_2) = \delta\overline{t}_1/c(\overline{t}_1)$. Since $\delta t = \delta\tau$,

$$\omega_2/\omega_1 = \delta\tau_1/\delta\tau_2 = c(\overline{t}_1)/c(\overline{t}_2) = \exp(-l/R) \tag{26}$$

where $l = c(\overline{t}_2 - \overline{t}_1)$. This result agrees with (15).

Because the metric (23) is invariant under

$$\overline{t} \rightarrow \overline{t} + \overline{t}_o, \qquad \overline{r} \rightarrow \overline{r} \exp(- c\overline{t}_o/R) \tag{27}$$

where $\overline{t}_o$ is a constant, the metric is unchanged by continuously varying time origin whilst appropriatly scaling radial distance. Redshift (27) can therefore be interpreted as continuous change of frequency at constant light velocity, which is the interpretation of (15).

UNIT FIELDS AND ARBITRARINESS OF GEOMETRY

Units for the three components of length and for time are of necessity arbitrary. By permitting each unit to become a field it is always possible to flatten space-time. Mathematically, the coordinate transformation is anholonomic (Browne, 1976b), and can be interpreted as projection from a curved space-time onto a flat space-time via coordinate curves in a 5-dimensional space in which both space-times are embedded. It happens that both de Sitter and Minkowski space-times can be embedded in a 5-dimensional Euclidean space.

It is possible to derive de Sitter metric by transformation from inhomogeneous unit fields in flat space-time to homgeneous unit fields (constant units) in de Sitter space-time (Browne, 1976b). The inhomogeneous unit fields are obtained by applying Lorentz contraction to measuring rods and time dilation to clocks appropriate for the velocity field of the cosmological fluid, which is energy density of the electromagnetic field, both radiation and matter. The unit fields become the description of gravitation in flat space-time.

REFERENCES

Arnowitt, R., Deser, S. and Misner, C.W., 1960, "Gravitational-electromagnetic coupling and the classical self-energy problem", Phys. Rev. 120:313.

Arp, H., 1987, "Quasars, Redshifts and Controversies", Interstellar Media, Berkeley, Ca.).

Boyer, T.H., 1975, "Random Electrodynamics: The theory of classical electrodynamics with classical electromagnetic zero-point radiation", Phys. Rev. D11:790.

Boyer, T.H., 1978, "Statistical equilibrium of nonrelativistic multiply periodic classical systems and random clasical electromagnetic radiation", Phys. Rev. A18:1228.

Browne, P.F., 1962, "The case for an exponential redshift law", Nature 193:1019.

Browne, P.F., 1976a, "A hierarchy hypothesis", Intern. J. Theor. Phys. 15:73.

Browne, P.F., 1976b, "Arbitrariness of Geometry and the Aether", Found. Phys. 6:457.

Browne, P.F., 1977, "Complementary aspects of gravitation and electromagnetism", Found. Phys. 7: 165.

Browne,P.F., 1979, "Universe expansion or reference system contraction", Phys. Lett., 73A:91.

Browne, P.F., 1985, "Magnetic vortex tubes, jets and nonthermal sources", Astron. Astrophys. 144:298.

Browne, P.F., 1993, "Active galactic nuclei: their synchrotron and Cerenkov radiations, "Progress in new cosmologies: beyond the big bang", ed. H.C. Arp et al., Plenum Press, New York, p.205.

De la Pena-Auerbach, L. and Garcia-Colin, L.S., "Possible interpretation of quantum mechanics", J. Math. Phys. 9:916

Kunieda, H. et al., 1990, "Rapid variability of the iron fluorescence line from the Seyfert 1 galaxy NGC6814", Nature 345:786.

Marshall, T.W., 1963, "Random electrodynamics", Proc. Roy. Soc. A276: 475.

Power, E.A., 1966, "Zero-point energy and the Lamb shift", Amer. J. Phys. 34:516.

Tolman, R.C., 1934. "Relativity, Thermodynamics and Cosmology", Clarendon Press, Oxford, p. 271.

Welton, T.A., 1948, "Some observable effects of the quantum-mechanical fluctuations of the electromagnetic field", Phys. Rev. 74:1157.

THEORETICAL BASIS FOR A NON-EXPANDING
AND EUCLIDEAN UNIVERSE

Thomas B. Andrews

3828 Atlantic Avenue
Brooklyn, N.Y. 11224

INTRODUCTION

LaViolette[1] has summarized observational data supporting a non-expanding universe. More recently, Sandage and J-M. Perelmuter[2] have concluded (with reservations) that the universe is expanding based on the analysis of the surface brightness of large elliptical galaxies. These are conflicting studies which do not prove conclusively that the universe is expanding or non-expanding.

To show that the universe is non-expanding, both a viable theory and observational data supporting the theory must exist. Any non-expanding universe theory or model must include, at the very least, a physical process for the Hubble red-shift and a process which prevents the gravitational collapse of the universe. This approach is almost self evident. However, a basic problem appears to exist. There is a long history of futile attempts to explain the Hubble red-shift in a non-expanding universe. If the universe is really not expanding, what do these attempts indicate? I believe these attempts show that current physical concepts are inadequate. Therefore, new physics is needed to develop a non-expanding universe model.

Consequently, a new and fundamental approach to physics is proposed based on the following paradigm: *The physical universe is a pure wave system consisting of a large number of wave modes.* This paradigm was inferred from the universal occurence of wave phenomena in physics. Note that a wave system has theoretical advantages as a basis for physics since it is Lorentz invariant and conserves energy.

As evidence for the validity of the paradigm, the paradigm has been used to derive the time-independent Schrödinger equation[3]. A symmetry argument of H. Giorgi[4] is also pertinent. He has proved a wave system must exist, given an infinite linear system in which the observed laws of physics are both translation and time symmetric. The proof is as follows: The modes of oscillation of a linear system with the above symmeties are governed by the representations of both the space and time translation groups. The solution of each representation is a complex exponential in space and time respectively. Combining these solutions results in left and right progressive waves, given by $\exp i(\omega t \pm kx)$, which form a standing wave system.

This symmetry argument must apply to the universe since it is a very large, if not

Frontiers of Fundamental Physics, Edited by M. Barone
and F. Selleri, Plenum Press, New York, 1994

infinite, system with linear responses at the level of the basic laws and exact time and translational symmetry of the laws of nature.

Two new theoretical results in cosmology are derived from the paradigm: First, a derivation of a new process for the Hubble red-shift which does not blur images over large distances. Second, an explanation of the "missing mass" problem.

Observationally, magnitudes and metric angular radii data for 1st rank elliptical galaxies are shown to be consistent with a non-expanding and euclidean universe.

WAVE SYSTEM THEORY

Consider a single wave mode which interacts with the other modes parametrically. For simplicity, one-dimensional wave motion equivalent to waves on a string will be considered. Assume the classical wave equation with variable parameters applies, given by

$$\frac{\partial}{\partial x}\left[T\frac{\partial \theta}{\partial x}\right] = \sigma\frac{\partial^2 \theta}{\partial t^2},\tag{1}$$

where T and σ are the tension and mass density.

The stability of the wave system is based on the following two principles: (1) The frequency is reduced if the mass density and tension are larger at the wave mode peaks; (2) As the number of particles mutually interacting increases, the wave system frequency increases.

To apply the first principle, assume the mass density and tension are proportional to the local energy density of the wave system. Then, the frequency of a wave mode will decrease when the constructive interference of the wave modes produces high energy concentrations at the peaks of the wave mode. Since the energy density at a peak is proportional to the square of the number of wave modes (10^{45}) constructively interfering, the decrease in the frequency of the wave mode is very large.

A lower bound on the frequency is given by the eigenvalue[5] equation

$$f = \frac{1}{2\pi}\sqrt{n}\sqrt{k/m}\tag{2}$$

where n is the effective number of particles interacting and k/m is the average interaction constant between any two particles. This vibrating system is remarkable since it has only two discrete frequencies, one degenerate with $n-1$ modes given by f and the other equal to $\frac{1}{2\pi}\sqrt{k/m}$, the independent frequency of a single element.

A stable frequency of the wave system is attained when the wave mode frequency and the frequency given by equation 2 are the same. This stable frequency is about 10^{23} hertz. Furthermore, it is proposed that the peaks are the elementary particles.

GENERAL FORCE EQUATION

Forces are the result of changes in the parameters. To derive a general force equation, set $\theta\left(x,t\right) = Y\left(x\right)T\left(t\right)$ in equation 1. Then the frequency is determined by the space dependent equation

$$\frac{d}{dx}\left[T\frac{dY}{dx}\right] + 4\pi^2\sigma f^2 Y = 0.\tag{3}$$

For small spatial variations in the values of the parameters, the perturbative solution[6] is

$$f^2 = f_o^2 \left[1 - \frac{2}{L\sigma_o} \int_0^L (\sigma - \sigma_o) \sin^2 (kx) \, dx \right.$$
$$\left. - \frac{2}{kLT_o} \int_o^L \frac{\partial T}{\partial x} \cos (kx) \sin (kx) \, dx \right] \tag{4}$$

where f_o is the stable frequency, f is the perturbed frequency and L is the length of the system. The system frequency is reduced when σ and T are larger at the constructive interference peaks than their average values, σ_o and T_o. Simplifying equation 4 by assuming the last two terms are equal, setting $\sin^2(kx)$ and $\sin(kx)\cos(kx) = 1/2$ and taking the square root, f is given by

$$f = f_o[1 - \frac{1}{L\sigma_o} \int_0^L (\sigma_\delta - \sigma_o) dx] \tag{5}$$

where the subscript δ indicates a constructive interference peak.

Assume there is only a single particle in the system and the particle or constructive interference peak has a linear width equal to the wavelength, λ. Then, $\sigma_\delta = m/\lambda$ and $\sigma_o = m/L$ where m is the mass of the particle. Integrating equation 5,

$$f = f_o[2 - \lambda\sigma_\delta/(L\sigma_o)]. \tag{6}$$

Without the constructive interference peak, $f = 2f_o$. With the constructive interference peak, $f = f_o$, the stable frequency of the wave system.

Setting $E = hf$ and $mc^2 = hf_o$ in equation 6, the force required to move a mass particle is given by

$$F = \frac{dE}{ds} = -\frac{mc^2\lambda}{L\sigma_o} d\sigma_\delta/ds = -c^2\lambda d\sigma_\delta/ds = -\lambda dI_\delta/ds \tag{7}$$

where dI_δ/ds is the change in the energy density of the particle. The type of force depends upon the process that changes I_δ.

DERIVATION OF THE HUBBLE RED-SHIFT

Consider the case of x-rays incident on a perfect crystal. Experimentally, the x-ray energy density inside the crystal decreases exponentially with distance due to reflections (interactions) from particles at lattice positions. By analogy, it is proposed that the individual wave mode energy densities are reduced exponentially by reflections from mass particles. Since the product of the average number density and effective area of mass particles is very small, the mean reflection distance, R, is very large, on the order of 4×10^{26} meters.

Due to the reflections, mass particles at different locations will tend to be partially uncoupled from the vibrations of the other. This leads to different wave mode sets associated with each particle. It is assumed that the particle only vibrates in-phase with its own wave mode set. *Consequently, as a particle moves away from its equilibrium position, the number of wave modes in-phase with the particle decreases and the energy density of the particle decreases.* This is the basic Hubble red-shift process which applies to both mass particles and photons.

Mass Particle Red-Shift

To calculate the decrease in the constructive interference energy density as a particle moves, define s as the distance the particle moves away from the equilibrium position. The number of modes interacting with the particle at s is reduced by the factor $\exp(-s/R)$. Since the constructive interference energy density at a peak is proportional to the square of the number of modes, the energy density at s is given by

$$I(s) = \frac{mc^2}{\lambda} \exp(-2s/R). \tag{8}$$

Then the decrease in energy density with distance is

$$\frac{dI(s)}{ds} = -\frac{2mc^2}{\lambda R} \exp(-2s/R). \tag{9}$$

Using equation 7, the force on the particle is

$$F = \frac{2mc^2}{R} \exp(-2s/R). \tag{10}$$

For $R = 4 \times 10^{26}$ meters, the force required to move a proton is approximately 10^{-36} newtons. I have named this new force the "Hubble force."

However, the increase in frequency of the mass particle introduces a new problem. At each new position, the particle frequency will be larger than the equilibrium frequency of the wave system. *Consequently, the particle will not be vibrating in-phase with the equilibrium wave system as it moves. Since very large forces will oppose out-of phase vibrations, this can not occur.* Instead the size of the particle and the velocity of light within the particle is reduced. until the particle is vibrating in-phase at the equilibrium frequency.

The equilibrium frequency relations then are such that

$$f_o = \frac{c_o}{\lambda_o} = \frac{c}{\lambda} \tag{11}$$

where $c < c_o$ and $\lambda < \lambda_o$. The energy density of the particle must increase so that it is greater than the energy density of the equilibrium particles. Consequently, c, the velocity of light, will be less in the red-shifted particles than in the equilibrium particles. Photons originating from particles with a lower c will be red-shifted. Thus, the particles will have an intrinsic red-shift similar to particles located in a gravitational field.

Paradoxically, particles with an intrinsic red-shift have more energy than equilibrium particles would have at the same position. To understand this, consider that equilibrium particles have N_o modes constructively interferring, resulting in a minimum energy system. If a particle is in-phase with only $N < N_o$ modes, the energy must be greater. Over time, as more wave modes constructively interfere with the red-shifted particles, the energy of the particles is reduced, releasing energy.

Alternatively, mass particles can initially have a large kinetic energy and move "up" in a Hubble field until they come to rest. These mass particles will also have an intrinsic red-shift.

This intrinsic red-shift may be the same effect as observed by H. Arp[7] in companion galaxies, assuming the companion galaxies were ejected from the dominant galaxy in the cluster.

Missing Mass Problem

Assuming all the stars in a galaxy have an equilibrium point near the center of a galaxy, then the stars will be accelerated by the Hubble red-shift effect towards the center of a galaxy. The Hubble acceleration has the nearly constant value, $2c^2/R = 5 \times 10^{-10} \ m/sec^2$, which explains quantitatively the higher than expected orbital velocities of stars in galaxies. *Therefore, the Hubble acceleration is proposed as the solution of the "missing mass problem."* Another effect of the Hubble acceleration is to stabilize the universe against gravitational collapse.

Photon Red-Shift

The Hubble potential energy for a mass particle is given by

$$\Delta P = \frac{2mc^2}{R} \int_0^s \exp\left(-2s/R\right)ds = mc^2[1 - \exp\left(-2s/R\right)]. \tag{12}$$

The Hubble potential has the same effect on a photon which moves away from an equilibrium position. As an analogy, consider the effect of a gravitational field on a photon moving away from the center of a finite spherical mass. As the distance from the center increases, the gravitational potential increases and the frequency of the photon decreases. Similarly, the frequency decreases for a photon as the Hubble potential increases.

Setting $\Delta P = h(f_o - f)$ and $mc^2 = hf_o$, the Hubble red-shift of a photon is given by

$$f = f_o \exp\left(-2s/R\right). \tag{13}$$

This red-shift law is exact and, to first order, equivalent to Hubble's velocity law, $v = Hs$ where $H = 2c/R$. Since the Hubble red-shift is due to a physical process similar to gravitation, all frequencies are equally affected and there is no blurring of images.

The above, however, is not a good physical explanation of the observed reduction in frequency. The physical explanation requires new gravitational concepts developed by R.A. Vera[8] which conflict with the usual concepts of gravitation.

Vera distinguishes between local and non-local quantities in a gravitational field. Local quantities are the locally measured values of physical quantities using local standards. An example of a local quantity is the rest mass of a proton measured on the surface of the Earth. Non-local quantities are local quantities referred to another location in the gravitational field or, equivalently, to a standard potential.

Vera has rigorously proved that no exchange of energy occurs between a freely falling mass particle or photon and the field. This is a result of the simple physical principle that the net number of signals sent by electro-magnetic radiation is conserved in a static and conservative field. Consequently, the photon's non-local frequency is conserved. But, the measured or local frequency of a photon decreases as the gravitational or Hubble potential increases although the non-local frequency or energy of the photon remains the same. This is a result of measuring the frequency in a different environment, one in which the potential is higher and the units of measrement are different.

The reason gravitation and Hubble red-shift effects do not blur observations of distant objects is that no energy loss to the field occurs for photons.

Size of Universe

Given a non-expanding universe, the determination of the size and age of the universe may have a solution. A possible starting point is the experimental evidence that the physical constants are exactly the same in very distant galaxies. Then, assuming the physical constants are a function of the stable frequency of the universe, given by equation 2, the distant galaxies must interact with the same effective number of particles as the local galaxies. However, these particles must be a different set of particles than the local set since all interactions are reduced exponentially with distance. From this argument, I conclude that the universe must be very much larger than R or even infinite since distant galaxies interact with parts of the universe with which we do not interact. Finally, an infinite universe implies an infinitely old universe.

If the non-expanding universe is infinite, some consequences are: (1) The gravitational potential must be finite and the same at all points (except near large masses) since each point interacts with the same number of particles. The universe is then euclidean (flat); (2) In an infinitely old universe, the universe must be stable on a large scale. Therefore, there is no evolution and processes must exist which maintain the stable state.

APPLICATION OF THEORY TO DATA

To test the non-expanding universe model, two data sets known to be very good were used. The first data set is from Kristian, Sandage and Westphal[9] and contains magnitudes for 1st rank elliptical galaxies to $z = 0.75$. The second data set is from Djorgovski and Spinrad[10] and contains metric angular radii of 1st and 2nd rank elliptical galaxies to $z = 1.05$.

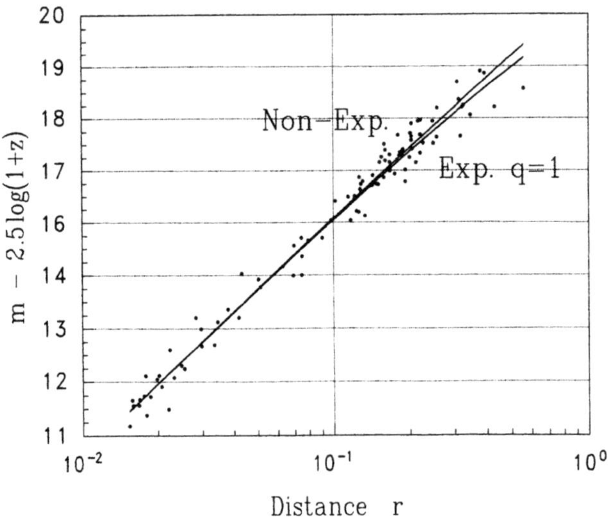

Figure 1 Aperture corrected magnitudes less $2.5 \log(1 + z)$ versus the euclidean distance r for 1st rank elliptical galaxies from data of Kristian, Sandage and Westphal

The luminosity, l, of a galaxy in the non-expanding universe model is

$$l = \frac{L}{4\pi r^2(1+z)} \tag{14}$$

where L is the absolute luminosity, r is the euclidean distance and $(1+z)$ is the reduction in luminosity due to the Hubble red-shift. The distance, r, is determined from the relation $f = f_o \exp(-2r/R)$. In terms of the red-shift, z,

$$r = \ln(1+z). \tag{15}$$

The angular radii of objects in a non-expanding euclidean universe theoretically vary as

$$\theta = \frac{S}{r} \tag{16}$$

where S is the metric radius (in the same units as r). For 1st rank elliptical galaxies, S is assumed constant since these galaxies have very nearly the same luminosity.

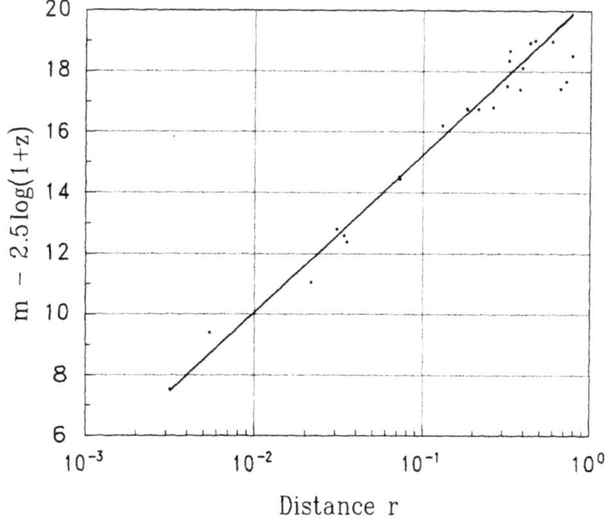

Figure 2. Magnitude less $2.5\log(1+z)$ versus the euclidean distance r for 1st and 2nd rank elliptical galaxies (some radio galaxies) from data of Djorgovski and Spinrad.

The theoretical relationships in equations 14 and 16 are very specific. Thus, the non-expanding universe model can be disproved by observations. In contrast, the expanding universe model can not easily be disproved since either the parameters can be changed or evolutionary effects can be cited to make the model fit the data.

Figure 1 is a plot of the magnitudes less $2.5\log(1+z)$ versus r. In this plot, aperture corrections to the magnitudes were made corresponding to a non-expanding universe, thus making the galaxies appear slightly less bright than in the $q = 1$ expanding universe model.

The linear regression equation, calculated using data points for which $z < 0.4$, is

$$m - 2.5 \log (1 + z) = 20.62 + 5.02 \log r. \qquad (17)$$

The slope is 5.02 which shows the inverse square law holds almost exactly. For comparison, the expected regression line for a $q = 1$ expanding universe model is the slightly curved line.

Figure 2 is a plot of the apparent magnitudes (K corrected) of the Djorgovski and Spinrad data. The regression line is based on all except the furthest three galaxies and is given by

$$m - 2.5 \log (1 + z) = 19.91 + 4.82 \log r. \qquad (18)$$

The three furthest galaxies (two are radio galaxies) are approximately 1.5 magnitudes

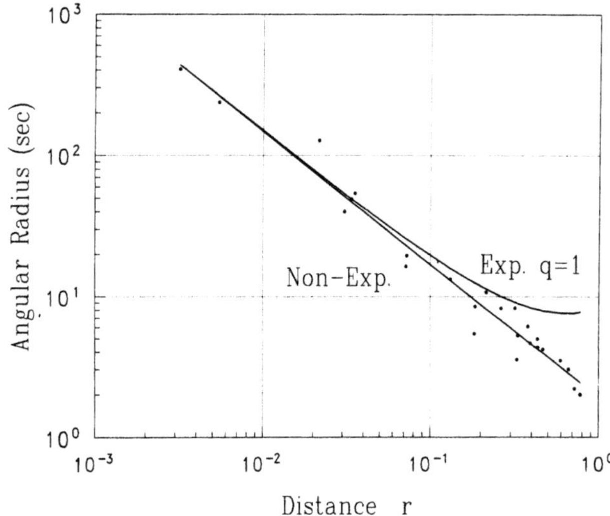

Figure 3. Metric angular radius versus the euclidean distance r of 1st and 2nd rank elliptical galaxies (some radio galaxies) from data of Djorgovski and Spinrad.

brighter than the other galaxies. It is difficult to explain why they are so much brighter. One possible explanation is that these galaxies are large spirals. These would appear brighter than 1st rank elliptical galaxies at large distances.

No aperture corrections in this case are necessary since measurements were made using a metric measurement method. The regression line (not shown) for a $q = 1$ expanding universe model is almost identical (only 0.5 magnitude less bright at $z = 1$).

The almost identical regression lines for the non-expanding and expanding universe models are explained as follows: For a Friedmann $q = 1$ expanding universe model, the usual relation is $m = A + 5 \log z$. The corresponding relationship for the non-expanding universe model is $m = A + 5 \log r + 2.5 \log (1 + z)$. Coincidently, these equations result in almost identical values of m because the values of z^2 and $r^2(1 + z)$ are very nearly the same for $z < 1$.

Figure 3 is a plot of the metric angular radius versus r of 1st and 2nd rank elliptical galaxies. The linear regression equation is

$$\log \theta = 0.28 - 0.94 \log r. \tag{19}$$

The slope is close to the theoretical slope of -1.

The curved line is the expected regression for a $q = 1$ expanding universe model. The fit to the data is obviously poor for the expanding universe model.

SUMMARY

As shown by the graphs, the non-expanding universe model is consistent with the observations. The expanding universe model is inconsistent since it fits the magnitude data but not the angular radius data.

Therefore, based on the new Hubble red-shift process proposed in this paper and the observational data, I conclude the universe is almost certainly non-expanding and euclidean. This conclusion is important since it will unify theoretical and observational cosmology and thus lead to greater knowledge about the universe.

Finally, note that the new physics based on the wave system paradigm was essential to the derivation of the Hubble red-shift process.

REFERENCES

1. P.A. LaViolette, Is the universe really expanding?, Ap. J. 301:544 (1986).

2. A. Sandage and J-M. Perelmuter, The surface brightness test for the expansion of the universe. 3. Reduction of data for the several brightest galaxies in clusters to standard conditions and a first indication that the expansion is real, Ap. J. 370:455 (1991).

3. T.B. Andrews, A wave system theory of quantum mechanics, Physica B 151:351 (1988).

4. H. Giorgi, An overview of symmetry groups in physics, Talk presented at 1990 fall meeting of the New England Section of the American Physical Society at Yale University, Harvard preprint number HUTP-90/A065 (1990).

5. F.Y. Chen, Similarity transformations and the eigenvibration problem of certain far-coupled systems, Am. J. Phys. 38:1036 (1970).

6. J.C. Slater, *Mechanics*, McGraw-Hill Book Co., p. 188-192 (1947).

7. H. Arp, *Quasars, redshifts and controversies*, Interstellar Media, p. 107-132 (1987).

8. R.A. Vera, "A dilemma in the physics of gravitational fields," Int. J. Theor. Phys. 20, No.1:19 (1981).

9. J. Kristian, A. Sandage and J.A. Westphal, The extension of the Hubble diagram, 2. New red-shifts and photometry of very distant galaxy clusters: First indication of a deviation of the Hubble diagram from a straight line. Ap. J., 221:383 (1978).

10. S. Djorgovski and H. Spinrad, Toward the application of a metric size function in galactic evolution and cosmology, Ap. J. 251:417 (1981).

LIGHT PROPAGATION
IN AN EXPANDING UNIVERSE

A. Paparodopoulos

3 Perikeleous Ave., GR-155 61 Athens, Greece

In an expanding space, propagating light waves/particles participate in the expansion as originally proposed by W.H. McCrea (1962) and S.J. Prokhovnik (1963). The resulting mode of light propagation is expressed by a light "kinematic equation", which is a direct consequence of the cosmological principle and can be mathematically deduced from the Robertson-Walker metric. This leads to an important consequence, that the Minkowski light cone is transformed, in expanding space, to a curved space-time surface. Our new approach offers a more physically intelligible interpretation of the observational astronomy and theoretical cosmology and provides simple and direct answers to many abstract expositions of the conventional theory.

ASSUMPTIONS AND DEFINITIONS

Our approach is based on the generally accepted assumptions of the Cosmological Principle (CP) and of the Hubble expansion of the universe. However, it involves also the notion of an observable privileged Fundamental Reference Frame (FRM), associated with the distribution of the expanding matter (galaxies, cluster) of the universe, in respect to which light waves/particles propagate, as implied by the Robertson-Walker metric and originally proposed by G.Builder[1] and S.Prokhovnik[2].

This notion is ignored by most physisists, since it seems to violate the basic principles of SRT. Yet, it is now further supported by the discovery of the Cosmic Background Radiation (CBR) and of its thermal anisotropy, through which the peculiar motion of our Galaxy was detected and measured; and it would be utterly absurd to contend that this frame is uniquely associated only with this specific m-w radiation and not with the whole range of the e-m radiation. There is further evidence of the association of the FRF with the distribution of distant galaxies. Over the last 20 years, the peculiar velocity of our Galaxy, as assessed from the systematic analysis of the motion of nearby galaxies[3], is in good agreement with the peculiar velocity estimates obtained from the background radiation thermal anisotropy criterion, so that we can confidently claim that the frame through which this radiation propagates, has all the properties of the privileged F.R. Frame.

With respect to the FRF, we shall distinguish **Fundamental Systems** (bodies-observers), at rest with it, and **peculiar systems**, moving relative to it. Fundamental systems, which in practice are identified with "ideal" galaxies (i.e. having zero peculiar velocity), are kinematically indistinguishable from each other, since any galaxy in accordance with the CP, can be considered as the centre of the universe and hence as the origin of a private privileged fundamental coordinate system. It follows that Fundamental systems share common length and time units. In particular, the existence of a "cosmic time"[4], common to all Fundamental observers is a direct consequence of the CP and of the Hubble expansion, whereby cosmic

Frontiers of Fundamental Physics, Edited by M. Barone
and F. Selleri, Plenum Press, New York, 1994

time may be measured by the expanding distance between any pair of ideal galaxies. On the other hand, relative to peculiar systems, the universal expansion and light propagation are not isotropic and hence, they are affected by a complex of anisotropy effects such that their length and time units are Lorentz transformed as described by S. Prokhovnik[5].

Fundamental Systems posses common time and length units, so their coordinate measures are related by the Galilean transformation, and the Galilean law of addition of velocities holds with respect to their view of peculiarly moving bodies. Thus, if o,o' ($\underline{oo'} = \underline{r}$) are two Fundamental observers, and P is a body moving along $\underline{r}$ with peculiar velocity u' with respect to o', its velocity u_p relative to o will be :

$$u_p = u' + \frac{dr}{dt} = u' + Hr \tag{1}$$

since $\qquad \dfrac{dr}{dt} = Hr \quad$ according to the Hubble Law.

THE ROBERTSON-WALKER METRIC AND THE LIGHT PROPAGATION IN EXPANDING SPACE

The RW metric is the most general metric satisfying the cosmological principle in expanding space. In the case of a light ray (ds=o) moving radially in Euclidian space, it is reduced, in spherical-polar coordinates, to the null metric form :

$$0 = -c^2 dt^2 + R_{(t)}^2 d\underline{r}^2 \tag{2}$$

where $\underline{r}$ is the dimensionless radial fixed comoving coordinate of the light wave front relative to an observer at rest in the origin O, and $R_{(t)}$ is the cosmic scale factor which describes the expansion of the universe. If the coordinate $\underline{r}$ is fixed as the distance r_o from the origin at the present epoch t_o, it can be labelled with the same dimensional figure as the distance r_o , so that $\underline{r} \equiv r_o$. Then the actual expanding distance of the light signal from the origin at epoch t, will be

$$r_{(t)} = r_o \frac{R_{(t)}}{R_o} = \underline{r}\frac{R_{(t)}}{R_o}$$

where $R_{(t)}, R_o$ are the scale factors at epochs t and t_o respectively; taking R_o as unit the above becomes :

$$r_{(t)} = \underline{r}R_{(t)} \tag{3}$$

relation which transforms the dimensionless index $\underline{r}$ to a dimensional expanding distance r .
We shall now try to transform the RW null metric to a dimensional form.

The metric (2) can be written as :

$$R_{(t)}\frac{d}{dt}\frac{r}{R} = \pm c$$

or invoking (3) and omitting suffixes :

$$R\frac{d}{dt}\frac{r}{R} = \pm c \tag{2a}$$

We may define by s the distance of the light signal from O when it reaches a fundamental point distant r from O at time t, so that $s \equiv r \equiv \underline{r}R_{(t)}$. Then (2a) can be written as :

$$R \frac{d}{dt}\frac{s}{R} = \pm c$$

$$\text{or} \quad R \frac{\dot{s}R - s\dot{R}}{R^2} = \dot{s} - s\frac{\dot{R}}{R} = \pm c$$

and since $\quad \dfrac{\dot{R}}{R} = H \quad$ (the Hubble Constant), we finally find :

$$\dot{s} = \frac{ds}{dt} = \pm c + Hs \tag{4}$$

This is the light propagation equation in an expanding space, of Big Bang or steady state model origin. The plus/minus sign refers to light rays moving towards or against the expansion, that is on whether the origin is associated with the source or the observer. The physical meaning of this equation is clearly that the speed of light $\frac{ds}{dt}$, relative to the receiver or emitter of the origin, is the sum of its local speed $\pm c$ and of the recession speed (Hs) of the locality which the light has reached. We might therefore say that the light propagates "riding" on the expanding "substratum", thereby expanding its wavelength as it passes all F. particles/observers at the same speed c, fact which satisfies the light principle of SRT. The equation (4) deduced from the RW metric, is identical to the equation (1) deduced from the Galilean addition law applied to Fund. Systems, (suffice is to substitute u by c). This identity is not surprising since both results are inferred from the same source, the Cosmological Principle.

EQUATION OF THE LIGHT PROPAGATION PATH IN THE EINSTEIN DE SITTER MODEL

In the E-de Sitter cosmological model, the scale factor $R_{(t)}$ is proportional to $t^{2/3}$, hence the Humble constant is :

$$H_{(t)} = \frac{\dot{R}}{R} = \frac{2}{3}\frac{1}{t}$$

Invoking the above in (4), we find the light propagation equation in the E-de Sitter model :

$$\frac{ds}{dt} = \pm c + \frac{2}{3}\frac{1}{t} \tag{5}$$

Using s=o at t=t_o (for light received at the present epoch t_o) we obtain as particular solution of (5) :

$$s_{(t)} = \pm 3ct \left[1 - \left(\frac{t_o}{t}\right)^{1/3} \right] \qquad \text{(m)}$$

Taking c as the unit speed and t in years we can express the above in light years, thus,

$$s_{(t)} = \pm 3t \left[1 - \left(\frac{t_o}{t}\right)^{1/3} \right] \qquad \text{[Ly]} \tag{6}$$

We shall maintain this notation, since it simplifies calculations and space-time depictions, wherein the velocity $\underline{c}$ is represented by a line of 45° gradient.

The result (6) describes the space-time paths of all light rays in expanding space, those received (-) and those emitted (+) from the Fund. Observer of the origin, at the present epoch t_o *(Fig.1)*; thus it corresponds precisely to Minkowski double light cone for static space. For receding light-rays (associated with the plus sign emitted from F_o, (6) represents the future branch (i) of the cone $(t > t_o)$ and for approaching light-rays (associated with the minus sign) received at F_o, it represents the past branch of the curve (ii) :

$$s_e = 3t_e \left[\left(\frac{t_o}{t_e}\right)^{1/3} - 1\right] \qquad (t_e < t_o) \tag{6a}$$

where t_e is the emission epoch and s_e the distance of the source at t_e. It is understood that the application of this formula for deep extrapolations into the past, in the case of the Big-Bang model, presupposes the Euclidean nature of the early universe, which holds only for the inflationary models.

Thus, in E-de Sitter universe, the double Minkowski cone of the static space, is transformed into a double curved, deformed cone (*fig.1*), closed at t=o of the hypothetical singularity since (6a) gives s=o at t_e=o. In the case of the steady state model, where H is time invariant, the equation of light path, corresponding to (6a), is given as:

$$s_e = \frac{1}{H}\left(1 - e^{Ht}\right) \qquad (Ly) \qquad \text{where } t < o,$$

which represents the generatrix of a deformed cone, having an assymptotic at the "horizon" distance 1/H (Ly), associated with a Hubble recession speed equal to c.

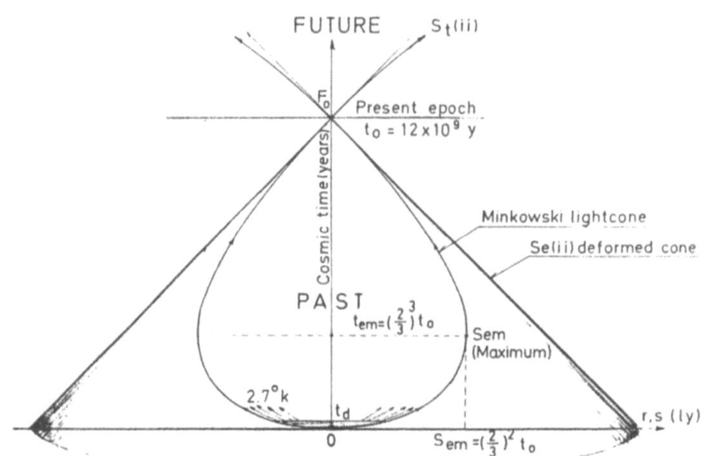

Figure 1. In the Einstein de Sitter model, the Minkowski's double light cone is transformed to the deformed double conelike surface $s_{(i)}$ and $s_{(ii)}$ which closes to the hypothetical big-bang event o. The observer F_o of the origin has access with light sources after the decoupling era t_d 10^5y.

Through this past branch (ii) we communicate with the "fireball event of the decoupling epoch, via the C.B. radiation which reaches us from every part of the sky. The equation (6a) describes the space-time path followed by photons emitted in the past from sources (galaxies) G_1, G_2, ... whose light is received simultaneously by us at the present epoch (t_o). These light sources, in their 3-dimensional spatial projection, are associated with the galaxies, quazars and other light sources, observable at present throughout the night sky. Their observed luminosity and redshift, relates directly to their present distances r_{o1}, r_{o2}, ... (*Fig.2*), which are given by as follows:

$$r_o = r_e \frac{R_o}{R_e} \equiv s_e \frac{R_o}{R_e} = s_e\left(\frac{t_o}{t_e}\right)^{2/3} = 3t_o\left[1 - \left(\frac{t_o}{t_e}\right)^{1/3}\right] \qquad (Ly) \tag{7}$$

where: $t_o = \dfrac{2}{3}\dfrac{1}{H_o} \cong 12 \times 10^9$ [Ly]

The emission epoch of observed light reaching us now, is assessed from (7)

$$t_e = \left(1 - \frac{r_o}{3t_o}\right)^3 t_o \qquad (y) \tag{8}$$

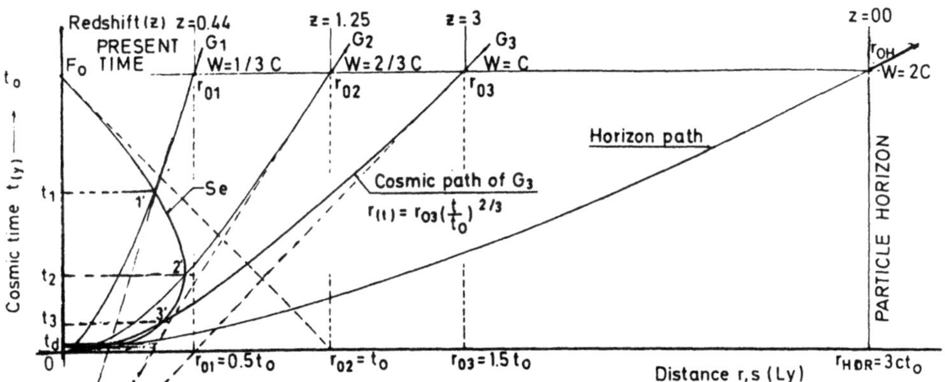

Figure 2. S_e deformed light cone; G_1, G_2, G_3 galaxies observed now in the sky at 1, 2, 3 at distances r_{01}, r_{02}, r_{03}; 1', 2', 3' their space-time position at their emission epochs t_1, t_2, t_3.

and the corresponding spectral ratio is given by :

$$\zeta = \frac{R_o}{R_e} = \left(\frac{t_o}{t_e}\right)^{2/3} = \left(1 - \frac{r_o}{3t_o}\right)^{-2} \tag{9}$$

from which we can also obtain the present distance as a function of the redshift z $\left(z = \zeta - 1\right)$,

$$r_o = 3\left(1 - \zeta^{-1/2}\right)t_o \quad (Ly), \tag{10}$$

and also for $\zeta \to \infty$, the event horizon distance for the E-de Sitter model :

$$r_{HOR} = 3t_o = 2H_o \qquad (Ly),$$

or $r_{HOR} = 3ct_o$, in metric units. This result is in full accordance with the conventional approach[7], for which there are also counterparts to our results (8), (9) and (10).

A very important feature (Fig.1) of the light path curve (6a) corresponding to the deformed light cone in the E-de Sitter space, is that it has a maximum value when $\dfrac{ds}{dt} = 0$, at the emission event, G_{em} having as coordinates :

$$t_{em} = \left(\frac{2}{3}\right)^3 t_o \quad \text{(y)} \quad , \qquad\qquad s_{em} = \left(\frac{2}{3}\right)^2 t_o \quad \text{(Ly)}$$

It follows that a galaxy of standard diameter, emitting from this epoch, will present to the observer F_o, a minimum angular diameter. (We remind that, according to the theory of astronomical observation, the distance of a source, assessed from its apparent diameter, is its distance at emission epoch.) The spectral ratio of the light received from this event will be :

$$\zeta_m = \left(\frac{t_o}{t_{em}}\right)^{2/3} = \frac{9}{4} = 2,25$$

exactly as suggested by the conventional theory[7], after long and complicated calculations. Our approach reveals the solution and meaning of this apparently inexplicable problem of the minimum angular diameter, posed by F. Hoyle in 1958, referred by J. Heidman[7] as a "very bizarre behaviour of light" and rendered by J. Silk[8] to the focusing of the light by the universal gravitational field, "which acts as a gigantic lens", a wild guess without any quantitative basis.

CONCLUSION

Our approach, based on the notion of a Privileged Fundamental Reference frame in an expanding Euclidean space and on the use of the light kinematic equation, is applied to observational astronomy and cosmology, equally to Big Bang and steady state models. This paper is limited to a brief outline of the basic consequences of the application of RW null geodesic. However further applications of our approach offer simple explanations to other paradoxes based by the conventional theory, while the employment of the light kinematic equation to the behaviour of peculiar massive bodies, leads to the Neuton's law of universal gravitation and to an understanding of the equilibrium of galactic formations despite the universal expansion. Finally, our approach is the only one whose results can be depicted by credible space-time diagrams, representing even the evolution of a Big Bang or steady state model of universe as a whole.

A great part of this paper is based on the essay "Cosmic Expansion Mechanics"[9] which offers further elaboration of the theory and of its consequences.

REFERENCES

1. G. Builder. "Austr. Journal of Physics", Vol. II, No.3 (1958).
2. S.J. Prokhovnik. "Proc. Cambr. Phil. Soc." 60,265 (1964).
3. M. Robinson. "The cosmological distance ladder", W.H. Freeman (1985).
4. M. Robinson. "Cosmology", Oxford Physic series (1977).
5. S.J. Prokhovnik. "Light in Einstein universe", Reidel (1985).
6. W. Rindler. "Essential Relativity". Springler-Verlag (1969).
7. J. Heidman. "Relativistic cosmology", Springler-Verlag (1980).
8. J. Silk. "The Big Bang", W.H. Freeman (1980).
9. A. Paparodopoulos. "Cosmic expansion Mechanics", Athens (1992).

FORNAX – THE COMPANION OF THE MILKY WAY
AND THE QUESTION OF ITS STANDARD MOTION

M. Zabierowski

Wrocław Technical University
50-370 Wrocław, Poland

The past encounter of Fornax and its perigalaction had as consequence the changing of the Fornax structure.

1. According to Hodge (1966) restructuralization of the Fornax density profiles proves that galaxies go round in an elliptical (Keplerian) motion. I am afraid that this vision of the Keplerian motion had been wrongly – but widely! – accepted, particulary at the decades when the short scale of the Universe age ($H = 100$ kms^{-1}Mpc^{-1}) was generally submitted.

2. The aim of the Keplerian-like conceptualization is to obtain the agreement between the computed tidally limited radius r_t and the observed tidally limited radius r_l of the Fornax system of star collection. Very much conflict is observed because $r_t = 7.0$ kpc is two times greater than $r_l = 3.1$ kpc. This noticeable divergence had been discussed by Hodge and he claimed that the observed radius of the Fornax system was the tidal radius imposed at perigalaction meeting. The new value of r_t is much reduced, just by the factor 2.3.

3. The Fornax density profiles form a structure which is frozen over the period t_f. The time taken for a star system to fill out the tidally allowed volume again is of the order of the relaxation time, t_r. We can consider the age of the frozen structure of the Fornax density profiles t_f as the age of an extremely old population: ergo as the age of the oldest system, t_h. Let's take the halo population, then $t_h = 4.73 \cdot 10^{17}$ s, which is greater than the Friedman's age of the dusty Universe for $H = 75$ kms^{-1} Mpc^{-1}.

4. This value of t_h seems to be very conservative, and is in agreement with my attitude: I try to reduce the importance of my deduction. Among others I reject $H = 100$ and even $H = 75$ kms^{-1} Mpc^{-1}. To decrease the influence of the free parameters I am taking Sandage's $H = 50.3$, the Friedman's age of the dusty Universe $t_F = 6.13 \cdot 10^{17}$ s.

5a. t_f (the age of the Fornax system as reconstructed from the Fornax'density profiles) involved by tidal forces cannot exceed the age of the stars (extremely old population), t_h, reduced by the disruption time for globular clusters t_d; $t_f \leq t_h - t_d$, where t_d is carefully taken from Burbidge (1960) and is consistent with Ogorodnikov's considerations. The Burbidge's t_d value minimizes the strength of our assumptions, hence our conlusion is particularly valuable. Thus t_f is not greater than $t_h - t_d$. If not, then Fornax system should have relaxed to a volume other than the volume organized by tidal action at the Hodge's (close) perigalacticon encounter.

5b. I doubt that the perigalacticon–Fornax meeting had been realized at the cloudy stage of this dwarf. In all probability, the relaxation time of the Fornax-from-gas-clouds was very short (instantaneous).

6. $t - \tau$, where t is the present moment and τ is the moment at which the Fornax (not cloudy-proto-Fornax) encountered its perigalaction point, is exactly $9.2 \cdot 10^{17}$ s. The calculated value of $t - \tau$ is 2.9 times larger than t_f. During the decades astronomers had not recognized that dwarfs could not quickly revolve, as postulated by Hodge.

7. $t - \tau$ is 1.51 times greater than the so-called global cosmological age of the dusty Friedman Universe, t_F. The Hubble constant H may be biased, however the all possible (considered in astronomy) errors cannot

change this conflict situation. If only 0.1% of the Fornax stellar matters was in the form of globular clusters, then t_r could not execeed the reaction time for tidal stripping. Keplerian motion refused to become reconciled with the calculated f, E, M, n, P. Can we bring into harmony two sets of notions: $(t_f, t_h, t_d, t_r, t_F)$ and (f, E, M, n, P)?

8. Hodge's explanation is popular and treated as realistic. Such a realism requires realistic evaluation. It brings us toward the Arp and Tifft hypotheses; cf. „Theory, model, reality" of T. Grabińska (Ch. 'Problems of Galactic Redshift Interpretation', §§ 'Tifft's Discovery', 'Arp's Hypothesis', 'Assessment of Tifft's Discovery', and 'How to Understand Tifft's Discovery').

References

G.R. Burbidge: In Die Entstehung von Sternen, Berlin, Göttingen, and Heidelberg, 1960.
P.W. Hodge: Astrophys. J., **144**, 869 (1966).
F. Hoyle: Proc. R. Soc. London, **260**, 201 (1961).
A.R. Sandage: Astrophys. J., **178**, 1 (1972).

COSMOLOGICAL REDSHIFTS AND THE LAW OF CORRESPONDING STATES

S.V.M. Clube

Department of Physics
University of Oxford
Keble Road
Oxford, OX1 3RH, UK

INTRODUCTION

"Those who are attached, as they may be with the greatest justice, to every doctrine which is stamped with the Newtonian approbation, will probably be disposed to bestow on these considerations so much the more of their attention as they appear to coincide more readily with Newton's opinion".

Thomas Young, 1802.

In the past, it evidently needed a physicist of uncommon politeness to confront and overcome that tendency within the breed to sanctify its great and its corpuscular. Physicists of course do not readily perceive themselves as being on probation. And yet, as providers of truth and the means to survive, they are usually so perceived in the public eye and do not always give a good account of themselves in the presence of scientists from other disciplines. As far back as the end of the nineteenth century, for example, there were distinguished authorities upholding a physical timescale for an obvious astronomical system (the Sun and Solar System) which could not be sustained in the face of the much preferred evidence from the ground for long term geological and biological calm. Admittedly this physical timescale was quite an advance on the position taken up during the middle years of the nineteenth century when the biblical timescale, requiring short term astronomical effects of a catastrophic nature, also proved disastrously wanting in the presence of geological and biological uniformitarianism. However, the fact that the exponents of a science which claimed its direct descent from Copernicus, Galileo and Newton could prove themselves so seriously mistaken in the face of demonstrable natural facts did not fail to have its effect; and by the beginning of the twentieth century, it is clear that physics and, to a greater extent, astrophysics were being conducted in an atmosphere of very low public and academic esteem. These conditions, which were not necessarily favourable to the maintenance of the highest intellectual standards, continued to the end of the Second World War, by which time some of the achievements of physical science had been such as to restore the discipline's lost esteem. Now, however, as the twentieth century itself is drawing to a close, this esteem is seen as having been built on fear as much as on respect and it

Frontiers of Fundamental Physics, Edited by M. Barone
and F. Selleri, Plenum Press, New York, 1994

is again an interesting question whether a new generation of physicists will repeat the failings of its predecessors and seek to force the cosmos into a preconceived mould.

THE PRESENT SITUATION

Physicists at the beginning of the twentieth century essentially pre-determined the nature of the universe as it has been perceived the last thirty to forty years. Thus, by preferring a model of the universe which comprised a stationary vacuum filled with transmitted photons (Einstein 1905) rather than one which comprised an inertial aether traversed by electromagnetic waves (Larmor 1900a), they took a gamble in respect of the cosmological redshift. In the former case, no cosmological redshift was evidently expected whereas in the latter case, the motions of the intervening aether, reflecting those of the imposed gravitational fields along any line of sight, were such that a cumulative lowering of the transmitted frequency at the second order in w/c was clearly anticipated. The picture here is that of typical relative motions w between successive regions of the aether supposedly "dragged" by matter (Lorentz 1892, Larmor 1897), and of secondary wavelets therefore which are unavoidably additive in respect of the quadratic doppler effect. It is true of course that the new hypothesis of photons, by reason of supposed economies in conceptualization, seemed to have a certain advantage over the aether model which was necessary if its capacity to transmit electromagnetic waves and to exhibit the property of drag were both to be explained. But one can hardly suppose that this reasoning would have been sufficient in itself to settle the issue unless there had also been a general conviction, however justified this may have been, that celestial space did not influence electromagnetic radiation eg so as to produce a cosmological redshift.

In the event, the presence of cosmological redshifts was established thirty years later (Hubble 1929). But instead of admitting the gamble and accepting the possible wave nature of electromagnetic radiation, the weight of opinion amongst physicists was still such as to endorse the 1905 preference and propel astrophysicists along their subsequent ptolemaic path. Thus, astrophysicists proceeded to invent a twentieth century "epicycle" to save the phenomena (Eddington 1931), suggesting for the vacuum an underlying state of universal expansion to which the material world was also tied. The invention was a rational response to the astrophysical facts in the presence of an imposed physical constraint; but one must take note of the timing of the inference for its epoch was also the period when astrophysicists and the cosmos were held in low esteem by scientists generally, including physicists, and the development was also widely seen as an unimportant aberration which could not be taken seriously. It would be to mock the role of scientific opinion in the progress of scientific knowledge, for example, not to allow Rutherford's well known view of Eddington to enter the statement of account.

In due course, the inferred expansion came to be perceived as a natural corollary of a "big bang" whose signatures were the microwave background and the observed distribution of nucleosynthetic species, but the fundamentally "epicyclic" character of the invention and its superstructure were still recognized and astrophysicists continued to seek more convincing "proof". To the extent that the universe is apparently uniform on the largest observed scale and dominated on a finer scale by islands of matter (such as our own) known as galaxies, there was evidently a *prima facie* expectation that the cosmological redshift due to a cumulative lowering of the transmitted frequency would be made up of characteristic units $\sim$ 30-50 kms^{-1} representing the separate, typical influence of interposed galactic environments. With the discovery eventually that the cosmological redshift is in fact so "quantized" (Tifft 1977, Guthrie & Napier 1991), there is understandable embarrassment for late twentieth century gamblers who may now be reluctant to add yet another cosmological "epicycle" to the adopted theory in order once again to save the phenomena. Not surprisingly perhaps, the existence of the cosmological " quantum", like the redshift itself for a while seven decades ago, is still largely denied!

THE INERTIAL AETHER

When the aether was first conceived, its function was that of a simple plenum at rest which filled the apparent vacuum. Direct observation seemed to imply an invisible, inviscid fluid medium in which the visible material of the universe was immersed and hence an aether wind in association with absolute motion. But with the development and recognition of an effective wave theory of light during the early nineteenth century, requiring an aether concept more in keeping with an elastic solid, the physical modelling of the luminiferous-cum-inertial aether was perceived to be a good deal more subtle than it first appeared (Schaffner 1972). The way forward, as it turned out, was to picture the material aether as a frictionless fluid in an essentially cellular configuration, supposing there to be a basic array of vortices whose precise structure and motion would then be arrived at through the construction of suitable physical models to be validated through experimentation and observation. A basic structure of the supposed kind evidently allowed the possibility of more complex configurations in practice, appropriate to the more elaborate angular momentum/energy states of both the invisible medium and visible matter, while the condition of visible matter in motion, treated as a sequence of displaced complex configurations, seemed to imply a succession of "corresponding states". The perception of the aether that emerged therefore was one which made it possible to combine locally and vectorially the circulatory motion of the vortex material (c say) and the translatory motion of a passing potential field (v say). In other words, it came to be seen at an early stage of the enquiry that the passing potential fields, treated as added patterns of inertial motion, basically determined the visible and invisible states of the underlying medium, so that the vacuum fields of visible material were in essence instantaneously co-moving or dragged aspects of the surrounding aether. By the end of the nineteenth century, a variety of analyses was tending to convince physicists that aether models involving spirally wrapped filamentary vortices (*ie* hollow vortices, not at all unlike the complex life-giving structures now known to exist at the molecular level), systematically disposed but nevertheless homogeneous on a larger scale (*ie* vortex sponge), offered the best prospect for eventual success (Whittaker 1951).

The idea of an inertial medium traversed by moving potential fields (gravitational, electric *etc)* received perhaps its most important endorsement when Maxwell successfully correlated the transverse wave properties of electromagnetic and luminiferous radiation with those of the electric and magnetic field bearing vortices of the aether. But the deepest level of understanding was probably achieved by Kelvin when he used a particular rigid array of gyrostats to demonstrate the action of a possible mechanical model of the vortices representing the boundless aether. Thus he proposed for the aether a so-called quasi-labile mechanism (Larmor 1894) according to which the medium could alternate between finitely short intervals of time in an incompressible state characterized by its inability to transmit condensational-rarefactional waves (*ie* in the state of a so-called MacCullagh aether) and infinitely short intervals of time in a compressible, or jelly-like, condition characterized by its ability to transmit condensational-rarefactional waves of exceptionally high speed ($ie >> c$ and in the state of a so-called Green aether). Such a perception may now be seen as marking the introduction of a phase-locked or universally coordinated aether whose successive (stable) vortex states are punctuated by regularly intervening (unstable) transitions during which the unsustainable condensations and rarefactions appearing in the medium are more or less instantaneously dispersed. These universally coordinated transitions can evidently be expected to include those changes of state that occur in the medium at large in which various pre-existing and post-existing transverse waves - the secondary wavelets of an earlier epoch - are, in effect, respectively dissolved and generated. The idea that visible matter occupies and jumps between various angular momentum/energy states is now of course part of the general currency of twentieth century physics though it appears not to be commonly appreciated that Larmor (1894) set the scene for this development by demonstrating that the Green and MacCullagh aethers, comprising the Kelvin aether, were in essence intertransferable. But having interposed the faceless vacuum and rendered photons paramount, any notion of an underlying mechanism appears then to have been lost and such transitions have disparagingly entered the realm of "hidden variables" !

It is well known of course that, during the last twenty years of the nineteenth century, it also came to be understood that the fundamental equations of the aether retained their form unaltered (to the second order in v/c) when the 'local' coordinate variables of moving reference frames (t, x, y, z) are subject to the Lorentz transformation:

$$
\begin{aligned}
ct &= ct' \cosh \alpha + x' \sinh \alpha \\
x &= x' \cosh \alpha + ct' \sinh \alpha \\
y &= y' \\
z &= z' \\
v &= c \tanh \alpha,
\end{aligned}
$$

and at the same time the electric and magnetic intensities are subject to the transformation:

$$
\begin{aligned}
E_x = E'_x, \quad E_y = E'_y \cosh \alpha + H'_z \sinh \alpha, \quad E_z = E'_z \cosh \alpha - H'_y \sinh \alpha \\
H_x = H'_x, \quad H_y = H'_y \cosh \alpha - E'_z \sinh \alpha, \quad H_z = H'_z \cosh \alpha + E'_y \sinh \alpha
\end{aligned}
$$

These effects were understood directly in terms of a physical contraction of material in the line of motion with absolute velocity v and a physical slowing of moving electromagnetic oscillators in the same proportion. However, it was also recognized that these compensatory physical effects influencing the form of the Maxwell equations were such as to render absolute motion through the aether and hence the latter's location in velocity space unobservable in a wide variety of physical experiments.This became something of an embarrassment for those who would have preferred to have the inferred aethereal mechanism more openly revealed.That the inertial aether had an observable influence on the passage of electromagnetic radiation was nevertheless clearly perceived and we cannot suppose therefore that the essential observation here was simply set aside.

According to Larmor (1900a) for example, by way of illustrating his general molecular theory of the aether, it was possible to consider the group formed of a pair of electrons of opposite sign describing circular orbits round each other in a position of rest: thus "we can assert for the [above correlations] that when this pair is moving through the aether with velocity v in a direction lying in the plane of their orbits, these orbits relative to the translatory motion will be flattened along the the direction of v to ellipticity $1 - \frac{1}{2}v^2/c^2$ while there will be a first order retardation of phase in each orbital motion when the electron is in front of the mean position combined with acceleration when behind it so that on the whole the period will be changed only in the second order ratio $1 + \frac{1}{2}v^2/c^2$. [.....]. The circumstance that the changes of their free periods, arising from convection of the molecules through the aether, are of the second order in v/c, is of course vital for the theory of the spectroscopic measurement of celestial velocities in the line of sight. That conclusion would however still hold good if we imagined the molecule to have inertia and potential energy extraneous to (*ie* unconnected with) the aether of optical and electrical phenomena, provided these properties are not affected by the uniform motion...". In other words, Larmor was clearly aware that the moving gravitational fields superimposed upon a celestial light path would inevitably result in a celestial redshift. Thus, when it is supposed that "the pressures and thrusts of the engineer, and the strains and stresses in the material structures by which he transmits them from one place to another, [are the] archetype of the processes by which all mechanical effect is transmitted in nature" and stated that " this doctrine implies an expectation that we may ultimately discover something analogous to structure *in the celestial space*, by means of which the transmission of physical effect will be brought into line with the transmission of mechanical effect by material framework" (Larmor 1900b), we can hardly be in doubt as to the observational effect which Larmor expected whilst concluding also that "we should not be tempted towards explaining the simple group of relations which have been found to define the activity of the aether by treating them as mechanical consequences of a concealed structure in that medium; we should rather rest satisfied with having attained to their exact dynamical correlation, just as geometry explores or correlates, without explaining, the description and metric properties of space". Whittaker (*loc cit*) in his famous review of aether theories unreservedly identifies Larmor, well before Einstein and Minkowski, as the one principally responsible for recognizing the aether as an immaterial medium not composed of identifiable elements having definite locations in absolute space. On the available evidence, there has to be some concern that he took this course to far reaching effect in the belief that the cosmological redshift did not exist. To what extent, then, did astrophysicists address the physical evidence thirty years later when the cosmological redshift was found, after all, to exist ? To what extent, indeed, is it recognized that Einstein may have introduced Mach's principle *post facto* to eliminate the possibility of observations that would reveal the aether's presence ?

THE COSMOLOGICAL REDSHIFT

The predicted doppler effect based on the Lorentz transformation for radiation passing between emitters and absorbers associated with relatively moving gravitational fields is

$$\nu' = \nu\beta(1 + \cos\theta\frac{w}{c})$$

where $\beta = (1 - w^2/c^2)^{-1/2}$, θ is the angle between the line of transmission and the line of relative motion, and ν is the oscillator frequency considered. It follows that the dispersion free spectral line shifts along and perpendicular to the direction of transmission (with velocity component u), assuming $w/c \ll 1$, are respectively given by

$$cz_{\parallel} = u + \frac{1}{2}c\frac{w^2}{c^2}$$

$$cz_\perp = \frac{1}{2}c\frac{w^2}{c^2}$$

the so-called longitudinal and transverse quadratic terms being experimentally confirmed utilizing positive rays (Ives & Stilwell 1938, 1941). In the absence of a material aether, photons are generally envisaged as travelling unaltered between the source and distant receiver with the result that the cosmological redshift is given by

$$cz_\parallel \text{ (observed)} = cz_\parallel + cz_\parallel \text{ (expansion)}$$

in which $(u, w) \ll c$ and $cz_\parallel$ (expansion) relates to the underlying zero mass substratum. In the presence of a material aether however, we expect a sequence of absorbers and emitters associated with the dominant intervening gravitational fields in average relative motion w, having components u along the radiation path, whence the observed spectral shift is given by

$$cz_\parallel \text{ (observed)} = \sum u + \frac{1}{2}c\sum \frac{w^2}{c^2}$$

Correcting for the relative motion of the Sun with respect to our Galaxy and summing over cosmological paths dominated by a sufficiently large number of intervening fields in random relative motion ($ie \sum u \to 0$), the spectral displacement is given by

$$c\bar{z}_\parallel = \frac{1}{2}c\sum \frac{w^2}{c^2}$$

For the purposes of discussion, we may assume that the successive relative velocities along any line of sight "between galaxies" and "between stars" are of comparable magnitude in which case $c\bar{z}_\parallel$ is dominated by the stellar component alone. To this degree of approximation therefore, it follows that the quadratic doppler redshift reduces to

$$cz_\parallel = Hr$$

the expression for the Hubble constant H being

$$cH = n_g \left(\tfrac{3}{4}n_s d_g w_s^2\right)$$

in which n_g is the number of typical (spiral) galaxies per megaparsec, d_g is a representative depth for a typical galaxy halo intercepted by the cosmological line of sight, n_s is a representative number density of stars per unit halo distance and w_s is one component of the typical "between stars" relative velocity. Evidently for a cosmological distribution of intervening gravitational fields of similar dimensions and mass which is isotropic, we anticipate a common value $Q = \tfrac{3}{4}n_s d_g w_s^2$ and the Hubble law is appropriately "quantized". This quantization is meaningful in the sense that individual galaxies have redshifts of the form

$$cz_\parallel = kQ + u_G$$

where k is an integer and u_G is the uncompensated line of sight velocity component of the particular observed galaxy relative to our own. In general this may be expressed in the form

$$cz_\parallel = (k + k')Q + \epsilon$$

in which k' is a (relatively small) integer and $|\epsilon| < Q$, with the awkward consequence that dynamical effects (reflecting the influence of both ordinary and dark matter) may be substantially masked by quantization.

In addition to a fundamental cosmological quantum, the quadratic doppler effect also produces halo-disc redshift discrepancies $\Delta(\frac{3}{4}n_s d_g w_s^2)$ for near parallel lines of sight originating from within the same galactic system. These depend on typical near-halo and near-disc values of (n_s, w_s) respectively. A preponderance of red rather than blue shifts is consistent with a stellar relative velocity effect which is stronger amongst older stars in accordance with the known galatic dynamical properties of the variously aged stellar populations generally. This effect is also broadly in accordance with estimates of the Sun's outward motion in the Galactic disc ($\sim 20 - 25$ kms^{-1} say) based on (1) the quantized cosmological redshifts of isotropically distributed nearby galaxies, (2) the detected north-south assymetry in the HI circular motion derived from material in orbits relatively nearer the Galactic centre than the Sun, (3) the motion of the (spiral arm) Stream I relative to the (older disc) Stream II in the Solar neighbourhood and (4) the pattern of interstellar cloud motions in orbits closer to the Galactic anticentre; for this motion amounts to roughly half the redshift of $Sgr\ A^*$ (~ 40 kms^{-1}) supposedly at a comparatively low velocity with respect to the Galactic nucleus (Blitz $et\ al$ 1993).

CONCLUSION

Reasons are given here for believing the cosmological redshift and quantum are added epicycles too many to justify the present course of physics and astrophysics which presupposes an absolute epoch BC for the origin of the universe. Rather it is suggested that we should restore the inertial aether as being fundamentally in accordance with the principal observed cosmological facts, recognizing also a material state for the universe which is both stationary and infinite in space and time. By so doing, we necessarily infer a strong radial motion for the local standard of rest in the Galaxy, thereby imposing upon the youngest spiral arm material in the solar neighbourhood and hence elsewhere in the Galactic disc, even extending to the central region itself, such an outward flow of cold gas and recently formed stars that we may also suspect that the nucleus of our Galaxy is the seat of repeated violent events of the kind seen in other galactic nuclei throughout the universe. This indicates that we should perhaps take seriously the modification of Lorentz's theory proposed by Dicke (1961) to explain the dynamical effects of gravitational fields. Thus an enhancement of mass (at the expense of c) in deep potential wells is predicted such as temporarily arise in highly evolved stars and superstars, an effect which in recurring galactic nuclei results in large gravitational redshifts, the release of spiral arms and the presence of dark matter. Quite apart from the implications of Lorentzian theory in general so far as fundamental physics and the infinite, stationary universe are concerned, the consequences for astrophysics are profound. Thus, we envisage spiral arms in the form of rapidly cooled plasma jets and a star, planet and comet formation regime therein which is of the kind originally envisaged by Jeans (1928).

ACKNOWLEDGEMENTS

I thank Franco Selleri for the invitation to speak at Olympia and his forbearance as the written version emerged; also Bill Napier for keeping me apprised of developments on the cosmological "quantum" front.

REFERENCES

Blitz, L., Binney, J., Lo, K.Y., Bally, J. & Ho, P.T.P., 1993, *Nature* 361:417.
Dicke, R.H., 1961, *Enrico Fermi Course XX* (ed., Möller, C., Academic Press).
Eddington, A.S., 1931, *Mon. Not. R. Astr. Soc.* 91:412.
Einstein, A., 1905, *Ann. der Phys.* 17:132.
Guthrie, B.N.G. & Napier, W.M., 1991, *Mon. Not. R. Astr. Soc.* 253:533.
Hubble, E., 1929, *Proc. Nat. Acad. Sci.* (USA) 15:168.
Ives, H.E. & Stilwell, G.R., 1941, *J. Opt. Soc. Amer.* 28:215.
Ives, H.E. & Stilwell, G.R., 1941, *Proc. Nat. Acad. Sci* 31:369.
Jeans, J.H., 1928, "Astronomy and Cosmogony", C.U.P.
Larmor, J., 1894, *Phil. Trans. Roy. Soc.* 185:719.
Larmor, J., 1897, *Phil. Trans. Roy. Soc.* 190:205.
Larmor, J., 1900a, "Aether and Matter", C.U.P.
Larmor, J., 1900b, *Brit. Assoc. Report.* 618.
Lorentz, H.A., 1892, *Versl. Kon. Akad. Wetensch. Amsterdam* 1:74.
Schaffer, K.F., 1972, "Nineteenth Century Aether Theories", Pergamon.
Tifft, W.G., 1977, *Astrophys. J.* 211:31.
Whittaker, E., 1951, "A History of the Theories of Aether and Electricity", Nelson.
Young, T., 1802, *Phil. Trans. Roy. Soc.* 92:387.

DID THE APPLE FALL?

Hüseyin Yılmaz
Hamamatsu Photonics K.K., Hamamatsu City, 435 Japan
Electro-Optics Technology Center, Tufts University
Medford, MA 02155 USA

INTRODUCTION

This paper is essentially a sequel to the Il Nuovo Cimento article (**107B**, 941, 1992) by the author on the new theory of gravity. Our aim is to discuss some of the points which were not possible to tend to in the original article and some that came to our attention since its publication. However, efforts are made to make the present discussion self-contained by briefly restating the general features of the theory in appropriate places. Professor Carroll O. Alley will present a concise summary and some of the recent results in these proceedings.

There exists an important mathematical fact in gravity theory of which most relativists seem to be unaware. This is the surprising truth that Einstein's field equations do not lead to interactive N-body *(N > 1)* solutions, hence being only a 1-body theory it has no correspondence to the Newtonian theory in the interactive N-body sense. [1a] It can at most be considered as a test-particle theory whereby it can be made to predict the three classical tests [a central body (sun) in the solution as in the Schwarzschild case, and a number of test bodies (planets) not in the solution but put by hand]. But such a test-particle theory makes also a number of false predictions. For example, test bodies (planets) cannot interact with each other so the planetary-perturbative (*532"* per century) part of the advance of the perihelion of Mercury would be missing, and, everything except the sun being a test particle, an apple detached from its branch would not fall to the ground. Also, most relevantly to the problem of motion, such a theory cannot consistently be applied to two masses of comparable size (Binary Systems) since one of the masses is in the solution (has a field) and the other is not (has no field) violating the universal N-body symmetry of gravity. [2]

COMPUTER CALCULATIONS

Outrageous as it may seem the truth of these statements can be proved in a general way, and special cases where exact solutions of Einstein's equations are obtainable can be exhibited in detail by computer calculations. [3] Gottfried Leibniz had a method of settling scientific disputes and had only one word to describe it – *Calculemus*– (let us calculate). So to settle the issue here considered we calculate, in each case, everything needed to all necessary detail. In the past this may have been difficult since some of the calculations are prohibitively long. But now there can be no excuse since the necessary calculations can be done by computer (for example, by MATHEMATICA, MACSYMA, REDUCE, etc.) in a matter of minutes and without mistake. Calculations involve finding solutions satisfying *conservation laws* and *appropriate boundary conditions* to Einstein's equations $^{1}/_{2}G_\mu{}^\nu = \tau_\mu{}^\nu$ where $\tau_\mu{}^\nu$ is the "matter" term alone (no gravity part $t_\mu{}^\nu$) and then evaluating $\sigma du_\mu/ds = {}^{1}/_{2}\,\partial_\mu g_{\alpha\beta}\,\tau^{\alpha\beta}$ which is *assumed* to lead to the geodesic equations of motion. But for *any such* solution of Einstein's field equations (including those of Schwarzschild type) the right-hand side, $^{1}/_{2}\,\partial_\mu g_{\alpha\beta}\tau^{\alpha\beta}$, when integrated over a body actually vanishes, implying that $m du_\mu/ds = 0$, hence there is no force (no interaction) on any mass m due to any other mass element m' in the (mass) distribution. To those of us who habitually write $\tau^{\alpha\beta} = \sigma\,u^\alpha u^\beta$ and expect the geodesic equations of motion $du_\mu/ds = {}^{1}/_{2}\,\partial_\mu g_{\alpha\beta}\,u^\alpha u^\beta$ this is quite a shock. The problem is that Einstein's field equations do not give $g_{\alpha\beta}$ and $\tau^{\alpha\beta}$ so as to lead to the geodesic equations of motion, nor to its N-body interactive Newtonian limit; they give something else such that, when calculated, $^{1}/_{2}\,\partial_\mu g_{\alpha\beta}\tau^{\alpha\beta}$ turns out to be zero. The situation is utmost serious because Einstein's theory thereby loses its presumed geodesic equations of motion on which practically all of its prediction procedures are based.

A SIMPLE EXAMPLE

We shall shortly give a general proof of this statement but it is instructive to first see how it works out in a simple specific case. For this purpose we consider the case of two parallel slabs S and S' of uniform densities σ and σ' placed parallel to the x-y plane. The form of the metric is found to be [4]

$$ds^2 = e^{-2\phi}\,dt^2 - e^{2\phi}\,[\,e^{2\varepsilon\phi}(dx^2 + dy^2) + e^{4\varepsilon\phi}\,dz^2]$$

where $\varepsilon = \pm 1$ in general relativity. (Note that the solution is not unique.) Here $\phi = Az + {}^{1}/_{2}\,\sigma z^2 + C$, and A and C are determined, in each of the five regions (three space and two matter regions), from the regularity condition that ϕ and its first derivatives be continuous. Calculating $\tau_\mu{}^\nu$ in S one finds $\tau_\mu{}^\nu = diag[\,(1+\varepsilon,\ {}^{1}/_{2}\varepsilon,\ {}^{1}/_{2}\varepsilon,\ 0)\,\sigma]$, hence

$$^{1}/_{2}\,\partial_\mu g_{\alpha\beta}\,\tau^{\alpha\beta} = -\,[\,1+\varepsilon - {}^{1}/_{2}\varepsilon(1+\varepsilon) - {}^{1}/_{2}\varepsilon(1+\varepsilon)\,]\,\sigma\partial_z\phi$$

$$= -\,(1 - \varepsilon^2)\,\sigma\partial_z\phi = 0$$

since $\varepsilon = \pm 1$ in general relativity. The $\partial_z\phi$ contains the σ' of the other slab. (Note that σ is the Laplacian of ϕ, hence $-\sigma\partial_z\phi = \nabla^2\phi\,\partial_z\phi$, so the equations of motion are "second-order" consequences of the field equations.) Something similar occurs in all cases calculated (uniform slabs, concentric shells, point particles, [4] etc.). In other words, in general relativity there is a conflict between the field equations and the presumed geodesic equations of motion. That it is so can, however, be more generally proved in a Lagrangian treatment embodying the conservation laws of energy-momentum and angular momentum.

GENERAL PROOF

That Einstein's theory is not an interactive N-body theory can be proved generally *if* an important but little known fact about Riemannian geometry is taken into account. This is the fact that in Riemannian geometry there is a second differential identity (Freud's identity) besides the usual Bianchi identity.[1a] In a conservative Lagrangian theory $\partial_\nu(\sqrt{-g}\tau_\mu{}^\nu) \equiv 0$, $\tau^{\mu\nu} = \tau^{\nu\mu}$ the two identities plus Einstein's field equations can be written as

$$\tfrac{1}{2}G_\mu{}^\nu = \tau_\mu{}^\nu \tag{1}$$

$$D_\nu G_\mu{}^\nu \equiv 0 \tag{2}$$

$$\partial_\nu(\sqrt{-g}\tau_\mu{}^\nu) \equiv 0 \tag{3}$$

Taking the covariant divergence of *(1)* and using *(2)* and *(3)* one finds

$$\tfrac{1}{2}D_\nu G_\mu{}^\nu = \partial_\nu(\sqrt{-g}\tau_\mu{}^\nu)/\sqrt{-g} - \tfrac{1}{2}\partial_\mu g_{\alpha\beta}\tau^{\alpha\beta} = 0 \tag{4}$$

$$\tfrac{1}{2}\partial_\mu g_{\alpha\beta}\tau^{\alpha\beta} = 0 \tag{5}$$

(that is, $\alpha du_\mu/ds = 0$, no acceleration) which can be taken as a general proof.

IMPORTANCE OF COMPUTER CALCULATIONS

Note that in the computer calculations the two identities are not explicitly used, nor are they even mentioned. This is (of course) because in a conservative system the identities assert themselves even if we are not aware of their existence. This points to the importance of computer calculations because a sufficient number of such calculations can indicate the presence of these identities. For example, given Einstein's equations plus the Bianchi identities the existence and form of the Freud identity (including the fact that in a consrvative theory an additive coordinate artifact $z_\mu{}^\nu$ must have zero density divergence) could be inferred from such calculations. Finally, the computer calculations seem to reliably single out the culprit as Einstein's field equations – the only equations used – for the failure of not recovering the geodesic equations of motion.

A FIRST ACQUAINTANCE

Having come this far with the simple example we may also indicate, in this simple case, how the new theory deals with the problem. It does so by modifying Einstein's field equations into $\tfrac{1}{2}G_\mu{}^\nu = \tau_\mu{}^\nu + t_\mu{}^\nu$ (where $t_\mu{}^\nu$ is the gravitational field stress-energy which, given $\tau_\mu{}^\nu$, is found via the two identities up to a tensor of zero divergence) and by requiring

$\tau_\mu{}^\nu = \sigma u_\mu u^\nu$. The latter **a)** is the appropriate generalization of the matter tensor of special relativity, and **b)** it recovers correctly the Poisson density of matter in the Newtonian limit. The Newtonian limit of the *2-slab* case now gives $\varepsilon = 0$ (note that the solution now is unique) and gives simple exponential metric $g_{00} \doteq e^{-2\phi}$, $-g_{ik} = \delta_{ik} e^{2\phi}$), hence

$$\sigma du_\mu/ds = {}^1/_2 \partial_\mu g_{\alpha\beta} \tau^{\alpha\beta} = -\sigma \partial_z \phi$$

Thus in this case the solution is 2-body (in general N-body by $\sigma \Rightarrow \sigma_1 + \sigma_2 \dots \sigma_N$) interactive in exact correspondence to the Newtonian theory and to the geodesic equations of motion; the difficulties encountered above do not occur (the apple falls). The pattern repeats for other cases as a general theorem at the Newtonian limit of the theory. Note that here initial velocities need not be zero. Motion is via the Hamilton-Jacobi equation $g^{\mu\nu} p_\mu p_\nu = m^2$, $H = p_0 = e^{-\phi}\sqrt{(m^2 + e^{-2\phi} p^2)}$, leading to the geodesic equations of motion. [In the new theory $\sigma_i du_\mu/ds = {}^1/_2 \partial_\mu g_{\alpha\beta}(\tau^{\alpha\beta} + t^{\alpha\beta})$ where σ_i is the inertial mass but the second term on the right does not contribute in the time-independent limit]. The new theory predicts correctly all known effects of gravity. (For example, it has energy-carrying gravity waves and allows calculation of the mass-quadrupole.) [5] It explains the apparent success of many calculations in general relativity by noting that in such cases the N-body interactive solutions are implicitly assumed without proof, whereby general relativity mimics the results of the new theory. This too can be verified with computer calculations.

THE GENERAL CASE

How the new theory proposes to generally overcome the exhibited defects arising from the absence of $t_\mu{}^\nu$ in Einstein' theory is given in the Il Nuovo Cimento article in terms of a variational principle. Although the variational treatment is probably most basic in terms of clarity and mathematical consistency, it is not the most intuitively accessible so we shall here adopt a more intuitive approach. In a conservative Lagrangian theory ${}^1/_2 G_\mu{}^\nu$ is composed of two parts $\tau_\mu{}^\nu$ and $t_\mu{}^\nu$ such that

$$\tfrac{1}{2}G_\mu{}^\nu = \tau_\mu{}^\nu + t_\mu{}^\nu \tag{1'}$$

$$D_\nu G_\mu{}^\nu \equiv 0 \tag{2'}$$

$$\partial_\nu(\sqrt{-g}\,\tau_\mu{}^\nu) \equiv 0 \tag{3'}$$

where, given $\tau_\mu{}^\nu$, the form of $t_\mu{}^\nu$ is determined by the two identities. Choosing the coordinates Cartesian and harmonic, one has

$$\tau_\mu{}^\nu = \square^2 \phi_\mu{}^\nu - \partial_\alpha(\sqrt{-g}\,\partial^\nu \phi_\mu{}^\alpha)/\sqrt{-g} \tag{4'}$$

$$t_\mu{}^\nu = -2(\partial_\mu \phi_\alpha{}^\beta \partial^\nu \phi_\beta{}^\alpha - \tfrac{1}{2}\delta_\mu{}^\nu \partial^\lambda \phi_\beta{}^\alpha \partial_\lambda \phi_\alpha{}^\beta) + \partial_\mu \phi \partial^\nu \phi - \tfrac{1}{2}\delta_\mu{}^\nu \partial^\lambda \phi \partial_\lambda \phi \tag{5'}$$

$$\sigma_i du_\mu/ds = {}^1/_2 \partial_\mu g_{\alpha\beta}(\tau^{\alpha\beta} + t^{\alpha\beta}) \tag{6'}$$

It is clear from the electromagnetic theory analogy that in the case of "hydrodynamic matter" we may choose the "matter" equations as

$$\tau_\mu{}^\nu = \Box^2\phi_\mu{}^\nu - \partial_\alpha(\sqrt{-g}\,\partial^\nu\phi_\mu{}^\alpha)/\sqrt{-g} \;\Rightarrow\; \sigma\, u_\mu u^\nu \tag{7'}$$

where, given $\tau_\mu{}^\nu$, the $t_\mu{}^\nu$ and its form is determined by the two identities. Strictly speaking, the above form is written in Lorentz-Harmonic (dx, dy, dz, dt, $\partial_\nu(\sqrt{-g}\,g^{\mu\nu}) = 0$, but not necessarily in freely fall) coordinates, that is, with the Lorentz gauge $\partial_\nu\phi_\mu{}^\nu = 0$.

$$\partial_\nu(\sqrt{-g}\,g^{\mu\nu}) \;=\; \partial_\nu\phi_\mu{}^\nu = 0.$$

Transformation to other coordinates and gauges is, however, possible since in the local Lorentz frames from which we start there is no $z_\mu{}^\nu$. [1a] (Other *matter* sources, scalar, vector, etc. field stress-energies, are introducible and treatable by perturbative methods.) The theory thus automatically satisfies both the Newtonian and the special relativistic limits correctly.

INTERACTIVE N-PARTICLE SOLUTIONS

In the time-independent limit the expressions of $\tau_\mu{}^\nu$ and $t_\mu{}^\nu$ reduce to

$$\tau_\mu{}^\nu \to \tau_0{}^0 = \nabla^2\phi = \Delta\phi\,/\,\sqrt{-g} = \sigma$$

$$t_\mu{}^\nu = -\,\partial^\nu\phi\,\partial_\mu\phi + \tfrac{1}{2}\delta_\mu{}^\nu\,\partial^\lambda\phi\,\partial_\lambda\phi$$

It can easily be shown (by computer or by hand) that there are N-body interactive solutions

$$\sqrt{-g}\,\sigma = \Delta\phi = \Sigma_A\, m_A\,\delta(x - x_A)$$

$$ds^2 = e^{-2\phi}\,dt^2 - e^{2\phi}\,(dx^2 + dy^2 + dz^2)$$

$$\phi = \Sigma_A\, m_A\,/|x - x_A| - C$$

$$\partial_\nu(\sqrt{-g}\,g^{\mu\nu}) = 0$$

Continuous matter can be expressed as $\phi = \int(\sqrt{-g}\,\sigma)'\,dV'/r' - C$. They are exact solutions.

A SURPRISING THEOREM

Theorem: If, in the time-independent slow motion limit of any matter distribution, the potential found in the Newtonian manner is substituted into the metric, one gets an exact solution to the corresponding curved space equations. The motion of such distributions can then be studied via the Hamilton-Jacobi method as mentioned. That is, for any matter distribution in the slow motion limit, *no matter how complicated,* we can immediately write down the corresponding exact curved space-time solution. The above slow motion solutions are special cases. (An analogous possibility with special relativistic fields replacing the Newtonian field is conjectured but so far it has not been proved.) One can see from the equations that the problem with Einstein's theory is one of mathematical overdetermination. For in the absence of $t_\mu{}^\nu$ the two identities apply at once on $G_\mu{}^\nu$ (or $\tau_\mu{}^\nu$). Since these are two different identities, a single quantity cannot sustain them mathematically. In the new theory there is the extra quantity $t_\mu{}^\nu$ which absorbs the conflict (in fact, by so doing gets determined up to an arbitrary tensor or nontensor with zero divergence), so there is no overdetermination. The $t_\mu{}^\nu$ represents the interaction energy (stress-energy), hence no $t_\mu{}^\nu$, no interaction.

THE PPN EXPANSION

In the above we have talked about exact solutions at the slow motion limit limit. We also have exact gravity wave solutions of $T\text{-}T$ (transverse-traceless) type [1b,c] which are crucial to the theory of the binary pulsar's gravity radiation. Between these two extremes we have slow motion N-body solutions of the type

$$\phi_\mu{}^\nu = \Sigma_A \, (mu_\mu u^\nu)_A / |x - x_A| - C_\mu{}^\nu$$

and these are useful in constructing Parametric Post-Newtonian (PPN) expansions. Such expansions are not legitimately constructible in Einstein's theory as it has no N-body interactive solutions even in the zero velocity limit. If nevertheless we construct them by implicitly assuming the N-body potentials, we are effectively working with the new theory.

NATURE OF THE POTENTIALS

As to the nature of the potentials, they are the same as the ones introduced by Einstein himself in the linear approximation and the gauge condition is the same as the one introduced by Hilbert in the same approximation. (Since $t_\mu{}^\nu$ is a second order quantity, the two theories are the same in first order.) But the equations of motion are second order consequences of the field equations, hence the linearized solution in terms of these potentials is not meaningful. One has to find a meaningful extension in second order. To this end we define the potentials directly in terms of metric connections (Christoffel symbols) as

$$\partial_\lambda \phi_\mu{}^\nu = -1/4 \, g^{\nu\rho}(\Gamma_{\rho\mu\lambda} + \Gamma_{\mu\rho\lambda}) + 1/4 \, \Gamma^\alpha{}_{\alpha\lambda}$$

This is equivalent to writing $\partial_\lambda \phi_\mu{}^\nu = (1/4) \, \mathbf{g}_{\mu\rho} \partial_\lambda \mathbf{g}^{\rho\nu}$ but it clearly exhibits the nature of the potentials. Their absolute values are not meaningful locally but their derivatives behave like Christoffel symbols. They arise from what is technically called tangent bundles, are invertible as [1b] $dg_{\mu\nu} = 2(g_{\mu\nu}d\phi - g_{\mu\alpha}d\phi_\nu{}^\alpha - g_{\nu\alpha}d\phi_\mu{}^\alpha)$. They are similar to the electromagnetic gauge potential A^ν. They seem to be integrable locally, but nonlocally this is not necessarily so if the path contains a flux tube as in the electromagnetic case. Such topological considerations are beyond the present experimental content. As noncommuting terms, do not contribute for existing tests (which are at most second order) we have a theory which is, at present, compatible with all experimental data available about gravity.

TWO GENERAL REMARKS

a) We have given at least one clear example of a complete solution to the problem in which we see exactly how the principles are working and we can calculate anything of relevance rigorously and to any desired accuracy. These include all Newtonian effects and the computation of the mass quadrupole and its derivatives. This is similar to the solution of the Quantum Mechanical harmonic oscillator problem. Although Heisenberg solved a simple (in fact the simplest) quantum mechanical problem, this solution indicated how other problems would work out since the principles were clearly elucidated.

b) In the other extreme we have shown that this theory gives exact gravity waves with positive energy, carrying energy-momentum. [1b,c] The relation between gravity radiation and the third derivative of the mass quadrupole is well known and leads to the period decay of the Binary pulsar as observed. These calculations cannot be carried out legitimately in general relativity. What happens in the intermediate cases of particles with velocities comparable to the velocity of light is at present accessible only via approximation methods. However, irrespective of whether the new theory is eventually right or wrong, the general relativity is clearly inadequate and it must be recognized as such to ensure healthy progress in the theory of gravity and its possible quantization.

SOME SUBTLE POINTS

We will now also clarify a few points where some readers might have uneasy feelings. They appear as legitimate problems at first but, when understood, there is no real problem:

1) The Freud identity is not a consequence of the Bianchi identity. The two are independent. For example, Landau pointed out a long time ago that the conservation law of energy-momentum is related to $\partial_\nu(\sqrt{-g}\tau_\mu{}^\nu) \equiv 0$ and not to $D_\nu G_\mu{}^\nu \equiv 0$. This is because Gauss' theorem (by which the conservation laws are implemented) works with the ordinary divergence and not with the covariant divergence. As a result the field equations allow a true tensor $t_\mu{}^\nu$. It turns out that it is exactly the stress-energy tensor of the new theory.

2) In any field theory (for example, electrodynamics) the field equations give time-independent solutions in the limit $c \to \infty$. One must not interpret such solutions as static since particles can have velocities as part of their initial conditions and the solution is still valid since the interaction is, in this limit, instantaneous. The Lagrangian further gives the equations of motion by a separate variation and by using this information one obtains the evolution of the system. In other words, in this theory static solutions are not to be interpreted as permanently stationary since there are forces between objects. They would be permanently static if there were no interaction between objects, which is what happens in Einstein's theory.

3) Mathematically $\tau_\mu{}^\nu$ is a second-order differential operation on $\phi_\mu{}^\nu$ (a gauge d'Alembertian satisfying Freud's identity). Physically it represents the matter stress-energy analogous to σ in Poisson's equation and reduces to it in the limit of slow-motion. However, matter generates a gravitational field $t_\mu{}^\nu$ which also has a stress-energy. As all energy is on equal footing $(E = mc^2)$, it is the *sum* of the two stress-energies that forms the source of the geometric curvatures. This is the a most basic motivation for the new theory.

4) The transition from a noninertial frame to a local inertial frame) should not be thought of as a coordinate transformation but as a local *compensation* or *balancing* of accelerations. This is because such transformations change the curvature quantities which are tensors, hence they cannot be coordinate transformations.

NATURE OF THE AFFINE TENSORS

We now mention also a few points that have come to our attention since the publication of the paper. In the new theory $U_\mu{}^\nu$, $u_\mu{}^\nu$ decompose as $U_\mu{}^\nu = \tau_\mu{}^\nu + z_\mu{}^\nu$, $u_\mu{}^\nu = - t_\mu{}^\nu + z_\mu{}^\nu$, but in general $^1/_2 G_\mu{}^\nu = U_\mu{}^\nu - u_\mu{}^\nu = \tau_\mu{}^\nu + \lambda t_\mu{}^\nu$ can be of the form $U_\mu{}^\nu = \tau_\mu{}^\nu + \alpha t_\mu{}^\nu + z_\mu{}^\nu$, $u_\mu{}^\nu = - \beta t_\mu{}^\nu + z_\mu{}^\nu$, $\alpha + \beta = \lambda$ where α depends on the choice of the coordinates. For example, in isotropic Cartesian coordinates $\alpha = - \varepsilon^2$, $\beta = 1$ so that λ is still $\lambda = 1 - \varepsilon^2$, although β is not equal to λ. Therefore it is better to characterize Einstein's theory directly through the overdetermination, that is, [6]

$$D_\nu G_\mu{}^\nu \equiv 0, \quad \partial_\nu(\sqrt{-g}\, G_\mu{}^\nu) = 0.$$

where $^1/_2 G_\mu{}^\nu = \tau_\mu{}^\nu$, If these relations hold, the solution belongs to Einstein's theory. Conversely, in order to be a solution to Einstein's theory these relations must be satisfied everywhere including the interior of matter distributions.

POSITIVE ENERGY AND THE METRIC

In order to have both the matter energy and the field energy positive, one can reinterpret the principle of equivalence as $- m_a = - m_p = m_i = m > 0$. This requires the reversing of sign of $G_\mu{}^\nu$ and $\phi_\mu{}^\nu$ but without a change in the form of the metric. This is actually the proper form, as can be seen by considering the following two facts: a) first is the observation that inertial and gravitational forces oppose each other and this leads correctly to the process of compensation in the free fall. b) secondly, the time-gain experiment of C. O. Alley and East-West West-East circumnavigation data of Hafele-Keating are so far calculated only by a convention since controvariant and covariant distinction of differentials is a convention. The positive energy requirement seems to remove this ambiguity by requiring a definite choice. That choice is counter intuitive and opposite to what one has become accustomed to. Rather than change all of our notations and conventions we may live with it as a historical accident as in the case of the sign of electronic charge.

ON GENERAL COVARIANCE

Finally, in what sense is this theory or any theory generally covariant? It seems that there is not a single experiment so far which requires the extent of generality required by Einstein's theory of gravitation. On the contrary there seems to be evidence that such unbounded generality is contrary to experiment. Take for example, a Lorentz frame in flat space-time and first transform it into spherical polar coordinates and then to $r \rightarrow r + K$ where K is an arbitrary number in the unit of length. [7] Calculating the time-delay for a ray passing a distance $a \geq K$ as if K represented a mass-point at the origin, we get a time-delay $\Delta t = 4K/c$. Is this a coordinate transformation in the same sense as going from the Lorentz frame to spherical coordinates which does not do anything like this? It seems not, because the first transformation does not stretch or tear the space whereas the second tears and stretches (here compresses) the space despite the fact that it is still flat. Perhaps the latter should not be called a coordinate transformation and should not be included in the repertoire of bona fide coordinate transformations (measuring rod is flexible but not compressible or stretchable) The new theory seems to be formulable in a local Lorentz covariant manner such that it can automatically disallow such transformations. This would be an additional hypothesis independent of the rest of the theory and testable by experiment. [1b, 7]

LOCAL LORENTZ COVARIANCE

Local Lorentz covariant formulation has the following three postulates: [7]

1) Physical observation is a *local* Process,
2) *Local* signal velocity is a universal constant,
3) Physical laws are *local* Lorentz-covariant.

The first of these implies that $\phi_\mu{}^\nu \rightarrow \phi_\mu{}^\nu - K_\mu{}^\nu$ where $K_\mu{}^\nu$ are constants and are the valuse of $\phi_\mu{}^\nu$ at the observation point. The second requires that we use local Lorentz variables where there is no $z_\mu{}^\nu$. The third requires the gauge condition $\partial_\nu\phi_\mu{}^\nu = 0$ in local Lorentz coordinates. The latter is equivalent to harmonicity in the new theory. These postulates can be used to identify $\tau_\mu{}^\nu$, $t_\mu{}^\nu$ and the field equations in the local Lorentz coordinates. An important point about the local Lorentz covariant formulation is that it takes over the measurement (operational) procedure of special relativity locally, hence it predicts that the locally measured velocity of light would be c in all directions even in a *noninertial* frame. In contrast general relativity does not have an operational procedure of measurement comparable to that of special relativity. A measurement of East-West and West-East one-way velocity of light on earth would settle whether the extra assumption is valid. It is interesting that Lorentz and Einstein have predicted the velocities to be different (that is, $c \pm v$) in the two directions. [9] Professor C. O. Alley and the author have proposed experiments to test the hypothesis. These are currently being pursued. [7, 8]

ACKNOWLEDGEMENTS

Deep gratitudes are due to Professor Carroll O. Alley and his associates (Ching Yun Ren, R. Kirk Burrows and others) at the University of Maryland, College Park, Maryland and to President Teruo Hiruma of Hamamatsu Photonics, K. K., and his staff (late Sakio Suzuki, Yoshiji Suzuki, Yutaka Mizobuchi and others) at Hamamatsu City, Japan. It is also a pleasure to thank Professor Willis E. Lamb and Mr. George W. Keros for their genuine interest and encouragement.

REFERENCES

1. H. Yılmaz, a) *"Towards a field theory of gravity"*, *Nu. Cim.* **107B**: 941 (1992); b) *Int. J. Theor. Phys.* **21**: 871 (1982); c) New theory of gravitation, *in: "M. Grossmann Meeting on General Relativity,"* Rome, Italy, Ed. R. Ruffini, North-Holland (1986).

2. H. Yılmaz, *in: "Wigner Symposium on Space-time Symmetries,"* Univ. Of Maryland, College Park, S. Y. Kim and W. W. Zachary, Eds., North-Holland (1988).

3. Computer calculations were directed by H. Yilmaz and C. O. Alley and performed by Dr. Ching Yun Ren and R. Kirk Burrows at the University of Maryland.

4. H. Yılmaz, Unpublished (1953); *Hadronic Journal*, **2**: 997 (1979). If the interaction is absent Einstein's equations do have N-body solutions. Thus, if in isotropic Cartesian Schwarzschild metric $\phi = \Sigma_k \alpha_k m_k / r_k$, $\alpha_k \alpha_i = \alpha_i \alpha_k = 0$, $\alpha_k^2 = 1$, the Einstein field equations are satisfied since the interactions (mutual terms) are removed.

5. Since $t_\mu{}^\nu$ and the field lagrangian is consistently recovered as in a local field theory, the new theory turns out to be perturbatively quantizable (H. Yılmaz, Unpublished 1992).

6. A. Einstein, *"The meaning of Relativity"* Princeton Univ. Press, p. 85 (1953).

7. H. Yılmaz, *"Maryland University Lectures"* Unpublished,(1988). Note that $s = \sqrt{(x_\mu x^\mu)}$, $\partial s / \partial x^\mu = x_\mu / s$ reduce to $ds = \sqrt{(dx_\mu dx^\mu)}$, $\partial s / \partial x^\mu = dx_\mu / ds = u_\mu$ when s is small, so that the form $\sigma d u_\mu / ds = \frac{1}{2} \sigma \partial_\lambda g_{\alpha\beta} u^\lambda u_\mu u^\alpha u^\beta$ is still a test-particle geodesic.

8. H. Yılmaz, *"One-Way velocity of light"*, in: *"Precision Measurements and Gravity Experiment"* International Conference Taipei Taiwan, ROC, p. 432, T. W. Ni, ed., (1983); See also C. O. Alley, et. al., *"PTTI Applications and Planning Committee, NASA"* Conference Publications 2265 (1982).

9. H. A. Lorentz, *"Problems of Modern Physics"* p. 218-21, Dover, (1967), Lectures originally delivered (1922).

INVESTIGATIONS WITH LASERS, ATOMIC CLOCKS AND

COMPUTER CALCULATIONS OF CURVED SPACETIME AND

OF THE DIFFERENCES BETWEEN THE GRAVITATION

THEORIES OF YILMAZ AND OF EINSTEIN

Carroll O. Alley

Department of Physics, University of Maryland at College Park,
College Park, Maryland 20742

INTRODUCTION

The description of gravitation by curved spacetime is a grand concept due to Albert Einstein. The form of his field equations for the determination of the metric coefficients for a given distribution of matter and field stress-energy is an assumption. The source term (right hand side) of these equations for the Einstein-Hilbert curvature tensor is taken to be the stress-energy tensor $\tau_\mu{}^\nu$ of all matter and fields except that of the gravitational field itself

$$(1/2) \ G_\mu{}^\nu \ = \ \tau_\mu{}^\nu \ .\tag{1}$$

The theory of Hüseyin Yilmaz[1] explicitly includes as an additional source term the stress-energy tensor $t_\mu{}^\nu$ of a gauge field which is a relativistic generalization of the Newton-Poisson potential field in a conservative system

$$(1/2) \ G_\mu{}^\nu \ = \ \tau_\mu{}^\nu \ + \ t_\mu{}^\nu \ ,\tag{2}$$

where $G_\mu{}^\nu$ and $\tau_\mu{}^\nu$ have the usual meanings. The expression for $t_\mu{}^\nu$ in the low velocity limit is given in eq. (17) below, and more generally, in eq. (5') of ref. 1.

It is the purpose of this paper to provide a brief physical and mathematical introduction to the new theory (complementing the exposition by Prof. Yilmaz in these proceedings) and to exhibit many of the differences between the Yilmaz theory and Einstein's legacy. The comparisons reveal many serious problems with general relativity in contrast to many positive features of the new theory. It is the opinion of the author that the explicit inclusion of the gravitational field stress-energy tensor in eq. (2) is as important for our understanding of physics as Maxwell's addition of the displacement current in his equations for the electromagnetic field.

The major new comparisons come from symbolic computer calculations carried out in the spirit of G.W.Leibniz[2] for deciding scientific questions: "let us calculate" ("calculemus" in the Latin original).

Modern computer workstations can evaluate, for a given metric, the symbolic expressions for

Frontiers of Fundamental Physics, Edited by M. Barone
and F. Selleri, Plenum Press, New York, 1994

the curvature tensors, connection coefficients, stress-energy expressions, etc. Such calculations done by hand could take months or years (and would almost certainly contain errors). It is possible to study in this way parameterized metrics describing simple physical arrangements (for example a single concentrated mass, two parallel slabs or two concentric shells) with some values of the parameter satisfying the Einstein field equations and another value satisfying the Yilmaz field equations. In the case of two slabs (Cavendish experiment) there is no interaction between them in Einstein's theory whereas the new theory has the correct Newtonian correspondence.

The new theory is in agreement with all of the known relativistic gravitational observations and experimental effects. For the ongoing experiments concerning the equality of light propagation times eastward and westward, the new theory predicts that these times should be the same, since the metric is locally Minkowskian even for accelerated observers. These experiments are therefore crucial tests of the Yilmaz theory in its local Lorentz covariant form.

EXPERIMENTAL RIEMANNIAN SPACETIME CHRONOMETRY

The imaginative concepts of a valid physical theory must be disciplined by agreement with experiment and observation. Fortunately, modern quantum electronics with the capabilities of atomic clocks and lasers now provides the means to perform new investigations of "Riemannian Spacetime Chronometry" in the phrasing of John Leighton Synge.[3]

Some of these experiments performed by the author in collaboration with others, both completed and ongoing, were reviewed in the talk given at the conference and are briefly described below as a prelude to the theoretical considerations and computer calculations presented later. The experiments include: laser ranging to corner reflectors on the Moon; proper time experiments with atomic clocks on aircraft with laser pulse time comparison; investigations of the isotropy of laser pulse propagation times between East $\rightarrow$ West and West $\rightarrow$ East on the Rotating Earth; examination of relativistic effects in the Global Positioning System (USA) and the Global Navigation Satellite System (Russia). Such experiments can serve to broaden and strengthen our base of knowledge from which the appropriate theoretical concepts must come and with which they must agree.

Laser Ranging to Corner Reflectors on the Moon

This experiment has been ongoing since the first manned landing on the Moon by Apollo 11 astronauts in July 1969. The first Laser Ranging Retro-Reflectors (LR[3]) array was deployed by Neil Armstrong and is the only experiment still working from that mission. Subsequent U.S. LR[3]'s were deployed by the Apollo 14 and Apollo 15 missions and French built reflectors were carried on two Soviet Lunar roving vehicles. There is now over 24 years of ranging data whose precision has steadily improved from $\sim$30 cm to the present $\sim$2 - 3 cm. Techniques to achieve measurement uncertainty of $\sim$2 mm are being developed (e.g. single photo-electron sensitive, circular scan streak tubes).[13]

The concept originated in the research group of Professor Robert Henry Dicke at Princeton University in the late 50's and early 60's during discussions of how to use the capabilities of lasers and of space flights to further our knowledge of gravity. A paper[4] on the idea was published in 1965 and a proposal prepared to the U.S. National Aeronautics and Space Administration by a team of investigators including P.L.Bender, R.H.Dicke, J.E.Faller, W.Kaula, G.MacDonald, H.H.Plotkin, and D.T.Wilkinson with the author serving as principal investigator. An extensive review and history of the Lunar Laser Ranging (LLR) experiment was published by the author[5] in 1983.

The primary purpose of the experiment was to investigate whether a prediction of the Brans-Dicke Scalar-Tensor theory of gravity[6] had experimental support. This prediction was that the gravitational self-energy of a body has a different effect from the mc^2 self-energy, leading to different accelerations toward the Sun for the Moon and for the Earth, thereby producing a polarization of the Lunar orbit about the Earth. (This was independently studied by K.Nordtvedt and is generally called the Nordtvedt effect). It was originally considered that the polarization could be as large as 100 cm. The latest analysis of the ranging data[7] has shown that it is -0.5 $\pm$ 1.3 cm. This makes the possible admixture of a scalar field so small as to exclude the Brans-Dicke theory as an interesting alternative theory of gravity.

The LLR measurements have been used by D.F.Bartlett and D.Van Buren[8] to show the equivalence of active and passive gravitational mass, a result present in Newtonian gravity theory but not deducible in Einstein's theory. It is exactly predicted by the Yilmaz theory.

Proper Time Experiments With Atomic Clocks in Air Craft and Laser Pulse Time Comparison

These experiments were conducted by the author's quantum electronics research group during 1975-1977 in collaboration with Leonard Cutler of the Hewlett-Packard Company, chief designer of the cesium beam atomic clocks which were used, and with Gernot Winkler, chief of the Time Services Department of the U.S. Naval Observatory. The work was financially supported by the U.S. Navy and the U.S. Air Force because of its relation to the Global Positioning System (see below). There is no space in the present paper to give some of the details presented during the talk at the conference. The interested reader may consult an earlier review[9]. Even more details are given in the Maryland Ph.D. theses[10] of Robert Reisse and Ralph Williams..

In a series of five 15 hour duration flights circling over the Chesapeake Bay in a Navy P3C antisubmarine plane, a typical time difference of about 50 ns was observed with the airborne clocks recording more elapsed time than the ground clocks. Laser pulse time comparison provided a record during the flights of the increasing time difference. Satisfactory agreement was found with the expected gravitational potential difference effect $\Delta\varphi/c^2 \sim 10^{-12}$ combined with the motional effect $-(1/2)v^2/c^2 \sim -10^{-13}$. The plane was flown as slow as possible to accentuate the former. Using an ensemble of three environmentally protected airborne clocks and a similar set of clocks on the ground, measurement precision between one and two percent was achieved (standard deviation of the mean for the five flights), in agreement with the calculated relativistic effects using the record of altitude and speed aquired by radar tracking.

In another experiment the clock set was transported to Thule, Greenland, from Washington D.C. in an Air Force C141 transport, left there for several days, and returned. This allowed a study of the combined effect of the gravitational potential and the centrifugal kinematic potential on the clocks. The results showed that the two effects cancel one another along the geoid. That is, the effect of decreased gravitational potential at Thule with respect to Washington due to the oblateness of the Earth is just compensated by the decreased Earth's surface rotational velocity at Thule with respect to Washington.

In a third type of experiment, the clocks were transported by the C141 aircraft from Washington in the northern hemisphere to Christchurch, New Zealand, in the southern hemisphere, left for several days, and returned. This was repeated a week later, both trials taking place near the time of the summer solstice when Christchurch was further from the Sun than Washington due to the tilt of the Earth's spin axis with respect to its orbital plane. In a sense, the clocks were carried from the floor to the ceiling of a freely falling elevator (the tilted Earth falling toward the Sun), kept there for a time, and returned. The results showed that the gravitational potential difference between Christchurch and Washington due to the Sun was compensated by the kinematic potential due to the Earth's acceleration towards the Sun. The effect of the Sagnac terms in the transport of the clocks, first westward and then eastward on the rotating Earth, had to be carefully considered. The plane's speed, altitude, and position were carefully measured using inertial navigation systems and plane to ground radar altimetry.

One-Way Light Propagation Times Eastward Versus Westward Over the Same Path

This experiment[11] makes direct comparison of the difference between the one-way propagation times of a $\sim$100 ps laser light pulse from E $\rightarrow$ W and W $\rightarrow$ E over the path across the city of Washington shown in Figure 1. It is the first such measurement of one-way times ever to be attempted.

Mirrors on a water tower and on the National Cathedral tower provide the changes in direction. A hydrogen maser atomic clock, whose relative rate and reading compared to an identical clock at the NASA Goddard Optical Research Facility (GORF) had been carefully measured in side by side comparison, is transported slowly in a carefully controlled environment to the U.S. Naval Observatory (USNO). The epoch t_1 of pulse emission from the 1.2 m telescope at the GORF is recorded by the non-travelling clock there. The epoch t_2 of pulse arrival at the USNO is recorded on the transported clock. The epoch t_3 of arrival of the reflected pulse back at GORF is also registered by the stationary clock.

The transported clock is kept at the USNO for an hour or two while numerous (t_1, t_2, t_3) measurements are made and then carefully moved back to GORF where it is again compared with the stationary clock. Optical calibrations and time comparison with the (t_1, t_2, t_3) method are made before and after the clock trip. (We note that the atmospheric path does not change appreciably during the 174 μs round trip time, so that this is not a source of error). The question is : Does $t_2 - t_1 = t_3 - t_2$? Rearranging, the question could be put: is $t_2 = (t_1 + t_3)/2$? This is the expected result for Einstein clock synchronization in an inertial frame (which we used in the laser pulse time time comparison for the aircraft flights described above). So the question can be rephrased: Is Einstein light pulse time comparison the same as clock transport comparison for an E - W path on the rotating Earth?

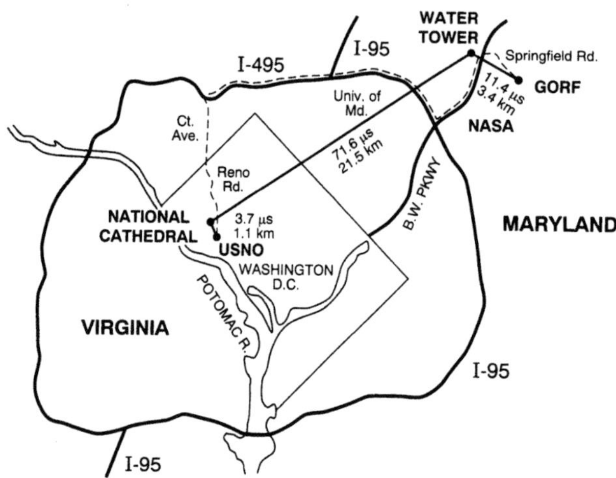

Figure 1. Paths for light pulses and clock transport. (North is at the top of the drawing).

If the speed of light is affected by the surface velocity v of the rotating Earth, being $c + v$ in one direction and $c - v$ in the other, the difference $t_2 - (t_1 + t_3)/2$ would be 80 ps. Results[12] to date (interrupted in the spring of 1989 for lack of funding) have had an uncertainty of 100 ps for this quantity because of incompletely identified systematic errors. A new series of experiments is in preparation to use improved optical detection timing and better hydrogen masers which may reduce the uncertainty to 10 ps.[13] This would allow a conclusive answer to the isotropy question.

The result will be important for our studies of possible systematic errors in the GPS (see below) which seems to use $c \pm v$ in its algorithms. The Yilmaz theory which has an observer dependent metric that is locally Minkowskian for observers even in accelerated frames of reference predicts isotropy for the local one-way propagation times in this experiment.

Relativistic Effects in the Global Positioning System (GPS) and Global Navigation Satellite System (GLONASS)

Each of these systems in its operational configuration will consist of 24 satellites in circular orbits around the Earth with periods of 12 hours (GPS) and 11 hours 15 minutes (GLONASS). The satellites for the GPS are distributed in six equally spaced orbital planes while those for GLONASS are arranged in three equally spaced planes. Every space vehicle carries an operating atomic clock (along with several back-up clocks). The space vehicle (SV) clock time is transmitted by a pair of L band carrier frequencies ($\sim$20 cm wavelength), each being phase reversal modulated by a "pseudorandom" bit string code whose bit spacing is controlled by the clock. Each SV transmits a different code in the GPS, while in the GLONASS the codes are the same but the carrier frequencies are different. Receivers possessing the codes can determine the time. (For the GPS the pseudorandom code starts anew at midnight Saturday/Sunday each week).

If all the SV clocks and receiver clocks were synchronized the shift in the pseudorandom code at a receiver would measure the propagation time of the signal from the SV and thus its range by using the speed of light. This is called the "pseudorange" (PR). In the operational systems the

SV's also transmit their precise location in space (orbital elements) at the time of transmission. By measuring the PR's from at least four SV's and using the orbital information, a receiver can use a microprocessor to solve for its location and time.

These systems constitute a grand scale laboratory for the application and study of relativistic effects on clocks. If the modeling of the known gravitational potential and motional effects is not applied correctly, systematic errors will result and the system performance will be degraded. For example, the combined average effect of these factors for a GPS SV is $\sim 4.5\times10^{-10}$, which if not allowed for, would lead in one day to an error of about 39,000 ns in time or 12 km in distance. This expensive ($\sim\$12$ Billion for GPS alone) technique, of great proven commercial and military benefit, can work properly only by using our scientific knowledge of relativistic gravity. It is a major practical application of gravitational theory.

In 1976 it was discovered by the author and his collaborators in the proper time experiments with aircraft, when presenting their results to the GPS program office, that it was planned to correct twice for the gravitational potential effect on the clocks (ref.8 p.421). After two years of discussion and argument the planned procedure was corrected. About six years ago the persistent appearance of unexplained residuals in the comparison of the SV clocks with the master clocks at the USNO as the GPS developed was called to the author's attention. The magnitude of these "bowing" and "hooking" deviations from the expected linear relation can be ten's of nanoseconds. After looking into the actual algorithms used in the GPS control segment, it seems that at least two mistakes are being made: (1) The relativity of simultaneity between receivers and SV's is being incompletely modeled in that the changing relative velocity between the two is not considered. This can readily lead to many ten's of nanoseconds during the transit of a SV. The actual signature will depend on the particular SV/monitor station pair. (2) The speed of light with respect to the receiver is treated as dependent on the Earth's rotation: $c \pm v$ (as discussed above). These errors seem to result from effectively considering all receiver measurements to be made by fictitious observers in a non-rotating Earth-centered inertial coordinate system, rather than the real monitor stations and users on the rotating Earth.

We are now engaged in an extensive study to identify these suspected problems. This involves the analysis of real unprocessed GPS data archived by the five Air Force monitor stations and the comparison of pseudorange data with actual laser range data. Each of the GLONASS satellites already carries a laser ranging retro-reflector. Recently a new GPS SV launched in August 1993 carried an LR^3 array and another GPS SV to be launched in March 1994 will carry a similar array. These LR^3 panels for the GPS satellites were purchased from the Russian Institute for Space Device Engineering by the University of Maryland with financial support from the U.S.Navy and Air Force. This may be the first instance of Russian-built equipment flying on U.S. military satellites.

The signals from the GPS SV's are one-way and hence can be readily misinterpreted as clock variations or orbital variations rather than the actual relativity effects. This seems to be what the GPS is doing, assigning by a Kalman filter statistical estimator procedure frequently changing values for the "clock states" and orbital parameters. The Kalman filter is actually applied to the monitor station data every 15 minutes.

As these investigations proceed we shall examine the consequences for the GPS of the observer dependent local Minkowski metric of the Yilmaz theory. The desire to identify the systematic errors of presently unknown origin in the GPS may help in clarifying the concepts of space, time and gravitation.

DIFFERENCES BETWEEN THE GRAVITATIONAL THEORIES OF YILMAZ AND OF EINSTEIN

This section is intended as a brief physical, conceptual and mathematical introduction to the new theory which may help to illuminate the accompanying exposition by Prof. Yilmaz,[1] and to assist in the understanding of his most recent publication.[14]

Physical and Conceptual Differences

The new theory emphasizes potentials in a successful relativistic generalization of the Newton-Poisson field theory. The metric coefficients $g_{\mu\nu}$ thus become functionals of this gravitational field $\varphi_\mu{}^\nu$ and are relieved of the double burden of serving both as potentials and as spacetime metric. The relation is formally exponential:

$$g_{\mu\nu} = (\eta e^{2(\varphi\hat{\imath} - 2\hat{\varphi})})_{\mu\nu} \, , \tag{3}$$

where $\eta = \mathrm{diag}(1,-1,-1,-1)$, $\varphi = \mathrm{tr}(\varphi_\mu{}^\nu)$, $\hat{1}$ is the unit matrix, and $\hat{\varphi}$ is the matrix array $\varphi_\mu{}^\nu$.

The quantity $\varphi_\mu{}^\nu$ is analogous to the four-potential A^ν in electrodynamics in that the $g_{\mu\nu}$ are determined from it, as the electromagnetic field $F^{\mu\nu}$ is determined from A^ν. With suitable gauge and/or coordinate conditions $\varphi_\mu{}^\nu$ satisfies the covariant d'Alambert equation with the matter tensor as source

$$\Box^2 \varphi_\mu{}^\nu = \tau_\mu{}^\nu \; . \tag{4}$$

Exact solutions are known in the low velocity limit where $\varphi_\mu{}^\nu \to \varphi_0{}^0 \to \varphi$ and (4) becomes the Poisson equation, and for traceless transverse gravity waves.

Many strong physical results are obtained by the new theory, some of which are briefly stated and discussed below.

Newtonian Correspondence in Second Order. This allows N-body interactive solutions including concentrated "point particles". General relativity seems to have no interactive N-body solutions at all.

Local Correspondence to Special Relativity. The metric becomes locally Minkowskian for any observer, even when in accelerated motion. The exponential metric allows the subtraction of a constant from the potential (including the kinematic potential). There is an unambiguous prediction for the local isotropy of the speed of light even for accelerated observers (e.g. an observer on the rotating Earth).

Principle of Equivalence by Local Kinematic Potential Compensation. The transition to a freely falling system is achieved by the addition of a kinematic potential to the gravitational potential. This is a more physical way of describing the equivalence of gravitational and kinematical accelerations than a coordinate transformation.

Localized Stress-Energy of the Gravitational Field. This exists in the new theory as the tensor $t_\mu{}^\nu$ and plays a central role. In general relativity it is argued[15] that there can be no localized stress-energy tensor since, in that theory, local free fall is described by a coordinate transformation to a local Riemannian normal coordinate system where the first derivatives of the metric coefficients vanish. The stress-energy tensor is a quadratic expression in the first derivatives and vanishes in that system, hence vanishes in all coordinate systems. Since the new theory describes the vanishing of $t_\mu{}^\nu$ at a point in a freely-falling frame by a compensation, or balancing, of the gravitational potential with a kinematic potential, **not by a coordinate transformation**, the argument does not apply. Local energy-momentum conservation between matter and gravitational field is consistently treated in the new theory. The serious problems in general relativity with local energy-momentum conservation are well known. (A recent review of the difficulties is given by Carmeli et al.[16] See also an early critical comment by Weyl.[17])

Exact Gravity Wave Solutions of Arbitrary Strength. These waves actually carry localized energy-momentum, with an analogous Poynting vector, between quadrupole sources and sinks. In general relativity gravity wave treatments require the limitation to 1st order expansions of the metric tensor. A localized stress-energy tensor for the gravity wave is assumed even though this is inconsistent with the argument against it given in reference 15.

The Strong Principle of Equivalence. This is the name often given to the equivalence of inertial mass, active gravitational mass and passive gravitational mass. It is readily deduced in the new theory from its second order Newtonian correspondence. It seems not to be deducible in Einstein's theory.

Quantum Theory Compatibility. Quantization using conventional methods of gauge field theory may be possible for the new theory with the $\varphi_\mu{}^\nu$ field.

Mathematical Differences

The above stated physical differences between the two theories can, of course, only be fully understood and comprehended in mathematical form. A brief exposition of some important results will be given in this section.

Exponential Metric. This is most readily discussed for the low velocity limit where $\varphi_\mu{}^\nu \to \varphi_0{}^0 \to \varphi$. The solution of (4) for $\tau_\mu{}^\nu = \sigma u_\mu u^\nu$, $\sqrt{-g}\,\sigma = \sum m_A\,\delta(x - x_A)$ becomes

$$\varphi(x) = \sum_A \frac{m_A}{|x - x_A|} + \text{constant} , \qquad (5)$$

the exponent in (3) evaluates as

$$2\begin{pmatrix}\varphi & & & \\ & \varphi & & \\ & & \varphi & \\ & & & \varphi\end{pmatrix} - 4\begin{pmatrix}\varphi & & & \\ & 0 & & \\ & & 0 & \\ & & & 0\end{pmatrix} \Rightarrow g_{00} = e^{-2\varphi},\ g_{ii} = e^{2\varphi} ,$$

and the metric becomes

$$ds^2 = e^{-2\varphi}\,dt^2 - e^{2\varphi}\,(dx^2 + dy^2 + dz^2) . \qquad (6)$$

It is most interesting and significant that an exact special relativistic treatment of the accelerated elevator with the use of the Principle of Equivalence leads directly to the exponential form for g_{00}. Recall that the relativistic Doppler factor k (k-calculus of Hermann Bondi[18]; see also Yilmaz[19]) is

$$k = \sqrt{\frac{1 + v/c}{1 - v/c}} = \frac{1 + v/c}{\sqrt{1 - (v/c)^2}} = \frac{1 + (\tanh\theta)}{\sqrt{1-(\tanh\theta)^2}} = \cosh\theta + \sinh\theta = e^\theta , \qquad (7)$$

where θ is the additive rapidity group parameter. In the usual manner of treating accelerated motion with a succession of instantaneously co-moving inertial frames, one finds that $\theta = \alpha\tau/c$ where α is the constant proper acceleration and τ is the accumulated proper time of a clock in the accelerated system. If one considers one clock on the floor of an accelerated elevator transmitting light pulses to another one fixed to the ceiling, the factor k is the ratio of the received interval between two light pulses to the emission interval between the same two pulses. Invoking the principle of equivalence one considers the elevator to be stationary in a gravitational field and replaces α by the local acceleration of gravity γ and the time τ by the transit time z/c of the light pulses from floor to ceiling at height z. One obtains

$$k = e^\theta = e^{\gamma z/c^2} = e^{\varphi/c^2} , \qquad (8)$$

where $\varphi = \gamma z$ is the Newtonian potential of the ceiling with respect to the floor. A metric describing the curvature of time is thus suggested:

$$d\tau^2 = e^{2\varphi}\,dt^2 , \qquad (9)$$

when we denote the received interval between pulses by $d\tau$ and the corresponding emitted interval by dt, with $d\tau = k\,dt$.

In his first writing[20] on the extension of relativity to gravity Einstein considered an exponential relation equivalent to (8) but chose instead to use the approximate expression

$$k = 1 + \gamma z/c^2 , \qquad (10)$$

with the words:

> "From the fact that the choice of the coordinate origin must not affect the relation, one must conclude that, strictly speaking, equation (30) [equiv. to our eq. (10)] should be replaced by the equation [equiv. to our eq. (8)]. Nevertheless we shall maintain formula (30)."

Einstein's reason for using (10) rather than (8), which he clearly recognized in the above passage to be the correct relation, seems to be his fixation on coordinate transformations in the spirit of the linear Lorentz transformations, focusing on the coordinate z which occurs linearly in (10) rather than on the potential φ. This attitude seems to have been maintained throughout the development of his theory.

It should be noted that the beginning of the new theory occurred when Yilmaz[21] found the equivalent of eq.(9) for the exact solution of the accelerated elevator, substituted the metric into the Einstein field equations and found, to his great surprise, that they were not satisfied! The development of the new theory has centered on the potential field.

[The difference in the sign of φ between (6) and (9) for the coefficient of dt^2 was chosen for reasons of convenience in the development of the new theory.]

Free Fall as Compensation by Kinematic Potential. Consider an elevator in free fall in the gravitational field of the Earth. Let the magnitude of its downward acceleration be denoted by γ. If, in the absence of the Earth's gravitational field, there were a kinematic acceleration γ in the same downward direction, the kinematic potential due to this acceleration would be $\varphi_{kin} = -\gamma z$ with respect to the floor of the elevator, so that $-\partial_z(\varphi_{kin}) = \gamma$, describing acceleration of objects toward the ceiling ($+z$ direction in the elevator). The exponential metric of the new theory admits a superposition of gravitational potentials, so the kinematic potential φ_{kin}, in a thorough acceptance of the principle of equivalence (it would be the gravitational potential of a large slab of matter) can be added to the gravitational potential of the Earth.

$$\varphi = \varphi_{grav} + \varphi_{kin} = -\frac{GM}{r} - \gamma z + \frac{GM}{r_0} + \gamma z_0 \tag{11}$$

$$-\frac{\partial \varphi}{\partial z} = -\frac{GM}{r^3} z + \gamma \tag{12}$$

The gradient will vanish at the point chosen ($z = r = r_0$, $x = y = 0$) because the local value of γ is given by GM/r_0^2. The constant $GM/r_0 + \gamma z_0$ is included in (11) so that at the point chosen $\varphi = 0$ and the metric becomes Minkowskian. $\nabla^2 \varphi = 0$ at r_0 which is a matter-free point in the interior of the elevator.

This method of treating the physical experience of weightlessness in a freely falling frame by compensating the gravitational potential with a kinematic potential related to the real inertial force experienced by objects in an accelerated frame is not a coordinate transformation, but a physical transformation. (Cutting the cable supporting the elevator can cause the death of its occupants when it hits bottom!) Indeed the kinematic potential $(M/2r_0^3)[2(z - z_0)^2 - x^2 - y^2]$ added to the expression (11) causes the full Riemann tensor to vanish at the chosen point in addition to the vanishing of the Christoffel symbols achieved by (11). This cannot be done with a coordinate transformation. These results have been demonstrated by explicit computer calculations.

Energy-Momentum Conservation. Perhaps the major result of the theory of special relativity is the possibility of transforming rest mass into other forms of energy. The general conservation law which encompasses this possibility is the vanishing divergence of the matter stress-energy tensor $\tau_\mu{}^\nu = \sigma u_\mu u^\nu$,

$$\partial_\nu (\sqrt{-g}\,\sigma u_\mu u^\nu) = 0 \ . \tag{13}$$

However, in general relativity this conservation law is replaced by the conservation of rest mass[21] in order to get the equations of motion from the field equations. The argument runs: take the covariant divergence of the field equations, which is zero by the Bianchi identity

$$\tfrac{1}{2} D_\nu G_\mu{}^\nu = D_\nu(\sigma u_\mu u^\nu) = \frac{1}{\sqrt{-g}} \partial_\nu(\sqrt{-g}\,u_\mu u^\nu) - \tfrac{1}{2} \partial_\mu g_{\alpha\beta}\, \sigma u^\alpha u^\beta \equiv 0 \ . \tag{14}$$

Differentiate the first term on the right as a product to yield

$$\frac{1}{\sqrt{-g}} \partial_\nu(\sqrt{-g}\,\sigma u^\nu)u_\mu + \sigma u^\nu \partial_\nu u_\mu - \tfrac{1}{2} \partial_\mu g_{\alpha\beta}\, \sigma u^\alpha u^\beta \equiv 0 \ . \tag{15}$$

The second term is $\sigma(\partial x^\nu/\partial\tau)\partial_\nu u_\mu = \sigma du_\mu/d\tau$ which would produce the geodesic equations of motion if the first term were zero. It is argued that one should require $\partial_\nu(\sqrt{-g}\sigma u^\nu) = 0$ as a conservation law. But this is the conservation of rest mass, not the conservation of energy-momentum! This whole procedure is illigitimate in view of the Freud identity, to be discussed in the next section, which requires the identical vanishing of expression (13). If the first term in (15) is set equal to zero, the second must also be zero, and one does not get the geodesic equation of motion.

In the new theory, one has

$$\tfrac{1}{2}\, D_\nu G_\mu{}^\nu = \tfrac{1}{\sqrt{-g}}\, \partial_\nu(\sqrt{-g}u_\mu u^\nu) - \tfrac{1}{2}\, \partial_\mu g_{\alpha\beta}\, \sigma u^\alpha u^\beta + D_\nu t_\mu{}^\nu \equiv 0\ . \tag{16}$$

Now one can require the first term in (16) to vanish, expressing the desired conservation law of energy-momentum and get the equation of motion since the divergence of the gravitational field stress-energy tensor $t_\mu{}^\nu$ is the force $\sigma du_\mu/d\tau$. In the low velocity limit, $t_\mu{}^\nu$ is given by

$$t_\mu{}^\nu = -\, \partial_\mu\varphi\partial^\nu\varphi + \tfrac{1}{2}\, \delta_\mu{}^\nu\partial_\lambda\varphi\partial^\lambda\varphi\ . \tag{17}$$

Taking the divergence, one obtains $-\nabla^2\varphi\partial_\mu\varphi$ which is just $\sigma_a\partial_\mu\varphi$ by the Poisson equation where σ_a is the active gravitational mass. This is instrumental in establishing the strong principle of equivalence in the new theory.

The Freud Identity. It has become clear recently that there is a mathematical requirement which forces the first term of eq (16) to vanish identically. In 1939 P.Freud published[23] a decomposition of the Einstein-Hilbert tensor $(1/2)G_\mu{}^\nu = U_\mu{}^\nu - u_\mu{}^\nu$ where $u_\mu{}^\nu$ is the quantity introduced by Einstein to describe the stress-energy of the gravitational field. He transformed the "obstreperous term" in eq (14) (phrase used by Schrödinger[24]) as follows:

$$-\tfrac{1}{2}\, \partial_\mu g_{\alpha\beta}\, \sigma u^\alpha u^\beta = \tfrac{1}{\sqrt{-g}}\, \partial_\nu(\sqrt{-g}u_\mu{}^\nu)\ . \tag{18}$$

$U_\mu{}^\nu$ is an expression satisfying the identity $\partial_\nu(\sqrt{-g}U_\mu{}^\nu) \equiv 0$. This is an indicial identity resulting from the antisymmetry of the superpotential $H_\mu{}^{\nu\alpha}$ given by the determinant

$$H_\mu{}^{\nu\alpha} = \begin{vmatrix} \delta_\mu^\nu & \delta_\mu^\alpha & \delta_\mu^\sigma \\ \sqrt{-g}g^{\nu\rho} & \sqrt{-g}g^{\alpha\rho} & \sqrt{-g}g^{\sigma\rho} \\ \Gamma^\nu_{\rho\sigma} & \Gamma^\alpha_{\rho\sigma} & \Gamma^\sigma_{\rho\sigma} \end{vmatrix}\ , \tag{19}$$

and the relation $\sqrt{-g}U_\mu{}^\nu = \partial_\alpha H_\mu{}^{\nu\alpha}$. It has been shown more recently that such an identity is true for a Riemannian geometry of any dimension and arbitrary signature for all symmetric nonsingular metrics.[25] The similarity of the Freud identity as discussed by Pauli[26], to a simpler expression used by Yilmaz to formulate the conservation laws was noted by the author. Their equivalence was established by Yilmaz who has shown that the Freud identity is nested so that the anharmonic part (the last two terms of $U_\mu{}^\nu$ as given in ref.14, Appendix B, p.959; in harmonic coordinates, they themselves vanish) and the coordinate dependent part separately satisfy the same identity. Freud had actually written $U_\mu{}^\nu = (1/2)G_\mu{}^\nu + u_\mu{}^\nu$ and used the identity to produce the energy-momentum conservation law of general relativity $\partial_\nu(\sqrt{-g}\tau_\mu{}^\nu + \sqrt{-g}u_\mu{}^\nu) \equiv 0$. However, writing the decomposition as $(1/2)G_\mu{}^\nu = U_\mu{}^\nu - u_\mu{}^\nu$ shows that the difference of the two "pseudotenors" must be the true tensor $(1/2)G_\mu{}^\nu$. There must be a common non-tensor part $z_\mu{}^\nu$ for each of $U_\mu{}^\nu$ and $u_\mu{}^\nu$ in addition to their tensor parts, and there must exist coordinate systems in which

$$U_\mu{}^\nu = \tau_\mu{}^\nu + z_\mu{}^\nu \quad\text{and}\quad u_\mu{}^\nu = -\, t_\mu{}^\nu + z_\mu{}^\nu\ . \tag{20}$$

In Einstein's theory the field stress-energy tensor $t_\mu{}^\nu$ is required to be zero in eq. (1), and therefore also in eq. (20), forcing the identification of $u_\mu{}^\nu$ with the non-tensor or coordinate artifact $z_\mu{}^\nu$. This has led to all of the difficulties with the energy momentum concept in general relativity discussed in refs. 16 and 17. It was already noted in 1918 by Schrodinger[27] that $u_\mu{}^\nu$ evaluated to zero for the Schwarzschild solution when expressed in Cartesian coordinates, and by Bauer[28] that $u_\mu{}^\nu$ evaluated to non-zero expressions for the flat space Minkowski metric expressed in polar coordinates. Schrodinger's criticism was replied to by Einstein[29] with the additional remark that $u_{\mu\nu} \neq u_{\nu\mu}$, leading to problems with angular momentum conservation. However he stated that the strange properties of $u_\mu{}^\nu$ in the Schwarzschild solution were due to its one-body nature and that $u_\mu{}^\nu$ would have appropriate physical properties as soon as the two-body solution to his equations was found. There seems to be no two-body (or N-body) solution for Einstein's equation with interaction between the bodies (See below and ref.1).

In a conservative system the non-tensor $z_\mu{}^\nu$ can be identified by starting with a coordinate system in which it is zero and transforming to another coordinate system where it is not.[1,14] From eq. (18) it can be seen that $z_\mu{}^\nu$ arises from the non-tensor part of the transformation of Christoffel symbols. From the indicial nature of the Freud identity, it seems that both $\tau_\mu\nu$ and $z_\mu{}^\nu$ have separately vanishing density divergences.[1,14] Therefore in Einstein's theory, $\partial_\nu(\sqrt{-g}u_\mu{}^\nu) \equiv 0$.

COMPUTER CALCULATIONS

It is important to provide concrete examples of the foregoing assertions. Using Mathematica (from Wolfram Research) and MathTensor (from Math Solutions) running on Digital Equipment Corporation 5000/240 workstations operating under Ultrix and on a DEC 3000/400AXP workstation under OSF-1, many different metrics for solutions of the new and old theories have been used to evaluate the curvature tensors, Christoffel symbols, etc. but emphasizing the study of $U_\mu{}^\nu$, $u_\mu{}^\nu$ and their density divergences. There is space for only a very limited presentation of some significant results.

Parameterized Schwarzschild Solution

The following parameterized metric satisfies eq. (1) for the values of $\varepsilon = \pm 1$ and eq. (2) for $\varepsilon = 0$. (the two general relativity solutions are related[30] by the transformation $r \Rightarrow r-2M$).

$$ ds^2 = (1 + \frac{2\varepsilon M}{r})^{-1/\varepsilon} dt^2 - (1 + \frac{2\varepsilon M}{r})^{1+1/\varepsilon}(r^2 d\theta^2 + r^2\sin^2\theta d\varphi^2) - (1 + \frac{2\varepsilon M}{r})^{1/\varepsilon} dr^2 \ . \quad (21) $$

If one sets $M/r = \varphi$ in (21) and evaluates the Einstein-Hilbert tensor, one finds

$$ \frac{1}{2} G_\mu{}^\nu = \begin{matrix} t & \theta & \varphi & r \end{matrix} \begin{pmatrix} 1+\varepsilon & & & \\ & \varepsilon/2 & & \\ & & \varepsilon/2 & \\ & & & 0 \end{pmatrix} \nabla^2\varphi + (1-\varepsilon^2)t_\mu{}^\nu \quad (22) $$

where $\nabla^2\varphi$ is the covariant Laplacian and $t_\mu{}^\nu$ is given by eq. (17). By correspondence with Newtonian theory $\nabla^2\varphi$ should be the mass density σ in the matter tensor $\tau_\mu{}^\nu = \sigma u_\mu u^\nu$, which in the low velocity limit should be $\tau_0{}^0$. But for $\varepsilon = -1$, $G_0{}^0 = 0$ and for $\varepsilon = +1$, $(1/2)G_0{}^0 = 2\nabla^2\varphi$! Also $\nabla^2\varphi$ appears in the $G_\theta{}^\theta$ and $G_\varphi{}^\varphi$ positions. These properties seem physically wrong. In the Yilmaz theory $\varepsilon = 0$ leads to the exponential metric eq. (6) expressed in spherical polar coordinates and the Laplacian occurs as expected only in the $G_0{}^0$ position.

The Schwarzschild solution in general relativity clearly seems not to describe a physical mass concentrated at the origin. Does it properly describe any real physical situation in nature?

Parameterized Two Slab Metric. Consider the configuration of two plane parallel slabs whose separation d is in the z direction between their central planes is small compared to their finite transverse x, y dimensions so that one can ignore edge effects — the situation often used in capacitor problems in electrostatics. Denote the uniform mass densities of the slabs by σ_1 and σ_2 and their thicknesses by w_1 and w_2, as shown in Figure 2.

134

The Newtonian potentials in regions I through V are sketched and their expressions written beside the drawing. The potentials and their slopes are continuous at the boundaries. The quantities γ_1 and γ_2 are the magnitudes of the acceleration of gravity in the external regions produced by the respective slabs. They are related to the density and thickness by $\gamma = \sigma w/2$.

The metric[31]

$$ds^2 = e^{-2\varphi}dt^2 - e^{2\varphi(1+\varepsilon)}(dx^2 + dy^2) - e^{2\varphi(1+2\varepsilon)}dz^2 \tag{23}$$

has been used to evaluate all of the relevant quantities for each of the five regions. It provides an exact solution of the new theory for $\varepsilon = 0$ in each region. However, both $\varepsilon = +1$ and $\varepsilon = -1$ provide solutions for Einstein's theory. Thus there is no unique solution even for this simple case. The two solutions cannot be related by a coordinate transformation since they have different curvatures. The evaluation of $G_\mu{}^\nu$ yields a general expression in each region which has exactly the same structure as for the parameterized Schwarzschild solution, eq. (22) with $(t, \theta, \varphi, r) \Rightarrow (t, x, y, z)$.

In this two body case one can evaluate the right hand side of the geodesic equations which by

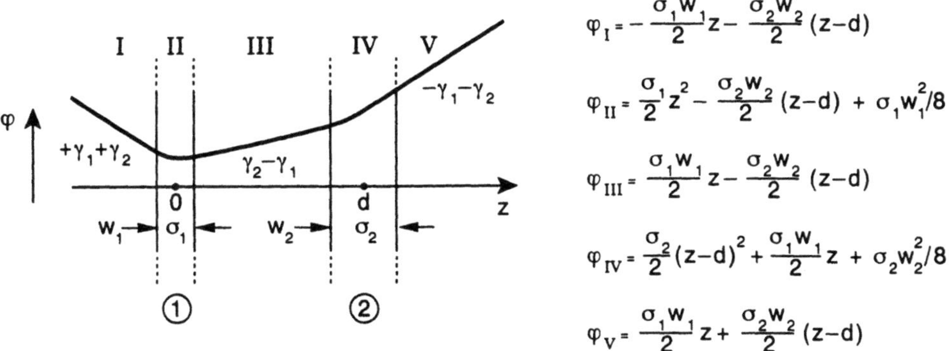

Figure 2. Two slabs, neglecting edge effects.

Einstein's introduction of $u_\mu{}^\nu$ is given by eq. (18). In reference 1 the left hand side of (18) is evaluated for the slabs. Here we give the results of the computer calculation in region II of $\partial_\nu(\sqrt{-g}u_\mu{}^\nu)$. The only non-zero component is

$$\partial_\nu(\sqrt{-g}u_z{}^\nu) = (1 - \varepsilon^2)[(1/2)\sigma_1\sigma_2 w_2 - \sigma_1^2 z] \tag{24}$$

The presence of $(1 - \varepsilon^2)$ in (24) makes it zero for the Einstein case! For the new theory it has the expected Newtonian value. Integrating over the thickness of the slab (The Jacobian $\sqrt{-g}$ in the integration cancels the $\sqrt{-g}$ in the denominator of the right hand side of eq. (18)) the self-force $\sigma_1^2 z$ vanishes and the force per unit area $(1/2)(\sigma_1 w_1)(\sigma_2 w_2)$ to the right remains. For slab 2, one finds the same expression but with opposite sign, giving a force to the left.

In this simplest of all gravitational problems the old theory predicts no interaction: the slabs would remain fixed, contradicting the results of Cavendish type experiments. The new theory, in strong contrast, describes the expected Newtonian interaction and predicts that the slabs will accelerate toward each other. Einstein often wrote that "above all else" his theory should have the appropriate Newtonian correspondence. In this concrete example it is shown not to be the case.

CONCLUDING REMARKS

In the implementation of Einstein's grand design of gravitation as curved spacetime serious problems have appeared. The theory does not possess N-body interactive solutions, nor does it generalize correctly the energy-momentum conservation law of special relativity and relativistic field theories, to mention only two major problems. The computer calculations of simple examples supporting these conclusions seem incontestable. A theory with such defects can not serve as the foundation on which to build our understanding of the physical universe.

There is a consistent curved spacetime theory of gravity which does not have these problems and which seems compatible with quantum theory. The author urges that the physics and astronomy communities shift the paradigm from general relativity and concentrate instead on the development and application of the theory of Yilmaz.

ACKNOWLEDGMENTS

The author wishes to express his profound gratitude to Professor Huseyin Yilmaz for the great pleasure of many illuminating and clarifying discussions about physics and its foundations.

Dr. Ching Yun Ren (now chairman of the Physics Department, Soochow University in Taipei, Taiwan) developed the computer programs for these investigations and applied them with great knowledge of physics to many different metrics. These programs have been extended with great skill by Robert Kirk Burrows and used to study many additional metrics. Our computer systems have been managed by Steve Sabean.

The endorsement by Professor Willis Lamb of the new theory of gravity and his encouragement of this research is greatly appreciated.

Professor Robert Henry Dicke's example of combining experimental and theoretical investigations of gravity has been a continuing inspiration.

We thank Mr. George Keros for his active interest and encouragement during a critical period of the investigations.

Assistance in the ongoing GPS related investigations is being provided by: Ron Beard and the Space Applications Branch of the Naval Research Laboratory; John Degnan and the Space Geodesy Branch of the Goddard Space Flight Center; Gernot Winkler and the Time Services Department of the U.S. Naval Observatory and Glen Spalding of the Office of Naval Research. The investigations are being carried out by members of the author's quantum electronics research group at the University of Maryland: R.K.Burrows, T.E.Kiess, A.Krikoriantz, S.W.Sabean, A.V.Sergienko (who also has given very able help in the technical composition of this paper), Y.H.Shih, T.Van Flandern, B.C.Wang, and M.Z.Zhang. Financial support for the studies of the GPS is being provided by the Federal Aviation Administration and the Air Force Space Command under FAA Grant 92-G-0025.

REFERENCES

1. H. Yilmaz, Did the apple fall?, these proceedings.
2. G.W. Leibniz, Untitled essay *in his*: "Philosophischen Schriften", vol. 7,
   C.J.Gerhardt, ed., Olms Verlagsbuchhandlung, Hildesheim, p.200 (1961).
3. J.L. Synge. "Relativity, The General Theory", North-Holland, Amsterdam, (1960).
4. C.O.Alley, P.L.Bender, R.H.Dicke, J.E.Faller, P.A.Franken, H.H.Plotkin, and D.T.Wilkinson,
   Optical radar using a corner reflectors on the moon, *J.Geophys.Res.*, 70 (1965).
5. C.O.Alley, "Laser ranging to retro-reflectors on the moon as a test of theories of
   gravity, *in*: "Quantum Optics, Experimental Gravitation and Measurement Theory",
   P.Meystre and M.Scully ed., Plenum Press pp.429-495 (1983).
6. C.Brans and R.H.Dicke, Mach's principle and a relativistic theory of gravity, *Phys.Rev.*,
   124:925 (1961).
7. J.O.Dickey, Jet Propulsion Laboratory, private communication, 1994. The last published
   relativity results are in: J.O.Dickey, X X Newhall, and J.G.Williams, Investigating
   relativity using lunar laser ranging: geodetic precession and the Nordtvedt effect,
   Advances in Space Research, 9:75 (1989). A more recent analysis emphasizing geophysics
   is: J.G.Williams, X X Newhall, and J.O.Dickey, Lunar laser ranging: geophysical results
   and reference frames, *in*: "Contributions of Space Geodesy to Geodynamics", v.24,
   D.E.Smith and D.L Turcotte eds., American Geophysical Union, Washington, D.C. (1993).

8. D.F.Bartlett and D.Van Buren, Equivalence of active and passive mass using the moon, *Phys.Rev.Lett.*, 57:21 (1980).

9. C.O.Alley, Proper time experiments in gravitational fields with atomic clocks, aircraft and laser light pulses, *loc.cit.ref.5*, pp.363-427.

10. R.A.Reisse, "The Effect of Gravitational Potential on Atomic clocks as Observed with a Laser Pulse Time Transfer System", University of Maryland, Ph.D. Thesis, May (1976). R.E.Williams, "A Direct Measurement of the Relativistic Effects of Gravitational Potential on the Rates of Atomic Clocks Flown in Aircraft", University of Maryland Ph.D. Thesis, May (1976).

11. C.O.Alley, R.A.Nelson, Y.H.Shih et al, Differential comparison of the one-way speed of light in the east-west and west-east directions on the rotating Earth", *in*: Proceedings of the Twentieth Annual Precise Time and Time Interval (PTTI) Applications and Planning Meeting, pp. 261-285, Nov. 29 - Dec. 1, (1988). [Available from U.S. Naval Observatory Time Services Departent Washington, D.C. 20392]

12. R.A.Nelson, C.O.Alley, J.D.Rayner, et al, Experimental comparison of time synchronization techniques by means of light signals and clock transport on the rotating Earth, *in*: Proceedings of the Twentieth Annual Precise Time and Time Interval (PTTI) Applications and Planning Meeting, pp.87-104, December (1992) [NASA Conference Pablication No.3218. Available from U.S. Naval Observatory Time Services Departent Washington, D.C. 20392]

13. C.O.Alley, T.E.Kiess, R.A.Nelson, A.V.Sergienko, Y.H.Shih, B.C.Wang and F.M.Yang, Plans to improve the experimental limit in the comparison of the East-West and West-East one-way light propagation times on the rotating Earth, loc.cit. ref 12, pp. 105-112 (1993).

14. H.Yilmaz, Towards a field theory of gravity, *Nuovo Cimento.* 107B:941 (1992).

15. C.W.Misner, K.S.Thorne, and J.H.Wheeler. "Gravitation", W.H.Freeman and Co., pp.466-468 (1973).

16. M. Carmeli, E.Leibnitz, and N.Nissani. "Gravitation: SL(2,C) Gauge Theory and Conservation Laws", ch.4 , World Scientific (1990).

17. H. Weyl. "Space, Time, Matter", Dover reprint of fourth edition, p. 270 (1950).

18. H.Bondi. "Assumption and Myth in Physical Theory", Cambrige University Press (1967).

19. H. Yilmaz, The new theory of gravitation, *Nuovo Cimento.* 10B:79 (1972).

20. A. Einstein, On the relativity principle and the conclusions drawn from it", *Jahrbuch der Radioactivitat und Elektronik* 4:411 (1907); Translation *in*: "The Collected Papers of Albert Einstein, v.2, The Swiss Years: Writings 1900-1909", p.305, Princeton University Press (1989).

21. H. Yilmaz, private communication.

22. H. Yilmaz, Correspondence paradox in general relativity, *Nuovo Cimento Lett.* 7:337 (1973).

23. P. Freud, On the expression of the total energy and the total momentum of a material system in the theory of general relativity [tr.], *Ann. Math.* 40:417 (1939).

24. E. Schrödinger. "Space-Time Structure", Cambridge University Press, p.102 (1950).

25. H. Rund, The Freud identity in Riemannian geometry, *Algebras, Groups and Geometries*, 8:267 (1991).

26. W. Pauli. "Theory of Relativity", Supplementary Note 8a, p.215, Pergamon Press (1958). Pauli acknowledges being informed of Freud's result by V. Bargmann. The present author attended lectures by Prof. Bargmann at Princeton in the 1950's on general relativity which included the Freud work. Notes on these lectures were made and retained by D.Brill.

27. E. Schrödinger, The energy components of the gravitational field [tr.], *Phys.Zeits.*, 19:4 (1918).

28. H. Bauer, On the energy components of the gravitational field [tr.], *Phys.Zeits.*, 19:163 (1918).

29. A. Einstein, Note on E.Schrödinger's paper [tr.], *Phys.Zeits.*, 19:115 (1918).

30. This was pointed out to the author by D. Brill.

31. H. Yilmaz, Einstein, the exponential metric, and a proposed gravitational Michelson-Morley experiment, *Hadronic Journal*, 2:997 (1979).

GRAVITY IS THE SIMPLEST THING!

David F. Roscoe

Department of Applied Mathematics
University of Sheffield, Sheffield S10 2TN, UK.
Fax: 742-739826, Email: D.Roscoe@uk.ac.shef

INTRODUCTION

The General Theory of Relativity (GR) is one of the best tested of modern physical theories, and has proved its superiority over rival theories many times during the past seventy years in the arena of the critical observation; in particular, the binary pulsar observations have recently proved decisive in this context, leaving GR as almost the only credible theory available. See Will[1] for a comprehensive discussion. As a consequence of these circumstances, it is easy to understand why, within the narrow context of purely gravitational phenomena, GR is considered to be securely founded. The only obvious problem facing the theory is the apparent difficulty of reconciling the gravitational force (as understood within the context of GR) with the three other forces of nature, as these are understood within their respective theoretical contexts; as a measure of this difficulty, we can reflect that, although a huge effort has been expended in the attempt to effect such a reconciliation - particularly over the past fifteen or twenty years - it is still not clear that it is possible *even in principle*. In view of this uncertainty, it would seem prudent to invest a modest effort towards a continued questioning of the theoretical basis of gravitation theory, and this paper is intended to represent such an effort.

As a result of analysing the properties of the various theories which have failed one test, or another, and comparing these with those of the one theory, GR, to have past all tests to date, a general concensus has arisen concerning what constitutes the *essence* of a 'good' theory of gravitation; this *essence* is the Strong Equivalence Principle (SEP), Will[1], which is satisfied only by GR amongst all extant theories, and which states

- *The Weak Equivalence Principle (WEP) is valid for self-gravitating bodies, as well as for test bodies;*

- *The outcome of any local test experiment is independent of the velocity of the freely falling test apparatus;*

Frontiers of Fundamental Physics, Edited by M. Barone
and F. Selleri, Plenum Press, New York, 1994

- *The outcome of any local test experiment is independent of where and when in the Universe the experiment is performed.*

The basic result of this paper is to show how the physics of four-momentum conservation provides all the structure that is required to arrive at a description of gravitation which is distinct from GR, and which satisfies the SEP. This latter quality is sufficient to guarantee the theory as a 'good' theory of gravitation - up to and including the binary pulsar - whilst the fact that four-momentum is globally conserved is sufficient to guarantee that the gravitational force as described by this theory is conceptually in the same category as that occupied by the other forces of nature.

A LORENTZIAN MODEL UNIVERSE

The presented theory is developed by arguing for the local observer who takes the philosophical position that global physics is a simple extension of local physics; consequently, for example, this observer would argue that since physics appears to be locally Lorentz invariant then, in the absence of empirically informed arguments to the contrary, there is justification in persisting with the assumption that physics is globally Lorentz invariant for as long as such an assumption leads to realistic descriptions of observed physics.

Following the local observer's philosophical position, this paper starts by setting up an N-particle model Universe in which Lorentzian physics is assumed to have a global validity; in particular, suppose that this model Universe is defined as an interaction event, $\mathbf{e}$, involving a total of N particles through the event, where by 'particle' is meant any entity which is governed by the laws of relativistic physics. Consequently, if there are n particles into $\mathbf{e}$, having respective four velocities $\mathbf{U}^r, r = 1..n$, then there will be $N-n$ particles out of $\mathbf{e}$, having respective four-velocities $\mathbf{U}^r, r = n+1..N$. Now suppose the particles have rest masses given by $m_1..m_N$ respectively then, if four-momentum is to be conserved, which it must be in a model Universe which is globally Lorentz invariant, we have

$$\sum_{r=1}^{n} m_r \mathbf{U}^r - \sum_{r=n+1}^{N} m_r \mathbf{U}^r = 0. \tag{1}$$

Further suppose that the first sum of (1), which represents the collection of all universal particles which enter into $\mathbf{e}$, is arbitrarily partitioned into two parts, $m_1 \mathbf{U}^1 + ... + m_k \mathbf{U}^k$ and $m_{k+1} \mathbf{U}^{k+1} + ... + m_n \mathbf{U}^n$ respectively; given these partitions, then the first part can be written *uniquely* as $M_1 \mathbf{V}^1$, where $\mathbf{V}^1$ has the structure of an ordinary four-velocity vector, whilst the second part can be written *uniquely* as $M_2 \mathbf{V}^2$ where $\mathbf{V}^2$ similarly has the structure of an ordinary four-velocity vector. It follows that M_i, $\mathbf{V}^i$, $i = 1, 2$, can be considered as the notional rest-masses and four-velocities respectively of two 'conglomerated particles' which go into $\mathbf{e}$; the second sum of (1) can be treated in a similar way to give two 'conglomerated particles', having notional rest-masses and four-velocities M_i, $\mathbf{V}^i$, $i = 3, 4$ respectively, coming out of $\mathbf{e}$. Given this arbitrary four-fold partitioning, then the global four-momentum conservation statement (1) can be rewritten uniquely in the form

$$M_1 \mathbf{V}^1 + M_2 \mathbf{V}^2 - M_3 \mathbf{V}^3 - M_4 \mathbf{V}^4 = 0, \tag{2}$$

where it is important to recognize: (i) that the partition corresponding to M_1, for example, could be the collection of particles which form a particular star, or a galaxy, or any structure whatever, short of the whole Universe; (ii) statement (2) is a true statement about the Universe at *any* instant, given only that we are willing to accept Lorentzian physics as a globally invariant physics; (iii) the *total* of the theoretical content of (2) and therefore, as we shall see, of all that follows, is this latter assumption that physics is globally Lorentz invariant. This assumption is based on the 'orthodox truth' that physics is locally Lorentz invariant, and the fact that there are no *empirically* informed reasons to suppose that this local truth is not also a global truth.

The representation (2) of global four-momentum conservation in the Lorentzian model Universe is the primary form upon which the following analysis is predicated.

THE LORENTZIAN REFERENCE SCALAR

According to (2), the trajectories, represented by the four-velocity vectors $\mathbf{V}^1$, $\mathbf{V}^2$, $\mathbf{V}^3$, $\mathbf{V}^4$ of the four partitions involved in the global four-momentum conserving interaction, must lie in some Lorentzian invariant 3-dimensional subspace of the four-velocity space. Any such subspace can, in turn, be considered to define a tangent plane to a level-surface of some Lorentzian scalar function, U say, defined on the coordinate space, $\mathbf{x}$. In this way, it is clearly seen that the level-surfaces of the scalar function U are abstract representations of the agglomeration of trajectories into, and out of, the global four-momentum conserving interaction. In other words, the interactive structure of the Lorentzian model Universe can be modelled in terms of an unknown Lorentzian scalar function and, in the following, it will be shown how the level-surfaces of this scalar function provide a natural means for referencing the motion, under gravitational action, of any test particle chosen arbitrarily from any of the four partitions; for this reason, the scalar U will be referred to as the 'reference scalar'.

GRAVITATION, GEOMETRY AND THE REFERENCE SCALAR

From the point of view of local physics, gravitational interactions could be defined as that class of interactions through which:

- *four-momentum is conserved;*

- *test particle trajectories are independent of the specific mass values of the test particles concerned.*

The first of these conditions is the basic statement of relativistic mechanics and is already satisfied in the Lorentzian model Universe by virtue of (2). The second of these conditions is a statement of the WEP which simply reflects that characteristic gravitational phenomonology which is discernible to a local observer; in the following, it will be shown how the reference scalar can be used to define a class of trajectories which satisfies this latter condition.

Consider the trajectory, $\mathbf{x}$, of a single particle chosen arbitrarily from one of the four partitions of all the particles; in general, such a trajectory will not be confined to any particular level-surface of U, but will cut across such surfaces with the consequence that it can be parameterized by changes in U. To see how this can be done, first note that, since $n_a \equiv \nabla_a U$, $a = 1, 2, 3, 4$, defines normal vectors to the level-surfaces of the reference scalar, U, then

$$dn_a = \nabla_j(\nabla_a U)dx^j \equiv g_{ja}dx^j \quad \text{where} \quad g_{ab} \equiv \nabla_a \nabla_b U \tag{3}$$

gives the covariant change in n_a along an arbitrary infinitessimal arc $d\mathbf{x}$. Effectively, therefore, if $d\mathbf{x}$ is confined to the particle trajectory, $\mathbf{x}$, then dn_a provides a vector measure of reference-scalar change along an infinitessimal section of trajectory and, in particular, the scalar product

$$dn_i dx^i \equiv g_{ij} dx^i dx^j \equiv ds^2$$

will give a scalar measure of the invariant resolution of this change along $d\mathbf{x}$.

Now suppose p and q are two arbitrarily chosen point coordinates, and suppose the latter expression is integrated between these points to give the scalar invariant

$$I(p,q) = \int_p^q \sqrt{dn_i dx^i} \equiv \int_p^q \sqrt{g_{ij} dx^i dx^j} \equiv \int_p^q \mathcal{L} d\tau,$$

where $\mathcal{L} \equiv (g_{ij}\dot{x}^i\dot{x}^j)^{1/2}$ for arbitrary parameter τ, and further suppose that $I(p,q)$ is minimized with respect to choice of the trajectory connecting p and q; this minimizing trajectory is then geodesic in the Riemannian space which has g_{ab} as its metric tensor. Since the existence of this metric space - and hence of geodesics within it - depends only on the *general fact* of momentum-conservation and is explicitly independent of specific inertial-mass values, then this latter step makes particle trajectories *independent* of particle inertial masses, and is the modelling step defining that general subclass of trajectories which conform to the characteristic phenomonology of gravitation. If, finally, the connections in this metric space are taken to be the metrical connections then g_{ab} given at (3) can be written explicitly as

$$g_{ab} \equiv \nabla_a \nabla_b U \equiv \frac{\partial^2 U}{\partial x^a \partial x^b} - \Gamma^k_{ab} \frac{\partial U}{\partial x^k}, \tag{4}$$

where Γ^k_{ab} are the Christoffel symbols, and given by

$$\Gamma^k_{ab} \equiv \frac{1}{2} g^{kj} \left(\frac{\partial g_{bj}}{\partial x^a} + \frac{\partial g_{ja}}{\partial x^b} - \frac{\partial g_{ab}}{\partial x^j} \right).$$

To summarize, the four-momentum conserving interactive distribution of all material in a Lorentzian model Universe can be represented as an abstract metric space within which geodesic trajectories serve as natural models for the trajectories of arbitrary particles which are gravitationally interacting with all Universal material. **In effect, this means that the collection of all material in the Universe acts as the natural frame of reference for describing gravitational action.**

THE SEP IS SATISFIED

The WEP states that test-particle trajectories are independent of the specific mass values of the test particles concerned whilst the SEP states:

- *The WEP is valid for self-gravitating bodies, as well as for test bodies;*

- *The outcome of any local test experiment is independent of the velocity of the freely falling test apparatus;*

- *The outcome of any local test experiment is independent of where and when in the Universe the experiment is performed.*

The first of these conditions is satisfied by the presented theory because the conservation statement (2), upon which the whole analysis is predicated, is apriori a statement of four-momentum conservation between two *arbitrarily constituted* conglomerated particles going into an interaction, and two similarly arbitrarily constituted conglomerated particles coming out of the same interaction and, in general, such conglomerated particles will be self-gravitating bodies, such as stars, galaxies etc. The second and third of these conditions are automatically guaranteed in the theory simply because it is predicated upon the assumption of a global Lorentz invariance.

THE BASIC CASE OF INERTIAL MOTION

It is generally perceived that any material body which is in a state of free-fall and remote from any other substantial material body is in an (almost) perfectly inertial state even though *in principle* it must be presumed the body is interacting with the remainder of the Universe. In this case, the metric tensor, g_{ab}, of the general formalism, specified at (4), which is 'seen' by the freely falling material body can be reasonably approximated by

$$g_{ab} = \gamma_{ab} = \frac{\partial^2 U}{\partial x^a \partial x^b},$$

and the most general solution of this equation is easily shown to be given by

$$U \equiv U^{(0)} = \frac{1}{2}(x^i - x_0^i)(x^j - x_0^j)\gamma_{ij},$$

where $\mathbf{x}_0$ is an arbitrarily chosen origin. Consequently, the class of all possible trajectories of inertial particles is represented by that whole class of reference scalars characterized by $U^{(0)}(\mathbf{x})$. For convenience, any member of this class will be referred to as an 'inertial field'.

A SIMPLE GRAVITATING SYSTEM

The simple gravitating system to be considered is a perturbation of the inertial system discussed in the previous section. The reference scalar which is 'seen' by any body, mass M say, which is remote from any other substantial material body, is one of the inertial fields, $U^{(0)}(\mathbf{x})$, of the previous section. Now suppose that a particle, mass $m << M$ say, which is a subordinate component of the 'rest of the Universe', is allowed into close proximity to M; since $m << M$, the trajectory of m will generally not be inertial, but the trajectory of M will be almost unaffected so that it can still be considered to be in an inertial state. Now imagine M did not exist - it is manifestly the case that, because of m's proximity to the original position of M, it would simply adopt the inertial state that was previously held by M. Consequently, it can be concluded that the non-inertial motion of m in the presence of M is due entirely to the proximity of this latter body which can therefore be reasonably designated as the 'gravitating source'.

The problem now is to identify the reference scalar which is 'seen' by m, and which enables a description of its motion to be given. To do this, the model Universe is redefined to consist of the particle m into and out of an interaction event together with the 'rest of the Universe' into and out of the same interaction event, where the 'rest of the Universe' now includes M. In the absence of M, the corresponding reference scalar would simply be one of the inertial fields, $U^0(\mathbf{x})$, of the previous section; however, from the point of view of m, the presence of M disturbs this inertial field in some fashion and, in the following, the modelling assumption is made that this disturbance has the structure of a spherically symmetric retarded wave generated by a point source which represents M.

THE REFERENCE SCALAR AND THE LINE ELEMENT

The most general form of reference scalar consisting of the inertial field perturbed by a spherically symmetric relativistic wave is given by

$$U(\mathbf{x}) = U^{(0)}(\mathbf{x}) + \frac{H(R - ct)}{R},$$

where $U^{(0)}$ is the inertial field, H is any twice differentiable function, and R is the radial displacement from the point-source. Using (4), it can be seen that the metric tensor describing this perturbation of the inertial field is given, *to first order*, by

$$g_{ab} = \frac{\partial^2 U}{\partial x^a \partial x^b} - \Gamma^k_{ab} \frac{\partial U}{\partial x^k} \approx \gamma_{ab} + \frac{\partial^2}{\partial x^a \partial x^b} \left(\frac{H}{R} \right). \tag{5}$$

Expanding this expression, forming the corresponding proper time element, and transforming to spherical polar coordinates leads to

$$d\tau^2 = \Delta_0 + \Delta_1 + \Delta_2 \tag{6}$$

$$\Delta_0 \equiv c^2 dt^2 - dR^2 - R^2 d\Phi^2,$$

$$\Delta_1 \equiv \frac{\ddot{H}}{R} \left(dR^2 - 2cdRdt + c^2 dt^2 \right),$$

$$\Delta_2 \equiv \frac{\dot{H}}{R^2} \left(-2dR^2 + R^2 d\phi^2 + 2cdRdt \right) - \frac{H}{R^3} \left(-2dR^2 + R^2 d\phi^2 \right).$$

The Δ_0 component represents the infinitessimal proper time registered by a test particle moving in the undisturbed inertial field, whilst Δ_1 and Δ_2 describe that part of the disturbance of this proper time which is due to the $O(1/R)$ approximation to g_{ab} given at (5). In the following two sections, it is shown how this part of the total proper time disturbance - the total has not been calculated owing to algebraic complexity - describes a gravitational structure which has an event horizon at the usual Schwarzschild radius, and displays the usual Schwarzschild-type physics outside of this radius, but which *has no singularity* inside this radius.

SCHWARZSCHILD PHYSICS AND THE SCALAR WAVE

Now suppose that the disturbance which gives rise to (6) is simply one of a continual train of identical disturbances passing through the particle with a regular frequency measured in the rest frame of the source (the source is considered as an oscillator), then $\ddot{H}$, $\dot{H}$ and H can be replaced by their mean values between successive minima in the train of disturbances, and $d\tau^2$ of (6) can be redefined as a measure of the mean proper time registered by the test particle between these successive minima.

With this understanding, and writing the mean value of $\ddot{H}$ as λ, then, to $O(1/R)$, a spherically symmetric disturbance in the inertial field passing through the test particle causes the particle to register a mean infinitessimal proper time, given by

$$d\tau^2 \approx \Delta_0 + \Delta_1 = c^2\left(1 + \frac{\lambda}{R}\right)dt^2 - \frac{2c\lambda}{R}dRdt - \left(1 - \frac{\lambda}{R}\right)dR^2 - R^2d\phi^2,$$

measured over the interval for which the disturbance can be said to be passing through the particle. Now, since the integrated magnitude of the deviation of the disturbed proper time from the undisturbed proper time, measured over any finite interval, will depend on the frequency with which the disturbances arrive at the test particle then it can be deduced that the undetermined parameter λ must be a measure of the frequency, and hence rest mass, of the disturbing source. Consequently, $\lambda = \beta M$, for constant β and rest mass M, and the proper time element approximation $d\tau^2 \approx \Delta_0 + \Delta_1$ is reduced to the form of that arising in any conventional metric theory.

Finally, it is noticed that the general form of the latter proper time element approximation is identical to that of the Eddington form of the Schwarzschild proper time element of General Relativity, and that the two forms match exactly if $\beta = -2\gamma/c^2$; consequently, under the transformation,

$$cdt \rightarrow cdt + \left(1 - \frac{c^2R}{2\gamma M}\right)^{-1}dR$$

then

$$d\tau^2 \approx \Delta_0 + \Delta_1 \equiv c^2\left(1 - \frac{2\gamma M}{c^2R}\right)dt^2 - R^2d\phi^2 - \left(1 - \frac{2\gamma M}{c^2R}\right)^{-1}dR^2,$$

and this form is identical to the Schwarzschild proper time element, as required.

The proper time element, (6), is only fully determined when the magnitudes of the $O(1/R^2)$ and $O(1/R^3)$ perturbations respectively, are given. It is readily shown that both the $\dot{H}/R^2$ and H/R^3 terms contribute effects which disturb the classical orbital equations at the same order as the $\ddot{H}/R$ term; but, since the proper time element, determined up to the $\ddot{H}/R$ term, is identical to the Eddington form of the Schwarzschild proper time element - and therefore giving 'perfect' predictions, at least for the orbit of Mercury and for the bending of light rays - it can be concluded that the effects of the $\dot{H}/R^2$ and H/R^3 terms are very small for R of planetary orbital dimensions, at least so far as the sun, as the gravitating source, is concerned.

The quantitative differences between GR and the presented formalism emerge when the neglected $\dot{H}/R^2$ and H/R^3 terms are included at small R: It is found that the Schwarzschild boundary at $R_s = 2\gamma M/c^2$ still exists as a 'one way membrane' for test particles, but the essential singularity which exists in the GR model at $R = 0$ is absent here. Instead, what happens is that at some $R = R^*$, satisfying $0 < R^* < R_s$, gravitational attraction 'turns off', and becomes a repulsion for $0 < R < R^*$, and the origin, $R = 0$, becomes the top of a 'potential hill'. The effect of this on test particles is that, instead of collapsing into a singularity at $R = 0$, they orbit the origin inside $R = R_s$, oscillating between $R = 0$ and $R = R_s$. This behaviour on the interior on the Schwarzschild boundary clearly has profound implications for the physics of gravitational collapse.

CONCLUSIONS

The foregoing discussion has shown how global four-momentum conserving inter-actions in a finite model Universe can be used to define an abstract metric space which, in turn, can be interpreted as a formal representation of the interactions concerned. The existence of this metric space, and hence of geodesics within it, is independent of the specific mass-values of all the particles concerned, and so can be used to model those interactions which are specifically gravitational in origin. In this way, it has been shown that 'metric gravity', as a general concept within the context of relativistic phys-ics, is implicitly contained within the concept of global four-momentum conservation, and is completely independent of the curved space-time concepts which characterize most other modern theories. **In effect, this means that the collection of all material in the Universe acts as the natural frame of reference for describing gravitational action.**

The theory cannot be distinguished from General Relativity for all local tests and, at the order of approximation calculated, the essential singularities at gravitational origins, which are features of both GR and Newtonian theory, do not exist. Furthermore, unlike all other competing theories, except for General Relativity, the presented theory satisfies the Strong Equivalence Principle which, it is conjectured (eg Will[1]), is a necessary and sufficient condition for a theory to pass the test of the 'binary pulsar'. The truth of this latter conjecture receives support in the present case because, since the theory is globally Lorentz invariant and is predicated of the global statements (1) and (2), mass-energy and momentum/angular momentum are automatically conserved through gravitational interactions; consequently, the absence of monopole and dipole components to gravitational radiation is guaranteed (eg Thorne[2]), and these absences are the precise conditions imposed by the binary pulsar (eg Will[1]). It follows that the theory is indistinguishable from General Relativity at the level of the binary pulsar also.

Finally, because the model Universe in which this theory is developed is globally Lorentz invariant, then gravitational processes are finally reduced to particle/particle interactions of a kind familiar to the rest of physics. The wider implication of this is simply that the conceptual barriers which have previously prevented the assimilation of gravitational physics into quantum physics do not exist here.

REFERENCES

1. Will, C.M., 'Theory and Experiment in Gravitational Physics', CUP (1981).
2. Thorne, K.S., 'Gravitational Wave Research: Current Status and Future Prospects', Rev. Mod. Phys., 52, 285-297 (1980).

FOURDIMENSIONAL ELASTICITY: IS IT GENERAL RELATIVITY?

Angelo Tartaglia

Dipartimento di Fisica, Politecnico
Corso Duca degli Abruzzi 24
I-10129 Torino, Italy

INTRODUCTION

The purpose of this paper is to show that it is possible to describe the space-time as a four dimensional elastic medium. This, when unstressed, has a Euclidean metric; when stressed however it may change its metric, acquire curvature and in general display features typical of the general relativistic space-time with and without gravitational interaction. In particular we shall see that under a uniaxial stress it is possible to recover the Minkowski metric, provided the elastic parameters of the medium satisfy appropriate but rather loose conditions. Similarly we shall show that a simple stress pattern produces the Friedman-Robertson-Walker universe.

The behaviour of the medium will be studied using, for simplicity, the linear theory of elasticity in n dimensions and considering only a static situation. This treatment, though yielding some results coincident with those of special and general relativity, is not equivalent to those theories in particular because of the linearity of the equations. It is especially fit for a new interpretation of the singularities, including the big bang.

Some consequences of the theory will be discussed in the conclusion.

N-DIMENSIONAL ELASTICITY

We suppose that in the unstrained medium the geometry is Euclidean and consequently the line element, in any given coordinate system, is expressed by the formula

$$dl_o^{\,2} = \varepsilon_{\alpha\beta}dx^{\alpha}dx^{\beta} \tag{1}$$

where $\varepsilon_{\mu\nu}$ is the Euclidean metric tensor.

Frontiers of Fundamental Physics, Edited by M. Barone
and F. Selleri, Plenum Press, New York, 1994

The introduction of strain may be described by a displacement vector field **w**, such that eq.(1) is transformed into

$$dl^2 = \left(\varepsilon_{\alpha\beta} + d\varepsilon_{\alpha\beta}\right)\left(dx^\alpha + dw^\alpha\right)\left(dx^\beta + dw^\beta\right) \tag{2}$$

The coordinates in use are still the unperturbed ones. It will be assumed that the w's are small enough to allow

$$\delta\varepsilon_{\alpha\beta} = \varepsilon_{\alpha\beta,\mu} w^\mu \tag{3a}$$

Furthermore one has

$$\delta w^\alpha = w^\alpha_{,\mu} dx^\mu \tag{3b}$$

Introducing (3a) and (3b) into eq.(2) and developing yields

$$dl^2 = dx^\alpha dx^\beta\Big(\varepsilon_{\alpha\beta} + \varepsilon_{\alpha\mu} w^\mu_{,\beta} + \varepsilon_{\mu\beta} w^\mu_{,\alpha} + \varepsilon_{\mu\nu} w^\mu_{,\alpha} w^\nu_{,\beta} + \varepsilon_{\alpha\beta,\mu} w^\mu +$$

$$+\varepsilon_{\alpha\mu,\nu} w^\nu w^\mu_{,\beta} + \varepsilon_{\mu\beta,\nu} w^\nu w^\mu_{,\alpha} + \varepsilon_{\mu\nu,\lambda} w^\lambda w^\mu_{,\alpha} w^\nu_{,\beta}\Big) \tag{4}$$

which may be written synthetically as

$$dl^2 = \left(\varepsilon_{\alpha\beta} + u_{\alpha\beta}\right) dx^\alpha dx^\beta \tag{5}$$

where $u_{\mu\nu}$ is a symmetric tensor identifiable as the strain tensor.

In eq.(5) we recognise

$$g_{\mu\nu} \equiv \varepsilon_{\mu\nu} + u_{\mu\nu} \tag{6}$$

as the new metric tensor of the strained medium.

The next step in ordinary elasticity theory is to write the Helmholtz free energy of the medium in terms of the strain tensor and to develop it in powers of the u's. The lowest order non trivial approximation gives

$$F = F_o + \frac{\lambda}{2}\left(u_\alpha^{\ \alpha}\right)^2 + \mu u_{\alpha\nu} u^{\nu\alpha} + \dots \tag{7}$$

where F_O is independent from the strain and λ and μ are the first and second Lamé coefficients.

Differentiating F with respect to the u's produces the stress tensor $\sigma_{\mu\nu}$. The explicit expression in the lowest approximation, together with the equivalent interchanging u's and σ's, is:

$$\sigma_{\alpha\beta} = \left(K - \frac{2\mu}{n}\right)\varepsilon_{\alpha\beta}\varepsilon^{\nu\lambda} u_{\lambda\nu} + 2\mu u_{\alpha\beta}$$

$$u_{\alpha\beta} = \left(\frac{1}{n^2 K} - \frac{1}{2\mu n}\right)\varepsilon_{\alpha\beta}\varepsilon^{\nu\lambda}\sigma_{\nu\lambda} + \frac{1}{2\mu}\sigma_{\alpha\beta} \tag{8}$$

Now μ, the second Lamé coefficient, is the shear modulus, $K = \lambda + \frac{2}{n}\mu$ is the uniform compression modulus, and n is the number of dimensions. To this approximation the raising and lowering of the indices of the tensors may be achieved using simply the ε's rather than the full g's.

The first linear equation in the (8)'s is the Hooke's law.

STATIC EQUILIBRIUM AND FIELD EQUATIONS

The equilibrium conditions of a stressed elastic medium are written as

$$\sigma_{\alpha}^{\ \beta}{}_{,\beta} + f_{\alpha} = 0 \tag{9}$$

where $\sigma_{\alpha\beta}$ may be interpreted as the α-component of the elastic force per unit surface acting

on a surface element orthogonal to the β-direction. f_{α} is the α-component of any force per unit volume applied to the medium. Eq.(9) as well as the elements of the linear elasticity theory may be found in any text book such as [1].

Combining now the Hooke's law, the definition of the strain tensor, eq.(6) and eq.(9) we obtain an equation for the metric tensor:

$$\left[\left(K - \frac{2\mu}{n}\right)\left(\varepsilon^{\nu\lambda}g_{\lambda\nu} - n\right) - 2\mu\right]\varepsilon_{\alpha\beta}\left(\varepsilon^{\mu\nu}g_{\mu\nu}\right)_{,\beta} + 2\mu g_{\alpha\beta,\beta} = -f_{\alpha} \tag{10}$$

Repeated indices imply summation. Using Cartesian coordinates, (10) simplifies to:

$$\left(K - \frac{2\mu}{n}\right)g_{\alpha\alpha,\beta} + 2\mu g_{\beta\nu,\nu} = -f_{\beta} \tag{11}$$

Eq.'s(10) or (11) may be thought of as the field equations of the theory. They are n equations for n(n+1)/2 unknowns. The problem of equilibrium is in general underdetermined unless suitable boundary conditions are provided.

APPLICATIONS

Uniaxial stress

As a first application of the present theory let us consider an homogeneous n-dimensional elastic medium and apply to it a uniform stress along any axis; let us call τ the coordinate along this axis. The stress tensor in Cartesian coordinates may be written as

$$\begin{aligned}
\sigma_{\tau\tau} &= p \\
\sigma_{\alpha\beta} &= 0 \qquad\qquad \alpha \neq \beta \\
\sigma_{ii} &= \Sigma
\end{aligned} \tag{12}$$

Greek indices run from 0 to n-1, assuming the 0-th axis be the τ one; Latin indices run from 1 to n-1. Σ and p are constants and in particular p>0 implies traction and p<0 compression along the τ axis.

In this case eq.'s(11) are of little use for us, because any constant g tensor is a solution for them, provided no volume force is present.

To solve the problem we can have recourse to the second equation in the (8)'s, directly computing the strain tensor:

$$u_{rr} = \frac{1}{n}\left[\left(\frac{1}{nK}+\frac{n-1}{2\mu}\right)p+(n-1)\left(\frac{1}{nK}-\frac{1}{2\mu}\right)\Sigma\right]$$

$$u_{\alpha\beta} = 0 \qquad\qquad\qquad \alpha \neq \beta \tag{13}$$

$$u_{ii} = \frac{1}{n}\left[\left(\frac{1}{nK}-\frac{1}{2\mu}\right)p+\left(\frac{n-1}{nK}+\frac{1}{2\mu}\right)\Sigma\right]$$

Recalling (6) and imposing the coincidence $g_{\mu\nu}=\eta_{\mu\nu}$, where $\eta_{\mu\nu}$ is the Minkowski metric tensor, leads to the conditions

$$p = \frac{n-1}{n}(2\mu-nK)$$

$$\Sigma = -\frac{2\mu+nK(n-1)}{n} \tag{14}$$

In four dimensions these become:

$$p = \tfrac{3}{2}(\mu-2K)$$

$$\Sigma = -\tfrac{1}{2}(\mu+6K) \tag{15}$$

For any reasonable elastic medium it is

$$\mu > 0, \qquad K > 0 \tag{16}$$

which implies, in our case, $\Sigma < 0$. Adding the condition

$$\Sigma > 2K \tag{17}$$

it is also p>0. What is being described is a medium stretched along one axis ("time" axis) and compressed along the others. Under the given conditions the original GL(4,R) symmetry has been reduced, by the stress, to the Lorentzian SO(3,1) symmetry: space-time looks like such an elastic medium.

Spherical symmetry and the expanding universe

Let us now look for a spherically symmetric solution to the problem of equilibrium in four dimensions. The centre of symmetry, that we use as origin of the coordinates, will be singular and we shall be lead to introduce a radial coordinate which we call τ.

Let us try a line element like this:

$$ds^2 = a^2(\tau)\left(d\tau^2 - dx^2 - dy^2 - dz^2\right) \tag{18}$$

where the directions orthogonal to the τ axis are spanned by three-dimensional Cartesian coordinates.

In general relativity (18) is the typical line element of a spatially flat expanding (or contracting) universe.

Introducing this metric into eq.(11) we find solutions if $f_i=0$. This means that any volume field must be radial (as it was obvious).

Assuming then $f_\tau=F=$constant and solving for $a^2(\tau)$ the result is

$$a^2(\tau) = \frac{F}{2K-3\mu}\,\tau+constant \tag{19}$$

Now rescaling the radial coordinate according to

$$\alpha(\tau)d\tau = dt$$

we introduce "time".

The line element, with the new variable, assumes its synchronous form

$$ds^2 = dt^2 - a^2(t)(dx^2 + dy^2 + dz^2) \tag{20}$$

with

$$a^2(t) = \left|\frac{F}{3\mu - 2K}\right|^{2/3} t^{2/3} \tag{21}$$

The scale factor displays the typical time dependence of a matter dominated Friedman universe [2].

More general Robertson-Walker metrics with positive or negative space curvature are in general no solution to eq.(11).

CONCLUSION

We have seen that space-time can behave as an elastic medium in four dimensions, at least in a couple of examples, but, since the problem of equilibrium in such a medium is largely underdetermined, it is easy to argue that other similar correspondences must exist. It is important to stress that this concerns the space-time in itself, without answering the question: what is matter? Actually matter is the source of internal and local symmetries or, otherwise stated, of internal stress. A point massive object is for example a line (or tube) inducing around it a cylindrical pattern of strains and stresses. The big bang is the central point of a spherically symmetric field of strain.

There are however some problems in interpreting our model. One is related to signature. In the uniaxial symmetry case the explicit expression of the elements of the strain tensor in four dimensions may be obtained from (13) and (15). It is:

$$u_{\tau\tau} = 0$$
$$u_{ii} = -2 \tag{22}$$

This is a well behaved real tensor, however it should be related to the displacements field $\mathbf{w}$ by the expression included in brackets in (4) (excluding only the first term). It should be

$$u_{\mu\nu} = w_{\mu,\nu} + w_{\nu,\mu} + w^{\alpha}{}_{,\mu}w_{\alpha,\nu} \tag{23}$$

Solving (23) under the conditions (22) gives

$$w^{\tau} = constant - 2\tau$$
$$w^{r} = (-1 \mp i)r \tag{24}$$

Now r is the distance from any arbitrary space origin. The price to pay for the Minkowski signature is a complex displacement vector field.

Another problem for the treatment in the preceding sections is that it is the result of a linearization of the equations. This means that the theory could describe weak field regions only. To include stronger field regions one has to pass to higher order approximations: this has the same difficulties as in non linear classical elasticity. The further approximation following the one already used is achieved introducing into the development (7) the third order terms:

$$\frac{v}{3}\left(u_{\alpha}{}^{\alpha}\right)^3 + \pi u_{\alpha}{}^{\alpha}u_{\mu}{}^{\nu}u_{\nu}{}^{\mu} + \rho u_{\alpha}{}^{\beta}u_{\beta}{}^{\mu}u_{\mu}{}^{\alpha} \tag{25}$$

where ν, π, ρ are three new parameters characterising the properties of the medium. Now everything is more complicated; raising and lowering indices, for instance, must be performed using the g's developed up to first order in the u's.

The last thing to be stressed is that this approach is entirely static. There is nothing like propagating signals in this picture: in four dimensions a light ray or any other moving object is as static as a line drawn on a piece of paper. However if this picture of the elastic medium has to be taken seriously dynamics cannot be excluded. Of course it needs an evolution parameter that is not internal to the medium: it cannot be what we called τ or t. Let us call T this parameter and treat it in a sense like the good old Newtonian time. T will not be observable by any four-dimensional observer inside the medium, but nonetheless it will preside to the dynamical behaviour of the system.

Allowing space-time to perform elastic vibrations what will imply for four-dimensional observers? Consider for instance a strain line joining two fixed points and suppose it is oscillating. An observer along it not able to perceive T will notice around him the coexistence of different possible curvatures and histories. The evolution in T will be transformed in different probabilities to be attached to the alternative histories. This is a hint to a possible reinterpretation of the quantum mechanics. All is of course rather simple and crude but maybe it deserves further investigation.

REFERENCES

1. L. Landau and E. Lifshitz. "Théorie de l'Elasticité", Editions Mir, Moscow (1967)
2. W. Misner, K.S. Thorne. J.A. Wheeler. The elementary Friedman cosmology of a closed universe, *in*: "Gravitation", Freeman and Co., San Francisco (1973).

UNIVERSALITY OF THE LIE-ISOTOPIC SYMMETRIES FOR DEFORMED MINKOWSKIAN METRICS

A.K.Aringazin[1,2] and K.M.Aringazin[1]

[1]Department of Theoretical Physics
Karaganda State University, Karaganda 470074, Kazakhstan
[2]Institute for Basic Research
P.O. Box 1577, Palm Harbor, FL 34682, U.S.A.

INTRODUCTION

Lie-isotopic and Lie-admissible theories are based on non-trivial realisation and generalisation of the conventional product and Lie algebra. Various studies are now performed in applying this formalism to metric spaces, gauge theory, classical and quantum mechanics, field theory, and quantum groups. Lie-isotopic construction provides consistent generalisations of Hamiltonian mechanics refered to as Birkhoffian mechanics and Birkhoff-Santilli mechanics.

In metric spaces, the application of the Lie-isotopic approach gives rise to generalisations of Minkowskian space-time called Minkowski-isotopic space-time. This generalisation can be treated as a deformation of the Minkowski space-time providing both gravitational and non-gravitational effects. In the non-gravitational sector, Minkowski-isotopic space-time metric has been studied, and various generalisations of the Minkowski metric, which were proposed in the context of particle physics, are shown to fit the Minkowski-isotopic metric. This is due to a general nature of the Lie-isotopic element, which may depend on parameters characterising geometrically a medium, such as anisotropy, velocity-dependence (Finslerian) characteristics.

Lorentz invariance is replaced by more general Lorentz-isotopic invariance. Particlularly, specific effects of the Lorentz-non-invariance (LNI) models may be all interpreted in the context of Minkowski-isotopic approach. Specifically, these models have been suggested to explain anomalous energy dependence of the life-times and other fundamental parameters of unstable particles which has been indicated to not fit the Einsteinian law, in high energy region (above 10 GeV). As a conclusion, we can state that the Lie-isotopic approach gives a natural anzatz to investigate the LNI effects.

In the absence of gravity, space-time is determined as a smooth flat manifold endowed with the Minkowski metric $\eta_{ij} = \mathrm{diag}(+1, -1, -1, -1)$. Transformation group of the space-time which leaves the space-time interval, $ds^2 = dx^i \eta_{ij} dx^j$, invariant is the Lorentz group. The Lorentz symmetry is one of the fundamental symmetries of physical theories and various experiments verify it to a high accuracy. The Lorentz symmetry

Frontiers of Fundamental Physics, Edited by M. Barone
and F. Selleri, Plenum Press, New York, 1994

seems to be exact. Evidently, this is correct when one deals with test particles and ordinary conditions. Hovewer, when, for instance, extended particles or high energies, or unusual physical conditions leading to a new reality are considered conventional formalism fails to describe associated relativistic effects.

Now, there are both theoretical and experimental arguments to treat the proper Lorentz symmetry as an approximate symmetry of physical processes. Various models have been proposed and experimental results are obtained to verify this conjecture.

Blokhintsev[1] and Redei[2] suggested the modification of the conventional relativistic life-time formula for unstable particles, $\tau = \tau_0\gamma(1 + 10^{25}\gamma^2 a_0^2)$, where a_0 is a "fundamental length", $[a_0] = cm$. According to the original treatment, the length a_0 plays an universal role in the sense that all conventional fundamental theories must be modified at the a_0 distance scale.

Nielsen and Picek[3] have developed the Lorentz-non-invariant (LNI) model based on the "minimally" generalised Minkowski metric $g_{ij} = \eta_{ij} + \chi_{ij}$, $\chi_{ij} = \mathrm{diag}(\alpha, \alpha/3, \alpha/3, \alpha/3)$, with α being small Lorentz symmetry breaking parameter, which suggested to be tangible on the scale of electroweak unification, $a \sim 1/M_W \sim 10^{-16} cm$. The traceless additional tensor χ_{ij} provides a residual $SO(3)$ symmetry. The model yeilds, in particular, the life-time high-energy formula, $\tau = \tau_0\gamma(1 + 4\alpha\gamma^2/3)^{-1}$. From consideration of the π- , μ- , and K-meson data they obtained the following estimation of the average of α, $<\alpha> = (0.54 \pm 0.17) \times 10^{-3}$. It should be stressed that this LNI model does not lead to CP-violating physics[3].

Also, series of K-meson regeneration experiments at Fermilab[4], in the energy range $E_K \sim 30\text{-}130$ GeV, display an anomalous energy dependence of the fundamental parameters of the $K^0 - \bar{K}^0$ system . Aronson $et\ al.$[5] have shown that the eventual anomalous behavior can not be attributed to an electromagnetic or hypercharge field, or to the scattering of kaons from stray charges or cosmological neutrinos, and that it can not arise from gravitational interaction either. They arrived at the conclusion that the anomalous energy dependence of the $K^0 - \bar{K}^0$ parameters may be the signature of a new interaction. In course of the work, they supposed that it is due to an interaction of the $K^0 - \bar{K}^0$ system with an external field or medium. To describe the anomalous behavior, Aronson $et\ al.$[5] have made the modification of the quantum equation of proper time evolution of the $K^0 - \bar{K}^0$ system by adding to it velocity-dependent terms and, after ultrarelativistic expansion, obtained the values of the slope parameters, $b_x^{(N)}$, defined by the following formula: $x = x_0(1 + b_x^{(N)}\gamma^N), \gamma = E_K/m, N = 1, 2$. Here, x denotes the life-time τ_S, the mass difference $\Delta m = m_L - m_S$, and the CP-violating parameters $|\eta_\pm|$ and $tan\phi_\pm$. However, it was emphasised that this treatment is purely phenomenological in that it makes no assumption concerning the origin of the additional velocity-dependent terms. An attemption to treat the origin of such terms on the basis of Lie-isotopic Finslerian lifting of the Lorentz group has been made in [6]. We will discuss this problem below.

Also, in a recent paper by Cardone, Mignani and Santilli[7], the K_S^0 life-time data reported in [5] have been re-analysed, within the framework of Lie-isotopic approach. They argued that a *non-linear* dependence of the life-time on energy is needed, in the energy range above 10 GeV, rather than the linear one provided by conventional special relativity. The fit parameter is found to be approximately constant at intermediate energies, 30-100 GeV. Another set of data on the K_S^0 life-time, in the energy range 100-350 GeV, reported by Grossman $et\ al.$[8] shows no evidence for an energy dependence of the life-time that is in contradiction with the data reported by Aronson $et\ al.$[4], in the range 100-130 GeV. So, more experiments are needed to solve this contradiction.

Gasperini[9] has considered ultrarelativistic particle motion in the model with broken local Lorentz gauge symmetry. In a cosmological aspect, the broken Lorentz symmetry in the very early universe has been considered[10]. The possibility that the Lorentz non-invariance can be considered as an effect produced by strong gravity has been shown[11]. Also, various gravitational consequences of the theory with broken local Lorentz symmetry were studied[12]. As to a Lie-admissible formalism, Gasperini[13, 14] formulated a Lie-admissible (Lie-isotopic) theory of gravity using the extended geometrical framework based on Lie-admissible (Lie-isotopic) underlying algebra.

Also, Ellis *et al.*[15] and Zee[16] have discussed the possibility that the proton decay may violate Lorentz invariance in the context of grand unified theories. Zee[16] mentioned, particularly, that in such a small region as an interior of the proton "anything can happen" and that perhaps it is not totally far fetched and outregious idea to allow violation of the Lorentz symmetry at some distance scale, for example, on the scale of superheavy X, Y-bosons, $a \sim 1/M_{GUT} \sim 10^{-23} cm$. Lorentz non-invariance of the primodial fluid was suggested by Rosen[17].

For a review of gravitational consequences of an eventual local Lorentz non-invariance and Finslerian approach to relativity and gravitation providing naturally a velocity-dependence framework see[18, 19] . For a comprehensive introduction to Finsler geometry, we refer the reader to monographes by Rund[20], Asanov[21], Matsumoto[22], and Asanov and Ponomarenko[23]. Also, for non-metric effects in flat space-time see[24] and references therein.

Original investigations by Santilli on the Lie-isotopic generalisation of Galilei's and Einstein's relativities has been presented in his monographes[25, 26]. Mathematical foundations of the Lie-isotopic generalisation has been reviewed in a monograph by Kadeisvili[27]. For a comprehensive review of the Santilli's Lie-isotopic generalisation of the relativities we refer the reader to a recent monograph by Aringazin, Jannussis, Lopez, Nishioka, and Veljanoski[28]; see also [29].

In the next section, we present a brief introduction to the Lie-isotopic generalisation of the Lorentz symmetry with special reference to its continuous part.

We discuss in detail problem of internal conditions. Non-Minkowskian part of the metric is assumed to describe geometrically local physical properties of the space-time such as nonhomogeneity, deformations, resistance, anisotropy, and velocity-dependence.

The origin of these properties in a microscopic region is perhaps due to some quantum effects, which can lead to small deviations from the conventional pseudo-Euclidean structure in four dimensions, and may be revealed at high energies. It is worthwhile to note here that de Brogle in his theory of double solution emphasised that even if a particle is not subjected to any gravitational or electromagnetic field, its possible trajectories are the same as if space-time possessed the non-Euclidean metrics. So, in view of this theory, the deviations may be sought to be caused by subquantum vacuum fluctuations. However, we shall not discuss this deep issue here.

Discussion in the last section serves both to illustrate more general arguments, and to set the stage for a subsequent description of the Lorentz-non-invariance effects within the framework of strict classical algebraic Lie-isotopic generalisation.

LORENTZ-ISOTOPIC SYMMETRY

Lie-isotopic approach in connection with the Lorentz symmetry generalisation problem has been originated by Santilli[30, 31]. Let us recall main aspects of the Lie-isotopic generalisation of the Lorentz transformations.

The associative enveloping algebra E equipped with associative product, AB, and the unit I can be isotopically lifted to the algebra $\hat{E}$ with the generalised product

$A * B = ATB$ and the new unit $\hat{I} = T^{-1}$, where T is a fixed nonsingular element of E. The isotopic lifting of the Lie transformation group G then reads, $\hat{G} : x' = g * x = exp(Xu)_{|\hat{E}} * x = exp(XTu)_{|E}x$, where X denote generators of the original Lie group G and u are parameters. The isotopic lifting of commutator is defined accordingly by

$$[X_i, X_j]^* = X_i * X_j - X_i * X_j = \hat{c}_{ij}^k * X_k \tag{1}$$

where $\hat{c} = c * \hat{I}$ are structural constants.

Let $E(4, \eta, R)$ be Minkowski space-time. Define isotopic lifting of the Minkowski metric

$$\eta \to g = \hat{\eta}(t, x, \dot{x}, \ldots). \tag{2}$$

According to Theorem 1 of [30], the isotopic lifting of the Lie group of transformation of the Minkowski-isotopic space $E(4, g, R)$ leaves invariant the metrical form defined by $x * x = x^i g_{ij} x^j$. Thus, when the isotopic element T is given by the new metric g, the new unit is $\hat{I} = g^{-1}$.

Then, by suitable generalisation of the Minkowski metric (2) providing preservation, under the lifting, connectivity properties of the proper Lorentz group one can construct step by step the isotopic liftings of the enveloping algebra of the Lorentz group, the Lorentz group, and the Lie algebra of the Lorentz group. In the limit

$$\hat{I} \to I \quad or \quad g \to \eta \tag{3}$$

Lie-isotopic theory covers conventional one.

Lorentz-isotopic transformations can then be explicitly computed when the new metric g is defined. We note here that in spite of the local isomorphism between the lifted Lorentz group and proper Lorentz group the Lorentz-isotopic transformations may be different from the conventional ones. Clearly, it is worthwhile investigate implications of this generalisation in particle physics. For more complete and precise development of the Lie-isotopy on metric spaces, we refer the reader to [30, 31].

As to examples, explicit calculations by Santilli[30] for locally deformed Minkowski metric
$$x * x = x^i g_{ij} x^j = x^1 b_1^2 x^1 + x^2 b_2^2 x^2 + x^3 b_3^2 x^3 + x^4 c^2 x^4 \tag{4}$$

where $b = b(t, x, \dot{x}, \ldots), c = c(t, x, \dot{x}, \ldots)$, have been made, and the associated *Lorentz-isotopic* transformations found to be in the form

$$t' = \hat{\gamma}(t - V b_3^2 z/c), \quad z' = \hat{\gamma}(z - Vt), \quad \hat{\gamma} = (1 - V b_3^2 V/c^2)^{-1/2} \tag{5}$$

The metric (4) with the associated generalised Lorentz transformations (5) have been used in [6, 7] to describe anomalous energy dependence of the parameters of the $K^0 - \bar{K}^0$ system[4].

Another example is given by anisotropic Finslerian metric by Bogoslovski[32]

$$x * x = x^i g_{ij} x^j = x^i \left[\left(\frac{\nu^k \eta_{kl} \nu^l}{x^k \eta_{kl} x^l} \right)^{r/2} \eta_{ij} \right] x^j \tag{6}$$

where ν^i is an anisotropy vector, and r is a scale parameter. Associated generalised Lorentz transformations are

$$t' = \hat{\gamma}(t - Vz/c), \quad z' = \hat{\gamma}(z - Vt), \quad \hat{\gamma} = \left[\frac{1 - V/c}{1 + V/c} \right]^{r/2} \left(1 - \frac{V^2}{c^2} \right)^{-1/2} \tag{7}$$

LIE-ISOTOPIC LIFTING

Examples presented in previous Section show that the Lie-isotopy provides a useful method to derive generalised Lorentz transformations while in the proper geometrical approach, for example, within Finslerian framework, one meets difficulties in constructing an explicit form of generalised Lorentz transformations leaving generalised metric form invariant (see, *e.g.,* discussions in [19, 21]).

Note that there are no theoretical constraints for dependence of the metric g on local variables except for general demands such as non-singularity. So, the question how one can determine the *non-Minkowskian* dependence of g for a specific application arises. Equivalently, the question is how one can choose specific element T of the isotopic class of generalisation. Fixing the isotopic element T of the algebra E means that one choose fixed generalised metric and, accordingly, fixed type of generalised Lorentz transformations associated to this algebra.

Phenomenological approaches

One way to pick out isotopic element T among possible ones is to suppose that the generalised metric should be derived uniquely from given internal physical conditions. Conventional theory and experiments of special relativity learns us that the ordinary conditions, *i.e.,* the case of "empty" space-time and point-like particles, should lead to conventional Minkowski metric. If this is not the case - complicated internal conditions - one can appropriately generalise the metric, and then derive associated generalised Lorentz transformations.

Washing out such complicated conditions should yield contunious reducing of the isotopic unit, $\hat{I} = g^{-1}$, to the ordinary one due to (3). This obvious requirement put strong limits on possible generalisations and does not admit exotic metric structures.

However, in general case a way to derive the generalised metric from (suppose known) internal conditions is far from being strightforward. One can try axiomatic way defining some specific type of the metric *a priori,* and then verifying the associated Lorentz-isotopic symmetry.

Once more way to specify the metric g is to account for the internal Lorentz-non-invariance effects on a phenomenological level. This method, albeit in somewhat *ad hoc* fashion, enables one to apply the generalised theory when internal conditions are not presented by well defined (Lagrangian-based) equations for the metric g.

It is highly remarkable that any phenomenological deviation from the Minkowski metric finds its counterpart in the Lie-isotopic formalism so that all such deviations can be treated, in effect, as the Lie-isotopic extension of the conventional underlying algebra.

In terms of *small* deviations from conventional Lorentz symmetry, one can proceed as follows. General expansion of the metric g on the Minkowskian background can be performed to set a parametrical representation of the lowest approximation. Then, one need to establish for what physical effects the parameters entering the metric are responsible. The sizes of the effects, if measured, will put upper limits or estimations on these parameters. Known example is the simplest LNI parametrisation by Nielsen and Picek[3] mentioned in Introduction where meson life-times data had been used to estimate value of the LNI parameter α. Another example[6] concerns to an ultrarelativistic expansion of the velocity-dependent metric (4), with estimations on λ's in series, $b_3^2 = 1 + \lambda_0 + \lambda_1 \gamma + \lambda_2 \gamma^2 + \ldots$, being derived from data of the anomalous energy behavior of the $K^0 - \bar{K}^0$ system parameters[4, 5]. In a recent paper by Cardone *et al.*[7], these data were used to estimate the fit parameter $a = b_3^2 / b_4^2$.

As it was stressed[7, 30], the LNI parameters of the metric may vary not only from weak interaction to strong one but also from reaction to reaction within each type of interactions. In fact, analysis on meson life-times data[3] showed that the LNI parameter α even has different sign for different meson decays. Thus, dependence of g on parameters is not universal, and should be varied when different physical conditions occur.

We should note that this LNI anzatz is in close analogy with the one of the parametrised post-Newtonian (PPN) framework where all possible self-consistent metrical theories of gravity can be tested on the basis of the "fiducial" PPN metric with experimentally determined PPN parameters (see, for a review, [19, 33]).

Using the metric g one can also try to construct Lagrangian of the theory and then derive equations for internal terms which will impose restrictions on g in a dynamical way. As an example, we refer to the Finslerian theory[19, 21] where dependence of the Finslerian metric tensor $g_{ij}(x, y)$ on the "internal" tangent vector $y \in TM$ can be determined from equations for curvature tensor of the tangent bundle TM. It should be noted that at the same time basic manifold M is characterised by its own curvature tensor, and can be taken flat in the absence of gravity. In a physical context, this means that we have a locally curved momenta space that implies a non-quadratic relation between energy and momentum of a free particle (such a relation arising from supposing that space-time is really a lattice has been discussed by Aronson et $al.$[5]; see also [14]). On the other hand, one can define a section $\sigma : M \to TM$ to treat vector y as an auxiliary vector field on M due to the concept of oscullation[21].

In the context of strict geometrical framework, the intrinsic behavior of the internal variable y causing some microscopic non-linear effects has been considered in detail by Ikeda[34]. In particular, Ikeda established that the intrinsic behavior of y is reflected in a whole spatial structure. This observation may have a direct use in the case of algebraic extension of the metrical structure provided by the Lie-isotopic approach.

Also, in Kaluza-Klein anzatz (see, for example, [35]) internal variables (extra dimensions) are assumed to be curled up to a small compact manifold C to be not visible at low energies. Internal fibers are sought to carry a Lie-group srtucture which defines an internal symmetry of the system at each point of space-time M_4; see also Weinberg's insight[36] and[37] for a recent development of the standard Kaluza-Klein technique. Despite of the relevance of the usual compactification scheme to get a stable vacuum solution for which a vielbein and connections obey the isometry on $M_4 \times C$ it would be very instructive to construct a dynamical framework for compactification (see discussion in [14]). Also, recent development of superstring theory[38, 39, 40] revealed that the compactification to Calabi-Yau manifold reducing ten-dimensional theory to four-dimensional one can be provided in enormous numbers of ways. This indicates a non-triviality of the internal condition problem on a fundamental level.

Path integral approach

Another way to solve the problem of the isotopic metrical degrees of freedom is to require something like an invariance under the Lie-isotopic lifting. Namely, one can treat a set of all Minkowski-isotopic metrics g as a functional space H, $g \in H$, and introduce a covariant measure in H. The theory then will contain integral, $S = \int dg F[g]$, over all possible Minkowski-isotopic metrics defined on a manifold with fixed topology. To define volume element in the space of all metrics g one first need to define the "distance" $\|g\|$ between infinitesimally different metrics, g and $g + \delta g$. The metric variation δg can be appropriately parametrised to adapt symmetries of the theory. For example, general coordinate transformation parameters, rotational transformation parameters of the internal vector y, and the conformal transformation parameter σ, $g_{ij}(x,y) \to exp[2\sigma(x,y)]g_{ij}(x,y)$, can collectively constitute possible set of variables, u^a, in respect of which the variation δg can be formulated. The measure will be of the following bilinear form:

$$||\delta g||^2 = P_{ab}\delta u^a \delta u^b \qquad (8)$$

So, the integration will be made over these independent parameters u^a,

$$S = \int du\sqrt{Det(P)}F[g] \qquad (9)$$

We refer the reader to [40] to consult for measures in space of metrics in the context of string theories. It should be noted that the rotational invariance does not eliminate the dependence of the metric on y so that the generalised character of the associated Lorentz-isotopic transformations still remains in this case.

Another choice is to require complete isotopic invariance of the theory under certain class of transformations of the isotopic unit, $\hat{I} \to \hat{I}'$, which includes conventional Minkowski one, $\hat{I} \to I$. Then, all the Lorentz-isotopic transformations will be equivalent to the proper Lorentz one so that the theory can be formulated in the usual terms, with the only difference of a volume element of H in the action. Introducing of dependence of g on local coordinates x^i leads one to consideration of the Lie-isotopic theory of gravity [13, 14, 26, 28], which appears to be more general than general relativity because of the manifested local Lorentz-isotopic symmetry instead of the conventional Lorentz one. However, we do not elaborate further on this topic here restricting our consideration by non-gravitational sector of the Lie-isotopic lifting of the Lorentz symmetry. So, dependence on coordinates x^i in the intergral (9) may be treated as a formal one, and one can omit it to avoid additional complications arising in part from topology of basic manifold M. We wish to note, incidentally, that the Lie-isotopic gravity, especially in view of the proposed invariance in respect to Lie-isotopic lifting, may be of interest in low dimensions, 2+1 or 1+1, where one deals with Riemannian surfaces[38, 39, 40, 41].

Perhaps, these suggestions seem to be speculative at this stage, and its realisation will be complicated. However, it can be considered as a very attractive alternative way which enables us to exploit the idea of a symmetry in the space of Lie-isotopically lifted metrics, on a high-energy scale.

CONCLUDING REMARKS

Although we have phrased our discussion in the context of LNI effects, the Lie-isotopic lifting of the Lorentz group may have use as a framework to generalise other fundamental theories such as quantum field theory and quantum gravity. It should be noted also that the Lie-isotopic lifting of a continuous group of transformations provides a Lie-isotopic generalisation of conventional gauge theories[42] which seems to have far reaching implications. Further generalisation of the Lie-isotopic theory is a Lie-admissible approach, within which one can, particularly, describe an evolution of quantum group systems[43], and we refer the interested reader to [44] for a review on Lie-admissible theory and its applications.

References

[1] D.I.Blokhintsev, *Phys. Lett.* 12: 272 (1964).

[2] L.B.Redei, *Phys. Rev.* 145: 999 (1966).

[3] H.B.Nielsen and I.Picek, *Phys. Lett.* B144: 141 (1982).

[4] B.H.Aronson, G.J.Bock, H-Y.Cheng, and E.Fischbach, *Phys. Rev.* D28: 476 (1983).

[5] B.H.Aronson, G.J.Bock, H-Y.Cheng, and E.Fischbach, *Phys. Rev.* D28: 495 (1983).

[6] A.K.Aringazin, *Hadronic J.* 12: 71 (1989).

[7] F.Cardone, R.Mignani, and R.M.Santilli, *J. Phys.* G18: L61 (1992).

[8] N.Grossman *et al.*, *Phys. Rev. Lett.* 59: 18 (1987).

[9] M.Gasperini, *Phys. Lett.* B177: 51 (1986).

[10] M.Gasperini, *Phys. Rev.* D33: 3594 (1986); *Phys. Lett.* B163: 84 (1985).

[11] M.Gasperini, *Phys. Lett.* B141: 364 (1984).

[12] M.Gasperini, *Phys. Rev.* D33: 3594 (1986); *Mod. Phys. Lett.* A2: 385 (1987); *Class. Quantum Grav.* 4: 485 (1987).

[13] M.Gasperini, *Nuovo Cimento* B81: 7 (1984); *Hadronic J.* 7: 234, 650, 971 (1984).

[14] M.Gasperini, *Nuovo Cimento* A83: 309 (1984).

[15] J.Ellis, M.Gaillard, D.Nanopoulos, and S.Rudaz, *Nucl. Phys.* B176: 61 (1980).

[16] A.Zee, *Phys. Rev.* D25: 1864 (1982).

[17] N.Rosen, *Astrophys. J.* 297: 347 (1985).

[18] A.K.Aringazin and G.S.Asanov, *Gen. Rel. Grav.* 17: 1153 (1985).

[19] A.K.Aringazin and G.S.Asanov, *Rep. Math. Phys.* 25: 183 (1988).

[20] H.Rund, "The Differential Geometry of Finsler Spaces," Springer, Berlin (1959).

[21] G.S.Asanov, "Finsler Geometry, Relativity and Gauge Theories," D.Reidel, Dordrecht (1985).

[22] M.Matsumoto, "Foundations of Finsler Geometry and Special Finsler Spaces," Kaiseisha Press, Kaiseisha (1986).

[23] G.S.Asanov and S.P.Ponomarenko, "Finsler Fiber Bundle Over Space-Time, Associated Gauge Fields and Connections," Stiinca, Kishinev (1989).

[24] A.K.Aringazin and A.L.Mikhailov, *Class. and Quantum Grav.* 8: 1685 (1991).

[25] R.M.Santilli, "Foundations of Theoretical Mechanics," Vol.II, Springer, New York (1983).

[26] R.M.Santilli, "Isotopic Generalization of Galilei's and Einstein's Relativities," Vols. I,II, Hadronic Press, Palm Harbor (1991).

[27] J.Kadeisvili, "Santilli's Isotopies of Contemporary Algebras, Geometries and Relativities," Hadronic Press, Palm Harbor (1992).

[28] A.K.Aringazin, A.Jannussis, D.F.Lopez, M.Nishioka, and B.Veljanoski, "Santilli's Lie-Isotopic Generalization of Galilei's and Einstein's Relativities," Kostarakis Publishers, Athens (1991).

[29] A.K.Aringazin, A.Jannussis, D.F.Lopez, M.Nishioka, and B.Veljanoski, *Algebras, Groups and Geometries* 7: 211 (1990).

[30] R.M.Santilli, *Lett. Nuovo Cimento* 37: 545 (1983).

[31] R.M.Santilli, *Hadronic J.* 8: 25, 36 (1985).

[32] G.Yu.Bogoslovski, *Nuovo Cimento* B40: 116 (1977); B43: 377 (1978).

[33] C.M.Will, "Theory and Experiment in Gravitational Physics," Cambridge Univ. Press, London (1981).

[34] S.Ikeda, *J. Math. Phys.* 26: 958 (1985); *Nuovo Cimento* B100: 493 (1987).

[35] E.Witten, *Nucl. Phys.* B186: 412 (1981); S.Randjbar-Daemi, A.Salam, and J.Strathdee, *Nucl. Phys.* B214: 491 (1983).

[36] S.Weinberg, *Phys. Lett.* B125: 265 (1983).

[37] M.Chaichian, A.P.Demichev, and N.F.Nelipa, *Phys. Lett.* B169: 327 (1986).

[38] M.B.Green, J.H.Schwarz, and E.Witten, "Superstring Theory," Cambridge Univ. Press, Cambridge (1987).

[39] W.Siegel, "Introduction to String Theory," World Sci., Singapore (1988).

[40] A.M.Polyakov, "Gauge Fields and Strings," Harwood Academic Publ., Chur (1987).

[41] E.Witten, Preprint IASSNS-HEP-88/32 (1988).

[42] M.Gasperini, *Hadronic J.* 6: 935, 1462 (1983); M.Nishioka, *Hadronic J.* 7: 1636 (1984).

[43] A.Jannussis, G.Brodimas, and R.Mignani, *J. Phys.* A24: L775 (1991).

[44] "Proc. of the Third Workshop on Lie-Admissible Formulations," Hadronic Press, Nonantum (1981); "Proc. of the First Intern. Confer. on Non-Potential Interactions and Their Lie-Admissible Treatment," Hadronic Press, Nonantum (1982).

HERTZ'S SPECIAL RELATIVITY AND PHYSICAL REALITY

Constantin I. Mocanu

Department of Electrical Engineering
Polytechnica University of Bucharest
Bucharest 1, ROMANIA

> *Motto: Special Relativity is like the Moon which shows us only
> one of her faces: it is Einstein's SR. The hidden face is Hertz's SR.*

INTRODUCTION

Einstein's Special Relativity (ESR), in its original formulation[1], is limited to inertial
motions only, while an insight into real - world shows that the motions are non - inertial. It
is hard to accept that the launching of a rocket implies a succession of inertial motions.
Similalry, during the re - entry of a satellite in the Earth's atmosphere, its continuously
decaying angular momentum and the violations of the Lorentz - Poincaré symmetries, are
then evident. From the view - point of kinematics, defined as the science of pure motion,
apart from causes, the motion may be either inertial or non - inertial. Since a motion
cannot be, at the same time, either a uniform translation or a non - uniform motion, then
they form *a pair of complementary kinematic concepts*, mutually exclusive. If the
gravitational effects are neglected, ESR assumes uniform translations, while under similar
circumstances (regarding the neglecting of the space curvature) to the best of our
knowledge, nothing is known in the literature to take into consideration non - inertial
motions. As a consequence, we shall denote by *Special Relativity* that branch of physics
where by neglecting the gravitational effects, the motion may be either inertial or non -
inertial, called in this paper *permissible motions*. As a consequence, the following chart of
complementarities:

Uniform translations	←	complementary	→	Non - uniform motions
(inertial motions)		to		(non - inertial motions)
↓				↓
Einstein's Special	←	complementary	→	? (An unknown SR)
Relativity		to		

holds, where to the *pair of complementary motions*, the *pair of complementary special
relativities*, corresponds; ESR based on inertial motions, respectively an unknown SR

which must be valid for permissible non - inertial motions. The problem of a SR involving non - inertial permissible motions has not drawn the attention of the earlier investigators.

A possibility to fill the gap in the above chart consists in the extension of ESR to non - uniform motions, which in turn implies the extension of Lorentz transformation from 1 + 1 to 1 + 3 dimensions. In order to impart group properties to the Lorentz transformation in higher space dimensions, Thomas[2] proposed that a twist had to occur supplementary to uniform linear translations and, in this way, to form a motional group. As a consequence of Thomas rotation, the canonical method of handling accelerated motions in ESR is to say that the theory is competent to describe the motions of accelerated particles, but that it cannot describe the view - point of accelerated observers or frames of reference. In a series of papers[3-7], the author proves that Thomas rotation leads to conflicts with the concepts of inertial motion and inertial reference frames.

However, fifteen years before the appearance of Einstein's paper[1], Heinrich Rudolph Hertz elaborated an electromagnetic theory[8], called Maxwell - Hertz Electrodynamics (MHE), valid for all kind of motions, either inertial or non - inertial, but at small velocities only. Critical comments were made by Poincaré, Weil, Lorentz, Pauli, Bohr, etc., to the effect that Hertz's theory gave an ether related interpretation that brought the theory into conflict with observations. Also, since it is based on classical kinematics, MHE leads to certain results not confirmed by experiments, fully clarified by Minkowsky's theory. Nevertheless, MHE still provides the theoretical basis of applications of electrodynamics of moving bodies, which in most cases occur at non - relativistic velocities. This theory, susceptible to be extended at higher velocities, has indisputable advantages, such as its operations with quantities attached to the medium (called proper quantities) and applicability to all kind of motions.

The first step in the construction of Hertz's Special Relativity (HSR), which has to cover the hole in the chart of complementarities (presented in this work), consists in the elaboration, beforehand, of a relativistic electrodynamics, by an extension of MHE at relativistic velocities. The new proposed theory, called Hertz's Relativistic Electrodynamics (HRE), does not require postulates. In the derivation of this theory an iterative scheme is conceived, based on Poincaré's induction principle and topologically founded by a repetitive application of Helmholtz's calculus. According to this procedure, Maxwell's theory of motionless media represents the approximation of order zero (the trivial case for $v = 0$) and the MHE theory that of order one of HRE. By successive iterations, from small to large velocities, an infinite set of equations is obtained and its convergence is fulfilled if and only if the velocity of the particle is less than the speed of light in vacuum ($v < c$). In this way, the Einstein's principle of constancy of light speed in vaccum is confirmed. The experiments that MHE was unable to explain (Röntgen - Eichenwald, Wilson, Fizeau - Fresnel, etc.) are correctly interpreted within the framework of the new theory.

The second step in the construction of HSR (which makes the object of another work[9-11]) takes into account that the iteration scheme of HRE implies no discussion of the nature of space and time, and does not require the specification of the reference frame, which may be inertial or not. But once the derivation of the theory is carried out, a justified question arises: to what extent the new theory requires modifications of the concepts of space and time, suggested by the convergence condition $v < c$? As a consequence, a kinematic analysis of the new theory is necessary, in order to see whether it is possible to determine a coordinate transformation, from one reference frame to another, so as to deduce the transformation equations of electric and magnetic quantities, in the same way, as in the Einstein - Minkowsky theory. On exclusively kinematic considerations, a Hertz's Relativistic Mechanics (HRM) is built up[9-11], able to specify the modifications of the classical concepts of space and time, of length contraction and time dilation, of the variation of mass with the velocity and the Einstein's formula of rest energy.

The first Section is devoted to the statement of MHE, with the definition of terms and the physical motivation of the theory. In order to make the development of the subject, as simple as possible, we shall restrict the presentation of the HRE to the nonpolarizable and nonconducting media. The iterative scheme makes the object of the second Section and the Einstein's composition law of colinear velocities is derived in the third Section. Comments on HSR are presented in the last Section.

1. STATEMENT OF MHE

Hertz's nonrelativistic electrodynamics is an attempt to extend Maxwell's theory for stationary media to moving bodies at relativistic velocities, on the basis of the following premisses:

(i), To each point of the space, a point mass with a state of motion, unambiguously determined by a given velocity $\mathbf{v}$ and, consequently, a proper reference frame $S_0(\mathbf{r}_0,t_0)$, which moves with respect to the reference frame at relative rest $S(\mathbf{r},t)$, can be associated. To $S_0(\mathbf{r}_0,t_0)$ $[S(\mathbf{r},t)]$ the position vector $\mathbf{r}_0(\mathbf{r})$ at time $t_0(t)$ is referred;

(ii), The moving media carry with them the curve Γ, the open (closed) surface $S_\Gamma(\Sigma)$ and the volume v_Σ, on which the line, surface and volume integrals of the electromagnetic quantities are written;

(iii), The local and instantaneous electric and magnetic states are described by the proper quantities: the electric field strength $\mathbf{E}_0$, the magnetic flux density $\mathbf{B}_0$, and the free charge density ρ_0, defined with respect to the proper reference frame $S_0(\mathbf{r}_0,t_0)$. In the case of nonpolarizable and nonconducting media, from the integral equations,

$$\oint_\Sigma \mathbf{E}_0 \cdot \mathbf{n}_\Sigma dA = \frac{1}{\varepsilon_0}\int_{v_\Sigma} \rho_0 dv, \qquad \oint_\Sigma \mathbf{B}_0 \cdot \mathbf{n}_\Sigma dA = 0, \tag{1}$$

$$\oint_\Gamma \mathbf{E}_0 \mathbf{ds} = -\frac{d}{dt}\int_{S_\Gamma} \mathbf{B}_0 \cdot \mathbf{n}_{S_\Gamma} dA, \qquad \oint_\Gamma \mathbf{B}_0 \mathbf{ds} = \varepsilon_0\mu_0 \frac{d}{dt}\int_{S_\Gamma} \mathbf{E}_0 \cdot \mathbf{n}_{S_\Gamma} dA, \tag{2}$$

the corresponding local equations are derived,

$$\mathrm{div}\mathbf{E}_0 = \rho_0/\varepsilon_0, \qquad \mathrm{div}\mathbf{B}_0 = 0, \tag{3}$$

$$\mathrm{curl}\,\mathbf{E}_0 = -d_s\mathbf{B}_0/dt, \qquad \mathrm{curl}\,\mathbf{B}_0 = c^{-2}d_s\mathbf{E}_0/dt, \tag{4}$$

where $d_s\mathbf{F}/dt$ defined by

$$\frac{d}{dt}\int_{S_\Gamma} \mathbf{F} \cdot \mathbf{n}_{S_\Gamma} dA = \int_{S_\Gamma} \frac{d_s\mathbf{F}}{dt} \cdot \mathbf{n}_{S_\Gamma} dA, \tag{5}$$

is the Helmholtz's total derivative, given by [10],

$$\frac{d_s\mathbf{F}}{dt} = \frac{\partial\mathbf{F}}{\partial t} + \mathbf{v}\,\mathrm{div}\,\mathbf{F} + \mathrm{curl}(\mathbf{F}\times\mathbf{v}). \tag{6}$$

Taking into account Eq. (6), Eqs. (4) may be written as follows

$$\text{curl } \mathbf{E}_1 = -\partial \mathbf{B}_0 / \partial t, \quad \text{curl } \mathbf{B}_1 = \mu_0 \rho_0 \mathbf{v} + c^{-2} \partial \mathbf{E}_0 / \partial t \tag{7}$$

where $\mathbf{E}_1$ and $\mathbf{B}_1$ are given by

$$\mathbf{E}_1 = \mathbf{E}_0 - \mathbf{v} \times \mathbf{B}_0, \quad \mathbf{B}_1 = \mathbf{B}_0 + c^{-2} \mathbf{v} \times \mathbf{E}_0. \tag{8}$$

2. DERIVATION OF THE ITERATIVE SCHEME OF HRE

As Rosser[12] pointed out, the uniqueness theorem of a vector field in terms of its sources (UTVF) plays an essential role in the analysis of the electromagnetic field equations. According to this theorem, any time - varying vector field $\mathbf{F}(\mathbf{r},t)$ is uniquely determined at each point within the v_Σ if its div $\mathbf{F}$ and curl $\mathbf{F}$ are given, and the initial and boundary conditions are prescribed. The compatibility of Maxwell's equations with UTVF is evident.

Now, let us analyze the MHE Eqs. (3) and (7), on the view - point of UTVF. Although Eqs. (7) determine the curls of $\mathbf{E}_1$ and $\mathbf{B}_1$ (8), Eqs. (3) determine the divergences of $\mathbf{E}_0$ and $\mathbf{B}_0$, instead of $\mathbf{E}_1$ and $\mathbf{B}_1$. As a consequence, Eqs. (3) and (7) do not represent the conditions of local formulation of UTVF in moving media. We say that the structure of MHE equations is defective and it suggests their approximate character. Consequently, the defective structure of MHE equations may be corrected by a reformulation of Eqs. (4) in terms of $d_s\mathbf{B}_1/dt$ and $d_s\mathbf{E}_1/dt$, instead of $d_s\mathbf{B}_0/dt$ and $d_s\mathbf{E}_0/dt$. In this way, a better approximation of MHE is gven by Eqs.

$$\text{div } \mathbf{E}_1 = \rho_1/\varepsilon_0, \quad \text{div } \mathbf{B}_1 = 0, \tag{9}$$

$$\text{curl } \mathbf{E}_0 = -d_s\mathbf{B}_1/dt, \quad \text{curl } \mathbf{B}_0 = c^{-2}d_s\mathbf{E}_1/dt. \tag{10}$$

With the notation $\mathbf{E}_2 = \mathbf{E}_0 - \mathbf{v} \times \mathbf{B}_0, \quad \mathbf{B}_2 = \mathbf{B}_0 + c^{-2}\mathbf{v} \times \mathbf{E}_0$, Eqs. (10) may be written as follows

$$\text{curl } \mathbf{E}_2 = -\partial \mathbf{B}_1/\partial t, \quad \text{curl } \mathbf{B}_2 = \mu_0 \rho_1 \mathbf{v} + c^{-2}\partial \mathbf{E}_1/\partial t. \tag{11}$$

A similar analysis shows the defective structure of Eqs. (9) and (11). The attempt to correct the defective structure, by using higher iterations, leads to the n-th order Hertzian iterative system of equations

$$\text{div } \mathbf{E}_{n-1} = \rho_{n-1}/\varepsilon_0, \quad \text{div } \mathbf{B}_{n-1} = 0, \tag{12}$$

$$\text{curl } \mathbf{E}_n = -\partial \mathbf{B}_{n-1}/\partial t, \quad \text{curl } \mathbf{B}_n = \mu_0 \rho_{n-1} \mathbf{v} + c^{-2}\partial \mathbf{E}_{n-1}/\partial t, \tag{13}$$

where,

$$\mathbf{E}_n = \mathbf{E}_0 - \mathbf{v} \times \mathbf{B}_{n-1}, \quad \mathbf{B}_n = \mathbf{B}_0 + c^{-2}\mathbf{v} \times \mathbf{E}_{n-1}. \tag{14}$$

From the recurrence relations (14), a step by step calculation of $\mathbf{E}_n$ and $\mathbf{B}_n$ as functions of $\mathbf{E}_0$ and $\mathbf{B}_0$ leads to the expressions, according to whether n is even or odd:

Even iteration order, 2n

$$\mathbf{E}_{2n} = \gamma_H^{(n)}\mathbf{E}_0 - \gamma_H^{(n-1)}\left[\mathbf{v}\times\mathbf{B}_0 + c^{-2}\mathbf{v}(\mathbf{E}_0\cdot\mathbf{v})\right], \tag{15}$$

$$\mathbf{B}_{2n} = \gamma_H^{(n)}\mathbf{B}_0 - c^{-2}\gamma_H^{(n-1)}\left[\mathbf{v}\times\mathbf{E}_0 - \mathbf{v}(\mathbf{B}_0\cdot\mathbf{v})\right]; \tag{16}$$

Odd iteration order, 2n-1

$$\mathbf{E}_{2n-1} = \gamma_H^{(n-1)}\left(\mathbf{E}_0 - \mathbf{v}\times\mathbf{B}_0\right) - c^{-2}\gamma_H^{(n-2)}\mathbf{v}\left(\mathbf{E}_0\cdot\mathbf{v}\right), \tag{17}$$

$$\mathbf{B}_{2n-1} = \gamma_H^{(n-1)}\left(\mathbf{B}_0 + c^{-2}\mathbf{v}\times\mathbf{E}_0\right) - c^{-2}\gamma_H^{(n-2)}\mathbf{v}\left(\mathbf{B}_0\cdot\mathbf{v}\right), \tag{18}$$

where,

$$\gamma_H^{(k)} = 1 + \frac{v^2}{c^2} + \frac{v^4}{c^4} + \cdots + \frac{v^{2k}}{c^{2k}} = \sum_{j=0}^{k}\frac{v^{2j}}{c^{2j}}. \tag{19}$$

Similarly[9-11],

$$\rho_{2n} = \gamma_H^{(n)}\rho_0, \qquad \rho_{2n-1} = \gamma_H^{(n-1)}\rho_0. \tag{20}$$

Since Eqs. (12) and (13), are linear with respect to the vector fields $\mathbf{E}_n$, $\mathbf{B}_n$ and the scalar ρ_n, their convergence requires the convergence of the sequences $\mathbf{E}_{2n}$, $\mathbf{E}_{2n-1}$, $\mathbf{B}_{2n}$, $\mathbf{B}_{2n-1}$, ρ_n, and ρ_{2n-1}. Only if these sequences have, in the limit as $n\to\infty$, finite nonvanishing values $\mathbf{E}$, $\mathbf{B}$ and ρ,

$$\lim_{n\to\infty}\mathbf{E}_{2n} = \lim_{n\to\infty}\mathbf{E}_{2n-1} = \mathbf{E}, \qquad \lim_{n\to\infty}\mathbf{B}_{2n} = \lim_{n\to\infty}\mathbf{B}_{2n-1} = \mathbf{B}, \qquad \lim_{n\to\infty}\rho_{2n} = \lim_{n\to\infty}\rho_{2n-1} = \rho, \tag{21}$$

do Eqs. (12) and (13) tend to the equations of HRE with respect to the reference $S(r,t)$

$$\mathrm{div}\,\mathbf{E} = \rho/\varepsilon_0, \qquad \mathrm{div}\,\mathbf{B} = 0, \tag{22}$$

$$\mathrm{curl}\,\mathbf{E} = -\partial\mathbf{B}/\partial t, \qquad \mathrm{curl}\,\mathbf{B} = \mu_0\rho\mathbf{v} + c^{-2}\partial\mathbf{E}/\partial t. \tag{23}$$

The convergence of sequences (15 - 18) is conditioned by the value of the series

$$\lim_{j\to\infty}\gamma_H^{(j)} = \lim_{j\to\infty}\sum_{k=0}^{j}\frac{v^{2k}}{c^{2k}}. \tag{24}$$

If and only if the condition,

$$v/c < 1 \tag{25}$$

is fulfield, the series (24) is convergent and equal to

$$\gamma_H = \left(1 - \frac{v^2}{c^2}\right)^{-1}. \tag{26}$$

In terms of γ_H, the vector fields $\mathbf{E}$, $\mathbf{B}$ and the scalar ρ (15 - 18) become

$$\mathbf{E} = \gamma_H\left(\mathbf{E}_0 - \mathbf{v} \times \mathbf{B}_0\right) - v^{-2}(\gamma_H - 1)(\mathbf{E}_0 \cdot \mathbf{v})\mathbf{v}, \tag{27}$$

$$\mathbf{B} = \gamma_H\left(\mathbf{B}_0 + c^{-2}\mathbf{v} \times \mathbf{E}_0\right) - v^{-2}(\gamma_H - 1)(\mathbf{B}_0 \cdot \mathbf{v})\mathbf{v}, \quad \rho = \gamma_H\rho_0. \tag{28}$$

Eqs. (27) and (28) represent the Hertzian relativistic transformations of the electric field strenght and magnetic flux density, from the proper reference frame $S_0(r_0,t_0)$ to the reference frame $S(r,t)$ with respect to which S_0 is moving with the velocity $\mathbf{v}$ of any kind of motion. As a consequence, the nth-order Hertzian iterative system of Eqs. (12) and (13) becomes, in the limit as $n \to \infty$, the system of HRE equations (22) and (23). Denoting by $\mathbf{E}^{(E)}$, $\mathbf{B}^{(E)}$, $\rho^{(E)}$ respectively $\mathbf{E}_0^{(E)}$, $\mathbf{B}_0^{(E)}$, $\rho_0^{(E)}$ the quantities within the framework of ESR, the corresponding transformation from the moving inertial reference frame $S_0(r_0,t_0)$, to the inertial reference frame $S(r,t)$, at relative rest, are given by [14],

$$\mathbf{E}^{(E)} = \gamma_L\left(\mathbf{E}_0^{(E)} - \mathbf{v} \times \mathbf{B}_0^{(E)}\right) - v^{-2}(\gamma_L - 1)(\mathbf{E}^{(E)} \cdot \mathbf{v})\mathbf{v}, \tag{29}$$

$$\mathbf{B}^{(E)} = \gamma_L\left(\mathbf{B}_0^{(E)} + c^{-2}\mathbf{v} \times \mathbf{E}_0^{(E)}\right) - v^{-2}(\gamma_L - 1)(\mathbf{B}_0^{(E)} \cdot \mathbf{v})\mathbf{v}, \quad \rho^{(E)} = \gamma_L\rho_0^{(E)}. \tag{30}$$

Comparison of the HRE Eqs. (27 - 28) with the Eqs. (29 - 30) shows that they have the same form. with the difference that instead of the Lorentz's relativistic factor γ_L,

$$\gamma_L = \left(1 - \frac{v^2}{c^2}\right)^{-1/2}, \tag{31}$$

Hertz's relativistic factor γ_H appears. However, to this formal distinction, correspond important consequences, discussed in the Comments.

3. DERIVATION OF EINSTEIN'S COMPOSITION LAW OF COLINEAR VELOCITIES

Let us consider three reference frames S_0, S_1, S_2, with parallel homologous axes, such that $\mathbf{v}_1$ is the velocity of S_0 with respect to S_1, $\mathbf{v}_2$ colinear with $\mathbf{v}_1$, is the velocity of S_1 with respect to S_2 and $\mathbf{u}$ is the velocity of S_0 with respect to S_2. Writing the Eqs. (27) and (28) for the vector fields $\mathbf{E}_k$ and $\mathbf{B}_k$ with respect to the reference frames S_k, $(k=1,2)$ and identifying the coefficients, we have

$$\gamma_{H1}\gamma_{H2}\left(1 + v_1 v_2 c^{-2}\right) = \gamma_{Hu}, \quad \gamma_{H1}\gamma_{H2}(v_1 + v_1) = \gamma_{Hu}u, \tag{32}$$

$$\gamma_{Hk} = \left(1 + v_k^2 c^{-2}\right)^{-1}, \quad \gamma_{Hu} = \left(1 + u_u^2 c^{-2}\right)^{-1}, \quad k = 1, 2. \tag{33}$$

From the ratio of Eqs. (32), the composed velocity u

$$u = \frac{v_1 + v_2}{1 + v_1 v_2 / c^2} , \qquad (34)$$

identical with Einstein's composition law of colinear velocities is obtained. Since no restriction has been imposed on the motion of S_0 and S_1, the new Hertzian theory yields that Einstein's formula (34) is valid for accelerated rectilinear motions as well.

4. COMMENTS

Before confining our discussion to the consequences of HRE, it is of interest to know, beforehand, how the problem of the extension of MHE at relativistic velocities appears. Since we have at our disposal the relativistic electrodynamics, even if it is restricted to only inertial motions, and on the other side the general theory of relativity solves the problem implying non - inertial motions, naturally the question arises: is it justified the elaboration of a new relativistic theory? For the sake of simplicity, let us consider a gedanken experiment, or an electromagnetic device, involving a non - inertial motion at a velocity v, which is not small enough such that v^{2k}/c^{2k} might be neglected, at least for $k = 2$. In order to circumvent the complexities involved by the general theory of relativity, the temptation to apply the MHE, suitably extended at those velocities, is evident. In fact, in such circumstances, we have to think in terms of two electrodynamics, resting on totally different bases; the Einstein's relativistic theory, valid for all permissible velocities (v<c) but implying only uniform tranlations, and MHE valid for nonrelativistic velocities but for all kinds of motions, as they occur in the physical reality and hold for the non - inertial reference frame of our laboratory. At least, for those small relativistic velocities of any kind of motions, the new theory appears justified.

Now let us return to the aim of this Section. The Hertzian relativistic theory, proposed in this work, is completely independent and built - up in a completely different way, as regards the Einstein's theory. HSR is neither a challenge nor an alternative to ESR. At a first sight many discrepancies appears between the two theories; perhaps the most puzzlling is that related to the replacemnet of the Lorentz's factor γ_L in Eqs. (29) and (30) with Hertz's factor γ_H. The correct interpretation of these apparent conflicts implies the principle of complementarity, involved not only in Quantum Mechanics but also in the theory of relativity. With respect to the behaviour of a quantum object, Selleri[15] writes: "Bohr believed in a kind of double faced physical being which appears to us, in certain circumstances as a particle and in others as a wave; on the contrary, de Broglie considered that there is only one thing which is at the same time a particle and a wave". On the view - point of the principle of conplementarity, and under the assumption of the neglecting the gravitational effects, *Special Relativity is a double faced theory which appears, either as ESR when the motion is inertial, or as the HSR when the motion is noninertial.* Both theories are complementary in the sense that the two descriptions cannot be employed for the same motion, being mutually exclusive. As a consequence, to every statement of one of the SR, a complementary statement of the other SR corresponds. Due to their comnplementarity, there are no disputes between them; on the contrary, by adding the HSR to ESR a complete view over the relativity in our physical world is realized.

ACKNOWLEDGEMENTS

The author wishes to express the debt he owes to Prof. Franco Selleri for his support in publishing this paper. A particular expression of thanks is due to Dr. Alexandru Morega and Doctorate Christian Boghiu for their help in the editing of this work.

REFERENCES

1. A. Einstein "Zur Electrodynamik bewegter Körper," *Ann. der Phys.* 17, 810-921 (1905).
2. L. H. Thomas, "The kinematics of an electron with an axis," *Phil. Mag.* 3, 1-22 (1927).
3. C. I. Mocanu, "The paradox of Thomas rotation," *Galilean Electrodynamics*, 2, 67-74 (1991).
4. C. I. Mocanu, "The relativistic kinematic and electromagnetic problems of Thomas rotation," *Physics Essays*, 4, 238 - 243 (1991).
5. C. I. Mocanu, "On the relativistic velocity composition paradox and the Thomas rotation, *Found. Phys. Lett.* 5, 443 - 456 (1992).
6. C. I. Mocanu, "Kinematic confirmation of Thomas paradox," *Galilean Electrodynamics*, 4, 23 - 28 (1993).
7. C. I. Mocanu, "Is Thomas rotation a paradox?", Apeyron, 16, 1 - 8 (1993).
8. H. R. Hertz, "Über die Grundleichungen der Electrodynamik für bewegten Körper," *Ann. der Phys.* 41, 369 - (1981).
9. C. I. Mocanu, "Hertzian alternative to the special theory of relativity," *Hadronic J.* A series of five papers, **I.** Qualitative analysis of Maxwell's equations for motionless media, 10,   61 - 74 (1987). **II.** Qualitative analysis of Maxwell - Hertz and Einstein - Minkowski equations, 10, 153 - 166 (1987). **III.** Iterative system of equations and transformation formulae for electromagnetic quantities, 10, 231 - 247 (1987).**IV.** Kinematic analysis of the iterative process, 11, 35 - 53 (1988). **V.** Elements of Hertzian mechanics, 11, 55 - 69 (1988).
10. C. I. Mocanu, "Hertzian Relativistic Electrodynamics of Moving Bodies," Publ. House of Roum. Acad. Bucharest - Romania (1985).
11. C. I. Mocanu, "Hertzian Relativistic Electrodynamics and its Associated Mechanics," Hadronic Press, Palm Harbour Fl. (1991).
12. W. G. V. Rosser, "Electromagnetism via Relativity. An Alternative Approach to Maxwell's Equations," Butterworths, London (1968).
13. P. Hammond, "Applied Electromagnetism," Library of Sci. Techn. London (1971).
14. J. D. Jackson, "Classical Electrodynamics," J. Wiley, NY (1975).
15. F. Selleri, "Quantum Paradoxes and Physical Reality," Kluver Acad. Pub. Dordrecht (1989).

FROM RELATIVISTIC PARADOXES TO ABSOLUTE SPACE AND TIME PHYSICS

H.E. Wilhelm

Department of Materials Science and Engineering
The University of Utah, Salt Lake City, Utah   84112, U.S.A.

INTRODUCTION

Contradicting Maxwell, Larmor, Heaviside, Hertz, and others, Voigt (1887), Lorentz (1904), and Einstein (1905) introduced the hypothesis that Maxwell's equations and the electromagnetic (EM) wave equations hold in this form not only in the ether frame $S°(0)$ but in all other inertial frames (IF) $S(\mathbf{w})$ with ether velocities $\mathbf{w} \neq \mathbf{0}$, too.[1] This proposition is physically equivalent to the questionable assumptions: (i) an EM wave carrier or ether (vacuum substratum) does not exist, (j) the velocity of a light signal has the same value c in all IFs, and (k) electrodynamic phenomena are relative to the observer (nonexistence of a preferred or substratum frame $S°$).

The special theory of relativity (STR) and Lorentz (L) covariant physics in general are formalisms *relative to the observer,* which predict that the physical state of any material body depends on its velocity $+\mathbf{v}$ relative to the observer or the velocity $-\mathbf{v}$ of the observer relative to the body.[1] Accordingly, modern physics is no longer unique but many-valued, since $n = 1,2,3, \ldots< \infty$ observers with different velocities $-\mathbf{v}_n$ relative to a body allegedly see this body in n different physical states. If a body's state is being measured simultaneously by n observers with different relative velocities $-\mathbf{v}_n$, the body would have to assume n different states during the common observation period. Most paradoxes, physical impossibilities, and contradictions of the STR and L-covariant theories are caused by the concept of the relativity with regard to the observer.

In Galilei (G) covariant electrodynamics and quantum mechanics,[2-7] the physical state of material bodies is unique, since it depends on the G-invariant body velocity $\mathbf{v} - \mathbf{w} = \mathbf{v}°$ relative to the vacuum substratum, which is the carrier of all elementary force interactions, e.g. EM, gravitational, or nuclear. G-invariance of a field infers its independence of the IF $S(\mathbf{w})$ of observation with substratum velocity $\mathbf{w}$. Measurements are assumed to perturb not significantly the state of a macroscopic body. Even in quantum physics, the indeterminacy phenomenon is not caused by perturbations of the (noninteracting) observer but by fluctuations of the vacuum substratum.[8,9]

Since the vacuum appears to have hardly any weight and can be penetrated by (subluminal) particles and bodies without experiencing friction forces,[5] the substratum probably consists of particles with positive and negative *gravitational* masses condensed to a superfluid state. Experiments indicate that the substratum has physical properties, e.g. a magnetic permeability $\mu_0 \simeq 4\pi\times10^{-7}$ Vs/Am, dielectric permittivity $\varepsilon_0 \simeq 10^{-9}/36\pi$ As/Vm, EM wave speed $c = 1/\sqrt{(\mu_0\varepsilon_0)} \simeq 3\times10^8$ m/s, and EM wave resistance $z_0 =\sqrt{(\mu_0/\varepsilon_0)} \simeq 120\pi$ V/A. The empty STR vacuum (no substratum) can not have physical properties, nor can it transmit EM, gravitational, or other physical waves.

The concepts from Maxwell to Hertz concerning the propagation of EM waves in the "luminiferous ether" of the vacuum[1] have been vindicated through recent discoveries. The experiment of Penzias & Wilson demonstrates that the universe is filled with a uniform and

Frontiers of Fundamental Physics, Edited by M. Barone
and F. Selleri, Plenum Press, New York, 1994

isotropic (at large) 2.7°K microwave background.[10] Measurement of this background EM radiation (excitation of vacuum substratum) permits localization of an absolute reference frame S° everywhere in space.[11] From the measurement of the velocity of the Earth relative to the cosmic frame S° the terrestrial ether velocity is obtained as $\mathbf{w} \sim 3\times10^5$ m/s.[11] Alone these experimental facts refute the STR, which is based, inter alia, on the assumption of the physical equivalence of all inertial frames.[1]

THEORETICAL FOUNDATIONS OF STR

The STR is essentially Einstein's physical interpretation of the L-transformation which is deduced from the experimentally unsupported assumption that Maxwell's equations hold in their ordinary form in all IFs.[10,11] The STR is usually based on the L-transformation or the Minkowski space-time invariance, its *mathematical foundations.*

Subluminal and Superluminal L-Transformations. The wave equation in the 2-dimensional space, $\partial^2\Psi/\partial\tau^2 = \partial^2\Psi/\partial x^2$, where $\tau = ct$, is form-invariant in the transformations $\xi = \xi(x,\tau)$, $\eta = \eta(x,\tau)$, $\Psi(x,\tau) \rightarrow \Psi(\xi,\eta)$, if the transformations of the independent variables satisfy the fundamental equations[12]

$$\partial\xi/\partial x = \pm \, \partial\eta/\partial\tau, \qquad \partial\xi/\partial\tau = \pm \, \partial\eta/\partial x \tag{1}$$

Hence, $\partial^2\xi/\partial x^2 = \partial^2\xi/\partial\tau^2$ and $\partial^2\eta/\partial x^2 = \partial^2\eta/\partial\tau^2$. The partial differential equations (1) have an infinite number of linear and nonlinear solutions (depending on the boundary and initial conditions, or constraints). Hence, a large number of *mathematical* relativity theories could be proposed based on the L-covariance of the wave equation.

Subluminal L-transformation: By (1) with upper signs +, this transformation from the frame $S(x,\tau)$ to the frame $S'(\xi,\eta)$ is for subluminal velocities $|u|/c < 1$:

$$\xi = \gamma(x - \beta\tau), \qquad \eta = \gamma(\tau - \beta x), \qquad \beta = u/c, \qquad \gamma = 1/(1 - \beta^2)^{1/2} \tag{2}$$

Superluminal L-transformation: By (1) with upper signs +, this transformation from the frame $S(x,\tau)$ to the frame $S'(\xi,\eta)$ is for superluminal velocities $|u|/c > 1$:

$$\xi = \gamma^*(\beta x - \tau), \qquad \eta = \gamma^*(\beta\tau - x), \qquad \beta = u/c, \qquad \gamma^* = 1/(\beta^2 - 1)^{1/2} \tag{3}$$

The subluminal and superluminal L-transformations (2) and (3) satisfy the space-time invariance,[1] which determines the space and time scales γ and γ^* in (2) and (3),

$$x^2 - \tau^2 = \xi^2 - \eta^2 = \text{L-inv}, \qquad |u| \lessgtr c \tag{4}$$

$S(x,\tau)$ and $S'(\xi,\eta)$ are IFs by implication since the ordinary wave equation holds in non-accelerated reference systems. S' is assumed to move with velocity $\mathbf{u} = u\hat{x}$ relative to S.

One of the main predictions of the STR is the nonexistence of superluminal velocities. The superluminal L-transformation (3) invalidates, within the relativistic formalism, the STR claim of the nonexistence of superluminal velocities.

The subluminal L-transformation (2) gives a nonlinear velocity addition theorem, which is not commutative (physically untenable), $R(u_1+u_2) \neq R(u_2+u_1)$ for *noncollinear* velocities.[13] Similarly, the superluminal L-transformation (3) gives a nonlinear, noncommutative velocity addition theorem, which is commutative for *collinear* velocities $u_{1,2}$

$$R^*(u_1 + u_2) = (c^2 + u_1 u_2)/(u_1 + u_2) > c, \qquad |u_{1,2}| > c \tag{5}$$

It is obvious that the subluminal and superluminal L-transformations (2) and (3) are nonphysical, mathematical space-time solutions of (1).

Minkowski Space-Time Invariance. If the vacuum were a space without substratum, two light flashes originating at the times $t = 0$ and $t' = 0$ from the sources Q and Q' at the origins O and O' of the IFs $S(r,t)$ and $S'(r',t')$ would be given by $r = ct$ and $r' = ct'$ (STR notation), respectively. Minkowski's space-time invariance goes far beyond these

172

equations by asserting that *one single light source* Q fixed to the origin O of the IF $S(r,t)$
generates not only a light flash about O of S but also light flashes $r' = ct'$ about the
origins O' of an *unlimited* number of other IFs S', if the axes x',y',z' are parallel to those
of x,y,z, and the origins O' of the IFs S' coincide with the origin O of S at $t = t' = 0$:[1]

$$x^2 + y^2 + z^2 - c^2t^2 = x'^2 + y'^2 + z'^2 - c^2t'^2 = \text{L-inv} \tag{6}$$

The discovery of this *interrelation of 'space and time'* seduced Minkowski (1909) to the
prophesy:[1] *"Henceforth space by itself, and time by itself, are doomed to fade away into
mere shadows, and only a kind of union of the two shall preserve an independent reality."*
In reality, these novel space-time concepts are based on a confusion of the *Lagrangian*
coordinates $X(t),Y(t),Z(t)$ and $X'(t'),Y'(t'),Z'(t')$ of the spherical wave fronts in $S(x,y,z,t)$
and $S'(x',y',z',t')$ with the *independent* space and time variables x,y,z,t and x',y',z',t' in the
Eulerian field theory of Maxwell's equations ($\partial r/\partial t = 0$, $\partial t/\partial r = 0$). Hence, space (x,y,z)
and time (t) in S or space (x',y',z') and time (t') in S' are independent.

In Newton's experiment with the falling apple of mass m_0 in the gravitational field $\mathbf{g} =
-g\hat{z}$, the Lagrangian position coordinate $Z(t)$ of the apple is determined by $m_0 d^2 Z(t)/dt^2 =
- m_0 g$. From the solution, $Z(t) = Z(0) - \frac{1}{2}gt^2$ for $dZ(0)/dt = 0$, Newton (1687) concluded
that space (x,y,z) and time (t) are independent.[1]

In summary, existence of EM wave fronts $R(t)$ and spatial positions $Z(t)$ of falling
bodies does not prove that space (x,y,z) and time (t) are interrelated. Hence, the space-
time (6) has no physical meaning and, therefore, no bearing on experiments.

Einsteinian Light Signals in Arbitrary Inertial Frames. Consider a light source Q fixed
to the origin O of the IF $S(r,t)$ which emits a light flash $r = ct$ for $t > 0$. By the space-
time invariance (6), light signals $r' = ct'$ would be observed for $t' > 0$ about the *origins* O'
of an *unlimited* number of other IFs $S'(r',t')$ (coinciding with S for $t = t' = 0$), in spite of
the facts that (i) the origins O' of the IFs S' move away from the light source Q at O of S
for $t' > 0$ and (ii) the IFs S' have *no* light sources Q' attached to their origins O' !

The space-time invariance (6) is not satisfied by the G-transformation $x = x' + ut'$, $y
= y'$, $z = z'$, $t = t'$ between the IFs $S(r,t)$ and $S'(r',t')$, where S' moves with a velocity $\mathbf{u} =
u\hat{x}$ along the x-axis of S. For this reason, consider a modified G-transformation, in which
the length and time scales are "corrected" by the dimensionless factors α,κ,σ and δ, and
time is shifted by a function $T = T(?)$,

$$x = \alpha(x' + ut'), \qquad y = \kappa y', \qquad z = \sigma z', \qquad t = \delta(t' + T) \tag{7}$$

in order to create *mathematically* the *appearance* of a light signal $r' = ct'$ about the origin
O' of the IF S' as S' moves away with velocity $\mathbf{u}$ from the light source Q fixed to the
origin O of the IF S. Substitution of (7) into (6) gives an identity in which the coefficients
of x'^2,y'^2,z'^2 and $-c^2t'^2$ must be 1, and the coefficient of t' must be 0. By these conditions: $\alpha = 1/(1 - u^2/c^2)^{1/2}$, $\kappa = 1$, $\sigma = 1$, $\delta = 1/(1 - u^2/c^2)^{1/2}$, and $T = ux'/c^2$.
Substitution of $\alpha,\kappa,\sigma,\delta$, and $T(x')$ in (7) shows that the (subluminal) L-transformation is
the distorted G-transformation (7), which satisfies the space-time invariance (6),

$$x = (x' + ut')/(1 - \mathbf{u}^2/c^2)^{1/2}, \quad y = y', \quad z = z', \quad t = (t' + ux'/c^2)/(1 - \mathbf{u}^2/c^2)^{1/2} \tag{8}$$

Accordingly, the L-transformation (8) is a falsified G-transformation (7) which fakes a
light flash $r' = ct'$ about the *origin* O' (without attached light source Q') in any IF $S'(r',t')$,
although there is only one real light source Q, which is fixed to the origin of $S(r,t)$.

The physically inconceivable light flash $r' = ct'$ about the *origin* O' of any IF S' is a
hallucination created by the space-dependent time shift $T = ux'/c^2$ and the rescaling of the
space and time measures by $\alpha = \delta = 1/(1 - \mathbf{u}^2/c^2)^{1/2}$. It is *kinematically* and *energeti-
cally impossible* that one single real light flash $r = ct$ about the origin O of an IF S
produces light flashes $r' = ct'$ about the *origins* O' of an *unlimited* number of other IFs S'.
The light flashes $r' = ct'$ about the origins O' of the IFs S' are *mathematical illusions*
without EM momentum and energy in the complex Minkowski space. Hence, the L-
transformation (8) is not a *physical* space-time transformation and the Minkowski invariance (6) does not represent a *measurable* interrelation of space and time.

The space-time invariance (6) contradicts the STR principle of the physical equivalence of the IFs S and S', since S has a real light source Q attached at O whereas no

light source Q' is fixed to S'. The space-time (6) gives the most condensed representation of the *nonphysical* foundations hidden in the STR: *Special relativity is not based on physics or experimental facts but on the 'surreal' assumption that a single light source Q fixed to the origin O of an IF S produces a light flash r = ct not only about the origin O of S but also light flashes r' = ct' about the origins O' of all other conceivable IFs S' initially (t = t' = 0) coincident with S.* Einstein on space-time (1910):[1] *"After reading Minkowski's paper, I do no longer feel that I understand my theory."*

Galilean Light Signals in Arbitrary Inertial Frames. Since EM fields are excitations of the substratum, vacuum light signals always propagate with isotropic velocity $|c^\circ(\theta^\circ)| = c$ in the ether which is at rest in S°, but are *observed* to be propagating with anisotropic velocity $c(\theta)$ in all other IFs $S(w) \neq S^\circ$, in which the substratum streams with a velocity $w \neq 0$. E.g., a light flash originating from a point source at the origin O° of the substratum frame $S^\circ(r^\circ,t^\circ;0)$ expands there with a spherical wave front $R^\circ(t^\circ) = ct^\circ$ where $R^\circ(t^\circ) = \sqrt{[X^\circ(t^\circ)^2 + Y^\circ(t^\circ)^2 + Z^\circ(t^\circ)^2]}$. An observer in an arbitrary IF $S(r,t,w)$ moving with a velocity $u = u\hat{x}^\circ$ relative to S° sees a receding light flash of radius

$$R_c(t) = ct, \qquad R_c(t) = [(X(t) + ut)^2 + Y(t)^2 + Z(t)^2]^{1/2} \tag{9}$$

since $X(t) = X^\circ(t^\circ) - ut^\circ$, $Y(t) = Y^\circ(t^\circ)$, $Z(t) = Z^\circ(t^\circ)$ and $t = t^\circ$ in the G-transformation $S^\circ(0) \to S(w)$. By (9), the light flash is (i) centered in S about the instantaneous position $C(t)$ given by $X(t) = -ut$, $Y(t) = 0$, $Z(t) = 0$, and (ii) convected with the velocity $-u = w$ downstream the ether flow w in S. There is only one spherical light flash $R^\circ(t^\circ) = ct^\circ$ in the substratum (S°), and the observers in all conceivable IFs S moving with velocities $u = -w$ relative to S° *see* this *very same* spherical light flash. Accordingly, there are no light flashes about the *origins* O of all observation frames $S \neq S^\circ$ as postulated by Minkowski's space-time invariance or Einstein's relativity principle.[1]

In the IF $S(w)$, the velocity of the spherical light flash is $c = dR(t)/dt = d\{X(t), Y(t), Z(t)\}/dt = \{c^\circ - u, c^\circ, c^\circ\} = \{c^\circ + w, c^\circ, c^\circ\}$. Hence, $c(\theta) = w + c^\circ(\theta^\circ)$, where θ is the angle of the light velocity with respect to the direction of the ether velocity w and $c^\circ(\theta^\circ) = c$ (isotropic light propagation in S°). Thus, we find for the light speed $c(\theta)$ in the polar direction θ with w in an arbitrary IF $S(r,t,w)$[7]

$$c(\theta) = (c^2 - w^2\sin^2\theta)^{1/2} + |w|\cos\theta, \qquad 0 \leq \theta \leq \pi \tag{10}$$

with $c(0,\pi) = c \pm |w|$ and $c(\pi/2) = \sqrt{(c^2 - w^2)}$. Accordingly, light propagates with anisotropic velocity $c(\theta)$ in all IFs $S(w) \neq S^\circ$. The relativity principle of the constancy "$c(\theta) = c$" of the velocity of a light signal in all IFs S is *physically* untenable.

GALILEI COVARIANT ELECTRODYNAMICS

The generalized, G-covariant Maxwell equations follow from the ordinary Maxwell equations by means of two hypotheses: (i) the Maxwell equations hold in the preferred IF $S^\circ(r^\circ,t^\circ;0)$ in which the vacuum substratum is at rest $(w = 0)$; (ii) the interrelation between the coordinates r°,t° of S° and the coordinates r,t of an arbitrary IF $S(r,t,w)$, which moves with a velocity $u = -w$ relative to S°, is given by the absolute space and time transformation of Galilei: $r = r^\circ - ut^\circ$ and $t = t^\circ$ with $\nabla = \nabla^\circ$ and $\partial/\partial t^\circ = \partial/\partial t - u \cdot \nabla$.[2] The EM field equations obtained by transforming Maxwell's equations from the substratum frame S° to the IF S hold in this form for arbitrary IFs since they are form-invariant in subsequent G-transformations $r,t \to r',t'$ to other IFs $S'(r',t';w')$.[2]

Galilei Covariant EM Field Equations. For isotropic media (dielectric permittivity ε, magnetic permeability μ, electrical conductivity σ) with velocity field $v(r,t)$ in an arbitrary IF $S(r,t,w)$, the generalized, G-covariant Maxwell equations are in MKS notation:[2]

$$\nabla \times (E + w \times B) = -(\partial/\partial t + w \cdot \nabla)B \tag{11}$$

$$\nabla \times H = (\partial/\partial t + w \cdot \nabla)(D + \varepsilon w \times B) + j - \rho w \tag{12}$$

$$\nabla \cdot (D + \varepsilon w \times B) = \rho, \qquad \nabla \cdot B = 0 \tag{13}$$

where

$$D = \varepsilon E , \qquad B = \mu H , \qquad j - \rho v = \sigma[(E + w \times B) + (v - w) \times B] \qquad (14)$$

are the constitutive relations for the EM field E,H and Ohm's law, for isotropic media $(\varepsilon, \mu, \sigma)$. By means of the G-covariant relations for the EM fields $E = -\nabla\Phi - \partial A/\partial t$, $B = \nabla \times A$, and the generalized L-gauge $\nabla \cdot A = -\mu\varepsilon(\partial/\partial t + w \cdot \nabla)(\Phi - w \cdot A)$ for the EM potentials Φ and A, (11)-(13) are reduced to the G-covariant EM wave equations:[2]

$$[\mu\varepsilon(\partial/\partial t + w \cdot \nabla)^2 - \nabla^2]A = \mu(j - \rho w), \quad [\mu\varepsilon(\partial/\partial t + w \cdot \nabla)^2 - \nabla^2](\Phi - w \cdot A) = \rho/\varepsilon \quad (15)$$

The G-covariance of the generalized Maxwell (11)-(13) and EM wave (15) equations follows from the field and operator invariants: $E + w \times B = E' + w' \times B' = E^\circ$, $B = B' = B^\circ$, $j - \rho w = j' - \rho'w' = j^\circ$, $\rho = \rho' = \rho^\circ$, $A = A' = A^\circ$, $\Phi - w \cdot A = \Phi' - w' \cdot A' = \Phi^\circ$, $v - w = v' - w' = v^\circ$, and $\partial/\partial t + w \cdot \nabla = \partial/\partial t' + w' \cdot \nabla' = \partial/\partial t^\circ$, $\nabla = \nabla' = \nabla^\circ$.[2]

Since the densities, temperatures, and particle masses of the medium components are G-invariants, the medium properties $\varepsilon = \varepsilon' = \varepsilon^\circ$, $\mu = \mu' = \mu^\circ$, $\sigma = \sigma' = \sigma^\circ$ are G-invariants, too. All G-invariant field expressions are ether excitations ($^\circ$). The negligible effects of induced polarization in moving dielectrics ($\varepsilon > \varepsilon_0$) are treated elsewhere.[6]

Invariants of Rod Length, Clock Frequency, Particle Mass, Momentum, and Energy. The G-covariant Maxwell and EM wave equations permit calculation of the substratum effects on electrodynamic phenomena.[2-7] Contraction of a moving rod is a real effect caused by the convective squeezing of the electric equipotential surfaces of its electrons and nuclei in the direction of their motion $v - w = v^\circ$ relative to the substratum.[4] Retardation of a moving atomic or light clock, using a light signal reflected forth and back between mirrors held apart by a rod, is a real effect caused by anisotropic light propagation between the mirrors and contraction of the supporting rod in the ether flow w.[7] By the electron-positron annihilation reaction, $m_+ + m_- \leftrightarrow 2\gamma$, the EM energy of the γ quanta equals the energy of the particle masses, $m_\pm c^2 = \hbar\omega$. For this reason, mass, momentum, and kinetic energy of a charged or neutral particle with charge structure are G-invariant functions of its velocity v° relative to the substratum.[7]

The length $\ell(v^\circ)$ of a rod parallel to its absolute velocity $v^\circ = v' - w$, the rate $\upsilon(v^\circ)$ of a clock, the mass $m(v^\circ)$, momentum $p(v^\circ)$, and kinetic energy $K(v^\circ)$ of a particle, moving with a velocity v in an arbitrary IF $S(w)$, are given by:[7]

$$\ell = \ell_0[1 - (v - w)^2/c^2]^{1/2} = \ell_0(1 - v^{\circ 2}/c^2)^{1/2} \qquad (16)$$

$$\upsilon = \upsilon_0[1 - (v - w)^2/c^2]^{1/2} = \upsilon_0(1 - v^{\circ 2}/c^2)^{1/2} \qquad (17)$$

$$m = m_0/[1 - (v - w)^2/c^2]^{1/2} = m_0/(1 - v^{\circ 2}/c^2)^{1/2} \qquad (18)$$

$$p = m(v - w) = mv^\circ \qquad (19)$$

$$K = m_0c^2\{[1 - (v - w)^2/c^2]^{-1/2} - 1\} = m_0c^2[(1 - v^{\circ 2}/c^2)^{-1/2} - 1] \qquad (20)$$

or

$$K = mc^2 - E_0, \qquad E_0 = m_0c^2 \qquad (21)$$

where ℓ_0, υ_0, m_0 are the corresponding values when the rod, clock, or particle is at rest in the substratum frame S°, $v^\circ = 0$, or $v = w$ in the observation frame S. Equations (16)-(21) are G-invariants since they depend on the system velocity $v - w = v^\circ$ relative to the substratum. They reduce for $w = 0$ to the corresponding relativistic formulas, which hold in the ether frame S° where relative and absolute velocities are equal, $v \equiv v^\circ$ in S°.

ABSOLUTE SPACE AND TIME VERSUS RELATIVISTIC SPACE-TIME

Rod Length Contraction. Consider two *identical* rods, rod "1" is attached to the origin O' of an IF S' and rod "2" to the origin O" of an IF S", where S" moves relative to S' with velocity u. S' moves with velocity u' and S" with velocity $u"$ relative to an IF S.

By the STR, the observer in S' measures for (i) his rod the largest ("proper") length $\ell_1' = \ell_0$ and (ii) the rod of S" the shorter length $\ell_2' = \ell_0(1 - u^2/c^2)^{1/2}$. Whereas the observer in S" measures for (i) his rod the largest length $\ell_2'' = \ell_0$ and (ii) the rod of S' the shorter length $\ell_1'' = \ell_0(1 - u^2/c^2)^{1/2}$. In the IF S, the lengths of the two *identical* rods are $\ell_1 = \ell_0(1 - u'^2/c^2)^{1/2}$ and $\ell_2 = \ell_0(1 - u''^2/c^2)^{1/2}$, i.e: (i) $\ell_1 \gtrless \ell_2$ for $|u'| \lessgtr |u''|$ and (ii) $\ell_1 = \ell_2$ if $|u'| = |u''|$. The measurements in S contradict those in S' and S".

Since in S' rod "1" is longer than rod "2", in S" rod "2" must be longer than rod "1", in view of the equivalence (STR) of the IFs S' and S" and symmetry of the system S'-S". However, the identical rods "1" and "2" can *not* be *both* longer and shorter than the other. Hence, the rod lengths relative to the observer have no physical meaning.

It is apparent that the subjective, multivalued relativistic length measurement can not serve as a basis for interpretation of experiments and comparison of data measured in different IFs. The existence of the preferred frame $S°$ refutes the STR assertion that n identical rods attached to different IFs S^n have there the same proper length ℓ_0.[10,11]

By G-variant electrodynamics, the length of a rod $\ell = \ell_0[1 - (v - w)^2/c^2]^{1/2}$ is the same in different IFs of observation since it is independent of the velocity v of the observer, due to sole dependence on its G-invariant velocity $v° = v - w$ relative to the substratum. Since the relative velocity of the two identical rods is $u \neq 0$, they have different absolute velocities $v_{1,2}°$ and, hence, different unique lengths $\ell_{1,2} = \ell_0(1 - v_{1,2}°^2/c^2)^{1/2}$.

Clock Rate Retardation. Consider two *identical* clocks, clock "1" is fixed to the origin O' of an IF S' and clock "2" to the origin O" of an IF S", where S" moves relative to S' with velocity u. By the STR, the observer in S' sees his clock running at the fastest ("proper") rate $\nu_1' = \nu_0$ and the clock of S" at the slower rate $\nu_2' = \nu_0(1 - u^2/c^2)^{1/2}$. Whereas the observer in S" sees his clock running at the fastest rate $\nu_2'' = \nu_0$ and the clock of S' at the slower rate $\nu_1'' = \nu_0(1 - u^2/c^2)^{1/2}$. As each of the (identical) clocks "1" and "2" can *not* run *both* faster and slower than the other, the clock rate retardation relative to the observer can not be a physical effect.

The experiments of Ives–Stilwell indeed demonstrate that there is no relativistic clock rate retardation.[14] Their results with atomic clocks confirm the absolute clock rate $\nu = \nu_0[1 - (v - w)^2/c^2]^{1/2}$, where $v - w = v° = $ G-inv.[14] The increased life-time of mesons confirms absolute clock period dilation $\tau = \tau_0/[1 - (v - w)^2/c^2]^{1/2}$, since the velocity of cosmic mesons $v°$ is not relative to the observer (S) but relative to the substratum ($S°$). The existence of the preferred frame $S°$ refutes the STR concept that an observer attached to the IF of a clock (moving relative to $S°$) measures its proper rate ν_0.[10,11]

Velocity Dependence of Mass. Consider two *identical* particles (same mass constant, charge, spin), particle "1" is fixed to the origin O' of an IF S' and particle "2" to the origin O" of an IF S", where S" moves relative to S' with velocity u. By the STR, in S' the particle "1" has the smallest or "proper" mass $m_1' = m_0$ and particle "2" the larger mass $m_2' = m_0/(1 - u^2/c^2)^{1/2}$. Whereas in S" the particle "2" has the smallest mass $m_2'' = m_0$ and particle "1" has the larger mass $m_1'' = m_0/(1 - u^2/c^2)^{1/2}$. Since each of the (identical) particles "1" and "2" can *not* have *both* a smaller and larger mass than the other, a physical meaning can not be attributed to the relativistic mass.

The untenability of the mass dependence on the velocity v relative to the observer is an empirical fact. In S', the mass of particle "1" is measured by means of an inertia balance as m' kg. The observer in S" measures the same mass m' kg for particle "1" of S' by inertial reflectometry. Similarly, the mass of particle "2" of S" is measured as m'' kg by the observers in S" and S'. The mass m of a particle in a state of uniform motion $v°$ relative to $S°$ is a G-invariant, i.e. observers in all IFs measure the same mass for it.

These empirical facts support the mass formula $m = m_0/[1 - (v - w)^2/c^2]^{1/2}$ of G-covariant electrodynamics, where $v - w = v° = $ G-inv. Accordingly, $m_1' = m_0/(1 - v°'^2/c^2)^{1/2}$ and $m_2'' = m_0/(1 - v°''^2/c^2)^{1/2}$ are the masses of the particles "1" (resting in S') and "2" (resting in S"), no matter in or from which IF these masses are measured ($v°'$ and $v°''$ are the velocities of the particles "1" and "2" relative to $S°$).

Velocity Dependence of Temperature. The impossibility of the "physics relative to the observer" is most apparent in the case of the relativistic temperature of a body moving with a velocity **v** relative to the observer, given by $T = T_0(1 - v^2/c^2)^{1/2}$ according to Planck (1908)[16] or $T = T_0/(1 - v^2/c^2)^{1/2}$ according to Ott (1963).[17] The peaceful co-existence of these contradicting formulas in the literature is remarkable.

Consider a glass of water, fixed to an IF S, which has a temperature $T_0 = 300°K$ in S. An observer moving with a speed $|-v| \sim c$ relative to the water (S) should observe (i) ice at $T << 300 °K$ according to Planck and (ii) a fully ionized hydrogen-oxygen plasma at $T >> 300°K$ according to Ott ! Experiments with a dilute, close to luminal neutron beam ("observer" generating different interaction pictures with matter in different states) traversing a C2 emulsion plate initially at $T = 300 °K$ show recoil proton tracks and O,N,C nuclear reaction stars as typical for a nondeteriorated emulsion at room temperature. No transformation of the emulsion into the frozen or plasma state is observed in the emulsion (S) or neutron (S') frames. Thus, the STR prediction that the temperature of a body varies with its velocity **v** relative to the observer is experimentally refuted.

The experimental independence of the temperature from the body velocity (+**v**) relative to the observer or the observer velocity (-**v**) relative to the body is in agreement with G-covariant statistical mechanics, which gives a G-invariant temperature for a macroscopic body moving with a velocity **v** in an arbitrary IF S(**w**):

$$T = T_0/[1 - (v - w)^2/c^2]^{1/2} = T_0/(1 - v°^2/c^2)^{1/2} \tag{22}$$

Accordingly, the temperature of a body depends on its velocity $v - w = v°$ relative to the substratum frame S°. Inertial observers with different velocities -**v** relative to the body measure the same temperature since T is independent of **v**. As the water glass is fixed to an IF S, its state of motion $v°$ relative to the substratum frame S° does not change. Hence, its temperature is the same, no matter from which IF it is measured.

Simultaneity of Events. Let the events "1" at x_1 and "2" at x_2 be simultaneous in the IF S: $t_1 = t_2$. By the inverse L-transformation of (8), these events are *nonsimultaneous* in all other IFs S' (moving with velocities $u = u\hat{x}$ relative to S): $t_2' - t_1' = -\gamma(x_2 - x_1)u/c^2$. This relativistic prediction is an illusion produced by the nonphysical time-shift function $T(x) = ux/c^2$ and the rescaling factor $\gamma = 1/(1 - u^2/c^2)^{1/2}$.

The relativistic nonsimultaneity is unobservable in experiments, in accord with absolute space-time physics. By the G-invariance of time, two events that are simultaneous in an IF S are simultaneous in all other IFs S', too: $t = t' = $ G-inv, $t_1 = t_2 \Leftrightarrow t_1' = t_2'$.

Causality Conservation. In an IF S, consider an event at (x_1, t_1) which is the cause of an event at (x_2, t_2), i.e. in S: $t_1 < t_2$. By the inverse L-transformation of (8), the time – distance between these events is in all other IFs S' (moving with velocity u along the x-axis of S): $t_2' - t_1' = \gamma(t_2 - t_1)(1 - uV/c^2) < 0$ for $V > (c/u)c$, where $V = (x_2 - x_1)/(t_2 - t_1)$. Hence, for an adequately superluminal action transfer speed V, the event (x_2', t_2') would occur earlier than its cause (x_1', t_1') in any IF S' $\neq$ S. This unrealistic inversion of the sequence of cause and event is again created by the nonphysical time-shift $T(x) = ux/c^2$ and the relativistic rescaling factor $\gamma = 1/(1 - u^2/c^2)^{1/2}$.

Violation of causality has not been observed in experiments, in accord with absolute space-time physics. By the G-invariance of time, a cause (x_1, t_1), producing an event (x_2, t_2) in an IF S, occurs also in all other IFs S' $\neq$ S earlier than its consequence: $t = t' = $ G-inv: $t_2 - t_1 = t_2' - t_1'$, $t_2 > t_1 \Rightarrow t_2' > t_1'$.

Nonrelativity of Motions. The relativity of velocity was conceived through a misinterpretation of EM induction experiments.[1] EM interactions between particles (charges, electric and magnetic dipoles) are not determined by their velocities relative to the observer (**v**) but relative to absolute space $(v - w = v°)$.[4] The absolute nature of velocities appears to be generally valid in physics, e.g. the physical effects discussed in (17)-(21) and (22) depend on the velocity relative to the substratum, $v - w = v°$.

Although we can observe the *appearances* of physical phenomena in an infinite number of IFs, the physics occurs in the preferred frame $S°$ because of the G-invariance of physical laws. E.g., by (19)-(21) the conservation laws for momentum and energy in the elastic collision of two particles with masses $m(v°)$ and $M(V°)$ (rest masses in $S°$: m_0 and M_0), momenta $p(v°)$ and $p(V°)$, and energies $K(v°)$ and $K(V°)$ are for an arbitrary IF of observation $S(r,t,w)$ (* designates velocities after impact in S):

$$m(v°)v° + M(V°)V° = m(v°*)v°* + M(V°*)V°* \qquad (23)$$

$$[m(v°) - m_0]c^2 + [M(V°) - M_0]c^2 = [m(v°*) - m_0]c^2 + [M(V°*) - M_0]c^2 \qquad (24)$$

where

$$v° = v - w, \quad V° = V - w, \quad v°* = v* - w, \quad V°* = V* - w \qquad (25)$$

are G-invariants. For infraluminal particle velocities, (23)-(25) reduce to the classical momentum and energy conservation equations [n = unit vector in direction of $(v-v*)$ or $-(V-V*)$]: $m_0(v - v*) = -M_0(V - V*)$, and $(v + v*) \cdot n = (V + V*) \cdot n$, for $|v°,V°,v°*.V°*| << c$. In this *approximation,* the conservation equations become relative to the observer (continuity of the general, G-covariant dynamics to low-velocity classical dynamics).

Already Lenard (1922) objected to the postulate of the relativity of velocities arguing that in the collision between a train and the ramp of a station (attached to the Earth) the relative velocity between train and station is not relevant.[1] Since the passengers in the train are killed, whereas those in the station remain unharmed, the velocity of the station can not be relative to the train. Hence, the velocity of the train is not relative to the station either. *By (19)-(21), the velocity, momentum, and kinetic energy are relative to absolute space ($S°$), both for the train and the station.*

CONCLUSION

We have shown that the STR and L-covariant theories relative to the observer in general are mathematical formalisms, for which mathematical but no physical or experimental foundations exist. In an other paper, we will demonstrate that G-covariant electrodynamics and absolute space and time physics explain, and are confirmed by, all crucial electrodynamic and optical experiments.

The commitment of the *international* physics establishment to the preservation of the mystified absurdities of sanctified relativity is a betrayal of science, the public interest, and Einstein.[17-18] From the very beginning, Einstein questioned accepted theories, including his own contributions. E.g., in 1949 he anticipated the fate of his relativity theories: *"There is not a single concept, of which I am convinced that it will survive, and I am not sure whether I am on the right way at all."*

REFERENCES

1.   E. Whittaker. "A History of the Theories of Aether and Electricity," Nelson & Sons, London (1951).

2.   H.E. Wilhelm, *Fusion Techn.* **6**:174 (1984); *Radio Sci.* **20**:1006 (1985).

3.   H.E. Wilhelm, *Int. J. Math. & Math. Sci.* **8**:589 (1985).

4.   H.E. Wilhelm, *Z. Naturforsch.* **45a**:736 (1990); *Z. Naturforsch.* **45a**:749 (1990).

5.   H.E. Wilhelm, *Int. J. Math. & Math. Sci.* **14**:769 (1991); *Gal. Electrodyn.* **3**:28 (1992).

6.   H.E. Wilhelm, *Apeiron* **15**:1 (1993); *Apeiron* **13**:17 (1992).

7.   H.E. Wilhelm, *Hadronic J.,Suppl.* **8**:441 (1993). F.Selleri. "Fundamental Questions in Quantum Physics and Relativity," Hadronic Press, Palm Harbor (1994). L. Janossy, *Acta Phys.Hung.* **17**:421 (1964).

8.   F. Selleri. "Wave Particle Duality," Plenum Press, London (1992); "The Double Nature of Subatomic Objects," Hadronic Press, Palm Harbor (1993).

9.   H.E. Wilhelm, *Phys. Rev.* **D1**:2278 (1970); *Progr. Theor. Phys.* **43**:861 (1970).

10.   A.A. Penzias and R.W. Wilson, *Astrophys. J.* **142**:419 (1965).

11.   E.K. Conklin, *Nature* **222**:971 (1969). R.B. Partridge, *Rep. Prog. Phys.* **51**:647 (1988).

12.   H.E. Wilhelm, *J. Appl. Phys.* **64**:1652 (1988); *Arch. Elektrotech.* **72**:165 (1989).

13.   C.C. Mocanu, *Arch. Elektrotech.* **69**:97 (1986); *Phys. Essays* **4**:238 (1991); *Apeiron* **16**:1 (1993).

14. H.E. Ives and G.R. Stilwell, *J. Opt. Soc. Am.* **28**:215 (1938); *J. Opt. Soc. Am.* **31**:369 (1941).

15. M. Planck, *Ann. Physik* **26**:1 (1908).

16. H. Ott, *Z. Physik* **175**:70 (1963).

17. R.M. Santilli. "Ethical Probe of Einstein's Followers in the USA.", Alpha Publ.,Tarpon Springs (1984)

18. G.F. Weiss, *Inediti* (Soc. Ed. Andromeda) **51**:1 (1992).

THEORIES EQUIVALENT TO SPECIAL RELATIVITY

Franco Selleri

Dipartimento di Fisica
Università di Bari
INFN - Sezione di Bari
Via Amendola 173
I-70126 Bari

INTRODUCTION

The purpose of this paper is twofold: (1) to discuss the basis of the Lorentz transformations showing that the invariance of the velocity of light has in them a role even more important than usually believed, and (2) to find the complete set of theories empirically equivalent to the special theory of relativity (STR) under the assumption that the one-way velocity of light is not measurable.

In particular it will be shown that any modification of the coefficients of the Lorentz transformations, however small, gives rise to an ether theory, in the sense that the modified theory necessarily predicts the existence of a privileged frame that in principle can be detected experimentally. Therefore all the theories equivalent to STR but based on different transformation laws, must necessarily negate the validity of the relativity principle. We will come thus to the surprising conclusion that <u>if the one-way velocity of light is not measurable</u>, the content of the relativity principle is entirely conventional, since it can be affirmed or negated without any practical change in the predictive power of the theory.

SPACE-TIME TRANSFORMATIONS

The task of the present section is to study once more an old problem, how to set up the most general transformation laws of Cartesian coordinates and of time between two different inertial systems S and S'. We will show that these transformations must be linear if the following conditions hold:

"Empty" space is homogeneous, that is, it has the same properties in all points. Also time is homogeneous, that is, all properties of space remain the same with passing time.

We suppose that there is at least one inertial reference frame in which Maxwell's equations are valid, and that it coincides with S ("stationary system"). With respect to S the velocity of light is c in all directions, a well known consequence of Maxwell's equations. Therefore <u>in S</u> clocks can be synchronized by using Einstein's procedure.

Let us consider the most general form of space-time transformations:

$$x' = f(x, y, z, t)$$
$$y' = g(x, y, z, t)$$
$$z' = h(x, y, z, t) \tag{1}$$
$$t' = e(x, y, z, t)$$

where f, g, h, e are four functions of the space-time coordinates of the system S. Particularly interesting is the function e giving the time t' of S'. There is of course a considerable arbitrariness in the operative definition of simultaneity for two clocks placed in different points of a moving inertial system [1], and one can therefore define a "time" which is not the same in the two inertial systems S(x, y, z, t) and S'(x', y', z', t') considered in (1), and such that the "delay" t' - t (which a priori can be positive, zero, or negative) depends not only on the velocity of S' with respect to S, but also on the considered geometrical point. In other words a clock W' of S' can be retarded with respect to the clock W of S passing near it by a quantity depending not only on t, but also on the coordinates x', y', z' of the point where W' is placed. One can write therefore:

$$t' = \varepsilon(x', y', z', t) \tag{2}$$

where ε can be called "synchronisation function" and informs about how t' depends on t and on the point of S' where the clock W' is placed. Given a function of several variables it is in general enough that some of them are changed in order to obtain a different value of the function. Therefore, if we consider two events simultaneous in S (same t), but taking place in different points of space, Eq. (2) implies that they will in general not be simultaneous in t'. This is the <u>relativity of simultaneity,</u> of course. There is a considerable arbitrariness in the choice of ε which is largely conventional since it depends on the procedures used for synchronising clocks in S' [2]. The function ε is however not <u>totally</u> conventional, since its dependence on t gives rise to well known phenomena, e.g. to the positive result found by Hafele-Keating [3]. In any case ε is equivalent to e, since if one substitutes in (2) the first three Eq.s (1) one obtains a function depending on x, y, z, t, just like e. The simplest synchronisation in S' is of course the one implying no dependence of time t' on the space variables [4], but such a choice in not compatible with the relativity principle. Nevertheless it retains a great physical interest, as we will see.

From space-time homogeneity it follows that the variation of position generated by the addition in S of a rod of length Δx parallel to the x axis has the same effect in S' on the co-ordinates x', y', z' and on the time t', whichever be the point x, y, z and the time t where the rod is added. The functions (1) are very general and a priori it is of course not necessary that the considered Cartesian coordinates of S and of S' be orthogonal. One can write:

$$x' + \Delta x' = f(x + \Delta x, y, z, t)$$
$$y' + \Delta y' = g(x + \Delta x, y, z, t) \tag{3}$$

$$z' + \Delta z' = h(x + \Delta x, y, z, t)$$
$$t' + \Delta t' = e(x + \Delta x, y, z, t)$$

There is naturally an effect of Δx also on t' since we saw that by changing the position at fixed t in general the time t' changes. In agreement with the previous considerations we assume that all the variations of the primed variables in the left-hand sides of (3) are independent of x, y, z, t but depend only on Δx. By subtracting (1) from the previous relations one has:

$$\Delta x' = f(x + \Delta x, y, z, t) - f(x, y, z, t) \tag{4}$$

and so on. The left-hand side, and therefore also the right-hand side, must be independent of x, y, z, t. This will remain true if everything is divided by Δx and the limit $\Delta x \to 0$ is considered. It follows that

$$a_1 \equiv \frac{\partial f}{\partial x} \quad ; \quad b_1 \equiv \frac{\partial g}{\partial x} \quad ; \quad c_1 \equiv \frac{\partial h}{\partial x} \quad ; \quad d_1 \equiv \frac{\partial e}{\partial x}$$

will be independent of x, y, z, t, and therefore constant.

We can make analogous considerations with rods of length Δy and Δz, respectively parallel to the axes y and z, and with a time interval Δt. In the latter case we consider a fixed point x, y, z of S and <u>in it</u> an increase Δt of time t, and assume that the latter gives rise to variations of t' and of x', y', z' which are independent both of x, y, z and of the time t at which Δt is assumed to start. From these new conditions one obtains that also

$$a_2 \equiv \frac{\partial f}{\partial y} \quad ; \quad a_3 \equiv \frac{\partial f}{\partial z} \quad ; \quad a_4 \equiv \frac{\partial f}{\partial t}$$

must be constant (that is, independent of x, y, z, t). Analogous constants can obviously be found from the functions g, h, e, but we do not write them all down. As far as the function f is concerned one obtains by integration:

$$\frac{\partial f}{\partial x} = a_1 \quad \Rightarrow \quad x' = a_1 x + \alpha_1(yzt)$$

$$\frac{\partial f}{\partial y} = a_2 \quad \Rightarrow \quad x' = a_2 y + \alpha_2(xzt)$$

$$\frac{\partial f}{\partial z} = a_3 \quad \Rightarrow \quad x' = a_3 z + \alpha_3(xyt)$$

$$\frac{\partial f}{\partial t} = a_4 \quad \Rightarrow \quad x' = a_4 t + \alpha_4(xyz)$$

where the first result is obtained by integrating over x, the second one over y, and so on. The functions $\alpha_1, \alpha_2, \alpha_3,$ and α_4 are integration "constants": that is, constant with respect to the integration variable, but of course dependent on the other three variables. The previous four relations must be simultaneously valid, but <u>can be compatible with one another only if x' is linear in all the space-time variables of S</u>. Analogous conclusions can obviously be obtained for y', z', t' and one can write:

$$
\begin{aligned}
x' &= a_1 x + a_2 y + a_3 z + a_4 t + a_5 \\
y' &= b_1 x + b_2 y + b_3 z + b_4 t + b_5 \\
z' &= c_1 x + c_2 y + c_3 z + c_4 t + c_5 \\
t' &= d_1 x + d_2 y + d_3 z + d_4 t + d_5
\end{aligned}
\tag{5}
$$

The transformation laws (5) will be our starting point in the coming sections and will be used , in particular, for evaluating the velocity of light in theories more general than the STR.

SIMPLIFIED CHOICE OF AXES

We have so obtained a very general set of transformations containing twenty coefficients, which are constant with respect to x, y, z, t, but can a priori depend on the particular system S' considered, and especially on its velocity v relative to the stationary system S. It is of course possible to reduce strongly the number of free coefficients by considering that Cartesian coordinates are perfectly arbitrary, and thus to be chosen on the basis of convenience criteria. In particular one can choose the axes in S and in S' in such a way that the straight line joining their origins be parallel to their relative velocity.   This implies that there is a time at which the two origins coincide, and this is assumed to happen at time zero both in S and in S'. In other words we write:

$$[x = y = z = t = 0] \quad \Rightarrow \quad [x' = y' = z' = t' = 0]$$

where "$\Rightarrow$" is the symbol of implication. This condition used in (5) gives:

$$a_5 = b_5 = c_5 = d_5 = 0 \tag{6}$$

Let us next assume that the plane (x, y) coincides with the plane (x', y') for all times t. One must then have for all x, y, t:

$$[z = 0] \quad \Rightarrow \quad [z' = 0]$$

From the third Eq. (5) one gets immediately:

$$c_1 = c_2 = c_4 = 0 \tag{7}$$

Now we assume that also the plane (x, z) coincides with the plane (x', z') at all times t. One must then have for all x, z, t:

$$[y = 0] \quad \Rightarrow \quad [y' = 0]$$

From the second Eq. (5) one obtains:

$$b_1 = b_3 = b_4 = 0 \tag{8}$$

We saw that our Cartesian coordinates are not necessarily orthogonal: therefore the coincidence of two planes of coordinates has in general no implication for the third plane. We assume however that at time $t = 0$ also the plane (y, z) coincides with the plane (y', z'): this is like saying that the two systems of coordinates overlap exactly at time zero. Therefore:

$$[t = x = 0] \quad \Rightarrow \quad [x' = 0]$$

From the first Eq. (5) one has:

$$a_2 = a_3 = 0 \tag{9}$$

We consider now the condition arising from velocity: let the origin of S' (equation $x' = 0$) be seen from S to move with velocity v parallel to the x axis, that is with equation $x = vt$. In other words:

$$[x = vt] \quad \Rightarrow \quad [x' = 0]$$

From the first Eq. (5) one has:

$$a_4 = -a_1 v \tag{10}$$

We can now rewrite the transformation laws (5) by keeping into account (6), (7), (8), (9), and (10). They become:

$$
\begin{aligned}
x' &= a_1(x - v\,t) \\
y' &= b_2\,y \\
z' &= c_3\,z \\
t' &= d_1\,x + d_2\,y + d_3\,z + d_4\,t
\end{aligned}
\tag{11}
$$

These are the most general transformation laws for two systems of Cartesian coordinates, in general non orthogonal, perfectly overlapping at times $t = t' = 0$ and with the velocity v of S' parallel to the axes x and x'.

If we specify <u>at this point</u> that the Cartesian coordinates are orthogonal, we can also assume a complete equivalence of the axes y and z. In fact no physical phenomenon can distinguish them, if space is isotropic. Therefore:

$$b_2 = c_3 \quad ; \quad d_2 = d_3 \tag{12}$$

It is now necessary to invert the system (11), of course after keeping into account (12). It is a simple matter to obtain:

$$x = \frac{1}{a_1}\,x' + v\,Q$$

$$y = \frac{1}{b_2}\,y' \tag{13}$$

$$z = \frac{1}{b_2} z'$$

$$t = Q$$

where

$$Q = \frac{1}{R} \left[t' - \frac{d_1}{a_1} x' - \frac{d_2}{b_2} (y' + z') \right] \tag{14}$$

with

$$R = d_4 + d_1 v \tag{15}$$

These results are not valid if $a_1, b_2, R = 0$, cases devoid of any physical interest, since they respectively imply that x' vanishes for all x, that y' vanishes for all y, and that t' does not depend on t in the x'y'z' space, as one can see from (11). Eq.s (11)-(15) will allow us to calculate the velocity of light in S'. The same velocity in S is c by hypothesis. By assuming that c holds also in S' it is of course possible to move towards the Lorentz transformations. The generality of our results will make it easy to study the empirical consequences of theories equivalent to SRT, but not based on the relativity principle.

THE ONE-WAY VELOCITY OF LIGHT

In the inertial system S consider two points $P_1(x_1, y_1, z_1)$ and $P_2(x_2, y_2, z_2)$ and suppose that a light signal leaves P_1 at time t_1 and arrives in P_2 at time t_2. Since in S Maxwell's equations have an unlimited validity one must have:

$$c^2 \Delta t^2 = \Delta x^2 + \Delta y^2 + \Delta z^2 \tag{16}$$

where

$$\Delta x = x_2 - x_1 \; ; \; \Delta y = y_2 - y_1 \; ; \; \Delta z = z_2 - z_1 \; ; \; \Delta t = t_2 - t_1$$

The events "<u>departure of the light signal from P_1 at time t_1</u>" and "<u>arrival of the light signal in P_2 at time t_2</u>" will be described in S' respectively as "<u>departure of the light signal from P_1' at time t_1'</u> " and "<u>arrival of the light signal in P_2' at time t_2'</u> ", where primed and unprimed variables are related via (13) as follows

$$x_1 = \frac{1}{a_1} x_1' + v Q_1 \qquad ; \qquad x_2 = \frac{1}{a_1} x_2' + v Q_2$$

$$y_1 = \frac{1}{b_2} y_1' \qquad ; \qquad y_2 = \frac{1}{b_2} y_2'$$

$$z_1 = \frac{1}{b_2} z_1' \qquad ; \qquad z_2 = \frac{1}{b_2} z_2'$$

$$t_1 = Q_1 \qquad ; \qquad t_2 = Q_2$$

Here Q_1 and Q_2 have expressions similar to (14) with the space-time variables bearing the index "1" and "2", respectively. By subtracting from one another the relations of every line one gets equations strictly similar to (13), but written in terms of space-time intervals:

$$\Delta x = \frac{1}{a_1} \Delta x' + v \, \Delta Q$$

$$\Delta y = \frac{1}{b_2} \Delta y'$$

$$\Delta z = \frac{1}{b_2} \Delta z' \tag{17}$$

$$\Delta t = \Delta Q$$

where

$$\Delta x' = x_2' - x_1' \; ; \quad \Delta y' = y_2' - y_1' \; ; \quad \Delta z' = z_2' - z_1' \; ; \quad \Delta t' = t_2' - t_1'$$

and

$$\Delta Q = \frac{1}{R} \left[\Delta t' - \frac{d_1}{a_1} \Delta x' - \frac{d_2}{b_2} (\Delta y' + \Delta z') \right] \tag{18}$$

with R given by (15) as before. From (16) it follows:

$$c^2 \Delta Q^2 = \left[\frac{1}{a_1} \Delta x' + v \, \Delta Q \right]^2 + \left(\frac{1}{b_2} \right)^2 (\Delta y'^2 + \Delta z'^2) \tag{19}$$

which can be solved as a second degree equation in ΔQ and gives:

$$\Delta Q = \frac{\beta \Delta x'}{c a_1 (1 - \beta^2)} \pm \frac{1}{c} \left[\frac{\Delta x'^2}{a_1^2 (1 - \beta^2)^2} + \frac{\Delta y'^2 + \Delta z'^2}{b_2^2 (1 - \beta^2)} \right]^{1/2} \tag{20}$$

where, as usual, $\beta = v/c$. By inserting (18) in (20) one can obtain $\Delta t'$, the time of propagation in S' of the light signal from $P_1'(x_1', y_1', z_1')$ to $P_2(x_2', y_2', z_2')$. The result is

$$\Delta t' = \frac{d_1}{a_1} \Delta x' + \frac{d_2}{b_2} (\Delta y' + \Delta z') + \frac{R \beta \Delta x'}{c a_1 (1 - \beta^2)} \pm \frac{R}{c} \left[\frac{\Delta x'^2}{a_1^2 (1 - \beta^2)^2} + \frac{\Delta y'^2 + \Delta z'^2}{b_2^2 (1 - \beta^2)} \right]^{1/2} \tag{21}$$

Of course $\Delta t'$ is positive also in the case $\Delta x' = 0$ and $\Delta y' + \Delta z' = 0$ (but with $\Delta y'^2 + \Delta z'^2 \neq 0$). Therefore the minus sign in (21) must be discarded. Consider now in S' a suitable system of polar coordinates with centre on the straight line joining P_1' and P_2', which are of course <u>fixed points</u> of S':

$$\Delta x' = \Delta r' \cos\theta' \; ; \quad \Delta y' = \Delta r' \sin\theta' \cos\varphi' \; ; \quad \Delta z' = \Delta r' \sin\theta' \sin\varphi' \tag{22}$$

By inserting (22) in (21) one obtains:

$$\frac{\Delta t'}{\Delta r'} = \left[\frac{d_1}{a_1} + \frac{R\beta}{ca_1(1-\beta^2)} \right] \cos\theta' + \frac{d_2}{b_2} \sin\theta'(\sin\varphi' + \cos\varphi') +$$

$$+ \frac{R}{c} \left[\frac{\cos^2\theta'}{a_1^2(1-\beta^2)^2} + \frac{\sin^2\theta'}{b_2^2(1-\beta^2)} \right]^{1/2} \tag{23}$$

This ratio between time and distance represents of course the inverse velocity of the light signal with respect to S'. Its dependence on θ' and φ' implies that in general it will not equal c^{-1}. The task of the following section is to study the case in which (23) is not only independent of the angles, but also exactly equal to the known value of the inverse velocity of light. These requirements are of course consequences of the relativity principle. In the last section we will however also consider points of view not compatible with relativity.

INVARIANCE OF THE VELOCITY OF LIGHT

The condition of isotropy of the velocity of light in S' can easily be seen from (23) to be equivalent to the following three requirements:

1) <u>Independence of the azimuthal angle φ'</u>:

$$d_2 = 0 \tag{24}$$

2) <u>Disappearance of the term linear in $\cos\theta'$</u>:

$$d_1 = -\frac{\beta}{c} d_4 \tag{25}$$

3) <u>Disappearance of the residual dependence on θ'</u>:

$$b_2 = a_1\sqrt{1-\beta^2} \tag{26}$$

The three previous conditions are clearly sufficient for the isotropy of the velocity of light in S', but they are also necessary as can easily be shown by requiring that the derivatives of (23) vanish for all angles. From (24)-(26) one easily obtains:

$$\frac{\Delta t'}{\Delta r'} = \frac{R}{ca_1(1-\beta^2)} \tag{27}$$

Eq. (27) implies that the velocity of light is isotropic, but it does not yet establish that it equals c. This can be imposed as a new condition and one then obtains:

$$d_4 = a_1 \tag{28}$$

From (25) and (28) one also gets d_1 in terms of a_1. <u>One should notice the great theoretical power of Einstein's condition on the velocity of light, that has given four relations for the five coefficients</u> entering in (11) [after taking into account

(12)]. The most general transformation laws between inertial systems leaving the velocity of light isotropic and equal to c are thus:

$$x' = a_1 (x - \beta ct)$$

$$y' = a_1 \sqrt{1 - \beta^2}\ y$$

$$z' = a_1 \sqrt{1 - \beta^2}\ z$$

$$t' = a_1 (t - \beta x/c)$$

(29)

where only one undetermined coefficient is left, namely a_1. Nothing more can be said if only the invariance of the velocity of light is assumed, but a_1 can be fixed by using the relativity principle in a different way.

LAST CONSEQUENCE OF RELATIVITY

The simplest way to determine a_1 is to observe that by inverting the system (29) one has

$$y = \frac{1}{a_1 \sqrt{1 - \beta^2}}\ y'$$

The only difference between the previous relation and the second of (29) is in the factor multiplying the space variable, which is here inverted. Such a coefficient gives the "contraction" of a rod put on the y axis of S and seen from S', while its inverse gives the "contraction" of the same rod put on the y' axis of S' and seen from S. Obviously, if in one case there is a real contraction (i.e., if the coefficient is less than unity), in the other case there is an expansion, and vice versa. The principle of relativity requires however that S and S' should observe the same effect, and the only possibility to achieve this is clearly to require that the said coefficient has value unity, i.e. that

$$a_1 = \frac{1}{\sqrt{1 - \beta^2}}$$

(30)

Obviously (29) and (30) together are equivalent to the Lorentz transformations. Initially it had been assumed that the velocity of light was c only in S, but later this requirement has been extended also to S' and a complete symmetry between the two systems has been introduced. Given the arbitrariness of S and S' we can conclude that the Lorentz transformations hold for any pair of inertial systems. The particular system S from which our reasoning started was called "stationary", but <u>in the relativistic line of thought</u> it loses at the end every peculiarity and becomes any one of the infinitely many equivalent inertial systems that can be conceived.

A different form of the transformation of time can be obtained by substituting the first Eq. (29) into the fourth one, and by using (30). One gets so:

$$t' = \sqrt{1 - \beta^2}\ t - \frac{\beta}{c} x'$$

(31)

from which one sees that the "delay" of the time t' of S' with respect to the time t of S has a double origin since it arises both from the factor multiplying t and from the presence of the term proportional to x'. In order to obtain the velocity c for the light signals in all inertial frames a very particular clock synchronisation was needed, which generated the precise dependence (31) of time on space. Recent research has led to the conclusion that such a synchronization is basically conventional and does not necessarily reflect an objective property of physical reality[5]. Einstein himself was aware of this aspect of clock synchronization[6].

THE TWO-WAY VELOCITY OF LIGHT

It is important to stress that Eq.s (24)-(30) are all necessary consequences of the relativity principle. Their eventual violation implies thus that relativity itself does not hold as a description of nature. It can therefore be said that the STR is "unstable", in the sense that any shift, however small, of any one of the five coefficients a_1, b_2, d_4, d_1, and d_2 away from their relativistic values [given respectively by (30), (26), (28), (25), and (24)] implies necessarily the existence of a privileged system[7]. In other words, either Lorentz has given mankind a final truth with his transformations, or some kind of ether shall have to be accepted in the future[8]. After all Einstein modified his negative opinion about ether and after 1916 reverted to acceptance of this conception[9], and today there are even proposals of detecting it with suitable new experiments[10].

The problem is not only that experiments can never check a mathematical expression with infinite precision, so that it is always possible to conceive small deviations of the coefficients from their relativistic values; the problem is also that arguments have recently been advanced[11] in favour of the thesis that the one-way velocity of light is measurable neither directly nor indirectly, with the consequence that d_1 and d_2 are essentially arbitrary and only dependent on the synchronization procedure chosen in the moving reference frame.

In fact we have seen that the left-hand side of (23) is the inverse <u>one-way</u> velocity of light in the direction specified by θ' and φ'. In practically all the performed measurements only the two-way velocity of light has been measured in a system S' instantaneously at rest with respect to the Earth, for example by sending a light pulse from a point P_1' to a point P_2' in which a mirror was placed that reflected the light back to P_1'. In such experiments the velocity of light was measured as the ratio between twice the distance $\Delta r'$ from P_1' to P_2' and the time necessary for the $P_1'P_2'P_1'$ round trip. Introducing the symbols F and B to specify the forward (from P_1' to P_2') and backward (from P_2' to P_1') trip, respectively, one must have for the time intervals:

$$\Delta t'_{FB} = \Delta t'_F + \Delta t'_B$$

which can also be written in terms of velocities:

$$\frac{2\Delta r'}{c'_{FB}(\theta')} = \frac{\Delta r'}{c'_F(\theta',\varphi')} + \frac{\Delta r'}{c'_B(\theta',\varphi')} \tag{32}$$

where c' indicates the velocity of light in S'. By eliminating $\Delta r'$ from (32) we see that <u>the inverse two-way velocity of light is the average between the inverse forward and backward one-way velocities.</u> Given the obvious fact that

$$c'_B(\theta',\varphi') \;=\; c'_F(\pi-\theta',\varphi'+\pi)$$

one has from (23)

$$\frac{1}{c'_{FB}(\theta')} \;=\; \frac{R}{c} \left[\frac{\cos^2\theta'}{a_1^2(1-\beta^2)^2} + \frac{\sin^2\theta'}{b_2^2(1-\beta^2)} \right]^{1/2} \tag{33}$$

which justifies our notation $c'_{FB}(\theta')$, since all dependence on φ' has disappeared. As one can see the terms proportional to d_1 and d_2 are indeed absent in (33).

If we now impose only the condition that in S' the two-way velocity of light equals c, from (33) we get the results:

$$b_2 \;=\; a_1 \sqrt{1-\beta^2} \;\;;\;\;\;\; R \;=\; a_1 (1-\beta^2)$$

Because of (15) the last equation is equivalent to:

$$d_4 \;=\; a_1 (1-\beta^2) - d_1 v$$

Therefore the most general transformation laws giving a two-way velocity of light equal to c in all inertial frames are

$$x' \;=\; a_1 (x - \beta ct)$$

$$y' \;=\; a_1 \sqrt{1-\beta^2} \; y$$

$$z' \;=\; a_1 \sqrt{1-\beta^2} \; z \tag{34}$$

$$t' \;=\; a_1 (1-\beta^2) t + d_1(x - \beta ct) + d_2 (y + z)$$

These transformations represent a good approximation to reality, since the two-way velocity of light is known with a precision better than 10^{-11}. It is possible to introduce in (34) some further information, by considering that time dilation is also a well established phenomenon[12], even though the numerical precision is not as good as that concerning the two-way velocity of light. A clock at rest in the origin of S' satisfies $y = z = 0$ and $x = \beta ct$ and the last Eq. (34) gives the well known time-dilation effect only if a_1 satisfies (30). This can be inserted in (34) and the first three equations become identical with the corresponding ones of the Lorentz transformations, but the last one becomes:

$$t' \;=\; \sqrt{1-\beta^2} \; t + d_1(x - \beta ct) + d_2 (y + z)$$

At this point length contraction by the usual factor $\sqrt{1-\beta^2}$ is a consequence of the transformations (34). As is well known this phenomenon has never been directly verified[13], in spite of opposite claims[14]. Length contraction has been shown to find a natural explanation in the frame of prerelativistic physics, due to deformation of the electromagnetic field of moving charges[15]. There is of course a considerable freedom left in the transformations (34), but nevertheless not enough to accept easily the idea that the Galilei transformations should be preferred[16]. This can still be done only by invoking the strange idea that Earth is at rest in ether. In all cases (34) represent the complete set of theories equivalent to special relativity: if d_1 and d_2 are varied, different theories are obtained which are all equivalent to STR as far as the explanation of experimental results is concerned. In all cases but that of STR such theories negate the relativity principle which becomes thus a disposable convention. In the case $d_1 = d_2 = 0$, corresponding to the so-called absolute synchronization[1], the Tangherlini transformations[4] are obtained, which are of course incompatible with the relativity principle, but are nevertheless particularly simple and elegant.

If the one-way velocity of light should turn out to be measurable, contrary to expectations, the previous results would anyway imply that today there is still a large set of theories logically possible, because compatible with empirical evidence, and that only future experiments will choose the right one.

REFERENCES

1. R. Mansouri and R.U. Sexl, *Gen. Relativity Grav.* , **7**: 497 (1977); *ibid.,* **8**:515 (1977); *ibid.,* **8**:809 (1977).
2. R.A. Nelson, et al., Experimental comparison of time synchronization techniques by means of light signals and clock transport on rotating earth, *in*: Proceedings of 24th Precise Time and Time Interval Meeting (1993).
3. J. Hafele and R. Keating, *Science* , **177**:166 (1972).
4. F.R. Tangherlini, *Nuovo Cim. Suppl.* , **20**:351 (1961).
5. G. Cavalleri and C. Bernasconi, *Nuovo Cim.* , **104B**:545 (1989).
6. A. Einstein, "Über die spezielle und die allgemeine Relativitätstheorie", Vieweg, Braunschweig (1988).
7. F. Selleri, *Z. Naturforsch.* **46a**:419 (1990).
8. T.E. Phipps Jr.,*Physics Essays*, **1**:150 (1988); G. Spavieri,*Found. Phys. Lett.* , **1**, 373 (1988); F. Winterberg, *Z. Naturforsch.* , **41a**:1261 (1986); J.P. Wesley, "Selected topics in fundamental physics", B.Wesley, Blumberg (1991).
9. M. Sachs, *Ann. Fond. Louis de Broglie,* **4**:85 (1979);
    L. Kostro, *Studies Hist. Gen Rel.* , **3**:260 (1992).
10. E.W. Silvertooth, *Nature* , **322**:590 (1986); *Spec. Science Techn.* , **10**:3 (1986); H. Aspden, *Spec. Science Techn.* , **10**:9 (1986).
11. T. Sjödin and M.F. Podlaha, *Lett. Nuovo Cim.* , **31**:433 (1981).
12. J. Bailey, et al.,*Nature* , **268**:301 (1977).
13. T.E. Phipps, *Apeiron* , **14**:5 (1992).
14. D. Hils and J.L. Hall, *Phys. Rev. Letters* , **64**:1697 (1990).
15. S.J. Prokhovnik, "Light in Einstein's Universe", Reidel, Dordrecht (1985); J.S. Bell, How to teach Special Relativity, in "Speakable and Unspeakable in Quantum Theory", Cambridge Univ. Press, (1988).
16. H.E. Wilhelm, *Z. Naturforsch.* , **45a**:736 (1989); *Apeiron* , **15**:1 (1993); C.I. Mocanu, *Hadronic Journal* , **10**:61, 153, 231 (1987); *ibid* ., **11**:35 (1988); U. Bartocci, private communication.

THE PHYSICAL MEANING OF ALBERT EINSTEIN'S RELATIVISTIC ETHER CONCEPT

Ludwik Kostro

Institute of Experimental
University of Gdańsk
Wita Stwosza 57
80-953 Gdańsk, Poland

INTRODUCTION

In 1905, as is well known A. Einstein began, to deny the existence of an ether as it was concieved in 19th-century physics, in particular of Lorentz's ether, which was in the first place a privileged reference frame. He denied its existence because it violated his principle of relativity, according to which there is no privileged reference frame for the formulation of the laws of nature. Nevertheless, in 1916, after the definitive formulation of the general theory of relativity, Einstein prosposed a completely new conception of the ether. In this conception, the new ether does not violate the principle of relativity because the space-time of the theory of relativity is conceived, in it, as a material medium sui generis that can in no way constitute a frame of reference.

In Einstein's letter to Lorentz of June 17, 1916, in which he introduced for the first time his new notion of ether, we read:

> This new ether theory, however, would not violate the principle of relativity, because the state of this $g_{\mu\nu}$ = ether would not be that of a rigid body in an independent state of motion, but every state of motion would be a function of position, determined trough the material proceses[1].

As we can see, physical space (intimately connected with time) the local state of which is discribed by the components $g_{\mu\nu}$ of the metrical tensor g was regarded by Einstein as the new ether. In this new conception, the physical space (identified with the new ether) is no longer considered as a rigid quasi-object to which we can apply the notion of motion and velocity (and the structure of which is independent of the presence and motion of material bodies) but it is conceived as a field

Frontiers of Fundamental Physics, Edited by M. Barone
and F. Selleri, Plenum Press, New York, 1994

the structure of which depends on the material processes and determines the motion of ponderable bodies.

Einstein denied the existence of the ether only 11 years from 1905 do 1916. Then he began to consider this denial as an opinion which was too radical. In the so-called "Morgan Manuscript", in 1919, he wrote:

> ...in 1905, I was of the opinion that it was no longer allowed to speak about the ether in physics. This opinion, however, was too radical[2].

In a letter to Lorentz, written also in 1919, he regreted even his denial of the ether existence.

> It would have been more right, if I had limited myself in my earlier publications to emphasizing only the nonexistence of an ether velocity, instead of arguing the total noneexistence of the ether, for I can see that with the word ether we say nothing else than that space has to be viewed as a carrier of physical qualities[3].

On the basis of Einstein's papers and letters we can say that, in 1916, Einstein's physics of space-time became a physics of a new ether. Thus, the notion of ether found, in the theory of relativity, a new interesting application. Einstein himself wrote in 1938:

> We may still use the word ether, but only to express the physical properties of space. This word ether has changed its meaning many times in the development of science. At the moment in no longer stands for a medium built up of particles. Its story, by no means finished, is continued by the relativity theory[4].

The history of Einstein's ether concept is very interesting. I presented it in several papers[5,6,7] and in my book[8]. In the present paper I would like to show especially its physical meaning.

I. EINSTEIN IDENTIFIED ETHER WITH THE PHYSICAL SPACE

Einstein never considered ether as something in space. He always identified it with the physical space. Hi did it when he denied its existence and also when he introduced his new ether. We see here P. Drude's great influence on Eistein that with great interest studied Drude's textbooks and papers. For Einstein like for Drude ether meant physical space ednowed with real physical properties.

Einstein was also under a great influence of E. Mach. According to Mach space (especially absolute space) constitutes a metaphysical intercalation or foreign matter which must be removed from physics which constitutes an experimental science. Absolute space must be removed from physics because it has not experimentally accessible properties. Until 1915, Einstein was convinced that Mach was right and therefore, in 1905, when he formulated his special relativity, in which he rejected the existence of the absolute space which he identified with the old ether, he was convinced that he was realising Mach's programme.

In his 1905 paper we read:

> the unsuccessful investigations, the purpose of which was the ascertainment of the earth motion with respect to the "luminifrous medium", lead to the supposition that not only in mechanics but also electromagnetism to the notion of absolute rest do not correspond properties typical of physical phenomena[9].

As we can see, according to Einstein like to Mach "to the notion of absolute rest do not correspond properties typical of physical phenomena" and therefore it must be removed from physics. Since Einstein identifies the absolute space at absolute rest with old ether the latter proves, according to him, to be superfluous.

> The introduction of a "luminiferous ether" will prove to be superfluous inasmuch as the view here to be developed will not require an "absolutely stationary space" provided with special properties[9].

This quotation taken also from Einstein's 1905 paper testifies that he identified the old luminiferous ether with the absolute space.

In Einstein's mind the notion of reference frame was closely connected with the notion of space. He identified the notion of privileged reference frame with the notion of absolute space. And therefore when he removed, in special relativity, the privilegeded reference frame he was convinced that he removed from physics the absolute space. Subsequently, when he arrived at the conclusion that in his general relativity he succeeded to remove the privileged set of inertial reference frames and when he became aware that, in this new theory of gravitation, the coordinate systems lost physical meaning he began to be convinced that the general theory of relativity achieved Mach's goal and removed from physics space and time as methaphysical intercalation. According to him the only thing which remained were the space-time coincidences of events.

In 1916, in his general relativity paper he wrote:

> That this requirement of general covariance, which takes away from space and time the last remnant of physical objectivity, is a natural one, will be seen from the following reflection. All our space-time verifications invariably amount to a determination of space-time coincidences[10].

In a letter to Moritz Schlik, on December 14, 1915, he stated:

> Thereby [through the general covariance of the field equations] time and space lose the last remnent of physical reality[11].

In a letter to Ehrenfest, on December 26, 195, he emphasized:

> The physically real in what happens in the world (as opposed to what depends on the choice of the reference system) consists of spatio-temporal coincidences. (In a footnote, Einstein added "and nothing else!")[12].

In the period from 1913 to 1916 Einstein did not believe in the existence of the physical space endowed with real physical properties. In a letter to Ernest Mach, in late 1913 or early 1914 he wrote:

> For me it is absurd to attribute physical properties to "space"[13].

In a paper of 1914 he stated:

> As much I am not disposed to believe in ghosts so I do not believe in the enormous thing about which you are me talking and which you call space[14].

Since for Einstein ether meant "physical space with real properties" therefore, at that time, he solidified his disbelief in the existence of the ether.

In June 1916 Einstein changed his ideas. In his mind the notion of space broke off from the notion of reference frame. The physical space ceased to be, in his opinion, a reference frame. And so he could recognize the reality of space. It did no longer violeted his principle of relativity. Under the influence of a letter written by Lorentz[1] he arrived at the conclusion that physical space does really exist and is described by the components $g_{\mu\nu}$ of the mertical tensor g but it can no longer be considered as an enormous quasi-body composed of points which could serve as reference frame[1]. In such a way the notion of ether was resurrected in the relativity theory. Einstein emphazised this fact, in 1919, in the so-called "Morgan Manuscript":

> Thus, once again "empty" space appears as endowed with physical properties,
> i. e., no longer as physical empty, as seemed to be the case according to special
> relativity. One can thus say that the ether is resurrected in the general theory
> of relativity, though in a more sublimated form[2].

According to Einstein, the notion of ether was resurrected not only in the general theory of relativity but also in the special one. Therefore, in 1920, in his Leiden lecture Einstein said:

> More careful reflection teaches us, however, that the restricted principle of rela-
> tivity does not compel us to deny ether. We may assume the existence of an ether;
> only we must give up ascribing a definite state of motion to it...[15].

Let's note that Einstein, when reincorporating the notion of ether into the special and general relativity theory, identified always the ether with the physical space. He did not consider the new ether as something in space but as space itself endowed with physical properties. In the above mentioned lecture he said:

> ...there is a weighty argument to be adduced in favor of the ether hypothesis. To
> deny the ether is ultimately to assume that empty space has no physical qualities
> whatsoever...space is endowed with physical qualities, therefore, there exists an
> ether[15].

We find the same identification of ether with space in Einstein's attempts to construct a relativistic unified field theory in which he considered the gravitational field and the electromagnetic field as states of the same physical space. In 1934 he wrote:

> Physical space and the ether are only different terms for the same thing; fields
> are physical states of space. If no particular state of motion can be ascribed to
> the ether, there does not seem to be any ground for introducing it as an entity
> of special sort alongside of space[16].

When Einstein is speaking about "physical space" he means the space closely connected with time i. e. the space-time. Therefore, as he emphazised it, in 1924 in his paper "Über den Äther" his new *"ether became, to some extent, four-dimensional"* [17].

II. THREE MODELS OF THE RELATIVISTIC ETHER

After 1916, Einstein published several papers[2,5,16,17,18,19,20] in which he interpreted his models of space-time as models of the new ether. In these papers, one can make a distinction between three models of the relativistic ether:

1. The first one is the ether of the special theory of relativity. In this model the ether is identified with the flat space-time which, possesses a pseudo-Euclidean metric. Since in the flat space-time of special relativity there are coordinate systems in which the 10 components $g_{\mu\nu}$ of the tensor g are constans and presented by the symbol $\eta_{\mu\nu}$, therefore Einstein considered the components $\eta_{\mu\nu}$ as the mathematical tool to describe the metrical behavior of the special relativity ether. In connexion with the fact that in reference frames in which $g_{\mu\nu} = \eta_{\mu\nu}$ the freely moving test particles behave according to the inertia principle, i. e. are at rest or move with constant velocity on straight lines, Einstein called the ether of special relatiwity like the ether of Newton mechanics "inertial ether"[21].

Because of its flatness the inertial ether is extended in infinity. According to Einstein it is also rigid and absolute (i. e. the presence of matter and the matter movement do not exert an influence on its structure) like Newton's absolute space and the three-dimensional Lorentz's ether.

> The four-dimensional space of special theory of relatiwity is just as rigid and absolute as Newton's space[16].
> The rigid four-dimensional space of the special theory of relativity is to some extent a four-dimensional analogue of H.A. Lorentz's ether[22].

In brief, using the present-day terminology and symbols, we can say: The pair (M, η), where M is the four-dimensional differential manifold and η- the Minkowski metric on M, represents the ether of the special theory of relativity.

2. The second model of the ether is that of the general theory or relativity. Einstein identifies this ether with the space-time which possesses a pseudo-Riemannian metric and the metrical behavior of which is described by the 10 components $g_{\mu\nu}$ of the symmetrical tensor g. The components $g_{\mu\nu}$ represent mathematically the physical properties of the new ether i. e. the gravitational potentials. Einstein, therefore, called this ether "gravitational ether"[15]. According to Einstein the general relativity theory ether is no longer rigid and absolute in the above mentioned sense because it *"not only conditions the behavior of inert masses, but it is also conditioned in its state by them"* [15].

We can say, using the present day terminology and symbols, that the general relativity theory ether is mathematically represented by the four-dimensional differential manifold M together with the imposed on it differeniable field of the symmetrical tensor g (i.e. together with Lorentz metric called also pseudo-Riemanian). In brief, the pair (M, g) designates the new ether in the general theory of relativity. Note, however, that although both of them (i.e. M and g) represent the new ether, only the components $g_{\mu\nu}$ of the tensor g represent its physical disinctiveness i.e. the gravitational potentials.

3. The third model of the new ether is the one that appears in Einstein's attempts to construct a relativistic unified theory. This model has as many versions as there are Einstein's attempts of unification of the elecrtomagnetic field with the gravitational one. Their common characteristic is Einstein's supposition that the 10 components $g_{\mu\nu}$ of the symmetrical tensor g (where $g_{\mu\nu} = g_{\nu\mu}$) no longer describe complitly the structure of real space-time. According to Einstein, this structure must be richer than Riemannian because, in reality, we are dealing in space-time not only with gravitational field but also with the electromagnetic one.

Therefore, Einstein looked for:

> a theory of the continuum in which a new structural element appears side by side with metric such that it forms a single whole together with the metric[16].

In Einstein's general theory of relativity, as well as in special relativity the electromagnetic field still appears as something that "fills space"[10] i. e. as something that does not belong to the structure of the space-time as described by the components $g_{\mu\nu}$ of the tensor g. Therefore Einstein looked for another mathematical entity which could mathematically reperesent the "total or entire field" ("Gesamtfeld"). In different versions of such an unification Einstein used different mathematical entities to express mathematically the "Gesamtfeld". For instance, in the version of the unsymmetrical unified field theory, the unsymmetrical tensor with 16 components ($g_{\mu\nu} \neq g_{\nu\mu}$) constituted such an entity. 10 of them reperesented the gravitational field and 6 - the electromagnetic one. While, in the version of bivector fields the mathematical entity called symmertical bivector $g^{\mu\,\nu}_{\alpha\beta}$ represented both of them. And still, in another version, the Hermitian metrical tensor $g_{\mu\nu} = \bar{g}_{\nu\mu}$ did it.

Briefly, the manifold M together with the different geometrical structures imposed on it constituted the different versions of the third model of the relativistic ether. The imposed structures were conceived as mathematical represetatives of physical properties of space-time i.e. of the new ether.

III. FROM A RIGID BODY CONCEPTION OF SPACE TO A FIELD CONCEPTION OF IT

According to Einstein, the transition from the old ether to the new consisted in the transition from a rigid body conception of space to a field conception of it. According to him, we are inclined to conceive the physical space as an unique infinite enormous all permeating rigid body to which we relate the position of all physical bodies. This all permeating body is inaccessible to our senses but was invented by mankind for the convenience of our thinking.

> When considering the mutual relation of the location of bodies, the human mind
> finds it much simpler to relate the locations of all bodies to that of a single one
> rather than to grasp mentally the confusing complexity of the relations of every
> body to all others. This *one* body, which is everywhere and must be capable of
> being penetrated by all others in order to be in contact with all, is not given to
> us by the senses, but we devise it as a fiction for convenience in thought[18].

According to Einstein also in the Newtonian mechanics the rigid body became the prototype of the notion of absolute space. The absolute space of Newtonian mechanics constitutes an idealisation of an enormous rigid body composed of particles. Points of the absolute space are idealisation of these particles. Space, therefore, is there conceived as an infinite flat quasi-object composed of points. These points do not change their position with respect to each other and therefore the absolute space is rigid.

Also the interial reference spaces which are at rest or move with respect to the absolute space were conceived in the same way. They are infinite flat rigid quasi-objects that move with respect to the absolute space and with respect to each other. Thus, the rigid body remained the archetype for the notion of space in Newtonian physics.

The rigid body remained also as prototype in several conceptions of the ether. E.g. in Lorentz's ether concept.

According to Einstein, a new prototype of the notion of space begin to play its part when Faraday and Maxwell introduced into physics the notion of field. First the field was conceived as a state of a mechanical medium. Therefore, the luminiforous ether was invented. But step by step such a mechanical carrier of the field was no longer needed. All mechanical interpretations of electromagnetic waves failed. According to Einstein:

> the emancipation of the field concept from the assumption of its association with
> a mechanical carrier finds a place among the psychological most interesting events
> in the devolopment of physical thought[22].

In the relativity theory, in which, according to Einstein, we are dealing with the victory over the concept of absolute space or over that of the inertial system, the new prototype of space, i.e. the field, plays, its fundamental part.

> The victory over the concept of absolute space or over that of the inertial system became possible only because the concept of the material object was gradually replaced as the fundamental concept of physics by that of the field. Under the influence of the ideas of Faraday and Maxwell the notion developed that the whole of physical reality could perhaps be presented as a field whose components depend on four space-time parameters. If the laws of this field are in general covariant, that is, are not dependent on a particular choice of coordante system, then the introduction of an independent (absolute) space is no longer necessary. That which constitutes the spatial character of reality is then simply the four - dimensionality of the field. There is then no "empty" space, that is there in no space without a field[23].

When Einstein identifies his new ether with the physical space-time he does it in the framework of the new field conception of the space. The physical properties of the "inertial ether" which determine the inertial behevior of test particles in the special theory of relativity are represented mathematically by the field of the tensor η. The physical properties of the "gravitational ether" which determine the inertio-gravitational behavior of the test particles in the general theory of relativity are represented mathematically by the field of the tensor g. And the physical properties of the "Gesamtfeld ether" that determine the inertio-gravitational and electromagnetic behavior of test particles in Einstein's attempts to construct an unified theory are represented mathematicaly by respective fields of respective mathematical entities like e.g. by the field of unsymmetrical tensor (where $g_{\mu\nu} \neq g_{\nu\mu}$) with 16 components.

IV. THE RELATIVISTIC ETHER CONSTITUTED AN ULTRAREFERENTIAL FUNDAMENTAL REALITY

In identifying the new ether with the physical space, Einstein made a very clear distinction between space as such (*"Der Raum als solche"*) conceived as it was indicated above and the reference spaces (*"Bezugsräume"*). According to Einstein there is only one single physical space as such which physically manifests itself trought field properties which are mathematically represented by the components $\eta_{\mu\nu}$ of the tensor η in special relativity, by the components $g_{\mu\nu}$ of the tensor g in general relativity and by the components of respective mathematical entities in the respective versions of Einstein's unified theory. The physical space as such is not composed either of particles or of points and is indivisible in parts. The new ether has to be identified with this space.

There is also an infinite number of reference spaces which are artificial extensions of reference bodies. We introduce a reference space through an infinite number of points that we connect with a reference body. Therefore, every reference space is composed of points like every material medium is composed of particles. Every reference space like every material medium can serve as reference frame. If we move with respect to a material medium we feel a wind or a change of temperature. When we move with respect to a reference space we "feel" a wind of points. The particles of material medium or points of a reference space can be followed in time. Therefore, the notion of motion is, in the full sense of the word, applicable to material media and to reference spaces. The reference spaces move with respect to each other.

The notion of motion and velocity, however, cannot be applied at all to the new ether i.e. to the physical space as such because it constitutes un ultrareferential fundamental reality which is not composed either of points, particles or of parts the motion of which can be followed in time.

But, this ether may not be thought of as andowed with the properties characteristic of ponderable media, as consisting of parts that may be tracted through
time. The idea of motion may not be applied to it[15].

According to Einstein, the new ether cannot be identified with any of the reference spaces because that would mean that one of them is favorated or privileged with respect to others. This contradicts the principle of relativity. The ether, in such a case, would be connected only with one reference space and would be reduced to a simple usual material medium. The new ether constitutes a material medium but in another sense. It is material in the sense in which we attribute materiality to a field. With the gravitational and electromagnetic field is connected a certain density of energy and in this sense they are material.

Physical space as such, indified with the relativistic ether, constitutes an ultrareferential reality. It means that physical space is over or behind all reference spaces. With respect to these, it constitutes a more fundamental bakcground reality, which makes possible the existence and the motion of reference spaces, although it is not in motion or at rest itself. The ultrareferential physical space manifests itself through real field properties. Its structure determines the behaviour of free moving bodies.

ACKNOWLEDGMENT

The author would like to express his sincere thanks to the Volkswagen-stiftung for a scholarship that made also this research possible and to the Hebrew University of Jerusalem for the kind permission to quote from Einstein's papers and letters.

REFERENCES

The abbreviation EA and the numbers in brackets refer to the control index of the Einstein Archive at the Hebrew University of Jerusalem.

1. A. Einstein, Letter do H.A. Lorentz, June 17, 1916 (EA 16-453).

2. A. Einstein, "Grundgedanken und Methoden der Relativitätstheorie in iher Entwicklung dargestellt", Unpublished manuscript called "Morgan Manuscript" (EA 2-070).

3. A. Einstein, Letter to H.A. Lorentz, November 15, 1919 (EA 16-494).

4. A. Einstein and L. Infeld, "The Evolution of Physics: The Growth of Ideas from Early Concepts to Relativity and Quanta" (Simon & Schuster, New York, 1938).

5. L. Kostro, "Einstein's Conception of the Ether and its up-to-dated Applications in the Relativistic Wave Mechanics", in: *Quantum Uncertaines*, eds. W. Honing and E. Panarella, (Plenum Press, New York 1987), pp. 435-449.

6. L. Kostro, "Einstein's New Conception of Ehter", in: "Physical Inerpretations of Relativity Theory", ed. M.C. Duffy, (Brithis Society for the Philosophy of Science, London, 1988), pp. 55-64.

7. L. Kostro, "An Outline of the History of Einstein's Relativistic Ether Concept", in: "Einstein Studies" Vol. 3, Studies in the History of General Relativity, eds. J. Eisenstaedt and A.J. Kox, (Birkhäuser, Boston, Basel, Berlin 1992), pp. 260-280.

8. L. Kostro, "Alberta Einsteina koncepcja eteru relatiwistycznego. Jej historia, sens fizyczny i uwarunkowania filozoficzne", (Wydawnictwo Uniwersytetu Gdańskiego, Gdańsk 1992).

9. A. Einstein, *Annalen der Physik*, 17, 891-921 (1905).

10. A. Einstein, *Annalen der Physik*, 49, 769-822 (1916).

11. A. Einstein, Letter to M. Schlick, December 14th, 1915 (EA 21-610).

12. A. Einstein, Letter to P. Ehrenfest, December 26th, 1915 (EA 9-363).

13. A. Einstein, Letter to E. Mach, December 1913 (EA 16-454).

14. A. Einstein, *Scientia*, 15, 337-348 (1914).

15. A. Einstein, "Äther und Relativitätstheorie", (Springer, Berlin, 1920).

16. A. Einstein, "Das Raum - Feld - und Äther - problem der Physik", in: "A. Einstein Mein Weltbild", (Querido Verlag, Amsterdam, 1934).

17. A. Einstein, *Schweizerische naturforschende Gesellschaft. Verhandlungen*, 105, 85-93 (1924).

18. A. Einstein, *Forum Philosophicum*, 1, 173-184 (1930).

19. A. Einstein, *Transactions of the second world power conference*, 29, 1-5, (1930).

20. A. Einstein, *Die Koralle*, 5, 486-487, (1930).

21. A. Einstein, "Fundamental ideas and problems of the theory of relativity, in: Nobel lectures, Published for the Nobel Foundation" (Elsevier Publishing Company, Amsterdam-London-New York, 1967).

22. A. Einstein, "Relativity and the Problem of Space", in: "A. Einstein, Ideas and Opinions" (Crown Publishers, Inc., New York, 1960) pp. 360-377.

25. A. Einstein, "Foreword", in: M. Jammer, "Concepts of Space. The History of Theories of Space in Physics" (Harward University Press, Camridge-Massachsetts, 1954) pp. 13-14.

THE LIMITING NATURE OF LIGHT-VELOCITY AS THE CAUSAL FACTOR UNDERLYING RELATIVITY

Trevor Morris

Teddington, TW11 0BH, UK

THE ETHER: *AD HOC* OR SUPERFLUOUS?

It is commonly accepted that Einstein's 1905 paper setting out the Special Theory of Relativity was a turning-point, and his approach has entirely displaced the earlier one developed by Lorentz (1892, 1904) and Poincaré (1904). Two main reasons for this preference are usually given:

1: Einstein's version is more economical because it makes a preferred frame of reference, the ether, "superfluous" (Einstein, 1905; Jeans, 1925).

2: The Lorentz-Poincaré approach is not satisfactory anyway because its basic postulation of the FitzGerald-Lorentz contraction was *ad hoc*.

Reason (1) can be shown to be based on faulty logic (and Einstein later revised his view accordingly). Reason (2) is negated because although the original suggestion of the contraction effect by FitzGerald (1889) may indeed have been *ad hoc*, it need not have been so. The contraction effect follows directly as a result of the retardation of the field-potentials of moving charges in Maxwell's electromagnetic theory, as shown by Voigt (1887), and Heaviside in a series of papers in 1888-89 (Heaviside, 1892), for example. Feynman (1964) gives a clear outline of the procedure "...*to show how naturally the Maxwell equations lead to the Lorentz transformation*". The slowing of moving clocks was also shown to follow from this retarded-potential effect by Larmor (1900) (though he did not appear to recognise the significance of the result) and so is not a separate, arbitrary assumption, as was noted by Builder (1958a). These retarded potential effects are a result only of the finite propagation velocity in the ether of changes in the field (that of light). The answer to the question at the head of this introduction is, therefore, "neither". A basically similar position has been taken by many others, including Ives and Janossy, and especially Builder and Prokhovnik.

Frontiers of Fundamental Physics, Edited by M. Barone
and F. Selleri, Plenum Press, New York, 1994

LOGICAL PROGRESSIONS

The list of statements below is intended to set out the logical elements in the derivation of Special Relativity.

    1 - Unique reference frame for all energy transfer
    ↓
    2 - No energy transfer faster than light in a vacuum
    ↓
    3 - Lorentz contractions etc. occur in moving systems
    ↓
    4 - Lorentz transformations are valid between inertial frames
    ↓
    5 - Measured speed of light is constant for all inertial frames
    ↓
    6 - Principle of relativity is universally valid

It is emphasised that this list is a reconstruction, with hindsight: the development of the theory was never expressed in this form by Lorentz or Einstein, or anyone else, as far as I am aware. However, it contains all of the essential features and results of either approach. It serves to emphasise the similarities, and clarify the differences.

The Lorentz-Poincaré approach proceeds following the arrows from top to bottom in the list: each step is a causal basis for the next. Einstein, on the other hand, took steps 5 and 6 as postulates and deduced step 4. Step 3 can then be deduced from 4. Unfortunately, Einstein's expression of this step was ambiguous:

> "Thus, whereas the Y and Z dimensions (of every rigid body) do not appear modified by the motion, the X dimension appears shortened in the ratio $1 : (1 - v^2/c^2)^{1/2}$, *i.e.* the greater the value of v, the greater the shortening."

This ambiguity has resulted in continuing confusion as to whether the Lorentz transformations describe real, physical processes or mere artefacts of observation, quite different in nature from the physical contractions proposed by Lorentz and FitzGerald (*e.g.* Reichenbach, 1928; Pais, 1982). However, there can now be no doubt about the physical reality of the entailed clock-slowing effects (*e.g.* Newman *et al.*, 1972), and it would therefore be absurd to maintain that the length contractions are only "apparent".

Step 2 was also shown by Einstein to follow from the Lorentz transformations, but he then denied the necessity of step 1 by declaring the ether to be "superfluous", without suggesting any alternative. Physical causality was thus replaced by a "universal principle", in a process somewhat analogous to affirming Ohm's Law and denying the need for any conduction mechanisms to exist. This denial is clearly an error of logic. Although the converse "2 implies 1" of the Lorentzian "1 implies 2" is not *necessarily true*, it is certainly not *necessarily false*. To falsify step 1, we would need to show either that it does *not* imply step 2 (so that 1 could be false even if 2 were true), or that it implied some other result which could be shown to be false. This is possibly the motive behind the common, but incorrect, statements to the effect that the null result of the Michelson-Morley experiment proves that the ether does not exist. In some later, little-known "second thoughts" on the subject, Einstein (1920) seems to have recognised this error:

> "More careful reflection teaches us, however, that the special theory of relativity does not compel us to deny ether...But on the other hand, there is weighty argument to be adduced in favour of the ether hypothesis. To deny the ether is ultimately to assume that empty space has no physical qualities whatever. The fundamental facts of mechanics do not harmonise with this view."

It seems to have been already too late for such a radical revision, however, and both this paper, and a later elaboration of the links between General Relativity and the notion of a preferred frame (Einstein, 1924), are rarely quoted (see also Janossy, 1971). In particular, it is clear that Einstein's initial assertion that the ether was superfluous can not be taken as definitive proof that a unique, fundamental reference frame is incompatible with Special Relativity. While the theory can be derived by starting from the sole relevant property of a unique frame - an upper limit on the speed of propagation of causal influences - the fact that it can also be derived by a reversal of the logical order of the argument does not negate the possibility of existence of a unique frame. Conversely, of course, the ability to derive the theory *assuming* a preferred frame does not *prove* its existence, though in the absence of any other proposed physical basis it must be regarded as at least highly plausible.

RETARDED POTENTIALS AND MATTER

How, then, might the retarded Liénard-Wiechert potential formula of Voigt (1887), Heaviside (1892) and Lorentz (1904) lead to relativistic contraction of moving material bodies? The only relevant property of the forces determining and maintaining the dimensions of solids is that the exchanges of energy involved should not be able to overtake a light-signal in the fundamental reference frame. Since, following Poincaré (1904), this is also a necessary condition for the validity of the Principle of Relativity, the many experimental demonstrations of that Principle may be taken to support the limiting nature of the speed of light for energy transmission. Given this limitation, any fields around and between charged particles and atoms in solids will, in general, be characterised by a series of ellipsoidal equipotential surfaces, with a ratio of minor to major axes given by the Lorentz contraction formula in which the value of the speed v is that of the body with respect to the fundamental reference frame. The mean equilibrium separation of the atoms (about which thermal and other vibrations will occur) will be determined by the pattern of these equipotential surfaces. There will be a position where the field-energy has a minimum value, and where the coulomb-derived attractive and repulsive fields balance each other (*e.g.* Dekker, 1958). Any change in the shape of the equipotential ellipsoids will thus necessarily lead to a corresponding change in the equilibrium distribution of equipotentials, and hence to a change in the equilibrium separation both of the atoms and their internal constituent parts.

This conclusion applies regardless of the nature of the bonding within and between atoms or molecules, *given only that changes in all such influences can not overtake light in a vacuum.*

It might be argued that if all the results of the Lorentzian theory can be derived (as by Einstein) without explicit reference to a preferred reference frame, the concept is surely redundant even if it can not be logically excluded. However, this approach fails to provide the necessary physical basis. For example, as shown by Builder (1958a) and Prokhovnik (1967, 1985), the concept has to be invoked in order to provide the necessary physical asymmetry needed to resolve the paradox of the travelling twins, or clocks. The kinematic symmetry has to be broken by ascribing a unique status to the acceleration of the one who goes on a high-speed journey, and returns the younger. Many authors make statements like that of Whitrow (1961):

> "The essential difference between the two clocks concerns their relations to the Universe as a whole."

(*e.g.* Rosser, 1964), implicitly (only) acknowledging the essential role of a preferred reference frame.

We should note that the Lorentz transformations themselves contain only the mutually-measured relative velocity of the twins: this symmetrical quantity alone does not reveal which of them has accelerated. The fact that Einstein's initial rejection of the relevance of the ether to his derivation of the Lorentz transformations is taken as meaning that a preferred frame for energy propagation is impossible (or, at least, unnecessary), therefore, for example, makes the usual textbook resolutions of the twins paradox seem very unconvincing. On the other hand, under the Lorentzian approach, the travelling twin is automatically the younger on return to home, because his mean squared velocity in the unique frame is necessarily the greater during the period of the journey. Any underlying velocity of their common (arbitrary) inertial reference frame relative to the preferred frame does not affect the final difference between the ages of the twins.

THE PREFERRED REFERENCE FRAME: PHYSICAL FACT OR HEURISTIC AID?

Thus, not only is the existence of a unique, physical frame of reference compatible with Special Relativity, but it also provides a necessary and sufficient basis for the principle of relativity itself if nothing can exceed the speed of light in it. (See Builder, 1958a,b; though it should be noted that Builder invoked the principle of relativity as a separate hypothesis, equivalent in effect to ascribing to the unique frame the necessary physical constraint on the speed of energy transfer). The satisfying way in which the physically anisotropic energy-exchanges in arbitrary inertial frames, moving relative to a unique, basic reference frame, obey the laws of relativity is described in detail by Prokhovnik (1967, 1985). Prokhovnik thus extends the intuitive notion (expressed by Whitrow (1961), for example) that the distributed matter of the Universe and the associated 2.7K black-body background radiation delineate this reference frame. It must be emphasised that the only relevant property of this reference frame is that no energy exchange can overtake light in it, and that this is consistent with all the known phenomena of energy exchanges.

than, the views of those who, like Bell (1976), acknowledge the heuristic superiority of the Lorentzian approach to relativity, but still maintain that the choice of "preferred" reference frame remains a matter of philosophical taste, and without special physical significance. This latter position, however, robs the preferred frame of the essential property of providing a (yet unknown) physical mechanism for the propagation of electromagnetic and other fields. In effect, it tries to maintain the ether as merely a convenient heuristic concept without the uniqueness and physical significance envisaged by Lorentz *et al.* and even Einstein, at least for a period in the 1920s. There have been occasional attempts to revive the concept in connection with Quantum Theory, but these have been overshadowed by the mistaken belief that "ether" and relativity are mutually exclusive (*e.g.* Dirac, 1951; Ives, 1953; Sciama, 1978).

CONCLUSION

The results of Special Relativity can be derived from the single postulate:

"No causal influence can be mediated at a velocity greater than that of light *in vacuo* relative to the inertial reference frame defined by the observable Universe."

I contend that this formulation is more "simple" than the conventional derivation by Einstein (1905). It brings together the recognition by Poincaré (1904) that the limiting nature of light-velocity for all interactions is the necessary and sufficient basis for the universal operation of the Principle of Relativity, and the physical mechanism of retarded potential effects on matter moving in a preferred frame of reference for energy propagation, as set out tentatively by Larmor (1900) and Lorentz (1904). The association of the preferred frame with the distributed matter and background radiation of the Universe is consistent with the General Theory of Relativity, and with the most basic cosmological principles (Einstein, 1920, 1924; Prokhovnik, 1985). This approach provides also a physical basis for relativity, and thus a clear physical explanation of the asymmetries of clock-rates and synchronisation observed between inertial reference frames, without any "paradoxes" (Builder, 1958a,b). It also makes explicit the formal links between relativity and the notion of causal connection already noted by Zeeman (1964).

At the very least, the existence of a unique inertial frame of reference which imposes an upper speed limit for any physical interaction equal to that of light *in vacuo*, should be recognised as a necessary and sufficient physical basis for relativity, rather than its antithesis, as is now commonly believed. If an upper speed limit for causal connection between events is problematic in Quantum Theory, it seems more likely that the latter will need to be adjusted than that the principle of relativity will need to be qualified.

REFERENCES

Bell, J.S., 1976, How to teach special relativity, in "Speakable and Unspeakable in Quantum Mechanics", Cambridge University Press, Cambridge, UK (1987).

Builder, G., 1958a, Ether and relativity, *Austr. J. Phys.*, 11:279-297.

Builder, G., 1958b, The constancy of the velocity of light, *Austr. J. Phys.*, 11:457-480 (Other refs. to Builder in Prokhovnik, 1967.)

Dekker, A.J., 1958, "Solid State Physics", MacMillan, London.

Dirac, P.A.M., 1951, Is there an ether?, *Nature*, 168:906.

Einstein, A., 1905, Ann. d. Phys. 17:891-921. (English trans. "The Principle of Relativity", Methuen, London (1923) and Dover, New York, (1952).)

Einstein, A., 1920, Lecture at University of Leyden. (English trans. by G. B. Jeffery and W. Perrett: 1922, "Sidelights on Relativity", Methuen, London.)

Einstein, A., 1924, Über den Äther, *Vehr. d. Schweiz. Nat. Ges.*, 105, Teil II:85-93.

Feynman, R.P. et al., 1964, "The Feynman Lectures on Physics", Addison-Wesley, New York, Vol. 2, 21.6.

FitzGerald, G.F., 1889, The Ether and the Earth's Atmosphere, *Science*, 13:390 (See Pais, 1982).

Heaviside, O., 1892, "Electrical Papers", MacMillan, London, Vol. 2, 490-499 and 504-518.

Ives, H.E., 1953, Genesis of the query "Is there an ether?", *J. Opt. Soc. Amer.*, 43:217-218. (This and all of Ives's published work between 1937 and 1953 in support of an ether-based interpretation of relativity, together with highly polemical material by the editors, is contained in "The Einstein Myth and the Ives Papers", Eds. Turner and Hazelett, 1978, Devin-Adair, Connecticut.)

Janossy, L., 1971, "Theory of Relativity Based on Physical Reality", Akademiai Kiado, Budapest.

Jeans, J.H., 1925, "The Mathematical Theory of Electricity and Magnetism", Cambridge University Press, Cambridge, 5th Edn., Chap.20, Sec.701.

Larmor, J., 1900, "Aether and Matter", Cambridge University Press, Cambridge, Chap.11, 173-193.

Lorentz, H.A., 1892, in "Collected Papers", Nijhof, The Hague, Vol.4:219-223 (1936).

Lorentz, H.A., 1904, ibid., Vol.5:172.

Newman, D., Ford, G.W., Rich, A. and Sweetman, E., 1978, Precision experimental verification of Special Relativity, *Phys. Rev. Lett.* 40:1355-58.

Pais, A., 1982, "Subtle is the Lord...", Oxford University Press, Oxford/New York.

Poincaré, H., 1904, *Bull. Sci. Math.* 28:302-324 (English trans., 1905, The principles of Mathematical Physics, *The Monist*, 15:1-24).

Prokhovnik, S.J., 1967, "The Logic of Special Relativity", Cambridge University Press, London and New York.

Prokhovnik, S.J., 1985, "Light in Einstein's Universe", Reidel/Kluwer, Dordrecht.

Reichenbach, H., 1928, "Philosophie der Raum-Zeit-Lehre". (English trans. "The Philosophy of Space and Time", Dover, New York (1958), Chap. III, Para. 31.)

Rosser, W.G.V., 1964, "An Introduction to the Theory of Relativity", Butterworths, London.

Sciama, D.W., 1978, The ether transmogrified, *New Scientist*, 2 February, p.298-300.

Voigt, W., 1887, Über das Döppler'sche Prinzip, *Goett. Nachr.*, March 10, p.41.

Whitrow, G.J., 1961, "The Natural Philosophy of Time", Nelson, London.

Zeeman, E.C., 1964, Causality implies the Lorentz Group, *J. Math. Phys.* 5:490-493.

THE ETHER REVISITED

Adolphe Martin

2235 Brébeuf, Apt. 3
Longueuil, Quebec J4J 3P9
Canada

C. Roy Keys

4405 St. Dominique
Montreal, Quebec H2W 2B2
Canada

INTRODUCTION

We will show that the Lorentz transformation applies in Galilean space-time, such that the laws of electromagnetism and classical mechanics become invariant. Assuming the existence of a gas permeating all space and matter, we conclude that the mechanical properties of gases, known for over a century, are sufficient to explain the known physical phenomena such as electromagnetism, light propagation, gravitation, quantum mechanics and the structure of elementary particles, including the photon.

HISTORICAL BACKGROUND

The history of the ether may be traced back to the pre Socratic Greek philosophers, who held that corporeal matter arises from an infinite substratum. Anaximander of Miletus called this substratum the *apeiron*. Descartes, in the 17th century, was the first to attribute definite properties to the ether, which he imagined to be comprised of minute particles engaged in constant rotational motion within an all-pervading system of vortices. Newton, meanwhile, believed that the ether was responsible for deflecting and directing light, though he did not conceive of light as a vibration of the ether.

The nineteenth century spawned a plethora of ether models, many of them relying on fluid analogies. Cauchy, for example, proposed a stationary luminiferous medium which possessed elastic solid characteristics for high-velocity perturbations (and hence could transmit transversal waves, *i.e.* light). The medium acquired gas properties when perturbations propagated at low velocity, since macroscopic motions necessitated a low viscosity coefficient, in agreement with the assumption that the medium remains stationary and is not entrained, as Bradley's discovery of aberration seemed to require.

By the mid-nineteenth century, Maxwell was in a position to formulate a full theory of the propagation of electromagnetic waves by a "magnetic medium" in which displacements of rolling particles arranged on a lattice of rotating cells represented electric currents. The stresses developed in this system of vortices may be likened to a hydrostatic pressure, which would be accompanied by a tension along the axes of rotation. Thus, it may be said that the canonical laws of electromagnetism were borne of the ether, and that this ether passed on to electromagnetism many traits of fluid dynamics.

Frontiers of Fundamental Physics, Edited by M. Barone
and F. Selleri, Plenum Press, New York, 1994

In subsequent years, investigations by W. Thomson, Kirchhoff, Bjerkness, Leahy, Lorentz, Larmor, Helmholtz, Hicks and FitzGerald invoked different forms of hydrodynamic models as a foundation for an ether theory. Of these models, the one proposed by Larmor was perhaps the most elaborate. He imagined that electrons and particles were structures in the ether, which he thought of as a perfect fluid medium whose flow determined the motions of particles (in a manner that anticipated the deBroglie pilot wave).

Toward the end of the 19th century, a crisis swept over physics, largely instigated by the electron theory devised by Lorentz. One perplexing implication of Lorentz's research was that the rest mass of the electron, and consequently of matter in general, appeared to be produced only by the electromagnetic field. As a result, it seemed impossible to explain the properties of the ether and electromagnetism by means of mechanical models. Accordingly, Lorentz proposed an entirely immobile, force-free ether lacking any mechanical properties. In his conception, the ether's various properties were reduced to essentially one: the ability to serve as a medium for light waves. The elements of the Lorentz ether were assumed to be at rest with respect to one another and this state of internal immobility was attributed to the ether as a whole.

The history of the ether virtually comes to a close with Lorentz's stationary ether. The new electron theory and the "dematerialization" of the electron's mass seemed to suggest that there was no further place for a mechanical representation of the ether, while contemporary fluid dynamics were then too little advanced to offer arguments to the contrary.

THE ETHER AND RELATIVITY

At the beginning of this century, Einstein (1905), adopting Poincare's idea of relativity, developed his theory of special relativity. Einstein observed that "the phenomena of electrodynamics as well as of mechanics possess no properties corresponding to the idea of absolute rest." Einstein could then criticize Lorentz's immobile ether as a throwback to absolute rest, concluding that the ether concept had outlived its necessity.

In the conventional view, then, the question of a medium for the propagation of light was dismissed from the agenda of physics by Einstein's synthesis of space and time. However, by eliminating the physical basis for the transmission of wave phenomena, special relativity created the impossible situation that effects must somehow propagate through empty space. Physics thus found itself back in the clutches of "spooky" action-at-a-distance.

As Kostro (1992), Janossy (1962) and others have shown, shortly after the formulation of the general theory of relativity, Einstein was persuaded, chiefly by Lorentz, to revive the notion of an ether as a transmitter for physical interactions. Einstein wrote to Lorentz in 1919:

> It would have been more correct if I had limited myself, in my earlier publications, to emphasizing only the nonexistence of an ether velocity, instead of arguing the total nonexistence of the ether, for I can see that with the word ether, we say nothing else than that space has to be viewed as a carrier of physical qualities. (Einstein to H.A. Lorentz, November 15, 1919, in Kostro 1992)

The ether here is equated with space, a space which acts as a medium for physical processes. In another text, written a few years earlier, Einstein distinguished between a special relativistic ether (defined as a rigid space-time analogous to Lorentz's rigid 3-dimensional ether) and a general relativistic ether, portraying the latter as a field whose distribution could vary in space.

The ether of the general theory of relativity differs from that of classical mechanics or from that of the special theory of relativity in so far as it is not 'absolute' but its spatial distribution is determined by that of matter... We shall not be able to dispense, in the field of theoretical physics, with the ether, *i.e.* a medium which possesses physical properties; indeed, the general theory of relativity... excludes any direct distant action. Every theory based on close action supposes the existence of continuous fields; thus they also presuppose the existence of an 'ether'. (Einstein 1924, translated in Janossy 1962)

The use of the term "ether" remains ambiguous in Einstein's writings, and it is almost completely absent from the literature on relativity—save for the many accounts of its demise. Einstein appears to have used the terms "physical space", "ether" and "total field" almost interchangeably (Kostro 1992). In some instances, space is referred to as "the sole medium of reality" and it appears to be conceived as a replacement for the ether. In Einsteinian relativity, concrete properties are ascribed to a geometric concept, and this new hybrid concept is then manipulated as if it were fully plastic and capable of variation.

If this confusion is to be overcome, physics must return to the materialist conception of the objective world. The reality of space and time is inescapable, inasmuch as there can be no existence outside of space and time. However space and time *as such*—since they are abstractions from spatial and temporal *relations* established by experience and measurement—are incapable of assuming any kind of state or structure. Space-time cannot serve as a material support, and, consequently, the existence of a medium for the transmission of actions must has to be posited. The classical fields can, then, be treated as states of this medium, whose properties should explain the Maxwell equations, the Lorentz transformations, the fields of electromagnetism and gravitation, and quantum mechanics while it must also account for the structure of elementary particles, including the photon.

THE LORENTZ TRANSFORMATION

The special theory of relativity was erected upon two postulates: the relativity principle of Poincaré and the postulate of the constancy of light velocity in all reference frames. The Lorentz transformation may be derived from these two postulates, but the reverse is not true.

Lévy-Leblond (1976) has demonstrated that two—and only two—kinds of space-time can exist based on the requirements set by the principle of relativity, the homogeneity of space and time, the isotropy of space, the principle of causality and the need for group properties for coordinate transformations. Einstein special relativity adds the requirement of constant velocity of light, while Galilean relativity postulates invariance of space and time. These last two requirements are mutually exclusive in any coherent theory.

If the postulate of constant light velocity is dropped, the only recourse is to a Galilean system of coordinates. We must therefore ask: Are there grounds for a kinematical theory in Galilean space and time possessing Lorentz invariance?

In the Einstein interpretation, where light velocity is normalized to c_o in all frames, *i.e.* relative to all observers, it is implicit that the reception of a light flash by a receiver in motion and by a receiver fixed relative to the source is the same event. It is this assumption that leads to the many paradoxes that have animated the literature for so many decades. Moreover as Einstein noted in 1905, normalization of light velocity to a constant value in all frames is tantamount to transforming all reference frames into a frame that is stationary relative to the light source.

One of us (A.M.), using the Doppler effect in Galilean space-time, showed that the time and position of reception of a light flash by an observer moving relative to the light source are not the same as the time and position of reception of the same light flash by an observer fixed relative to the source. Distances and time intervals for the two events, measured from the instant the moving observer passed at the coordinate of the source, are related by the Lorentz factor. The velocities of light signals to and from a moving receiver differ from the standard c_o and are functions of the relative velocity. Light returned to the source from the two reception events arrives at the same time in both theories, as observed in experiment.

In 1970, Jacques Trempe (1990) formulated a kinematics with Galilean coordinates and velocities. The relation between Galilean coordinates is given by the same Lorentz transformation as for the Einstein coordinates, except that Galilean speed V replaces Einstein speed v and light speeds C and C' (with the same Galilean time T) replace the normalized light speed c_o (with the different times t and t'). For a given equivalent relative velocity $v = \tanh(B) = \tanh(V/c_o)$, all distances in Galilean space-time have the same ratio $B/\sinh B$ to their equivalent distances in Einstein space-time. All corresponding angles are the same, thus making Galilean space-time an isomorphism of the Einsteinian space-time for each pair of uniformly moving reference frames. Galilean space-time retains vectorial composition of velocities.

We can say that Einstein space-time, where the speed of light is normalized to the light speed between points relatively at rest, permits an easier resolution of high relativistic speed problems. The derived coordinates and speeds can then be easily translated into the corresponding Galilean values. This means that all problems presently solved in Einstein space-time represent physical phenomena in Galilean space-time.

This Galilean viewpoint of relativity establishes a full symmetry, not found in Einstein space-time, in the description of phenomena as seen by two observers in relative uniform motion, thus respecting the relativity principle. It eliminates the paradoxes of length contraction, time dilation and of differently aging twins. However it introduces more natural Galilean speeds varying from zero to infinity. The Lorentz transformation in both space-times relates the actual coordinates of a uniformly moving object to the coordinates of its image as seen by an observer at the same instant. It simply gives what physicists call the retarded apparent position of moving objects due to the finite velocity of light.

In a detailed analysis, Trempe (1992) demonstrated that the Lorentz transformation is, in fact, the equation of an ellipse or ellipsoid of revolution, which

> ... transforms the coordinates from one reference frame with its origin at one focus to another reference frame with its origin at the other focus. The Lorentz transformation is therefore a purely geometrical relation between lengths in two or three dimensions.

Because the Lorentz transformation is fully valid in Galilean space-time, the electromagnetic equations become invariant in Galilean space-time, while classical dynamics are invariant under the Galilean transformation. We may thus assume that it is possible to envision an explanation of electromagnetism by means of a mechanical ether.

A GAS ETHER MODEL

Despite over one hundred years of research into the nature of the luminiferous medium, the concept of a gas with all its presently known properties has not been properly investigated. The assumption by one of us (A.M.) that all of space and matter are permeated by a gas seems to account for all known physical phenomena, which may be modeled by known gas mechanical properties.

The gas kinetic theory teaches us that a gas consists of particles agitated in all directions. Since the ether particles are considered the smallest entities in the Universe, they are also the only substantial entities. These grains of cosmos will be referred to as "cosmons".

Cosmons are individual spheres of a definite diameter and volume. Between cosmons we assume that there is an absolute void which cannot transmit any signal. Thus at the cosmon level there are no fields or forces. As the cosmons have no moving parts, they possess no internal energy and, according to Einstein, no rest mass (inertial or gravitational), no charge (electric or color) and no spin, since friction does not occur at this level. Hence, a cosmon is a boson.

Due to their agitation and lack of rest mass, cosmon velocities vary from zero to indefinitely high values. Interchange of velocity components when cosmons encounter other cosmons produces a velocity distribution similar to Maxwell's. However, at the cosmonic gas level, to account for our usual concepts of gas properties we have to assign energy to a moving cosmon. The cosmonic gas has mean energy and mass densities, producing the temperature, internal heat i and static pressure p of the gas. A general movement of cosmons produces gas flow and gas kinetic energy.

The ideal gas equation is sufficient to account for QED. We have:

$$\frac{\rho v^2}{2} + p + \frac{3}{2} p = \frac{\rho v^2}{2} + NkT + \frac{3}{2} p$$

$$i = p + \frac{3}{2} p = NkT + \frac{3}{2} p = \frac{5}{2} p$$

$$P = \frac{\rho v^2}{2} + p = \frac{\rho v^2}{2} + NkT$$

A very important notion when interpreting physical phenomena with gas properties is the "total pressure", P, which is the sum of static pressure and dynamic pressure due to the gas kinetic energy density. When the gas properties are uniform everywhere, we see and measure nothing; *i.e.* no variations are detectable and vacuum conditions obtain. However this vacuum (space evacuated of elementary particles and fields) is not an absolute void, since it is full of cosmonic gas. Values of vacuum properties are taken as our datum of physical measurements. This is the reason for the impossibility to detect the vacuum directly.

In a gas, two kinds of flow must be considered. Irrotational flows occur when viscosity is negligible: the total pressure is uniform, although the static pressure and the kinetic energy density may vary in complementary fashion as in the Bernouilli effect. Pure classical mechanical phenomena are then observed. Rotational flows occur when gradients of total pressure are present. They are always accompanied by vorticity, circulation and viscosity effects due to rotation of the gas around vortex lines. Viscosity effects in gases, ordinarily small compared to pressure effects, produce shear phenomena which allow propagation of transversal shear waves.

Gradients of total pressure parallel to streamlines control variations of velocity. Gradients of total pressure normal to streamlines control their curvature. These phenomena are quantitatively described by the right-hand (three-finger) rule in hydrodynamics, which has an exact counterpart in electromagnetism. Accordingly, total gas pressure represents the product of electric charge density by the electric potential: $P = \rho_e \psi$. The total pressure gradient (with a change of sign) corresponds to the product of electric charge density by the electric field, the source of electrostatic forces: $-\nabla P = \rho_e E$. The vortex lines of the gas in the direction of vorticity give the magnetic field lines of force. Both lines respond to the same mathematics, while a tension is produced along the vortex lines and a pressure arises normal to the lines. The total pressure gradient component normal to streamlines, expressed as the product of mass flow density by the vorticity in a gas, corresponds to the cross-product of electric current density by the magnetic field, the source of induction forces: $-\nabla P_n = \rho v \times 2\omega = \rho_e v \times B$. Given the similarity between the gas part and the electromagnetic part in this equation, a relation between ρ and ρ_e might be expected. This relation is supplied by Newton gravitational constant G, the square root of which is exactly ρ_e/ρ electrostatic units per gram. Other non-dimensional factors may be involved. In statics, Newton's G multiplies the grams^2 of scalar gravity, while a $G = 1$ multiplies the e.s.u.^{2} of scalar electrostatics.

The phenomenon of gravitation is accounted for by the Clausius term added in the equation of state of a non-ideal gas. It is a negative dimensionless term, equal to the ratio of half the volume of a cosmon's sphere of exclusion b to the total volume V, or the ratio of cosmon diameter (σ) over mean free path L. These ratios depend exclusively on the ratio of cosmon number density N over the absolute maximum value $N_{\max}$. This term multiplies the pressure p of the ideal gas equation to become the gravitational term, which is subtracted from p.

$$p(1-b/V) = NkT = p(1-\sigma/L) = p\left[1 - 2\pi N/(N_{\max} - N)\right]$$

Dividing by the cosmon number density, we obtain a mean energy and mass per cosmon, only due to their velocity relative to a particular inertial frame of reference. Separating parameters, dividing by N, and including all energy terms, we obtain the mean energy and mass per cosmon:

$$\frac{\mu v^2}{2} + \frac{5}{2}p\left(\frac{1}{N} - \frac{2\pi}{N_{\max} - N}\right) = \frac{\mu v^2}{2} + \frac{5}{2}kT = \mu c_o^2$$

$$\mu = \frac{5}{2}\frac{kT}{c_o^2\left(1 - v^2/2c_o^2\right)}$$

In these formulas v is the Galilean velocity, giving the classical kinetic energy $v^2/2$ per unit mass.

The basic equations of micro- and macrophysics can be derived from the viscosity coefficient formula for gases. In terms of gas dynamics, Planck's constant is interpreted as the viscosity coefficient per particle of ether gas, *i.e.*, per cosmon. The mean free path of the cosmon corresponds to its deBroglie and Compton wavelengths, which implies that the mean energy speed of cosmons in the vacuum is the standard speed of light. Through the mean frequency of encounters between cosmons, we obtain the Einstein (mc^2) and Planck (hf) formulae for the mean cosmon energy. Consequently, the basic equations of physics apply even at the cosmonic gas level, a fact which explains the universality of these equations.

This gas ether model has an analogy in engineering practices used in high speed aerodynamics. Airfoil design under conditions of near sonic speed, first contemplated in the years around 1920, was advanced by the discovery of the Prandtl-Glauert transformation factor, which is identical in form to the Lorentz factor (the same symbols are used in both!). Aeronautical engineers apply this transformation to obtain optimum performance for conditions near the speed of sound where compressibility is important. We could use this analogy to say that the Lorentz transformation accounts for the compressibility of the cosmonic gas. If the ether medium were incompressible and non-viscous, the Lorentz transformation would not be required.

As mentioned previously, since cosmons have zero spin, they should be considered as bosons and their energy distribution in vacuum should follow Planck's energy distribution function for radiation in thermal equilibrium with the cosmonic gas, as observed in the background radiation. Because of its very low vacuum temperature (2.75° K), the cosmonic gas acts as a superfluid medium which neutralizes viscosity effects via mechanical resonance (quantum conditions), such that quasi-permanent concentrations and movements are possible. All elementary particles may be treated as polytropic gas spheres described by the same Lane-Emden function used for stellar gas models (Chandrasekkar 1938). The Lane-Emden function corresponds to the spherical coordinate function (Kompaneyets 1961), and provides the values of pressure, density and temperature at all radii of a concentration or particle.

The weak interaction is due to the superimposed spinning vortex velocity patterns of all elementary particles, which are minute gyroscopic magnets subject to precession in uniform magnetic fields and to attraction or repulsion in varying magnetic fields. This spinning vortex produces mechanical spin in all particles, as well as the magnetic moment and isospin of charged particles.

According to Maxwell, a charge density arises whenever there is a divergence (sum of partial differentials on three axes) of the electric field. Since in a spherical concentration the electric field varies from the center to infinity, electric charge density should be extended in space and is not infinitely concentrated at a point, as assumed ordinarily. The Lane-Emden function eliminates the need to normalize the infinities that emerge from standard model calculations based on the sacrosanct Coulomb law. It introduces a modification of the Coulomb law, whereby the invariant charge is replaced by a charge that varies with the radius from zero at the center to the standard value at infinity. This variation of charge is due to what is normally called polarization of the vacuum.

Since a concentration of a single charge would explode by repulsion, the light particles (leptons), especially, must be composed of two concentric charges of opposite sign. It is the attraction between these two charges which builds up the total pressure or electrical potential within the particle. Using the Gell-Mann formula for a particle charge $Q = t + Y/2$ modified to $Q = (t + Y) - Y/2$ we obtain for the central charge $(t + Y)$ an unbroken series from -13 to $+13$ when counted in $e/6$ units, covering all the known particles. Charge Q and hypercharge Y vary by $e/3$ units; isotopic spin t varies by $e/2$ units. Each elementary particle may be interpreted as a composite of two concentric charges of opposite sign totaling the particle charge. Only the down quark has two charges of the same sign that have to be balanced by a gravitational black hole. A black hole exists within each quark, giving a baryon number 1/3. The ratio of the two concentric charges identifies the particle. It is usually negative except for the down quark (two $-e/6$ charges) and the photon (two zero charges) where the ratio is $+1$.

Because the gravitational term is a product of the same pressure that accounts for all the effects observed in electromagnetism, similar effects are to be expected at the gravitational scale of high energy and mass densities. When this gravitational term exceeds the total pressure (static pressure plus dynamic pressure due to spin or orbiting speed), the cosmon number density increases continuously until it reaches its maximum value at the center of a concentration, which brings the temperature to absolute zero, producing what has been called the gravitational potential well of a black hole. But there are no singularities (infinite densities) and the strong force may be understood as an extreme value of the gravitational force occurring in the high density gradients of quarks, required to counterbalance the centrifugal forces due to spin and orbital motion and electrical repulsion between quarks within hadrons. A similar effect is to be expected in the vicinity of the massive black holes of astronomy.

The reverse is found in anti-quarks where the cosmon number density falls to zero at the center, producing an anti-black hole. Anti-quarks can only exist in the presence of a high quark density, as is observed in mesons. The ratio of electrical to gravitational forces varies greatly from one kind of particle to another. Gravity is negligible in leptons, while it is overwhelming in the black holes of quarks and in nuclei. An effect similar to electromagnetic induction and the weak interaction should be observed. Such effects are observed in atomic nuclei and inside hadrons, under the name "color charge". The three colors of quantum chromodynamics (QCD) seem to be associated with three permissible orthogonal spin axes inside hadrons.

All the mathematics for these effects is presently available, while the vacuum quantization of the cosmonic gas model elucidates how high-density phenomena associated with the strong force are actually produced by gravitation. This reduces the four forces of nature to three: electromagnetic, weak-magnetic and gravitational.

All moving elementary particles, including the photon, are accompanied by what is called in fluid dynamics a spherical vortex composed of an infinity of circular vortices, all centered on the trajectory of the particle. This velocity pattern is similar to a smoke ring. Not surprisingly, then, stationary and moving particles behave as waves in the gas.

Because of charge symmetry between particles and their anti-particles, the calculated values at the center of the electron are taken to be the same as for the vacuum properties. It is possible then, to estimate the vacuum mass density, pressure, cosmon number density, diameter and maximum number density. The same procedure can be applied to all other particles with rest mass to obtain variations of their gas properties along the radius.

CONCLUSION

An ether gas model, we may conclude, provides rational mechanical explanations for electromagnetism, gravitation, the general applicability of quantum or wave mechanics, the apparent non-locality of elementary particles including the photon—and the diffraction phenomenon—from the quantization of the ether medium into its ultimate entity, the cosmon.

REFERENCES

Chandrasekkar, S., 1938, "An Introduction to the Study of Stellar Structure," Dover, New York.

Einstein, A., 1905, On the electrodynamics of moving bodies [*Annalen der Physik* 17 (1905)], in: "The Collected Papers of Albert Einstein," Vol. 2, Princeton University Press.

Einstein, A., 1924, Über den Äther, *Verh. d. Schweizer. Nat. Ges.* 105(II):85-93.

Janossy, L., "Relativity Theory based on Physical Reality," Adad. Kiado, Budapest.

Kompaneyets, A.S., 1961, "Theoretical Physics," Dover, New York.

Kostro, L., 1992, An outline of the history of Einstein's relativistic ether conception, in: "Studies in the History of General Relativity," J. Eisenstaedt, A.J. Kox, eds., Birkhäuser, Boston.

Lévi-Leblond, Jean-Marc, 1976, One more derivation of the Lorentz transformation *Am. J. Phys.* 44:3.

Trempe, Jacques, 1990, Laws of light propagation in Galilean space-time, *Apeiron* 8:1.

Trempe, Jacques, 1992, Light kinematics in Galilean space-time, *Physics Essays* 5:1.

WHAT IS AND WHAT IS NOT ESSENTIAL IN LORENTZ'S RELATIVITY

Jan Czerniawski

Institute of Philosophy
Jagellonian University
Kraków, PL-31-044

The special theory of relativity has already existed for nearly a century, and the controversy between the adherents of its two versions is almost as old. While Einstein's version is widely accepted by the scientific community, its Lorentz's alternative is even far from being widely known. Worse still, its partisans are often accused of lack of competence. Of course, this strange state of affairs may be partly explained by various "external factors" – psychological, sociological, historical and the like. It is, however, more interesting to ask about possible "internal factors". Maybe, the followers of Lorentz's relativity are guilty to some extent.

First, they lay too much stress on their opposition to the orthodox Einstein's relativity. Some of them even consider Lorentz's relativity to be an alternative to the special theory of relativity as such. In fact, these theories are indiscernible with respect to empirical predictions. Thus, it would be better, if they were treated consistently as two different *interpretations* of the same physical theory.

Secondly, the Lorentzians dislike some Einstein-relativistic concepts too much. *Spacetime* is one of these concepts.[1] It has proved, however, very useful as a mathematical tool and it would be unreasonable to reject it too hastily. One should be only warned against ascribing too much ontological significance to this concept.

The *principle of relativity* is another such concept. Negative attitude of some Lorentzians with respect to it[1] is the result of confusing of this physical principle with the philosophical principle of *relativity of motion.*[2] While the first one, stating that the laws of nature have the same form in all *inertial frames of reference,*[3] is quite sound and empirically well confirmed, the second one, stating that it only makes sense to speak about motions of bodies relative to other bodies, is not only wrong, but also incompatible with the very theory of relativity!

Under the influence of Mach's philosophy, Einstein aimed at elimination of *absolute motions* from physics. This program has failed. Although there is no *absolute rest* in Einstein's special relativistic world, there are absolute motions in it, and not only the accelerated ones, but non-accelerated as well. All *inertial motions*, which determine inertial frames of reference, are absolute in the sense that the class of such motions is defined by the spacetime geometry alone,[3] with no reference to any bodies; what more, the differences between them are also absolute, as they may be described by suitable invariants.[4] No wonder that they may cause absolute effects, as in the case of the so-called "*clock paradox*". Finally, a serious blow at

Einstein's program was struck by his general relativity theory,[3] which even reintroduces (local) absolute rest frames, at least in standard models.[5-7]

Thirdly, almost all Lorentzians reject Einstein's *second postulate*.[1,7,8] Some points need to be clarified in this context. First of all, literally understood, this postulate states nothing more than the independence of light velocity of the motion of its source, which is common to the relativity theory and all ether theories, contrary to "ballistic" theories, like the one of Ritz. Unfortunately, this rather weak assumption is often confused with the so-called "*constancy of light velocity* hypothesis" (CLV),[9] according to which the value of light velocity is always equal to the universal constant c in any inertial frame of reference. It is CLV that is usually contested under the misnomer "Einstein's *light postulate*". In fact, it is not identical with the second postulate, but it is a consequence of both postulates, taken in conjunction.

The Lorentzians often take up the problem of the *one-way velocity* of light. They hold that only the *measure* of light velocity is equal to c in all inertial frames.[1,7,8] The light velocity itself is in general direction-dependent in consequence of the motion of the reference frame with respect to the ether, i.e., in the moving frames light propagation is *anisotropic*, contrary to CLV. The apparent *isotropy* of light propagation in moving frames is a matter of convention, depending on the choice of synchronization. Only the average *round-trip velocity* of light has a direct empirical sense, i.e. is non-conventional.

This last claim is, however, not very consequent. If the synchronization of clocks is conventional, why the length and time units are not? They are at least as important components of the frame-dependent perspective as the synchronization.

The Lorentzians are often strongly attached to the idea of *conventionality of synchronization*. They are probably afraid of being compelled to accept Einstein's point of view once rejecting this idea. But is the synchronization conventional any more than the units of length and time? In the Lorentzian world, as well as in the Einsteinian world, the so-called "absolute synchronization" may be achieved only as an *external* synchronization.[10] This in turn depends on some *internal* synchronization in a chosen *base frame*,[11] being thus secondary to this internal synchronization (and deserving rather the name "quasi-absolute synchronization"). Moreover, such quasi-absolute synchroniation introduces abitrary preference of the base frame among physically equivalent inertial frames.[10,11]

On the other hand, among possible internal synchronizations the *standard* synchronization is the only one consistent with the assumption of homogeneity and isotropy of space. Any non-standard synchronization must break at least one of these symmetries, which requires some reason to prefer rather this than that place or direction in space. For the non-standard synchronizations considered by the Lorentzians the second is the case.

The reason needed, however, cannot be given, unless the direction of the frame's absolute motion is known, which is impossible. With the standard synchronization there is no such trouble, so only it deserves the name of "the" synchronization in a given frame of reference. Furthermore, the adoption of non-standard synchronization would preclude the use of Lorentz transformations, which are strongly preferred over other possible kinematic transformations because of the well-established invariance of physical laws.[12] The synchronization is, therefore, conventional only in some trivial sense.[12,13]

Synchronization defines some relation between events. As this relation is *relative*, i.e. frame-dependent, the Lorentzians hesitate to call it "simultaneity", reserving this term for the *absolute simultaneity*.[8] But the term "absolute simultaneity" is as good for the last and so there is no reason for such terminological preference. Why not to standardize the terminology as much as possible? It might prevent many misunderstandings, and this would be in favour of the Lorentzians, who are in minority. On the other hand, it would by no means hinder them to express their point of view.

Is then the one-way velocity of light measurable? In principle it is - if this term means the light velocity in a given inertial frame. Still more, its value is always equal to c ,   in full accordance with CLV, so that the distinction between the one-way and round-trip velocities of light, as well as the distinction between velocity and its measure, prove redundant. But another distinction may be of some use, namely between the *velocity* of an object in a given reference frame and its *velocity relative to* another object in this frame. The second may be defined as the result of subtraction of velocities of the objects in question in the first sense.[14] Then, although the light propagation in any inertial frame is isotropic and its velocity in any such frame is always equal to c ,   in accordance with CLV, the propagation of light relative to any moving body (in particular: to its moving source) is not.   It is so, because its velocity is direction-dependent, as a result of subtraction of the body's velocity from the velocity of light, both in the same inertial frame. The so-called *relativistic composition* (or: *addition*) *rule* for velocities is in fact a *transformation rule* for them.

Is this the anisotropy the Lorentzians mean? Certainly not, as this is a relative, frame-dependent effect, which disappears in the body's rest frame. What do they mean, then? Of course some absolute effect. It could be the anisotropy of the *absolute velocity* of light relative to *absolutely moving* bodies. The definitions of these terms must be postponed at this moment. Nevertheless, it may be already said that it would be better, if the Lorentzians used the above expression, instead of speaking about the alleged anisotropy of light velocity in moving frames. Similarly, it would be better if the Einsteinians did not speak about the isotropy of light propagation with respect to any observer[15] without adding that they mean this propagation as viewed by this observer, i.e. in his rest frame.

The relativistic effects of *length contraction* and *time dilation* are sometimes held to be only apparent.[16,17] This opinion, wrong irrespective of interpretation one adopts,[13,18,19] is a result of misidentifying of *reality* with *absoluteness*, i.e. with frame-independence. The *rest length* of a body, equal to its length as determined in its *rest frame*, is no more real than its length as determined in any other inertial frame of reference, at least unless this rest frame is the preferred one. The same holds for the *proper time* as contrasted with time as determined in some inertial frame. The claim that physically identical bodies have "in reality" the same length, irrespective of their relative motion, results from equivocation, since the expression "length in the body's rest frame" does not mean the same for two bodies, unless they rest relative to each other. If that claim were true, it would follow that no body "in reality" moves, because it is at rest in its rest frame. The rest length of a moving body is not its actual length, but only the length the body would have if it were at rest.

At most the Lorentzians might be allowed to hold e.g. that the rest length of a body *may* be somehow more real than its lengths in the frames moving relative to it - namely in the case when this body is *absolutely at rest*. It would be better, however, if they adopted more neutral terminology, looking for as much agreement with the most competent Einsteinians as possible. And the latter ones will never insist on the claim that the relativistic effects are apparent or unreal.[19,20] There are also apparent effects in relativity - e.g. the *visual appearance* of rapidly moving objects and the *relativistic Doppler effect* may be viewed as composed of the real part, due to length contraction and time dilation, respectively, and the apparent part, due to finiteness of the velocity of light as the information carrier.[21]

On the other hand, the Einsteinians deny, as a rule, meaningfulness of the problem of causal explanation of these effects.[20] They regard them as mere results of the observer's perspective, or as *purely kinematical* effects. Is this claim legitimate?

It may be maintained as far as only qualitative aspect of the problem is considered. Different perspective can, of course, account for the fact that the same things look different. It cannot, however, explain the concrete quantitative differences.

Now, even if the latter are explained as effects of spacetime geometry, it does not preclude further explanation, if someone is not satisfied with the bare fact that material things behave in accordance with the Minkowskian spacetime geometry. This further explanation may well have dynamical character;[13] it may be even held that only such explanation is the genuine one, the kinematical explanation being a pseudo-explanation, which "explains" *idem per idem*. The old dynamic explanation of relativistic effects, given by Lorentz and others, may be almost unchanged adopted in the framework of the Einstein's relativity,[22] the only difference from the original version being the substitution for the ether rest frame some freely chosen inertial reference frame. The same is true about many other explanations, e.g. the explanations of the null results of *ether drift* experiments, like the Michelson-Morley experiment.[23]

Some peculiarities of the Lorentzians' terminology which cause misunderstandings were discussed above. The Einsteinians are also partly responsible for the misunderstandings. They usually confuse *absolute* with *invariant*. A quantity may be absolute, i.e. frame-independent, in two ways: either it is the same in all inertial frames, or it is defined with no reference to any frame. Thus invariance implies absoluteness, but not *vice versa*.

Now, let us assume that any body might have, apart from its different lengths in various frames of reference and apart from its invariant rest length, some "true" length, which would be a non-relational property, while the length in some frame of reference is a relational property, based on certain relation between the body and the frame. Such "true" length might be properly called *absolute length*. Similarly, a physical process might occupy some interval of the "true" time, which would be in the same sense absolute and then might be called *absolute time*. The absolute length and time in the above sense are not physical, but ontological concepts.

In a special-relativistic world the absolute length of a body would be affected by motion in the manner its (relative) length in any inertial frame is. The same would be true about the interval of the absolute time occupied by some physical process in moving matter. If so, then a preferred frame might be defined as the one, in which the effects of *absolute length contraction* and *absolute time dilation* disappear for the matter being at rest. The state of motion of this frame might be properly called *absolute rest*. The assumption of existence of *absolute rest frame* in this sense, as related to the ontological concepts of absolute length and time, operates thus on the ontological level[13] and cannot be rejected on the basis of the physical principle of relativity, which excludes only the possibility of determining the absolute rest frame by means of local physical experiments.[6,18,24]

It would be reasonable to assume that this absolute rest frame would coincide with some inertial frame of reference. By suitable choice of length and time units the (relative) length in this frame, measured by absolutely resting measuring rods, and the (relative) time, measured by absolutely resting clocks, could be made to coincide with the absolute length and time, respectively, thus becoming their measures. It may be shown that if an absolutely resting observer would introduce Cartesian space coordinates and a time coordinate in the standard way, and an observer in some other inertial frame would do the same, the respective coordinates would be connected by the formulae of the appropriate Lorentz transformation. Now, since any observer "sees" the physical phenomena through his own coordinates, the (relative) length and time in his rest frame would play in his description of the world strictly the same role as the absolute length and time for the observer being absolutely at rest.[18]

In terms of absolute length and time measures the *absolute velocity* of an object may be defined. It will, of course, coincide with the (relative) velocity in the absolute rest frame, which is thus its measure. Since the light velocity CLV speaks about is always determined in some inertial frame, it is not an absolute velocity in the above sense. Moreover, if there is no absolute rest frame, as the Einsteinians assume, then there is no absolute light velocity in this sense. It would be then safer for them to speak about invariance of light velocity[25] than about its absoluteness.

Let us assume that the absolute length contraction and time dilation are caused by the motion relative to the ether. It means that the rest frame of the ether must coincide with the frame, in which these absolute effects disappear for the resting matter. This would be, of course, the absolute rest frame. The terms "absolute rest" and "absolute motion" may then be redefined to mean "rest relative to ether" and "motion relative to ether", respectively. This will make them a little less mysterious.

The coincidence of the ether rest frame with the absolute rest frame does not mean that the ether, or the *cosmological substratum*, may be identified with the preferred inertial frame - even if it makes sense to speak of the rest relative to the ether, which is not the case for the so-called *Einstein's ether*.[26] The ether and the absolute rest frame are separate concepts. It is the ether that may serve as the *source of inertia*[7] - even in the absence of absolute rest.

If there is absolute rest, the absolute length of any body coincides with its length in the privileged frame and then in general does not coincide with its rest length. The absolute length contraction is simply identical with the difference between the absolute length and the rest length of a body; other absolute relativistic effects may be interpreted in a similar way. In contrast with the absolute length and with the rest length, the length of a body is *relative*, i.e. frame-dependent. The length contraction as a result of its (relative) motion in a given inertial frame is in the same sense relative. It may be explained as a common effect of its (eventual) absolute contraction and of (eventual) influence of the absolute motion of the frame in question on the length unit, time unit and simultaneity standards in this frame. It is, however, by no means apparent,[6,13,18] since it may be investigated in this frame by means of strictly the same techniques as the absolute contraction in the absolute rest frame, which is assumed to be real.

The same is true about other relativistic effects like the *mass increase* and the time dilation. Then, they are real irrespective of the existence of the absolute rest frame. Consequently, they may result in real effects like the one considered in connection with the clock paradox. The presence of such velocity-dependent effects cannot then serve as a basis for the proof of existence of the absolute rest.[7,8]

This alleged proof is sometimes stated otherwise: it is claimed that the existence of absolute velocity-dependent effects implies the existence of an absolute inertial system[7] (or: frame). This conclusion is quite legitimate as far as the expression "absolute inertial frame" does *not* mean "absolute rest frame", or "the privileged inertial frame". Moreover, it is a mathematical fact that some combinations of relative quantities can be invariants of Lorentz transformations and thus be in this sense absolute. Any effect that may be described in terms of such invariants, is in the same sense absolute. Thus, even if all velocities are relative, i.e. there are no absolute velocities (which is the case if there is no absolute rest frame), there may be absolute velocity-dependent effects.

What is then the difference between the Lorentzians' and Einsteinians' positions? They disagree mainly with respect to the assumption of existence of the privileged inertial frame, i.e. of the absolute rest frame. There is a deeper reason for this point of disagreement. Whereas the Lorentzians prefer the space-and-time picture of the world, the Einsteinians are attached to the spacetime four-dimensional picture. The controversy thus reduces to some ontological preference.[13,24] All other points of disagreement either result from this one, or are products of misunderstandings. Some of these misunderstandings may be eliminated by clarifying and conforming terminology, others - by avoiding invalid inferences.

REFERENCES

1. H.E. Ives, Derivations of the Lorentz transformations, *Philosophical Magazine* 36:392 (1945).
2. F.A. Muller, On the principle of relativity, *Found. Phys. Lett.* 5:591 (1992).
3. M. Friedmann, Relativity principles, absolute objects and symmetry groups, *in*: "Space, Time and Geometry," P. Suppes, ed., Reidel, Dordrecht (1973).
4. A. Aurilia and F. Rohrlich, Invariant relative velocity, *Am. J. Phys.* 43:261 (1975).
5. S. Weinberg, The cosmological principle, *in*: "Gravitation and Cosmology," Wiley, New York (1972).
6. W.L. Craig, The elimination of Newton's absolute time by relativity theory, *in*: "Physical Interpretations of Relativity Theory," British Society for the Philosophy of Science, London (1990).
7. S.J. Prokhovnik, The empty ghost of Michelson and Morley: a critique of the Marinov coupled-mirrors experiment, *Found. Phys.* 9:883 (1979).
8. P.-A. Ivert and T. Sjödin, Poincaré's principle determines the behaviour of moving particles and clocks, *Acta Phys. Hung.* 48:439 (1980).
9. K.F. Schaffner, Einstein versus Lorentz: research programmes and the logic of comparative theory evaluation, *Brit. J. Phil. Sci.* 25:45 (1974).
10. R. Mansouri and R.U. Sexl, A test theory of special relativity: I. Simultaneity and clock synchronization, *Gen. Rel. Gravit.* 8:497 (1977).
11. A.P. Stone, Non-standard clock synchronization in special relativity and the hypothetical ether frame, *Found. Phys. Lett.* 4:581 (1991).
12. P. Mittelstaedt, Conventionalism in special relativity, *Found. Phys.* 7:573 (1977).
13. D. Dieks, The "reality" of the Lorentz contraction, *Zeitschrift für allgemeine Wissenschaftstheorie* 15/2:330 (1984).
14. G. Cavalleri and Ø. Grøn, On the experimental indistinguibility between Lorentz theory of electrons and special relativity, *Lett. Nuovo Cim.* 18:508 (1977).
15. A. Eddington, The velocity of light, *in*: "The Mathematical Theory of Relativity," Cambridge Univ. Press, Cambridge (1960).
16. M. Born, Schein und Wirklichkeit, *in*: "Die Relativitätstheorie Einsteins," Springer, Berlin, (1920).
17. J.L. Synge, Apparent contraction of a moving body and apparent retardation of a moving clock, *in*: "Relativity: The Special Theory," ** Amsterdam (1965).
18. H.A. Lorentz, The principle of relativity for uniform translations, *in*: "Lectures on Theoretical Physics" Vol. 3, MacMillan, London (1931).
19. A. Einstein, Zum Ehrenfestschen Paradoxon, *Phys. Z.* 12:509 (1911).
20. C. Møller, Contraction of bodies in motion. The retardation of moving clocks. The clock paradox, *in*: "The Theory of Relativity," Clarendon Press, Oxford (1952).
21. V.F. Weisskopf, The visual appearance of rapidly moving objects, *Physics Today* 13/9:24 (1960); J. Terrell, Invisibility of the Lorentz contraction, *Phys. Rev.* 116:1041 (1959).
22. J.S. Bell, How to teach special relativity, in: "Speakable and Unspeakable in Quantum Mechanics," Cambridge Univ. Press, Cambridge (1987).
23. C. Møller, The Doppler effect, the aberration of light, and the dragging phenomenon according to the theory of relativity, *in*: "The Theory of Relativity," Clarendon Press, Oxford (1952).
24. N. Maxwell, Are probabilism and special relativity incompatible?, *Phil. Sci.* 52:23 (1985).
25. D. Bohm, The Lorentz transformation in Einstein's point of view, *in*: "The Special Theory of Relativity," Benjamin, New York (1965).
26. L. Kostro, Einstein's relativistic ether, its history, physical meaning and updated applications, *Organon* 24:219 (1988).

VACUUM SUBSTRATUM IN ELECTRODYNAMICS AND QUANTUM MECHANICS - THEORY AND EXPERIMENT

H.E. Wilhelm

Department of Materials Science and Engineering
The University of Utah, Salt Lake City, Utah   84112, U.S.A.

INTRODUCTION

The substratum of the vacuum is the carrier of the elementary force interactions, such as electromagnetic (EM), gravitational, or nuclear.[1] These basic fields occur as *excitations* whereas elementary particles are probably *defects* in the substratum. Experiments show that the substratum has physical properties, e.g. a magnetic permeability $\mu_0 \simeq 4\pi \times 10^{-7}$ Vs/Am, dielectric permittivity $\varepsilon_0 \simeq 10^{-9}/36\pi$ As/Vm, EM wave speed $c = 1/\sqrt{(\mu_0\varepsilon_0)} \simeq 3\times 10^8$ m/s, and EM wave resistance $z_0 = \sqrt{(\mu_0/\varepsilon_0)} \simeq 120\pi$ V/A.

The substratum has either very small or no *gravitational* mass density, and consists probably of positive and negative *gravitational* mass particles with positive *inertial* mass (confirmation of negative g-masses would invalidate equivalence principle). The substratum appears to be a superfluid since (subluminal) particles move in vacuum without experiencing retarding forces.[2,7]

Maxwell's equations for the *ideal vacuum* (no substratum defects or particles) are deductable from the wave equation for an isotropic, linear, elastic substratum:[1] $\rho \partial^2 s/\partial t^2 = (G + L)\nabla\nabla\cdot s + G\nabla\cdot\nabla s$. From the displacement field $s = s(r,t)$, the velocity $\upsilon = \partial s(r,t)/\partial t$ and rotation $\pmb{\varphi} = \frac{1}{2}\nabla\times s(r,t)$ fields of the substratum follow. The substratum has positive *inertial* mass density ρ, shear module G, and Lamé module L.

Transverse Substratum Waves. For transverse waves with $k \perp s$, i.e. $\nabla\cdot s(r,t) = 0$ and $\nabla\cdot\upsilon(r,t) = 0$, the elastic wave equation reduces to $\partial^2(\rho s)/\partial t^2 = - G\nabla\times\nabla\times s$. Hence: (i) $\partial(\rho\upsilon)/\partial t = - 2G\nabla\times\pmb{\varphi}$ and (j) $\nabla\cdot(\rho\upsilon) = 0$, with the integrals (k) $\rho\upsilon = \nabla\times A$ and (l) $2G\pmb{\varphi} = - \partial A/\partial t - \nabla\Phi$, and the kinematic equations (m) $\partial\pmb{\varphi}/\partial t = \frac{1}{2}\nabla\times\upsilon$ and (n) $\nabla\cdot\pmb{\varphi} = 0$. With the Heaviside identification,[1] $B = \alpha(\rho\upsilon)$, $H = \mu_0^{-1}\alpha(\rho\upsilon)$, $D = \beta\pmb{\varphi}$, $E = \varepsilon_0^{-1}\beta\pmb{\varphi}$, Maxwell's equations for the EM field E,B in the substratum frame S° follow from (i), (m), (n), (j) as: $\nabla\times E = - \partial B/\partial t$, $\nabla\times H = \partial D/\partial t$, $\nabla\cdot D = 0$, $\nabla\cdot B = 0$, where $D = \varepsilon_0 E$, $B = \mu_0 H$ by definition, and $\alpha = 1/(\varepsilon_0 G)^{1/2} = (\mu_0/\rho)^{1/2}$, $\beta = 2(\varepsilon_0 G)^{1/2} = 2(\rho/\mu_0)^{1/2}$ are constants [velocity of light in vacuum, $c = \sqrt{(G/\rho)} = 1/\sqrt{(\mu_0\varepsilon_0)}$].

The wave equations for the potentials $A(r,t)$ and $\Phi(r,t)$ are obtained, under consideration of the gauge, $\nabla\cdot A = - c^{-2}\partial\Phi/\partial t$, by elimination of υ and $\partial\pmb{\varphi}/\partial t$ from (k), (m) & (l), and $\nabla\cdot A$ from the divergence of (l). Since $A = \alpha A$ and $\Phi = \alpha\Phi$, the wave equations for the EM potentials $A(r,t)$ and $\Phi(r,t)$ in the substratum frame S° result as: $\partial^2 A/\partial t^2 = c^2\nabla^2 A$, $\partial^2\Phi/\partial t^2 = c^2\nabla^2\Phi$, where $\nabla\cdot A = - c^{-2}\partial\Phi/\partial t$ (Lorentz gauge).

In the substratum picture, magnetic $B = \alpha\rho\upsilon$ and electric $E = (\beta/\varepsilon_0)\pmb{\varphi}$ fields are essentially velocity $\upsilon(r,t)$ and rotation $\pmb{\varphi}(r,t)$ perturbations of the ether. These identifications permit a "fluid dynamic" explanation of EM forces. $G = \rho c^2$ (energy of *inertial* substratum mass density), α, and β are given in terms of ρ.

Frontiers of Fundamental Physics, Edited by M. Barone
and F. Selleri, Plenum Press, New York, 1994

Longitudinal Substratum Waves. Introduction of the substratum compression field $\Pi = \nabla \cdot s(r,t)$ into the elastic wave equation gives the longitudinal wave equation $\partial^2\Pi/\partial t^2 = C^2\nabla^2\Pi$ with *superluminal* wave speed $C = \sqrt{[(2G + L)/\rho]} > c\sqrt{2}$ (refutation of relativity theories). It is conceivable that these longitudinal substratum waves are gravitation waves with potential $V(r,t) = u^2\Pi(r,t)$ where u [m/s] is a constant.

The elastic substratum theory can be extended to consider elementary particles as substratum defects with finite life-times (as in solids). The observed uncertainty $\Delta p_i \Delta x_i > \tfrac{1}{2}h$ of microscopic systems is explainable by fluctuations of the substratum. Photons $\hbar\omega$ are excitations of the substratum (similar to phonons as excitations of ordinary matter). The vacuum substratum provides a physical basis for a unified field theory of electrodynamics, quantum dynamics, and gravitation.[3]

EM DOPPLER EFFECT IN SUBSTRATUM

The special relativity theory (STR) and Lorentz (L) covariant theories predict the physical state of a material body to depend on its velocity **v** relative to the observer.[1] Galilei (G) covariant electrodynamics shows that all known electrodynamic and optical phenomena are relative to the substratum frame $S°$.[4-9] In observation frames $S \neq S°$, we see the *appearances* of the physical processes which *actually* occur in the substratum.[4-9] As an example, the substratum nature of the EM Doppler effect is discussed.

Consider a light source Q fixed to the origin O of an inertial frame (IF) $S(r,t,w)$ with substratum velocity **w**, moving with a velocity **u** = −**w** relative to the substratum frame $S°(r°,t°,0)$. The EM wave frequency (ω) emitted by the source Q is measured (ω') by an observer fixed to the origin O' of an IF $S'(r',t',w')$, moving with a velocity **u'** = −**w'** relative to $S°$. Hence, the velocity **v** of the observer O' in S' *relative* to the light source Q in S is the *difference* **v** = **u'** − **u** = **w** − **w'** = **G-inv** of *absolute* (**u'**,**u**) or ether (**w**,**w'**) velocities. The G-covariant EM wave equations for the vector potentials[4] A,A' give the EM wave dispersions in the IFs $S(r,t,w)$ and $S'(r',t',w')$:

$$[(\partial/\partial t + \mathbf{w}\cdot\nabla)^2 - c^2\nabla^2]A(r,t) = 0: \quad A = A_0 e^{i(\mathbf{k}\cdot\mathbf{r} - \omega t)} \quad \Rightarrow \quad \omega = kc + \mathbf{k}\cdot\mathbf{w} \quad (1)$$

$$[(\partial/\partial t' + \mathbf{w'}\cdot\nabla')^2 - c^2\nabla'^2]A'(r',t') = 0: \quad A' = A'_0 e^{i(\mathbf{k'}\cdot\mathbf{r'} - \omega't')} \quad \Rightarrow \quad \omega' = k'c + \mathbf{k'}\cdot\mathbf{w'} \quad (2)$$

In the G-transformation, **r** = **r'** + **v**t', t = t', from IF $S(r,t,w)$ to IF $S'(r',t',w')$, the vector potential is an invariant, $A(r,t) = A'(r',t')$.[4] Hence, $\mathbf{k}\cdot\mathbf{r} - \omega t = \mathbf{k'}\cdot\mathbf{r'} - \omega't'$, or

$$\mathbf{k}\cdot(\mathbf{r'} + \mathbf{v}t') - \omega t' = \mathbf{k'}\cdot\mathbf{r'} - \omega't' \quad \Rightarrow \quad \omega' = \omega - \mathbf{k}\cdot\mathbf{v}, \quad \mathbf{k'} = \mathbf{k} \quad (3)$$

With **n** = **k**/k, **k'** = **k**, **n'** = **n** (k = $2\pi/\lambda$), combining of (1) to (3) gives the EM Doppler frequency ω' measured by the observer O' in S'(**w'**) for a light source Q(ω) in S(**w**):

$$\omega' = \omega[(1 + \mathbf{n}\cdot\mathbf{w'}/c)/(1 + \mathbf{n}\cdot\mathbf{w}/c)], \qquad \omega' \simeq \omega(1 - \mathbf{n}\cdot\mathbf{v}/c), \quad |\mathbf{w}| << c \qquad (4)$$

$$\omega' = \omega[1 - (\mathbf{n}\cdot\mathbf{v}/c)/(1 + \mathbf{n}\cdot\mathbf{w}/c)], \qquad \omega' \simeq \omega(1 - \mathbf{n}\cdot\mathbf{v}/c), \quad |\mathbf{w}| << c \qquad (5)$$

$$\omega' = \omega/[1 + (\mathbf{n}\cdot\mathbf{v}/c)/(1 + \mathbf{n}\cdot\mathbf{w'}/c)], \qquad \omega' \simeq \omega(1 - \mathbf{n}\cdot\mathbf{v}/c), \quad |\mathbf{w}| << c \qquad (6)$$

These formulas are identical (**v** = **w** − **w'**) and reduce for infraluminal velocities $|\mathbf{w}| << c$ to the classical formula.[1] The substratum nature of the EM Doppler effect is obvious in (4), and in (5) and (6) since **v** = **w** − **w'**. Since measuring clocks (EM oscillators) are retarded in the IFs S(**w**) and S'(**w'**) with substratum flow, the frequencies ω and ω' are related to their *uncorrected* measured (m) values by[9]

$$\omega = \omega_m(1 - \mathbf{w}^2/c^2)^{1/2}, \qquad \omega' = \omega_m'(1 - \mathbf{w'}^2/c^2)^{1/2} \qquad (7)$$

As a generalization, consider a light source Q moving with a velocity **U** in an IF S(**w**), i.e. a velocity **U** − **w** = **U°** relative to $S°$, and an observer O moving with a velocity **V** in S(**w**), i.e. a velocity **V** − **w** = **V°** relative to $S°$. In this case, the frequency ω' measured by the observer O is related to the frequency ω of the source Q by:

$$\omega' = \omega[(1 - \mathbf{n}\cdot\mathbf{V}^\circ/c)/(1 - \mathbf{n}\cdot\mathbf{U}^\circ/c)], \qquad \omega' \simeq \omega(1 - \mathbf{n}\cdot\mathbf{v}/c), \quad |\mathbf{U}^\circ| \ll c \tag{8}$$

$$\omega' = \omega[1 - (\mathbf{n}\cdot\mathbf{v}/c)/(1 - \mathbf{n}\cdot\mathbf{U}^\circ/c)], \qquad \omega' \simeq \omega(1 - \mathbf{n}\cdot\mathbf{v}/c), \quad |\mathbf{U}^\circ| \ll c \tag{9}$$

$$\omega' = \omega/[1 + (\mathbf{n}\cdot\mathbf{v}/c)/(1 - \mathbf{n}\cdot\mathbf{V}^\circ/c)], \qquad \omega' \simeq \omega(1 - \mathbf{n}\cdot\mathbf{v}/c), \quad |\mathbf{U}^\circ| \ll c \tag{10}$$

where

$$\mathbf{v} = \mathbf{V}-\mathbf{U} = \mathbf{V}^\circ-\mathbf{U}^\circ = \text{G-inv}, \quad \mathbf{V}^\circ = \mathbf{V}-\mathbf{w} = \text{G-inv}, \quad \mathbf{U}^\circ = \mathbf{U}-\mathbf{w} = \text{G-inv} \tag{11}$$

The velocity $\mathbf{v}$ between observer O and light source Q is G-invariant since $\mathbf{V}^\circ$ and $\mathbf{U}^\circ$ are the absolute velocities of O and Q in $S^\circ(0)$. The absolute nature of the Doppler effect in the formulas (8),(9) and (10) is obvious, which are identical by (11). In view of the retardation of the measuring clocks, the frequencies ω and ω' and their *uncorrected* measured (m) values are interrelated by[9]

$$\omega = \omega_m(1 - U^{\circ 2}/c^2)^{1/2}, \qquad \omega' = \omega_m(1 - V^{\circ 2}/c^2)^{1/2} \tag{12}$$

Thus, the EM Doppler effect is shown to be a substratum or absolute space and time effect. This theory is strictly applicable only to the vacuum, in the absence of ordinary material media. The relativistic Doppler effect, in particular the quadratic one, is unobservable in experiments.[1] In the presence of nonuniform gases or hyperdense plasmas, isoredshifts or isoblueshifts occur which may be dominant.[10]

ELECTRODYNAMICS OF MEDIA MOVING IN SUBSTRATUM

In order to explain the (usually negligible) speeding up or slowing down of EM waves in nonmagnetic dielectrics ($\varepsilon > \varepsilon_0$, $\mu = \mu_0$) observed in the Fizeau interferometer,[11] induced polarization $\mathbf{P} = (\varepsilon - \varepsilon_0)[(\mathbf{E} + \mathbf{w}\times\mathbf{B}) + (\mathbf{v} - \mathbf{w})\times\mathbf{B})]$ has to be added to the generalized, G-covariant Maxwell equations,[4] which generates a G-invariant polarization current $\mathbf{j}_P = \nabla\times[\mathbf{P}\times(\mathbf{v} - \mathbf{w})]$. Thus, the G-covariant electrodynamic equations for dielectric (ε), conducting (σ) media moving with a velocity field $\mathbf{v}(\mathbf{r},t)$ in the substratum are obtained for an arbitrary IF $S(\mathbf{r},t,\mathbf{w})$ with substratum velocity $\mathbf{w}$:[4,8]

$$\nabla\times(\mathbf{E} + \mathbf{w}\times\mathbf{B}) = - (\partial/\partial t + \mathbf{w}\cdot\nabla)\mathbf{B} \tag{13}$$

$$\nabla\times\mathbf{H} = (\partial/\partial t + \mathbf{w}\cdot\nabla)[\varepsilon_0(\mathbf{E} + \mathbf{w}\times\mathbf{B}) + \mathbf{P}] + \mathbf{j} - \rho\mathbf{w} + \nabla\times[\mathbf{P}\times(\mathbf{v} -\mathbf{w})] \tag{14}$$

$$\nabla\cdot[\varepsilon_0(\mathbf{E} + \mathbf{w}\times\mathbf{B}) + \mathbf{P}] = \rho, \qquad \nabla\cdot\mathbf{B} = 0 \tag{15}$$

where

$$\mathbf{j} - \rho\mathbf{v} = \sigma[(\mathbf{E} + \mathbf{w}\times\mathbf{B}) + (\mathbf{v} - \mathbf{w})\times\mathbf{B}], \qquad \mathbf{D} = \varepsilon\mathbf{E}, \quad \mathbf{B} = \mu\mathbf{H} \tag{16}$$

are Ohm's law and the constitutive EM relations (standard MKS notation). The Minkowski electrodynamics is flawed by an nonsymmetric stress-energy tensor (predicting unobservable torques on ponderable bodies) and the contradictions of L-covariance.[1]

Equations (13)-(16) simplify for nonconducting media ($\sigma = 0$, $\rho = 0$) with uniform velocity field $\mathbf{v}$ in an IF $S(\mathbf{r},t,\mathbf{w})$ with substratum velocity $\mathbf{w}$ to: $\nabla\times\mathbf{E}^\circ = - d\mathbf{B}/dt$, $\nabla\times\mathbf{H}$ $= d(\varepsilon_0\mathbf{E}^\circ + \mathbf{P})/dt + \nabla\times(\mathbf{P}\times\mathbf{v}^\circ)$, $\nabla\cdot(\varepsilon_0\mathbf{E}^\circ + \mathbf{P}) = 0$, $\nabla\cdot\mathbf{B} = 0$, $\mathbf{P} = (\varepsilon - \varepsilon_0)(\mathbf{E}^\circ + \mathbf{v}^\circ\times\mathbf{B})$, where $d/dt = \partial/\partial t + \mathbf{w}\cdot\nabla$, $\mathbf{E} + \mathbf{w}\times\mathbf{B} = \mathbf{E}^\circ$, and $\mathbf{v} - \mathbf{w} = \mathbf{v}^\circ$. By elimination, $\nabla\times\mathbf{B}/\varepsilon^*c^{-2}$ $= \{d\mathbf{E}^\circ/dt + (1 - \varepsilon^{*-1})[2\mathbf{v}^\circ\cdot\nabla\mathbf{E}^\circ - \nabla(\mathbf{v}^\circ\cdot\mathbf{E}^\circ)]\}$ for $v^{\circ 2} \ll c^2/\varepsilon^*$, where $\varepsilon^* = \varepsilon/\varepsilon_0$. Multiplication by $\nabla\times$ results in the fundamental equation for the magnetic field $\mathbf{B}$ in a moving ($\mathbf{v}$) uniform dielectric (ε^*) in an IF $S(\mathbf{r},t,\mathbf{w})$:

$$\varepsilon^{*-1}c^2\nabla^2\mathbf{B} = (\partial/\partial t + \mathbf{w}\cdot\nabla)^2\mathbf{B} + 2(1-\varepsilon^{*-1})(\mathbf{v}-\mathbf{w})\cdot\nabla(\partial/\partial t + \mathbf{w}\cdot\nabla)\mathbf{B}, \quad v^{\circ 2}\ll c^2/\varepsilon^* \tag{17}$$

Hence, plane EM waves $\mathbf{B} = \mathbf{B}_0 e^{i(\omega t - \mathbf{k}\cdot\mathbf{r})}$ show in a dielectric, with uniform phase speed $c(\omega) = c/\sqrt{\varepsilon^*(\omega)}$ and velocity $\mathbf{v}$ in an IF $S(\mathbf{w})$, the dispersion $\omega = \omega(\mathbf{k})$:

$$(\omega - \mathbf{k}\cdot\mathbf{w})^2 - 2[1-\varepsilon^*(\omega)^{-1}][(\mathbf{v} - \mathbf{w})\cdot\mathbf{k}](\omega - \mathbf{k}\cdot\mathbf{w}) - k^2c(\omega)^2 = 0, \quad v^{\circ 2} \ll c(\omega)^2 \tag{18}$$

Since $|(\mathbf{v} - \mathbf{w})\cdot\mathbf{k}|^2 << k^2 c(\omega)^2$ for $(\mathbf{v} - \mathbf{w})^2 << c(\omega)^2$, (18) has the simple solution $(\omega - \mathbf{w}\cdot\mathbf{k}) \simeq kc(\omega) + [1 - \varepsilon^*(\omega)^{-1}](\mathbf{v} - \mathbf{w})\cdot\mathbf{k}$ for $(\mathbf{v} - \mathbf{w})^2 << c(\omega)^2$. Hence, the phase velocity $V = \omega/k$ of the EM waves in the moving $(\mathbf{v})$ dielectric is in $S(\mathbf{w})$ $(\mathbf{n} = \mathbf{k}/k)$:

$$V(\omega) = c(\omega) + \mathbf{w}\cdot\mathbf{n} + [1 - n(\omega)^{-2}](\mathbf{v} - \mathbf{w})\cdot\mathbf{n}, \quad (\mathbf{v} - \mathbf{w})^2 << c(\omega)^2 \quad (19)$$

where $n(\omega) = \sqrt{\varepsilon^*(\omega)}$ is the refractive index. Due to the Doppler effect (9), each atom of the moving $(\mathbf{v})$ dielectric "sees" an EM wave of apparent frequency $\omega' \simeq \omega - \mathbf{k}\cdot\mathbf{v} = \omega[1 - (k/\omega)v\cos\theta] \simeq \omega\{1 - [v/c(\omega)]\cos\theta\}$, $|\mathbf{w},\mathbf{v}| << c(\omega)$. By Taylor expansion, $n(\omega') \simeq n(\omega) + [dn(\omega)/d\omega](\omega' - \omega) = n(\omega)\{1 - (v/c(\omega))\cos\theta[d\ell nn(\omega)/d\ell n\omega]\}$, $|\mathbf{w},\mathbf{v}| << c(\omega)$, where $c(\omega) = c/n(\omega)$ and $\theta = \angle(\mathbf{k},\mathbf{v})$. In the same approximation, the Doppler shifted phase velocity (19) of the EM wave in the moving $(\mathbf{v})$ dielectric is in $S(\mathbf{w})$:

$$V(\omega) = c(\omega) + [1 - n(\omega)^{-2} + d\ell nn(\omega)/d\ell n\omega]|\mathbf{v}|\cos\theta + |\mathbf{w}|\cos\phi/n(\omega)^2, \quad |\mathbf{w},\mathbf{v}| << c(\omega) \quad (20)$$

where $\phi = \angle(\mathbf{k},\mathbf{w})$. This fundamental equation shows that an EM wave $(\omega,\mathbf{k})$ is speeded up or slowed down (depending on the angles θ and ϕ) by the moving dielectric $(\mathbf{v}$, Fizeau experiment[1]) and ether flow $(\mathbf{w}$, Hoek experiment[1]) in the observation frame $S(\mathbf{w})$. *L-covariant electrodynamics can not explain the substratum $(\mathbf{w})$ effect in (20).*

CRUCIAL SUBSTRATUM EXPERIMENTS

In experiments, the substratum effects either (i) compensate and are hidden or (ii) are observable by measurement. G-covariant electrodynamics explains all crucial electrodynamic and optical experiments, but only a few can be discussed here.

Michelson-Morley Experiment. On the optical table of the MM-interferometer, a light ray from a monochromatic source is split by a semitransparent mirror P into two perpendicular, coherent beams of lengths $L_1 = PM_1$ and $L_2 = PM_2$, which are reflected by the mirrors M_1 and M_2, and then pass through P to a telescope T.[1] The fringes observed in T permit to calculate the difference ΔT of the *2-way* light travel times of the interfering beams "1,2", $T_1 = L_1(\theta)/c(\theta) + L_1(\pi-\theta)/c(\pi-\theta)$ and $T_2 = L_2(\pi/2+\theta)/c(\pi/2+\theta) + L_2(\pi/2-\theta)/c(\pi/2-\theta)$, where θ is the angle the path $P\rightarrow M_1$ forms with the substratum velocity $\mathbf{w}$. With the 1-way light velocity $c(\theta) = (c^2 - w^2\sin^2\theta)^{1/2} + |\mathbf{w}|\cos\theta$, and length of a rod $L(\theta) = L_0(1 - w^2/c^2)^{1/2}/[1 - (w/c)^2\sin^2\theta]^{1/2}$, in the θ-direction,[9] one finds the θ-independent results:

$$T_1 = 2L_1(\theta)[1-(w/c)^2\sin^2\theta]^{1/2}/c(1-w^2/c^2): \quad T_1 = (2L_{10}/c)/(1 - w^2/c^2)^{1/2} \quad (21)$$

$$T_2 = 2L_2(\theta+\pi/2)[1-(w/c)^2\cos^2\theta]^{1/2}/c(1-w^2/c^2): \quad T_2 = (2L_{20}/c)/(1 - w^2/c^2)^{1/2} \quad (22)$$

Due to the compensation of the *anisotropy* of the $L(\theta)$ and $T(\theta)$ ether effects, the 2-way light travel times T_1 and T_2 are independent of the anisotropy in the IF $S(\mathbf{w})$: $\Delta T = T_1 - T_2 = \Delta T_0/(1 - w^2/c^2)^{1/2}$. Hence, the interference pattern can not change when the optical table, on which the interferometer is rigidly mounted, is turned into any direction of space. *This is exactly what the MM-experiment shows.*[1]

The STR claims that the MM-experiment proves *isotropic* light propagation in vacuum and nonexistence of ether. *This experiment does not measure the 1-way velocity of light nor does nonobservation (compensation) of an effect (ether) in one experiment prove the effect (ether) not to exist.*

Ives-Stilwell Experiment. This experiment uses hydrogen (H_2,H_3) canal rays in a Dempster vacuum tube.[11] Hydrogen ions are passed from an arc through a hole in an electrode "1" of zero potential, and accelerated by an electrode "2" at high negative potential ($\sim$ -20,000 to -30,000 V) with a hole, through which the accelerated particles move with a velocity $\mathbf{v} \sim c/10$ to the end of the tube which faces a spectrometer slit SS. The latter is illuminated by photons of frequencies (i) υ emitted by the stationary particles ($H_{2,3}$) from the arc, and (ii) $\nu_\pm$ (Doppler shifted satellite lines) emitted by the accelerated particles of velocity $\mathbf{v}$ in the directions of $\pm \mathbf{v}$ (a lateral mirror just in front of the hole of electrode "2" reflects the ν_- photons towards SS).

The spectrometer S measures wavelengths (by comparing them with the grating's spacing). The wavelengths λ and $\lambda_\pm$ corresponding to ν and $\nu_\pm$ are $\lambda = c/\nu$ and $\lambda_\pm = c/\nu_\pm$, where (i) $\nu = \nu_0(1 - \mathbf{w}^2/c^2)^{1/2} \simeq \nu_0$ since the *stationary* atomic clocks (arc) are retarded by moving with velocity $-\mathbf{w} \sim 3\times10^5$ m/s $<< c$ relative to S°, and (ii) $\nu_\pm \simeq \nu_0(1 - \mathbf{v}^{\circ 2}/c)^{1/2}(1 \pm |\mathbf{v}|/c)$ by the Doppler effect (9) for the atomic clocks *moving* with velocity $\mathbf{v}^\circ = \mathbf{v} - \mathbf{w}$ relative to S°. Expansion gives for $|\mathbf{w},\mathbf{v},\mathbf{v}^\circ| < c$:

$$\lambda \simeq \lambda_0(1 + \tfrac{1}{2}\mathbf{w}^2/c^2) \simeq \lambda_0, \quad \lambda_\pm \simeq \lambda_0[1 - (\pm)|\mathbf{v}|/c + \tfrac{1}{2}\mathbf{v}^{\circ 2}/c^2], \quad \lambda_0 = c/\nu_0 \quad (23)$$

where λ_0 and ν_0 are wavelength and frequency of the atomic clocks when at rest in the substratum (S°). Accordingly, the center of gravity $\lambda_c = \lambda_0 + \Delta\lambda$ of the satellite lines $\lambda_\pm$ is displaced from λ_0 by $\Delta\lambda = \tfrac{1}{2}\lambda_0 \mathbf{v}^{\circ 2}/c^2$. *These results are in full agreement with Ives-Stilwell who measured $\Delta\lambda$ and, thus, the velocity $\mathbf{v}^\circ = \mathbf{v} - \mathbf{w} \simeq \mathbf{v}$ of the atomic clocks relative to the substratum.*[11] The widespread opinion that the Ives-Stilwell experiment ($\theta=0$) proves the transverse ($\theta=\pi/2$) Doppler effect (STR) is inapplicable.

Wilson Experiment. This experiment confirms the ether effect $\varepsilon\mathbf{w}\times\mathbf{B}$ in the G-covariant EM field equations.[4] A hollow dielectric cylinder ($\varepsilon > \varepsilon_0$) of height d, inner radius R_1 and outer radius R_2, rotates uniformly with angular velocity $\mathbf{\Omega}$ about its symmetry axis, in a homogeneous magnetic field $\mathbf{B}_0$ in the axial direction. The dielectric cylinder surfaces $R_{1,2} \pm 0$ are surrounded by thin metal coverings M ($\varepsilon_M \simeq \varepsilon_0$) of height d. Wilson showed that rotation of the dielectric cylinder in $\mathbf{B}_0$ charges the capacitor formed by the metal coverings, and measured their surface charges.[1]

Since rotation is G-invariant motion in S°, we evaluate the G-invariant surface charges in the quasi-IF ($\Omega R_{1,2} << c$) $S(r,\phi,z,t,\mathbf{w})$ rotating with the dielectric cylinder. The substratum velocity and the magnetic field are $\mathbf{w}(r) = -\mathbf{\Omega}\times\mathbf{r} = -\Omega r \mathbf{a}_\phi$ and $\mathbf{B} = B_0 \mathbf{a}_z$ in S for $0 \leq r < \infty$, $|z| < \infty$ ($\mathbf{B} = $ G-inv). The G-covariant equation $\nabla \cdot \varepsilon(\mathbf{E} + \mathbf{w}\times\mathbf{B}) = \rho$ determines the surface charge densities at $r = R_{1,2} \pm 0$ ($\mathbf{n} = \mathbf{a}_r$):[4] $\rho^* = \mathbf{n}\cdot[\varepsilon(\mathbf{E} + \mathbf{w}\times\mathbf{B})]$. The radial electric field is short-circuited, $\mathbf{E} = \mathbf{0}$, in the dielectric by a sliding circuit across the metal fittings.[1] Thus, we find for the surface charge densities:

$$\rho^*(r=R_{1,2}) = \pm[\varepsilon\mathbf{w}(r=R_{1,2}\pm 0) - \varepsilon_0\mathbf{w}(r=R_{1,2}\mp 0)]B_0: \ \rho^*(r=R_{1,2}) = \pm(\varepsilon-\varepsilon_0)(\Omega R_{1,2})B_0 \quad (24)$$

since $\mathbf{w}\times\mathbf{B} = \mathbf{w}(r)B_0 \mathbf{a}_z$ is continuous across the dielectric interfaces at $r = R_{1,2}$. The result (24) agrees with all aspects of the Wilson experiment,[1] which became famous since it is explainable (formally) by Minkowskian but not by Hertzian electrodynamics. *G-covariant electrodynamics shows that the surface charges are a substratum effect.*

Unipolar Induction Without Relative Motion. Already Faraday noted that a radial electric field $\mathbf{E} = - \mathbf{v}\times\mathbf{B}$ is induced in the iron of the pole areas of a cylindrical magnet rotating ($\mathbf{v}$) about its axis.[1] This shows that the magnetic field lines $\mathbf{B}$ do not rotate with the magnet but are fixed in the vacuum substratum (G-invariance). For analytical reasons, consider a unipolar generator consisting of a thin copper disc of radius a and height $2\delta << a$ in the narrow gap (a $>> 2\delta$) of the S and N poles of two coaxial cylindrical magnets, when conducting disc and magnet are rotating rigidly with the same angular velocity $\mathbf{\Omega}$ about their symmetry axis. In the quasi-IF $S(r,\phi,z,t,\mathbf{w})$ ($\Omega a << c$) rotating with this system, the substratum rotates with velocity $\mathbf{w} = - \mathbf{\Omega}\times\mathbf{r}$. Hence, the G-covariant electrodynamic equations for the stationary EM fields of the disc are in S: $\nabla\times(\mathbf{E} + \mathbf{w}\times\mathbf{B}) = 0$, $\nabla\times\mathbf{B} = \mu_0(\mathbf{j} - \rho\mathbf{w})$, $\nabla\cdot[\varepsilon_0(\mathbf{E} + \mathbf{w}\times\mathbf{B})] = \rho$, $\nabla\cdot\mathbf{B} = 0$, $\mathbf{j} = \sigma\mathbf{E}$, in the thin disc approximation ($\mathbf{w}\cdot\nabla\mathbf{B} = 0$, $\mathbf{w}\cdot\nabla(\mathbf{E} + \mathbf{w}\times\mathbf{B}) = 0$).[8] Since $B_\phi = 0$ and $(\nabla\times\mathbf{B})_{r,z} = 0$, also $(\mathbf{j},\mathbf{E})_{r,z} = 0$, while $E_\phi = 0$ in steady state. Thus, $\mathbf{j} = \mathbf{0}$ and $\mathbf{E} = \mathbf{0}$ in the disc, and the electrodynamic equations for the thin disc become in $S(\mathbf{w} = -\Omega r \mathbf{a}_\phi)$:[8]

$$\boldsymbol{w}\times\mathbf{B} = -\nabla\Phi, \quad \nabla\times\mathbf{B} = -\mu_0\rho\boldsymbol{w}, \quad \rho = \varepsilon_0\nabla\cdot(\boldsymbol{w}\times\mathbf{B}), \quad \nabla^2\Phi = -\rho/\varepsilon_0; \quad r \leq a, \ |z| \leq \delta \quad (25)$$

By elimination, $dB_z/B_z = [d(\Omega^2 r^2/c^2)]/(1-\Omega^2 r^2/c^2)$. Hence, the magnetic $\mathbf{B}(r)$, space charge $\rho(r)$, and ether potential $\Phi(r)$ fields are in the thin disc, $r \leq a$, $|z| \leq \delta$:

$$B_z(r) = B_0/(1-\Omega^2 r^2/c^2) \quad \Leftrightarrow \quad B(r) \simeq B_0, \quad \Omega a \ll c \qquad (26)$$

$$\rho(r) = -2\varepsilon_0\Omega B_0/(1 - \Omega^2 r^2/c^2)^2 \quad \Leftrightarrow \quad \rho(r) \simeq -2\varepsilon_0\Omega B_0, \quad \Omega a \ll c \qquad (27)$$

$$\Phi(r)-\Phi(0) = -\tfrac{1}{2}cB_0(c/\Omega)\ell n(1 - \Omega^2 r^2/c^2) \quad \Leftrightarrow \quad \Phi(r)-\Phi(0) \simeq \tfrac{1}{2}\Omega B_0 r^2, \quad \Omega a \ll c \qquad (28)$$

The EMF across the copper disc is $\Phi(a)-\Phi(0) \simeq \tfrac{1}{2}\Omega B_0 a^2$, in accord with the measurements of E.H. Kennard.[1] *This EM induction (i) is a substratum effect by the EM field equations in (25), which vanish for $\boldsymbol{w} = 0$, and (ii) can not be explained by L-covariant electrodynamics.* The theory of Landau-Lifshitz by which, in the corotating frame S, the external sliding circuit rotates *across* the $\mathbf{B}$-field *resting in* S, is false since the $\mathbf{B}$-lines do not corotate with the magnet and their integration $\oint\mathbf{E}\cdot d\mathbf{r}$ is incomplete.

Fizeau Experiment. A light ray from a monochromatic source is split into two coherent components "1" and "2" by a semi-transparent mirror M_0. Beam "1" is guided clockwise along the *rectangular* path $M_0{\rightarrow}M_3{\rightarrow}M_2{\rightarrow}M_1{\rightarrow}M_0{\rightarrow}T$, whereas beam "2" is guided counter-clockwise along the path $M_0{\rightarrow}M_1{\rightarrow}M_2{\rightarrow}M_3{\rightarrow}M_0{\rightarrow}T$, by the mirrors $M_{0,1,2,3}$. These beams pass through two water (ε) sections of length L *within* the light paths (i) M_3-M_2 with water velocity $+v$ in the direction $M_3{\rightarrow}M_2$ and (ii) M_1-M_0 with water velocity $-v$ in the direction $M_1{\rightarrow}M_0$, whereas the remaining light paths are in air (ε_0). Fizeau observed a shift of the interference pattern in the telescope T, after switching on the water flows ($\pm v$) in the connected tubes.[1]

By (19), the phase velocities $V^{\pm}_{1,2}$ of the beams "1" and "2" in the 'upper' ($+$) and 'lower' ($-$) tubes with flowing water ($\pm v$) are ($n = k/k$):

$$V^{+}_{1,2}(\omega) = c(\omega) \pm [1 - n(\omega)^{-2} + d\ell n n(\omega)/d\ell n\omega]|v| \pm \mathbf{w}\cdot\mathbf{n}/n(\omega)^2, \quad |w,v| \ll c(\omega) \qquad (29)$$

$$V^{-}_{1,2}(\omega) = c(\omega) \pm [1 - n(\omega)^{-2} + d\ell n n(\omega)/d\ell n\omega]|v| - (\pm)\mathbf{w}\cdot\mathbf{n}/n(\omega)^2, \quad |w,v| \ll c(\omega) \qquad (30)$$

Accordingly, the beams "1" and "2" interfere in T with the phase time difference $\Delta t = t_2 - t_1 = (L/V^{-}_2 - L/V^{+}_1) + (L/V^{+}_2 - L/V^{-}_1)$, where $V^{+}_1 V^{-}_2 \simeq c(\omega)^2 \simeq V^{-}_1 V^{+}_2$ for $|w,v| \ll c(\omega)$. Hence, the observed fringe shift is $Z = c\Delta t/\lambda$ where

$$\Delta t \simeq 4L[1 - n(\omega)^{-2} + d\ell n n(\omega)/d\ell n\omega]|v|/c(\omega)^2, \quad |w,v| \ll c(\omega) \qquad (31)$$

Note that in the nominator of (31) the ether terms $\pm\mathbf{w}\cdot\mathbf{n}/n(\omega)^2$ subtract out rigorously, but occur in the exact denominators $V^{+}_1 V^{-}_2$ and $V^{-}_1 V^{+}_2$ as small effects, which are probably unmeasurable. *The Fizeau experiment is in complete agreement with (31).*

Hoek Experiment. This uses the same rectangular, optical circuit as the Fizeau experiment, M_0-M_1-M_2-M_3-M_0-T, through which the monochromatic light beams "1" (clockwise) and "2" (counter-clockwise) are guided to T. However, *only* in the "upper" light path $M_3{\rightarrow}M_2$ the beams "1" and "2" travel through a tube insert of length L $<$ $M_3 M_2$, filled with *resting* water (ε, $v = 0$ in laboratory S), in the opposite directions $M_3{\rightarrow}M_2$ and $M_2{\rightarrow}M_3$, respectively. Hoek oriented the water tube parallel to the (then estimated) Earth velocity $\mathbf{V}° = -\mathbf{w}$ relative to the substratum (S°), so that the ether velocity $\mathbf{w}$ is in the direction $M_2{\rightarrow}M_3$.

Since the water is at rest ($v = 0$) in the tube (S), (19) gives for the phase velocity of light $V(\omega) = c(\omega) + |w|\cos\phi/n(\omega)^2$, $|w| \ll c(\omega)$. Hence, the travel times of the beams "1" and "2" are $t_1 = L/[c/n(\omega) - |w|/n(\omega)^2] + L/(c + |w|) + t_0$ and $t_2 = L/(c - |w|) +$

L/[c/n(ω) + |w|/n(ω)²] + t₀, |w| << c(ω) (t₀ = remaining travel time of *each* beam in air).
Thus, the phase-time difference of the interfering light beams is found to be

$$\Delta t = t_2 - t_1 = (2L/c)(|w|/c)^3[1 - n(\omega)^{-2}], \qquad |w| << c(\omega) \tag{32}$$

Hoek's interferometer was not accurate enough to make a quantitative determination of
the substratum effect (32), since $\Delta t \sim (2L/c)(|w|/c)^3)$ is of the order 10^{-35} s on the
Earth. *L-covariant electrodynamics* ($w \equiv 0$) *can not explain the Hoek experiment.*

G-COVARIANT QUANTUM MECHANICS IN EM FIELDS

Quantum mechanics rests on the de Broglie relation λ = h/mv, which explains phe-
nomenologically the particle-wave dualism observed in experiments,[3] e.g. the diffraction of an electron beam by a slit. Since the empty STR vacuum can not carry matter
waves, these have to be understood as excitations of the substratum.

G-Covariant Schroedinger Equation. The classical energy of a particle (rest mass
m_0 in substratum, charge e) in the presence of EM potentials $A^\circ(r^\circ,t^\circ)$ and $\Phi(r^\circ,t^\circ)$ is
$E^\circ = (p^\circ - eA^\circ)^2/2m_0 + e\Phi^\circ$ in S°. By the operator interpretations E° = ih∂/∂t° and p°
= –ih∇°, the Schroedinger equation is in the substratum frame $S^\circ(r^\circ,t^\circ,0)$:

$$ih\partial\psi^\circ/\partial t^\bullet = [(-ih\nabla^\circ - eA^\circ)^2/2m_0 + e\Phi^\bullet]\psi^\circ \tag{33}$$

Transformation of (33) from $S^\circ(r^\circ,t^\circ,0)$ to an IF S(r,t,w), moving with a velocity u = –
w relative to S°, by means of the G-transformation r° = r + ut, t° = t, ∂/∂t° = ∂/∂t
– u·∇, ∇° = ∇ yields the Schroedinger equation for an arbitrary IF S(r,t,w):

$$ih(\partial/\partial t + w\cdot\nabla)\psi = [(-ih\nabla - eA)^2/2m_0 + e(\Phi - w\cdot A)]\psi \tag{34}$$

with

$$[\mu_0\varepsilon_0(\partial/\partial t + w\cdot\nabla)^2 - \nabla^2]A = \mu_0(j - \rho w) \tag{35}$$

$$[\mu_0\varepsilon_0(\partial/\partial t + w\cdot\nabla)^2 - \nabla^2](\Phi - w\cdot A) = \rho/\varepsilon_0 \tag{36}$$

are the EM wave equations [gauge ∇·A = –$\mu_0\varepsilon_0$(∂/∂t + w·∇)(Φ – w·A)], and ρ(r,t)
and j(r,t) are the charge and current sources of the EM potentials A(r,t) and Φ(r,t).[4]
The covariance of (34)-(35) in G-transformations from S(w) to arbitrary IFs S'(w'),
moving with velocities u' = w – w' relative to S, follows from the G-invariants:[4]

$$\partial/\partial t + w\cdot\nabla = \partial/\partial t' + w'\cdot\nabla' = \partial/\partial t^\circ, \quad \nabla = \nabla' = \nabla^\circ, \quad \psi(r,t) = \psi'(r',t') = \psi^\circ(r^\circ,t^\circ) \tag{37}$$

$$A = A' = A^\circ, \quad \Phi - w\cdot A = \Phi' - w'\cdot A' = \Phi^\circ, \quad j - \rho w = j' - \rho'w' = j^\circ, \quad \rho = \rho' = \rho^\circ \tag{38}$$

*Equations (34)-(38) represent a formulation of parabolic quantum mechanics, in which
both the Schroedinger and EM wave equations are G-covariant.* They permit evaluation
of substratum effects (w) on microscopic quantum and wave phenomena in IFs S(w)
with substratum velocity w. E.g., in an IF S(w), the energy of a free particle matter
wave (frequency ω, wave number k) is Doppler shifted, hω = hk·w + $(h^2/2m_0)k^2$.

G-Covariant Dirac Equation. In the absence of EM potentials A° and Φ°, the
dynamics of a high velocity (p°= mv°) particle (e,m_0) with G-invariant mass m =
$m_0/\sqrt{(1-v^{\circ 2}/c^2)}$ is deductable in the substratum frame S° from a nonlinear Hamiltonian,
which can be linearized: $H^\circ = c(p^{\circ 2} + m_0^2c^2)^{1,2} = -c\Sigma_i\alpha_ip_i^\circ$, where $p_0^\circ = m_0c$, and p_1° =
p_x°, $p_2^\circ = p_y^\circ$, $p_3^\circ = p_z^\circ$ (i = o, 1, 2, 3). Since $H^{\circ 2} = c^2(p^{\circ 2} + m_0^2c^2) = \frac{1}{2}c^2\Sigma_i\Sigma_jp_i^\circ p_j^\circ(\alpha_i\alpha_j$
$+ \alpha_j\alpha_i)$, the 4×4 matrices α satisfy the conditions: $\alpha_i\alpha_j + \alpha_j\alpha_i = 2\delta_{ij}$ and $\alpha_i^2 = 1$ for
i = o,1,2,3. Hence $(\sigma_{1,2,3}, I = \begin{pmatrix}1 & 0\\0 & 1\end{pmatrix},$ and $O = \begin{pmatrix}0 & 0\\0 & 0\end{pmatrix})$ are 2×2 matrices):

$$\boldsymbol{\alpha} = \begin{pmatrix} 0 & \sigma \\ \sigma & 0 \end{pmatrix}, \quad \alpha_0 = \begin{pmatrix} 1 & 0 \\ 0 & -1 \end{pmatrix}; \quad \sigma_1 = \begin{pmatrix} 0 & 1 \\ 1 & 0 \end{pmatrix}, \quad \sigma_2 = \begin{pmatrix} 0 & -i \\ i & 0 \end{pmatrix}, \quad \sigma_3 = \begin{pmatrix} 1 & 0 \\ 0 & -1 \end{pmatrix} \tag{39}$$

Accordingly, in the presence of EM interaction potentials $\Phi°(\mathbf{r}°,t°)$ and $\mathbf{A}°(\mathbf{r}°,t°)$, the Hamiltonian is $H° = - c\boldsymbol{\alpha}\cdot(\mathbf{p}° - e\mathbf{A}°) + e\Phi° - \alpha_0 m_0 c^2$ in $S°$. With the operator identification, $E° = ih\partial/\partial t°$ and $\mathbf{p}° = - ih\nabla°$, the wave equation $H°\Psi° = E°\Psi°$ is explicitly $[(E° - e\Phi°) + c\boldsymbol{\alpha}\cdot(\mathbf{p}° - e\mathbf{A}°) + \alpha_0 m_0 c^2]\Psi° = O$ in $S°$, where $\Psi°$ is a *column*-matrix of 4 scalar wave functions $\Psi°_\alpha$ ($\alpha = 1,2,3,4$). Multiplying this equation *on the left* by the operator $[(E° - e°\Phi) - c\boldsymbol{\alpha}\cdot(\mathbf{p}° - e\mathbf{A}°) - \alpha_0 m_0 c^2]$ and considering that $[\boldsymbol{\alpha}\cdot(\mathbf{p}°- e\mathbf{A}°)]^2 = (\mathbf{p}° - e\mathbf{A}°)^2 - eh\boldsymbol{\sigma}'\cdot\nabla\times\mathbf{A}°$ $[\boldsymbol{\sigma}' = \begin{pmatrix} \sigma & 0 \\ 0 & \sigma \end{pmatrix}$ is a 4×4 matrix, σ_i = Pauli matrices], yields the Galilean Dirac equation in the substratum frame $S°(\mathbf{r}°,t°,0)$:

$$[(ih\partial/\partial t° - e°\Phi°)^2 - c^2(-ih\nabla° - e\mathbf{A}°)^2 - m_0^2 c^4 + ehc^2\boldsymbol{\sigma}'\cdot\mathbf{B}° + iehc\boldsymbol{\alpha}\cdot\mathbf{E}°]\Psi° = O \tag{40}$$

since the EM field is $\mathbf{B}° = \nabla°\times\mathbf{A}°$, $\mathbf{E}° = - \nabla°\Phi° - \partial\mathbf{A}°/\partial t°$ in $S°$.[4] Equation (40) indicates that the charged particle (e,m_0) has (i) a magnetic moment $\boldsymbol{\mu} = (eh/2m_0)\boldsymbol{\sigma}'$ and (ii) an electric moment $\boldsymbol{\varepsilon} = (eh/2m_0 c)\boldsymbol{\alpha}$, associated with its spin $h/2$.

Since (40) holds only in the substratum frame $S°$, it is transformed by means of the G-transformations (37)–(38) from $S°(\mathbf{r}°,t°,0)$ to an IF $S(\mathbf{r},t,\mathbf{w})$. Thus, the generalized, Galilean Dirac equation is found for an IF $S(\mathbf{r},t,\mathbf{w})$ with substratum velocity $\mathbf{w}$:

$$\{[ih(\partial/\partial t + \mathbf{w}\cdot\nabla) - e(\Phi - \mathbf{w}\cdot\mathbf{A})]^2 - c^2(- ih\nabla - e\mathbf{A})^2 - m_0^2 c^4$$

$$+ ehc^2\boldsymbol{\sigma}'\cdot\mathbf{B} + iehc\boldsymbol{\alpha}\cdot(\mathbf{E} + \mathbf{w}\times\mathbf{B})\}\Psi = O \tag{41}$$

Equation (41) holds in any IF, since it is form-invariant in G-transformations from the IF $S(\mathbf{r},t,\mathbf{w})$ to an arbitrary other IF $S'(\mathbf{r}',t',\mathbf{w}')$ by (37)–(38), which imply G-invariance of the EM field $\mathbf{E},\mathbf{B}$ and the wave function column-matrix Ψ:

$$\mathbf{B} = \mathbf{B}' = \mathbf{B}°, \quad \mathbf{E} + \mathbf{w}\times\mathbf{B} = \mathbf{E}' + \mathbf{w}'\times\mathbf{B}' = \mathbf{E}°, \quad \Psi(\mathbf{r},t) = \Psi'(\mathbf{r}',t') = \Psi°(\mathbf{r}°,t°) \tag{42}$$

Since $\mathbf{B}$ is proportional to a velocity perturbation υ of the substratum, the magnetic field is necessarily G-invariant. It is clear that the characteristic *scalar substratum* (i) wave *speed* $c = c'$ and (ii) properties $\mu_0 = \mu_0'$ and $\varepsilon_0 = \varepsilon_0'$ are G-invariants.[4]

Equation (41) permits calculation of the effects of the substratum velocity $\mathbf{w}$ on microscopic quantum phenomena in arbitrary IFs $S(\mathbf{w})$. The electron spin is not a "relativistic effect", as the derivation based on absolute space and time demonstrates. The appearance of magnetic $(\boldsymbol{\mu})$ and electric $(\boldsymbol{\varepsilon})$ particle moments in (41) is remarkable, since only the particle properties e,m_0 are assumed in the Galilean Hamiltonian $H°$.[9] The particle spin $h/2$ can be shown to be an EM momentum effect induced in the substratum by a charged particle (e,m_0).

The tragedy of the STR and L-covariant theories is that they hold strictly only in the substratum frame $S°$ where $\mathbf{w} = 0$, the existence of which is denied through their unrealistic foundations (subjective relativism and empty vacuum concept).

REFERENCES

1. E. Whittaker. "A History of the Theories of Aether and Electricity,"Nelson & Sons, London (1951).
2. K.P. Sinha, E.C.G. Sudarshan, and J.P. Vigier, *Phys. Lett.* **114a**:298 (1986).
3. F. Selleri. "Wave Particle Duality," Plenum Press, London (1992);
   "The Double Nature of Subatomic Objects," Hadronic Press, Palm Harbor (1993).
4. H.E. Wilhelm, *Fusion Techn.* **6**:174 (1984); *Radio Sci.* **20**:1006 (1985).
5. H.E. Wilhelm, *Int. J. Math. & Math. Sci.* **8**:859 (1985); *Phys. Rev.* **D1**:2278 (1970).
6. H.E. Wilhelm, *Z. Naturforsch.* **45a**:736 (1990); *Z. Naturforsch.* **45a**:749 (1990).
7. H.E. Wilhelm, *Int. J. Math. & Math. Sci.* **14**:769 (1991); *Gal. Electrodyn.* **3**:28 (1992).

8. H.E. Wilhelm, *Apeiron* **13**:17 (1992); *Apeiron* **15**:1 (1993).

9. H.E. Wilhelm, *Hadronic J.*, *Suppl.* **8**:441 (1993); L. Janossy, *Acta Phys. Hung.* **17**:421 (1964).

10. R.M. Santilli, Isominkowskian representation of quasar redshifts and blueshifts, this Volume; *Lett. Nuovo Cimento* **37**:545 (1983).

11. H.E. Ives and G.R. Stilwell, *J. Opt. Soc. Am.* **28**:215 (1938); *J. Opt. Soc. Am.* **41**:369 (1941).

THE INFLUENCE OF IDEALISM IN 20TH CENTURY PHYSICS

Heather McCouat[1] and Simon Prokhovnik[2]

[1]Centre for Liberal and General Studies
[2]School of Mathematics
University of New South Wales
Sydney, NSW 2052

INTRODUCTION: INTELLECTUALS AND IDEALISM

In a critique of idealism, it would be somewhat incongruous if one were seen to be tracing 'threads' or philosophical themes through history, as each historical culture has its own particular configuration; but an academic sojourn at Olympia, the meeting-place of Ancient Greece, prompts reflection on the elements of cultural continuity between some parts of that earlier world and that of Western culture today. In both, we find similar forms of idealism; we can also identify material and social conditions which, though unalike in many ways, are in some respects comparable.

Ancient Greece was not culturally homogeneous.[1] The expanse and diversification of Mediterranean economic life after the fall of the Bronze Age empires resulted in two quite different intellectual and practical responses within the Greek world. Both were based on money economies – one saw the emergence of the materialist world-view of the Ionians, reflecting their busy trading life involving merchants, craftsmen and free-holding peasants in an early form of political democracy. The other saw the emergence of an idealist intellectual production situated within a slave-based economy, in which there was a clear division between thought and action, between intellectual and manual labour. Everything, including human beings, could have their value expressed *quantitatively*. Pythagoras and his followers, including Plato, extrapolating from the dominance of number in everyday life, claimed that underlying reality is also structured according to mathematical principles. A religious, mystical status was accorded by these thinkers to mathematics. Slave-owners, the usual patrons of intellectual life, no doubt found this very reassuring – the dehumanising of their workers was lent some legitimacy by such an anti-humanist ethos. Moreover, the abstract, ethereal quality of mathematical discourse could provide for these slavemasters a degree of mental relief from the brutality of everyday life!

---

1. See E. Lerner, "The Big Bang Never Happened", p.60 ff. and V.G. Childe, "What Happened in History", Chapter 10. (For bibliographical details of works cited see Reference List.)

Frontiers of Fundamental Physics, Edited by M. Barone
and F. Selleri, Plenum Press, New York, 1994

In the 20th century, mathematics has once again been accorded a privileged position, this time in the physical sciences. Indeed, a quasi-*magical* quality is attributed to it in some quarters. It is very rare to hear physicists spell out the ontological status of mathematics as their Greek forerunners, such as Plato, so clearly did. Rather, if they are – in what seems to be the majority – instrumentalists, they are quite content if their equations form a deductive chain predicting and linking the 'appearance' of phenomena often in the form of instrument readings. The real *physical* processes which might *explain* the perceived events are either implicitly or explicitly waved aside as being unknowable, 'irrational', or non-existent.

Mathematical equations, matrices and the like are presented in lieu of a realist, physically based, causal explanation, and are apparently believed to render such an explanation unnecessary. In this 'intellectual' *milieu* just *why* such equations are empirically effective (or not, as the case may be) remains a mystery, as does the relationship between the *axioms* of the mathematical formalisms and the material world.

Whose interests are served by the placing of so much emphasis on an abstract mathematics within physical science today? Is the intellectual élite indulging in such fetishism to provide for itself some kind of comforting rationalisation of its generally privileged social position? That is, is there a belief afoot in academia that the rest of the population is simply incapable of understanding today's physics (inter alia)?

A deeper probe results in even more disturbing realisations: within capitalism, intellectual activity is sponsored by capitalists and the state. Neither has its interests served by the endorsement of *realism* in theory or practice – instrumentalism asks less tricky questions both about the structure of the economy and about the nature of physical reality. Everything is 'normal' as long as we can add up our income and expenditure and provide the military with what they need!

A PHILOSOPHICAL AND HISTORICAL SKETCH

It will be clear from what has already been said that an historical investigation of the philosophy of Graeco-Western physical science reveals that it is permeated with various forms of anti-realism. This infection can be seen as one of the principal dangers to and within science when its current state is examined from both political and educational perspectives. More will be said about these in the concluding section.

Plato is often considered to be one source of the problem.[2] While he postulated the existence of a primordial material substance, he claimed that the study of the ever-changing phenomena formed from this substance was of secondary value to the devotion to *reason*, principally that mode of reason inherent in number and geometry. For him, the principles according to which the cosmos is structured are mathematical, and the ideal mathematical Forms reflected in imperfect earthly objects had provided the 'blueprint' for the 'Craftsman' who had fashioned this material world. Underpinning the ideal numbers, lines, planes and solids, are the 'One' and the 'Indefinite Dyad', the former being the 'Good' and the foundation of the soul.[3] By our rational study of the earthy reflections of the Forms, we may discover 'windows' through which we are able to gain insight into

---

2. Discussed in G.E.R. Lloyd, "Early Greek Science ...", Chapters 3 and 6; J. Losee, "A Historical Introduction to the Philosophy of Science", pp.17,19; E. Lerner, op.cit., p.66 ff.
3. See discussion of Plato's lecture, 'On the Good', in "Lore and Science in Ancient Pythagoreanism", W. Burkert, Chapter 1.

those perfect mathematical, eternal Forms.   In this cosmology, mathematics is clearly accorded a metaphysical/ mythical/mystical status and nature.

I mention these details of Plato's philosophy because of its emergence, in various forms, time and again, within Western discourse.   Lloyd,[4] in a passage reflecting certain 20th century attitudes, claims that Plato's conception of an ideal mathematical physics is something that we today take 'so much for granted', as well as our belief in the mathematical structure of the universe.   Not only currently have such Pythagorean/Platonic ideas flourished: in medieval Europe, Plato's *'Timaeus'*, with its mathematical doctrines, was incorporated into biblical teaching, and in the amalgamation God was seen as the Great Mathematician; and later, Galileo, whilst espousing a critical empiricism, believed that, *a priori*, mathematical reasoning is crucial to the 'reading' of the mathematical characters of the 'Book of Nature'.   Today, and it bears repeating, such a Pythagorean orientation in the work of many theoretical physicists, while lacking an ontological claim that Nature *is* mathematical, presents itself as an over-indulgence in mathematical formalisms at the expense of serious investigation into real physical processes.

This modern form of mathematical idealism requires highlighting here so that its relationship to theories of knowledge can be made explicit.   Its character is *positivist*, or *instrumentalist*, and under its regime sets of equations and matrices are used to systematise phenomena (including instrument readings) and mathematical deductions are used to predict such phenomena, employing in the calculations unexamined axioms.   Physicists adopting these procedures should be reminded that Ptolemy's mathematical manipulations also produced 'correct' predictions!

The positivist approach, content with its formalisms linked to *observable* 'results' is a version of *phenomenalism*, the philosophy of many empiricist writers working within the scientific culture which emerged in Europe in the 17th to 19th centuries.   This culture developed within an expanding capitalism, whose ideology emphasised the *individual*'s immediate engagement with the sensory world, the *raison d'être* of which was *utility*.   All the individual could hope to 'know' was to be gained through experience of sense data; there are, it was claimed, no means by which one can hope to gain access to any kind of reality other than the phenomenal, whether that reality be ultimately material or God's mind!

A prominent philosopher in the phenomenalist mould was John Locke (1632-1704). For him our sciences consist in the compilation of extensive natural histories and the application to these of the methodology of correlation and exclusion.   Though an atomist, he could not see how we could ever hope to know the myriad of ways in which atomic motions produce 'secondary qualities' or effects in us.   All we have access to are *ideas* of nominal 'essences', i.e. *observed* properties of and relations amongst bodies.

David Hume (1711-1776) drew a demarcation line between the necessary statements of mathematics, e.g. Euclid's axioms and theorems, and the contingent statements of empirical science.   The latter are derived solely from sense impressions which allow us to know only that certain events are constantly conjoined with other specific events.   Our ideas can be 'traced back' as far as our sense impressions, but we have no access to underlying natural causes.[5]

A phenomenalist critique of Newtonian mechanics was provided by Berkeley (1685-1753).   He declared that 'forces' in these mechanics are analogous to epicycles in Ptolmeic astronomy: mathematical constructions are instrumental only and scientific laws are nothing but computational devices for the description and prediction of phenomena.   There are no

---

4. G.E.R. Lloyd, op.cit., p.79.
5. D. Hume, "Enquiries ...", Section 4, pp.25, 39.

primary qualities[6] of bodies; all qualities are sensible, and what is more, *minds* are the sole causal agents, ultimately dependant upon the first principle – the universal Mind. This last notion is comparable with Plato's idea of a 'divine' cause, the 'One'.[7]

The first writer to formulate the doctrine of positivism *per se* was Auguste Comte (1798-1857). This was in relation to the history of knowledge, the law of which, he claimed, had three stages: the first stage of human history was 'theological' where all causes were animistic and took on the forms of gods and spirit beings; the second stage was 'metaphysical', where explanation of phenomena was attempted through the invocation of the existence of entities, often indefinable, as in the case of Plato's 'the One'; thirdly, human history reaches a *positive stage*, in which aetiological questions are shelved and science limits itself solely to the 'straightforward' *description* of observable facts of experience.

Comte could not have hoped for a more devout exponent of positivism than Ernst Mach. A thorough-going phenomenalist, Mach refused to look for objective causes of our experiential data. An 'object' for him was constituted by a collection of ideas formed via our senses: an 'apple' is a certain colour, taste, smell, shape and consistency, all experienced correlatively. That this characterisation is Machian in nature is significant because, as Menger explains,[8] it is in fact also that of Berkeley; but, unlike the latter, Mach confined himself to phenomena, rejecting any form of metaphysics, spiritualist or otherwise. His anti-metaphysics incorporated a critique of Newton's mechanics: for Mach there was no absolute space, time or motion. All motion was seen as relative and science was viewed as being concerned only with relations between observations. To the end, he vehemently denied the existence of atoms. Empirical significance was seen to be dependant upon specification of procedures for measuring spatial and temporal intervals.

The early work of Einstein was similarly positivistic. Einstein's Special Theory of Relativity (1905) was founded on two empirically-based assumptions: (i) the principle of relativity; and (ii) the principle of the constancy of the speed of light. As well, Special Relativity was based on 'conventions' for determining synchronism of separated clocks and the 'distance' and 'time' of an event using reflecting light-signals, and an algorithm which assumes that the speed of light is constant and equal in all directions with regard to an inertial frame, e.g.

$$t = \frac{t_1 + t_2}{2} , \quad r = c/2 \, (t_2 - t_1), \text{ etc.}$$

Einstein deduced the *Lorenz transformation* from his assumptions and definitions, showing thereby that the co-ordinates of the transformation relate specifically to his light-signal *measures* of the time and distance co-ordinates of an event. The transformation and its associated composition of velocities formula confirmed the observational equivalence of all inertial frames and the constancy of the *observed* speed of light where the speed is measured using a reflecting light-signal over a measured distance, and one clock.

Many other results were deduced by Einstein: the reciprocity of observation between observers stationary in different inertial frames; length contraction; the relativity of simultaneity; time dilation, etc. He confirmed the Lorenz-invariance of Maxwell's

---

6. 'Primary qualities' are objective properties of bodies; 'secondary qualities' exist only in the perceptual experience of the subject.

7. See G. Berkeley, 'De Motu', in: "The Works of George Berkeley ...", A. Luce and T. Jessup, eds., pp.38, 52.

8. Discussion by K. Menger, 'Introduction ...', in "The Science of Mechanics", E. Mach, pp.(xii)-(xiii).

equations, and restated the laws of dynamics and optics in a Lorenz-invariant form, thereby revealing new relations between mass and energy. However, he offered no *physical* interpretation for these new results, leaving the way open for fanciful, esoteric 'explanations' (e.g. time-subjectivity; 'time' as having no meaning apart from 'space-time', etc.). The stage was also set for a long-standing controversy about the meaning of time-dilation and related issues. Einstein's positivist, *measurement*-oriented approach in his formulation of Special Relativity had important repercussions. Many theoreticians were attracted to it, e.g. Minkowski, who further reduced the physical content of the theory with his metrics which later became 'standard' in texts and courses on Special Relativity. Particularly after 1919, when Einstein's ideas became widely known, Special Relativity was hailed as a *model* for physical theories. Thus, in the years 1925-6, both Schrödinger's equation[9] and Heisenberg's matrix mechanics based on observed spectrum frequencies made a *virtue* out of their lack of clear physical meaning as well as of their avoidance of the wave-particle dilemma.

The absence of an unambiguous physical interpretation of quantum mechanics, quantum electrodynamics, and relativity, encouraged the emergency of a spate of esoteric notions. These included Feynman's time-reversal; the uncertainty characteristics of elementary particles; the 'collapse' of wave-function due to human observation; the claim that the world exists by virtue of its being observed; and the idea that all wave-function possibilities are supposedly fulfilled in an increasing number of different universes.

The quantum theories and Special Relativity have enjoyed considerable practical successes, despite the employment of sometimes dubious stratagems, e.g. the 'renormalisation' of divergent series results by Weinberg *et al.* Empirical 'confirmations' have brought for such theories the status of established authority, leaving, for example, the Copenhagen interpretation of quantum mechanics able to withstand the criticism of a few lonely voices, such as Einstein's and Bohm's. It is not surprising that apparently illogical, quasi-mystical processes, proclaimed to exist by some physicists, have become the subject of both philosophical and journalistic speculation in which the obscurities are promoted and Eastern mysticisms often invoked.

As a result of the quantum theories and Special Relativity becoming 'standard' their positivistic assumptions have *embedded* themselves in texts and University courses. As dogmas, they have proved highly resistant to criticism or re-interpretation, and have stood as an 'establishment' barrier to consideration of alternative views such as those of de Broglie, Vigier and Bohm.[10] The emphasis on the primacy of mathematics in these 'standard' theories has made them attractive only to those students who are prepared to forgo an understanding of an objective physical world.

CONCLUSION

Ironically, what started for Comte as the pursuit of *certain* knowledge of the physical world and a rebuff for metaphysics, has revealed itself as an anti-realist enterprise which yields a psychological 'certainty' in the minds of its followers at the expense of a

---

9. It must be noted that Schrödinger did show concern about the ontological status of his wave function and acknowledged the interpretation of a wave as an energy field.

10. Thankfully, weaknesses have been found in this barrier erected against a physical interpretation of these mathematical theories. The work of Builder (1958) and the findings of Modern Cosmology have resulted in a self-consistent physical interpretation of S.R.; and over the last few years there has been progress made towards achieving a physical interpretation of quantum mechanics, following the work of Bell, Selleri, et al., and to some extent, Bohm.

retreat to metaphysics. The positivists' faith in mathematics as a foundation for, and justification of, their scientific practices is achieved only at the expense of ignoring the fact that this useful descriptive and deductive symbolic system is based on *assumptions* and *definitions*. Likewise, the other 'foundation' of the positivist program, phenomena, the sense-data of observation, rely for their existence on the human *mind* – surely shaky ground on which to build a science!

If, as it seems, there is a minority only of physicists who are realists, the implications for education, political practice, and for science itself, are alarming. If students' world-views are constructed within educational institutions which fail to foster an appreciation of the existence of a *real material world* and the means by which to investigate it, then we are failing as educators. This is today of particular importance, as we live in an age where alienation from the 'dirty messy' world of matter is built into the very nature of our sophisticated equipment and instrumentation. Even more powerful 'pure' magic is provided within the computer confines of 'virtual reality'!

Where realism is denied we may find that mathematics is accorded, without question, a quasi-explanatory status; and in some quarters, it is apparently believed that it is unnecessary to even allude to explanation as a function of science at all. In such an educational environment, intellectual argumentation is discouraged, and the student is expected to submit to the authority of the teacher *and* of what is taught. Education becomes, therefore, mere transmission of orthodoxy, and *critical enquiry* is eschewed. This situation should cause great concern as today's students are required in almost every discipline to consume large amounts of specialised information, leaving almost no scope for debate. Relationships between social life and the physical world cannot be questioned by those who lack the skills of critique and a breadth of vision.

Some scientists discount the possibility of a physical interpretation of their theory; others go so far as to claim that a physical world apart from that created by observations does not exist. If it is denied that there is a knowable material reality existing independently of human thought and action, then there can be no rational demand for individuals or social groups to construct *mutually consistent views* of reality, scientific or otherwise. A mystical 'New Age' philosophy 'will do' just as well as an appeal to the causal efficacy of the gods. In the absence of a critical realism these sorts of views will be just as indisputable as any positivist science.

We surely do our students (and ourselves) a disservice to the extent that we fail to work with them to construct realist scientific and social discourses which are communicable amongst, and able to be rationally debated by, both 'specialists' *and* the wider community. We have bequeathed to our descendants a world riddled with irrationality and superficial ideologies, one possessed with the ethos of economic growth for the sake of the profit of the few. Sadly, we are leaving the task of social and environmental reconstruction to a generation which is being denied true understanding of the real structure of both social and physical reality.

ACKNOWLEDGMENT

Thanks are due to Dr Scott Mann, Centre for Liberal and General Studies, University of New South Wales, for many helpful discussions.

REFERENCES

Berkeley, G., 1951, 'De Motu', in: "The Works of George Berkeley, Bishop of Cloyne", Vol. IV, A. Luce and T. Jessop, eds., Thomas Nelson, London.

Burkert, W., 1972, "Lore and Science in Ancient Pythagoreanism", Harvard University Press, Mass.

Childe, V.G., 1954, "What Happened in History" Penguin Books Ltd., Harmondsworth.

Hume, D., 1975, "Enquiries Concerning Human Understanding and Concerning the Principles of Morals" (3rd edit.), O.U.P., Oxford.

Lerner, E.J., 1992, "The Big Bang Never Happened", Simon and Schuster, London.

Lloyd, G.E.R., 1982, "Early Greek Science: Thales to Aristotle", Chatto and Windus, London.

Losee, J., 1980, "A Historical Introduction to the Philosophy of Science", O.U.P., Oxford.

Menger, K., 1960, Introduction to the sixth American edition, in "The Science of Mechanics: A Critical and Historical Account of Its Development", E. Mach, The Open Court Publ. Co., La Salle.

CREEDS OF PHYSICS

S. Warren Carey

University of Tasmania
GPO Box 252C
Hobart, Tasmania, Australia 7001

Science started in the middle. Visions of the wise, and "revelations" from "rapid-eye-movement" sleep, when the mind is released from beliefs we think we know, were passed on and repeated indefinitely to become dogma. A gene is a self-replicating organic complex. A meme is a self-replicating concept, which likewise may propagate through generations.

Astrology starts with the time and date of birth. Calculations establish the relative positions of the planets and the zodiac constellations. So far so good. — impeccable science. The meme is to believe that the relative position of lordly Jupiter and voluptuous Venus, or of peripatetic Mercury and belligerent Mars, to Libra's scales of justice or the threat of Scorpio have profound influence on a specific billionth of humanity. Such anthropomorphism of stars and planets is utterly absurd fantasy. Yet more than a third of adult Americans believe it. Every newspaper and magazine has its astrology page. Nancy Reagan, wife of an American President is reported to have her personal astrologer. Kepler, great physicist though he was, earned his livelihood as an astrologer. The chant: Such precise mathematics could not be wrong.

The flat Earth was the universal meme — obviously true. Until Pythagoras saw Earth's shadow on the moon and concluded that Earth was round. Absurd! cried the establishment — men on the other side would have their feet in the air, rain would fall upward, water would drain from lakes, and so on. The debate was settled by Aristotle three generations later. Pythagoras' beautiful meme of the harmony of the spheres—the distances of the heavenly bodies were in proportion to his tone scale, and emitted melody heard only by maidens—was still the dogma of Shakespeare's day.

Ptolemy's meme, that all heavenly bodies were on crystal spheres rotating a stationary earth, had to be continually adjusted with epicycles to fit each new fact found in their motions, until Fracastero in the sixteenth century needed 79 cyclic and epicyclic motions to satisfy the data. Yet the patched and repatched meme survived for two millennia, ratified by prediction of eclipses and canonized by the church, until overturned by Copernicus. He substituted a daily rotating Earth, which orbited the Sun with the others, a model previously proposed by Aristarchus two centuries B.C., two centuries still earlier by Heraclides, and by the Sanskrit philosophers two millennia before that, but rejected. The Copernican system was rejected by the Church. Bruno was burned at the stake for teaching it, and Galileo was forced to recant. Entrenched memes prevail regardless of common sense. The chant: Ptolemy's elegant maths could not be wrong!

Frontiers of Fundamental Physics, Edited by M. Barone
and F. Selleri, Plenum Press, New York, 1994

Nowadays, many laugh at Archbishop Ussher, who dated the creation of Earth as 4004 B.C. at 9 in the morning, summer time. But Ussher was no fool; on the contrary he was most erudite, a scholar of the scriptures and in several relevant languages, and his timing was determined by eclipses. He was believed and accepted for two centuries. His meme was his belief in the absolute literal truth of the scriptures.

Meanwhile geologists were claiming longer and longer time for the seemingly endless succession of mountain-building, destruction, and resurrection. Then Kelvin, master physicist of his generation, scoffed at non-numerate geologists. The only ultimate source of energy was gravitational collapse. Knowing the present heat flux and allowing even a molten beginning, Earth could not be more than 20 million years old, and even Sun could not be more than 100 million! The chant: Such elegant mathematics could not be wrong. Many geologists wilted before the heat. But some held their ground. Something must have been left out which would explain their empiricism. Then radioactivity was discovered, and with it, adequate time. Far from cooling, Earth might even be getting hotter.

Diapirism

The meme of English-speaking geophysicists is that folding, thrusting, and mountain-building result from crustal compression, originally from a shrinking earth. In contrast, generations of European geologists had argued that the compression is superficial, through gravity spreading from tumour-like diapirs from the interior. Among the many leaders were Gillet-Laumont (France), Kuhn (Switzerland), Bombicci (Italy), Reyer (Germany), Steinmann (Germany), Van Bemmelen (Netherlands), Beloussov (Russia), Wegmann (Switzerland), and Ramberg (Sweden). But in recent decades most geologists have wilted before the English-language bandwagon, based though it is on two false memes — that folding implies tangential crustal shortening, and that Earth radius is constant. The latter I will discuss below. Folding implies flow, even perhaps with tangential *widening*. Strata in salt domes, which have moved only upward, usually with at least some tangential widening, are complexly contorted. Figure 1 shows how strata may be intricately folded only by flow parallel to the axial surfaces, while maintaining their volume and tangential width, because the whole may be analysed into segments, each a parallelogram with constant intercept on the parallel flow lines. The two beds, A, were first folded as in E, to produce B, and then flowed again

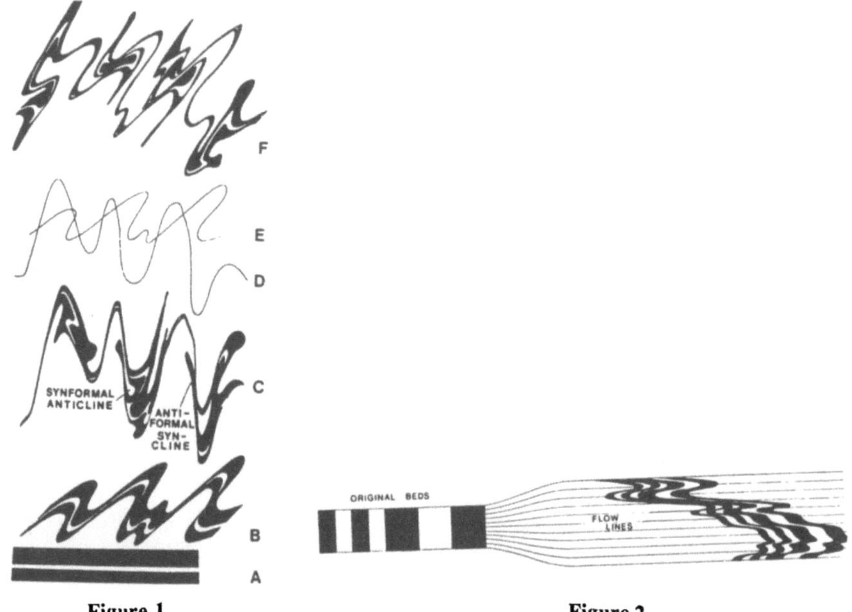

Figure 1

Figure 2

242

according to D, to produce C. If the order of flow were D then E, the result would be F. In Figure 2 the original beds have flowed upward between flow-lines which diverge to twice their spacing. The folded beds still have exactly the same area in cross-section as the original beds, because the intercept on the flow-lines is halved. Although the beds are intensely folded, their crustal width is doubled, although most field geologists would say they had been fore-shortened by compression.

At all stages the orogenic core zone moves upward. The top of the mantle, originally some tens of kilometres below sealevel, continually rises, gravity-driven by the weight deficiency of the orogenic zone below regional isostatic equilibrium. Mantle rock, now gabbroic or serpentinite, eventually becomes the ocean floor, then cores of fold mountains. Point P in the lower diagram of Fig. 3 is driven upward by the rising column below, but pressing down on it is the weight of the pile above. If the latter is less than the former, P continues to rise, but when the weight of the pile equals or exceeds the upward drive, rocks squeeze sideways. A stable height is reached where further upward drive causes continuing lateral spread as nappes overthrust the miogeosyncline, at a rate determined by the load excess and the viscosity ($10^{17\text{-}19}$ poises), but effectively much less where there is fluid over-pressure). Because of earth's rotation, continent-ocean borders, and other factors, orogens are rarely symmetrical but commonly flow to one side.

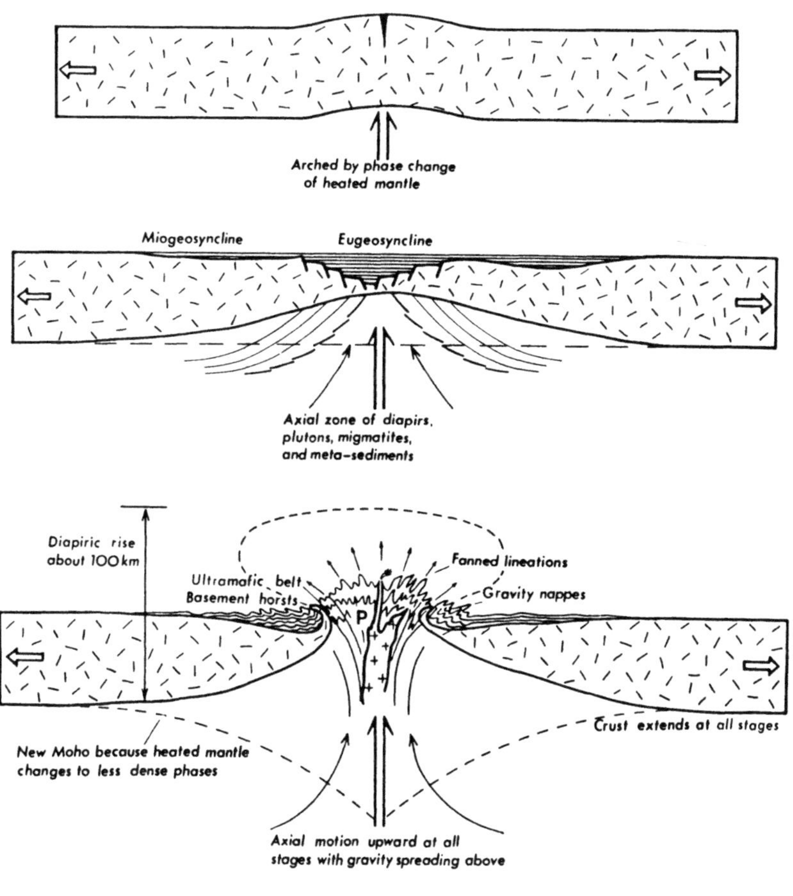

Figure 3

Numerically, an orogenic core, say 100 km wide, rises some 100 km, that is 10,000 sq. km in cross-section area. Let us be ultra-conservative and assume only 2500 sq. km, a nappe-sheet one km thick would be driven laterally 2500 km! But this cannot happen, because friction resistance on the base increases with the square of the area. Hence when a nappe has been driven out at most a few tens of km, the friction on the base exceeds the drive. The nappe halts, and a new nappe is driven over it until it too comes to a halt. But the cumulative outward flow of the nappe pile could still reach the 2500 km of our conservative example — all resulting from the diapiric drive of the orogen. Yet geologists, finding cumulative nappe travel of perhaps 700 km for the Alps or Himalayas, proclaim intense crustal shortening! Far from shortening, orogenic belts represent moderate crustal widening, both in the initial stretching to form the geosyncline, and in the subsequent orogenesis.

Permanence of Continents and Oceans

A century ago, the universal meme was that continents and oceans were permanent geographical fixtures, although shallow seas had flooded the continents from time to time. In 1912, Wegener argued that continents had moved relatively thousands of km, that the Atlantic Ocean opened as the Americas had split from Africa and Europe. In the last tenth of geological time, all the continents had been united as a single continent, Pangaea, the Pacific being the only original ocean. Absurd! cried the physics establishment. Continents could not possibly move across their base. The physicists were of course right. No continent has ever moved far over its base. Modern seismic analysis has shown that the roots of continents penetrate deeply into the mantle. Continents have dispersed, not by sliding tangentially over their bases, but by the growth of new oceanic crust between them.

Oroclines

There are many places where orogenic belts bend at a sharp angle. Forty years ago, I asked whether these were impressed strains, or the belts were born bent. I found that most were indeed impressed bends, and that unwinding them went far toward reproducing Wegener's Pangeia. My paper was rejected by the Geological Society of Australia, so I published it locally (Carey, 1954). During subsequent years many such rotations were tested by palaeomagnetic measurements —various rocks, different ages, on several continents, by numerous independent workers. In every case they confirmed my predictions.

	Carey	*Palaeomagnetic*		*Carey*	*Palaeomagnetic*
Alaskan orocline	28°	30°	Mendocino orocline	60°	63°
North America	30	30	Puerto Rico	45	53
Africa to S. America	45	45	Jamaica	42	50
Newfoundland	25	25	Hispaniola	39	40
Spain to Europe	35	35	Colombia	large	80
Italy to Europe	110	107	Appalachian arcs	20-40	29
Corsica and Sardinia	90	50	Malay Peninsula	about 70	70
Sicily to Africa	0	0	Seram	large	98
Arabia to Africa	4	7	Scotia arc	large	90
East New Guinea	35	40	India	70	70
Honshu, N to S	40	35			

The Alaskan orocline was one of the first I had described (Fig. 4). Runcorn sent S. B. Stone, then one of his graduate students, to check it out palaeomagnetically. But instead of the expected anticlockwise rotation of the Alaska Range against North America, he found clockwise rotation (Packer & Stone, 1972). For both the Arctic Ocean and the Gulf of Alaska to be opening at the same time by large angles about a common centre, is impossible — *except on an expanding earth* (Fig. 5). The total angles subtended at the centre has doubled, indicating a doubling of Earth radius. The geology and tectonics of this region is discussed in greater detail in Carey (1994).

Earth Expansion

After 50 years of scornful rejection all now agree that new crust is added at the mid-ocean spreading ridges causing dispersion of the continents in a logarithmic process, which has become very rapid during the last 100 million years. Current dogma has adopted two arbitrary memes — that orogenesis is a compressional phenomenon, and that the radius of the earth is essentially constant. Hence the added oceanic crust must be balanced by equivalent removal of crust by concurrent subduction down the oceanic trenches.

Permian reconstructions agree that all the continents were then assembled in a single continent—Pangaea, which filled one hemisphere and the Pacific Ocean filled the other. With the opening of the other oceans, the Pacific must have greatly reduced in area since the Permian, *unless the earth is expanding*. The Pacific is roughly circular, ringed by the continents—North and South America, Antarctica, Australia, and Asia. During the relevant time, the distance between North and South America has increased by 2500 km; the distance between South America and Antarctica by at least 3200 km; between Antarctica and Australia by 3500 km; between Australia and China by 3800 km; and between Asia and North America by 1000 km. (The basis for these estimates is discussed in full in Carey, 1994). Here is the dilemma: Current dogma demands that the Pacific area has reduced by at least 150 million sq. km since the Permian; but its perimeter increased by some 14,000 km— impossible, *except on an expanding earth.*

Palaeomagnetic data, confirmed by tropical faunal assemblages, indicate that the Permian equator passed through Texas and New York. The present equator is in Brazil. So North America is now some 35° nearer the North pole than in the Permian. Similarly the Permian equator was in the south of France. The present equator is in central Africa. So Europe is now 40° nearer the north pole than in the Permian. Likewise Siberia is now 20° nearer the north pole than it was in the Permian. Similar results came independently from Triassic and Jurassic paleomagnetic data, progressively less toward the present. Here is the paradox: America Europe and Siberia have converged on the Arctic by large angles, whereas throughout this time the Arctic region has widened by 30° — impossible *except on an expanding earth.*

Quaternary palaeomagnetic poles around the world for the last two million years fall in a 95% confidence oval centred a few degrees beyond the pole, but it does include the pole. The confidence oval for Plio-Pleistocene (from 2 to 7 million years) palaeomagnetic poles falls beyond the north pole. The confidence oval for the Neogene (7–25 million years ago) overshoots the north pole by 6° (Fig. 6). This paradox arises because, in plotting the

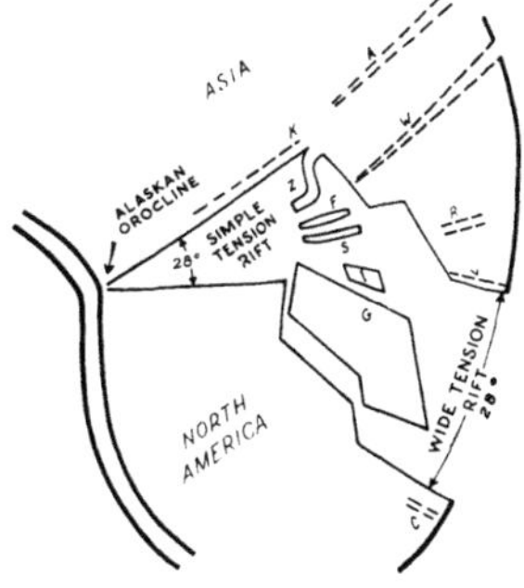

Figure 4 Alaskan Orocline.
A, Aral depression. C, Cabot trough.
F, Franz Josef Land. G, Greenland.
K, Katanga rift. L, Lisbon scarp.
R, Rhine graben. S, Spitzbergen.
W, White Sea depression.
Z, Novaya Zemlya.

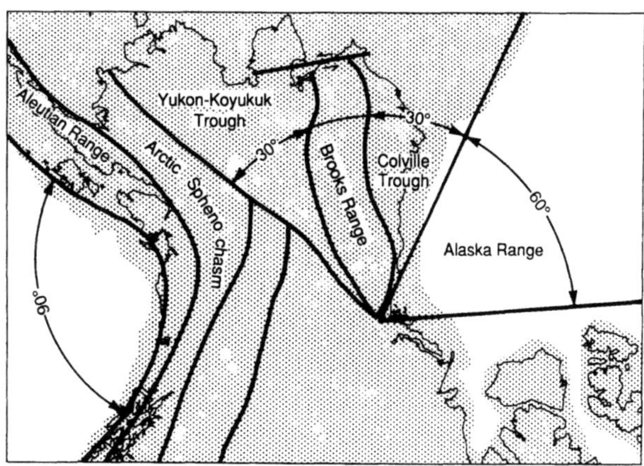

Figure 5

245

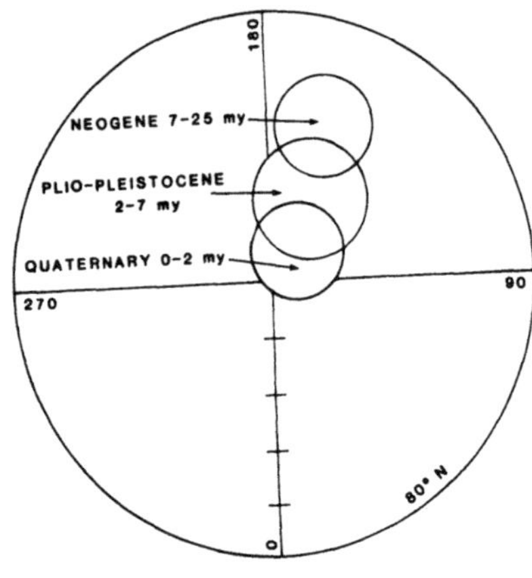

Figure 6

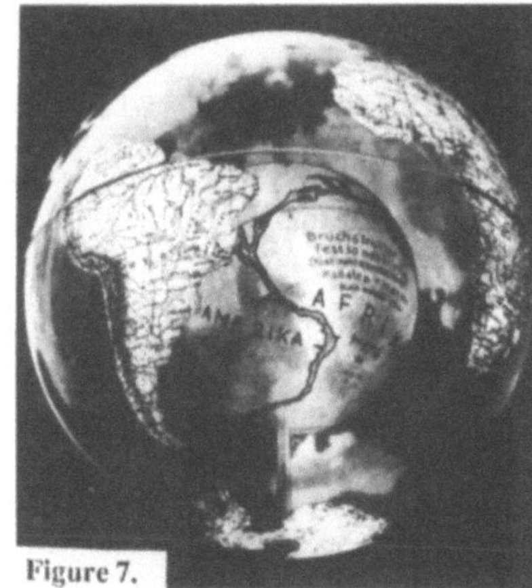

Figure 7.

palaeopoles, we have assumed the length of a palaeomagnetic degree from the present earth radius. With progressively smaller radius, each of the confidence ovals would include the pole. This extends to the present the results of the ancient equators described in the last paragraph. Impossible — *except on an expanding earth.*

Some thirty people have quite independently constructed globes, eliminating the oceans, and have all fitted the continents together on a globe some 60% of the present earth radius. But Vogel went further. He placed his small globe inside a transparent globe with the continents and oceans as at present. Whereas current tectonicists achieve the dispersion of the continents by shifting them tangentially (removing equivalent ocean crust down trenches) Vogel's model (Fig. 7) shows that the continents simply moved out radially with the oceanic crust appearing between them. Perry developed a computer algorithm to permit the continents to move radially, while preserving their areas. His print-out, Fig. 8, developed a series from an initial globe with all the continents in contact, to the present globe with the continents separated by the oceans, and developed the succession of palaeomagnetic growth lines observed in the oceans. Embleton and Schmidt, (1979), endeavouring to reconstruct Proterozoic polar wander paths, were astonished to find that that the polar wandering path for Africa, Australia, North America, South America, and Australia coincided when plotted on their present geographic positions, implying that these continents then occupied the same angular positions relative to the centre of the earth, notwithstanding the large angular separations since inserted between each pair, implying that the continents separated by radial outward movement, not tangential. Vogel, Perry and Embleton, in three different continents, have never met. But their quite different methods each indicate *gross earth expansion.*

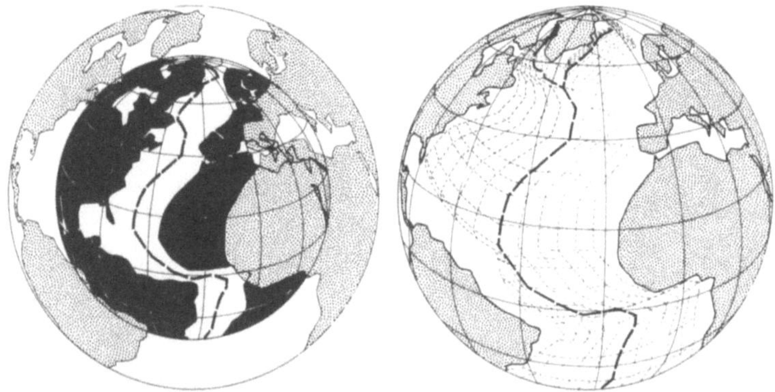

Figure 8. Perry's computer print (left) of the inner globe (black) expanded 25% inside the present globe (stippled). Right: expanded to present radius showing computer growth lines which match actual palaeomagnetic growth lines.

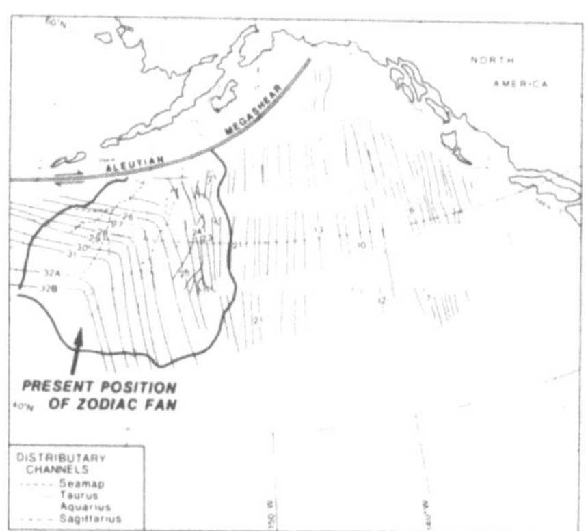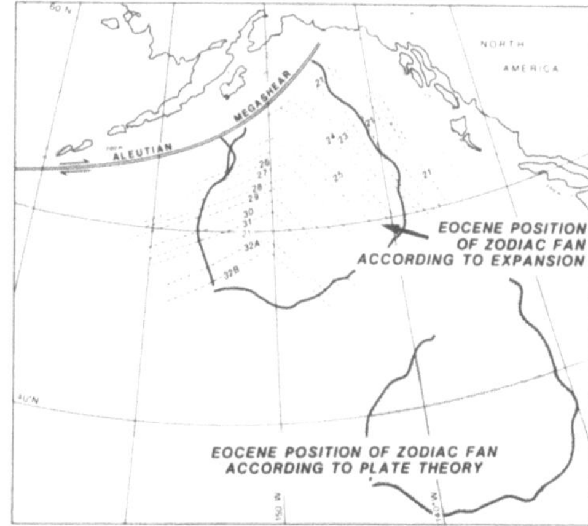

Figure 9

The Zodiac fan, Figure 9, is a million-sq. km submarine delta off the Gulf of Alaska which accumulated during the Eocene. Plate-tectonics workers assert that this region of the Pacific has been subducted down the Aleutian trench at 7 cm per year. Hence to reconstruct the late Eocene position of the fan (palaeomagnetic anomaly 17), the fan must have been formed more than 3000 km south in the Pacific Ocean, some 2500 km from the nearest landmass with no possibility of its silt reaching it (Stevenson, Scholl, & Vallier, 1983). But according to the earth expansion model, subduction is a myth, and seafloor inserted since anomaly 17 was not there in the Eocene; removing it brings the fan back to the Alaskan coast, with ample source for its three million cu. km of silt. The Zodiac fan *indicates earth expansion*.

Plate-tectonicists agree that Africa is surrounded by mid-oceanic spreading ridges where the new crust has been added continuously for the last hundred million years. According to the plate model, Africa's share of this growth should have been subducted within Africa — which is clearly not so, as Africa has suffered extensive rifting throughout the relevant time. So they assumed that Africa must have been stationary, and that its share of subduction must be accepted by the trenches flanking the Andes, implying 7000 km of thrusting there. But seismic profiling and some sampling, shows that some parts of the trench are empty, and other parts have accumulated as much as 6 km thickness of sediment, which should have been obviously crumpled and overthrust if 7000 km of foreshortening had occurred there. But they show no sign of this. Besides, the sediments are mainly turbidites, derived from the erosion of the Andes, not a vast accumulation of pelagic sediments which should have been scraped off the ocean crust if 7000 km of it had been dragged down the trench. *The subduction hypothesis is wrong.*

A million cu. km of Devonian micaceous sediments from Bolivia to Argentina imply an extensive continental source to the west where there is now the deep Pacific Ocean. The nearest such source is Australasia, now 7000 km away. The Triassic Kitikami conglomerates of Japan imply a source to the east where only the deep Pacific lies (Ichikawa, 1951, Choi, 1984). The nearest such source is North America, now 3000 km away, where suitable strata are truncated as they trend toward the Pacific. Several faunal distribution disjuncts (Ordovician trilobites, Devonian and Permian brachiopods and fusulinids, Cathaysian floras, and many others) are interrupted by the Pacific. Thus, contrary to the subduction hypothsis, the Pacific is a rift ocean, as Ager (1986) concluded on more general grounds. These facts indicate *gross expansion of the earth.*

In global perspective, what is the primary inhomogeneity of Earth? Clearly that Earth is a sphere with a fluid core, and a crystalline mantle some 3000 km thick. Surely such should

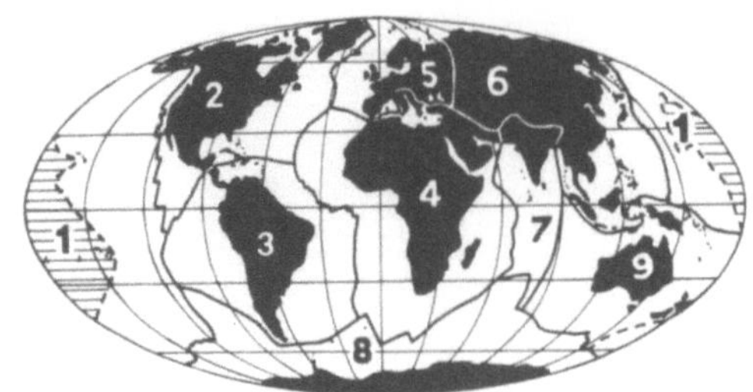

Figure 10

express itself as a major feature, and of course it does (Fig. 10). Earth's surface consists of nine major polygons, with growing boundaries a few thousand kilometres across, the loci of maximum seismicity, vulcanism, and heat flux, the latter causing ridge expansion. If the mantle were thinner, there would be more, smaller, polygons. The polygons have a central continent, surrounded by new crust which has been added during the last hundred million years. If this new crust be removed, the continents fit on a smaller globe, as shown previously, the mantle then being only half its thickness.

The second order inhomogeneity of Earth is a weak asthenosphere a few hundred kilometres down. This too is expressed as polygonal cracking a few hundred kilometres across (Fig. 11) where minor faulting, seismicity, and heat flux peaks, forming swells surrounding stable inactive basins. A critic suggested bias, Africa being a special case. Figure 12 shows Africa in global context. The second order polygons are indeed regional. Elsewhere I have described an expansion hierarchy to smaller scales, down to epeirogenic jointing which is found wherever rocks have suffered regional movements — two sets of joints nearly at right angles, with roughly vertical intersection.

NASA has been conducting some inter-continental measurements, using VLBI from remote radio galaxies, interferometry from moon reflectors, and from the LAGEOS satellite. As plate tectonics is the standard dogma, results favouring that theory are published, while others are treated as anomalous and withheld. A closer look is required. Quoted distances are

Figure 11

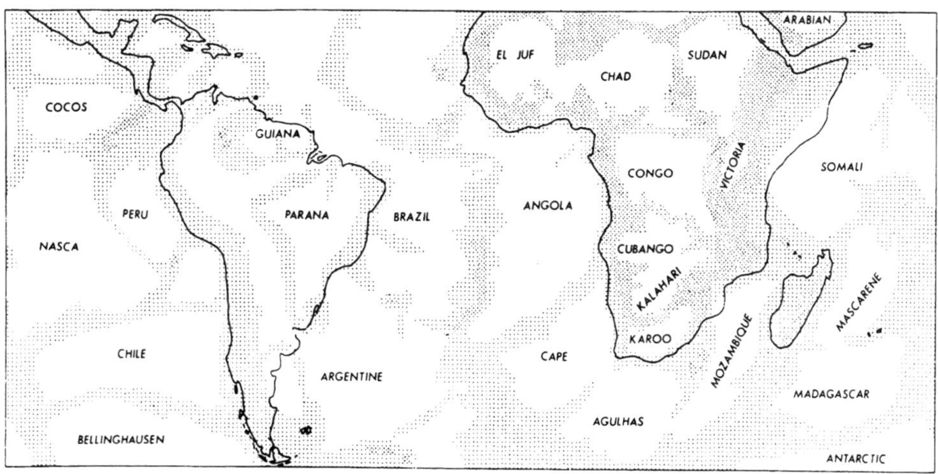

Figure 12

chords, not surface measurements. Where three stations fall on a great circle, a specific test of expansion can be made, as in Fig. 13, in which dR/dt is the rate of expansion of the earth. If the radius is constant, dR/dt should come out as zero. But it doesn't. No stations on a single great circle have been published, but Arizona–Hawaii–Canberra intersect at 159°, but the difference from 180° does not affect the result seriously. Parkinson, using NASA published data, found the rate of increase of radius was 2.8 ± 0.8 cm per year. This is pure geometry, irrespective of hypothesis or the presence or absence of spreading or subduction zones.

If this rate were constant, it implies circumference increase between 12,600 and 22,600 km since the mid Cretaceous, which agrees with the amount of new crust since that time, with no subduction. The increase in length of the equator is about 18,300 km, in the middle of the probability range. The arc through Antarctica and Australia is about one sixth of a great

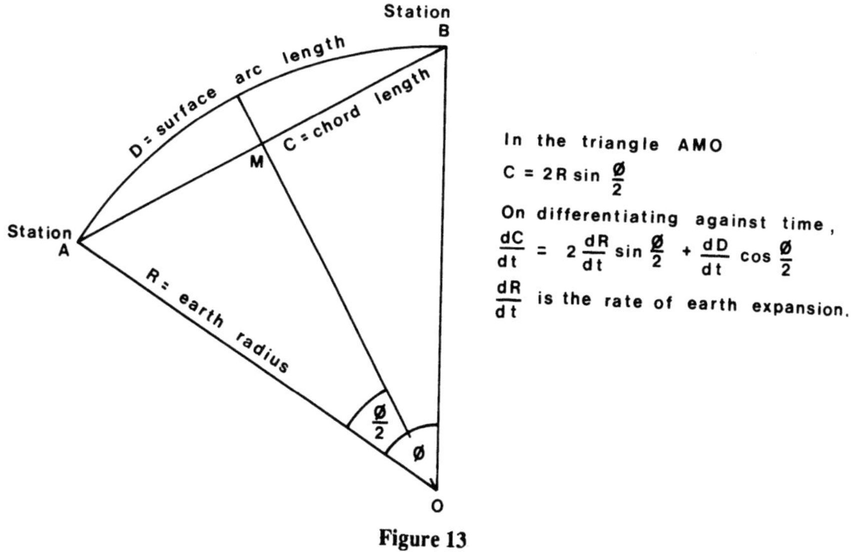

Figure 13

circle, so an increase between 2100 and 3800 might be expected. The southern ocean between Antarctica and Australia since the Cretaceous has opened by 3300 km, again in the middle of the range. India-Antarctica occupies four-tenths of a great circle, so an extension between 5000 and 9000 km might be expected. Their post-Cretaceous separation has been about 7000 km.

Where a slice of new crust has been inserted in a great circle, any segment with no new crust in it, will subtend a smaller angle at the centre of the Earth. No new crust has been inserted between Hawaii and Japan since the Jurassic, so this arc would appear to be shrinking at 6 cm per year, which is about what NASA finds. But they interpret it as subduction of crust, whereas I interpret it as caused by insertion of new crust between Hawaii and Peru and elsewhere within the Hawaii–Japan great circle. This matter has caused trouble for NASA because arcs between Denver and Connecticut, and between Canberra and Western Australia appear to be shrinking, although no subduction or other shortening has been observed or hypothesized.

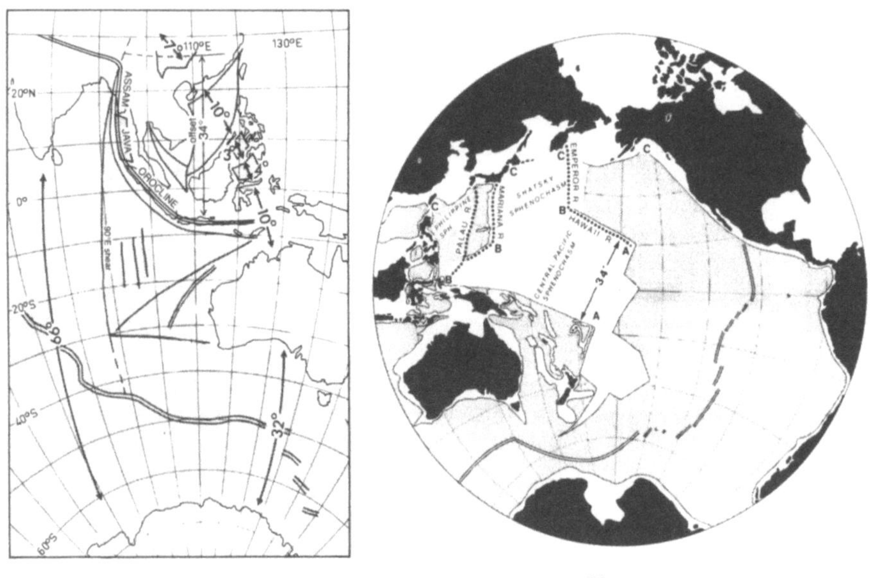

Figure 14
Figure 15

A crucial test is between Easter Island and the South American mainland, because Easter Island is east of the Pacific spreading ridge, and according to the standard meme only Andean subduction should occur within this arc, whereas I would look for minor diapiric extension. I knew that NASA had made some measurements, so I called at NASA geodynamics headquarters in Virginia seeking results. They were withheld, because they were regarded as 'anomalous' and a search was in progress for some hitherto unexpected spreading zone within the arc!

The new ocean crust between India and Antarctica is 66° (Fig. 14). The new crust between Australia and Antarctica is only 32°. The distance that Australasia, Indonesia, and East Asia has been moved south by the Nintyeast orocline is 34°, and the new crust in the several new small seas between Australia and China is also 34°. If this orocline is straightened so that Australia moves north 34°, the Melanesian plateau abuts the Hawaian ridge, all the pre-Cretaceous ocean crust is removed. Elimination of the crust inserted since, completely closes the Pacific and Indian Oceans (Fig. 15). *The earth has expanded.*

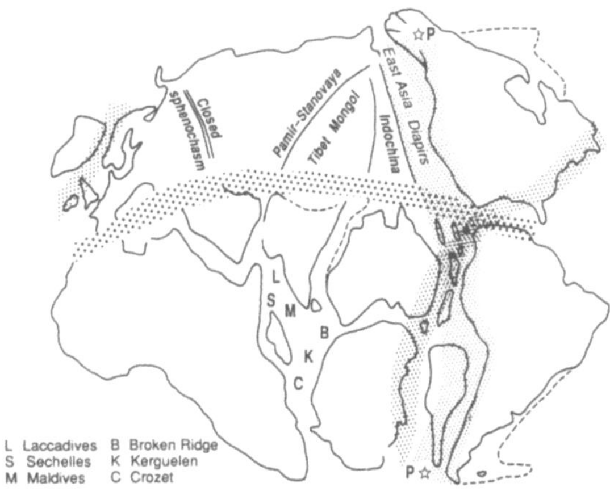

Figure 16

Figure 16 shows the Permian relations of the continents. The finest stipple is the Cordilleran orogen, the intermediate-sized stipple, the Caledonian–Appalachian–Tasmanide orogen, and the coarsest stipple shows the Tethyan orogen. Note that it occupies the whole globe — South America nestles into Africa, and the African bulge fits between North and South America, so this is the whole globe.

Newton and Hubble

When Kepler, after six years of analysis of Brahe's observations of Mars, found that the planets had elliptical orbits with Sun at one focus, Newton developed his d^2 law of gravitation. Newton's empiricism was on the scale of the solar system, and no greater when applied to binary stars. An empirical law cannot be safely extended beyond the scale of the empiricism, but astrophysicists have applied Newton's law to transgalactic distances, even to infinity. This *could* be valid, but risks the possibility of other terms, too small to detect on the scale of the original empiricism, but which might dominate on larger scales. Unfortunately, there is such a term.

When Hubble found that many of the "nebulae" were island universes outside our Milky Way galaxy, receding from us with a velocity which increased in proportion to distance from us, an extra term was needed:

$$\text{Force on unit mass} = mG \ (\underbrace{1/d^2}_{\text{Newton}} - \underbrace{ad^2 \ H^4/c^4}_{\text{Hubble}})$$

where a is a pure number scaling factor, G is the gravitation constant, H is Hubble's constant, and c the velocity of light. c^4/H^4 is the hypervolume of the Universe, and the volume of the solar system being insignificant in comparison. The Newton term diminishes with the square of distance, whereas the Hubble term increases with the square of distance. H/c is a very small faction, and when raised to the fourth power, is quite indetectable on the scale of the solar system; but it is real and positive, and when the scale is increased by 10^{10} to intergalactic distances, the Hubble term first equals the Newton term as distance increases, and thereafter becomes the only term that matters.

When real numbers are inserted, with 'a' set arbitrarily at 10^{20} (Dirac's big number), the Magellanic Clouds are just beyond the null, so, originally part of our galaxy, they are now hiving off, and will continue to recede. The M31 spiral galaxy in Andromeda, is ten times this distance, and parted from us, much earlier. The 10^{13} observable galaxies form a Gaussian distribution, a few very large and some very small, but no-one has suggested why this size — not millions of times larger nor smaller The reason is now obvious. Any matter within the Newton-Hubble null is attracted to the galaxy, but any matter beyond the null recedes. The so-called dark matter of the Universe is a myth, arising from the application of Newton's law to trans-galactic distances, without the Hubble term. Hence the amount of alleged dark matter increases with scale. For our own galaxy, it is trivial. Spiral galaxies generally appear to have up to five times their visual mass, elliptical galaxies up to twenty times, small galaxy clusters twenty to fifty times, and large galaxy clusters 100 times as much unseen matter as what is visible, indeed the greater part of the mass of the universe appears to be unseen dark matter, a startling conclusion — but false.

Olbers long ago pointed out that in an infinite universe, every ray should eventually reach a star, and that although intensity declines with distance their number exactly balances this, so that the whole sky should resemble the searing surface of Sun — Olbers' paradox. At the limiting distance of optical telescopes, some 10^{13} galaxies can be resolved. Beyond, they become fainter and vanish from sight. At this distance the angular separation of so many galaxies over the surface of a sphere is very small. But beyond, they still exist, at decreasing angular separation, and increasing redshift, until at the limit of resolution of radio telescopes (a function of aperutre and focal length) they fuse into continuous radiation, now near absolute zero. Thus Olbers' universal radiation dawns — but not at the searing heat of Sun's surface, but at 2.7 K. Such universal background radiation is inevitable in any infinite universe (Fig. 17).

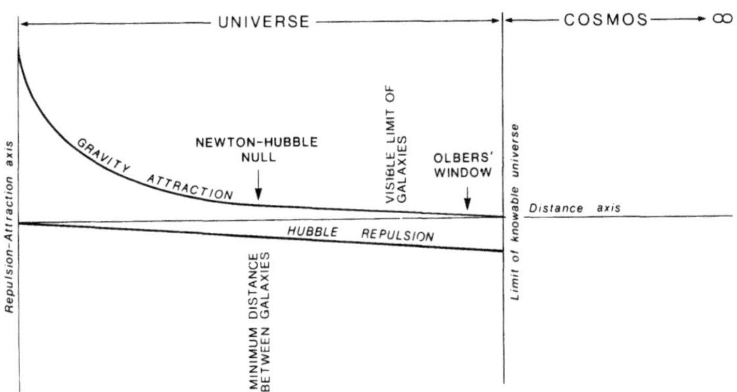

Figure 17

Gamov's big bang was based on Hubble's discovery that all galaxies are receding and the Penzias and Wilson accidental discovery of the universal micro-wave radiation. Both are inevitable in any infinite cosmos, hence the big-bang fantasy should be scrapped. The relative proportion of hydrogen and helium is not definitive, and could arise from other scenarios.

Universe and cosmos are used as synonyms. But the universe is the limit we can know physically, where galaxies appear to us to be retreating at the velocity of light, whereas an observer there would believe that he was at rest, and that we were receding from him at the velocity of light. Compare this with the receding spots on the surface of an inflating balloon. Each spot appears to be at rest while all other spots recede at speeds which increase with distance. A mariner sees as far as his horizon, but another mariner there sees the first on his horizon. There are as many horizons as there are mariners, and as many universes as there are observers — all parts of an infinite cosmos This is what Einstein called the cosmological principle, or the perfect cosmological principle when applied through infinite time past and future.

Newton's law created the problem of action-at-a-distance, and hence the concept of æther, abandoned by most since the Michelson-Morley experiment. But I still find need for such a universal medium, call it space-time, field, vacuum, or æther as you will. Each of these names except æther has a specific meaning, so I stick to æther, and proceed to define its properties. Æther is all pervasive, massless, is not affected by matter passing through it, or *vice versa*, transmits the energy of electromagnetic waves with the same mathematical form as though it were isotropically elastic (normal elasticity depends on the displacement of atoms from their equilibrium stress position, and æther is not made of atoms or of any other particles, on the contrary they themselves are strains of the æther).

252

Every galaxy in the cosmos exerts tension in the æther, which integrally creates a uniform pervasive tension through space. A single galaxy dominates locally but is surpassed at the Newton-Hubble null by the integrated tension of all the other galaxies. The pervasive tension determines the velocity of light, which is therefore a function of the mean density of the cosmos. In an empty universe, the velocity of light would be infinite. The velocity of light is reduced near matter where the tension is increased, hence gravitational lensing. Traversing a crystal, light passes very close to atoms, hence refractive index. In non-isotropic crystals, the closeness of atoms differs in specific directions, hence birefringence.

Matter and energy are equal and opposite, like the two faces of a coin. One cannot exist without the other. The tension applied to the æther by a mass is exactly equal to the mass — by Einstein's formula $E = mc^2$. The total mass-energy of the cosmos is zero — a universal null. It is like borrowing a large credit from a bank, acquiring a large sum with which to operate, but also an equal debt. Pay off the loan and the debt vanishes. The model of a particle being unaware of another particle until it creates and transmits to it a virtual particle, is fantasy. Every mass exerts a diminishing strain in the æther which affects every other particle out to infinite distance, unless swamped by strains from other masses.

As mass and energy are inseparable opposites, matter cannot be little hard balls, the meme handed down through the ages from Leucippus and his pupil Democritus. If energy is a strain in the æther, so also must be mass — I suggest as a soliton, a self-limiting wave, which maintains its amplitude and form as it propagates indefinitely. There could be as many kinds of soliton as there are particles, each with its specific radial æther strain, observed as mass. Such solitons might, or might not, carry charge, æther-strain tangential to radius, either dextral or sinistral, also diminishing as the square of distance. Æther has a specific magnitude of pervasive tension, which we may denote by h. At unit frequency, a soliton would only propagate when its maximum is in phase with h. Other phases would attenuate rapidly but multiples would propagate, their stable energy being nh, where n is the frequency. Thus Planck's constant seems to be a measure of the pervasive tension in the æther.

When a particle is accelerated to high velocity, the added kinetic energy appears as an increase in mass. Also, the mass-energy of an atom must be distinguished from its configuration energy. When an archer's bow is strained, the constituent atoms are displaced by the applied stress to a configuration different from their neutral rest position. When the arrow is released, the atoms of the bow return to their original state and the configuration energy is transferred to the arrow. When a lead battery is charged, the lead is raised to a higher chemical energy state. When the battery is discharged, the chemical configuration energy does external work, while the lead returns to its rest energy. In the construction of atoms, configuration energy is added to the mass-energy of the constituent particles. Large atoms like uranium have high configuration energy which may be released by fission to smaller atoms with less configuration energy with release of configuration energy and the reduction of mass. This process may theoretically continue as far as iron, which has the lowest energy per nucleon, but mass-energy of iron cannot be released by fission, because new configuration energy would have to be added. Similarly, configuration energy may be released in each fusion step from hydrogen towards iron, but no configuration energy could be released by fusion of iron, because each heavier atom requires the addition of configuration energy. Thus all current nuclear energy is configuration energy.

Solar System

The Hertzsprung-Russell sequence plots all stars according to their spectral class and luminosity magnitude, forming a sigmoid curve. Although this is called a sequence, current dogma assumes that stars do not move along the curve, but remain on a single ordinate. Stars at the blue end are believed to be the youngest, because they radiate such stupendous energy, that they could not last long. On the contrary, I suggest that they are the oldest, and that in due course Sun will attain that stage, and that the H-R sequence *is* a sequence, and stars begin at the low-mass low-luminosity end as brown dwarfs, then small red stars, then move along

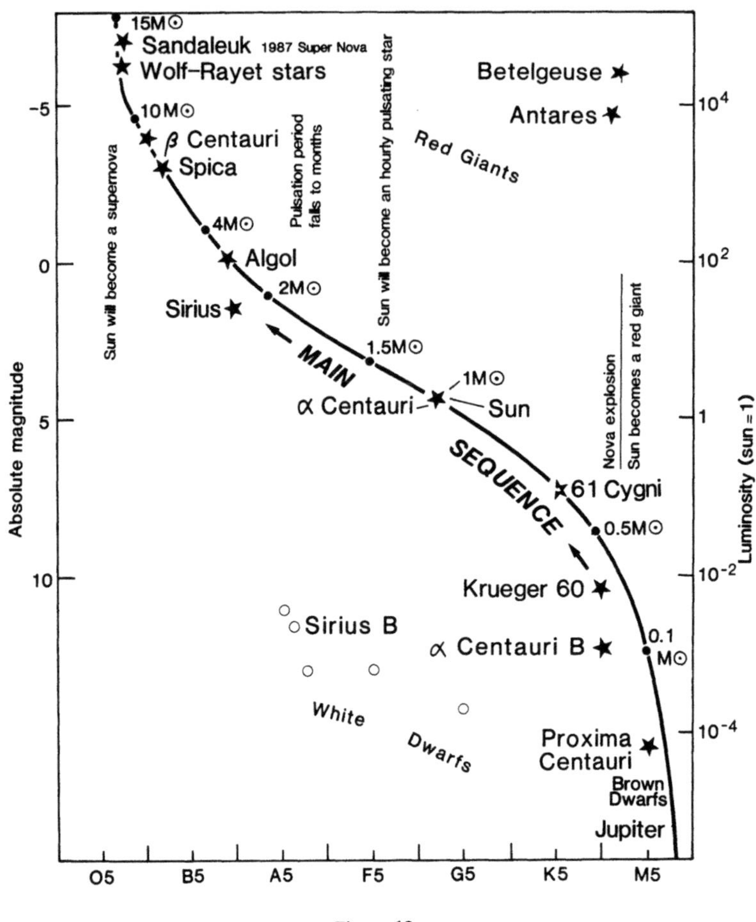

Figure 18

the sequence with increasing mass and progressive spectral type through yellow stars like Sun, to double its mass as a cepheid variable, to white stars like Sirius, bluish stars like β Centauri and Algol, to bright blue-white stars ten times the mass of Sun, and on to still more massive stars at the high end of the sequence, like the one that exploded as the 1987 supernova (Fig. 18).

During their journey along the sequence, stars pass through various stages of instability, when the rate of production of energy exceeds the rate of excretion by conduction, convection, radiation, or any other mode. Betelgeuse and Antares, only half the mass of Sun, are extremely tenuous, with, for comparison with Sun, radii extending beyond the orbit of Mars. I suggest that some five thousand million years ago Sun passed through this stage, and that was the origin of the solar system. Sun was then a binary as are Antares and Betelgeuse today, which accounted for the ecliptic plane and the anomalous distribution of angular momentum in the system. The binary companion might have been α Centauri, which on known proper motions could journey to its present distance in that time; this is unlikely because α Centauri is a large angle from the ecliptic; but this might have been explained by a gravitational instability which ejected it from the Sun's vicinity.

Such a stellar sequence would end in a neutron star, or as a black hole. What happens to black holes? Is that where the mass and energy fuse to zero to finally discharge the debt of their creation from the null cosmos? Black holes must be a temporary state, for otherwise they would be very numerous.

254

When we grade the bodies of the solar system by size, the 3000 asteroids less than 100 km in radius are irregular rocks. The 30 asteroids and 11 planetary satellites of 100–250 km radius are irregular in shape with density less than 2. With radius above 250 km and mass 10^{20-21} kg, the five asteroids and eight satellites have enough self-gravitation to pull themselves together and become spherical, they show polygonal surface fractures, and the first signs of vulcanism and outassing (although still on the flat part of the exponential growth curve. At 10^{22-23} kg, which includes Moon, Mercury, Mars, and the six largest planetary satellites, density exceeds 3, vulcanism becomes prominent, and there may be a tenuous atmosphere. At 10^{24} kg, which includes Earth and Venus (now on the steepening growth curve) density exceeds 5, with extensive vulcanism, a fluid core, rapid expansion disrupting the lithosphere,

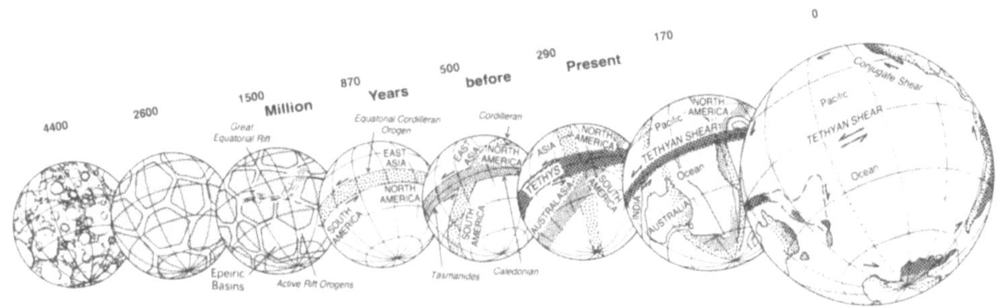

Figure 19

and a thick atmosphere. Above 10^{25} kg, which includes Uranus, Neptune and Saturn, a threshold of rapid expansion is passed, density drops below 2, and rings appear. At 10^{28} kg (Jupiter) already a net radiator, is more like an embryo star. At 10^{29} kg we enter the field of brown dwarf stars, like Van Briesbroek 8B, which radiates in the infra-red at 2000° K.

Figure 19 is a reconstruction of the evolution of Earth. The expansion is logarithmic. At 4400 my ago, it was like Moon of today. At 1500 my was like Mars with his great equatorial rift. The Cordilleran, Caledonian and Tethyan orogens were each equatorial, with a 60° pole migration between each and a sinistral torsion of some 10°. After the Paleogene Alpide orogeny, with a sinistral torsion, the axis shifted again to its present position. These matters are discussed in Carey (1988).

Physics now is in a similar position to that when Kelvin scoffed at geologists and ruled that the long time demanded for the age of the Earth was far more than physics could grant. Such mathematics could not be wrong. Physics is now mature, and only filling of minor gaps lies ahead. Now, I find empirically that the Earth has been expanding logarithmically since the beginning and has doubled in diameter and mass during the last hundred million years. No presently known physics can accommodate this. We must be on the brink of a fundamental new advance. Herein lies the opportunity for a new young genius.

References

Ager, D., 1986, Presidential address, *Brit. Assoc. Adv. Sci.*

Carey, S.W., 1954, The orocline concept in geotectonics, *Roy. Soc. Tasmania* 89:255-288.

Carey, S.W., 1962, Anniversary address: Folding, *Alberta Soc. Petrol. Geol.* 10 (2) 95-144.

Carey, S.W., 1988, Theories of the Earth and Universe. Stanford, 413 pp.

Carey, S.W., 1994, Earth, Universe, Cosmos. (in preparation).

Choi, D.R., 1984, Late Permian-Early Triassic paleogeography of northern Japan. *Geology* 12: 728-731.

Embleton, B.J.J. and Schmidt, P.W.,1979, Recognition of common Precambrian polar wandering r e v e a l s conflict with plate tectonics. *Nature* 282 (5740): 705-707.

Ichikawa, K., 1951, Note on the basal conglomerate of Triassic Inai Group in the Okazu district. *Minerals and Geology* 4: 17-19.

Parkinson, W.D., 1988, *in* Carey, 1988: 168–170.

Perry, K., 1986, *in* Carey, 1988: 267-9.

Stephenson, A.J., Scholl, D.W., and Vallier, T.L., 1983, Tectonic and geologic implications of the Zodiac fan, Aleutian abyssal plain, northeast Pacific. *Geol. Soc. America, Bull.* 94:259-273.

EARTH COMPLEXITY VS. PLATE TECTONIC SIMPLICITY

Giancarlo Scalera

Istituto Nazionale di Geofisica
Via di Vigna Murata 605
I-00143 Rome, Italy

INTRODUCTION

A strong impulse to the geosciences was due to the journey of Christopher Columbus and to the subsequent work of contemporary cartographers in mapping the new continents. The similarity of the Atlantic coastline shapes of South America and Africa become immediately evident, transporting the 16th century thought towards new "continents" of ideas and awarenesses. From the sometimes different and attenuated flat Earth dogmas (Randles, 1980) defended by the church in the age of Columbus, three centuries were needed to slowly develop the links with other fields which prompted the French naturalist Buffon to affirm that:

They (elephants, rhinoceroses, hippopotamuses) lived contemporaneusly in the northern region of Europe, Asia and America; a fact revealing that the two continents were once contiguous and that they separated in later epochs. (......) Maybe the separation of Europe from America happened 10000 years ago.

The above quotation, its precise formulation, bear witness to a previous circulation of mobilistic ideas which until today present hidden historical modalities and documents which would be very interesting to highlight. In the nineteenth century there is an increase in the number of documents containing mobilistic ideas (Green, Snider-Pellegrini), which corresponds to the evolutionism in biological sciences, but not even Wegener in the next century was able to accomplish the 'scientific revolution' because of the amount of data lacking from the ocean floors. Only in the 60s did, increasing amounts of geophysical measurements, better knowledge of the oceanic topography, the discovery of the symmetric magnetic anomalies and their correlation with the ocean floor age, allow the scientific community to bridge the gap from fixism to mobilism. To the very prolonged time revolution of thought (at least from Buffon to Wegener and beyond to the Carey meeting of 1956), was added a very quick and generalized conversion of geoscientists occurring from 1960 to 1970 on the new factual basis, but growing on a cultural ground already fertilized by two centuries of circulation of mobilistic ideas.

The same situation, in my opinion, may now be the case for the expanding Earth hypothesis. From the first written traces of this idea (Mantovani, Yarkowski) at the end of the last century, a long series of books dealing with the argument has been published without raising definite interest (Lindemann, Hingelberg, Keindle, Jordan, Carey). The reason for this theory's difficult dissemination is on the one hand similar to that encountered by mobilism in the first half of our century because of the lack of data (today the GPS and VLBI are still at their beginning era, and are not able to provide conclusive results), but on the other hand its acceptance is hindered by a difficoult explanation of the physical causes of such a large amount of expansion, which nearly necessarily needs an increase of the amount of the matter of the planet. On the contrary, very early on, plate tectonics have found in the convective fluxes in the mantle, the virtual engine which has been generally accepted because of its symplicity.

We actually can formulate only uncertain hypotheses on the physical-chemical processes active on the deep, pressure and temperature of the core, nor do we completely know the physics of high pressure, which is tested only in short fractions of seconds into micro high-pressurized environments. The possibility to choose the true mechanisms of the expansion among the few which have already been proposed (conversion of heter in matter, superdense earth core, dark matter, etc....) is very far today, albeit the Olympia Conference bear witness to an interest of a minority in the scientific community to return to a space with physical property, not only geometrical as in relativity, and to test with neutrino physics, the NESTOR Project, both the Earth's interior and the space surrounding galaxies.

The aim of this short contribution is then limited mainly to discussing not exhaustively some factual, experimental and logic evidence which strongly favours an expansion of the Earth, tryng at this stage to make this theory self-sustaining without need to propose a precise cause. The hope is that the entire set of these arguments (together with other arguments presented in the papers of different authors in these proceedings) could work as a solid and convincing basis for the physicists to start an exploration of the possibilities to theoretically explain expansion and to support the theory by specific experimental works.

HAVE WE TODAY NEW REAL DATA WHICH ALLOW US TO OVERRIDE PLATE TECTONICS?

New Perspectives from Real Data

Plate tectonics was born as a geometrical-kinematical formalism which had to be connected to geologic evidence, initially mainly surface geology evidence. The traces of the kinematics of the continents or of the plates (wider than the continents because of the surrounding ocean floor) were recognized in the oceanic fracture zones, generally ortogonal to the mid-oceanic ridge (Morgan, 1968). Albeit this is valid, at least to some degree, in the Atlantic; historically a first contraddiction was estabilished with the real data because of the discordant trend of the fracture zones in the North East Pacific, which extend from the North american Cordillera towards New Guinea. To overcome this - whose meaning is an expanding Earth - two solutions were proposed: the first stating that once the motion of the Pacific plate was in the direction traced by fracture zones while recently the direction has changed with a rotation of nearly 90° of the vector velocity (strangely, this solution, with the truth confined in the past, was more acceptable than the modern one in which the motion of the Pacific is traced by the Emperor Hawaii submarine chain of volcanos, never in agreement with fracture zone direction), and the second proposing that perhaps it was better to take the transform fault near mid oceanic ridges as indicator of kinematics, forgetting the embarrassing fracture zones. In this paper I will subseguently show that, examining without prejudices the real data, many other clues exist in favour of fracture zones as path indicators and in favour of the Cordilleran side of the Laurentian plate near North Australia and New Guinea, and of an asymmetrical growing of the Pacific basin.

The Pacific Ocean as Seat of Main Contraddictions

Coincidence of Maximum Ages. The map of Larson et al. (1985) shows the ages of ocean floors and of continental provinces. The maximum age of all the oceans is Jurassic, a fact that is suspicious because of the different status of the Pacific ocean which should be in contraction, having undergone the subduction process at its margins; in other words, nothing could have forbidden a different maximum exposed age (greater or lesser than the age of the other oceans) of the Pacific, but the data say that in this case Pacific is perfectly analogous to the other oceans. If differences did exist it should be easier to convince ourselves of the contraction status of the Pacific, but this strange coincidence of ages is a clue favouring an identical expansion status of all the oceans.

Accordance of Expansion Rates and Biologic Evolution. Another equivalent form of the preceding argument can be expressed observing the expansion rates of the Atlantic and Pacific mid-oceanic ridges. The rate of expansion of the mid-Pacific ridge is from three to four times that of the mid-Atlantic ridge. If the Pacific rate had been equal or only two times the Atlantic rate, today we would be able to observe conspicuous remains of the Triassic and Paleozoic ocean-floor in the North West Pacific. The expansion rates are in a sort of mutual accordance to prevent observation of oceanic crust earlier than Jurassic, and moreover in accordance with the biologic evolutionary process which only now make available the biologic organisms able to develop scientific observations. If this sufficient degree of biologic evolution had been reached 100 million years before or after the modern age, the intelligent organisms would have observed the Paleozoic ocean floor or the absence of Jurassic age in the Pacific, respectively. All these accordances are not acceptable from a philosophical point of view because the anthropic science is not a product of a special epoch. A change in the tectonic style in the Late Paleozoic and the contemporaneous opening of all the modern oceans in the Jurassic, with an asymmetric expansion of the Earth, more pronounced in the Pacific, could be a better explanation of the ocean age pattern without appeal to a large amount (more than a hemisphere) of Pacific crust recycling (subduction).

Existence of Conformities Stronger than the Atlantic Ones. Besides the accordance of ages, expansion rates and biologic evolution - in the time dominion -, which are completely disregarded by plate tectonics, even space dominion is richer with informations and clues than the ones considered by the theory of the plates. The conviction that shape similarities among coastal contours - which testify to a former contact among continents - exist only in the Atlantic, in the Antarctic Ocean between Australia and Antarctica and, less markedly, in the Indian, is widespread. Not even the supporters of the repeated Wilson cycle theory (or the "dance of the continents") have searched for the same kind of shape similarity in the Pacific. The factual reality is, even in this case, richer than we aspected. Similarities exist in the Pacific, and they are of an higher rank with respect to the Atlantic ones.

A different kind of similarities exists in the Pacific, which we can call "entire surface shape similarities". This kind of similarity is a further step in the rank of similarities because we have to consider them not matching portions of contours, as was done for the Atlantic similarities, but matching of shapes of entire areas. The first pair of this second kind of similarities is that between the entire Australian continent and the Nazca plate (Fig.1a). If we compare, by means of cartographical transformations, the shapes of the Australian continent to that of the Nazca plate we can see that there is remarkably good conformity between the two. Likewise, there is conformity between the shape of the Tasman and Coral sea basin and the whole South American continent (Fig.1a). In this case the match is not perfect regarding the northern part of the Coral sea basin due to the fact that it is cut by the Solomon Vityaz trenches. However this group of trenches corresponds to the setting of the Marajo Rift and the Amazon River in South America, and -- because rivers have a propensity to follow preexisting troughs or tectonic grabens -- this could suggest a record of the position of deep geofractures along which trenches and river preferentially tend to lie. If Greenland is excluded, a third example of conformity is the one between the North American continent and the northwestern Pacific basin (Fig.1b), which is bounded by the series of trenches of New Guinea, Manus, Salomon, Vityaz, by the East Asiatic arcs and by the Emperor-Hawaii volcanic chain. There is a good comparison between the two shapes: each convexity of an Asiatic arc corresponds in shape and position to each concavity of the North American artic margin. Moreover the small ovoidal Juan de Fuca plate corresponds to the ovoidal New Britain plate, and this constitutes a fourth example. The contemporaneous occurrence of four pairs of conformity is a clear sign that their emplacement happened not by chance (Scalera, 1991a,b, 1993) but following tridimensional decoupling processes in which not only the crust but also deeper lithospheric strata can be involved (a heuristic model has been constructed in Scalera, 1993). It is possible to provide logic proof that in a set of different tectonic theories, only the expansion of the Earth without subduction is favoured by this new kind of large scale morphologic conformities (Scalera 1993).

In Fig.2 a comparison between the actual position of the continents and their Permian position in the classic Pangea (Bullard, Everett and Smith, 1965; Smith and Hallam, 1970) is performed, and also the conformities are shown. Australia, South America and Laurentia should have converged towards the Pacific to reach their modern position, but if this is valid, no simple explanation can exist of the set of conformities which are, on the contrary, clues of former proximity of the involved continents in the Paleozoic. The concomitant well-known clues of proximity even in the Atlantic and Indian side of the Pangea suggests an Earth of smaller size and nearly completely covered by sialic crust.

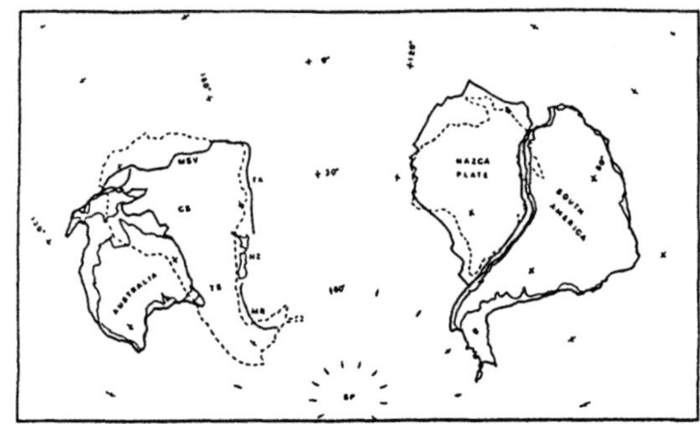

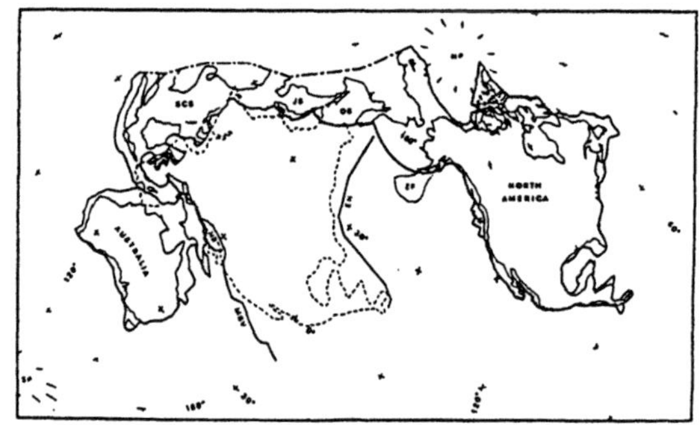

Fig. 1. The similarities in the Pacific. They are of higher order in respect to the Atlantic ones because they refer to entire surface shapes. In a) the Australia-Nazca and South America-Coral Tasman Sea conformities are shown. In b) the conformity between North America and Northwestern Pacific is shown. North America corresponds in shape to Western Pacific. The Juan de Fuca plate corresponds in shape to the New Britain ovoidal plate. The dotted lines represent computer-aided rotation of the continental contours. The broken bold line is an arbitrary boundary between Asia (not represented in figure) and the East Asiatic trench are backarc zones. MSV= Manus, Salomon, Vityaz trenches; TK= Tonga Kermadec trench; NZ= New Zealand; MR= Macquarie ridge; CS= Coral Sea; TS= Tasman Sea; SCS= South China Sea; JS= Japan Sea; OS= Ochotsk Sea; ZF= Zodiac Fan; EH= Emperor Hawaii volcanic chain; JF= Juan de Fuca plate; NB= The little New Britain ovoidal plate; NP= North Pole; SP= South Pole.

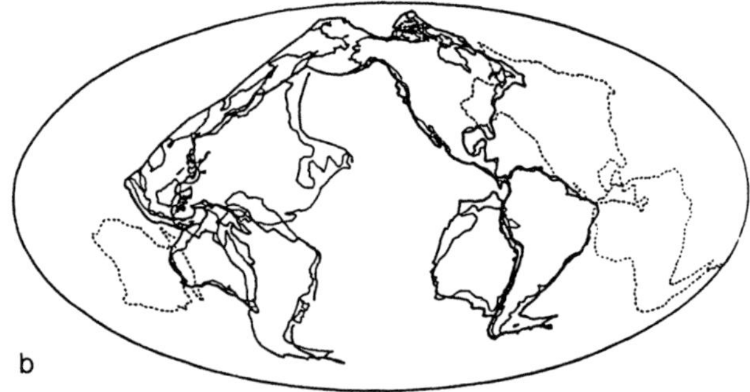

Fig.2. a) The reference Pangaea. The supercontinent have been reconstructed following the classic work of Bullard, Everett and Smith (1965) and Smith and Hallam (1970); b) All the conformities among continents and basins together with the (dotted) outlines of Australia Laurentia and South America in the positions which they assume in the reference Pangaea. It is hard to imagine that the conformities could be formed by the convergence of Laurentia, South America and Australia towards the Pacific.

The Emperor Hawaii Volcanic Chain and the Pacific Volcanism. The Emperor Hawaii chain is one of the most prominent structures of the oceans. No other oceanic chain has the same characteristics and such strong gravimetric anomalies(Haxby, 1987). Several different interpretations of this chain have been provided in the course of the last few decades, but the one accepted today by the geophysical community, in its acceptance of a rigid movement of the entire Pacific sea-floor, is completely ancillary to plate tectonics and prefers to hide the complexity of the situation surrounding this ridge.

A mantle plume rising from a source in a steady position in the mantle and a rigid mouvement of the Pacific plate would provide the right succession of age of the volcanic ridge. The age of the volcanoes decreases, going from Cretaceous to north-west, near the Aleutine arc, to the now active Kilaewa on the extreme east of Hawaii. As a consequence of their assumption and of those observed volcanism ages, we should be able to observe the same shape and pattern of ages of Emperor-Hawaii on other successions of volcanic islands and seamounts. In the box of Fig.3 is represented a typical example of what is searched for. The real age of the Pacific volcanism, if examined globally, leads to different conclusions. In Fig.3 I have plotted the data extracted from the Unesco Geologic Map (1976). Besides the Emperor-Hawaii progression of ages only two other long progressions of volcanic ages can be recognized in the Pacific basin (Fig.3).

The first starts with a largely dispersed distribution of volcanic apparatuses of Lower and Upper Cretaceous age, located north and east of Marianas, and ending, becoming progressively younger and more narrow, at the Austral Seamount (Isles Marotiri, east of Isles Cook). The second, also very dispersed, starts south of Hawaian chain with the same Lower and Upper Cretaceous age, and ends, becoming narrow, at the Isles Pitcairn, showing recent volcanism.

The different shapes, dimensions and caracteristics of the three volcanic lineaments are not compatible with a rigid and coherent movement of the Pacific crust on a motionless plume. The only thing which can be affirmed is a general tendency of the volcanic activity to follow, with a space-delay of thousands kilometers, the enlarging boundary of the plate on the mid-oceanic spreading ridge. A new working hypothesis could be the trapping of deep magmatic chambers on both sides of the mid oceanic ridge, in correspondence with deep discontinuity crossing the axis of the ridge, and their becoming active when they are sufficiently near to the surface and the state of the litospherc stress is changed because of the changing curvature on an expanding Earth.

The Emperor-Hawaii volcanism is distributed on a very narrow band, while the other two are wider, going back through time. The first fact can further support the role of the Emperor-Hawaii as a boundary between two plates (Scalera, 1991a,b, 1993) along which the volcanic activity is emplaced preferentially. In fact a great difference can be noted between the eastern and western side of Emperor, the first one characterized by transform faults and parallel fracture zones which trace the outlines of an environment produced by plate tectonic processes, and a western one characterized by a higher seismicity level (Walker, 1989),an uneven topography containing many seamounts, canyons, seamount chains and level differences.

The Magnetic Anomalies Around Emperor. The behaviour of the Pacific as a rigid plate can be excluded also on the basis of the detailed examination of the sea-floor magnetic anomaly. Besides the well-known dislocations of the magnetic anomalies along transform faults and fracture zones, a major dislocation of anomalies exists between the east and west side of the Emperor-Hawaii chain (Fig.4). The anomaly 32b (71 Ma) on the east of the Emperor is flanked by the anomaly M1 (122 Ma) on the west side. This means a difference in age of nearly fifty million years between the two zones and a transcurrence of about thirty geographical degrees. The existence of step faults along the chain is in agreement with this interpretation.

The mouvement responsible for the displacement of the anomalies could be happened mainly on the Emperor-Hawaii chain, but also other zones of intense deformation cannot be excluded on the basis of our present knowledge. The two other great lineaments described in the above section could have a role in the history of the Pacific basin deformations, and new lines of research are needed to ascertain these new possibilities.

The Sedimentary Fans in the North Pacific Boundaries. A higher degree of awareness can be gained about the real processes acting on the margins of the Pacific by considering the terranes near these margins, both oceanic terranes and land ones.

The huge Zodiac Fan is located near the east side of the Aleutian trench (Fig.4) in a position which, if taken back to the time of its formation ($\approx$40 Ma), would end up, according to plate tectonics, right in the middle of the Pacific Ocean (Carey, 1988; Harbert, 1987), separated from North America by the hypothetical (but necessary for the plate tectonics congruency) Kula--Pacific and Farallon--Pacific ancient spreading ridges, which being topographic reliefs should have kept the sediments from overleaping it. On the other hand, according to a version of the expanding earth hypothesis, the Zodiac fan never moved from its place but the North American continent was the one to move away from it, enlarging the Pacific, whose opening started from the mentioned match just north of New Guinea. The same kinematics is compatible with palaegeographic reconstructions constrained by palaeomagnetic data (Scalera, 1990).

On the western corner of the Aleutine trench another sedimentary fan exists, the Meiji sediment tongue, with a maximum age of 16 Ma (Fig.4). Also in this case, reconstructing the position of the fan to the epoch of its initial sedimentation, no simple reason can be found for its growing so far from the coasts. The streams system needed to explain the Meiji tongue is judged *ad hoc* by Scholl et al. (1977) which evaluate a maximum of 300-400 km of subduction which occurred due to the limiting presence of the tongue, and a maximum of 500 km due to the limit imposed by the Zodiac fan. Another line of research could be that to better ascertaining the maximum age of its first sedimentation, because if it resulted older than the presently accepted value, it could enhance the paradox to a level nearer to the Zodiac fan one.

Both these anomalies can be eliminated assuming a Pacific plate with no large amount of undertrust on its margins (no subduction) and this can also mitigate the problems deriving from the so called "accreted terranes". The accreted terranes of, say, Cretaceous age, must necessarily originate on the eastern side of the cetaceous mid-Pacific spreading ridge and they had to perform a long journey across the Pacific, which was wider than today with a symmetric pattern of ages on the two sides of the ridge axes. In the expanding Earth framework, the origin of the accreted terranes can always be very near to the Cordilleran margin, and they can be considered alloctonous because of their displacements along a trans-tensional dextral margin, of which the Cordillera, the Juan de Fuca ridge, the Basin and Ranges region and the San Andreas transform fault are the modern examples.

Problems with Convective Cells. The hypothesis of convective cells as engine for plate tectonics has always encounterd several objections. The main objection is that the horizontal size of the convective cells should be of the same magnitude of the vertical size, a condition not fulfilled in the Pacific. In the Atlantic, near the equatorial zone, the alignment of the long fracture zones, which cut the central ridge in many short segments oriented north-south, and the strong dislocation of the northern ridge axis with respect to the southern ridge one, are incompatible with the expected pattern of flux-lines produced by convection.

Another decisive argument can be considered that of the absence in the geoid anomalies of peak to peak amplitude and wavelength which we expect for the presence of convective motions (Sandwell and Renkin, 1988). Moreover, the cause of the previous appeerence of geoid anomalies, which was supposedly linked to the convection, is ascribed to the removing of the spherical harmonic coefficients through degree and order 10 without any computational precaution. This artifact is typical of the way a systematic error can invalidate a research field even for a long time, before awareness of it is reached. The same situation could also be in act today for the very long baseline geodetic measurements (VLBI, GPS, SRL).

PALEONTOLOGICAL CLUES

Paleontology provide some clues which suggest an analogous behaviour of the Pacific basin with respect to the other oceanic basins. I can here quote only a selected choice of few cases among the numerous which effectively exist. Stait and Burrett (1987) report a strong similarity of Ordovician nautiloids of Tasmania with Asiatic and Laurentian ones. In the Silurian, an anomalous correspondence can be recognized in the strong similarity between the planctonic floras (74 species) of the Canning Basin in Western Australia and the planctonic floras (41 species) of Yowa, USA (Colbath, 1990). Devonian vertebrate fossils of South America have been found displayng close similarities with correspondent faunas of Antarctica,

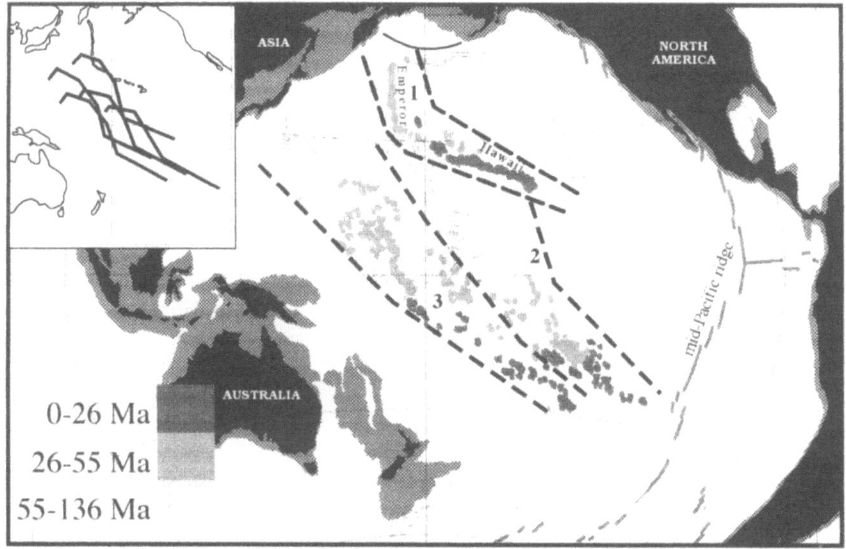

Fig.3. The ages of Pacific volcanism. Three major long volcanic provinces, 1,2,3, can be recognized in the Pacific Basin, and they have been divided by bold dotted lines. The shapes and positions of the provinces are incompatible with the rigid motion of the Pacific plate on hot spots, as alleged by plate tectonics. In the little box the shapes of volcanic chains which are sought by the theory of the plates.

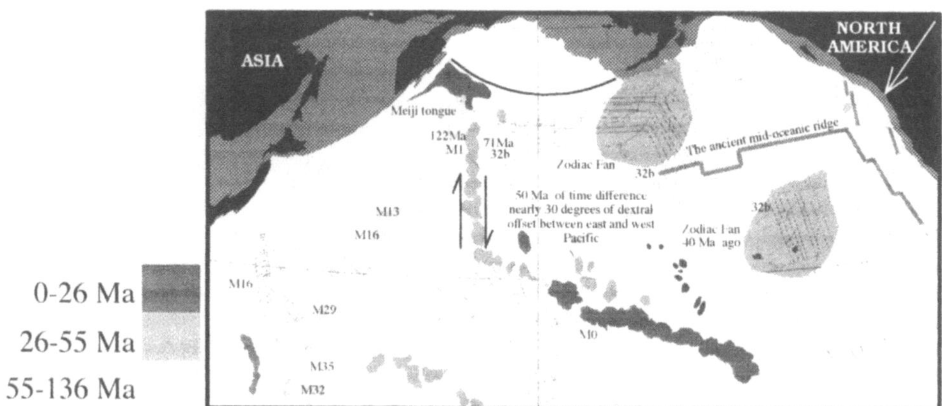

Fig.4. The offset of the magnetic anomalies on the Emperor Hawaii chain, and the position of the Meiji and Zodiac sedimentary fans. The paleo-position of the Zodiac fan is represented at nearly 40 Ma, following plate tectonics. The white arrow represents the possible ancient region of alimentation of the fan, and the flux direction. The paleo-mid-oceanic ridge was a topographic barrier to the alimentation of the fan.

Australia and South Cina (Goujet and Janvier, 1984). This fact testifies to an extremely narrow marine barrier between the two zones, and is in perfect agreement with my Silurian paleogeographical reconstruction (Earth's radius = 3500 km), were Australia is in contact with the Cordilleran zone of Laurentia (Scalera, 1990).

The biogeographic anomaly of the Paleozoic and Early Mesozoic Tethyan faunas consists in the presence of faunas which are typical of Tethys sea on rocks of the Cordilleran margins of North and South America (Newton, 1988). Plate tectonics explains the anomaly appealing to the eastward displacement of several micro-plates which carried (traveling through an entire hemisphere) the faunas from East Asia to the Western Americas. A different and more attenuated version of this solution appeals to micro-plates in the middle of the Pacific, colonized using the interposed islands as "stepping-stones" which, only after their colonization, started a journey towards the Cordilleras. The second solution proposes the existence of marine corridors between the Tethys and Pacific, from Iberia through the two Americas, which have never been found. Only terrigenous sediments exist where they have been proposed. The third solution, adopted by Newton (1988) is the possible West-East diffusion both of biotas along islands chains, and of larvae by marine streams. The greater size of the ancient Pacific is a strong limitation for this last proposal. All the problems inherent to these solutions are extensively discussed by Newton (1988) and in a debate with several authors (AA.VV. vs. Newton, 1988). A Pacific of a reduced extension in the Paleozoic and Early Mesozoic is the obvious alternative in the expanding Earth theory, and Davidson (1983, 1994) and Shields(1979, 1983) support this view through global paleontological considerations.

Many links of the antartic flora have been found, in the course of the exploration of this continent, with the flora of the Northern Hemisphere from Paleozoic to Mesozoic (Truswell, 1991). In particular numerous antarctic pollens and floras are similar to some European and North American ones, and strong affinity exists in the Jurassic flora with India and Europe. Very important is the paradox of the Cretaceous flora, fossil forests, which show indisputable signs of long growing seasons, equability of light and frost-free conditions, typical of temperate or low latitudes (Truswell, 1991), a fact incompatible with the classic Pangea reconstruction and with the paleomagnetic data. In a paleogeography based on expanding Earth (see the sections on the reconstructions) the paradox disappears, with the Antarctica located on intermediate latitudes.

The bones of the triassic reptile Lystrosaurus have been found in several places of Gondwana, in India, but also in many places of Eurasia, separated by the wide Tethys Sea from Gondwana in the Triassic. Colbert (1991), an eminent paleontologist, affirms that the global distribution of the Lystrosaurus fossil sites could be more easily understandable if the Eurasian sites were displaced near the Gondwana sites, with a closure of the interposed Tethys. The same conclusion, from a different point of view is reached by Ahmad (1983), Chatterjee (1984, 1987) Chatterjee and Hutton (1986), Sahni (1984), Sahni et al. (1987), Tripathi and Singh, 1987, Smith (1988) and Patterson and Owen (1991), concerning also the northern and western boundary of India. Similar situation results for several saurus of Indochina or, more generally, of north Gondwana of the Mesozoic age, which are strictly linked to central and northern Asia or European species, suggesting an active interchanging across a very narrow Tethys sea (Buffetaut, 1989a, 1989b; Buffetaut et al., 1989; Astibia et al., 1990). Ager (1986) develops analogous reasoning especially using the distribution of brachiopods, and Stöcklin (1984, 1989) reaches the same conclusion for the Afghanistan-Pamir-Pakistan region on the basis of geological evidence. All the quoted papers are in conflict both with the long and isolated journey of India before its hypothesized collision with Laurasia, and with the large extension of the Tethys sea which both are the result of paleogeographical reconstructions performed on an Earth of constant radius.

The East Pacific Barrier is a marine barrier extending from Hawaii, Line islands, Marquesas islands to the western shore of the Americas. It is the most effective barrier to dispersal of the modern shallow water fauna. Fossil records testify that this has been true throughout all the Cenozoic, but that the barrier was less effective in limiting the faunal dispersal during the Cretaceous (Grigg and Hey, 1992). This factual reality could be better explained in an expanding Earth framework in which the Cretaceous Pacific has a reduced extent with the exclusion of just the Cenozoic area occupied by the barrier. The remaining Cretaceous area today has a very uneven topography, numerous seamounts and islands, which suggest a similar topography even in the geological past. The faunal diffusion from island to island and finally to American shores should have been easier in Cretaceous with these conditions. The above examples (and many others) are tied together by showing a strong continuity through geological time, and the whole of it suggests a progressive opening of the Pacific Basin starting from a very narrow size.

A NEW CARTOGRAPHICAL EXPERIMENT USING THE LARSON MAP

The stimulating arguments above described claims for a practical proof of their simultaneous consistency. The main proof which is possible to produce is to perform paleogeographical reconstructions, at radii lesser then the modern one, constrained by different set of data.

After a first trial of paleogeografical reconstructions performed with the help of few paleopole and magnetic anomaly data (Scalera, 1990), a new project started with the aim to repeat the paleogeographic reconstruction with an updated set of geophysical and geological data. The digitization has been performed of all the ocean floor fragments of same age and continental shields from the map "Bedrock Geology of the World" (Larson et al. 1985), and of all the magnetic anomalies from the map "Identified magnetic sea-floor spreading anomalies" of Roeser and Rilat (1982) and the paper of Nakanishi et al.(1992). Software has been created to extract with a wide set of independent options the data from the updated version of the Global Paleomagnetic Database (Lock and McElhinny, 1991, McElhinny and Lock 1990a,1990b) filtering the paleopole on the basis of their quality (Florindo et al., in press; Sagnotti et al., in press; Scalera, et al., 1993). Finally a cartographic computer program (Scalera, 1988, 1990) has been implemented to allow the plotting of all the different kinds of data (with different transformation laws) on globes of any assigned radius. Then, a series of paleogeographic reconstructions has been started using all the above described tools. I show here only two examples of this reconstructions of which a more complete set is now in progress. It should be intended that the present day continental outlines are for reference only.

The Jurassic Reconstruction

All the Jurassic oceanic fragments from the map of Larson et al. (1985) have been plotted together with the continents. Each ocean floor fragment has been considered connected to the adjacent continent to constitute an entire plate. These obtained Jurassic plates have been matched on globes with radii lesser than the modern one to seek a solution of the simultaneously imposed boundary condition, paleogeographic, paleomagnetic, paleontologic.

In the Atlantic, the large Jurassic band near northwest Africa must be matched to the correspondent Jurassic band adjacent to North America. The Pacific becomes equal in size to the Jurassic ocean-floor today near the Asiatic trench-arc-backarc zones (Fig.5). The Australia paleoposition has been assigned with the help of paleopole data, and albeit many margins could be of slightly different form in the considered age, I have corrected only the New Guinea position taking into account the interposed rifting with Australia and cutting the little extensions of younger sea-floor between the main island and Australia. The submerged continental plateau surrounding New Zealand finds a right position interposed between Australia and South America (Fig.5).

A first impressive fact in the paleo-Indian ocean is the matching which results in this reconstruction between the Jurassic nucleus of the Pacific and the Jurassic of Wharton basin, northwest of the modern Australia, whose last basin has for a long time been seat of idle exploration in search for the Paleozoic oceanic crust there necessarily in the classical Pangea reconstructions (Dietz and Holden, 1971). The expanding Earth easily removes this difficulty of plate tectonics, requiring exactly a Jurassic age for the Wharton basin, which becomes simply a split fragment of the Pacific.

The position of Antarctica in this reconstruction has been assigned in strict consequence to the matching of its east continental margin, from Victoria Land to Wilkes Land, with the southern Australian margin. The second impressive fact is the matching of the Jurassic band adjacent to Somalia and Southern Arabian Peninsula with the Jurassic band adjacent to Antarctica from the Weddel sea to the Maud Queen. The two mentioned bands have never been posed in mutual relation in the plate tectonic framework, but in the expanding Earth they become analogous to the two jurassic bands of the Atlantic at its first opening, albeit in this case with a clear trans-tensional dextral component which is possible to observe by comparison of the map with the next one of Cretaceous age, and with modern geography. The

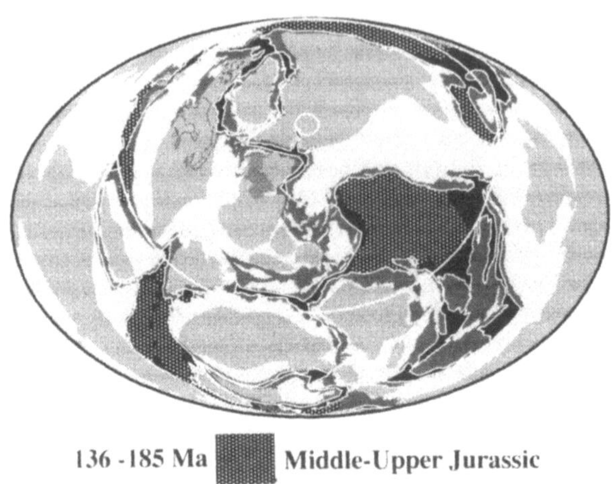

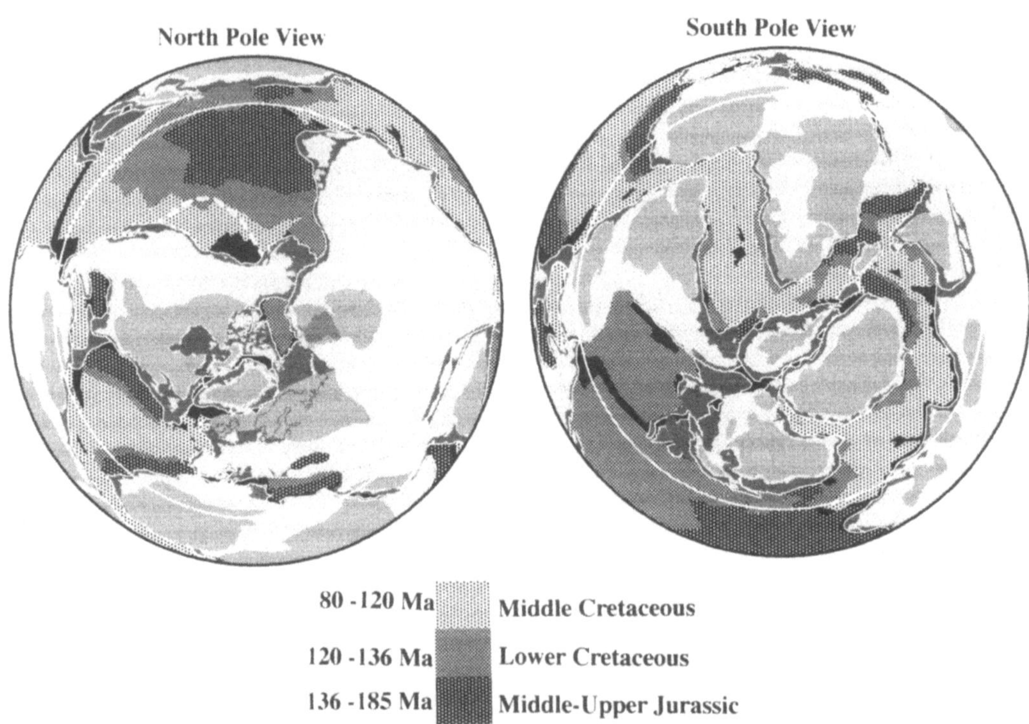

Fig.5. The Jurassic and Cretaceous paleogeographical reconstructions

position of the Antarctic Peninsula has been corrected for a trans-tensional rifting interposed with the eastern craton. The long segment of Antarctica, between India and Australia, in contact with the proto East-Asia is in agreement with the existence of a granulite facies Precambrian mobile belt (Katz, 1993) which can be better restored in its continuity from Sri Lanka, India, Antarctica, North China, Indochina, Australia, with respect to classic Pangea reconstructions. The absence of superposition of the shield zones is a further guaranty of coherence of the recontruction.

The Cretaceous Reconstruction

The Cretaceous reconstruction (Fig.5) has been performed by adding the respective oceanic areas up to a minimum age of 100 million of years. Paleopoles have been time-filtered by windows from 95 to 105 Ma or 90 to 110 Ma depending on the amount of the data available for each continent. The breakdown of the Pangea is now in an advanced state with a neat opening of all the oceans. India is virtually in contact with Eurasia and only a shallow sea with interposed islands has been traced south of Tibet, in agreement with geologic and paleontologic data. It is not possible to exclude the presence of large areas of trench-arc-backarc in eastern Asia (not traced in the reconstruction), in complete analogy to the modern situation. The areas surrounding the Eurasian cratons is now larger and can be the seat of the Alpine orogeny and of the Ob sphenochasm. A suitable space is found for the Bering Sea and the Artic Sea Cretaceous fragments.

Antarctica is now largely at intermediate latitudes, in agreement with the paleontologic data of vegetation (Truswell, 1991). Antarctic Peninsula has also in this reconstruction a non ambiguous position fixed by the neighbouring oceanic fragments and more close to the central craton. The northern and the sothern cratons of South America has been located independently, and the northern one shows a sinistral dislocation assigned by the extension and the matching of the Atlantic oceanic fragments. The line from India, Sri Lanka, Madagascar, Antarctic rift, Alpine fault of New Zealand, appears as an elongated line of tectonic instability of global extension. The Emperor Hawaii chain is indicated by a broken bold line, and the non perfect matching with the North American continent is in agreement with the strong dislocation revealed by the magnetic anomalies described in a preceding section. The limited extension of the Pacific is in agreement with the non existence of the "Pacific Barrier" in this period (Grigg and Hey, 1992).

THE EXPANDING EARTH TRUE MOBILISM AND NOT A COMPROMISE

Some conceptions describe expanding Earth as a fixistic view in which all the terrains, once created, remain in their site of emplacement without other displacements. Other conceptions prefer to consider expanding Earth a compromise between fixism and mobilism. Both this opinion is untenable because not grounded on factual experiences. In reality the expanding Earth is an evolution of mobilism, and it is possible to support this statement on the basis of several arguments.

The first argument is a very simple cartographic experience, in which a pure operation of variation of the Earth's radius is applied to a globe with the continents in their modern position. The continents pass from the modern radius (Fig.6a) to the lesser one, mantaining fixed the geographical coordinates of a point near their centroid. Fig.6b shows that no match is possible of the Pangea because rotations have occurred among the continental masses, which could be eliminated only by choosing the 'fixed point' of each continent very far from the respective centroid position, in an improbable position. This experience means that the continents have undergone true trans-tensional mutual displacements and not only purely radial ones. A second information from this experience is that by simply decreasing the radius it is impossible to completely close the Pacific without having a strong overlapping of the continents on the opposite hemisphere. This means that both true rotations and strong differences in mutual drift velocities are superimposed to the radial displacements, which suggest real mobilism even in this case.

Same suggestions came from the described paleogeographic reconstructions (Fig.5) in which strong trans-tensional events are recognizable (e.g. India-Africa; Antartica-Antartic peninsula; Antartica-Africa; India-Australia ocean floors and the creation of the 90° East Ridge; the strong dislocation of the magnetic anomalies across the Emperor chain; etc...), and some narrow zone of destruction of shallow marine basins (India-Tibet; North and South American Cordilleras; etc...). The latter is supported by overtrusts of hundreds of kilometers which are geologically recognizable in many orogenic zones. Finally the existence of the conformities 'basin-continent' in the Pacific is a clear sign of a more general and three-dimensional mobility of the major strata, with mechanisms still to be understood (Scalera, 1991b).

All this points to an expanding Earth which is a more extreme version of the classic mobilism and which could open new perspectives in the remodelization of several processes and structures. An example of a possible new modelization, which is necessary because of the described clues against the subduction concept, is given in a next section.

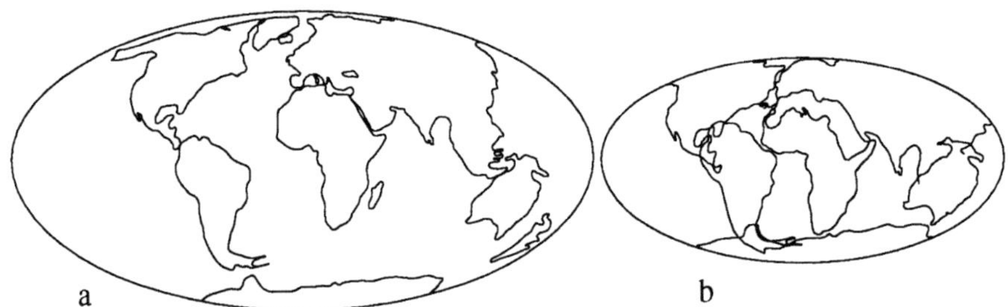

Fig.6.   In this experience precise computerized cartography is used and a simple operation of decreasing the Earth's radius is applied to a modern radius globe (a); the resulting halvened radius globe (b) shows that the continents do not match in a Pangea without mutual rotations and translations, and therefore the expansion was not simply radial but asymmetrical - more pronounced on the Pacific, which is impossible to close avoiding spurious superimposition of continental contours, and less pronounced on the Mediterranean, where a strong superimposition is evident. In applying the operation of radius decrease, the geographical coordinates of the continents' centers of mass have been maintained fixed. Expansion is not uniform nor isotropical.

A MODEL OF EVOLUTION OF THE TRENCH-ARC-BACKARC ZONES WITHOUT SUBDUCTION

All the considerations developed in the preceding sections point to a limited amount of crustal undertrust starting from Paleozoic. Therefore an obvious line of research is the definition of an alternative model of the trench-arc-backarc (TAB) zones in which subduction does not exist, at least in the strong sense (thousands of kilometers) used by plate tectonics.

The non-subductive model of the evolution of TAB zones has been developed using as key points the tensional state of stress of trenches and the presence of a volcanic zone on the arc and the backarc, which exert a rear push on the orogenic arc. The two cooperate with gravity and with high frequency tidal movements (Kosygin and Maslov, 1986) to favour gravitationl spreading of the arc toward the trench. The spreading is accompained by the folding and trusting of oceanic crust beneath the accretion complex. Once the trench is filled in by the advancing edge of the arc, a new tensional zone starts seaward from the old, creating a new trench. Both old vanishing and new born zones of tensional weakness are traced by earthquake foci explaining the presence of the double Benioff surface. The TABs initiate their evolution with uplift of a new orogen on the continental edge, subsequent subsidence of the orogen and open of backarc. Covering of the subsided orogen by pillow lavas and a cyclic sequence of uplift and exumation of deep crust could explain several typical features of the TABs, namely the obduction of ophiolitic fields, the high backarc heat flow, the special characteristics of the methamorphism (Fig.7).

In the presence of hot intrusive complexes, spreading occurs at all scales. The proposed model is then substantiated by both experimental works on an analogical scale model (Merle and Vendeville, 1992) and by observational evidence on different scale structures like trust fault on glacial morenes (Lehmann, 1992), the spreading of volcanic cones (Borgia et al., 1992; Borgia and Treves, 1992; Delaney, 1992) and the role of melt in the uplift and evolution of orogenic belts (Hollister, 1993; Merle et al. 1993). The model needs further toonings to be adapted to special regional cases. With no pretense to be a solution for all the complex problems at the TABs, the main advantage of the proposed model is to fit observation with simple mechanisms, while the physical difficulties encountered by subduction are very strong.

Another fundamental advantage is the possibility to consider the different situations present around the circum-Pacific margins as different phases of the same process, overcoming the difficulty of plate-tectonics to explain the strong differences among the opposite Pacific shore and the North and South American Cordillera.

COSMOLOGICAL LINKS

The preceding sections contain arguments favouring a manifold increase of size of the Earth, which cannot result from simple changes of phases or other very simple physical-chemical process. The conclusion is that an increase of mass could be the real process active in the interior of the planet. I consider an inpassable limit which I have to respect to remain neutral as regards the possible explanations of this phenomenon, which I consider at this moment a working hypothesis to be pursued only by searching for arguments which can support the increasing mass.

An argument from geology is the finding of fossil heaps of loose materials which show angle of repose higher than the present ones going back through geological time (Mann and Kanagy, 1990; see also russian quotations in Neiman, 1990). Maximum values recorded also exceed by 30°-40° (e.g. some Silurian values) the modern ones, while a conservative estimate of several effects, assuming the big-bang cosmology and plausible variations of the Earth spin, assigns only 10° of possible increase. This evidence seems to me not an artifact and more convincing because we have to espect the contrary bias of the compactation of the strata. Angle of repose up to 60° testifies to a gravity lesser than the modern one.

Neiman (1990) discussed the paleontological clue of the position of the heavy tail of the biggest quadruped dinosaurs, whose traces of dragging on the ground near the fossil footmarks are extremely rare. This can be undoubtedly interpreted as evidence of a Mesozoic lesser gravity, but in my opinion stronger paleontological evidence is the posture of biped dinosaurs whose weight was up to 7-8 tons. I have examined the complete catalogue of the known dinosaurs genera and I have verified that nearly 73% of their families are biped or biped-quadruped. Their posture was not plantigrade but tridigital (tridactyl) with a posterior digit typical of arboricol species. Also their shorth arms could have been suitable for arboricol life. The same posture has been adopted today only by birds, very light animals, and heavier species are now all plantigrade (canguroes and similar) or quadruped. It is possible that a progressive increasing of gravity has drived in this direction the biological evolution?

I have performed a simple computerized experiment in search of the macroscopical effect of a possible increase of gravity on characteristic doubling time of few hundreds of millions of years. A series of test masses have been thrown (all in the same plane) at increasing time in a central gravitational field which increases exponentially. I cannot throw the series starting from infinity but starting from a finite distance only imposing the condition that the masses spiralize with regularity without large differences among them. I also impose that observers are very far on the plane of the open orbits and can measure only the component of the velocity of the masses along the line mass-observer. The graphic part of the program plots this component of the velocity of each i-th mass (Fig.8a) and a curve envelope is traced because it more strictly represents the obseved "present time" of all the series (Fig.8b).

The important fact is that the envelope is similar to the experimental curves of the rotation of the galaxies (Fig.8c), most of which measured by Rubin. Another noteworthy fact is the dependence of the kind of spiralization from both the doubling time τ of the field and the "period" T of the open orbits (period is an improper word, T is the time needed to cover 360° of the open orbit) of a test mass. If $T \gg \tau$ then the orbit is very elongated and near a straight line, if $\tau \approx T$ the orbit is a neat spiral resemblig the galaxies arms, and if $T \ll \tau$ the orbit became virtually an elliptical one. All these conditions are experienced by each test mass,

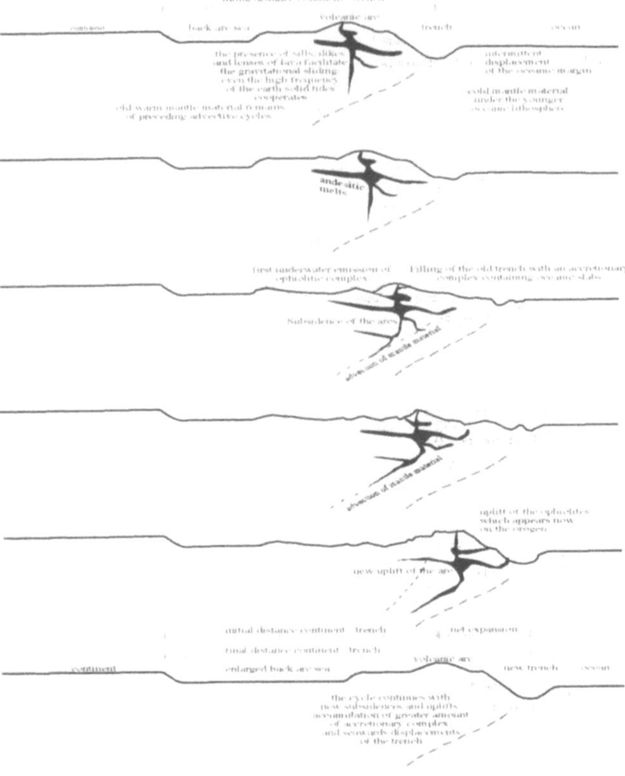

Fig.7. The model of evolution, without subduction, of the trench-arc-backarc zones.

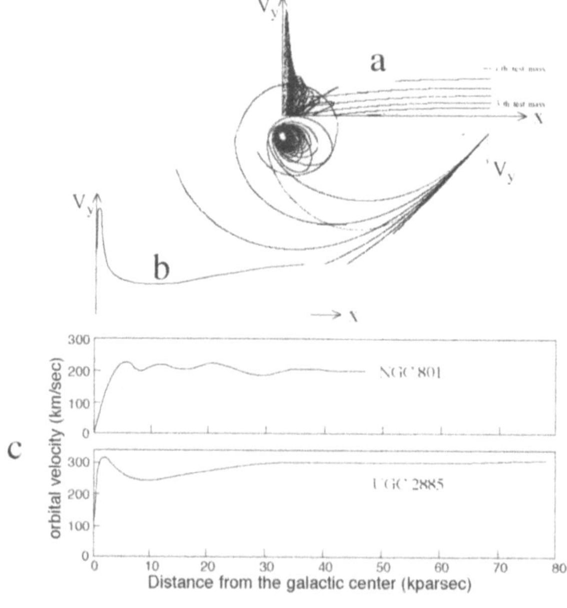

Fig.8. Astronomical simulation : gravitational field which increases exponentially; **b** is the envelope curve of the curves of the test masses in **a** ; in **c** NGC801 and UGC2885 are two examples of galactic rotational velocity measured by Rubin.

270

which after a sufficient time spiralizes elliptically in a limited region resembling the galactic nucleus. The argument is also in favour of a galactic evolutive sequence starting from open spirals and ending on the elliptical galaxies.

The results of this oversimplified computer simulation cannot have any pretense to be the solution of the intricate problem of astrophysics, but could be indicative of how the adherent to expansionism could consequently link Earth to cosmology, linking the morphology of the galaxies to the process of new mass creation because of yet unknown cosmological processes. This argument can also be a suggestion for a line of research. The cosmological scenario, starting from expanding Earth, coud become very different from the big-bang one, with several possible concomitant variations of physical "constants".

CONCLUSIONS

The presence of large undisturbed detrital fans along the Aleutine arc, the existence of astonishing conformities between the members of basin-continent pairs in the Pacific (and the consequent non-random emplacement of the trench arc zones), the progression of the ages of the Pacific volcanic provinces, the many new palaeontological and geological data, the complex distribution of the densities and seismic waves velocities revealed by the seismic global tomography, the seismic evidence of deep continental roots, and the existence of a large amount of data and regional situation which do not fit with the plate tectonics, are all signs that the factual content of the reality, which the collective scientific enterprise progressively discovers, has become extremely larger today than the one once covered by plate tectonics or by its successive adjustments. My personal convinction is that no kind of theory, in the present or in the future, will completely cover the infinite deep and wealth of nature, but through discontinuous advancements our objective knowledge can increse, overcoming old conceptions on philosophical and scientific ground. This is a favourable moment to present new ideas and models, and the more they are, the more probability we have in encountering the right ones. In any case the new models would take into account the fact that a hypothetical civilization confined only around the Pacific shores could have created a mobilistic theory on the basis of the numerous trans-Pacific clues, and therefore created a hypothetical Pangea by closing the Pacific instead of the Atlantic.

Expanding Earth is a framework which seems to better explain some general and regional features of global evolution and morphology, but we should be aware that its efforts to gain knowledge on earth's complexity are not yet over. I hope these efforts never find an end - even in the case of a generalized acceptance of the expansion ideas - because we need an open theory, very different from the rigid plate tectonics schemata, which could be able to incorporate different regional situations, each with a different modellization. In particular we should provide two different models, mutually non-contradictory, of emission of mantle material in the mid-oceanic ridges and in the trench-arc-backarc zones, and of the influence of continental roots on these fluxes and on the general lithospheric flux.

Some general information on the physical processes occurring in the Earth interior are still hidden, the role of the solid and liquid core in global geodynamics is almost completely unknown and the same uncertitude is present concerning the earth's water and atmospheric cycle which could be less closed then we imagine.This Conference could be the best occasion to propose new lines of research able to solve important points. NESTOR is certainly a key project in this direction.

REFERENCES

AA.VV. versus Newton, C.R., 1988, Paleobiogeography of the ancient Pacific, *Science* 249:680-683.

Ager, D.V., 1986, Migrating fossils, moving plates and an expanding Earth, *Modern Geology* 10: 377-390.

Ahmad, F., 1983, Late Palaeozoic to Early Mesozoic palaeogeography of the Tethys region, *in* :"Expanding Earth Symposium," S.W. Carey, ed., University of Tasmania, Sydney 1981,131-145.

Arason, P. and Levi, S., 1990, Compaction and inclination shallowing in deep-sea sediments from the Pacific Ocean, *Journ.Geoph.Res.* 95:4501-4510.

Astibia, H., Buffetaut, E. et al., 1990, The fossil vertebrates from Laño (Basque Country, Spain); new evidence on the composition and affinities of the Late Cretaceous continental faunas of Europe, *Terra Nova* 2:460-466.

Borgia, A., Ferrari, L., and Pasquarè, G., 1992, Importance of gravitational spreading in the tectonic and volcanic evolution of Mount Etna, *Nature* 357:231-235.

Borgia, A., and Treves, B., 1992, Volcanic plates overriding the ocean crust: structure and dynamics of Hawaiian volcanoes, *in*: "Ophiolites and their Modern Oceanic Analogues," L.M. Parson, B.J. Murton and P. Browning, eds., *Geol.Soc.Spec.Publ.* n°60:277-299.

Buffetaut, E. 1989a, The contribution of vertebrate palaeontology to the geodynamic history of South East Asia, *in*: "Tectonic Evolution of the Tethyan Region," A.M.C. Sengör, ed., Kluver Academic Publishers, 645-653.

Buffetaut, E., 1989b, Archosaurian reptiles with Gondwanan affinities in the Upper Cretaceous of Europe, *Terra Nova* 1:69-73.

Buffetaut, E., Sattayarak, N. and Suteethorn, V., 1989, A psittacosaurid dinosaur from the Cretaceous of Thailand and its implications for the palaeogeographical history of Asia, *Terra Nova* 1:370-373.

Buffon, G. 1780, "Les Époques de la Nature," *quoted in*: "La macchina della Terra," N. Morello, Loescher, Torino, pp. 231.

Bullard E.C., Everett J.E. and Smith A.G., 1965, Fit of continents around Atlantic, *Roy. Soc. London Phil. Trans.ser.A* 258:41-75.

Carey S.W., 1975, The Expanding Earth-an Essay Review, *Earth Science Reviews* 11:105-143.

Carey S.W., 1976, "The Expanding Earth," Elsevier, Amsterdam, pp.488.

Chatterjee, S., 1984, The drift of India: a conflict in plate tectonics, *Mém.Soc.géol.France N.S.* n°147:43-48.

Chatterjee, S., 1987, A new theropod dinosaur from India with remarks on the Gondwana-Laurasia connection in the Late Triassic, *in*: "Gondwana Six: Stratigraphy, Sedimentology, and Paleontology," G.D. McKenzie, ed, *AGU Geophys. Mon.* n°41:183-189.

Chatterjee, S. and Hotton, N., 1986, The paleoposition of India, *Jou.Southeast Asian Earth Sci.* 1:145-189.

Colbath, G.K., 1990, Palaeobiogeography of Middle Palaeozoic organic-walled phytoplankton, *in*: "Palaeozoic Palaeogeography and Biogeography," W.S. McKerrow and C.R. Scotese, eds., *Geological Society Memoir* n.12:207-213.

Colbert, E.H., 1991, Mesozoic and Cainozoic tetrapod fossils from Antarctica, *in*: "The Geology of Antarctica," R.J. Tingey, ed., Clarendon Press, Oxford, 568-587.

Davidson, J.K., 1983, Tethys and Pacific stratigraphic evidence for an expanding Earth, *in*: "Expanding Earth Symposium," S.W. Carey, ed., University of Tasmania, Sydney 1981, 191-197.

Davidson, J.K., 1994, *this volume*.

Delaney, P.T., 1992, You can pile it only so high, *Nature* 357:194-196.

Dietz, R.S. and Holden, J.C., 1971, Pre-Mesozoic oceanic crust in the eastern Indian Ocean (Wharton basin)? *Nature* 299, 309-312.

Florindo, F., Sagnotti, L. and Scalera, G., 1994, Complete automatic management of the ABASE ASCII version of the Global Paleomagnetic Database, *EOS* (in press).

Grigg, R.W. and Hey, R. 1992, Paleoceanography of the tropical eastern Pacific Ocean, *Science* 255, 172-178.

Goujet, D. and Janvier, P., 1984, Devonian vertebrates from South America, *Nature* 312:311-311.

Harrison C.G.A. and Watkins N.D., 1977, Shallow inclinations of remanent magnetism in Deep-Sea Drilling Project igneous cores: Geomagnetic field behavior or postemplacement effects? *Journ.Geoph.Res.* 82:4869-4877.

Haxby, W.F., 1987, "Gravity Field of the World's Oceans," map published by the National Geophysical Data Center, NOAA, Boulder.

Hilgenberg, O.C., 1933, "Vom Wachsenden Erdball," Giessmann & Bartach, Berlin, pp.56.

Hollister, L.S., 1993, The role of melt in the uplift and exumation of orogenic belts, *Chemical Geology* 108:31-48.

Hunt, C, Banerjee S.K. and Marvin J., 1986, Magnetic exsolution in oceanic basalts lead to CRM controlled by ambient field, not host-rock magnetization (abstract), *EOS* 67:923.

Jordan, P., 1971, "The Expanding Earth," Pergamon Press, Oxford, pp.202.

Katz, M.B., 1993, The Kannack complex of the Vietnam Kontum massif of the Indochina block: An exotic fragment of Precambrian Gondwanaland? *in*: "Gondwana Eight," R. Findlay et al., eds., Balkema, Rotterdam, 161-164.

Keindl, J., 1940, "Dehnt sich die Erde aus? Eine Geologische Studie," Herold-Verlag, München, pp.50.

Kosygin, Yu.A. and Maslov, L.A., 1986, The role of solid lunar tides in the tectonic process, *Geotectonics* 20:451-454.

Larson, R.L., Pitman III, W.C., Golovchenko, X., Cande, S.C., Dewey, J.F.,Haxby, W.F., La Brecque, J.L., (map's compilers) 1985, "The Bedrock Geology of the World," Freeman & Co.Inc., New York.

Lehmann, R., 1992, Artic push moraines, a case study of the Thompson Glacier moraine, Axel Heiberg Island, N.W.T., Canada, Z. *Geomorph. N.F. supp.* 86:161-171.

Levi, S. and Banerjee, S., 1990, On the origin of inclination shallowing in redeposited sediments, *J.Geoph. Res.* 95, 4383-4389.

Lindemann, B., 1927, "Kettengebirge, Kontinentale Zerspaltung und Erdexpansion," Verlag von G. Fischer, Jena, pp.186.

Lock, J. and McElhinny, M.W., 1991, The Global Paleomagnetic Database, *Surveys in Geophysics* 12:317-491.

Mann, C.J. and Kanagy, S.P., 1990, Angles of repose that exceed modern angles, *Geology* 18:358-361.

McElhinny, M.W. and Lock, J., 1990a, Global Paleomagnetic Database Project, *Phys. Earth Plan. Int.* 63:1-6.

McElhinny, M.W. and Lock, J., 1990b, IAGA Global Paleomagnetic Database, *Geophys. J. Int.* 101:763--766.

Merle, O.R., Davis, G.H., Nickelsen, R.P. and Gourlay, P.A., 1993, Relation of thin-skinned thrusting of Colorado Plateau strata in southwestern Utah to Cenozoic magmatism, *Geol.Soc.Amer.Bull.* 105:387-398.

Merle, O.R. and Vendeville, B., 1992, Modélisation analogique de chevauchements induits par des intrusions magmatiques, *C.R.Acad.Sci.Paris* 315 ser.II:1541-1547.

Morgan, W.J., 1968, Rises, trenches, great faults, and crustal blocks, *J.Geoph.Res.* 73:1959-1982.

Nakanishi, M., Tamaki, K. and Kobayashi, K., 1992, A new mesozoic isochron chart of the Northwestern Pacific Ocean: paleomagnetic and tectonic implications, *Geophys.Res.Lett.* 19:693-696.

Neiman, V.B., 1990, An alternative to Wegener's mobilism, in: "Critical Aspect of the Plate Tectonics Theory. Vol.II (Alternative Theories)," Beloussov et al, eds., Theophrastus Publications S.A., Athens, 3-18.

Newton, C.R.,1988, Significance of "Tethian" fossils in the American Cordillera, *Science* 242:385-391.

Patterson, C. and Owen, H.G., 1991, Indian Isolation or Contact? A response to Briggs, *Syst. Zool.* 40(1):96-100.

Randles, W.G.L., 1980,"De la Terre Plate au Globe Terrestre," Librarie Armand Colin, Paris, pp.172.

Roeser, H.A. and Rilat, M., 1982, "Identified Magnetic Sea-floor Spreading Anomalies (map)," Federal Institute for Geosciences and Natural Resources, Hannover, Germany, *Geologisches Jahrbuch E* 23:71-80.

Sagnotti, L., Scalera, G. and Florindo, F. 1994, Gestione automatica del database paleomagnetico globale, *Geoinformatica* (in press).

Sahni, A., 1984, Cretaceous-Paleocene Terrestrial Faunas of India: Lack of endemism during drifting of the Indian plate, *Science* 226,441-443.

Sahni, A., Rana, R.S. and Prasad, G.V.R., 1987, New evidence for paleobiogeographic intercontinental Gondwana relationships based on Late Cretaceous - Earliest Paleocene coastal faunas from peninsular India, in: "Gondwana Six: Stratigraphy, Sedimentology, and Paleontology," G.D. McKenzie, ed, *AGU Geophys. Mon.* n°41:207-218.

Sandwell, D.T. and Renkin, M.L., 1988, Compensation of swells and Plateaus in the North Pacific: No direct evidence for mantle convection, *Jou.Geoph.Res.* 93B:2775-2783.

Scalera, G., 1988, Nonconventional Pangaea reconstructions. New evidences for an expanding Earth, *Tectonophysics* 146:365-383.

Scalera, G., 1990, Palaeopoles on an expanding Earth: a comparison between synthetic and real data sets, *Phys.Earth.Plan.Int.* 62:126-140.

Scalera, G., 1991a, The significance of a new class of shape similarities for global tectonic theories, *Atti 8° Congresso Annuale GNGTS 1989*, Roma, 635-639.

Scalera, G., 1991b, Non-chaotic emplacements of arc-trench zones in the Pacific Hemisphere, *Pubblicazione I.N.G.* n°531, pp.19.

Scalera, G., 1993, Non-chaotic emplacements of trench-arc zones in the Pacific Hemisphere, *Annali di Geofisica* XXXVI, n°5-6:47-53.

Scalera, G., Favali, P. and Florindo, F. 1993, Use of the paleomagnetic databases for geodynamical studies: some examples from the mediterranean region, in: "Recent Evolution and Seismicity of the Mediterranean Region," E. Boschi et al., eds., Kluwer Acad.Pub., Netherlands, 403-422.

Scalera, G. and Meloni, A., 1991, "L'evoluzione della Terra," Dedalo Edizioni, Bari, pp.240.

Scholl, D.W., Hein, J.R., Marlow, M. and Buffington, E.C., 1977, Meiji sediment tongue: North Pacific evidence for limited movement between the Pacific and North American plates, *Geol.Soc.Am.Bull.* 88:1567-1576.

Shields, O., 1979, Evidence for initial opening of the Pacific Ocean in the Jurassic, *Palaeogeog. Palaeoclimat. Palaeoecol.* 26:181-220.

Shields, O., 1983, Trans-Pacific biotic links that suggest Earth expansion, in: "Expanding Earth Symposium," S.W. Carey, ed., University of Tasmania, Sydney 1981, 199-205.

Smith, A.B., 1988, Late Palaeozoic biogeography of East Asia and palaeontological constraints on plate tectonic reconstructions, *Phil.Trans.R.Soc.Lond.* A 326:189-227.

Smith, A.G. and Hallam, A., 1970, The fit of the southern continents, *Nature* 225:139-144.

Stait, B. and Burrett, C., 1987, Biogeography of Australian and southeast Asian Ordovician nautiloids. in: "Gondwana Six: Stratigraphy, Sedimentology, and Paleontology," G.D. McKenzie, ed., *AGU Geophys. Mon.* n°41:21-28.

Stöcklin, J., 1984, The Tethys paradox in plate tectonics, in: "Plate Reconstruction from Paleozoic Paleomagnetism," R. Van der Voo, C.R. Scotese and N. Bonhommet,eds., *AGU Geodyn. Ser.* n°12:27-28.

Stöcklin, J., 1989, Tethys evolution in the Afghanistan-Pamir-Pakistan region, in: "Tectonic Evolution of the Tethyan Region," A.M.C. Sengör, ed., Kluver Academic Publishers, 241-264.

Tripathi, C. and Singh, G., 1987, Gondwana and associated rocks of the himalaya and their significance, in: "Gondwana Six: Stratigraphy, Sedimentology, and Paleontology," G.D. McKenzie, ed., *AGU Geophys. Mon.* n°41:195-205.

Truswell, E.M., 1991, Antarctica: a history of terrestrial vegetation, in: "The Geology of Antarctica," R.J. Tingey, ed., Clarendon Press, Oxford, 499-537.

UNESCO, 1976, "Geological World Atlas (22 maps)," Commission for the Geological map of the World (C.G.M.W.), UNESCO, Paris.

Walker, D.A., 1989, Seismicity of the interiors of plates in the Pacific Basin, *EOS* 70:1543-1544.

AN EVOLUTIONARY EARTH EXPANSION HYPOTHESIS

Stavros T . Tassos

National Observatory of Athens
Seismological Institute
P.O. Box 20048
GR-118 10 Athens

INTRODUCTION

Although I am a research scientist in geophysics - seismology , I feel that an attempt to answer fundamental philosophical questions , like : is the universe finite or infinite ? is it static or dynamic ? is it an open or a closed system ? and so on , will provide some intuitive insights into the incomplete and scarce set of data that scientists have to work with .

Furthermore I've been impressed since my student years , by the dogmatic and simplistic approach of plate tectonics advocates . I rejected , on philosophical grounds , the perfect equilibrium between input and output of crust. I could not understand , how easily the energy deficit of convection cells , of plate movements and of subduction was overlooked .

In 1981 Professor Carey , the patriarch of earth expansion , invited me to the Expanding Earth Symposium , in Sydney where I had the chance to express , in an abstract form (Tassos ,1981), my first thoughts about these matters . Since then a lot of literature concerning criticism of plate tectonics and proposing earth expansion as an alternative theory has appeared , even though these authors were always treated as heretics by the mainstream scientists . Such a fideistic approach appreciates only the status of power and depreciates the joy of the game , thus depriving human behaviour and thought from their playful character .

For me, thinking , wondering and imagining is a personal need and I enjoy it perhaps the same way children enjoy playing . Being in Olympia this playful mood is enhanced , and I feel that this mental game of mine is part of the ancient Greek spirit heritage. This great universal heritage inspires us to look at , explore through and play with : The free , independent , original and incisive reasoning of Sophists . The knowledge as means of moral perfection of Socrates . The unchanging ideas and monism of Plato . The entelechy and the encyclopaedic wisdom of Aristotle . The chaos and the infinite air of Anaximandros . The fire and the continuous creation of Heracleitos . The opposing forces of Eros and Eris of Empedocles , the first leading to universal union and the later to universal separation through a constant organizational movement . The realistic thinking and the atomic theory of Democritos and Epicuros . The abstract, mathematical thinking and the numbers of Pythagoras . The heliocentricity of Aristarchos. The ... The ...

A paradigm that is the marriage , the synthesis of Science and Philosophy , of Observing and Acting , of Being and Becoming , of Necessity and Chance .

THE LAWS OF THERMODYNAMICS

Conservation of energy (rest, potential, kinetic), of momentum (linear, angular) and of electric charge of the first law, as well as the entropy increase and the thermal equilibrium of the second law refer to isolated , closed systems .

Looking at their implications we see that conservation of energy implies infinity, lack of beginning and end, symmetry and reversibility, the possibility of creation of "something" from "nothing" provided that the resulting "something" has zero net values for all conserved quantities (matter and equal amount of anti-matter, positive charges and equal amount of negative charges etc.). On the other hand conservation of angular   momentum   introduces   asymmetry, since only asymmetric   bodies can spin. Asymmetry is a characteristic of change and evolution. Also in quantum physics the energy conservation law is violated during the "tunnelling effect" or when a change in the number of existing particles occurs .

Irreversibility and increase in disorder are the direct implications of the second law of thermodynamics. Although direction in time is in accordance with the observed behaviour of complex far from equilibrium evolving systems, the increase in disorder in nature as a whole is not . All facts indicate an increase in order and complexity in all observable (directly or indirectly) natural systems that "survived" for a considerable length of time. There is an increase of matter and energy participating in the structures of the known world. An obvious question is : where these come from ? The answer lies in the marriage of the reversibility and of the infinite as an inexhaustible source of energy and matter of the first law to the irreversibility, the birth , growth and death implied by the second law .In other words the "new something" is a product of an "old something" .

The question boils down to: Are symmetry, stability , reversibility or asymmetry, instability, irreversibility the rule in nature? In other words, is nature a closed or an open system ?

In the next few lines I will try to sketch , since the space provided is objectively limited , why , I think , asymmetry , instability and irreversibility are the rule and why evolution , which leads to an increase of complexity and order , is a general natural process that applies to all observable world .

THE UNCERTAINTY PRINCIPLE

The Newtonian laws of relatively slow motion , are different from those of Einstein's for fast moving bodies that approach the limit of the speed of light . Both theories refer to the gravitational force and to relatively big bodies - above the atomic level - and to relatively long distances .

Heisenberg's quantum theory and his uncertainty principle refer to masses and distances at the atomic and subatomic level where the gravitational force is very weak compared to electromagnetic and nuclear forces . The accuracy of determination of the spin and the position of an atomic particle (eg. , an electron) are inversely proportional . The higher the energy oscillation , the higher the uncertainty of position and therefore

276

the higher the instability , the unpredictability and the indeterminability . The ultimate implication is that uncertainty relates to motion and since , according to Einstein's theory, a motionless frame of reference does not exist , chance is inherent in nature , is a property of matter .

Prigogine and Stengers (1984) , with their pioneer work in chaos theory have shown why uncertainty is not only an attribute of the microscopic high energy world , but also of the macroscopic low energy world . Instability seems to be the general mechanism of breaking the symmetry of an originally homogeneous chaotic "cold" state, or the asymmetry of an inhomogeneous ordered "hot" state , initiating a process of differentiation and self-organization . Nature despite the simple and reversible Newtonian laws of a static universe , is complex and dynamic . Random and irreversible events play the primary role in the course of its evolution. Asymmetry is the rule , symmetry is the exception . The universe is an "alive" not a "dead" system , and instability and asymmetry are the cause and the result of change.

THE CREATIVE NATURE

We start with chaos , a state of preexisting "something" which approaches the simplest possible and therefore lacks order , or with a state of another preexisting "something" which is extremely complex and therefore highly ordered . In the first case prevailing vector sum direction in the movement and communication between the elements of the system are lacking . Its elements ignore each other . On the contrary , in the second case , we have countless interactions between a great number of elements and subsystems , so that their mathematical description becomes impossible. Simplicity and complexity are the two sides of the same coin . Nature is simple in its complexity and complex in its simplicity.

The symmetry of the homogeneous chaotic state or the asymmetry of the inhomogeneous but also chaotic state , can break by a random perturbation which in both cases will introduce branching . At a bifurcation point , a system can choose among, potentially infinite , new regimes . The question is: Can all of them "survive"? The Darwinian principle of natural selection can give the answer . And the answer is , no. The ones that better "fit" to the demands of the surrounding environment will "survive" and grow. The other ones, even they are formed, they will soon "die". Relatively stable are those systems that better "adjust" to a given set of conditions and relatively unstable those that do not. Stability at low energy level is the simple and outward macroscopic manifestation of this "adjustment".

Another question comes next : The systems that survived are more or less complex compared to the systems they came from ? Concerning chemical and biological systems the answer is known . They will be more complex , and their increased complexity implies quantitative increase since more elements participate in their structure . I think none can argue that it is not an observable fact that the organic matter , the biomass has increased with time. Could things run in the opposite direction in the inorganic world ? All evidences from the known observable world indicate the tendency towards greater order and complexity is an inherent one, it is a property of matter . The continuous discovery in high energies of new "elementary particles" and antiparticles , some like the proton , the electron , the photon and the neutrino stable , others like the hyperon, the neutron , the meson , the pion and the muon unstable , shows that indivisible units do not exist . Collisions between smaller particles produce bigger particles electrons , protons , hydrogen, helium and so on . The Russian doll of nature has not a solid final doll.

The classical mechanical, Newtonian approach explains the entropy decrease in the open system of living matter, by an increase in entropy of the inorganic world, thus considering this world a closed system. Organic matter has the same relatively stable barionic structural units as inorganic matter and acquires its energy and matter from it, the preceding level of organization. Therefore with the same reasoning inorganic matter also behaves as an open system and acquires its information (energy and matter) from the subatomic, particle world, through successive chaotic situations, and so on "ad infinitum".

An influx of information leads to an increase of collisions and into two conflicting processes. One in which disorder increases as a result of velocity increase of each element, and another in which we have an increase in order due to the establishment of new relations between the elements of the system. The two phenomena could be mutually cancelled (annihilate) by acting in opposite directions and by being quantitatively equal. This though can only happen under conditions of absolute equilibrium and symmetry which only in theory exist. Opposites exist and they are the "sine qua non" of change and evolution, but they are not at a perfect balance. They can only momentarily reach such a state, when systems pass from one ordered to another ordered state. Disorder produces order and vice versa. For example if we accept the existence of matter and antimatter, the moment matter was formed the universe was out of balance and matter was more than antimatter.

According to Einstein's theory, if we exceed the speed of light, we can go back in time, we can "see" events of the past. In order to do that, we should be able to break up existing correlations in a whole in such a way to cancel the results of collisions that produced this whole of correlations. This though is practically impossible because, as the chaos theory implies, the whole contains more information than the sum of information of its components and breaking it up, can most likely give parts that contain different information and therefore they are different from the initial ones. The past is deterministic and is contained in the present, the future is not. Trying to go back in time, is like trying to go back into the future of the past, and as future cannot be predicted and cannot be duplicated, is uncertain.

In the organic world reproduction is the inherent tendency to self-organization. In life the necessary information is contained in the DNA. According to Cramer (1993), the true self-organization is neither inherent nor "preorganized". It is a property of an entire system which under precisely defined conditions and a high degree of complexity organizes itself, so it is a physical property.

The ability to reproduce is connected with open systems that exchange energy and matter with their surroundings. Systems that do not exchange information, behave like isolated, closed systems. They exhaust their information and finally die, decompose and acquire a chaotic symmetry. Although under the influence of Cramer`s evolutionary field, the building blocks of this chaotic state will become ordered once again at a later time and at a higher level.

What is important to mention here is that, although evolution is an endless process, does not take place at a constant rate and is not linear. There is a time of maturity, of integration between the initiation of a process and its outwardly manifestation. Qualitative jumps, and catastrophic events are the rule in the evolutionary process.

CONSERVATIVE OR EVOLUTIONARY EARTH EXPANSION ?

The earth has all the characteristics of an "alive", evolving, open system. It exchanges energy and matter with its surroundings, and all existing evidence indicate that it is at the stage of growth. In evolutionary terms it means quantitative and qualitative increase , and its expansion involves increase of its mass . On the contrary, expansion of the earth , according to the conservation laws , implies constant mass and increase in volume due to decrease in density .

In the context of an expanding earth, we accept that the radius close to its origin time, 4.5 billion years ago, was about 3500 km - 55% of its present size- with an all encompassing Pangean crust .Its composition was close to its present core composition, with iron being its main constituent. We must have an average density over 30 g/cm^3 , if conservation of mass is assumed . Gottfried (1990), proposes such values of density for the inner core , and accepts a hydridic gas giant primordial earth with 99.7% hydrogen and a small rocky core of highly compressed metallic elements in the form of hydrides and other highly reduced complexes . At elevated pressures and temperatures iron can accept hydrogen on a one-to-one basis (Antonov et al. , 1980 ; Fukai and Akimoto , 1983) . Hunt (1992), in his "Expanding Geospheres" proposes , in the context of conservation laws, a new earth theory based on the concept of phase changes of carbides and hydrides starting in the core and continuing in the mantle and the crust .

On the other hand if the density of the premordial earth was comparable to the one of the present core -about $10.7 g/cm^3$- then we have over a three fold increase of the mass of the earth. That amounts to $4-4.5 \times 10^{24}$ kg of matter with an energy equivalent of $35-40 \times 10^{40}J$. The total amount of lava added each year mainly at the mid-oceanic ridges, is in the order of 10^{13} kg . Assuming that this amount represents only about 1%, while the rest is trapped before it reaches the surface, then 4-4.5 billion years are needed , with the present rate of annual accretion, for the earth to reach its present size . This is though a highly speculative and linear assumption. Most likely expansion is not taking place at a constant rate and each episode is exponentially accelerated compared with the previous one . How does then an evolutionary expansion of the earth take place?

A possible process could be the following: Cosmic particles , with a very small mass , travel unimpeded through the crust and the mantle and are trapped in the much denser core . Through collisions and fusions they convert their energy into the masses of newly created bigger particles like electrons , protons and neutrons , which in turn collide to form hydrogen atoms . As it was mentioned before at the high temperatures and pressures iron can accept hydrogen almost on a mol-for-mol basis. Hydrogen also can form light metal hydrides with silicon and aluminium , non-metallic complexes with carbon, chlorine , fluorine as well as complexes with sodium , potassium and carbon (Gottfried,1990) . The presence of high amounts of hydrogen in the core may accelerate or slow down the expansion rate . Badding et al. (1991), have found in the laboratory , that at low pressures , near 3.5×10^3 atm, hydrogen atoms entering the crystalline lattices of iron, add to the weight but not to its volume . Above this pressure the metal lattices are destroyed and expansion occurs rapidly .

In the proposed model earth expansion is a result of the combined and competing effects of new mass creation due to capturing of cosmic particles by the core , and of phase changes in the core and the mantle . New mass in the form of hydrogen atoms tends to increase , with the reinforcing effect of high pressures , the volume of the core. On the contrary the exhalation of silanes and hydrocarbons , as it is proposed by Gottfried (1990), tends to contract the core and the lower mantle and to expand the upper mantle and the crust. The created new mass tends to increase the volume of the earth as a whole.

From the known barionic particles the one which is closer to the attributed characteristics is the neutrino. The neutrino, initially considered as massless and chargeless, is a product of beta decay and of the reaction of electron capture in the stars, where under gravitational pressure the electrons combine with protons into neutrons and neutrinos. Neutrinos could also emitted when charged pions decay into muons. Dart (1993), in his exponential decay of photons hypothesis proposes the conversion of the radiant energy of a photon into two neutrinos with a rest mass of 1.165×10^{-65} g and rest energy 1.05×10^{-44} erg. Neutrinos are not massless, contrary to what it was thought initially, and most likely not even chargeless, since they possess energy and momentum. Therefore they can't pass unimpeded through any matter, and in the high densities of the core the effect of their collisions with other particles could be profoundly important in the mass creation process.

Another possibility is that in the core we have some sort of controlled successive fusion reactions. After the formation of hydrogen atoms, the same way as mentioned before, there is the formation of helium, carbon, oxygen, silicon and finally iron atoms, in all cases except the last with energy release. The fusion process ends with the formation of iron, because it is the most stable element. For the same reason iron is the end product of fission.

Finally the particles could be non-barionic and in that case we have to refer to the unknown matter and process. Even if the nature of the matter and the kind of processes involved are presently unknown, it doesn't mean that earth expansion is not happening. Authors in this volume and elsewhere present convincing geological and geophysical evidences. Furthermore the works of nature and of the earth provide the most convincing ones.

Would the earth continue to expand? In the "near" future -and this could be the next second or billions of years- the answer is yes, as a result of statistical determinism. On the other hand it is also possible that the earth might either shrink, or explode, or may be destroyed by mankind, or even by a meteorite, or who knows what else.

CONCLUSION

In the proposed evolutionary earth expansion hypothesis the earth is expanding through a process of new mass creation. Processes controlled by conservation laws play a secondary role in phase changes. This is a speculative, but realistic hypothesis, based on an open and creative universe, where irregular series of symmetry breaks produce new matter, and new relations, none of which are predictable.

REFERENCES

Antonov, V.E., Belash, I.T., Degryareva, E.G., Ponyatovskii, E.G., and Shiraev, I., 1980, Obtaining iron hydrides under high hydrogen pressure, *Sov. Phys. Dokl.* 25.

Badding, J,V., Hemley, R.J., and Mao, H.K., 1991, High-pressure chemistry of hydrogen in metals. In situ study or iron hydride, *on press*.

Carey,S.W., 1981, "The Expanding Earth. A symposium," University of Tasmania, Sydney.

Cramer, F., 1993, "Chaos and Order," VCH, Weinheim.

Dart, H.P., 1993, A new alternative to the big-bang theory, *Apeiron.* 17:5

Fukai, Y., and Akinoto, S., 1983, Hydrogen in the earth's core, *Proc. Jap. Acad.* 59B:158.

Gottfried,R., 1990, Origin and evolution of the earth chemical and physical verification, in "Critical Aspects of the Plate Tectonics Theory," Barto, A. Kyriakidis, ed., Theophrastus, Athens.

Hunt, C.W., 1992, "Expanding Geospheres," Polar Publishing, Calgary.

Prigogine, I., and Stengers, I., 1984, "Order out of Chaos," Heinemann, London.

Tassos, S.T., 1981, Entropy and expansion of the earth, in "The Expanding Earth. A symposium," S.W. Carey, ed., Univ. of Tasmania, Sydney.

GLOBAL MODELS OF THE EXPANDING EARTH

Klaus Vogel

Gabelsbergerstr. 27
08412 Werdau
Germany

ABSTRACT

The theory of Earth expansion starts from the assumption that "Pangaea" covered completely the surface of an earth with approximately 55 % - 60 % of the present diameter. By growing volume of the earth due to endogenic processes this continental crust broke to pieces and the widening gaps developed to the oceans of today according to the pattern of seafloor spreading.

With the help of several global models a reconstruction of this small earth is presented. Globes with growing diameters are enclosed within transparent spheres of the modern earth (globe-in-globe model) to compare the starting positions with the present situation. In general the continents are fixed at the extensible substratum, maintaining their positions to each other. The movements are mainly determined by radial outward pressing of the continents.

Finally this process is shown at geological globes (diameters 85 cm and 54 cm) with more accuracy and clearness.

HISTORICAL REMARKS AND FUNDAMENTAL PRINCIPLES

In 1927 B. Lindemann in his book "Kettengebirge, kontinentale Zerspaltung und Erdexpansion" tried to develop the idea of an expanding earth due to rising temperatures as a working hypothesis of equal value to the assumption of a shrinking earth due to sinking temperatures. He considered extensional processes as a predominating element for the formation of the earth's surface, including the movements of the continents, which then were still discussed controversially.

Some years later O.C. Hilgenberg concluded from Alfred Wegener's theory of Continental Drift that the originally coherent supercontinent "Pangaea" was the sial shell which completely covered the surface of a relatively smaller earth. This shell was disrupted to pieces by overpressure (hypothesis of "Krustensprengung"), originated by endogenic forces. The material raising from deeper layers filled out the extensional gaps and produced the sima of the ocean floors. In this way the surface of the expanding earth increased. With his book "Vom wachsenden Erdball" (1933) Hilgenberg was the first who presented an Earth globe where all continents fitted together with somewhat more than 60 % of the present diameter.

In the following decades many scientists in several countries were engaged in the question of earth expansion and numerous papers and books were published. A comprehensive presentation of the history and consequences involved in this subject gives S.W. Carey's book "The Expanding Earth" (1976). In 1981 in Sydney Carey also organized the "International Expanding Earth Symposium" which may be regarded as a highlight in the endeavours of promoting the idea that the earth is expanding. Beside of these activities several reconstructions of smaller globes, similar to Hilgenberg's were produced by various authors.

In 1977 by change the author read a short remark to the possibility of a growing earth. At this time he was without any special knowledge to this subject, but he was fascinated by the picture of the disrupting continental crust in the light of the technological thinking of an engineer. This was the starting piont to make these several globes in the course of the years till 1992.

First scentific contacts developed to Prof. M. Schwab (Halle) and Prof. S.W. Carey (Hobart). Since numerous useful and friendly connection followed, especially after the opening of the wall between east and west, e.g. to Prof. G.O.W. Kremp (Tucson), Dr. H. Owen (London), Prof. J. Pfeufer (Erlangen), Dr. G. Scalera (Rome), C. Strutinski (Cluj-Napoca) and last but not least to Mag. J. Koziar and his colleagues at the University and the Polish Geological Institute of Wroclaw.

THE GLOBAL MODELS AND THEIR INTERPRETATIONS

The basic model is a globe with 55 % of the present diameter, the surface area of which is equal to the areas of all continents together. It was really possible to assemble all continents to a nearly closed surface by eliminating the oceans. Deciding points were the closure of the Indic with India bordering Madagascar in the southern position and the Pacific, bringing the northern coast of Sibiria against the western coast of North america. The eastern and western parts of Antarctica formed originally a coherent nearly elliptical continent (figure 1). A globe with a diameter of 60 % has less distorsions due to radius changes and local intracontinental stretchings and shearings. Therefore it was used for futher work.

To compare the starting postions with today's situation the small globe was fixed in the centre of a transparent sphere of the present diameter on which the continents were pasted in their real positions. It is visible that the continents have maintained their orientations to each other, so in principle no drift or rotations have occured. The model sets represented in the figures 2-4 with diameters of 46%, 55%, 60%, 66 %, 75 % and 85 % show the continuity of the movements in which the continents were pressed from the centre outward in radial direction. The asymmetric distribution of the continents and oceans originated by the earlier opening of the Pacific and a far higher rate of spreading compared with the rates in the other oceans.

The fit at the 60 % globe was not only made with the outlines of the continents but also by pasting over the cuttings of the continents taken from the prints of a physical globe and a tectonic one. In this presentation one can compare the mountain chains and the corresponding mobile belts of the different eras which originated between the fragments of the old tables, forming geosynclinals. These old tables represent the remnants of the relatively uniform crust of the precambrian earth with a diameter of less than 50 % of today before its disruption. This event may be the same as the "Krustensprengung" after Hilgenberg or the "Algonkische Umbruch" in Late Precambrian, postulated by Stille, however on the earth with constant radius.

Especially remarkable is the Cenozoic folding, going as a continuous belt around the earth near the equator. This is the Tethys belt (but not the Tethys Ocean which never existed),dividing the northern and southern hemispheres or Laurasia and Gondwanaland. By growing expansion rates the upper crust with its substratum was stretched an thinned (shelf formation) until rupture. After this the formation of the basaltic ocean floor followed, beginning in the early Northwestern Pacific.

The latest models were made with help of prints for geological globes (scale 1:15000000, Moscow Academy of sciences 1972), one in the original diameter of 85 cm and the other one in the reduced size of 63 % or 54 cm. Like already the fit of Wegener between Africa and South America the comparison of these two globes shows the very numerous agreements of the geological structures of all the bordering continents around the world which by no means can be accidental.

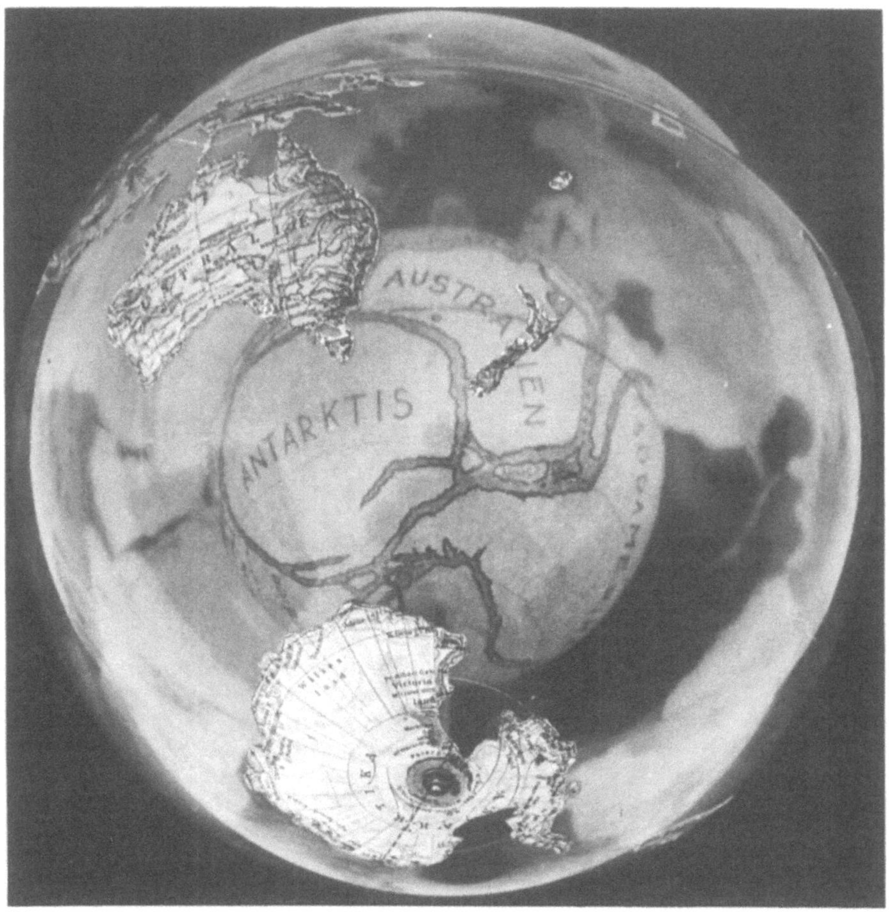

Figure 1. A globe with 55% of the present diameter inside a transparent globe of today, showing the connection of Austalia, Antarctica and South America before opening of the Pacific Ocean compared with the present distribution.

EXPANDING EARTH AND OTHER THEORIES

As explained above, Earth expansion carries on Wegener's theory of Continental Drift. The question arises about relations to Plate Tectonics which in the sixties also was developed by a modification of Wegener's theory and reached wide acceptance. This was the result of palaeomagnetic discoverings and the recognition of the young age of the growing ocean floors which lead to the term "seafloor spreading " and to a new causality for the movements of the continents. But the postulated convection cells which should carry like a conveyor belt the ocean floors from its origin in the middle of the oceans to the rims of the continents to be swallowed there down into the mantle do not fit with the geometric pattern of the continental movements. These conveyor belts would disturb each other and they would not have the power to push away additionally the continents. Whereas in the light of expansion the seafloor spreading is the increase of the growing surface of the earth which fits very well to the kinematics connected with this process.

With the postulation of the continental movements Wegener made the first important step to a new consideration of the development of the earth, but in this light the second and deciding step must follow to complete the Wegnerian revolution: The assumption of an **expanding** earth!

It remains the question wether Earth Expansion Theory is a mobilistic or a fixistic one. It is both! It is mobilistic because the forces of expansion push the continents outward in radial direction and simultaneously their horizontal distances are growing, but it is fixistic, because the continents are deeply rooted in the mantle and maintain their positions and orientations to each other during this process. In this light the long lasting controversy between Mobilism and Fixism is settled.

It may be the same with the dispute whether vertikal or horizontal movements are deciding for geological developments. The earth's crust is formed by the resultant of horizontal **and** vertical forces. Furthermore we find at the expanding earth phenomena of Uniformism **and** Catastrophism or Plutonism **and** Netunism. It is important to consider the necessity of continually producing water in the depth of the expanding earth. This explains the successions of transgressions and regressions in the course of the earth's history as a result of the changing proportion between the quantity of the existing water and the available capacity of the ocean basins. With all theses and still other arguments G.O.W. Kremp (1990) pointed out: "There is an abundance of evidence which validates the Earth Expansion Theory. This theory explains the earth's history in a clearer and simpler manner than any other".

PROBLEMS AND CONCLUSIONS

In his book "Die Expansion der Erde" (1966) the German physicist Pascual Jordan said: "The Expansion of the Earth belongs to the great problems of Natural Science". The proof of a valid reconstruction with all its details would need great efforts and a comprehensive programme of research. Althuogh these new reconstrutions in general may be convincing a lot of details are still waiting to be defined including the timescale of the growing diameters which is only based on raw approximations.

The further work requires the evaluation of very numerous data from geology and other branches of Earth sciences, e.g. to refine the structures of ocean floors, especially in the huge areas of the Pacific including cross sections of opposite continental rims. Within the continents areas of increments of the Precambrian crust and intracontinental deformations, not only due to diminishing of the curvature but also by stretchings, shearings or pushings must be localized and eliminated. Palaeomagnetic work is of high importance. Additionally the emerging results of intercontinental geodetic maesurements by satellite laser ranging and VLBI may be able to detect the trend of the growing circumference of the earth. Also the various extensional structures at the surfaces of the moons of the outer planets (e.g. Miranda, a moon of Uranus) may be signs of expansion within our planetary system.

But even the best results show only the outer development, the growing surface of the earth and not the causes which originate this process in the depth of our planet. They are still unknown, although many attempts of explanation were made. First Hilgenberg (1933) with his aether flux hypothesis postulated an increase of volume and mass ot the earth, its distance to the sun and the revolution period. Keindl (1940), Egyed (1957) and Mouritsen (1975) assumed a super-dense core and Jordan a diminishing of the gravitional constant G. By decrease of presure phase changes lead to a growing volume. Pfeufer (1981) supposed a similar effect by raising temperatures due to friction at the border between core and mantle. R. Gottfried (1990) and C.W. Hunt (1992) deduced the expansion from a primordial hydrdic earth, starting as gas giant from a protoplanetic cloud. But no one of these hypothesis now is generally accepted.

It remains the fundamental question to the nature of matter and gravitation. From this reason - despite of all indications - many geologists avoid Earth expansion as not consistent with the laws of physics. But should we discard therefore this theme with all its visible evidences? This is the question to the physicists and cosmologists (and even to the philosophers), this is the question at the FRONTIERS OF FUNDAMENTAL PHYSICS.

Finally again let me quote G.O.W. Kremp with his words:
"It is time to discard the notion that the earth could not have changed its size over time. It is time to acknowledge the evidence presented in this paper. It is time to explore, with open mind, the theory of Earth Expansion".

Figure 2-4. The model sets (upper row) show the disruption of the continental crust and the development of the oceans as a result of the growing diameter of the earth. Below this process is demonstrated by glass models and the corresponding reconstruction of the earth at a geological globe with 63% of the present diameter. The Atlantic (fig. 2) origineted similar to Wegener's postulation, the Indic (fig. 3) from a triangle between Africa, India and Antarctica and the Pacific (fig. 4) from a sickle-shaped gap around Australia.

REFERENCES

Carey,S.W.,1976,"The Expanding Earth." Development in Geotectonics 10, Elsevier,
     Amsterdam, 1-488
Carey,S.W.,1988,"Theories of the Earth and Universe" Stanford University Press,
     Stanford, California, 1-413
Egyed,L.,1957, A new dynamic conception of the internal constitution of the earth,
     Geologische Rundschau, 46, 101-121
Gottfried, R.,1990, Origin and evulution of the earth, chemical and phys.verifcations.
     in "Critical Aspects of Plate Tect. Theory II," Theophrsatus Publ.Athens, 115-140
Hunt, C.W.,Ed., 1992, "Expanding Geospheres."Polar Publishing,
     Calgary
Jordan, P.,1966,"Die Expansion der Erde",Verl. Viehweg & Sohn,
     Braunschweig, 1-182
Keindl, J., 1940, "Dehnt sich die Erde aus?" Herold- Verlag
     München Solln 1-50
Koziar, J..1991, Studies on the problems of the earth expans. carried in the
     Wroclaw-centre, Acta Univ.Wrocl.,1375,1-14
Kremp,G.O.W.,1991, Paleogeography of the last two cycles of earth expansion, in
     "Current prespectives in palynological research", Journ. of Palynol. New-Delhi,
     231-260
Kremp,G.O.W.,1992, Earth expansion theory versum statical earth assumption,
     in Chatterjee,S.; Hotton, N.III, eds., "New Concepts in Global Tectonics."
     Texas Tech. University Press, Lubbock, 279-307
Lindemann,B.,1927, "Kettengebirge, kontinentale Zerspaltung und Erdexpansion",
     Fischer-Verlag Jena, 1-183
Mouritsen,S.A., 1975, Introduction of planetary structure and terrestrial maria,
     Geologische Rundschau Bd.64 Heft 3, 899-915
Owen, H.G. 1983, "Atlas of Continental Displacement, 200 mill years to present"
     Cambridge Univ. Press 1-160
Pfeufer,J., 1981, "Die Gebirgsbildungsprozesse als Folge der Expansion der Erde ",
     Glückauf Verl. Essen 1-128
Pfeufer, J. 1992, Zur Hypothese der Expansion der Erde,
     Z.geol.Wiss., 20(5/5), Berlin, Dezember 1992, 527-546
Scalera, G. 1990, General glues favouring expanding earth theory, in: "Critical Aspects
     of the Plate Tectonics Theory", Theophrastus publ. Athens vol.2, 65-93
Strutinski,C., 1990, Tthe importance of transcurrence phenomena in mountain building
     in "Critical Aspects of the Plate Tectonics Theory, "Theophrastus publ.Athens,
     vol.2,141-166
Vogel,K., 1983, Global models and earth expansion, in:S.W.Carey, ed., "The Expansion
     of the Earth - a Symposium", Sydney 17-27
Vogel,K, Schwab,M. 1983, The position of Madagascar in Pangaea
     in: S.W.Carey (Ed.) "The Expanding Earth -a Symposium,"   Sydney , 73-76
Vogel,K.,1984, Beiträge zur Frage der Expansion der Erde auf der Grundlage
     von Globenmodellen, z.geol.Wiss. Berlin 12, 563-577
Vogel,K.,1989, Recent crustal movements in the light of earth expansion theory,
     in:"Geodesy   and Physics of the Earth" (Symposium),
     Veröffentl. Zentralinst. Physik der Erde
     Nr. 102, III; Potsdam, 284-288
Vogel,K., 1990, The expansion of the earth - an alternative model to the plate tectonics
     theory, in "Critical Aspects", Theophrastus publications Athens, vol.2, 19-34
Wegener,A.,1929, "Die Entstehung der Kontinente und Ozeane",4.Aufl.,Verl.
     Viehweg&Sohn Braunschweig, 1-231

AN OROGENIC MODEL CONSISTENT WITH
EARTH EXPANSION

Carol Strutinski

Romanian Geological Survey
Cluj Branch
POB 181
3400 Cluj 1, Romania

> *"Most heresy is false; yet latent*
> *within it are the gems of the age"*
> (S.W.Carey, 1983)

INTRODUCTION

Since the dawn of plate-tectonics theory, almost 30 years ago, a vaste amount of evidence has accumulated that has continuously necessitated correction of its paradigms. Anyhow, the basic principle - subduction on a constant-radius Earth - remained unaltered being assumed by the majority of Earth scientists. Still there is a little community of 'heretics' who question the unanimously accepted 'truth', the more so as , until now, there has been no peremptory proof either in favour or against a constant-radius Earth through geological time. It is beyond the scope of this paper to provide evidence in favour of Earth expansion. What we shall try to show is that orogeny may well be understood without any implication of subduction and with only subordinate participation of lithospheric compression. Yet, we are aware that beside direct evidence this is what the Earth expansion hypothesis needs most in order to increase its credibility.

STRIKE SLIP VERSUS SUBDUCTION

Notwithstanding the revolution in Earth sciences triggered in the early sixties by the evidence of ocean spreading, the general rules of orogeny are still poorly known. Plate-tectonics theory took over from older theories the crustal-shortening concept and elaborated a general model according to which spreading along ocean ridges is compensated by progressive closure of old oceanic basins, eventually followed by collision of continental blocks. Initially this process, whereby island arcs and/or orogenic belts are being created,was thought to take place by orthogonal plate convergence across subduction zones (Dewey and Bird,1970; Figure 1a).This scenario best accounted for some of the most obvious structural patterns (fold and thrust vergencies in particular) which imply tectonic transport *across* the strike of the orogen.However, it came out very soon that beside such

(Fitch, 1972), thus proving tectonic transport also *along* the orogen and forcing plate-tectonics theory to reconsider its earlier expectations. The first attempts to accommodate orogen-parallel displacements along strike-slip faults with the basic principle of plate tectonics assumed initiation of the faults in order to permit 'tectonic escape' in regions of strong convergence (Figure 1b). Under certain circumstances and over restricted areas such

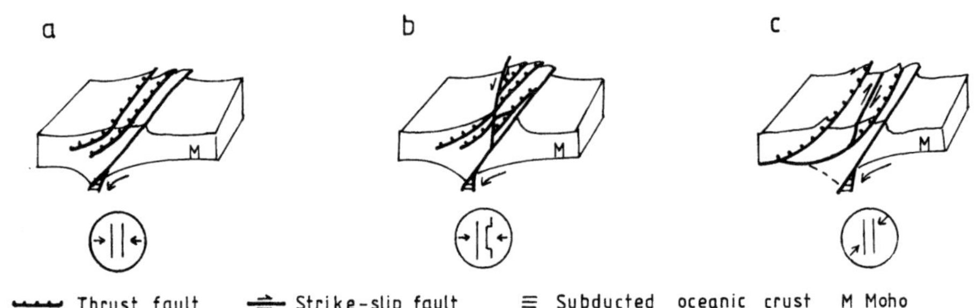

Figure 1. Three plate-tectonic models of orogeny successively developed in order to keep up with growing new evidence.

interpretations may actually apply. Yet, examples cited from Central Anatolia, south-eastern Asia and elsewhere clearly show that expulsion of continental blocks takes place along conjugated sets of faults, of which only some are sub-parallel to the strike of the orogen. The disposal reminds, however, of a pure shear mechanism, whereas orogen-parallel strike-slip faults that can be followed for thousands of kilometres along strike are in domains of simple shear.Moreover, the 'tectonic escape' model assumes nucleation of strike-slip faults at an advanced stage of crustal collision, in the late history of orogenic evolution. For true orogen-parallel strike-slip systems this assumption is contradicted by field evidence which suggests at least coeval evolution of contractional structures and deformation attributed to strike slip. As orthogonal-type subduction was unable to explain such coexisting features in the deformational field, the concept of 'oblique subduction' has been propounded. It assumes oblique plate convergence and its resolution into an orthogonal component of motion that creates the dipping faults "necessary for subduction" (Sleep,1992) and a parallel component, manifest as strike-slip faults. Searching for a unique transmission mechanism for such a combined motion, Oldow et al.(1990) have proposed that thrusting and strike slip occur along spatially segregated listric surfaces that merge downwards into a sole sub-horizontal decollement zone near the base of the crust (Figure 1c).In current plate-tectonics reasoning the corollary of 'oblique subduction' is 'terrane accretion'.It assumes that slivers of continental crust, so-called 'terranes',are separated by rifting from "mother" continents, incorporated into oceanic plates and successively 'accreted' by oblique subduction onto active continental margins where they are bounded by strike-slip faults. This interpretation best conforms with the observed ubiquity of longitudinal strike-slip faulting in 'collisional' orogens and island arcs, saving at the same time the alleged process of subduction.

Contrary to customary thinking in terms of collisional tectonics, we have supported the idea that megashears of the crust (i.e.strike-slip systems of continental or global

288

importance) may have caused "by their simple activity the formation of orogenic belts" (Strutinski,1987). This idea was later more extensively treated and opposed to subduction-related orogenic models (Strutinski,1990). There is now growing evidence in favour of our strike-slip model of orogeny coming especially from structural data that in plate-tectonics interpretations are regularly misinterpreted or ,at most, "not well understood".Thus stretching lineations in metamorphic rocks, almost unanimously accepted as markers of tectonic transport, clearly show that, without exception, orogen-parallel tectonic transport precedes, is coeval with and sometimes even outlasts tectonic transport across the orogen (Ellis and Watkinson,1987 and references cited herein; Piasecki,1988; Strachan et al.,1992), pointing to the *relatively late and episodic occurrence of a shortening component in mountain building.* This is in evident contradiction to plate-tectonic models that are tempted to ascribe orogen-parallel movements to a late phase of orogenic evolution, as it makes no sense to place subduction and shortening related to it at a moment when the greatest part of orogeny has already unfolded. Yet,seismic anisotropy patterns emphasize that during orogeny flow within the mantle must have been likewise parallel to the strike of folded belts (Ramananantoandro,1988; Vauchez and Nicolas,1991), being in perfect accordance with the findings of surface geology.

Space limitations do not permit an extensive treatment of the flaws of plate-tectonics theory regarding orogeny.It should, however, be mentioned that there are numerous aspects related to the inferred existence of palaeo-oceans, arc magmatism, timing of obduction, palaeo (bio) geography and to other topics that point to the debatable status of subduction as a fundamental mechanism of lithospheric motion and deformation.

OROGENS - MEGASHEARS OF THE CRUST

One of the most fertile papers in improving our ideas regarding orogeny was that of Tchalenko (1970) on similarities between shear zones of different magnitudes. It was really striking to learn that shear-zone structures are essentially the same on a regional scale and down to the magnification limit of the optical microscope. What was nearer as to suppose that even on continental or global scale things should not be very different?

Our model of orogeny predicts that differential movements in the asthenosphere induce shearing in the overlying lithosphere that proceeds in a similar manner to the shear-box experiments of Tchalenko (1970). The first stage of shearing, consisting of an almost homogeneous straining, produces stretching and a thinning out along the future shear zone. Other than in model experiments this stage lasts very long, in fact tens of millions of years, due to the very slow increment of shear stress and to the rheologic behaviour of the deeper crust and upper mantle. At the surface the thinning out implies subsidence and the formation of elongated troughs,i.e.geosynclines. The frictional drag exerted by the differential asthenospheric flow continually decreases upwards according to decreasing temperature and rheologic behaviour of rocks.This movement may best be visualized by two adjoining decks of cards progressively 'sheared' along their length in opposite directions (Figure 2). It is important to note that ,other than in plate-tectonic models, in our model deformation accompanied by metamorphism begins at this early stage of orogenic evolution, an assumption that is in full agreement with field evidence. It is obvious that foliation created during this stage must be flat-lying and more or less sub-parallel to the bedding within the geosynclinal pile. This is exactly what has been observed (e.g.Hanmer,1981; Piasecki,1988; Strachan et al.,1992).

In a second stage the increasing shear stress cannot be accommodated any further by primary homogeneous straining, so that some kind of a ductile 'rupture' must occur at depth. The first Riedel shears will appear, significantly increasing the stretching component

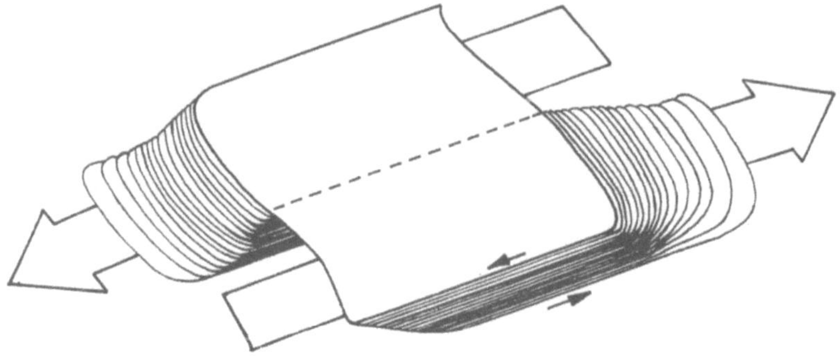

Figure 2. Shearing of the lithosphere above differential asthenospheric motion during the first stage of orogeny, as visualized by two adjoining decks of cards. The regime along the future zone of rupture is tensional.

of the motion. At surface level this means acceleration of subsidence in the geosynclinal furrows. The advanced stretching also produces tension gashes, the more so if there is also a tensional component beside simple shear, like in the present-day Gulf of California. These fissures may reach down to the updomed mantle, constituting conduits along which mantle material ascends giving birth to the ophiolite suite.At this stage metamorphism is (Riedel-) fault-bounded, meaning that it is restricted to steep-dipping zones and characterized by sub-vertical foliations, lineations remaining, however, sub-horizontal (compare the observations of Hanmer,1981; Piasecki,1988; Strachan et al.,1992).

The following stage in the evolution of an orogen is characterized by the initiation of principal displacement shears (Y-type shears, according to Tchalenko,1970) that most probably correspond to the 'late' orogen-parallel strike-slip faults visible at the surface. Shear-heating at depth begins to drive crustal anatexis (Sylvester,1988), and mixing-up of crustal melts with mantle material gives birth first to gabbro-dioritic and later on ,as melting proceeds, to granodioritic-tonalitic or granitic magmas. These will start to ascend diapirically, initiating the upheaval of orogens and eventually triggering folding and thrusting as well as, in the flanking basins, the formation of accretionary wedges. In plate-tectonics considerations these processes are conventionally ascribed to subduction-related compression, whereby 50 to 80% shortening is implied.Yet, one should recall that laboratory-model studies on transpression (reviewed by Sylvester,1988) all show that fault-bounded welts are forming above principal displacement zones,due to the accommodation of the component of shortening strain by uplift.Field relations observed along convergent strike-slip faults are consistent with these model experiments, evincing the upward flattening of fault planes and their conversion into oblique-slip thrust faults (Sylvester and Smith,1976; Strutinski,1987), along which crustal stacking occurs.We see no reason why this should not apply also to the orogen as a whole ,that, according to our model, is a shear system at least one order of magnitude greater than a common strike-slip fault.It should not be very difficult to prove that during the last stage of evolution shear heating not only triggers magmatism and related diapirism, but also implies crustal expansion ,thus constraining pure strike slip or transcurrency to turn into transpression. Swallowing of crust, be it oceanic or continental, is not required by our model. Instead, a steady transition from transtension to transpression is emphasized to be the main characteristic of evolving orogenic belts.

THE GEOTECTONIC FRAMEWORK OF OROGENS

In a recent paper Murphy and Nance (1991) distinguished between two different types of orogenic belts, so-called interior orogens, supposed to be the product of continent-continent collisions after 'contraction of interior oceans', and peripheral orogens regarded as being due to subduction of 'exterior oceans'. This duality of orogens seems to be a well established fact, being acknowledged also by our model of orogeny. The difference is, however, that, according to our view, peripheral orogens are only restricted to the actual border of the Pacific and did not occur before the breakup of Pangaea. Therefore we shall term them circum-pacific orogens. The other main type of orogens is not strictly restricted to a specific geological period, but its importance seems to be decreasing in recent times. According to Carey (1983) orogens of this type were born equatorially and have been, at least in part, globe-girdling. We shall refer to them as equatorial orogens.

a) **Equatorial orogens.** Their palaeomagnetically proven equatorial position has been regarded by Carey (1983) as being due to, or, at least, concordant with an equatorial sinistral torsion, thought to represent the combined effect of gravity and rotational inertia. Instead of a sinistral torsion we assume an easterly directed asthenospheric current, similar in direction, but opposed in sense, to the equatorial ocean currents.Such a current in a globe-enveloping asthenospheric layer may be inferred by analogy with the latitudinal disposal of the atmospheres on Jupiter and Saturn. This disposal is due to differential zonal motion that shows greater - in the case of Saturn significantly greater - wind velocities in the equatorial zone as compared to higher latitudes (Figure 3). As a consequence the equa-

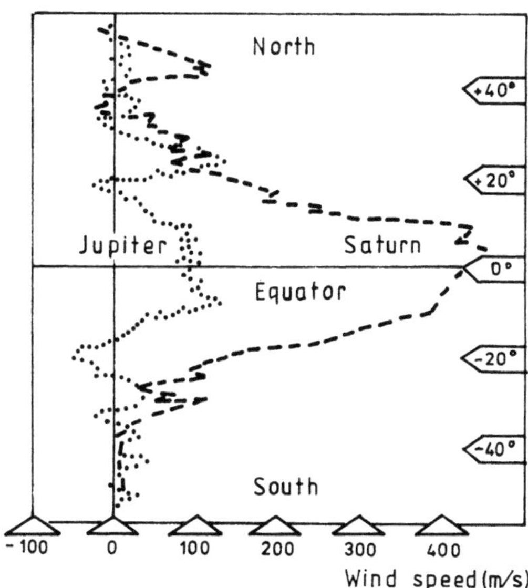

Figure 3. Wind pattern of the Jupiter and Saturn atmospheres (after Dorschner,1986).

torial zone must be "sheared off" sinistrally against the northern, and dextrally against the southern, tropical zones. On Earth shearing produced in the lithosphere above the implied asthenospheric current is inferred to be the motor that drives equatorial orogeny. Scattered palaeomagnetic data and shear sense indicators from literature accessible to the present author (Badham,1982; Bachtadse et al.,1983; Mawer and White,1987; Piasecki,1988; Crespo-Blanc,1992;Strachan et al.,1992) may indeed be cited in support of the interpretation of the Caledonian and Hercynian fold belts as originating above equatorial asthenospheric currents. At present only sinistral shear is acknowledged for the Caledonian orogen, whereas for the Hercynian belt some authors assume dextral shear against Gondwana (e.g.Badham,1982), while others substantiate sinistral shear against northern (stable) Europe(e.g.Bachtadse et al.,1983). The Alpine orogen is much more complicated due to intervening ocean spreading. Carey (1983) assumes only sinistral shear along it ('Tethyan torsion'). Yet, there are at lest some indications that dextral shear was likewise operative in the Dinaric (southern) branch of the Alpine orogen (Bébien et al.,1986). In chronological order the equatorial orogens all left their initial position, clearing the way for the next orogenic cycle. This appears to be due to the spinning of the asthenospheric current about the changing rotation axis of the Earth (Figure 4), a motion that cannot be obeyed by the relatively brittle lithosphere. There are some hints that zonal motion is still active, even if not precisely at 0° latitude but some 15° more to the north. Here we may observe the pronounced easterly bowing of the Mariana arc-trench system of the Pacific and the shear-bounded Caribbean 'plate' with its frontal arc-trench system of the Lesser Antilles. Both situations suggest the existence of an easterly flowing asthenospheric current beneath them.

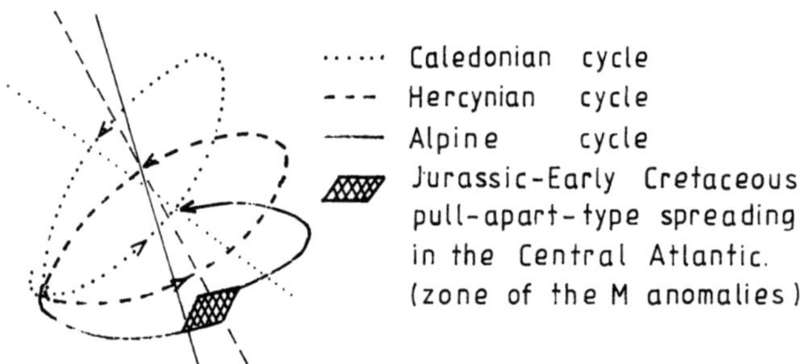

Figure 4. Inferred spinning of the asthenospheric equatorial current about the changing rotation axis of the Earth from Caledonian to early Alpine times.

    b) **Circum-pacific orogens**   may represent the alternative of orogeny to Earth expansion. However, here too shearing is the relevant factor. It seems that this is due to a counterclockwise rotation of the Pacific plate against surrounding continents and island arcs. Benioff zones ,trenches and volcanic arcs make up a trinity that characterizes circum-pacific orogens in their last (diapiric) stage of evolution that begins with the emplacement of large batholithic intrusions. The volcanic front is situated inland from the batholithic alignment (e.g. Peruvian Andes), that, in modern arc-trench systems is supposedly marked by the so-called non-volcanic outer arc or outer high characteristically evidenced by a positive gravity anomaly. Moreover, it may be observed that this outer high approximately halves the distance between the trench and the volcanic front. Other than in plate-tectonic models dimmed by the compression assumption,we herewith claim that both the trench and the volcanic arc are created essentially due to tensional stresses in the lithosphere that occur most probably as a rebound to the diapiric upheaval performed in between. That volcanic activity must be linked to normal faulting, and that, on the other side, trenches are as well characterized by normal faulting is all but new wisdom. Yet, in spite of having both a tensional character, volcanic arcs and trenches are very dissimilar, due to striking differences in their heat fluxes. Under the arc tensional faulting enables overheated (and decompressed) crust-mantle material to escape to the surface so that isostatic equilibrium is maintained.Under the cold trench there is nothing hot to escape, so the trench bottom sags down and the isostatic equilibrium is broken. However, recent data coming from submersible investigations of trench bottoms point to the presence of "overpressurized water oozing out from the zone of contact between plates" (Cadet et al.,1986; see also Mascle et al., 1986). Do we have to do here with a dewatering conduit of the mantle whose tap is switched on as soon as diapiric processes set in? And if so, might overpressurized waters have something to do with the so disputed topic of earthquake generation along Benioff zones?

CONCLUSIONS

Except for this earthquake generation along Benioff zones, that is taken by plate tectonicists as (circumstantial!) evidence in favour of subduction of oceanic lithosphere under continental lithosphere, and hence of an essentially compression-related evolution of orogens, geological field evidence as well as an increasing amount of geophysical data point to the fact that orogens are megashears of the lithosphere that have nothing to do with closure of oceanic basins. Thus closure may be an artefact as well as the alleged Tethys, Rheic, Iapetus and older oceans leading to the perspective that ocean spreading may not be compensated spatially, hence lending support to the Earth expansion hypothesis.

REFERENCES

Bachtadse,V.,Heller,F.,and Kröner,A.,1983,Palaeomagnetic investigations in the Hercynian mountain belt of Central Europe,*Tectonophys.*,91:285-299.

Badham,J.P.N.,1982,Strike-slip orogens - an explanation for the Hercynides,*J.geol.Soc.London*,139:493-504.

Bébien,J.,Dubois,R.,and Gauthier,A.,1986,Example of ensialic ophiolites emplaced in a wrench zone: Innermost Hellenic ophiolite belt (Greek Macedonia),*Geology*,14: 1016-1019.

Cadet,J.P.,Kobayashi,K.,Lallemand,S.,and Jolivet,L.,1986,Subduction in the Japan Trench:the Kaiko results,*in*:"The Origin of Arcs",F.-C.Wezel,ed.,Elsevier, Amsterdam.

Carey,S.W.,1983,Tethys and her forebears,*in*:"The Expanding Earth.A.Symposium",S.W.Carey,ed., University of Tasmania.

Crespo-Blanc,A.,1992,Structure and kinematics of a sinistral transpressive suture between the Ossa-Morena and the South Portuguese Zone,South Iberian Massif,*J.geol.Soc.London*,149: 401-411.

Dewey,J.F.,and Bird,J.M.,1970,Mountain belts and the new global tectonics,*J.Geophys.Res.*,75: 2625-2647.

Dorschner,J.,1986,"Planeten - Geschwister der Erde?", Urania Verlag, Leipzig.

Ellis,M.,and Watkinson,A.J.,1987,Orogen-parallel extension and oblique tectonics:the relation between stretching lineations and relative plate motions, *Geology*,15: 1022-1026.

Fitch,T.J.,1972,Plate convergence,transcurrent faults and internal deformation adjacent to southeast Asia and the western Pacific,*J.Geophys.Res.*,77:4432-4460.

Hanmer,S.,1981,Tectonic significance of the northeastern Gander Zone,Newfoundland: an Acadian ductile shear zone, *Can.J.Earth Sci.*,18: 120-135.

Mascle,A.,Biju-Duval,B.,de Clarens,P.,and Munsch,H.,1986,Growth of accretionary prisms: tectonic processes from Caribbean examples,*in*:"The Origin of Arcs",F.-C.Wezel,ed.,Elsevier,Amsterdam.

Mawer,C.K.,and White,J.C.,1987,Sense of displacement on the Cobequid-Chedabucto fault system, Nova Scotia,Canada,*Can.J.Earth Sci.*,24: 217-223.

Murphy,J.B.,and Nance,R.D.,1991,Supercontinent model for the contrasting character of Late Proterozoic orogenic belts,*Geology*,19:469-472.

Oldow,J.S.,Bally,A.W.,and Avé-Lallemant,H.G.,1990,Transpression,orogenic float,and lithospheric balance,*Geology*,18:991-994.

Piasecki,M.A.J.,1988,Strain-induced mineral growth in ductile shear zones and a preliminary study of ductile shearing in western Newfoundland,*Can.J.Earth Sci.*,25: 2118-2129.

Ramananantoandro,R.,1988,Seismic evidence for mantle flow beneath the Massif Central rift zone (France), *Can.J.Earth Sci.*,25:2139-2142.

Sleep,N.H.,1992,Archean plate tectonics:what can be learned from continental geology?,*Can.J.Earth Sci.*,29: 2066-2071.

Smith,M.,1988,Deformational geometry and tectonic significance of a portion of the Chilliwack Group, northwestern Cascades,Washington,*Can.J.Earth Sci.*,25: 433-441.

Strachan,R.A.,Holdsworth,R.E.,Friderichsen,J.D.and Jepsen,H.F.,1992,Regional Caledonian structure within   an oblique convergence zone,Dronning Louise Land,NE Greenland, *J.geol.Soc.London*, 142:359-371.

Strutinski,C.,1987,Strike-slip faults - what are they really standing for? General features with exemplifications from the Romanian Carpathians,*Studia Univ.Babeş-Bolyai*,Geol.-Geogr.,XXXII:47-59.

Strutinski, C.,1990,The importance of transcurrence phenomena in mountain building,*in*:"Critical Aspects of the Plate Tectonic Theory",vol.II,V.Beloussov a.o.,eds.,Theophrastus Publ.S.A.,Athens.

Sylvester,A.G.,1988,Strike-slip faults,*Geol.Soc.Am.Bull.*,100:1666-1703.

Sylvester,A.G.,and Smith,R.R.,1976,Tectonic transpression and basement-controlled deformation in San Andreas fault zone,Salton Trough,California,*Am.Assoc.Petrol.Geol.Bull.*,60:2081-2102.

Tchalenko,J.S.,1970,Similarities between shear zones of different magnitudes,*Geol.Soc.Am.Bull.*,81:1625-1640.

Vauchez,A.,and Nicolas,A.,1991,Mountain building:strike-parallel motion and mantle anisotropy, *Tectonophys.*,185:183-201.

EARTH EXPANSION REQUIRES INCREASE IN MASS

John K. Davidson

Petrecon Australia Pty Ltd
322 Liverpool Street
Hobart 7000. Australia

Research in one field of scientific endeavour can re-direct another. Geological evidence for an expanding earth is now at a point that a fundamental change in physics is imminent. Mass is being created within the Earth.

In 1915 Wegener (1966) postulated that the continents were originally a single unit, Pangaea. The scientific world incorrectly dismissed Wegener's hypothesis. In 1956 Carey (1958) rekindled continental drift using tectonic evidence. This resulted in the deep-sea drilling of the 1960's which led to the theory of "plate tectonics". That theory not only solved many problems in geology but also in several other branches of science.

Yet "plate tectonics" has a fundamental flaw; it assumes that the rate of new oceanic crustal creation at the spreading ridges is equal to the rate of oceanic crustal disappearance, or subduction, at the oceanic trenches. An excess of creation over subduction was postulated by Owen (1983) with his 180Ma (180 million years ago) Earth radius (R_{180}), 80% of the present, that is, $R_{180} = 0.8R_0$. This has been termed "slow" expansion compared with the "fast" expansion of Carey (1976) who had no subduction, that is, $R_{180} = 0.55R_0$.

This paper presents evidence for expansion at a little less than the "fast" rate and in which the mass (and volume) of the Earth has increased at an accelerating rate while average density has fluctuated.

ASYMMETRIC EARTH EXPANSION

The angle between the Earth's rotational axis and the plane of the ecliptic has usually been greater than at present. Increasing tilt from 45Ma to the present formed polar ice-caps and relatively well defined climatic zones. This situation last occurred in the Late Carboniferous/ Early Permian, approximately 280Ma. Four climatically controlled Early Permian floras were recognised by Wegener as important in continental reconstructions. These included the southern and northern cool-temperate to polar Glossopteris and Angaran floras and the tropical and subtropical Euramerian and Cathaysian floras (Figure 1a).

Frontiers of Fundamental Physics, Edited by M. Barone
and F. Selleri, Plenum Press, New York, 1994

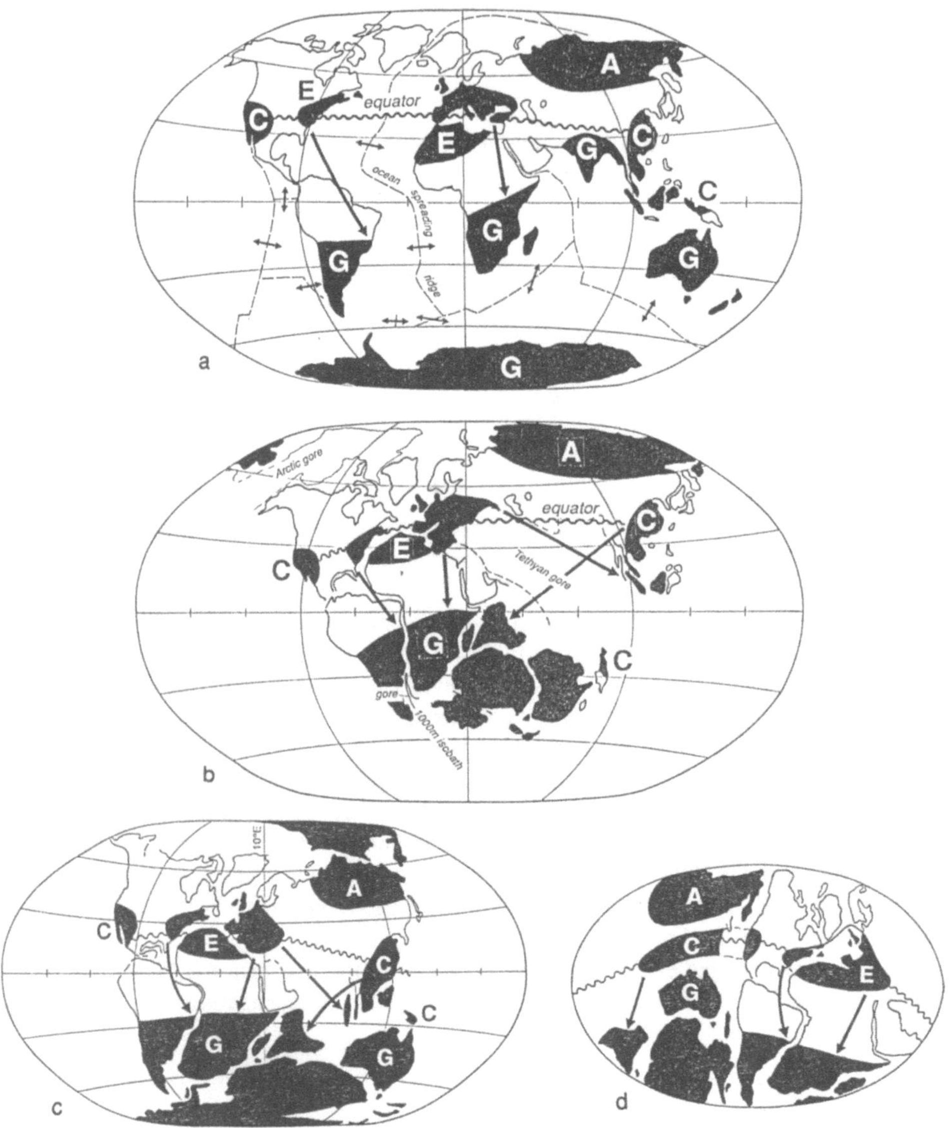

Figure 1. Distribution of Early Permian (280Ma) floras. (a) present Earth radius, $R_0 = 6400$ km; (b) Early Jurassic (180Ma) Earth radius, $R_{180} = R_0$; (c) Early Jurassic (180Ma) Earth radius, $R_{180} = 80\%R_0$ (a-c, after Owen, 1983); (d) Early Jurassic (180Ma) Earth radius, $R_{180} = 55\%R_0$ (after Carey 1988). Cool temperate to polar: G = Glossopteris, A = Angaran. Tropical: E = Euramerian. Sub-tropical: C = Cathaysian (after Davidson 1983).

Owen (1983) reconstructed the continents at 180Ma on a present radius Earth (Figure 1b). This reconstruction partly solved the problem of large oceanic expanses separating each non-marine flora imposed by the current distribution of the continents. Not only were the dispersed floras joined in a single continental mass (Pangaea) but the excursions of tropical forms into the southern cold flora were also partly explained, for example, Euramerian flora to Brazil and Zimbabwe. Owen used both dated oceanic crustal information and the edges of the continents in order to cartographically accurately re-assemble the globe. However, he found small oceanic "gores" for which there was no accounting, for example, between southern Africa and South America. He made six reconstructions using progressively smaller Earth radii (Figure 2a). The smallest was 80% of the present radius at 180Ma ($R_{180} = 0.8R_0$) which totally removed these small "gores" (Figure 1c). It is surprising that the removal of such relatively small gaps had the cumulative effect of removing the geologically unacceptable Tethyan and Arctic "oceanic" gores. His non-continent area or palaeo-Pacific Ocean was 50% of that in the present radius reconstruction (Figure 1b). Owen considered there were no oceans 700Ma on an Earth of 50% present radius (Figure 2a) because layered ophiolites (oceanic crust) are not known prior to that time.

The Figure 1c closure of "Tethys" further reduced the 280Ma floral distribution problem by placing the Cathaysian floras of China adjacent to those in Iryan Jaya. It also provided elements of the same flora with a land-bridge for mixing with the Glossopteris flora in India. However, the 80% radius is too high because the shape of each continent has changed; Owen did not take into account the tensional features within continents largely resulting from expansion-induced changes in radius of curvature. Further, his pre-100Ma reconstructions around Antarctica are interpretative due to lack of detailed sea-floor spreading data. He may have used too much subduction. Figure 1c shows the Euramerian and Cathaysian floras, 0° to 20° north of the 180Ma equator, about 30° from their current position 30° to 50° north of the present Equator. Therefore the 180Ma Earth circumference was approximately 240 present degrees, that is, $R_{180} = 0.67R_0$ (Figure 2a, Davidson) using Owen's reconstruction, not $R_{180} = 0.8R_0$. If the radius of curvature of the continents were increased the reconstruction would be on a smaller Earth, suggesting the 0.67 value might reduce further, the real value being nearer Carey's $R_{180} = 0.55R_0$ than Owen's $R_{180} = 0.8R_0$.

Davidson (1983) required the Cathaysian floras of China/SE Asia and western North America to be in communication, possibly as in Carey's (1988) draft reconstruction (Figure 1d). If the axis of the Euramerian and Cathaysian floras lay on the 280Ma equator, the 40° northwards migration of that axis, supported by palaeo-magnetic data (Carey, 1976), equates to a circumference of 200 present degrees, or Carey's $R_{180} = 0.55R_0$. However, the area of the cold Glossopteris flora is considerably greater than its northern equivalent, the Angaran flora. Hence the 280Ma equatorial floral axis may have been located slightly north of the geographic equator. The Earth may therefore have had a 280Ma radius, a little more than 0.55R, with the Cathaysian flora possibly encircling a small palaeo-Pacific, as in Vogel's (1983) $R = 0.6R_0$ reconstruction.

While Carey (1988) emphasised the preliminary nature of Figure 1d, the north/south asymmetry in the polar floras is not solved by other reconstructions, for example, Vogel's (1983) Pangaea differs little from Carey's in moving Australia/Antarctica south relative to South America and in rotating North America anticlockwise by moving Asia/Siberia southwestwards. The asymmetry is probably real, greater areal extent of the southern Glossopteris flora reflecting a broadening non-marine area due to the developing southern expansion bulge 280Ma. Davidson (1992) has related this expansion asymmetry, so clearly seen in the current distribution of spreading ridges (Figure 1a), to the 97%/3% northern/southern distribution of world oil reserves; an enormous asymmetry.

Figure 2a shows that in this paper $R_{180} = 0.67R_0$, possibly less than $0.67R_0$ and there were no oceans before 700Ma. Broad platformal depressions were present pre-1500Ma. From

1500Ma to 700Ma extensional grabens were common possibly representing slow expansion from $R_{1500} = 0.5R_0$ to $R_{700} = 0.55R_0$.

INCREASE IN MASS

The initial assumption of earth expansion is that increased volume is achieved by decreased density under constant mass and possibly with changes in G, the gravitational constant. The latter is very tightly constrained within a 10^{-11} variation per annum by astronomical observations (Napier, pers. comm.) which, over the last 180 million years in inadequate by three orders of magnitude in accounting for the rate of expansion of the Earth. The constancy of mass presents both chemical and physical problems. If the whole Earth were composed of dynamite its internal energy would be inadequate by an order of magnitude to generate the required expansion from $R_{180} = 4000km$ to $R_0 = 6000km$ (Napier, pers. comm.) suggesting chemical energy is inadequate. Even if this could be achieved, physical constraints are imposed by g, the acceleration due to gravity, variation of which over geological time has only recently been substantiated by Mann and Kanagy (1990). They noticed that the maximum angle of bedding

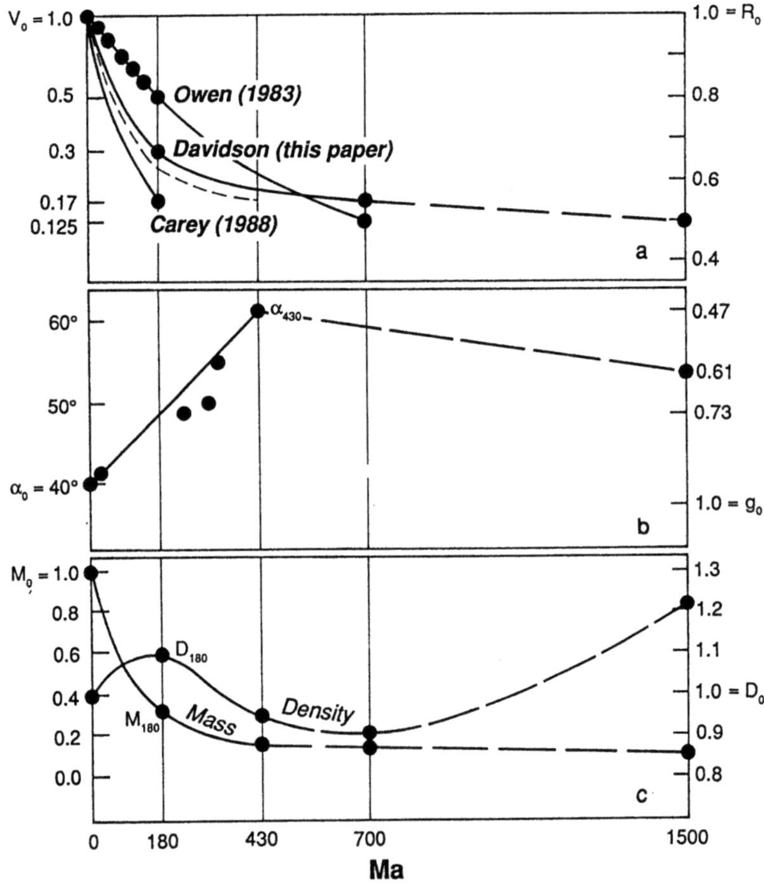

Figure 2. (a) volume (V) and radius (R) of the Earth against geological time (Ma = millions of years before present); (b) maximum angle of repose of cross-bedded sandstones (α) (after Mann and Kanagy, 1990) and acceleration due to gravity (g) against geological time; (c) mass (M) and density (D) of the Earth against geological time.

repose (α) increased from 40° on present sand-dune faces to 61° at 430Ma (Figure 2b). Intervening measurements indicate the angle of repose is almost linear from the present to 430Ma. Four very consistent maxima of 53° to 54° were reported at 1450Ma to 1500Ma.

The angle of internal friction (ϕ) of slope deposits is related to the shear stress on the potential sliding surface (s), the pressure exerted by the material (p) and the cohesion (c) by

$$s = c + p \tan \phi$$

For uncemented, dry sand grains and water-immersed sand grains c is zero. Therefore, for wind and water-deposited sands

$$s = (M/A) \, g_0 \tan \alpha_0$$

where M = mass, A = area and g_0 and α_0 are the present acceleration due to gravity and maximum angle of repose.

α_0 is less than 30° for rounded, well-sorted grains and increases to 40° for angular, poorly sorted grains. These variables in the Mann and Kanagy data sets can be eliminated in determining the variation in g with time by only considering the maximum value of the angle of repose in cross-bedded sandstones. Therefore, assuming the shear stress at the point of sliding has not changed since 430Ma, that is, the physical properties of sands have not changed

$$g_0 \tan \alpha_0 = g_{430} \tan \alpha_{430}$$
$$\text{or} \quad g_0 \tan 40° = g_{430} \tan 61°$$
$$\text{or} \quad g_{430} = 0.47 g_0$$

Note that this is conservative as 61° is the cross-bed angle after compaction.

Since force $F = (Gm_1 m_2)/R^2$, $g_0 = (GM_0)/R_0^2$ and $g_{180} = (GM_{180})/R_{180}^2$, where M_0, M_{180}, R_0 and R_{180} are the mass and radius of the Earth at the present and 180Ma, respectively.

From Figure 2b,

$$0.73 g_0 = g_{180},$$

therefore $(0.73 GM_0)/1^2 \ = \ (GM_{180})/0.67^2$

$$\text{or} \quad M_0 = 3 M_{180}.$$

Also, $M_0 = D_0 V_0$ or $M_0 = D_0 (3.3 V_{180})$, where D_0, V_0, V_{180} are the density and volume now and 180Ma. Therefore,

$$M_{180} = 1.1 D_0 V_{180}$$

Thus the mass of the Earth has increased three-fold since 180Ma and average density has decreased about 10% (Figure 2c).

The Earth mass/density relationships with time can be traced to 430Ma but prior to that are a little conjectural due to limited data (Figure 2c). Although Carey has no oceanic crust prior to 180Ma, the absence of ophiolites prior to 700Ma places a limit on the first appearance of oceanic crust. It seems likely that from 1500Ma to 700Ma the Earth's radius increased about 10% (Figure 2a) with a small increase in mass, but volume and mass increase began to accelerate exponentially from 430Ma. The current expansion rate is very rapid and gives rise to questions like, how is the extra mass being created (it seems to be occurring in the core as there is no evidence at the surface); will the Earth ultimately explode and form another asteroid belt or will it become a Jupiter, then a Sun and so on up the Main Sequence of stars?

REFERENCES

Carey, S.W., 1958, The tectonic approach to continental drift, *in* "Continental Drift, a Symposium," University of Tasmania, Hobart.

——, 1976, "The Expanding Earth", Elsevier, Amsterdam

——, 1988, "Theories of the Earth and the Universe", Stanford University Press, Stanford.

——, 1994, Creeds in Physics, this volume.

Davidson, J.K., 1983, Tethys and pacific evidence for an expanding earth, *in* "Expanding Earth Symposium", S.W. Carey, ed., University of Tasmania, Hobart.

————, 1992, Tectonic control of world oil reserves: Australia's position, *Aust. Petr. Explor. Assoc. Jour.* 1, 183.

Mann, C.J. and Kanagy II, S.P., 1990, Angles of repose that exceed modern angles, *Geology* 18, 358.

Owen, H.G., 1983, "Atlas of Continental Displacements", Cambridge University Press, Cambridge.

Vogel, Klaus, 1983, Global models and earth expansion, *in*"Expanding Earth Symposium," S.W. Carey, ed., University of Tasmania, Hobart.

Wegener, Alfred, 1966, "The Origin of Continents and Oceans", Trans. John Biram, Dover Publications, New York.

PRINCIPLES OF PLATE MOVEMENTS ON THE EXPANDING EARTH

Jan Koziar

Institute of Geological Sciences
Wrocław University, pl.M.Borna 9
50-204 Wrocław, Poland

INTRODUCTION

The existence of rigid lithospheric plates resting on plastic asthenosphere, allows one to build a quantitative model of their movement on the expanding earth. Such a model relates the kinetics with dynamics, and binds the lithosphere with its basement as a general reference frame. These features do not exist in plate tectonic model. The expanding earth model explains the observed scheme of development of the lithosphere, together with some relations incomprehensible within plate tectonics.

THE PINNED PLATE

Let us start with considering a rigid plate with fixed shape which rests on the basement with net of coordinates, being isotropically extended (Figure 1). Let us also assume that the

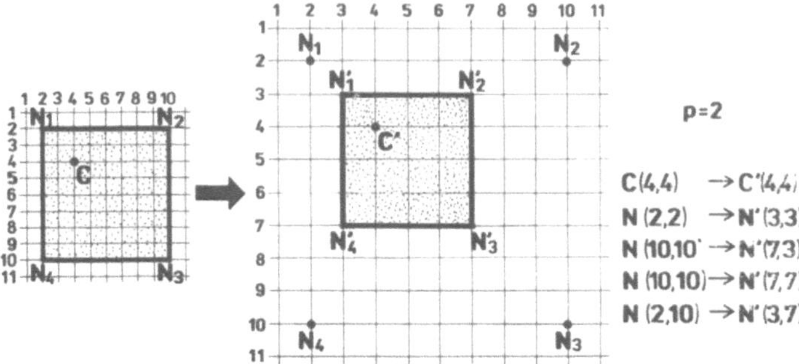

Figure 1. Transformation of plate corners' coordinates of the plate pinned to expanding basement at point C.

plate is pinned to the basement in point C, and that the basement enlarges its line dimension by extension in rate p = 2. The transformation of coordinates of the plate corners is shown in the table, of Figure 1. Among all points of the plate, only point C does not change its coordinates and then we may call it the "stable point of transformation" (SPT). General algebraic transformation of coordinates of any point of the plate is described by formulas

$$x' = x_0 + 1/p \, (x - x_0), \qquad\qquad y' = y_0 + 1/p \, (y - y_0) \qquad\qquad (1)$$

where (x_0, y_0) are coordinates of SPT, while (x, y) and (x', y') are coordinates of any point of the plate, respectively, before and after extension of the basement at linear rate p.

STABLE POINT OF TRANSFORMATION OF NONPINNED PLATE

Nonpinned plates (resting loosly on the basement), the lithospheric plates being like this, have their SPTs as well. In order to find such a point, we must analyse the friction forces between the plate and basement.

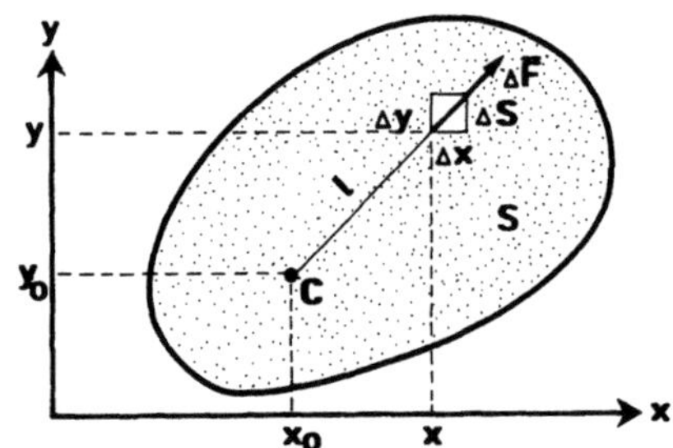

Figure 2. Friction force acting on an element of plate (area element) pinned to expanding basement in point C.

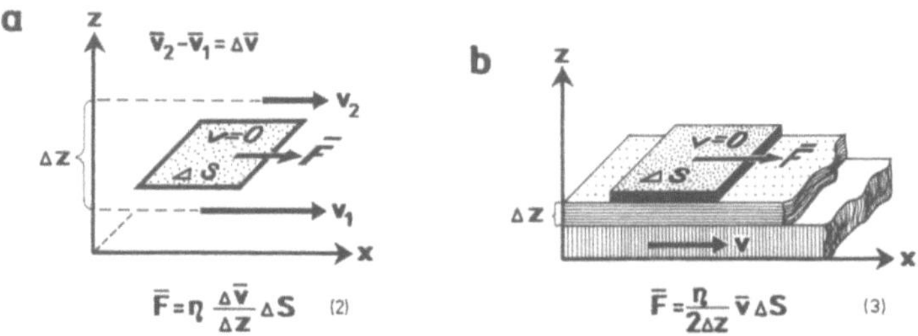

Figure 3. a). Friction exerted by flowing viscous fluid on fixed area element ΔS, where η- coefficient inner friction of fluid $\Delta v/\Delta z$ - velocity gradient of fluid's laminar movement. b). Single - sided friction force exerted by expanding basement on plate element ΔS.

Let us consider first the force acting on the area elements (ΔS) of a plate pinned in point C, (Figure 2). This force will be everywhere directed outwards from point C.

We can find its value from the formula (2) describing the friction of viscous fluid on area element ΔS, Figure 3a. Because friction acts only on the bottom side of area element, it is twice weaker, (Figure 3b), and since the layer of laminar movement has an unknown but constant thickness (Δz) then (2) takes the form (3). However basement velocity relative to element ΔS is expressed by formula

$$\bar{v} = \frac{\Delta \bar{l}}{\Delta t} = \frac{(p-1)\bar{l}}{\Delta t} = h\bar{l} \tag{4}$$

where (h) is the formal equivalent of known parameter of expanding Universe, which we may call, in general sense, the "Hubble factor" (in distinction from "Hubble constant"). This factor can change with time, but in a given time it is constatnt for the whole plate. In connection with the former statements, the formula for ΔF (Figure 2) appears as

$$\Delta \bar{F} = {}^{h\eta}\!/_{2\Delta z}\bar{l}\Delta S = k\bar{l}\Delta S \tag{5}$$

where k = $h\eta/2\Delta z$ is unknown but constatnt parameter for the whole plate.

Now we will calculate the resultant friction force acting on the whole plate. To do that, we must display the force element ΔF components:

$$\Delta F_x = k(x-x_0)\Delta S, \qquad \Delta F = k(y-y_0)\Delta S \tag{6}$$

and integrate them on the whole area (S) of the plate

$$F_x = k\iint\limits_S (x-x_0)dS, \qquad F_y = k\iint\limits_S (y-y_0)dS \tag{7}$$

The resultant friction force exerted by the basement on the plate will be, in general different from zero. Thereby, this force will tend to tear off the postulated physical connection between the plate and basement in point C. However, if this force equals zero, then the removing of the ties changes nothing, and point C will be a stable point of transformation as before, but this time - of loosly resting plate (nonpinned). Then, in order to find this point, we must compare the right sides of (7) to zero

$$k\iint\limits_S (x-x_0)dS = 0, \qquad k\iint\limits_S (y-y_0)dS = 0 \tag{8}$$

and by solving such equations we will find the coordinates (x_0, y_0). Since $k \neq 0$, (8) take the form

$$\iint\limits_S (x-x_0)dS = 0, \qquad \iint\limits_S (y-y_0)dS = 0 \tag{9}$$

Their solution is

$$x_0 = \frac{\iint\limits_S xdS}{S,} \qquad y_0 = \frac{\iint\limits_S ydS}{S} \tag{10}$$

These are the formulas for the coordinates of barycenter of the plate.

THE CRACKING PLATE

Now, let us demonstrate the case of a loosly resting plate which is cracking to smaller pieces during expansion of the basement, (Figure 4).

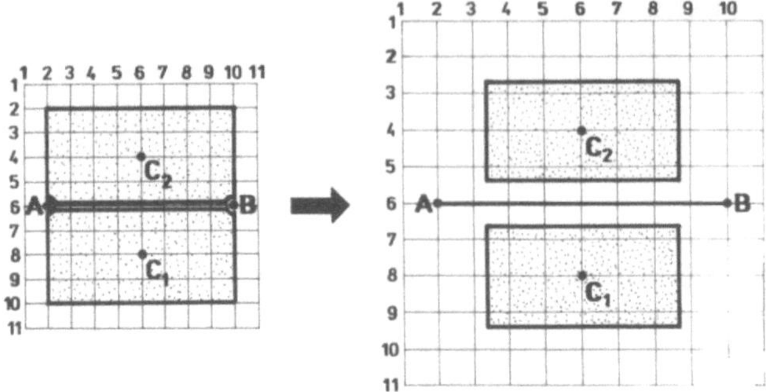

Figure 4. Cracking plate on the expanding basement.

As we see, the offshoot plates are carried away from themselves but, at the same time, they are tied to their basement in their SPTs. Then, the presented model solves the contradiction between fiksicism and mobilism, which is unsoluble on nonexpanding earth.

If the cracking plate leaves a split trace on the basement, then this trace undergoes enlargement relative to relevant edges of the offshoot plates (Figure 5). Such traces of plate splitting are in fact the oceanic ridges. Their enlargement relative to relevant continent's contour and the signs of their lengthwise extension are most important geotectonic features, nonexplained by plate tectonics.

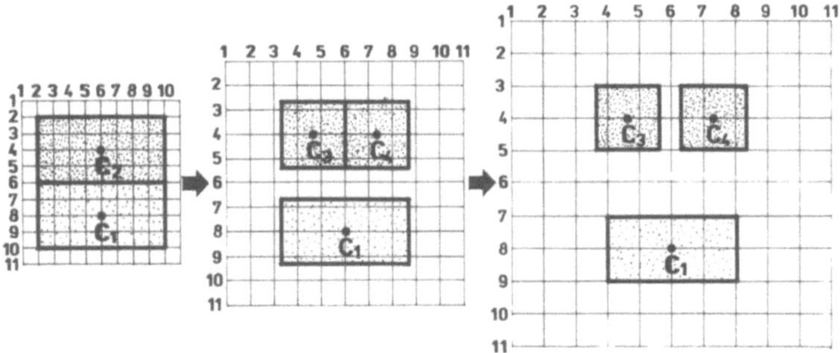

Figure 5. Trace of plate cracking on the expanding basement.

SPTs and traces of plate cracking (oceanic ridges) fix the general reference frame which is the deep basement. In plate tectonics, at most, one plate can be fixed, and it is not known which, at that. Recently, it was proved by seismic tomography (Woodhouse and Dziewoński, 1984) that plates and oceanic ridges are authochthonous, which corroborates the presented model.

THE GROWING PLATE

In reality, the lithospheric plates not only crack, but also grow as a result of spreading. We can prove, that if the growth of plates is regular i.e. their borders are motionless in relation to the basement (oceanic ridges are authochthonous), then their SPTs are stable.

At regular growth, the translation of every point of plate contour is proportional to its former distance from the SPT. Let us consider the area S and S' which has developed from S

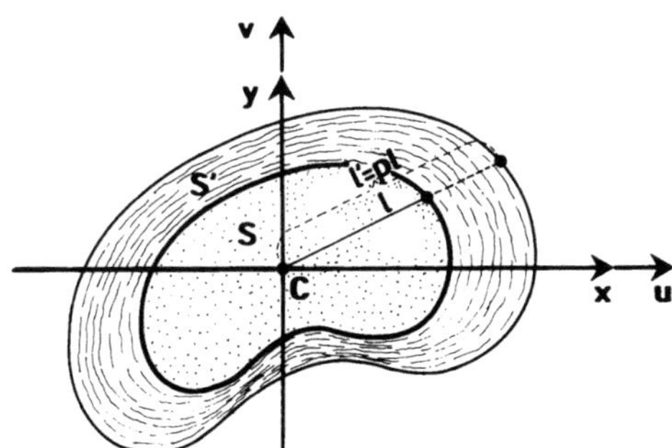

Figure 6. Regularly growing plate.

by regular growth, (Figure 6). For simplicity, let us assume that the center of reference frame lies in SPT of area S (point C). Then

$$\iint_S xdS = 0, \qquad \iint_S ydS = 0 \tag{11}$$

Then, let us transform the area S of (x, y) coordinates, to the S' area of (u, v) coordinates, according to the following formulas

$$u = px, \qquad v = py \tag{12}$$

This transformation corresponds with the radial enlargement of the area S relative to point $(0,0)$, at rate p. Then, it corresponds with the regular growth defined above. Let us now change the variables in formulas (11) according to transformation (12). Since the jacobian of this transformation is $1/p^2$, then equations (11) take the form

$$\tfrac{1}{p^3}\iint_S udS' = 0 \qquad\qquad \tfrac{1}{p^3}\iint_{S'} vdS' = 0 \tag{13}$$

Since $1/p^3 \neq 0$, then

$$\iint_{S'} udS' = 0, \qquad \iint_{S'} vdS' = 0 \tag{14}$$

We have proved in this way, that SPTs of areas S and S' are the same.

It is self-evident that the junction and disjunction of areas with the same SPTs, give areas also with the same SPTs (the analogy to physical barycenters). Then if we exlude area S from area S', the SPT of the obtained ring (S' - S) lies in SPT of area S. We have shown in this way, that if growing ring (S' - S) has different friction coefficient relative to the basement (e.g. because of smaller thickness of the lithosphere), then it has no influence on the position of SPT of the whole growing plate.

HOT SPOTS

Now let us think about the effects of hot spots activity. The volcanic chains produced by hot spots (mantle plumes) which are tied with the expanding basement, are divergent. We can demonstrate this first for the case of two intraplate hot spots (Figure 7).

A similar situation is in the case of interplate hot spots tied with oceanic ridges (Figure 8). Such hot spots do not pierce the plates but add swelling to their borders, which produces, in the course of spreading, also volcanic chains.

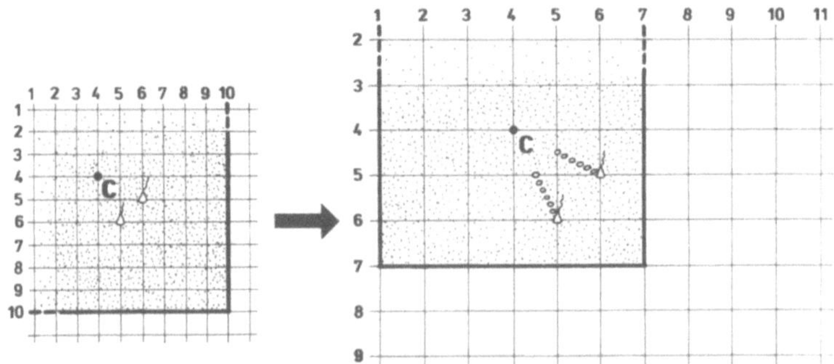

Figure 7. Divergence of volcanic chains produced by intraplate hot spots.

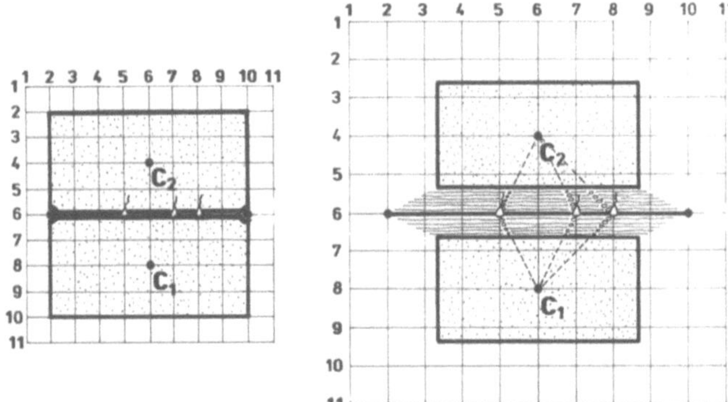

Figure 8. Divergence of volcanic chains produced by interplate hot spots.

Mutual moving away of the Mid-Atlantic Ridge hot spots (that is tantamount after all, with the lengthwise extension of the ridge) was pointed at already by Burke et al. (1973), but without connecting this phenomenon with earth expansion. On the contrary, the divergence of hot spot volcanic chains was noticed by Steward (1976) and interpreted as a manifestation of earth expansion.

DEVELOPMENT OF THE CENTRAL AND SOUTH ATLANTIC OCEAN

The described rules can be used to model the development of the Central and South Atlantic (Koziar 1985, 1993), as shown in Figure 9.

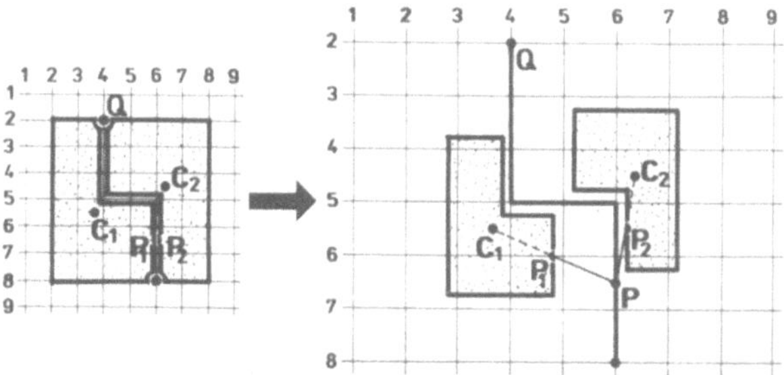

Figure 9. Development of the Central and South Atlantic Ocean on the expanding basement.

The two lines - P_1P and P_2P are, respectively, Rio Grande and Walvis ridges, produced by interplate hot spot placed near Tristan da Cunha island (point P). The arrangement of both ridges was noticed by Dietz and Holden (1970), who tried to explain the southwise translation of point P by the same translation of deep basement. However, such interpretation does not explain a similar, but northwise translation of the Azores region (Q), neither the general enlargement of the Mid Atlantic Ridge.

PHYSICAL MODEL

We can give a physical form for the geometrical model demonstrated above, in the shape of radially extended rubber slice (Koziar 1980, 1993). We can put different configurations of plates on it. Using such a device , it is possible to model (demonstrated above in a geometrical way) the development of the Central and South Atlantic (Koziar 1980, 1993), and the radial growth of oceanic lithosphere around Africa and Antarctica (Koziar 1980, 1993), which points streighforwards to earth expansion. For the first time this relationship was pointed out by Carey (1958), and latter by Heezen (1962) too. The devices can also be used to model the development of triple junctions and tear off processes of island arcs from continental margins (Koziar 1993). The tensional development of the latter structures is described elsewhere (Koziar and Jamrozik, 1991, 1994).

REFERENCES

Burke, K., Kidd, W.S.W., and Wilson, J.T., 1973, Relative and latitudinal motion of Atlantic hot spots, *Nature* 245:133.
Carey, S.W., 1958, A tectonic approach to continental drift, *Symp. Continental Drift Hobart*, 177.
Carey, S.W., 1976, "The Expanding Earth", Elsevier, Amsterdam - Oxford - New York.
Dietz, R.S., and Holden, J.C., Reconstruction of Pangea: breakup and dispersion of continents, Permian to present, *J. Geoph. Res.* 75:4939.
Heezen, B., 1962, The deep sea floor, *in:* "Continental Drift", S.K. Runcorn, ed., Academic Press, London.
Koziar, J., 1980, Ekspansja den oceanicznych i jej związek z hipotezą ekspansji Ziemi, *Sprawozdania Wrocławskiego Towarzystwa Naukowego* 35B:13.
Koziar, J., 1985, Rozwój oceanów jako przejaw ekspansji Ziemi, *Geologia (Uniw. Śl.)* 8:109.
Koziar J. 1993, Poster: The expanding earth, International Conference: Frontiers of Fundamental Physics, Olympia, Greece 27 - 30 September 1993.
Koziar, J. and Jamrozik, L. 1991, Tensyjno - grawitacyjny model subdukcji, PTG Oddz. Poznański, *Streszcz. referatów 1990-1991*:34.
Koziar, J. and Jamrozik, L., 1994, Tension - gravitational model of island arcs, (this volume).
Stewart, J.C.F, 1976, Mantle plume separation and the expanding Earth, *Geophys. J. R. Astr. Soc.* 46:505.
Woodhouse, J.H. and Dziewoński, A.M., 1984, Mapping the upper mantle: three-dimensional modeling of Earth structure by inversion of seismic waveforms, *J. Geophys. Res.* 89/B7:5953.

THE ORIGIN OF GRANITE AND CONTINENTAL MASSES IN AN EXPANDING EARTH

Lorence G. Collins

Geological Sciences Department
California State University Northridge
Northridge, CA 91330-8266 USA

INTRODUCTION

In the 1940s a major debate raged as to whether some granite bodies of plutonic dimensions were formed from magmas or by replacement processes (Grout, 1941; Read, 1948). Geologists favoring replacement, the "granitizers," suggested that Si and K are introduced into diorites and gabbros as Al, Fe, Mg, and Ca are subtracted from the mafic rocks to convert them to felsic granites (Table 1). "Magmatists," geologists supporting a melt-origin, believed that granitic magmas are created in "dry" rocks devoid of free water and at temperatures hot enough to cause thermal breakdown of water-bound minerals (Whitney, 1988; Leake, 1990). At these temperatures partial melting of granitic compositions occur and leave mafic restites. Experimental work in closed systems on melted natural and artificial granites convinced the magmatists that granites of plutonic dimensions formed from melts derived from mantle or lower crustal sources (Tuttle and Bowen, 1958; Luth et al., 1964). The experimental work showed that granitoid compositions plot near eutectics of phase diagrams. Lowest temperature points in these diagrams represent ultimate conditions for either (1) initial melting or (2) fractional crystallization of silica-saturated magmas. The magmatists concluded that because most granites plot near these low-temperature points, their bulk chemical content must be controlled by the sequence in which minerals first melt or the sequence in which minerals first crystallize and fractionally settle from a melt.

Magmatists also opposed the forming of granites by replacement processes on the basis that, (1) although many granitic terranes are heterogeneous, many large granite bodies are homogeneous, and (2) because replacing fluids are likely heterogeneous, the creation of uniform granite plutons from them is unlikely (Clarke, 1992).

Additional objections to a replacement origin include the problems of: (1) finding sufficient silica and potassium to produce the quartz and K-feldspar in granite, (2) providing large volumes of hydrous fluids necessary to transport the added and

Table 1. Chemical analyses, modal compositions, and densities of typical diorite-gabbro and granite in the continental crust.

	Diorite-Gabbro	Granite	Diorite-Gabbro	Granite
	$G = 3.1$	$G = 2.68$		
	wt %	wt %	vol. %	vol. %
SiO_2	54	75	70 (Na-Ca	30 (Na plag.)
Al_2O_3	17	13	plagioclase)	35 (K-feldspar)
$FeO-Fe_2O_3$	10	2	30 (olivine,	32 (quartz)
MgO	12	0.5	pyroxene,	3 (biotite)
CaO	8	0.5	hornblende,	
Na_2O	4	4	biotite)	
K_2O	≤ 1	5		
	100	100	100	100

subtracted elements, and (3) determining where the subtracted elements were deposited.

Magmatists believe that because these obstacles produce severe restraints on the replacement hypothesis, large granite plutons cannot be formed by replacement processes.

THE ROLE OF MYRMEKITE IN THE ORIGIN OF GRANITE

Contained within many large granite bodies is a tiny mineral texture called myrmekite, which consists of an intergrowth of vermicular quartz in plagioclase. The plagioclase of the myrmekite is optically continuous with adjacent quartz-free plagioclase, and the myrmekite projects as wartlike protrusions into adjacent K-feldspar (Figure 1). Myrmekite generally constitutes less than 0.5 vol % of the rock. It has never been produced experimentally, although other quartz-plagioclase intergrowths have. The melt-derived textures occur in the form of graphic or granophyric textures in which the quartz exists as triangular or runic shapes, whereas quartz in myrmekite is vermicular. Moreover, the plagioclase in graphic and granophyric textures has a uniform composition, unlike plagioclase in myrmekite which has a variable composition. The lack of being able to produce myrmekite in a melt provides permissive evidence that some granites are *not* formed by magmatic processes, but this characteristic is not conclusive evidence because myrmekite could have been formed following the crystallization of a granite from a melt (Collins, 1988ab; Hunt et al., 1992).

Myrmekite was first reported by Michel-Lévy (1874) and has been extensively studied (Phillips, 1974). Earlier hypotheses utilize a variety of methods to explain the origin of myrmekite, but all use equations that are balanced mass-for-mass (Table 2). In these equations myrmekite is suggested to be formed by either (1) exsolution of either Ca^{+2} and Na^{+1} from a high-temperature K-feldspar or the components of

Table 2. Mass-for-mass equations used to explain the origin of myrmekite.

Exsolution of myrmekite from high-T K-feldspar

$$xKAlSi_3O_8 = xKAlSi_3O_8 + NaAlSi_3O_8 + CaAl_2Si_2O_8 + 4SiO_2 \qquad (1)$$

$NaAlSi_3O_8$ plagioclase   quartz

$Ca(AlSi_3O_8)_2$ x = a large number

H-T K-feldspar   K-feldspar   myrmekite

Replacement by Ca^{+2} and Na^{+1} to form myrmekite

$$KAlSi_3O_8 + Na^{+1} = NaAlSi_3O_8 + K^{+1} \qquad (2)$$
$$2KAlSi_3O_8 + Ca^{+2} = CaAl_2Si_2O_8 + 4SiO_2 + 2K^{+1}$$

                            plagioclase   quartz

K-feldspar   myrmekite

Schwantke's molecule ($CaAl_2Si_2O_8$) or (2) the replacement of K-feldspar by plagioclase, Ca^{+2} and Na^{+1} (Phillips, 1974).

Equation (1) fails because Schwantke's molecule [$Ca(AlSi_3O_8)_2$] has not been found to exist, and some volumes of myrmekite require more than twice the amounts of CaO dissolved in K-feldspar than has been experimentally found to be possible (Carmen and Tuttle, 1964). Equation (2) fails because the introduction of Ca^{+2}- and Na^{+1}-bearing fluids should produce a concentration of myrmekite near veins or the complete replacement of K-feldspar by myrmekite, and neither has been found to occur.

Instead of the mass-for-mass equations (1) and (2), Collins (1988a) found that the origin of myrmekite can be best explained by volume-for-volume replacement processes. In cataclastically-broken mafic diorites and gabbros, strained and deformed primary plagioclase crystals are replaced from the interior outward by K-feldspar. Locally, where replacement is incomplete, islands of altered plagioclases lattices recrystallize to form the myrmekite. Not all relatively-calcic plagioclase grains are replaced by K-feldspar, because during the replacement, sodium displaced by incoming potassium also replaces other altered plagioclase grains to form sodic plagioclase. In this process about half of the original plagioclase is replaced by K-feldspar and about half by sodic plagioclase (Table 1). Where the original mafic rock was gabbro, the myrmekite in the replacing granite contains coarse quartz vermicules (Figure 1 a). Where the original mafic rock was a calcic diorite, the myrmekite in the replacing granite contains intermediate-sized quartz vermicules (Figure 1 b). And where the original mafic rock was a sodic diorite, the myrmekite in the replacing granite contains fine-textured quartz vermicules (Figures 1 c and d). In current hypotheses for the origin of granite and myrmekite, such a correlation between the composition of adjacent mafic rocks and thicknesses of quartz vermicules in myrmekite is not anticipated or predicted. These correlations are supported by cathodoluminescence microscopy, scanning-electron studies, and electron-probe chemical analyses of transition rocks between the two end members (Collins, 1988a; Hunt et al., 1992).

The replacement of plagioclase by K-feldspar and myrmekite is accompanied by the simultaneous replacement of ferromagnesian silicates by quartz. Because many mafic igneous rocks contain about 70 vol % relatively-calcic plagioclase, the replacements of the relatively-calcic plagioclase by two feldspars and of the 30 vol % mafic silicates by

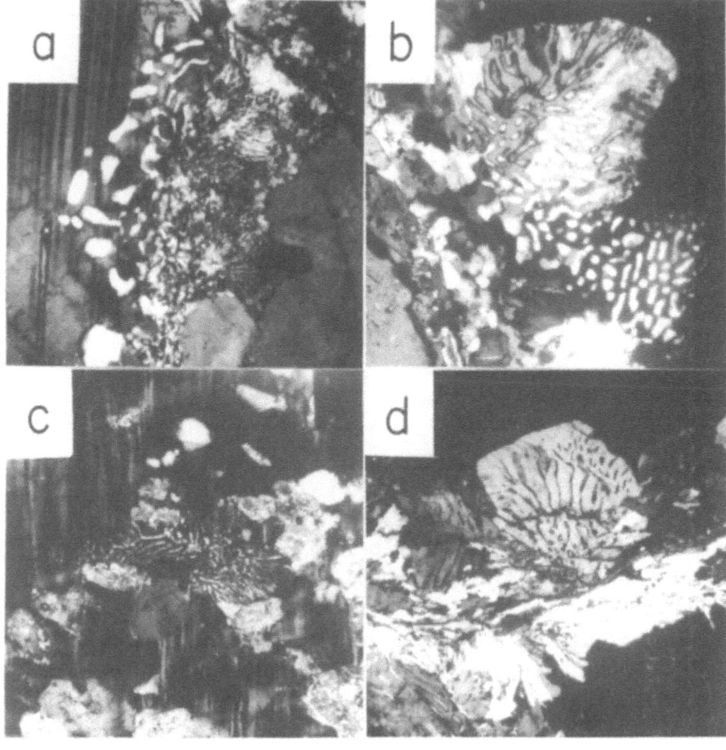

Figure 1. Photomicrographs of myrmekite, an intergrowth of quartz in plagioclase. Quartz occurs as white globs and vermicules in a, b, and c and as black vermicules in d. In photo a, vermicules are coarse; in b, intermediate; and in c and d, fine-textured. Myrmekite occurs as wartlike masses bordering K-feldspar (dark gray or black).

quartz produce a granitic rock containing about one-third sodic plagioclase, one-third K-feldspar, and one-third quartz (Table 1).

The source of silica is from silanes (SiH_4) that were generated by hydrogen reacting with silicon carbide deep in the Earth's mantle (Hunt, 1990; Hunt et al., 1992).

$$4H_2 + SiC \dashrightarrow SiH_4 + CH_4 \tag{3}$$

The source of water and hydrogen to facilitate movements of elements also comes from reactions of silanes with free oxygen, free water, or water in crystal structures.

$$SiH_4 + 2O_2 \dashrightarrow SiO_2 + 2H_2O \tag{4}$$

$$SiH_4 + 2H_2O \dashrightarrow SiO_2 + 4H_2 \tag{5}$$

$$SiH_4 + 2(OH)^1 \dashrightarrow SiO_2 + 3H_2 \tag{6}$$

The source of potassium comes from either deep sources in the mantle (Larin, in press) or from the breakdown of biotite in the mafic rocks into which the deeply-

312

sourced K was transferred. Many Precambrian diorites and gabbros contain as much as 30% biotite, which is a logical source of the K needed to produce the K-feldspar in the granite masses that replace these rocks (Collins and Davis, 1992).

THE ORIGIN OF GRANITIC CONTINENTAL MASSES

The sudden appearance of abundant anorogenic granite in the Earth's crust in the late Precambrian was facilitated by the depletion of K from the mantle at depths from 40 to 120 km (Larin, in press). The formation and degassing of water, as the Earth expanded, created acids which helped to transfer the K from the upper mantle to biotite in diorite and gabbro in the crust. Continued deformation of the crust as the Earth expanded later permitted silanes to replace the biotite with quartz. In that process K was released from the biotite to replace some of the primary, relatively-calcic plagioclase with K-feldspar and myrmekite. The remaining altered primary plagioclase was recrystallized as relatively sodic plagioclase. The recrystallizations and replacements progressively converted early-formed mafic rocks into rocks of more granitic composition. Increased temperatures created by the reactions of silanes with water and oxygen in the rocks locally caused melting that created new magmas that rose as plutons. Following the loss of heat and crystallization of these plutons, renewed deformation allowed the replacements to continue and progressively to convert the newly-formed, more-felsic magmatic rocks into rocks of still more granitic compositions. Thus, the replacement hypothesis does not deny that magmatic rocks exist, but shows how their compositions are progressively changed through time during volume-for-volume replacements prior to possible remelting. The volume-for-volume exchanges, however, are not necessarily at constant volume, because the granitic rocks that are produced have minerals in them of lower average density than were once in the original mafic rocks. Therefore, the granitic continental masses with lower densities rise as diapirs that expand, and their expansions may cause adjacent rocks to be thrust-faulted and piled up on adjacent oceanic crust in imbricate fashion.

ADDITIONAL EVIDENCE AND CONCLUSION

Additional support for a replacement origin of granite of plutonic dimensions and of granitic gneisses is provided by other studies (Collins, 1988a; Collins and Davis, 1992; Hunt et al., 1992). For example, ^{218}Po, ^{214}Po, and ^{210}Po halos in biotite in myrmekite-bearing granites and pegmatites indicate that these rocks cannot have crystallized from magmas because of the short half-lives of these polonium isotopes (Gentry, 1988). The 5 million years of cooling time that is needed before biotite can crystallize from a large magma body is too long to allow sufficient quantities of polonium to remain in the magma to form concentrations of 10^9 atoms that are needed to produce the halos in the biotite. The only way that these Po halos can form is by the decay of ^{222}Rn gas as it moves through cataclastically-broken mafic rocks being replaced by granite.

Second, oxygen isotopic studies show that volcanic rocks have lower $^{18}O/^{16}O$ ratios than occurs in plutonic rocks of equivalent compositions. This relationship is unexpected because magmatic plutonic and volcanic rocks of the same composition formed from melts at high temperatures should have the same $^{18}O/^{16}O$ ratios. Crystal settling of heavy mafic minerals to remove Ca, Mg, Fe, and Al from the melts should enrich the tops of magma chambers in felsic minerals rich in ^{18}O and thereby cause volcanic rocks emerging from tops of plutons to have higher $^{18}O/^{16}O$ ratios than the restite. The reversal of this pattern indicates that the Ca, Mg, Fe, and Al were removed from diorite and gabbro and transferred upward to volcanic rocks during replacing processes so that ^{18}O was concentrated in the silica-rich, granitic, plutonic residue.

Third, Rb-Sr isotopic studies confirm the upward movement of Ca because of the differential movements of Rb and Sr. Rb tends to remain behind in the granite with K in K-feldspar as Sr isotopes go up with removed Ca that was transferred to the mafic volcanic rocks. Because of the tendency for Rb to be retained in granite while Sr is extracted and moved to mafic volcanic rocks, some basalts have higher $^{87}Sr/^{86}Sr$ ratios than can be produced by their coexisting ^{87}Rb contents. The $^{87}Sr/^{86}Sr$ ratios of the basalts indicate impossible isochron ages greater than the age of the Earth. All these data support the hypothesis that large granite masses in the continental crust had histories of replacement during some time in their evolution (Collins, 1988a; Hunt et al., 1992).

Finally, although the sources of Si, K, and water for granitization were not known in the 1940s and their supposed absence was used by magmatists to reject a replacement origin of granite, these sources are now logically understood as part of an expanding Earth hypothesis (Hunt, 1990; Hunt et al., 1992; Larin, in press).

REFERENCES

Carmen, J.H., and Tuttle, O.F., 1964, Experimental study bearing on the origin of myrmekite. *Abstr., Geol. Soc. Amer. Spec. Paper,* 76:29.

Clarke, D.B., 1992, "Granitoid Rocks," Chapman & Hall, London, 283 pp.

Collins, L.G., 1988a, "Hydrothermal Differentiation And Myrmekite - A Clue To Many Geologic Puzzles." Theophrastus Publications S.A., Athens, 382 pp.

_____, 1988b, Myrmekite, a mystery solved near Temecula, Riverside County, California. *Calif. Geol.* 41:276-281.

_____, and Davis, T.E., 1992, Origin of high-grade biotite-sillimanite-garnet-cordierite gneisses by hydrothermal differentiation, Colorado, *in* "High Grade Metamorphic Rocks," S.S. Augustithis, ed., pp. 297-339, Theophrastus Publications, Athens.

Gentry, R.W., 1988, "Creation's Tiny Mystery," 2nd ed., Earth Sci. Assoc., Knoxville, 348 pp.

Grout, F.F., 1941, Formation of igneous-looking rocks by metasomatism: a critical review and suggested research, *Geol. Soc. Amer. Bull.,* 52:1525-1576.

Hunt, C.W., 1990, "Environment of Violence," Polar Publishing, Calgary, 211 pp.

_____, Collins, L.G., and Skobelin, E.A., 1992, "Expanding Geospheres, Energy and Mass Transfers from Earth's Interior," Polar Publishing, Calgary, 421 pp.

Larin, V.N., in press, "Hydridic Earth, the New Geology of our Primordially Hydrogen-rich Planet," C.W. Hunt, ed., Polar Publishing, Calgary.

Leake, B.E., 1990, Granite magmas: their sources, initiation and consequences of emplacement: *J. Geol. Soc. London,* 147:579-589.

Luth, W.C., Jahns, R.H., and Tuttle, O.F., 1964, The granite system at pressures of 4 to 10 kilobars, *J. Geophys. Res.,* 69:759-773.

Michel-Lévy, A.M., 1874, Structure microscopique des roches acides anciennes. *Soc. Fr. Mineral. Crystallogr. Bull.,* 3:201-222.

Phillips, E.R., 1974, Myrmekite - one hundred years later, *Lithos,* 7:181-194.

Read, H.H., 1948, Granites and granites, *in*: J. Gilluly, ed., "Origin Of Granite," Gilluly, J., ed., *Geol. Soc. Amer. Mem.,* 28:1-19.

Tuttle, O.F., and Bowen, N.L., 1958, Origin of granite in the light of experimental studies in the system $NaAlSi_3O_8$-$KAlSi_3O_8$-SiO_2-H_2O, *Geol. Soc. Amer. Mem.* 74, 153 pp.

Whitney, J.A., 1988, Origin of granite: the role and source of water in the evolution of granitic magmas, *Geol. Soc. Amer. Bull.,* 100:1886-1897.

THE PRIMORDIALLY HYDRIDIC CHARACTER OF OUR PLANET AND PROVING IT BY DEEP DRILLING

C. Warren Hunt

1119 Sydenham Road SW
Calgary, Alberta, Canada T2T 0T5

FAULTY FUNDAMENTALS: AN INTRODUCTION

Modern geoscience relies on fundamentals derived as much as a century ago under the severe constraints on observational knowledge of those early days. Brilliant geophysical and geochemical technology today depends for its application on 19th century geology, the principles of which fail to take into account new observational knowledge. From cold fusion to the red shift, much has been written opining the resistance to new ideas in science. Incongruent new data are ignored when theory fails to support observation. Instead of dealing with the fundamentals, scientists prefer to tinker with existing theory.

I will discuss two antique geological ideas that prevail despite evidence that they should be relegated to the archives: *first*, the concept that granitoid rocks have exclusively magmatic-intrusive origins and its corollary, the fiction that there is no possibility for sedimentary-type, clastic lithologies within or below a granitoid body unless tectonics has placed them there, and *second*, the attribution of hydrocarbon origins exclusively to biomass degradation, and its corollary, the notion that hydrocarbons only originate in sedimentary rock where biomass could have accumulated, never beneath it in the crystalline basement or plastic mantle of our planet.

The contrary observational evidence to the first mistaken theory is that unaltered sedimentary rock *does* occur enveloped within granitoids; and metasomatically-evolved "pseudo-metasedimentary" gneissic sequences *are* demonstrably in many cases, derived from mafic igneous host rocks. These phenomena, which are described in detail by Collins (1988) and Hunt et al. (1992), are very inconvenient observations for orthodoxy, and they are routinely ignored.

The contrary observational evidence to the second mistaken theory is the widespread occurrence of hydrocarbons in crystalline terranes worldwide. Methane, for example is a serious problem in mines where metal ores such as uranium, platinum, and gold are found in crystalline rocks. In many places the methane is accompanied by liquid hydrocarbons as well. Since hydrocarbons can certainly not be generated in granitoids and could not have migrated downward from sedimentary cover, they must instead rise from deeper levels of the planet.

THE NEW GEOLOGY OF HYDRIDES

Whereas hydrocarbons in crystalline rocks far removed from biomass sources may appear enigmatic at first glance, this perception is preconception, a refuge when no alternative is immediately apparent. The solution is, in fact, so evident that we must wonder why its discovery has taken so long.

Frontiers of Fundamental Physics, Edited by M. Barone
and F. Selleri, Plenum Press, New York, 1994

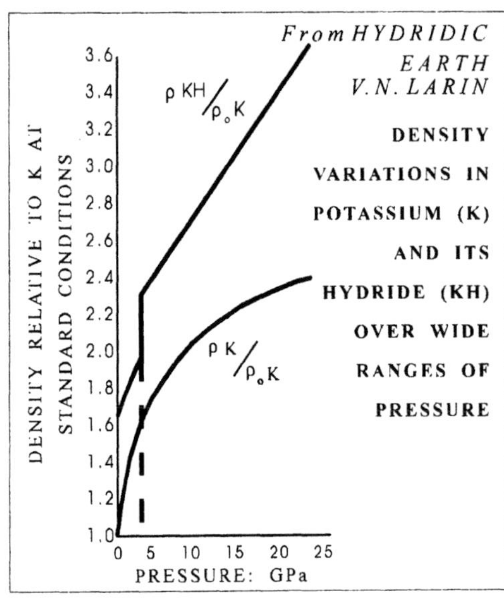

The problem is twofold: to find (1) *a source of hydrogen,* and (2) *a source of carbon* beneath the Earth's crust. Hydrogen in Earth's core has received much recent scientific attention as shown by a growing literature. Larin (1993) advances the proposition that hydrogen saturates Earth's metal core, which he shows is unlikely to be nearly pure iron, as orthodoxy would have it, but more probably, a dense mixture of metals. His high-pressure research demonstrates with impressive finality that *at high pressures the hydrogen proton is forced inside the first electron orbit of metals, where it takes on anion character (H⁻). In this way the metal atom is compacted and densified. It becomes a truly new substance. This is the meaning of "phase-change."* The above diagram illustrates the behavior of native potassium and contrasts it with that of potassium hydride.

The Consequences of Hydride Emanations

An hydridic planetary core is, manifestly, able to release hydrogen when pressure is decreased below thel level critical for retaining it. The critical levels, specific for individual metals, fall in the range of pressures that prevail in the liquid outer core. Thus, with any decompression, hydrogen is evolved and tends to rise into the mantle where it can react with mantle minerals.

The mantle mineral particularly interesting as a hydrocarbon source is silicon carbide, a mineral known to be prominent in kimberlites. The hydrogen reactions with SiC produce hydrides according to the following equations (Aston, 1983; germanium is included because it is next above carbon and silicon in the periodic table):

HYDROGEN REACTIONS WITH SILICIDES OF THE GROUP IVB ELEMENTS

$$4H_2 + SiC \longrightarrow SiH_4 + CH_4 \qquad (1)$$
$$4H_2 + SiSi \longrightarrow 2SiH_4 \qquad (2)$$
$$4H_2 + SiGe \longrightarrow SiH_4 + GeH_4 \qquad (3)$$

OXIDES PRODUCED BY REACTIONS OF SOME HYDRIDES WITH WATER

$$SiH_4 + 2H_2O \longrightarrow SiO_2 + 4H_2 \qquad (4)$$
$$GeH_4 + 2H_2O \longrightarrow GeO_2 + 4H_2 \qquad (5)$$

OXIDATION PRODUCTS OF HYDROCARBON, SILANE, AND GERMANE
UNDER CONDITIONS OF OXYGEN SCARCITY

$$CH_4 + 2O_2 \longrightarrow 2H_2O + C \qquad (6)$$
$$SiH_4 + O_2 \longrightarrow 2H_2 + SiO_2 \qquad (7)$$
$$2GeH_4 + 4O_2 \longrightarrow 2H_2GeO_3 + 2H_2O \qquad (8)$$

OXIDATION PRODUCTS OF HYDROCARBON, AND, SILANEUNDER
CONDITIONS OF OXYGEN ABUNDANCE

$$CH_4 + 2O_2 \longrightarrow 2H_2O + CO_2 \qquad (9)$$
$$SiH_4 + 2O_2 \longrightarrow 2H_2O + SiO_2 \qquad (10)$$

HEAT EVOLUTION FROM HYDROCARBON AND SILANE

$$CH_4 + 2O_2 \longrightarrow CO_2 + 2H_2O$$
Heat: $-19.1 + 94.4 + 2 \times 68.4 = 212.1$ kCal/g/80 mol-eq-wt
$$CH_4 + 2O_2 \longrightarrow CO_2 + 2H_2O$$
Heat: $-11.9 + 201.3 + 2 \times 68.4 = 326.2$ kCal/g/96 mol-eq-wt

When hydrogen mobilizes silicon and carbon from the silicides of Group IVB elements, the reaction products include alkane hydrocarbons, and their analogues, the silanes and

316

germanes (Eq.#1-3). These products, like the hydrocarbons, are all volatile and loaded with latent energy. At surface conditions the manufacture of silicon carbide requires more than 1000°C, the "carborundum" it yields then being a highly stable solid material, far different from the volatile hydrocarbons and silanes, which result after SiC is purged with hydrogen.

Silanes are extremely reactive as well as mobile. They are able to permeate some mineral lattices and substitute their silicon for heavier metals, thus transforming mafic [metal-rich] rock to more felsic [silicon-rich] lithologies by releasing the heavier metals. This process can be compared to the established industrial process of purging impurities from iron with hydrogen. It is known, for example, that most oxygen and carbon can be removed from iron in this way (99.5% and 73-87% respectively). It is less understood that 50% of magnesium can be removed, although this fact has important implications in geology, where a standing conundrum has been the source of magnesium for the dolomitization of carbonate rocks..

The first order of reaction of the silanes as they rise from deep in the mantle is, then, the *mineral transformation stage,* which is active under the anhydrous and anoxic conditions of the upper mantle and lower crust. The second order of reactions is the *hydrous reaction stage.* Silanes react vigorously with water on first encounter, which may be at the Moho or in mid-crustal levels near the ductile/brittle transition. Quartz, heat, and hydrogen [from water and silanes] are evolved (Eq.#4).

The hydrocarbons that accompany these silanes do not react with water and continue toward the surface with the newly-released hydrogen, thus setting up third stage reaction, *volcanism,* which occurs on first encounter with free oxygen in mid- to upper-crustal levels. Gaseous emissions comprise a mixture of H_2O, CO_2, CO, and C [as soot], depending on the amount of available oxygen, and carbon. More heat is released, and sand plus excavated cover rock is ejected.

Silane Systematics

The foregoing gives a picture of the emanation of silicon and carbon, and the release of the latent heat of the silicides from inner geospheres as the volatile hydrides rise through the outer geospheres. The process is one of ongoing creation of continental crust by the fixation of previously-mobilized silicon. This fixation of silicon generates crustal mass of greater bulk and less density than the preexisting mafic rock or overlying cover rock. Whether the less dense mass is a slurry [as in the case of a sand slurry under granite], or a gas-charged magma, any breach of the cover [as with tensional faulting, meteorite impact, or other environmental violence], a massive, *"convective overturning"* follows Hunt et al. 1992). This can take the volcanic forms of tephra ejection, basalt flooding, mud expulsion, or sand slurry emission; and it is followed by subsidence of the former superjacent surface. All such ejection amounts to a swap of masses, a *"convective overturning"* of solid Earth.

Massive *"convective overturns"* litter the planet. Conventional volcanism, tephra and lava, are the most obvious of these; and they will not be discussed here. Quartz sand produced in the *hydrous stage* of silane reactions will be our focus. Many sand deposits of pure quartz sand and silt of uniform grain size are known that have no apparent source but do have associations with volcanism and magmatism. It is my interpretation that this sand and silt first accumulated below the brittle crust by silane reactions with water. Silane infusion must have been restricted sufficiently that general meltdown and ensuing magmatism was avoided. In time and with ongoing infusion and high gas pressure the sands were ejected volcanically. It should be noted that silane-water reactivity that produces a sand slurry with dissolved hydrogen or a gasified magma, also dilates the brittle cover rock and creates the conditions for volcanic eruption.

Under a crustal dilation, *pressure buildup mitigates horizontal hydraulic breakout* (Bailey, 1990). This natural process is analogous to hydraulic fracturing in oilfield practice. The induced rupture propagates outward from an overpressured source with explosive rapidity, entraining entrapped sand or gassified magma as the case may be. At the surface the manifestation of the phenomenon is an earthquake. In the subterrane the manifestation is a permanent sill of magmatic rock, quartz sand, or quartzite. The great Siberian trapp sills probably result after silane melting of country rock and the subsequent injection of the melt into planar fractures that have this origin. This explains the seismically-visible sub-horizontal "discontinuities" that so often occur in granitoid terranes, usually radiating outward from volcanic centers. Sand sills of lateral expanse comparable to the trapp sills are not to be expected, because magma is naturally fluid and flows easily into open fractures, whereas sand injection is resisted by bridging. However, hydride systematics *do* predict sand injectites in crystalline terranes, where sand was forced by subcrustal injection preceding volcanism in Proterozoic time..

By contrast to the sand injectites, which at this writing are hard to prove, the field evidence for massive sand-slurry emissions on the surface is overwhelming. Pure quartz sand bodies [up to 50,000 km^3 (Hunt et al., 1992)] provide mute testimony. These sand deposits as well as their associated volcanic vents have escaped recognition by geologists up to now, because their crater terranes may be almost devoid of conventional volcanic features and because the ejected sand, once on the surface, is soon redistributed by erosional processes. In any case, volcanic features in association with sands may support but are not diagnostic of endogenic origins for the sands. The distinctive feature about silane-generated sands is their mineral purity. No other natural agency can produce that feature. ***Pure quartz clast character is diagnostic of silane activity.***

Hydrocarbon Systematics

Petroleum accumulations result from the carbide side of SiC in contrast to the silanes, which are products of the silicide component. Whereas there are many silicides of the major metals [Fe, Mg, Ni, etc.], there is only one silicide of carbon. Therefore, carbon minerals are intrinsically less abundant than silicon minerals. Petroleum, the hydride of carbon, is the mobile intermediary between the SiC of the inner geospheres and the crustal carbonatites, carbonates, and oxides of the outer geospheres. The mobile phase may accumulate if reservoir conditions occur along the way. Such conditions in granitoid rocks include injectite sands and fracture systems. Metal ores are often associated with the hydrides of carbon. Similar mobilities among metal hydrides and those of carbon lead to syndeposition. It is, in fact, a remarkable symbiosis that carbide/hydride systematics can produce not only metal orefields but oilfields and even the reservoir rock for the latter.

The Petroleum Prospect in Granite at Ft. McMurray, Alberta

I will discuss what I regard as the prime prospect for petroleum in granitoid rock anywhere in the world. The evidence starts with the 1.3 billion-year-old Athabasca sandstone, a quartz sand ejectite of $\approx$50,000 km^3 that followed a volcanic debris flow from a vent named the Carswell crater in northern Saskatchewan, Canada. The sand was expelled in a series of pulses to the surface, where it settled around the vent and slumped eastward by gravity. Its weight depressed the granite crust into a bowl approximately 1.4 km in depth. From the fact that the sand flowed mainly eastward, we can deduce that the eruption must have been on the east flank of the crustal dilation from which it burst.

These sands must have been generated under the brittle Crust over a long period of pre-eruption time with concomitant emplacement of hydraulic injectites around the crustal dilation on a grand scale The Carswell eruption would have emptied the sand chambers of the east flank of the welt, but sand accumulations farther west should have been unaffected and are, perhaps, to some extent still present to the west of the Carswell feature in the crustal granite after 1.3 billion years. To the west around Ft. McMurray, Alberta, the unroofed oilfield, [known generally as the "Athabasca tar sands"] occupies a huge area. The bitumens occur in early Cretaceous "McMurray sands," which resemble the Athabasca sands and likely represent a later eruption [~110 my ago] from the same crustal depth. This later sand was reworked after ejection in a north-south-trending estuary. The oval topology of the present sand distribution suggests the emission occurred through a north-south-trending tension fault [the *"tar sands fault"*] approximately on the present Athabasca River course north of the town of Ft. McMurray.

After the sands were reworked by surface waters, the welt beneath them subsided, and marine clastics buried them deeply. Volatile hydrocarbons entered the buried sand body in late Cretaceous time. Then, in Cenozoic time the crust dilated again and the oilfield was gradually exhumed and unroofed. Near-surface bacterial action turned the formerly hydrogen-rich petroleum into hydrogen-poor varieties, the "tars" we find today.

Injectite sands that were emplaced before the ejection of the Athabasca sands should still be under the former crustal welt on which the tar sands occur. Some of these were likely ejected to create the McMurray sands and some of the accumulated petroleum drained to provide the present bitumens. Sands and bitumens adjacent to the drained area should be unaffected. If 1.3 trillion barrels leaked out and large unleaked areas are still present, as seems to be the case, prospectiveness is still very great. High gravity petroleum and gas are prospective at deeper levels and lower gravities at shallower levels, depending on bacterial degradation factor.

In Conclusion

Hydrogen systematics is seen as the driver of the entire evolutionary process on our planet. The systematics of interactivity between hydrogen and silicides [including SiC] define Earth's endogeny. This includes the creation of oil- and gasfields and metal orefields. Because petroleum can survive under surface conditions where silanes and the hydrides of metals cannot, the proving of petroleum in granitoid rocks, far from biomass accessibility should prove beyond reasonable doubt that biomass is unnecessary for the generation of hydrocarbons. If quartz sands, quartzites, or very high-silica granites are associated with the petroleum, silane precursors will be emphatically indicated; and the case can be rested for the: hydridic Earth, for hydridic processes as the movers and shakers on our planet, and for the new geology they reveal.

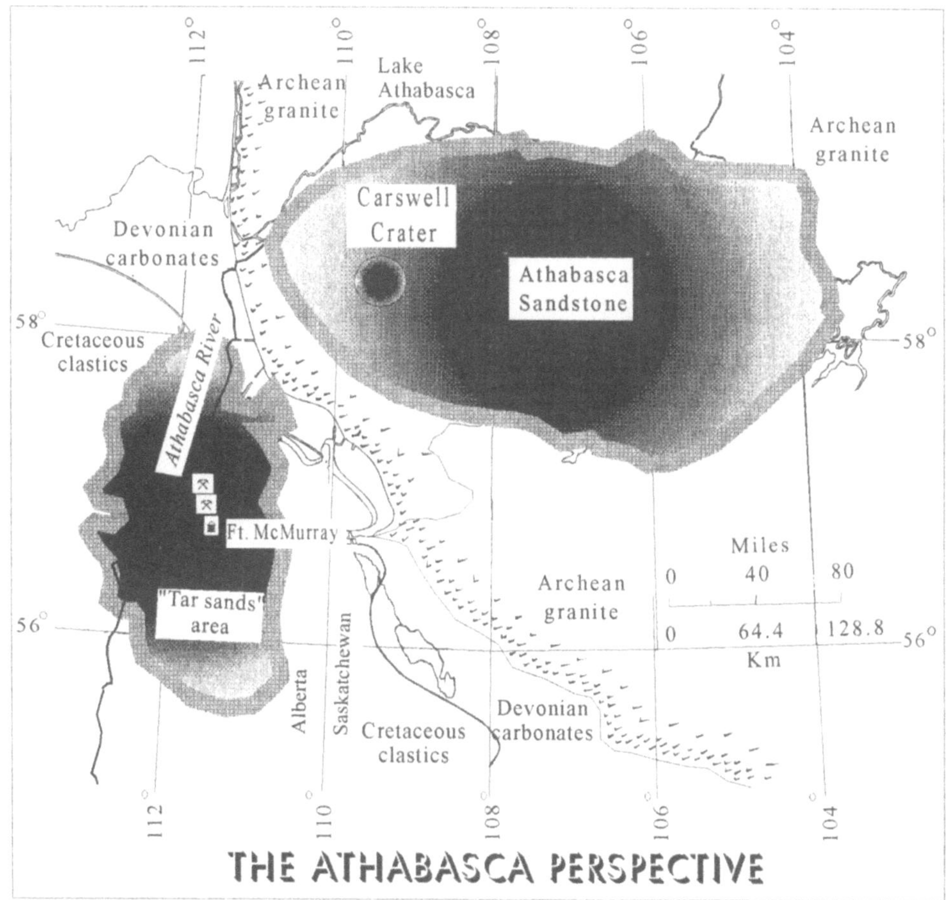

References

Aston, S.R., 1983, "Silicon geochemistry and biogeochemistry" Academic Press, London,

Bailey, R.C., 1990, "Trapping of aqueous fluids in the deep crust," *Geoph. Res. Lett. 27:191*

Collins, L.G., 1988, "Hydrothermal differentiation," Theophrastus Press, Athens

Hunt, C.W., 1992, Chap. VIII in "Expanding geospheres: energy and mass transfers from Earth's interior," C.W. Hunt, ed. Polar Publishing, Calgary

Larin, V.N., 1993, "Hydridic Earth: the new geology of our primordially hydridic planet," Polar Publishing,. Calgary

POSSIBLE RELATION BETWEEN EARTH EXPANSION
AND DARK MATTER

Stanisław Ciechanowicz[1], and Jan Koziar[2]

[1]Institute of Theoretical Physics
[2]Institute of Geological Sciences
Wrocław University, pl. M.Borna 9
50-204 Wrocław, Poland

INTRODUCTION

The theory of earth expansion has been developed since long from several basic observations. These are, first of all, mutual moving apart of all the continents, lenghtwise extension of the ocean ridges, radial growth of the oceanic lithosphere around Africa and Antarctica, extension and splitting of continental margins, increase of the Pacific Ocean (similar to other oceans), and deep rooting of the plates and ocean ridges which has recently been proved by seismic tomography.

The main objection which has been raised against earth expansion are unknown couses of this process. The objection is incorrect because we do not need to know a physical cause of a phenomenon, to be able to prove it as a fact. Neverthless, the recently developed theory of dark matter can give a casual explanation for earth expansion. What is more, this process appears as a being sought by physicists a tangible effect of the existence of dark matter.

GEOMETRICAL RANGE OF EARTH EXPANSION

Approximate magnitude and rate of earth expansion can be assumed from the extensional development of all oceans in the Meso-Cainozoic. This gives about double increase of the earth radius in the last 200 mln years. It was already supposed long ago, that the size of the earth is growing exponentially (Hilgenberg, 1933). Such a growth was supported by calculations based on paleomagnetic data (Hilgenberg, 1962; Neiman, 1962) and on isochrons of oceanic lithosphere (Koziar, 1980; Osipisin with Blinov, 1987). Furthermore the graph obtained by Koziar (Figure 1) is based on intracontinental reconstruction. It was confirmed by Vogel (1991) and we will use it for further calculations.

The graph is defined by the formula

$$R_t = R_1 + R_2\, e^{rt} \tag{1}$$

where: $R_1 = 2800$ km. $R_2 = 3570$ km. $r = 0.00725$ Ma^{-1}.

Frontiers of Fundamental Physics, Edited by M. Barone
and F. Selleri, Plenum Press, New York, 1994

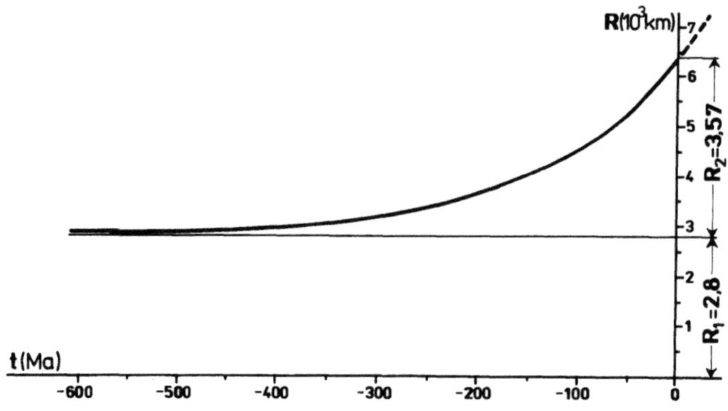

Figure 1. Growth of the earth radius

Differentiation of (1) gives the rate of earth expansion

$$dR/dt = r\,R_2\,e^{rt} \qquad (2)$$

Recent growth rate obtained from (2) is

$$r\,R_2 = 2,6\ cm/yr \qquad (3)$$

This value (Koziar, 1980) falls between the results of Blinov (1987) - 2,4 cm/yr and Parkinson - 2,8 ± 0,8 cm/yr (vide Carey, 1988).

GROWTH OF THE EARTH MASS

The growth of the earth mass is suggested by many authors (Jarkowski, 1888, 1889; Hilgenberg, 1933, 1974; Kirillov, 1958; Nejman, 1962; Wesson, 1973; Carey, 1976, 1983; Veselov, 1976, 1981; Blinov, 1983; Ivankin, 1990). The growth can be inferred from two empirical circumstances . First, the recorded increase of the earth radius assuming the constant earth mass, implies about four times greater gravitational surface acceleration at the turn of the Palaeozoic and Mesozoic. It would be too large for the tetrapods living at that time. Second, the earth growth is accelerated. A simple decrease of mass density, at the constant earth mass, would rather suggest slowing down the expansion rate.

The exponential growth of the earth size suggests an exponential growth of the earth mass (Hilgenberg, 1933, 1974; Carey, 1983; Blinov, 1983).

Recently the Czech geologist Hladil (1991) published the results of his detailed investigations on Upper-Ordovician impact structures made by dropstones from drifting icebergs Hladil gives the quanitative conclusions that 450 mln year ago the surface gravitationa' acceleration was equal to 15 ms^{-2}. Using this value and R = 2900 km - the Upper-Ordovician earth radius (1), G - the gravitational constant, and with the mass formula M = gR^2/G we obtain the mass of the earth 1.89 x 10^{27} g, i.e. 3.16 times lesser than the recent value (5.98 x 10^{27} g) and the mean density 18.15 g/cm^3, i.e. 3.3 times greater than the recent value (5.52 g/cm^3). Then, we record both, the growth of the earth mass and the decrease of its density.

322

The possible hyperbolic decline of the gravitational constant G, due to Universe expansion, as suggested by Dirac (1937), is relatively small over the period of the last 450 mln years, and then we assume G as constant in the calculations above.

Now we can try to find the exponential function of the earth mass growth for the whole Phanerozoic. First, we need to find the asymptote of this function. The initial earth volume defined by (1) is 92×10^9 km^3. Because the Ordovician mean density is close to the asymptotical value, we can (avoiding a greater error) multiply by it the above initial volume. We obtain in this way the initial mass of the earth equal to 1.67×10^{27}g.

Having this value, together with the Ordovician and recent earth mass, we are able to obtain the whole exponential function

$$M(t) = M_1 + M_2 e^{mt} \tag{4}$$

where $M_1 = 1.67 \times 10^{27}$g, $M_2 = 4.31 \times 10^{27}$g, m = 0.00652 Ma^{-1}. The function is shown in Figure 2.

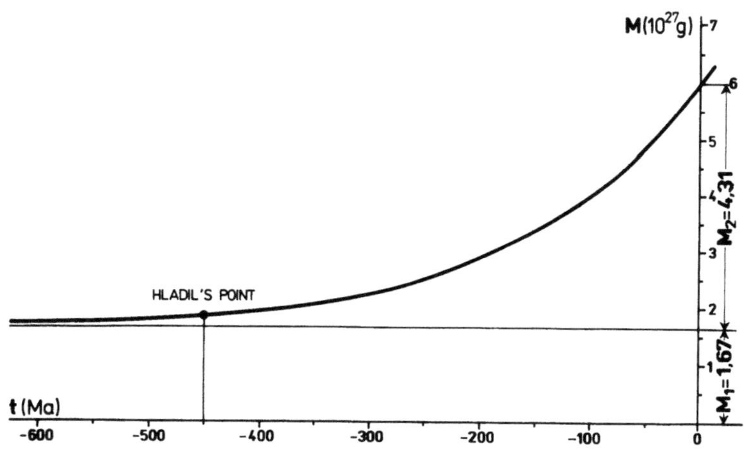

Figure 2. Growth of the earth mass.

Differentiation of (4) gives the earth mass growth rate

$$dM/dt = m\, M_2\, e^{mt} \tag{5}$$

Recent earth mass growth rate is

$$m\, M_2 = 2.82 \times 10^{19} g/yr \tag{6}$$

By reducing this value to 1 cm^2 of earth surface S, we obtain

$$m\, M_2/S = 5.5\; g/(cm^2\, yr) \tag{7}$$

The growth of the earth mass is possible to be measured by satellite geodesy. Using the method of SLR (Satellite Laser Ranging), the geocentric gravitational constant, GM, has already been observed with high accuracy, (Smith et al., 1985, 1990). The increase of the GM can be noticed in the data. We find its value corresponding to our results.

DARK MATTER AND EXPANDING EARTH

The majority of the authors mentioned above who are convinced about the earth mass growth, have postulated that the substance of yet unknown kind enters the earth interior, and converts into normal terrestrial matter. Here, we shall point out that the recent astrophysics concept of dark matter provides a candidate for this type of substance. Some physicists propose dark matter in the form of highly penetrating neutral particles to explain the well known problem with the more than 90% deficiency of the Universe mass. Until now, many objects of various kinds have been considered as possible components of penetrating dark matter. Among others, these include neutrinos and mini black holes with the mass of the order of 10^{12} kg.

Since 1986, a series of papers suggesting gravitation and collisional capture and accumulation of dark matter inside the cores of the sun and planets have been published. Here we mention only the papers dealing with the capture of dark matter by the core of the earth: Freeze, 1986; Krauss et al., 1986; Gould, 1987, 1988, 1991, 1992a, 1992b; Giudice and Roulet, 1989; Kawasaki et al., 1992. These authors treat the iron atom cores in the earth core as the main medium capturing by collision the so called WIMPs (Weakly Interacting Massive Particles). The main effect of the dark matter capture would be its anihilation and emission of neutrinos. Experimental results do not confirm the anticipated emission (Mori et al. 1992) but there is a possible different capture mechanism and its final effect.

None of the papers mentioned above do not take into account the possibility of the transistion of dark matter into baryonic matter in contact with the latter. None of these consider the earth expansion theory either. At the same time, the significant increase of the earth volume and mass found in geological data can be that, sought by physicists, empirical effect of interaction between dark and visible matter and another indicator of the existence of the former.

Assuming a connection between earth expansion and capture of dark matter by the earth core, we have in general two possibilities for the conversion mechanism. First, dark matter directly reacts with earth matter. Second, the earth matter only catalyses the change of dark matter into visible matter.

Previously, we presented the global annual amount of new earth matter (6), and now we shall discuss some qualitative effects. Considering more than the threefold growth of the earth mass since the Ordovician, we come to the conclusion that the main earth elements Fe, O, Si, Mg must be the products of dark matter conversion.

It is also possible to estimate the energy output due to the conversion process necessary to drive the earth expansion. We cut the radial cone of the earth with area of 1 cm^2 at its surface. The expanding earth needs to move the cone up by 2.6 cm every year (2). For the cone of differentiated density (4.2 g cm^{-3} - mean mantle density, and 12.2 g cm^{-3} - mean core density) we find the energy 5.6×10^{12} erg (1.34×10^5 cal) per 1 cm^2 of the earth surface per year. This amount of energy is much larger than the geothermic flux ($47.3 \text{ cal cm}^{-2} \text{yr}^{-1}$) and almost equals the mean flux of sun energy at the earth surface ($1.32 \times 10^5 \text{ cal cm}^{-2} \text{yr}^{-1}$).

Since there is 5,5 g mass increase by every square cm of the earth surface per year (7), then for each gram of the new matter the energy release should be about 2.45×10^4 cal g^{-1}. This is about 7 times more than in the most effective chemical reaction of oxygen with hydrogen (3.6×10^3 cal/1g H_2O). On atomic scale this ratio reads: $= 1.07$ eV per single nucleon or 59.9 eV per single atom of iron.

In these estimations, we have considered only the energy needed to overcome the gravity forces, and have neglected the energy needed to overcome the mechanical resistance of earth matter. We can expect that this energy which drives the endogenous geological processes is comparable to the value calculated above.

Indirectly, we have evaluated dark matter capture using the magnitude of the earth mass growth determined by geological methods. However, the resulting amount of dark matter mass is much larger than the theoretical estimates given in the papers cited above. For instance, Gould's analyses imply an annual global capture of about 30 kg. In these papers, the conjectured capture machanism resulted from the standard elementary particle interaction combined with the gravitational attraction. Then to explain the earth expansion, we admit for the earth much more effective capture mechanism, a probably connected with the earth magnetic field. This would explain an extraordinary behaviour of the earth among other near sun planets the expansion of which is insignificant (unknown exeption is Venus). So, all these planets have minute magnetic field compared with the earth. In this scenario, particles of dark matter must obviously have a tiny magnetic moment.

The proposed capture mechanism determines the optimal conditions for using the recently designed low-temperature particle detectors, see e.g. in Sadoulet (1988), for detection of dark matter.They should be inserted in traps of strong magnetic field deeply underground, in the neighbourhood of the earth magnetic poles.

Finally, we must point out that the described above possible conection between dark matter and earth expansion is not a necessary condition for the latter, which is evidenced independently by geological data.

REFERENCES

Blinov, V.F., 1983, Spreading rate and rate of expansion of the earth, *in:* "Expanding Earth Symposium", Carey S.W. (ed), Sydney.

Carey S.W., 1976, "The Expanding Earth", Elsevier, Amsterdam - Oxford - New York.

Carey, S.W., 1983, A null Universe, *in:* "Expanding Earth Symposium", Carey S.W. (ed), Sydney.

Carey, S.W., 1988, "Theories of the Earth and Universe", Stanford University Press, California.

Dirac, P.A.M., 1938, The cosmological constants, *Nature* 139:323.

Freese, K., 1986, Can scalar neutrinos or massive Dirac neutrinos be the missing mass?, *Phys. Lett.*, 167B:295.

Giudice, G.F., and Roulet, E., 1989, Energetic neutrinos from supersymetric dark matter, *Nu. P. B.* 316:429.

Gould, A., 1987, Resonant enhancements in weakly interacting massive particle captured, by the earth, *Astrophys. J.* 321:571.

Gould, A., 1988, Direct and indirect capture of weakly interacting massive particles by the earth, *Astrophys. J.* 328:919.

Gould, A., 1991, Gravitational diffusion of solar system WIMPs, *Astrophys. J.* 368:610.

Gould, A., 1992 a, Cosmological density of WIMPs from solar and terrestrial annihilations, *Astrophys. J.* 388:338.

Gould, A., 1992 b, Big bang archeology: WIMP capture by the earth at finite optical depth, *Astrophys. J.* 387:21.

Hilgenberg, O.C., 1933, "Vom wachsenden Erdball", Berlin - Charlottenburg (Publ. by the author).

Hilgenberg O.C., 1962, Palaopollagen der Erde, *Neues Jahrb. Geol. Palaontol. Abhandl.*, 116(1):1.

Hilgenberg, O.C., 1974, Debate about the earth. The question should not be: "drifters or fixists?" but instead: "earth expansion with or without creation of new matter?", *Geotekt. Forsch.*, 45:1959.

Hladil, J., 1991, The Upper Ordovician dropstones of Central Bohemia and their paleogravity significance, *Bull. Czech. Geol. Surv.*, 66/2:65.

Jarkowski, J., 1888, "Hipothese cinetique de la gravitation universelle en connexion avec la formation des elements chimiques", Moscow (chez l'author).

Jarkowski, J., 1889, "Vsemirnoje tjagotenije kak sledstvo obrazovanija vesomoj materii vnutri nebesnych tel", Moskwa (publ. by the author).

Kawasaki, M. et al.,1992, Can the strongly interacting dark matter be a heating source of Jupiter?, *Progress of Theoretical Physics*, 87/3:685.

Kiriłłow, I.V., 1958, Gipoteza razvitija Zemli, jeje materikov i okeaniceskich vpadin, Avtoreferat doklada, *Bjul. Mosk. Ob. Isp. Prirody Ot. Geol.* 2:142.

Koziar, J., 1980, Ekspansja den oceanicznych i jej związek z hipotezą ekspansji Ziemi, *Sprawozdania Wrocławskiego Towarzystwa Naukowego*, 35 B:13.

Krauss, L.M., Średnicki, M. and Wilczek, F., 1986, Solar system constrains and signatures for dark-matter candidates, *Phys. Rev.,* D33:1079.

Mori, M., Hikasa, K., Nojiri, M.N., Oyama, Y., Suzuki, A., Takahashi, K., Yamada, M., Takei, H., Miyano, K., Miyata, H., Hirata, K.S., Inoue, K., Ishida, T., Kajita, T., Kihara, K., Nakahata, M., Nakamura, K., Sakai, A., Sato, N., Suzuki, Y., Totsuka, Y., Koshiba, M., Nishijima, K., Kajimura, T., Suda, T., Fukuda, Y., Kodera, E., Nagashima, Y., Takita, M., Yakoyama, H., Keneyuki, K., Takeuchi, Y., Tanimori, T., Beier, E.W., Frank, E.D., Frati, W., Kim, S.B., Mann, A.K., Newcomer, F.M., Van Berg, R. and Zhang, W., 1992, A limit on massive neutrino dark matter from Kamiokande, *Physics Letters* B 289:463.

Nejman, W.B., 1962, "Razsirjajuscajasja Zemlja", Gosudarstvennoje Izdatielstvo Geograficeskoj Literatury, Moskva.

Osipisin, N.J. and Blinov V.F., 1987, Vozrastnaja zonalnost okeaniceskoj kory i jeje svjaz s razsirenijem Zemli, *Bjul. Mosk. O-va Isp. Prirod. Otd. Geol.* 62/4:18.

Sadoulet, B., 1988, Cryogenic detectors of particles; Hopes and chalenges. *Trans. Nucl. Sci.* 35:47.

Smith, D.E., Christodoulidis, D.C., Kolenkiewicz, R., Dunn, P.J., Klosko, S.M., Torrence, M.H., Fricke, S. and Blackwell, S., 1985, A global geodetic Reference Frame from LAGEOS Ranging (SL5. 1AP), *Journ. Geophys. Res.* 90/B11:9221.

Smith, D.E., Kolenkiewicz, R., Dunn, P.J., Robbins, J.W., Torrence, M.H., Klosko, S.M., Williamson, R.G., Paulise, E.C., Douglas, N.B. and Fricke, S.K., 1990, Tectonic motion and deformation from satelite laser ranging to LAGEOS, *Jour. Geophys. Res.* 95/B13:22013.

Vogel, K., 1991, Postertekst, 142 Hauptversamlung der Deutshen Geologische Gesellschaft von 3-6 October in Bremen.

Veselov K. E., 1976, Gravitacjonnyje pole i geologiceskoje razvitije Zemli, *Sov. Geol.,* 1976/5:70

Veselov K. E., 1981, Slucajnyje sovpadenija ili javlenija prirody, *Geof. Zurn.,* 1981/3:5.

Wesson,P.S., 1973, The implications for geophysics of modern cosmologies in which G is variable, *Q. Jl R.Soc.* 14:9.

EARTH EXPANSION AND THE PREDICTION OF
EARTHQUAKES AND VOLCANICISM

Martin Kokus

P.O.Box 119
Hopewell, PA. 16660, U.S.A.

Abstract. This paper argues that the accuracy of earthquake prediction would improve rapidly if earthquakes were analyzed with an expanding earth model instead of the prevailing one. It describes patterns of
seismic events that the present theory cannot begin to explain. These
patterns are that similar seismic events tend to reoccur at similar times
in the sunspot cycle and/or similar positions of the earth, moon and sun
which are not related to any maximum tidal stresses. It is then shown
that the most dominant pattern is a direct consequence of earth
expansion. This paper is not intended to be a rigorous treatment of
seismicity on an expanding earth, but a largely qualitative discussion
whose main purpose is to stimulate further research into methods of
seismic prediction that are presently being ignored.

INTRODUCTION

Within this century over 2,000,000 people died in earthquakes. This
exceeds all other causes of violent death during peacetime. Unlike other
sources of human suffering such as famine or war; its solution lies
largely within the scientific realm. But unfortunately, the recent
history of seismic prediction has not been encouraging.

With a few predictions in the early seventies that were accurate
enough to permit evacuations, it seemed that practical seismic prediction
would soon be commonplace. Why then have there been no practical
predictions made by official agencies in the last fifteen years? This
should be especialy curious considering the mushrooming of research
budgets and available data.

Perhaps it should be noticed that during this time plate tectonics[1]
has come to dominate the field. All of the ad hoc formulas that were
employed with some success now had to be reinterpreted to conform to the
new paradigm. Several empirical relationships were now in disfavor
because there was no way to incorporate them into plate tectonic theory.
Chief among the newly disgarded were many papers suggesting that some
particular volcano or earthquake region tended to erupt more frequently
at some combination of solar activity; time of year; or lunar phase,
distance, and declination. These correlations are loosely referred to as
earthquake signatures or cosmolocations.[2] If these conditions were such
that they approximately coincided with a high or low tide, they were
interpreted as tidal triggering and relegated to a minor role in the

plate tectonic explanation of seismicity. If the studies revealed a relationship that was not obvious, they were then treated as folklore regardless of how professionally they were done.

It is these disregarded studies that are the subject of this paper. It is my aim to show that they represent a cohesive body of research that is consistent with the earth expansion hypothesis.

The sun and moon do more than create tides on the earth. Their tidal bulges change the moments of inertia of the entire earth and create an even greater change in the moments of inertia of individual plates. These changes accelerate and decelerate the entire earth and try to accelerate and decelerate different plates and hemispheres creating major and minor torsions. Solar activity and weather can also effect the moments of inertia and rotation. Solar flares excite large scale turbulence increasing the atmosphere's moment of inertia which creates an atmospheric drag on the earth, slowing it measureably. Long term pressure patterns, such as *El Nino,* can lower sea level in one area for a long time changing its moment of inertia. Even though the area is small, the effect can be measured in global rotation.

All of the torsions caused by the above phenomena are small and insignificant with respect to the forces that are required to move the plates about in tectonic theory. But what if the plates are moved about by global torsions fueled by an expanding earth? The expansion produces a torsion about the equator because the northern hemisphere, containing more continents, has a higher moment of inertia than the southern hemisphere. The torsion would be modulated by the tidal bulges. The principle modulation is of the same period as and in phase withthat measured for equatorial volcanic eruptions.

Before we take an in depth look at the effects of tides and solar activity on an expanding earth, we will take a short look at an earthquake signature and the type of data that is being ignored.

AN EARTHQUAKE SIGNATURE: PARKFIELD, CALIFORNIA, USA

During the century before the advent of plate tectonics, it was quite fashionable for researchers in seismic prediction to look for a combination of lunar and solar variables that correlated strongly with seismic activity in a given area. Perhaps, the best way for the reader to understand the concept is to examine one of the classical examples. The authors choice not only illustrates the nature of the geologic phenomena, but it also illustrates the current nature of geology.

Parkfield is located on the San Andreas Fault midway between San Francisco and Los Angelas. The Parkfield segment of the fault was chosen by the United States Geologic Survey for in depth study because it was the location of five, maybe six, similar quakes which happen about every 20 years. This catalog was chosen collectively by a group appointed by the Geologic Survey, so there can be no charge of *aposteriori* statistics leveled against the statistical significances found in it.

The accepted explanation of the quasi-periodicity is that mantle convection is pushing the western side of the fault northward. The motion is not smooth because there is a kink at Parkfield and it takes about twenty years to build sufficient stress to break through the kink. The main objection to this is that out of five intervals between quakes, one is ten years short and another is ten years too long. Still, this was the basis for the only earthquake prediction sanctioned by the National Earthquake Prediction Council. The prediction was that a magnitude 6.0 earthquake would occur on this segment during a 10.4 year period, centered on January 1988 with a confidence level of 95%. The prediction failed.

But now, let us look at the earthquakes signature (Fig.1 and 2). If we plot the quakes by the lunar elongation (angle between the sun and moon

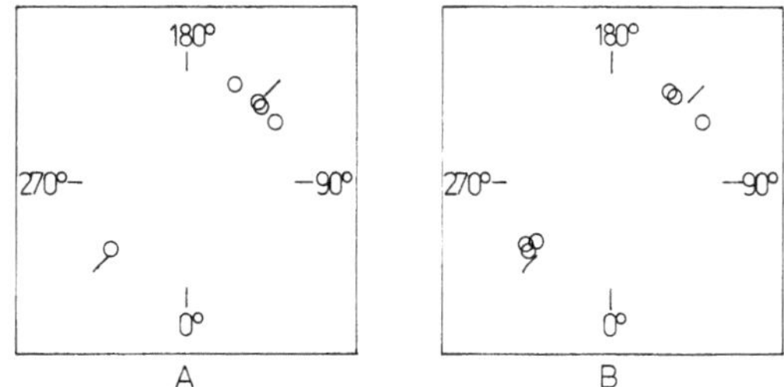

Figure 1. Lunar elongation (in degrees) at time of earthquakes on Parkfield segment of San Andreas Fault (A) and Imperial Fault (B).

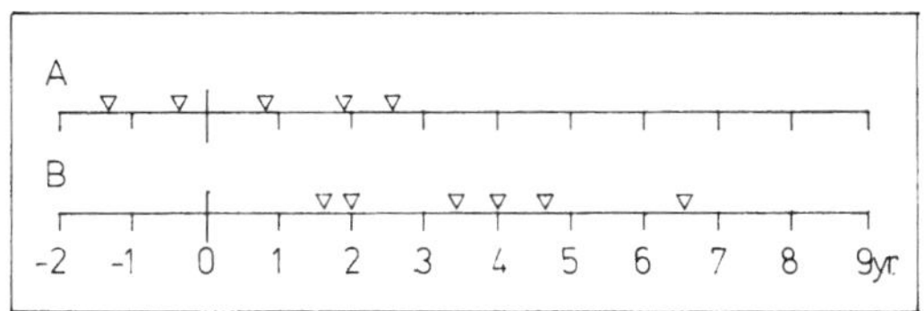

Figure 2. Years after minimum of sunspot cycle for earthquakes on Parkfield segment of San Andreas Fault (A) and Imperial Fault (B).

measured from the earth) when they occurred, we see a very tight clustering around 134.5° and 314.5° with random probability, Pr=0.005. On a yearly scale they cluster around Apr.23 with random probability less than 0.1; and they cluster 0.75 years after sunspot minimum with random probability less than 0.1.[3] These three probabilities could bracket the time of a recurring quake with greater accuracy than everything that has come out of the Parkfield project.[4] So far the only official comment on these relationships is that they have found no significant tidal correlation.

Also included is data[5] from the portion of the Imperial fault that is not under the Gulf of California (that part was excluded because tidal loading might dominate under water, so a comparison would be dubious). The Imperial Fault lies southeast of the San Andreas and has the same orientation and motion. These quakes cluster about the exact same phase angles with Pr=0.001 (Fig.1). They clustered 3.7 years after sunspot minimum with Pr= 0.04 (Fig.2).

These signatures were first discovered by Allen (1936). Since then they were rediscovered by Bagby (1975), Berkland (1988), Kokus (1988a) and an untold number of amateurs.

SEISMIC SIGNATURES: A GLOBAL SYNTHESIS.

A comprehensive review of seismic signatures has been undertaken elsewhere (Kokus 1989 and unpublished,a). This will be an attempt to seek a common thread through 184 claims of lunar and solar correlations that were done over a hundred year period in many countries by people with very different training.

Tidal triggerring and tidal loading. To be sure, a large number (35) of these studies are correlations (27) between tides and seismic activity or denials (8) of such correlations. The effect is easy to detect, especially along coastlines and mid-oceanic ridges. But what should have been curious is that while it is fairly easy to find a corelation for

diurnal and semidiurnal tides, the phase of tide that correlates varies
over the entire range as researchers move from region to region. This can
be easily explained if one looks at the wealth of accelerations as well
as displacements that the tides can produce.

But it gets more difficult to explain as we get to fortnightly tidal
cycles. While there are areas where seismic activity correlates with the
fortnightly tide maximums, there are correlations with the other phases
that make little sense. There appears to be one at fortnightly minimum
when the tides are essentially flat; that is, the difference between the
diurnal high and low tides is minimized. The situation referred to here
is when the lunar and solar tides partially cancel and a cross section of
the earth in the plane of the eccliptic is closest to a circle. While it
could be argued that this is the maximum tidal acceleration due to the
fortnightly term, this acceleration is dwarfed by the diurnal and
semidiurnal accelerations.

Correlations with long period tidal cycles such as the 18.6 year
declinational cycle or 8.85 anomalistic cycle rarely show correlations
between the tidal peaks and seismic activity. Periodograms for these
cycles often show well defined peeks but there are usually two of them,
neither one at the maximum tide.

Nontidal lunar, solar correlations. After the lunar phases that cause
tidal maximums (full and new moon), the most common phases associated
with seismic activity are the quarter phases. Wood (1918), the Hawaiian
Volcano Observatory (1927), Jaggar (1920-1945), Davidson (1938), Johnston
and Mauk (1972), Hamilton (1972), Sauers (1986a), Ritter (1987), Kokus
and Ritter (1988), and Kokus (1988b) have all demonstrated a seismic
preference for the quarter phases of the moon in either a limited area or
for a global sample.

In the definitive work on volcanism and fortnightly cycles, Mauk and
Johnston (1973) analyzed all 680 volcanic eruptions between 1900 and
1971. For the entire sample there were significant peeks at the maximum
and minimum of the fortnightly tidal cycle with the maximum peek being
stronger. The eruptions at minimum were more likely to occur in basaltic
volcanoes (as opposed to andesitic) and in the Tethyan Torsion Zone.[6]

Annual and semiannual tides are due to changes in the sun's declination and distance. Their stresses are small in comparison to those associated with the moon; but their periods show up quite strongly and unambiguously in earthquake statistics. Conrad (1933), Davison (1938), Morgan
et al (1961), Eggers and Decker (1969), Schneiderov (1973), DeSabbatta
and Rizzati (1977), McClellan (1984), and Stothers (1989) all showed
annual periods in seismic activity. Spalding (1915), Conrad (1933),
Davison (1938), and Schneiderov (1973) reported semiannual periods.

In studies of long lunar periods, we again meet the unexpected. In the
18.6 year lunar declinational cycle one would expect that if a triggerring effect would occur, it would be at maximum declination at the higher
latitudes where there is the greatest change in the tides. In one of the
most comprehensive studies, Hamilton (1973) showed that volcanic eruptions near the equator increased during years almost exactly between
maximum and minimum declination creating a 9.3 year period (when lunar
nodes are pointing toward the sun at summer solstice). Eruptions near the
poles occurred during years of minimum declination. Similar relationships
have been noticed by Jaggar (1945), Ward (1961), Lamakin (1966), Shirokov
(1973, 1983) and Kokus (1988b).

In the definitive study of long term periods in volcanic eruptions,
Stothers analyzed 380 events over a 400 year time span and found a 9.5
year period[7] with a confidence level of 99.5%. It should be noted that
Stothers was looking for an 11 year cycle related to sunspots.

Summary of global seismic signature. There are definite increases in
global seismic activity at certain configurations of the earth, sun and

moon which do not cause tidal extremes near the seismic activity. High
seismicity occurs when the moon is at the quarter phases, near the summer
solstice, with the nodes of the lunar orbit pointing towards the sun (and
to some extent, when the sunspot cycle is at a minimum). This configuration maximizes earth rotation rate (Kokus 1988c and 1989), and causes
the tides to partially cancel[8] in such a way that latitudinal cross
sections will most resemble a circle.

After a brief outline of earth expansion, we will attempt an
explanation.

EARTH EXPANSION AND SEISMICITY

The concept of earth expansion is old and there are many interpretations of it, but most attempts to apply it to seismic statistics have
been rather simplistic. An obvious effect of an expanding earth is that
as the core and mantle expand they will produce tension in the crust. If
the expansion were uniform, the fracture of the crust would be fairly
random. Conversely, if the expansion were uneven, the fractures would
tend to occur when it was most rapid. Certain early studies which assumed
that G, the universal gravitational constant, varied with a yearly cycle
looked for and found a yearly variation in seismicity.

In a better developed theory of earth expansion there is more to
crustal deformation than tension cracks. There are two distinct sources
of seismic energy and each will be modulated by tidal and rotation
effects in two distinct ways. One scource is due to a crust that was
formed over a mantle of small radius trying to fit over a mantle which
now has a larger radius of curvature. The other concerns the Tythean
Torsion which is due to the differences in moment of inertia between the
northern and southern hemispheres.

An orange peel trying to fit on a grapefruit. As the earth expands,
its crust is trying to fit over a mantle which is continually trying to
get larger. So the thicker, older parts of the crust, the continental
plates are slightly out of isostatic equilibrium with their underlying
mantle. Their radii of curvature will be continually less than that of
the mantle which is supporting them. The continental centers will be
slightly elevated over their equilibrium position and their margins will
be slightly depressed. This disequilibrium is the scouce of a great deal
of potential seismic energy. Chao and Gross (1987) showed that the net
effect of most earthquake displacements was to make the earth more round
(in direct contrast to the theory of plate tectonics where the number of
quakes that decrease the earth's moment of inertia should be about equal
to those that increase it). Kokus (unpub.,b) showed that the displacements responsible for the rounding of the earth in Chao and Gross's study
were partly due to the collapse of the excess roundness of the continents
(mostly Eurasia) and not just the excess equatorial bulges.

The strain due to the differential radii would not increase uniformly
because the radius of curvature of the mantle is modulated by the tidal
bulges. Tamrazyan (1962 to 1993)[9] has catalogued a great deal of
anecdotal evidence to suggest that seismic activity in very limited areas
is linked to a particular plate being at a tidal extreme.

The Tythean Torsion. In the most developed model of an expanding
earth, Carey (1976,1982,1988) argues that the primary scources of crustal
deformation on an expanding earth are not simple tensions but global
torsions. These torsions are the result of the assymetric distribution of
continents on the earth. The continental crust being thicker than ocean
crust, will have a higher inherent moment of inertia. As the earth
expands, the westward acceleration of the plates will be proportional to
their moments of inertia. Since the northern hemisphere is mostly
continent and the southern hemisphere is mostly ocean, the northern

hemisphere twists westward compared to the southern hemisphere. This is referred to as the Sinistral Tethyan Torsion. There are other inhomogeneities in the crust that give rise to lesser torsions.

Let us now look at all the things that can affect the Tethyan Torsion. Allow I to be the moment of inertia of a latitudinal cross section; L, its angular momentum; w, its angular velocity; and subscripts 1 and 2 represent adjacent cross sections. Then,

$$L_1 = I_1 w \quad \text{and} \quad L_2 = I_2 w \tag{1}$$

The torsion between the two cross sections is proportional to the difference between the time derivatives of the two angular momentums divided by the mean distance, s, between the two cross sections:

$$\text{Torsion} = [d(L_1 - L_2)/dt]/s = \{[d(I_1 - I_2)/dt]w + (I_1 - I_2)dw/dt\}/s \tag{2}$$

There are several ways that the tidal bulges and sunspots can affect the torsion. I_1, I_2, w, dw/dt, and s all vary and are interrelated. For our purposes, the important terms are the moments of inertia and angular velocity. The torsion will increase with increasing velocity, w, which increases with decreasing moments of inertia.

Torsion is maximized between two cross sections near the equator when the moon is at the quarter phases, the lunar node is pointing toward the sun and the earth is near summer solstice. This minimizes the moment of inertia of the equatorial cross sections and maximizes angular velocity. This is identical to the configuration during which volcanic eruptions peek near the equator.

A decent proposal

As plate tectonics matures, it requires that an increasing amount of data be disgarded to preserve the theory. No where is this more evident than in seismic prediction.

I am proposing that we construct a global model based on earth expansion that would calculate global torsions as a function of time. It would assign moments of inertia to the various plates and then allow the tidal bulges to perturb them as the earth expands. Periodic stresses could be calculated and calibrated with existing seismic catalogs. Besides being of theoretical value, it offers hope of an economical seismic forecast that would serve the entire globe; not just select real estate.

Acknowledgments

The author would like to thank the Foundation for the Study of Cycles and the Dewey Foundation for partially funding this research.

ENDNOTES

1. When I say plate tectonics, I mean the theory where continents drift about due to thermal convection in the mantle.
2. The term "earthquake signature" was, to the best of my knowledge, introduced by Bagby (1969). "Cosmolocation" is the literal translation of the term Tamrazyan (1992,1993) uses for the same phenomena.
3. The probabilities were calculated using Rayleigh's test of uniformity. My statistics include the quakes of April 10,1881; March 2,1901; March 10,1922; June 7,1934; and June 27,1966 but exclude the Fort Tejon quake of January 9, 1857. My motivation was that there was no strong evidence that a quake occurred near Parkfield on that day. Two quakes did happen in California that day, and it would enhance the accepted pattern

if one of those quakes were "moved" to Parkfield. If I include that quake
in my statistics, the random probability for the lunar phase drops to
0.015 and the sunspot probability jumps to 0.05.
4. While the value of these statistics may seem dubious, they can predict
when a quake will not happen with high confidence. This can be very
useful when scheduling hazardous tasks.
5. The Imperial Fault quakes used in the study were April 18,1906; June
22,1915; May 18,1940; March 4,1966; Oct.16,1979; and June 9,1980.
6. The Tethyan Torsion Zone circles the earth about the equator. It is
about 20^0 south of the equator in the Pacific and 20^0 north in the
Mediterranean. See Carey (1976,1988).
7. There are many claims of 9.3-9.5 year cycles in a variety of variables which seem totally unrelated. These cycles have been observed in
seismicity, the volcanic dust veil index, precipitation, tree rings going back 5000 years, grain yields, agricultural prices, and the general
economy. The ubiquity of this cycle was first noticed by E.R.Dewey and
was the main motivation behind the Foundation for the Study of Cycles.
8. If we look at the five largest tidal terms, S_2 partially cancels M_2,
and P_1 partially cancels O_1. K_1 is zero.
9. Contact author for bibliography and reprints of Tamrazyan's work.

REFERENCES

Allen, A.M.; 1936, *Bull. Seism. Soc. Am.* 26:147.
Bagby, J.P.; 1969, Icarus 10:1.
     1975a, *Nature* 253:482.
     1975b, Cornell Eng.41:6-12
Berkland, Jim; 1988, Personal correspondence.
Carey, S.W.; 1976, The Expanding Earth. Elsevoir, New York.
     1982, The Expanding Earth: A Symposium. Earth Resources Foundation,
     University of Sydney.
     1988, Theories of the Earth and Universe. Stanford U. Press.
Casetti,G.,Frazzetta,G.,Romano,R.,1981, *Bull.Volcanol.*, 44:283-284.
Chao and Gross; 1987, *Geophys. J.Roy. Astron. Soc.* 91:569.
Conrad, V.; 1933., *Gerlands Beitr.Z.Geophysik* 53:111-139.
     1934, *Nature*, Oct.20,p.631.
Davison, C.; 1893, *Phil. Mag.* 5:36,310.
     1896, *Phil.Mag.*5:42-463.
     1927, *Nature* 120:587.
     1928, *Bul.Seism.Soc.Am.*18:246
     1929, *Phil. Mag.* 7:580-586.
     1933, *Phil. Mag.* 15:1085-1091.
     1934a, *J.of Geology* 42:449.
     1934b, *Geological Magazine* 71:493.
     1935, *Nature* 135:76.
     1938, Studies on the Periodicity of Earthquakes. Thomas Murphy and
     Co., London.
De Mendoca Dias,A.A., 1962. *Bull. Volcanol.* 24:211-221.
De Sabbata, V.; Rizzati, P.; 1977, *Lettre al Nuovo Cimento* 20(4)117-120.
Dewey, E.R.; 1958, *Cycles* 9:86-87
Dzurisin,D., 1980, *Geophys. Res. Lett.* 7:925-928.
Eggers, A.A.; Decker, R.W.; 1969, *EOS. Trans. AGU* 50:343
Hamilton, W.L.; 1972, *Science* 176:1258.
     1973, *J. Geophys. Res.* 78:3356-3375.
Hawaiian Volcano Observatory;1927, *Volcano Let.*No.170,Mar.29.
Jaggar, T.A., 1920. *Bull.Seismol.Soc.Am.* 10:155-275.
     1931,*Volcano Lett.*326:1-3.
     1938, *EosTrans.AGU* 19:23-32
     1945. Volcanos Declare War: Logistics and Strategy of Pacific

Volcano Science. Paradise of the Pacific Ltd., Honolulu.
1947. Mem.21,p.39-318, Geolog.Soc. Amer., Boulder, CO.
Johnston, M.J.S.; Mauk, F.J.; 1972, *Nature* 234:266-267.
Kokus, M.; Ritter, D.; 1988, *Cycles* 39:56-57.
Kokus, M., 1988a, *Cycles* 39:230-233
1988b, *Cycles* 39:266-270
1988c, *Cycles* 39:288-291
1989,Seismic Periods That Can Be Related to Lunar and Solar Cycles: A Comprehensive Review. Found. for the Study of Cycles, Wayne,PA.
1990, *Cycles*,41:76-79.
1991a, *Cycles*, 42:189-195.
1991b, *Cycles*, 42:236.
unpublished,a, Seismic periods that can be related to lunar and solar cycles: A comprehensive review of the literature to 1991.
unpublished,b, Earthquakes, earth rotation and the excess elliptical bulge.
Lamakin,V.V.; 1966, *Doklady An SSSR* 170(2):410-413.
MacDonald,G.A., 1960, *Bull.Volcanol.* 23:211.
Machado,F.,1960, *Bull.Volcanol.* 23:101-107.
1967, *Bull.Volcanol.* 30:29-34
Mauk,F.J. 1979. *Eos Trans.AGU* 60:833.
Mauk, F.J.; Johnston, M.J.S.; 1973, *J. Geophys. Res.* 78:3356-3362.
Mauk,F.J.; Kienle,J.; 1973, *Science* 182:386-389.
McClellan, P.H.; 1984, *Nature* 307:153-156.
Morgan,W.J.;Stoner,J.O.;Dicke,R.;1961, *J.Geophys.Res.*66:3831-43.
Perret,F.A.,1908, *Science* 28:277-87.
Perrine,C.D.;1949, *Bull.Seism.Soc.Am.*39:109
Pines,D.; Shaham,J.; 1973, *Nature* 245:77-81.
Press,F.; Briggs,P.; 1975, *Nature* 256:270-273.
Rinehart,J.;1973, 7th International Symposium on Earth Tides,Sopron,Hun.
Ritter,D.C.;1987, unpub., A possible earthquake frequency exists which can be associated with a tidal cycle.
Roosen,R.;Harrington,R.;Giles,J.;Browning,I.;1976, *Nature* 261:680-681.
Sapper,K.,1930. *Vol. Let.*302:2-4.
Sauers, Jack; 1986a, *Cycles* 37:46-47.
1986b, *Cycles* 37:203-204.
Schuster, A.; 1897, *Proc. Roy. Soc. A* 61:455-465.
Shirokov, V.A.; 1973, The cosmos and volcanos. Man and the Elements; Leningrad, Gidrometeoizdat; p. 26-28.
1983, The Great Tolbachik Fissure Eruption; S.A. Fedotov and Ye. K. Markhinin, ed. Cambridge University Press. pp.232-242.
Shneiderov,A.;1973, *EOS,Trans AGU* 54:1138.
Spalding,W.;1915, *Bull.Seism.Soc.Am.* 5:30.
Stothers,R.D.,1989a, *Geophys.Res.Lett.* 16:453-455.
1989b, *J.Geophys.Res.* 94:17371-17381.
Tamrazyan, G.P., 1962, *Cycles* 18:232-241.
1966. *J.of Physics of the Earth (Japan)* 14:41-48.
1967. *Indian Geophysical Union J.* 4:131-141.
1968a. *Icarus* 9:574-592.
1968b. *J.Geophys.Res.*73:6013-6018.
1969a. *Intern.J.of the Solar System* 10:164-168.
1970a. *Geologische Rundschau (GFR)* 59:623-36
1972. *Nature* 240:406-408.
1974. *Geophys.J.of the Royal Astron.Soc.*38:423-9
1991. *Cycles* 42:93-99 and 134-140.
1993. *Cycles* 44:81-85.
Ward, L.; 1961, *Cycles* 12:317-319
Winkless, N.; Browning, I.; 1975, Climate and the Affairs of Men. Harpers Magazine Press, New York; p. 42-81.
Wood, H.O.; 1918, *Amer.J.Sci.* 4:45-146.

TENSION - GRAVITATIONAL MODEL OF ISLAND ARCS

Jan Koziar and Leszek Jamrozik

Institute of Geological Sciences
Wrocław University, pl.M.Borna 9
50-204 Wrocław, Poland

Active continental margins are the zones where, according to plate tectonics, the oceanic spreading is being compensated. The original model, based on this assumption, was created by Isacs et al.,(1968), being subsequently modified in different ways. Finally, it appeared as an artificial construction much less convincing than the model of spreading, and has been criticised by many authors (Tanner 1976, Carey 1977, Pfeufer 1981, Cudinov 1985, Koziar and Jamrozik 1991).

In clear disagreement with the plate tectonics model is a double seismic zone discovered beneath the Japanese Islands (Hasegawa et al., 1978) and the oppositely oriented tectonic regime in both planes of hypocenters: almost horizontal tension in the lower plane, and similarly oriented compression in the upper plane (Figure 1). The lower plane cuts the lower part of the horizontal oceanic plate. The distance between both zones is 30-40 km. The shear translations corresponding with the recorded tension and compression in relation to both plains of hypocenters means sliding down of the lithospheric material between these planes. This induces a scheme of the destruction of oceanic lithosphere shown in Figure 2a. This model is in accordance with the tension beneath oceanic trenches and with evidenced here stepwise lowering of oceanic lithosphere along gravitational faults. It comes out from this model that the tectonic frame is the opposite to the plate tectonics model (Figure 2b). Moreover, it agrees with diapirism beneath active continental margins, which similarly to diapirism beneath ocean ridges, indicates tension regime, and also with back arc spreading, as well as with the extensional development of marginal seas. The latter three processes have always been in strong contradiction with the plate tectonics model of plate collision.

The gravitational destruction of oceanic plate shown above explains the bended shape of island arcs and the dip of the Benioff zones always inwards the arcs (Figure 3). A similar gravitational interpretation was presented by Carey (1976).

The tensional development of the Benioff zone as a whole is shown in Figure 4. The sinking of segmented oceanic lithosphere is caused by heating and decreasing density of asthenosphere (which is well evidenced) beneath active continental margin. The thermal activation has the same cause as the gravitational destruction of lithosphere, i.e. extension.

The shallow part of the lithosphere between the oceanic trench and diapir top manifested as volcanic line, must be gravitationally sliding towards the trench (Figure 5). It is possible to prove this process.

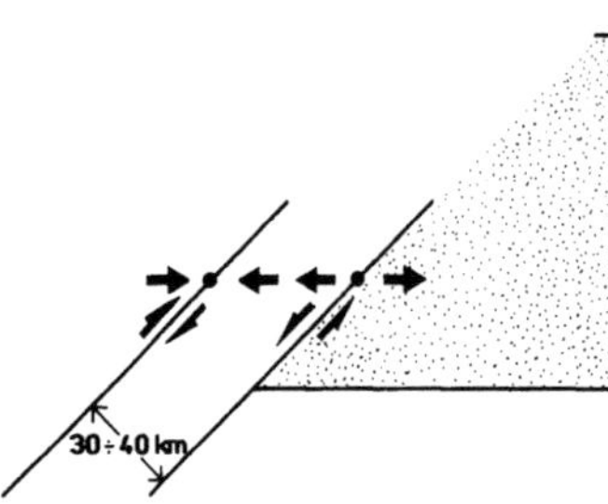

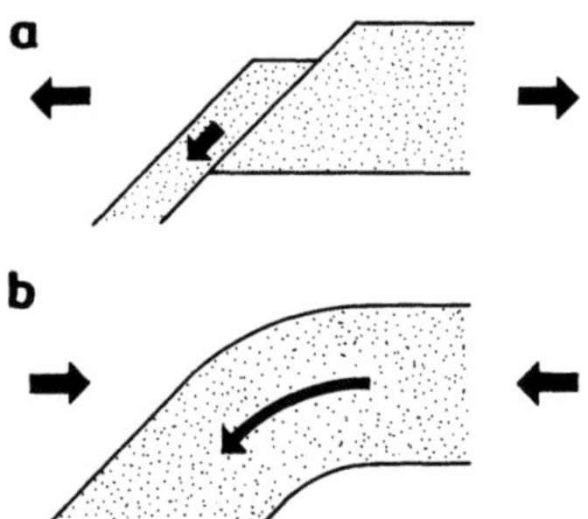

Figure 1. Double seismic zone and strain directions in the upper and lower planes of hypocenters.

Figure 2. a) Model of destruction of the lithosphere beneath island arcs induced from figure 1. b) Plate tectonics model of destruction of the lithosphere beneath island arcs.

The shear plane marked by shallow earthquakes beneath island arcs is not part of the Benioff zone as assumed in the plate tectonics model. It dips gently towards the line of volcanos (Plafker 1965). Then, this plane cannot be considered to support the scheme shown in Figure 2b. However, it can be interpreted as a plane of gravitational slide. The latter interpretation is confirmed by the horizontal displacement of rock masses associated with shallow earthquakes.

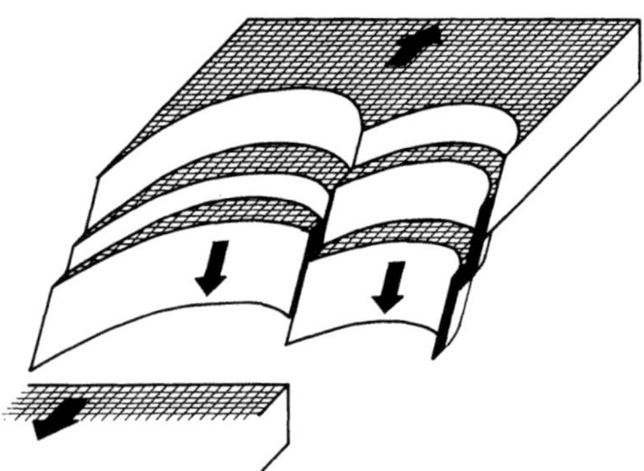

Figure 3. Gravitational destruction of oceanic plate determines the relationship between bending of island arc and a dip of seismic zone.

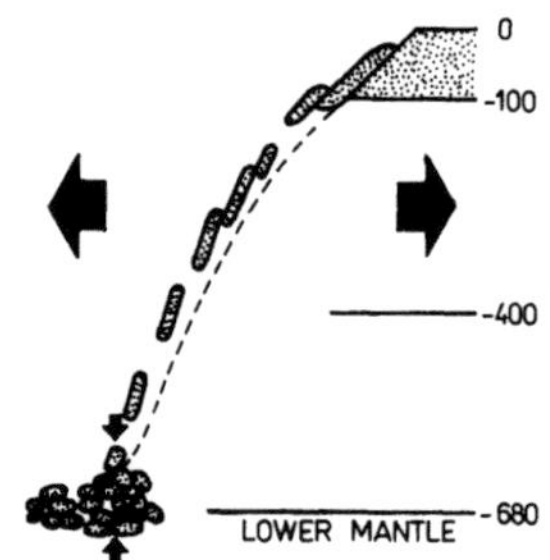

Figure 4. Tension · gravitational model of the whole seismic zone.

In gravitational slide, these masses should be transported trenchwards (Figure 6a); however, assuming oceanic lithosphere underthrusting (plate tectonics), they should be squeezed (Figure 6b). The first case is true as evidenced by the data presented by Parkin (1969), Plafker and Savage (1970), Fitch and Scholtz (1971).

Gravitational slide of island arcs is also reflected in the vertical displacement of rock masses. A characteristic feature of such slide is its rising frontal and sinking distal part. This is a rule in active continental margins (Plafker, 1965; Plafker and Savage, 1970; Fitch and Scholtz, 1971).

In the scheme of Figure 5, two separate, but causally connected mechanisms can be seen. The first, deep mechanism causes slow subsidence of the area near the trench and, on the other

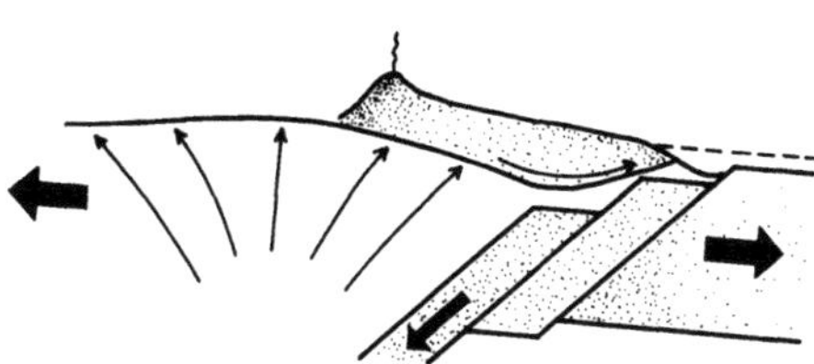

Figure 5. Gravitational sliding of island arc as a result of sinking of lithosphere in its front, and diapiric upheaval beneath the volcanic line.

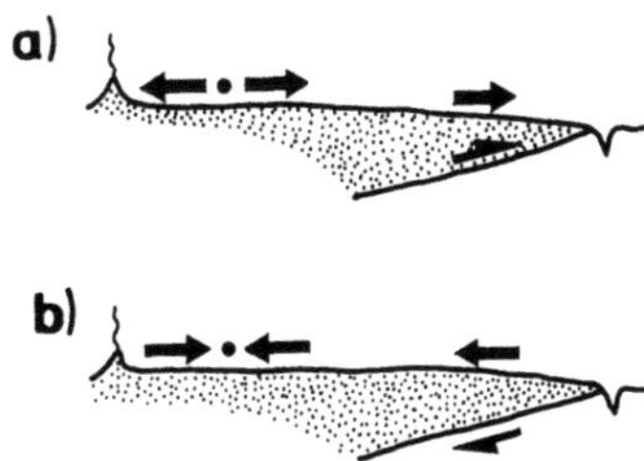

Figure 6. Horizontal displacemets of rock masses in island arc in the case: a) gravitational overthrusting of island arc, b) underthrusting of oceanic lithosphere (plate tectonic assumption).

hand, slow rising of the area near the volcanic line. The second mechanism is gravitational slide, leveling the growing vertical gradient of the earth surface. Many years ago both such mechanisms were recognized in gravitational tectonics by Haarmann (1926, 1930), who named them, respectively, primary and secondary tectogenesis. The character of the first mechanism has always been difficult to define. In active continental margins it can be deduced from recent well documented deep processes. The cause of the primary tectogenesis is evidently the break up of the lithosphere and extension of underlying mantle.

A similar model, though not reaching so deep into the mantle, was obtained by the authors for intracontinental orogens (Koziar, Jamrozik, 1985a,b).

From both models it comes out that the active continental margins and intracontinental orogens are not the place of oceanic spreading compensation. On the contrary, the lithosphere is being drawn aside also there, though to a less degree than at the mid-ocean ridges.

REFERENCES

Carey, S.W., 1976, "The Expanding Earth," Elseviere, Amsterdam - Oxford - New York.

Cudinov, J., 1985, "Geologija Aktivnych Okeaniceskich Okrain i Globalnaja Tektonika," Nedra, Moskva.

Fitch, T., and Scholtz, Ch., 1980, Mechanism of underthrusting in southwest Japan: A model of convergent plate interactions. *J.Geoph.Res.* 76:7260.

Haarmann, E., 1926, Uber die Kraftquelle der Tektogenese, *Zeitschr. Deut. Geol. Gesel.* 78:71.

Haarmann, E., 1930, "Die Oszillationstheorie", Ferdinand Enke Verlag, Stuttgart.

Hasegawa, A., Umino, N., and Takagi, A., 1978, Double-planed structure of the deep seismic zone in the Northwestern Japan Arc, *Tectonophysics* 47:43.

Isacks, B., Oliver, J., and Sykes L., 1968, Seismology and the New Global Tectonics, *J.Geoph.Res.* 73:5855.

Koziar, J. and Jamrozik, L., 1985a, Tension - gravitation model of tectogenesis, *in:* "Proceeding reports of the XIII-th Congress of KBGA", Polish Geological Institute, Cracov.

Koziar, J., and Jamrozik, L., 1985b, Application of the tension - gravitation model of tectogenesis to the Carpathian orogen reconstruction, *in:* "Proceeding reports of the XIII-th Congress of KBGA", Polish Geological Institute, Cracov.

Koziar, J. and Jamrozik, L., 1991, Tensyjno - grawitacyjny model subdukcji, *Polskie Towarzystwo Geologiczne, Oddz. Poznański, Str. referatów (1990-1991)*:34.

Parkin, E.J., 1964, Horizontal crustal movements determined from surveys after the Alaskan earthquake of 1964. The Prince William Sound Alaska earthquake of 1964 and aftershocks. *U.S.Dep. of Commer.* 3:35.

Pfeufer, J., 1981, "Die Gebirgsbildungsprocesse als Folge der Expansion der Erde", Verlag Gluckauf, Essen.

Plafker, G., 1965, Tectonic deformation associated with the 1964 Alascan earthquake, *Science* 148:1675.

Plafker, G., and Savage, J., 1960, Mechanism of the Chilean earthquake of May 21 and 22, 1960. *Geol. Soc. Am. Bull.* 81:1001.

Tanner, W., 1976, Deep-sea trenches and the compression assumption., *AAPG Bull.* 57:2199.

ELECTROMAGNETIC INTERACTIONS
AND PARTICLE PHYSICS

A. O. Barut*

International Center of Physics and Applied Mathematics
Trakya University
Edirne, Turkey

INTRODUCTION

The foundations of physics must be continuously reexamined. Not only the physical theories, but also the very basic notions and entities undergo changes. Physics is very subtle and is difficult to be completely axiomatized. It turns out that there are tacit assumptions, inprecise wordings, preconceived ideas or pictures that have been taken for granted. For example, there was a prejudice about the simultaneity of distant events which was removed by the special theory of relativity. In quantum theory, the notions of what we mean by electrons or light have undergone various paradigms, and there is still imprecision in the conceptualisation of a single individual electron or "photon" which is necessary for an understanding of quantum theory. After the successes of quantumelectrodynamics, the general concensus is that we understand almost completely the interaction of the electrons, positrons and light, even though this theory is, as it stands, necessarily a perturbative approach. Therefore, phenomena which do not fit into this perturbative picture were accounted for by the introduction of new forces and new basic fundamental particles. It is our contention here to show that elecrodynamics processes viewed nonperturbatively have a rich structure at short distances and may already unify these new phenomena of week and strong interactions.

FOUNDATIONS OF ELECTRODYNAMICS

We have been reexamining the foundations of quantumelectrodynamics (QED) for two reasons: (1°) In order to be able to extrapolate the electromagnetic forces to short distances; (2°) In order to be able to eliminate the infinities in the perturbation theory and to provide a nonperturbative approach. In perturbaton theory the short distance behavior of QED is completely unknown, for example, the forces between e^+ and e^- in very high energy colliding experiments. What is now used in practice from QED is essentially the Born approximation together with a few higher order radiative corrections to it. Although Born approximation is good at high energies for a given potential, the forces themselves, specially spin forces, change completely at high energies or short distances. In perturbation theory then one is confronted with summing up infinitely many terms, which is hardly possible, because one has reached a practical computational limit to order α^4 in the calculation of $(g-2)$ for example[1].

Frontiers of Fundamental Physics, Edited by M. Barone
and F. Selleri, Plenum Press, New York, 1994

Experimentally, starting with a purely electromagnetic process, e.g. electron-position scattering, a host of new particles or resonances are observed, as well as novel asymmetries in the angular distributions, which cannot be accounted for in the perturbation theory. Some simple models show however that these new phenomena may be understood by certain plausible behavior of electromagnetic interactions at short distances. This motivates that we must seek ways, other than perturbaton theory, to extrapolate quantumelectrodynamics to high energies. For such a successful and universal theory as electrodynamics, which accounts for all phenomena, except gravitation, from astronomical down to atomic distances, to remain incomplete and inconclusive is a real impediment to the development of fundamental physics. Otherwise we will be forced, and we are forced, to use phenomenological models with many parameters and many new entities. I cannot think of any more urgent and basic problem for theoretical physics than the completion of QED.

Our reexamination of QED begins with the postulate of separate second quantizations of both the electromagnetic field and the matter field. In perturbative QED the electromagnetic field and the matter field have separate degrees of freedom, and the corresponding quanta, after quantization, are the photons and the electrons. In contrast to this, one may express the electromagnetic field in terms of its sources, and therefore can take the point of view that every light has some source, thus the degrees of freedom of the field are those of the sources.

A theory is called *dualistic* if we take both fields and particles as primary fundamental objects as in QED. It is called *unitary* if either particles, or the fields alone are the only fundamental objects. Thus in a theory with electrons being fundamental, the fields are just mathematical constructions to describe their interactions; whereas in a theory with fundamental fields only, the particles would be realized as localized concentrations of fields.

I now present a formulation of electrodynamics in terms of particles only. First in classical electrodynamics, it can be shown that[2] the set of Maxwell's equations, the Lorentz equation, and the conservation laws, all follow from a particle Lagrangian

$$L = \frac{1}{2} \sum_k m_k \dot{x}_k^2 + \sum_{k<\ell} e_k e_\ell \frac{1}{r_{k\ell}} \left(1 - \frac{\dot{\boldsymbol{x}}_k, \cdot \dot{\boldsymbol{x}}_\ell}{c^2} \right) , \tag{1}$$

$$r_{k\ell} = |\boldsymbol{x}_k - \boldsymbol{x}_\ell|$$

The fields are defined quantities in terms of the sources. The fields at the position of the kth particle (thought to be a test charge) are

$$E(x_k) = -\nabla_k \Phi(x_k) - \frac{1}{d} \frac{d^*}{dt} A(x_k), \tag{2}$$

$$B(x_k) = \nabla_k \wedge A(x_a)$$

where $\Phi(x_k) \equiv \sum_m \frac{e_m}{r_{km}}$, $A(x_k) \equiv \sum_m \frac{e_m}{c} \frac{\dot{x}_m}{r_{km}}$ and $\frac{d^*}{dt}$ means differentiation of functions $x_m(t), \dot{x}_m(t)$; those of $x_k(t), \dot{x}_k(t)$ being separated and carried out. With these definitions we obtain from (1) the Lorentz-force

$$m_k \ddot{x}_k = e_k E(x_k) + e_k \frac{\dot{x}_k}{c} \wedge B(x_k) \tag{3}$$

as well as all the Maxwell's equations

$$\nabla \cdot B = 0, \quad \nabla \wedge E = -\frac{1}{c}\frac{d^*}{dt}B,$$

$$\nabla \wedge B = \frac{4\pi}{c}J + \frac{1}{c^2}\frac{d^*}{dt}D, \quad \nabla \cdot D = 4\pi\varrho \tag{4}$$

The magnetic effects are due to the second term in the potential of eq. (1), and the magnetic units also follow from it.

The relativistic covariant form of eq. (1) is

$$A = m_1 c^2 \int dt_1 \sqrt{1 - \beta_1^2} - m_2 c^2 \int dt_2 \sqrt{1 - \beta_2^2} - e_1 e_2 \int dt_1 dt_2 \left(1 - \bar{\beta}_1(t_1) \cdot \bar{\beta}_2(t_2)\right)$$

$$\times D\left(x_1(t_1) - x_2(t_2)\right) \tag{5}$$

where $D(x_1 - x_2)$ is the symmetric Green's function of the Laplacian and t_1 and t_2 are the retarded times relative to each other. The quantum correspondence of this is the Lagrangian density

$$\mathcal{L} = \sum \bar{\psi}_k \left(\gamma^\mu i \partial_\mu - m_k\right)\psi_k - \frac{1}{2}\sum_{km} e_k e_m \bar{\psi}_k \gamma^\mu \psi_k \int dy D(x - y)\bar{\psi}_m \gamma_\mu \psi_m(y) \tag{6}$$

Maxwell's equations follow from (5) and (6), respectively if we define

$$A_\mu(x) \equiv \sum_m \int dy D(x - y) j_\mu^m(y) \tag{7}$$

It follows then that

$$F^{,\nu}_{\mu\nu} = j_\mu = \sum_m e_m \bar{\psi}_m \gamma_\mu \psi_m$$

$$F^{*,\nu}_{\mu\nu} = 0 \tag{8}$$

and $A^\mu_{,\mu} = 0$, since the current j_μ is conserved. This derivation of field equations from particle interactions is in fact not surprising, because the particle action (6) follows from the "dualistic" Lagrangian density

$$\mathcal{L} = \mathcal{L}_{\text{matter}} + A_\mu(x)j^\mu(x) - \frac{1}{4}F_{\mu\nu}F^{\mu\nu} \tag{9}$$

by eliminating the potnetial A_μ in the second and third terms, in the Loretnz gauge, using the field equations $\Box A_\mu = j_\mu$.

EXTRAPOLATION OF ELECTRODYNAMICS

Thus a unitary point of view in quantumelectrodynamics is possible. It does not really contradict the dualistic $\mathcal{L}$ of eq. (9) on the classical level. But now we can study the particle action (6) by itself. It involves particle-particle interactions, as well as self-interactions, the latter in a nonlinear way. But now instead of quantizing both fields A_μ and ψ in (9), we have a possibility to formulate quantumelectrodynamics as a relativistic classical field theory. This has been carried out in great detail and applied to almost all radiative processes[3]. Since A_μ has been eliminated, its quantization does not arise. Furthermore, with a proper interpretation of the negative energy solutions of the Dirac equation as antiparticles, which is indeed possible, even in first-quantized Dirac theory, we do not need even to second quantize the matter field ψ.

The elimination of the electromagnetic field A_μ and the resultant form of the action (6) allows us further to obtain a nonperturbative covariant two- or many-body equation in closed form involving a single time t, so that we have now a possibility to extrapolate the two-body interactions to short distances.

If we start with a 2-body action (6) and introduce the composite field

$$\Phi(x,y) = \psi_1(x)\psi_2(y)|_{(x-y)^2=0} \tag{10}$$

i.e. the points x, y are space-like separated, it is possible to derive the 16-component two-body equation[4]

$$\left[(\gamma^\mu i\partial_\mu - m_1) \otimes \gamma \cdot n + \gamma \cdot n \otimes (\gamma^\mu i\partial_\mu - m_2) - e_1 e_2 \gamma^\mu \otimes \gamma_\mu / r + V^{\text{self}}\right] \Phi = 0 \tag{11}$$

This is a one-time equation, because of the condition $(x-y)^2 = 0$ in the definition of $\Phi(x-y)$, the relative time drops out in eq. (11). The equation (11) has been studied extensively, separation of radial and angular parts, separation of center of mass and relative coordinates, the solutions of the radial equations.[5]

Here we will only point out that the resulting structure of the interactions at short distances is very intricate and rich, whereas at large distance we get approximatively the Dirac or Schrödinger equations for the Coulomb problem.

More importantly, and perhaps unexpectedly, the self interactions lead to potentials which become important at short distances, and which have not been generally considered in QED. Among these is the one due to the induced anomalous magnetic moment. These effects are very small at the level of positronium, but because they induce potentials of the form $1/r^n$, n=3,4,5,...they become the dominant terms at short distances.

MAGNETIC INTERACTIONS AT SHORT DISTANCES

Based on these facts one can construct models of magnetic and anomalous magnetic moment interactions (short of an exact solution of (11) which at the moment is incomplete). These models show indeed that some new physics takes place at short distances, namely the occurence of narrow, massive resonances due to deep potential wells at short distances. Experimentally one observes that in the $e^+ - e^-$ scattering at high energies there are indeed sharp resonances superimposed on a background given by Born approximation. This is in fact typical for the the phenomenon of "barrier penetration" in the presence of deep potential wells followed by a barrier. Furthermore, the magnetic interactions are spin-dependent. The importance of highly spin-dependent interactions, notably in proton-proton interactions, is now recognized and poses a challenge to perturbative theories like QCD, in which they cannot be explained. For composite systems like proton-proton scattering, there are in addition magnetic exchange forces, exchange of charged constituents in the strong short distance magnetic fields. The spin effects are generally masked in unpolarized scattering experiments, but show up in experiments with polarized particles, because they average out if one sums over all polarizations. Therefore, polarization experiments are now very actively carried out and planned in most recent experiments. We see that the magnetic interactions at short distances have a very rich structure, and manifestations as "strong interactions," like the rich structure of particle physics itself. But ultimately they are all based on the simple basic equations of electrodynamics. We recall that, on the other end of the energy scale, the chemical forces between neutral atoms or alpha-decay are ultimately some weak manifestations of electromagnetic interactions. In fact the so-called "weak-interactions", e.g., beta-decay, maybe also a weak manifestation of electromagnetism as we shall try to show in the next section.

PARTICLE MODEL

The above considerations lead to the idea of constructing a model of particle physics on the basis of electromagnetic interactions and using as fundamental entities the absolutely stable particles only.[6] It turns out that the two absolutely stable particles that might be called truly elementary are sufficient for this purpose: the electron and the electron-neutrino, and their antiparticles. This is the simplest, most economic model that one can imagine, considering the fact that e and ν are very akin, and correspond to one state in the representation of the conformal group. This may come as a surprise, for all the other standard particle models have dozens of fundamental particles (e.g. 36 or more in the standard model). All we need is to endow the neutrino with an anomalous magnetic moment of the order of $10^{-9} - 10^{-10}$ Bobr magneton, which is is also the experimental limit put on this quantity. In terms of forces all we need to postulate is a short range strong magnetic force between leptons, as it seems indeed to follow from electrodynamics.

Examples of forerunners of this idea are very old. When Pauli introduced the concept of neutrino in β-decay in 1931 he envisaged the neutron to be a bound-state of $(pe\bar{\nu})$, bound by the magnetic moment of the neutrino. But this model was soon forgotten in favor of a phenomenological theory of beta decay by Fermi. So was also forgotten the early calculation of the neutral-current process $e + \nu \to e + \nu$ due to neutrino magnetic moment, until both were revived recently in the context of a more comprehensive model of particles and their interactions.[7]

The construction of particles, multiplets and currents proceeds from more stable states to less and less stable states according to an Aufbau principle. The ordering of particles according to their stability is very remarkable and shows a regularity so that I consider the stability criterion to be an immmportant element of understanding the particle states. The three levels of most stable particles and their lifetimes are

$$
\begin{array}{llll}
\text{level A} : & e & \nu_e & \text{lifetime } \tau = \infty \\
\text{level B} : & p = (e^+e^+e^-), & \nu_\mu = (\nu_e\nu_e\bar{\nu}_e) & \tau \text{ very large} \\
\text{level C} : & n = (pe^-\bar{\nu}_e), & \mu \equiv (\nu_\mu e^-\bar{\nu}_E), \tau_n = 888\text{sec.} & \tau_\mu = 2 \times 10^{-6}\text{sec}
\end{array}
$$

all other particles have shorter lifetimes (with the exception of perhaps ν_τ). If a particle is rather stable it can be used to build other particles in the subsequent level and the process can be continued. The quantum numbers of particles are obtained as follows. Since electron is absolutely conserved and all charge is due to electron, it provides an absolutely conserved quantum number Q=number of electrons − number of positrons, which explains why charge is both quantized and conserved. Similarly, the neutrino number $N_\nu = L$ is quantized and conserved. A metastable state of liftime τ provides a conserved quantum number for all processes whose duration is less than τ. Thus $N_p =$ Baryon number $B, N\nu_\mu = C, N_\mu = S, N_n$ are conserved in strong interactions, but the strangeness S (μ-number) is not conserved when μ-decays, i.e. in weak decays. The notation $p = (e^+e^+e^-)$ indicates the minimum number of constituents; a pair $(\ell\bar{\ell})$ can always be added without changing the quantum numbers. The general rule to obtain all other states is: Fermions $= (\ell \otimes \ell \otimes \ell)$, Mesons $= (\ell\otimes\bar{\ell})$. One can further systemetize this construction so that one obtains a one-to-one correspondence with the quark scheme. Taking the six leptons $(\nu_e, e^-, \mu^-, \nu_\mu, \tau^-, \nu_\tau)$ and assuming six baryons already made i.e. $b = (p^-, n^0, \Lambda_s^0, \Lambda_c^+, \Lambda_b^0, \Lambda_r^+)$ and all other baryon states obtained as $\ell \otimes \ell \otimes \bar{\ell}$, one obtains a model which has been shown to be equivalent to 3-quark states with three colours. And it turns out that the average charge in the $(b, \ell, \bar{\ell})$ system is equal to the fractional quark charges. How many states one can make depends on the dynamics and on stability. It is likely that very massive states corresponding to the so-called top quark are so unstable that they will not be found.

One support for this model comes from the simple fact that if one waits long enough all particle reactions eventually must end up with the absolutely stable particles, e and ν_e, although at present the decay of level B particles, p and ν_μ, have not yet been seen. Secondly, it is a theorem of S-matrix theory that if a particle decays into something, e.g. $n \to pe^-\bar{\nu}_e$, then conversely, this particle must be constructible from the decay products, e.g. the neutron is a resonance pole in the reaction channel $p + e^- + \bar{\nu}_e$.

With the two-basic particles, e and ν_e, we have now the following four currents:

$$j_{em}^\mu \overset{e^-}{\underset{e^-}{\diagdown\diagup}} \text{---} \quad A_\mu \; j_{NC}^\mu \overset{\nu}{\underset{\nu}{\diagdown\diagup}} \text{----} \quad Z_\mu \; j_-^\mu \overset{\bar{\nu}}{\underset{e^-}{\diagdown\diagup}} \text{---} \quad W^- \; j_+^\mu \overset{\nu}{\underset{e^+}{\diagdown\diagup}} \text{----} \quad W^+,$$

and if we introduce the four fields $A_\mu, Z_\mu, W_\mu^+, W_\mu^-$, generalizing electrodynamics, the Lagrangian of the electroweak theory can be obtained without using the idea of gauge theory, or the need of Higgs particles, because the gauge symmetry is completely broken anyway. In our case the new interaction of leptons, or the resonance states Z^0, W^+, W^- would be given by the very short range magnetic interactions and the corresponding deep potential wells, as we discussed earlier. The phenomenon barrier penetration is equivalent to forming an intermediate resonance states which eventually decays again. In fact experimentally, we do not see Z^0, W^+, W^- themselves, but only their decay products.

In conclusion, we have reviewed here that a very intuitive economic particle model in the tradition of atomic and molecular physics, without unobservable constituents and without confinement problems, which can account for the complexity of particle phenomena. We made it plausible that the dynamics might be given by magnetic interactions, although the precise calculations of bound states is incomplete. But this is also incomplete in quark models.

*Physics Department, University of Colorado, Boulder, CO, 80309-0390.

REFERENCES

1. "Quantumelectrodynamics," edited by T. Kinoshita, World Scientific, Singapore, 1990.
2. A. O. Barut, Formulation of electrodynamics from a single principle, *Found. Phys.* (in press).
3. For a review see A. O. Barut, "Brief History and Recent Developments in Electron Theory and Quantumelectrodynamics," in *The Electron*, edit. by D. Hestenes and A. Weingarshofer (Kluwer, 1991), p. 105.
4. A. O. Barut, The covariant many-body problem in quantumelectrodynamics," *J. Math. Phys.* **32**: 1091 (1991).
5. A. O. Barut, A. J. Bracken, S. Komy and N. Ünal, "New approach to the de-

termination of of eigenvalues and eigenfunctions for a relativistic two-fermion equation," *J. Math. Phys.* **34**: 2089 (1993), and references therein.

6. A. O. Barut, "How to build hadron multiplets from stable particles," *Lecture Notes in Physics*, Vol. **94**, 496 (Springer, Berlin, 1979) and "Stable particles as building blocks of matter," *Surv. High Energy Physics.* **1**: 113 (1980).

7. For a review see A. O. Barut, "Unification based on Electrodynamics," *Annal. der Phys.* **43**: 83 (1986), and references therein, and W. Grandy, Jr. "A nonstandard model," *Found. Phys.* **23**: 439 (1993).

ISOTOPIC AND GENOTOPIC RELATIVISTIC THEORY

A. Jannussis[1,2] and A. Sotiropoulou[1]

[1]Department of Physics, University of Patras, Greece
[2]I.B.R. The Institute of Basic Research, P.O. Box 1577,
 Palm Harbor, Florida 34682-1577, U.S.A.

SUMMARY

In the present paper we investigate the isotopic and genotopic structure in the relativistic theory, via the theory of isominkowskian spaces. According to Santilli the central assumption is that motion of particles and electromagnetic waves within inhomogeneous and anisotropic physical media implies an alteration of space-time representable via isominkowski spaces. For the isotopic case we obtain an interaction maximal speed of propagation of causal signals which is $\rightarrow c$ (speed of light in vacuum). For the genotopic case we introduce the simple Lie-admissible complex time model in which the interaction speed is complex and leads to complex values of mass. Also by using the small-distance derivative model we obtain the Lie-admissible Wheeler - DeWitt equation and new concepts concerning the connection between space-time and particles. Finally for the isotopic and genotopic cases we obtain the asymmetry of space-time and the inhomogeneity as well as the arrow of the cosmological time.

INTRODUCTION

As it is well known the Lie-admissible theory which has been developed in the pioneering work of Santilli[1],is divided in two large branches; the genotopic and the isotopic. The structure of the above branches has today a big importance, especially for Physics. Also the existing literature on this subject is very extended[2].

Basically the two branches arosed from the generalization of the Lie-commutator via the so-called Lie-admissible algebras.

According to Santilli[1,2] the genotopies, meaning "induce configuration", have been applied in the sense of altering the original axioms to induce more general axioms which however are restricted to admit the original ones as a particular case. For the Lie-algebras with product [A,B], the Lie-admissible genotopies are given by the broader mappings (genotopies).

$$[A,B] = AB - BA \quad \rightarrow \quad (A,B) = ARB - BSA \quad R \neq S \tag{1.1}$$

and the special case $R = S = T$ corresponds to the isotopies, e.g.

$$[A,B] = AB - BA \quad isotopies \quad (A,B) = ATB - BTA, \quad T = T^+ \tag{1.2}$$

The isotopies, meaning "preserve-configuration", are applied in the sense of preserving the original axioms.

Recently Santilli[3] has written a book with the title "Elements of Hadronic Mechanics" which belongs in the field of the isotopies of contemporary mathematical (integral) and nonhamiltonian generalizations of given mathematical or physical structures which are such to preserve the abstract axioms of the original theory. As it becomes evident from this book, every physical problem is faced from two points of view: i.e.

a)   The external dynamical problem, which is characterized by motion of point-like particles within the homogeneous and isotropic vacuum and

b)   The internal dynamical problem, which is characterized by motion of extended and therefore deformable particles within inhomogeneous and anisotropic physical media, resulting in the most general known dynamical equation of nonlinear, nonlocal and nonhamiltonian type.

Also in a new paper Santilli[4] has extended the isotopies into the relativistic physics, with considerable application in the cold fussion of particles.

The central assumption is that motion of particles and electromagnetic waves within inhomogeneous and anisotropic physical media implies an alteration of spacetime representable via isominkowski spaces[4]. As is well-known, the Minkowski space provides a geometrization of the homogeneity and isotropy of empty space (vacuum). Its preservation in internal physical problems then implies the suppression of the inhomogeneity and anisotropy of the medium, in which motion occur, that is, the elimination of the primary geometric characteristics to be represented.

In a gravitation theory the concepts of genotopy and isotopy are directly connected with the Riemannian metric, which as is well known has the form

$$ds^2 = g_{ik}dx_i dx_k \tag{1.3}$$

where g_{ik} are the metrical coefficients. The above metric (1.3) for the genotopic case is written[3]

$$ds^2 = g_{ik}dx_i T_{ik}dx_k \quad , \quad T_{ik} \neq T_{ki}^+ \tag{1.4}$$

and for the isotopic

$$ds^2 = g_{ik}dx_i T_{ik}dx_k \ , \ T_{ik} = \pm T_{ki} (symmetric \ and \ antisymmetric \ tensor) \tag{1.5}$$

In the present paper we shall study the genotopic case as well as the isotopic case as a special case of the former.

In section 2 we'll study the metric $ds^2 = dx^\mu \hat{\eta}_{\mu\nu} dx^\nu$ which corresponds to isotopic cases with $T = diag(b_1^2, b_2^2, b_3^2, b_4^2) = diag(n_1^{-2}, n_2^{-2}, n_3^{-2}, n_4^{-2}) > 0$ and

$$\hat{\eta}_{\mu\nu} = T\eta \ , \quad \eta = diag(1,1,1,-1)$$

and the b's (or the n's) are the relativistic quantities of the medium considered with the functional dependence

$$b_\alpha = \frac{1}{n_\alpha} = b_\alpha(x,\dot{x},\ddot{x},\psi,\psi^+,\partial\psi,\partial\psi^+) \tag{1.6}$$

In section 3 we will study the metric $ds^2 = dx_1^2+dx_2^2+dx_3^2-c_o^2(1+i\lambda)^2dt^2$ which corresponds to the genotopic case and is equivalent with the initiation of the Lie-admissible complex time model, i.e. $t \to t(1+i\lambda)$

In section 4 we will apply the Lie-admissible theory in the study of the Wheller-DeWitt equation according to the results of Gonzalez-Diaz[5] and Jannussis[6]. Finally in section 5 is devoted to concluding remarks.

RELATIVISTIC ISOTOPIC STRUCTURE

Recently Santilli[3,4] has studied the relativistic isotopic case on isominkowskian spaces which can be written in the form.

$$\hat{M}(x,\hat{\eta},\hat{R}), \ x =(r,x^4) = (r,c_ot), \ r\epsilon\hat{E}(r,\hat{\delta},\hat{R}), \tag{2.1}$$
$$\hat{E}(r,\hat{\delta},\hat{R}) \ is \ the \ isoeuclidian \ space$$

$$\eta = diag(1,1,1,-1)\epsilon \ M(x,\eta,R), \ \hat{\eta} = T\eta \tag{2.2}$$

$$\hat{R}(\hat{\eta},+,^\star)\approx R,$$

$$T = diag(b_1^2,b_2^2,b_3^2,b_4^2) = diag(n_1^{-2},n_2^{-2},n_3^{-2},n_4^{-2})>0 \tag{2.3}$$

$$\hat{R}\approx R\hat{I} \ , \ \hat{I}=T^{-1}$$

$$x^2 = x^\mu\hat{\eta}_{\mu\nu}x^\nu = x^1b_1^2x^1+x^2b_2^2x^2+x_3^3b_3^2x^3-x^4b_4^2x^4$$
$$= x^1\frac{1}{n_1^2}x^1+x^2\frac{1}{n_2^2}x^2+x^3\frac{1}{n_3^2}x^3-t\frac{c_o^2}{n_4^2}t \tag{2.4}$$

where c_o is the speed of light in vacuum and the invariant quantity s is defined by

$$ds^2 = dx^\mu\hat{\eta}_{\mu\nu}dx^\nu = dr^kb_k^2dr^k-dtc_o^2b_4^2dt \tag{2.5}$$

The b's (or the n's) are the relativistic characteristic quantities of the medium considered with the functional dependence

$$b_\alpha = \frac{1}{n_\alpha} = b_\alpha(x,\dot{x},\ddot{x},\psi,\psi^+,\partial\psi,\partial\psi^+...) \tag{2.6}$$

and the dual form $b_\mu = n_\mu^{-1}$ is introduced for physical interpretations connected with the index of refraction. As it is well known Riemannian Spaces geometrize the homogeneous and isotopic vacuum with a metric $g(x)$ solely dependent on the local coordinates x. Instead the isominkowskian spaces $\hat{M}(x,\hat{\eta},\hat{R})$ geometrize interior physical media and for $T>0$ are locally isomorphic to the conventional Minkowski space $M(x,\eta,R)$. Also isospace $\hat{M}(x,\hat{\eta},\hat{R})$ therefore permits a nonlinear - nonlocal-noncanonical generalization of the conventional Minkowski space by preserving its geometric axioms and topology, because all non-linear - non local - non canonical departures from the Minkowski space are embedded in the isounit $\hat{I} = T^{-1}$. In the final analysis the isospaces $\hat{M}(x,\hat{n},\hat{R})$ are a manifestation of the Lie-isotopic theory (for more details and properties of the isospace see ref.(3,4).

For the case of special relativity[3,4,7] the Lie-isotopic theory permits a non local relativistic kinematics. For the case $b_k > o$ constant (2.6), the infinitesimal interval ds^2 has the following form

$$ds^2 = b_k^2(dx^k)^2 - b_4^2 c_o^2 dt^2 \tag{2.7}$$

$(k=1,2,3)$. For an isotropic threedimential space we get for null separation $ds^2 = o$:

$$b^2(dx^2 + dy^2 + dz^2) = b_4^2 c_o^2 dt^2 \tag{2.8a}$$

or

$$u^2 = (\frac{1}{dt^2})(dx^2 + dy^2 + dz^2) = (\frac{b_4}{b})^2 c_o^2 \tag{2.8b}$$

From the above equation it easily follows that

$$u = (\frac{b_4}{b})c_o \tag{2.9}$$

i.e, a maximal casual speed u which depends on the physical system (and its interaction). Moreover, one has

$$u \to c_o \qquad according \; to \qquad \frac{b_4}{b} \to 1 \tag{2.10}$$

The maximal casual speed u can be interpreted, from a physical stand point as the speed of quanta of the interaction which requires a representation in terms of a nonlocal Minkowski space[3,4,7].

The nonlocal Lorentz transformation is the (isotopically linear) transformation Λ leaving (2.7) invariant[8]:

$$x' = \hat{\Lambda} \star x = \hat{\Lambda} T x \tag{2.11}$$

$$x^t \star x = x^t \star \hat{\Lambda}^t \star \hat{\Lambda} \star x = x^t \star x \tag{2.12}$$

(where the upper "t" denotes transpose). The nonlocal boosts (for motion, saying, along the x-axis) can be expressed as

$$x' = \hat{\gamma}(x - \beta x_4) \; , \; y' = y \; , \; z' = z \; , \; x_4' = \hat{\gamma}(x_4 - \frac{\hat{\beta}^2}{\beta}x) \tag{2.13}$$

where $\beta = \frac{v}{c_o}$ (v is the relative speed of the reference's frame) is the usual speed parameter and

$$\hat{\beta} = \beta \frac{b}{b_4} \quad , \quad \hat{\gamma} = (1 - \hat{\beta}^2)^{-1/2} \tag{2.14}$$

According to ref.(3,4,7) the main nonlocal kinematical laws (for an isotropic space) are summarized:

Minkowski space

$$\upsilon=(\upsilon_1+\upsilon_2)(1+\frac{\upsilon_1\upsilon_2}{c_o^2})^{-1} \ , \quad \Delta t=\Delta t_o(1-\frac{\upsilon^2}{c_o^2})^{-1/2} \ , \quad \Delta L=\Delta L_o(1-\frac{\upsilon^2}{c_o^2})^{1/2} \qquad (2.15)$$

Isominkowski space

$$\upsilon=(\upsilon_1+\upsilon_2)[1+(\frac{b}{b_4})^2\frac{\upsilon_1\upsilon_2}{c_o^2}]^{-1} \ , \Delta t=\Delta t_o[1-(\frac{b}{b_4})^2\frac{\upsilon^2}{c_o^2}]^{-1/2} \ ,$$

$$\Delta L=\Delta L_o[1-(\frac{b}{b_4})^2\frac{\upsilon^2}{c_o^2}]^{1/2} \qquad (2.15a)$$

Also the isotopic kinematical laws of time duration and length contraction for a particle of rest mass m can be expressed in terms of the usual energy E:

Minkowski space

$$\Delta t = \Delta t_o\frac{E}{m_o} \ , \quad \Delta L = \Delta L_o\frac{m_o}{E}$$

Isominkowski space

$$\Delta t = \Delta t_o[1-(\frac{b}{b_4})^2+(\frac{b}{b_4})^2(\frac{m_o}{E})^2]^{-1/2} \ , \quad \Delta L= \Delta L_o[1-(\frac{b}{b_4})^2+(\frac{b}{b_4})^2(\frac{m_o}{E})^2]^{1/2} \quad (2.16)$$

Clearly for $E>>m_oc^2$., E can be considered as the total energy of the particle.

Besides the corresponging kinematical quantities, i.e., iso-four-velocity

$u^\mu = \dfrac{dx^\mu}{ds}$, , iso-four acceleration $\alpha^\mu = \dfrac{du^\mu}{ds}$, Santilli has also introduced the iso-four momentun

$$p = (p^\mu) = (\hat{m}u^\mu) = (m_o\hat{\gamma}c_o\upsilon^\kappa \ ,m_o\hat{\gamma}\hat{c}_o) \ ,\hat{m} = m_o\hat{\gamma} \qquad (2.17)$$

where υ is the velocity in isoeuclidean isospace.

The isocasimir p^2 implies the fundamental isoinvariant of the isospecial relativity (see ref.(4), form (4.22)).

The isospecial relativity is then based on isotopy of all basic postulates of the special relativity studied in detail by Santilli[3]. (For more details and applications we suggest the ref.(4,7)).

RELATIVISTIC GENOTOPIC STRUCTURE

A general example of a relativistic genotopic structure is the case where ds^2 is given from the Riemann relation

$$ds^2 = dx^\mu\hat{\eta}_{\mu\nu}dx^\nu \qquad (3.1)$$

and the metric coefficients $\hat{\eta}_{\mu\nu}$ are complex.

For the above metric we see that this is not any more Riemanian. The metric which we could study here is exactly the relation (2.5) with the assumption that b_k are not all real. The special case $b_k^2 = 1$, $k =1,2,3$ and $b_4^2 = (1+i\lambda)^2$ has been studied by Jannussis and his collaborators[9] and it is the "Lie-admissible Complex Time

Model", for which the invariant quantity s is defined by

$$ds^2 = dx^2+dy^2+dz^2-c_o^2(1+i\lambda)^2dt^2 \qquad (3.2)$$

From the above metric the causal speed or interaction speed is complex, i.e.,

$$u = c_o(1+i\lambda) \qquad (3.3)$$

with

$$|u| = c_o\sqrt{1+\lambda^2} > c_o \qquad (3.4)$$

The introduction of the Lie-admissible complex time can be simply done by means of the transformations $t \rightarrow t(1+i\lambda), \lambda$ real. We remark that the metric is not Riemannian for the Special and General Relativity. In other words, the metric is complex. Consequently the introduction of the Lie-admissible complex time model leads exactly to generalized theories of gravitation[10-13]. Recently Marques[14] has derived the Dirac equation in a non-Riemannian manifold by using complex algebra and supposing that there exists a complex mass. Halliwell and Hartle[15] suggest complex metrics in their paper "Integration contours for the no-boundary wave function of the universe". Recently, Hawking et al.[16] considered a model in which the action is given by the Einstein-Hilbert action I_g plus the massive field action I_ϕ ref. (16) forms (1.1) and (1.2)). In essence, the no-boundary proposal, consists in trying to define a quantum state of the universe via a path integral over all metrics with action. In general, the metrics in the path integral are complex, rather than purely Lorentzian or purely Euclidean.

The metric (3.2) for $1+i\lambda = (1+\lambda^2)^{1/2} e^{i\theta}$ takes the form

$$ds^2 = dx^2+dy^2+dz^2-c_o^2(1+\lambda^2)e^{2i\theta}dt^2 \; , \quad \theta = \arctan\lambda \qquad (3.5)$$

Since in the special theory of relativity the time appears as polarized by the constant c_o (light velocity), metric (3.5) can be derived from the usual Minkowski metric by the equivalent transformaction $c_o \rightarrow c_o(1+i\lambda)$. Because of this, we can use in what follows the known formula of the masses putting instead c_o the value $c_o(1+i\lambda)$ and according to ref.9 we obtain

$$m = m_o\cos\phi.[1-2\frac{1-\lambda^2}{1+\lambda^2}\frac{w^2}{c_o^2}+\frac{w^4}{c_o^4}]^{-1/4} +i\; m_o\sin\phi.[1-2\frac{1-\lambda^2}{1+\lambda^2}\frac{w^2}{c_o^2}+\frac{w^4}{c_o^4}]^{-1/4} \qquad (3.6)$$

where $\phi = \dfrac{1}{2}\arctan\dfrac{w^2\sin2\theta}{c_o^2-w^2\cos2\theta}$, $w = \dfrac{v}{\sqrt{1+\lambda^2}}$ and v is the relative speed of the referance's frame.

The new formula of the complex mass (3.6) is of important physical interest, since for $\lambda\rightarrow0$ from positive values we obtain $\theta=0$, $\phi=0$ with the condition $c_o>w$ and (3.6) reduces to the known Einstein mass formula. For $\lambda\rightarrow0$ from negative values, we have $\theta=\pi,\phi=\pi/2$ with $c_o<w$. In this case formula (3.6) reduces to the one of the classical tachyons according to Recami's theory[17] i.e.

$$m = im_o(1-\frac{v^2}{c_o^2})^{-1/2} = m_o(\frac{v^2}{c_o^2}-1)^{-1/2} \qquad (3.7)$$

The absolute value of the mass is also of interest.
We have:

$$|m| = m_o[1 - 2\frac{1-\lambda^2}{1+\lambda^2}\frac{w^2}{c_o^2} + \frac{w^4}{c_o^4}]^{-1/4} \qquad (3.8)$$

The idea that mass can become a complex quantity, mainly in suitable particle's "resonances", has already been developed some years ago by several authors[18-20].

From the mass formula (3.6) we note that it depends on the velocities v and c, as it happens in the theory of Einstein and even on the parameter λ, which describes the interaction of a particle with the rest of the word[9]. According to ref.9 the determination of the parameter λ for different models can be done using the initial energy of particles, the Planck constant h, the speed c of light and the known "chronon" τ of Caldirola. For the case of Dirac energy operator $H = \pm c_o(p^2 + m_o^2 c_o^2)^{1/2}$ and according to ref.9 we take the form

$$\lambda = \pm \frac{\hbar}{\tau(p^2(o)+m_o^2c_o^2)^{1/2}}\ln\ [1+\frac{\tau^2c_o^2(p^2(o)+m_o^2c_o^2)}{\hbar^2}] \qquad (3.9)$$

The Caldirola "chronon" τ is of important physical significance, because it describes the time of interaction between two physical systems[21,22].

This result has been expected, because the use of the Lie-admissible complex time model is equivalent to the complex energy model.

The determination of λ in (3.9) gives in our opinion, an important significance in the formula of mass (3.6) not only in what concerns the theoretical physical interest but also for practical applications. According to Santilli[23], the Lie-admissible complex time model is a simple model for an open theory of gravitation, with new results particularly relevant to quantum gravity.

WHEELER-DEWITT EQUATION IN THE LIE-ADMISSIBLE THEORY

The quantum mechanical operator equation for the wave function of the universe leads to the Wheeler-DeWitt (WD) equation[24], which is a second order hyperbolic differential equation in the dynamical phase space variables and which possesses only stationary solutions. The wave function is time independent and there is temporal development in a spatially closed universe.

A few years ago, Gonzalez -Diaz[5] and Jannussis[6] applied the Lie-admissible theory to the Wheeler-DeWitt equation. Due to the fact that Lie-admissible structures are incompatible with the formulation of conservation law, such structures have to describe open systems. In order to give a Lie-admissible character to (WD) equation, the above authors have used the small distance derivative (SDD) model[25,26]. Gonzalez-Diaz[5] started from the functional integral proposed by Hartle and Hawking[27]

$$\psi[h_{ij}] \propto \int_c \delta g(x)e^{iS_E(N,g)} \qquad (4.1)$$

for a surface with space like three-metric h_{ij} . In eq. (4.1) S_E is the Euclidean classical action integral for gravity, N is the lapse function and the functional integral is taken over all four-geometries on a space-like boundary on which the induced metric is h_{ij}. Since (3.1) does not depend explicitly on time, there exists a Hamiltonian constraint with ^3R denoting the scalar curvature constructed of the form

$$H = \tilde{K}^2 - \tilde{K}_{ij}\tilde{K}^{ij} + {}^3R = 0 \qquad (4.2)$$

where the three-dimensonal metric h_{ij} and $\tilde{K}$ is an extrinsic curvature of the boundary three-surface, related to the conventional quantity by

$$\tilde{K} = K(1 - \frac{\eta^{\star 2}}{N^2})^{-1/2} \quad , \qquad (4.3)$$

$$\tilde{K}_{ij} = K_{ij}(1 - \frac{\eta^{\star 2}}{N^2})^{-1/2} \quad , \qquad (4.4)$$

and $\eta^\star$ is the value of the lapse function corresponding to a proper-time separation equal to the Planck time. Hence, the conventional Hamiltoniam of general relativity

$$H = K^2 - K_{ij} K^{ij} + {}^3R \qquad (4.5)$$

is no longer rigorously zero and the quantum state of a closed universe is described by a wave function ψ satisfying the Schrodinger equation $\tilde{H}\psi = 0$, , where the Hamiltonian $\tilde{H}$ is constructed form (4.2) by exressing the second fundamental form K_{ij} in terms of the momenta conjugate to the h_{ij}.

Using the concepts of the small-distance derivate model, Gonzalez-Dianz[5] obtained a new modified Lie-admissible Wheeler-DeWitt equation and calculated the energy of the ground state of the universe. Since we have given a Lie-admissible structure to the universe, we are forced to consider it as being an open system which is created by some kind of physical reality.

The Lie-admissible Wheeler-DeWitt equation is of the form

$$[-G_{ijkl}\frac{\tilde{\delta}^2}{\tilde{\delta}_{hij}\tilde{\delta}_{k\ell}} - h^{1/2}(1 - \frac{\eta^{\star 2}}{N^2}) + {}^3R] \; \psi \; [h_{ij}] = 0 \qquad (4.6)$$

where G_{ijkl} is the metric on the sphere formed by all three-metricts h_{ij} and a tilde over δ means small-distance variation. Using a mini superspace approximation consisting of a homogeneous isotropic universe with no matter fields and zero cosmological constant, we finally obtain the relation

$$\hat{E}\psi(\alpha) = \pi M^\star(n + 1/2)\psi(\alpha) \quad , \quad n = 0,1,2,.... \qquad (4.7)$$

where α is the radius of the three-sphere that bounds all compact metrics of the form $ds^2 = 2(-N^2dt^2 + \alpha^2 d\Omega_3^2)$, $M^\star$ is the Planck mass and $d\Omega_3^2$ demoting the metric on a three-sphere of unit radius.

The interpretation of equation (4.7) is that the considered universe has a nonzero total energy whose values coincide exactly with the corresponding ones of a harmonic oscillator with Planck mass $M^\star$, it is an open system which interacts with some sort of "exterior" world and it is created by a kind of physical reality. The wave function of the universe is now interpreted as giving the probability amplitude for the universe to have been created from some sort of additional physical reality.

From the above results we conclude that the universe is no longer considered to be closed and has a nonzero vacuum energy.

Recently many authors[16,24,28,29,30,31] study in several ways the (W.D) equation and almost all of them are trying to justify the origin of time asymmetry and the time arrow of the universe.

But the results of the Lie-admissible (W.D) equation (4.6), following the application of SDD model (actually this is a dissipative model) help us in the definition of the space-time asymmetry. According to Jannussis et al[6] this model is directly related to the Planck constants $L \sim 10^{-33} cm$, $\tau = 10^{-43}$ sec. The application of this model to the study of space-time regions close to Planck dimensions is expected to be relevant for quantum Gravity. The fluctuating topology of space-time appears in a simple and natural way using the SDD model, as we have demonstrated[6]. Another characteristic of this model is that it accommodates the existence of a constaint on our measurements (resolution limits) for any physical system of the form

$$\Delta q \geq =L = 10^{-33} \ cm, \quad \Delta t \geq \tau = 10^{-43} \ sec$$

as repeatedly pointed out in refs.(32,33). The SDD model is a special case of the Lie-admissible formulation. Jannussis et al[34] and Nishioka[35], by using the noncanonical commutation relations in the Lie-isotopic as well as in the Lie-genotopic formulation, showed that Hadronic Mechanics has, in general, a noncanonical character in time and space. According to ref.(34;35), by forming the commutator in the isotopic case of the operators p_μ and q_ν we obtain the following expression for the element T

$$T = 1 + \frac{q_\nu^{(o)}}{q_\nu} \ , \quad T(q_\nu) \neq T(-q_\nu) \ , \quad q_\nu \geq q_\nu^{(o)} \tag{4.8}$$

The element T is the only one which leaves the usual canonical commutator of p_μ and q_ν invariant. If q_ν, is large enough, then $T(q_\nu)$ approaches unity. In this case the Lie-isotopic commutator coincides exactly with the ordinary canonical commutator. It follows, from the standpoint of the space components[35], that in a microscopic region, such as the neighborhood of a hadron, we may use the Lie-admissible noncanonical, commutation relations, though, in the range of the atomic structure, we may use the cannonical commutation relations.

At this point we want to refer the results of sections 2 and 3 of this paper, which are derived via the Lie-admissible formulation and leads to a generalization of the Minkowski space to isominkowski space[3].

Relation (4.8) admits a general interpretation since it destroys the symmetry of space and time. The desctruction of the symmetry of time $(T \neq -T)$ is studied by physicists who are particularly concerned with the problem of strong interactions. The element T is also important because its interpretation is compatible with the interpretation of the quantum vacuum by Wheeler[36] and demostrates the time asymmetry. Consequently the formula (4.8) accounts not only for the asymmetry of space-time but also for its inhomojeneity.

CONCLUSION

In the present paper we have investigated the isotopic and genotopic structure in the relativistic theory. According to Santilli[3,4] the central assumption is that motion of particles and electromagnetic waves within inhomogeneous and anisotropic media implies an alteration of space-time representable via isominkowski spaces.

For the isotopic case we see from eq. (2.3) that the maximal speed of propagation of causal signals is not an absolute constant but depends on the local physical condition, being c in vacuum, a higher (or lower) value in the interior of hadronic (or nuclear) matter. For the genotopic case we have introduced the simple

Lie-admissible complex time model and obtained some intriguing results. The time evolution of the commutators [q(t),p(t)] is not anymore canonical even for the non-relativistic quantum physics[9] . The metric is not a Riemannian metric but a complex metric and the interaction velocity (3.3) is also complex. Due to the fact that the interaction speed is complex we obtain also complex values for the mass (3.6). From the quantum mechanical aspect and according to the results of ref.9 we conclude that the universe is open and we obtain a unified description of the effects of quantum gravity with respect to the Lie-admissible complex time, which contains as special cases the time theory of Hawking[37] and Gonzalez-Diaz[38]. Also the study of the Wheeler-DeWitt equation, via the (SDD) model, i.e. the Lie-admissible (W.D) equation (4.6) and the eigenvalues of the Max-Planck energy $\pi M^{\star}(n+{}^1/2)$ leads us to the conclusion that the universe can not be considered as closed and has a nonzero vacuum energy. Consequently, formula (4.8) accounts not only for asymmetry of space-time but also for its inhomogeneity and, therefore, for the cosmological arrow is evident. (For more details about the cosmological arrow see ref. (16,24)).

REFERENCES

1. R.M.Santilli: Hadr. Journ. 1, 574 (1978); Lie-Admissible Approach to Hadronic structure Vol. I (1978) Vol. II (Nonantum, Ma. Hadronic Press); Fountations of Theoretical Mechanics Vol.I (1978) Vol.II (1983) (Springer-Verlag, Heidelberg, N. York).
2. H.C. Myung and R.M. Santilli: Hadr. Journ. 5, 1267, 1377 (1982); R. Mignani, Myung and R.M. Santilli: Hadr. Journ. 6, 1878 (1983); A. Jannussis et al: Hadr. Journ. 5, 1923 (1982), 6, 1653 (1983); R.M. Santilli; Preprint I.B.R, Isolinear, Isolocal and Isocanonical, Axion-Preserving Formulation of Local Realism, to appear (1993), and references therein.
3. R.M. Santilli:Isotopic Generalization of Galilei's and Einstein's Relativities, Vols. I,II,Hadronic Press (1991);Elements of Hadronic Mechanics Vol's I,II and III, JINR, Dubna, in Press.
4. R.M. Santilli: Application of Isosymmetries to the Apparent Chemical, Synthesis of Hadrons (Inter. Workshop Symmetry Methods in Physics in honor of Prof. Yu.R. Smorodinsky, J.I. N.R. Dubna, Russia, July 6-10 (1993) and references therein.
5. P.F. Gonzalez-Diaz: Hadr. Journ. 9, 199 (1986); Hadr. Journ. Suppl. 2, 437 (1986); "A Heuristic Particle-Geometry Principle" (Preprint (1986), Dept. of Applied Math. and Theor. Physics, Univ. of Cambridge, Cambridge CB EW, U.K); Hadronic Mechanics and Non-potential Interactions, Ed. M. Mijatovic (Nova Science Publ. p. 77 (1990).
6. A. Jannussis: Hadr. Journ. Suppl. 2, 458 (1986); A. Jannussis and I. Tsohantzis: Hadr. Journ. 11, 1 (1988); Hadronic Mechanics and Nonpotential Interactions, Ed. Mijatovic (nova Science Pl. p. 1 (1990).
7. F. Cardone and R. Mignani: Nonlocal Relativistic Kinematics"Univ. of Rome preprint No 910 Nov. 1992.
8. R.M. Santilli: Lett. N. Cim. 37, 545 (1983).
9. A. Jannussis: Hadr. journ. 13, 399, 425 (1990); A. Jannussis D. Skaltsas and G. Brodimas: Hadr. Journ. 13, 415 (1990): Hadronic Mechanics and Nonopotential Interactions, H.C. myung, Ed. Nova Science Science Publ. N. Y. 1991, Part. II Physics.
10. K. Borchenius: phys. Rev. D13, 2707 (1976); D. Boal and J. Moffat: phys. Rev. D11, 2026 (1975); R. Mann J.Moffat: phys. Rev. D26, 1858 (1982).
11. J. Moffat: Phys. Rev. D19, 3554, 3562 (1979); journ. Math. phys. 21, 1978 (1980); J. Moffat and Boal: phys. Rev. D11, 1375 (1975).
12. S. Marques and C. Oliveira: phys. D36, 1716 (1987).

13. J. Moffat: phys. Rev. $\underline{D39}$, 479 (1989).
14. S.Marques: Journ. Math. Phys. $\underline{29}$, 2127 (1988).
15. J. Halliwell and J. Hartle: phys. Rev. $\underline{D41}$, 1815 (1990).
16. S. Hawking, R. Laflamme and G. Lyons: phys. Rev. $\underline{D47}$, 5342 (1993) and references therein.
17. E. Recami: Riv.N. Cim. $\underline{9}$ No 6 (1986) and references therein.
18. E. Recami: Possible causality effects and comments on tachyons, Virtual particles, resonances, report 1 FUM-088ISM (Univ. of Milan, Phys. Inst. August 1968); G. Fis.(Bologna) 10, 195 (1969).
19. E. Sundarshan: physics of complex mass Particles, report ORO-3992-5 (Texas Univ. Austin 1970).
20. J. Elmonds: Found. phys. $\underline{4}$, 473 (1974).
21. A. Jannussis: Lett. N. Cim. $\underline{39}$, 75 (1981).
22. R. Bonifacio and P. Caldirola: Lett. N. Cim. $\underline{33}$, 197 (1981); $\underline{38}$, 615 (1983); R. Bonifacio: Lett. N.Cim. $\underline{37}$, 48 (1983).
23. R.M. Santilli: Hadr. Journ. $\underline{2}$, 1460, 1764, 1773 (1979).
24. J. Moffat: Found. phys. $\underline{23}$, 411 (1993) and references therein.
25. P. Genzalez-Diaz: Lett. N. Cim. $\underline{41}$, 489 (1984).
26. A. Jannussis: N. Cim. $\underline{90B}$, 58 (1985); A. Jannussis, M. Symeonidis and G. Karayannis: Hadr. Journ. Suppl. $\underline{1}$, 239 (1985).
27. J. Hartle and S. Hawking: phys. Rev. $\underline{D28}$, 2960 (1983).
28. P.D. Collins and Euan J. Squires: Found. phys. $\underline{23}$, 913 (1993).
29. M. Castagnino and F. Lombardo: phys. Rev. $\underline{D48}$, 1722 (1993).
30. M. Makela: phys. Rev. $\underline{D48}$, 1679 (1993).
31. L. Gardy: phys. Rev. $\underline{D48}$, 1710 (1993).
32. T. Padmanabhan: Ann. phys. (N.Y). 165, 38 (1985).
33. C. Mead: phys. Rev. $\underline{135}$, 849 (1964); 143, 990 (1966).
34. A. Jannussis, G. Brodimas, A. Pierris and H. Ioannidou: Hadr. Journ. $\underline{6}$, 623 (1983).
35. M. Nishioka: Hadr. Journ. $\underline{7}$, 1636 (1984).
36. J. Wheeler: Geometrodynamic (Acad. Press, Gambridge 1979); C. Patton and J. Wheeler, in Quantum Gravity, an Oxford Symposium, Vol. 1.
37. S. Hawking: A Brief History of Time (Bandam Books (1988).
38. P. Gonzalez-Diaz: SIS preprint - RCG 189-01: Maximum Resolution and the Initial Conditions of the Universe (1987).

A LOOK AT FRONTIERS OF HIGH ENERGY PHYSICS: FROM THE GeV(10^9eV) to PeV(10^{15}eV) AND BEYOND

M. Barone*

National Center for Scientific Research"DEMOKRITOS"
15310 AG. Paraskevi-Athens, Greece

Since the talk is devoted to people not strictly involved in High Energy Physics, the formalism will be sacrified to the simplicity. We will review:

1. Some aspects of the Accelerator Physics
2. Some others of the Non-Accelerator Physics
3. The NESTOR neutrino telescope
4. Conclusions

1. THE ACCELERATOR PHYSICS

The years going from 1970 to the end of 80s led to development of the Standard Model which provides the present understanding of the particle physics. The Standard Model combines:
-The Quantum Chromodynamics (strong forces).
-The Electromagnetics and Weak forces
incorporating all the known quarks and leptons believed to be the basic building blocks of the matter.
The forces are transmitted trough bosons and the interaction modes are:
-Electromagnetic interaction where electrons exchange a photon.
-Strong interaction where quarks exchange a gluon and a color.
-Weak interaction where quarks and leptons exchange a W or a Z particle.
This is the way the Nature in the frame of such a theoretical model works down to a resolution of 10^{-18} m (sub nuclear level).
To test the Standard Model the physicists have build linear and circular particle accelerators[1,2,3],which allow them to probe the structure of the matter to increasing depth. Because of quantum rules wich govern the sub-atomic physics a price in form of energy has to be paid to go deep and deep and achieve higher resolution.

----

*work supported by the Attica Technology Park.

The Colliders based on the colliding beam technique have been so far the way to proceed see Fig1; therefore experiments were made either using fixed target either colliding beam devices and from the examination of the debris-released in the collision information is gained about the nature of the particles and the forces acting between them .

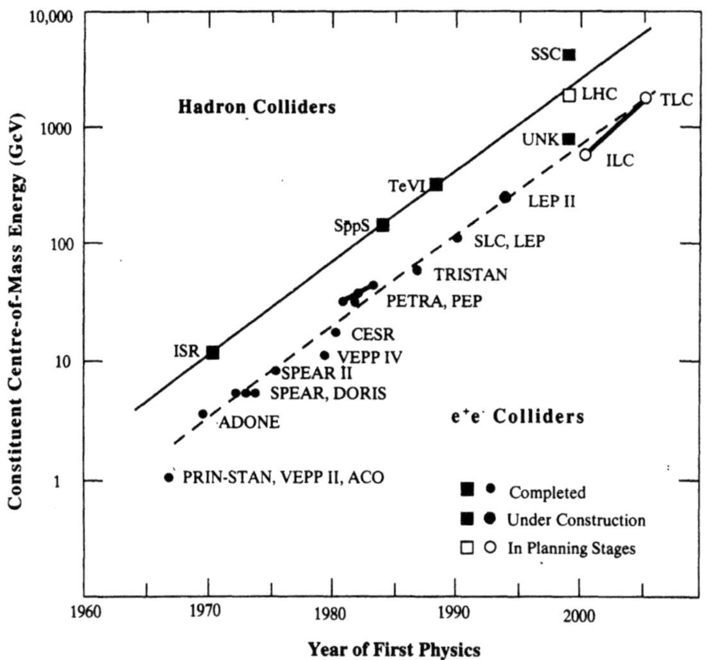

Fig. 1

One of the latest accelerator build is the LEP(electron-positron collider) which will explore the mass range of 100-200 GeV,corresponding to the dimension of 10^{-18} m.It is a very good tool to test the Standard Model.The physics which we meet at the energies mentioned before is also that which prevailed when our expanding Universe (according to the BIG-BANG model) was 10^{-10} sec. old.
The LEP with is 27 Km of circumference is presently the largest accelerator in the world and togheter with the SLAC collider in U.S.A, has given the contribution to discover the existence of only 3 neutrino particle families in the Universe confirming the BIG-BANG nucleosynthesis model which predicts aboundances of Deuterium, He-3, He-4 and Litium.

1.1 LIMIT OF THE STANDARD MODEL

The Standard Model is felt to be incomplete because:

- it requires about 20 apparently arbitrary constants (particles masses,coupling constant,mixing angles and so on).
-when it is applied at higher energies than those now accessible it gives solutions inconsistent with some laws conservation(CP violation,Baryon Number and so on). On the other hand the unification of the electromagnetic and weak nuclear

	TESLA	DLC (D-D)	NLC (SLAC)	JLC (KEK)	VLEPP (P/N)	CLIC (CERN)
Collision Energy (GeV)	500	500	500	500	500	500
Gradient (MV/m)	25	17	38	28	96	80
Active Length (km)	. 20	30	14	17	6.4	6.6
Frequency (GHz)	1.3	3	11.4	5.7	14	30
Pulse rate (Hz)	10	50	180	150	300	1700
Bunches per pulse	800	172	90	72	1	1-4
Luminosity (10^{33} cm^{-2} s^{-1})	8	4	9	9.7	12	8-3.2

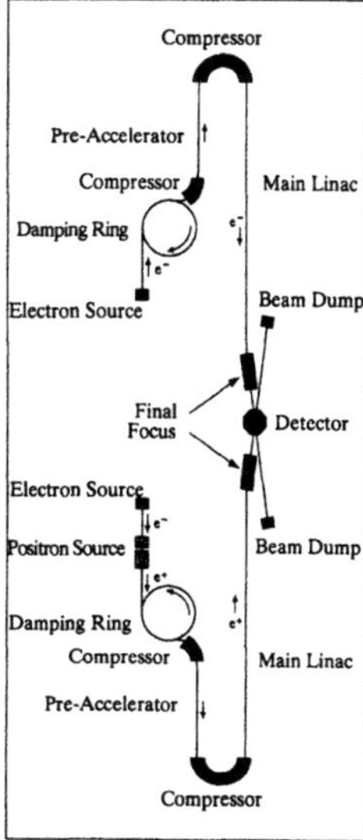

Fig. 2
Future linear colliders

forces in the electro-weak theory push people to envisage a concept of symmetry which would combine quarks and leptons coupled trough an unique interaction mode and a Grand Unified Theory could be open at Higher Energies (about 10^{15} eV). Then the unification of Strong,Weack and Electromagnetic forces could be achieved,therefore a further order of magnitude in resolution is needed to go to 10^{-19} m when the Universe was 10^{-12} sec old.
Because of that linear[4] and circular colliders in the TeV region are under construction. Among the linear colliders[5] should be mentioned the following shown in Fig.2

LEP/LHC
p (proton)
ion
e+ (positron)
e- (electron)
SPS
BOOSTER
PS
EPA
LIL
e+ e-
Linacs
Proton ion
linacs

Fig. 3

15 km
100 TeV main ring
10 TeV injector ring
500 GeV second
injector ring
86 km
300 km Circumference
Interaction regions

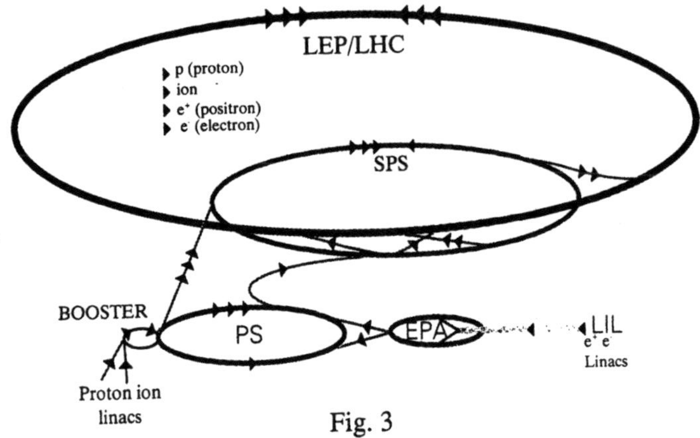

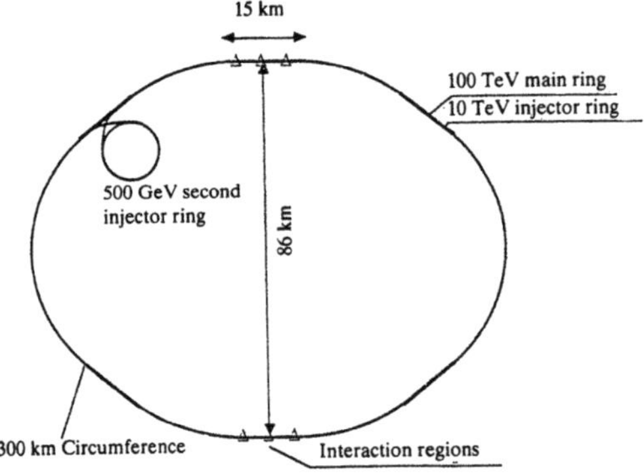

Fig. 4

The proton colliders seems to offer an easiest approach to the TeV region,using s.c. magnets as elements of the accelerator rings.The CERN is planning to build in the period 94-99 the Large Hadron Collider(LHC) which will achieve p-p collisions at 16 TeV in the center of mass; it will be installed in the same tunnel than the LEP see Fig. 3.
In the USA next to Dallas the Super conducting Super Collider(SSC) is also under construction.It will reach 20 TeV/beam/protons allowing to have 40 TeV in the center of mass.It is expected to be in operation during 98.*
In Italy a feasibilty study of an 100TeV+100TeV proton accelerator,the ELOISATRON[6] is also carried out See Fig. 4.

What next?.Will the physicists take into consideration the suggestion made by Fermi 40 years ago to construct an accelerator that would uncircle the world?

2. THE NON-ACCELERATOR PHYSICS

During the 80s the High Energy Physics community started to use large Non -Accelerator facilities to study the nature at sub-nuclear scale detecting cosmic rays,which up to the 50s were the main tool to study a large fraction of the "known" elementary particles,afterward the accelerators took over with their flexibility in providing high intense beams of predetermined energies and species.
Cosmic radiation contains particles with incredible energy;the high energy part of the spectrum is distinguished as :
-Very High Energy,VHE ~ 1   TeV $= 10^{12}$ eV
-Ultra High Energy,UHE ~ 1   PeV $= 10^{15}$ eV
-                      ,EeV ~ 10^{3} PeV $= 10^{18}$ eV

These energies are far beyond the range of the available accelerators.
The exploration of these energies generated a new discipline at the junction of High Energy Physics,Astrophysics and Cosmology:

"THE PARTICLE ASTROPHYSICS[7,8,9]"

2.1 THE PARTICLE ASTROPHYSICS

This new experimental,observational and theoretical field brings together the physics of the early Universe based on large scale to the physics of the very small at very highest energy and it is not possible to distinguish between them when we try to answer to fundamental questions like:

-The dark matter in the Universe and the origin of Galaxies.

For half century,Astrophysicists have suspected that most of the matter in the

---

* Just after the conference, the American Congress decided to stop the SSC.

Universe is non luminous and transparent.A close analysis of relative motion of individual galaxies or clusters of galaxies shows that the masses implied by gravitational forces are much bigger than the amount of visible matter.

Such dark matter seems to pervades the Universe and may not be made by the ordinary protons and neutrons (the baryonic matter) but by relic particles (like light neutrinos,axions and Weackly Interacting Massive Particle(i.e.WIMP's)) from the early Universe.

The combination of sub nuclear physics and relativistic cosmology generated the Theory of Cosmic Inflation, which push in the existence of hidden symmetries, no longer existing in the Universe,but which must have been manifested at high tempe-ratures($>>10^{12}$ K) of the first stages after the Big-Bang.

When the Universe started to cool it went trough a series of transition phases during which fundamental symmetries broke down; the Theory of Inflation suggest that during one of these phases ,one single low symmetry bubble grew so rapidly to provide a possible explanation of the galaxies formation.The only messengers of these are to be found in the fossil signal(the relic particles) such as the photons of the galactic background radiation discovered in 1964.

Another cosmic remnant is a Magnetic Monopole which also according to some cosmological theories was produced $\sim 10^{-34}$sec after the Big-Bang at the phase transition associated to the breaking of the Great Unification Symmetries.

In order to understand better the stabilty of the stars and the mechanism for the acceleration of the cosmic rays to energies as high as 10^{20} eV,physicists using a new generation of particle's detectors investigate about U.H.E. photons and neutrinos of Low,High and TeV energy,the last ones coming from point sources or Active Galactic Nuclei that led to the NEUTRINO ATRONOMY.

A possibile source of U.H.E photons and TeV neutrinos is shown in the Fig.5.

Particles of 10^{15} eV and beyond are produced and possible sources may be the CIGNUS X-3,VELA X-1,the CEN X-3 and to detect such particles underground and underwater laboratories have been made because the layers of rocks and water is used as screen to shield the detectors.

The table A shows the mayor underground and underwater installations of past and present; among them enphasis will be given hereinatfter to NESTOR.

2.2 THE NESTOR NEUTRINO UNDERWATER TELESCOPE

NESTOR[10] stands for Neutrinos from Supernovae and TeV sources Ocean Range and indicates an underwater neutrino astrophysics telescope to be located in the international waters off the Southwest of Greece, near the town of Pylos. The project has been funded for the initial stage and in the last 2 years a group of physicists from Greece and Russia have carried out demonstration experiments in 4Km deep water, counting muons and veryfying the adequacy of the deep sea site.

Future plans for a 100.000 m^2 high energy neutrino water Cerenkov detector have been made. It will be composed of hexagonal towers with 1176 optical modules.A single tower with 168 phototubes currently under construction is shown in Fig. 6.

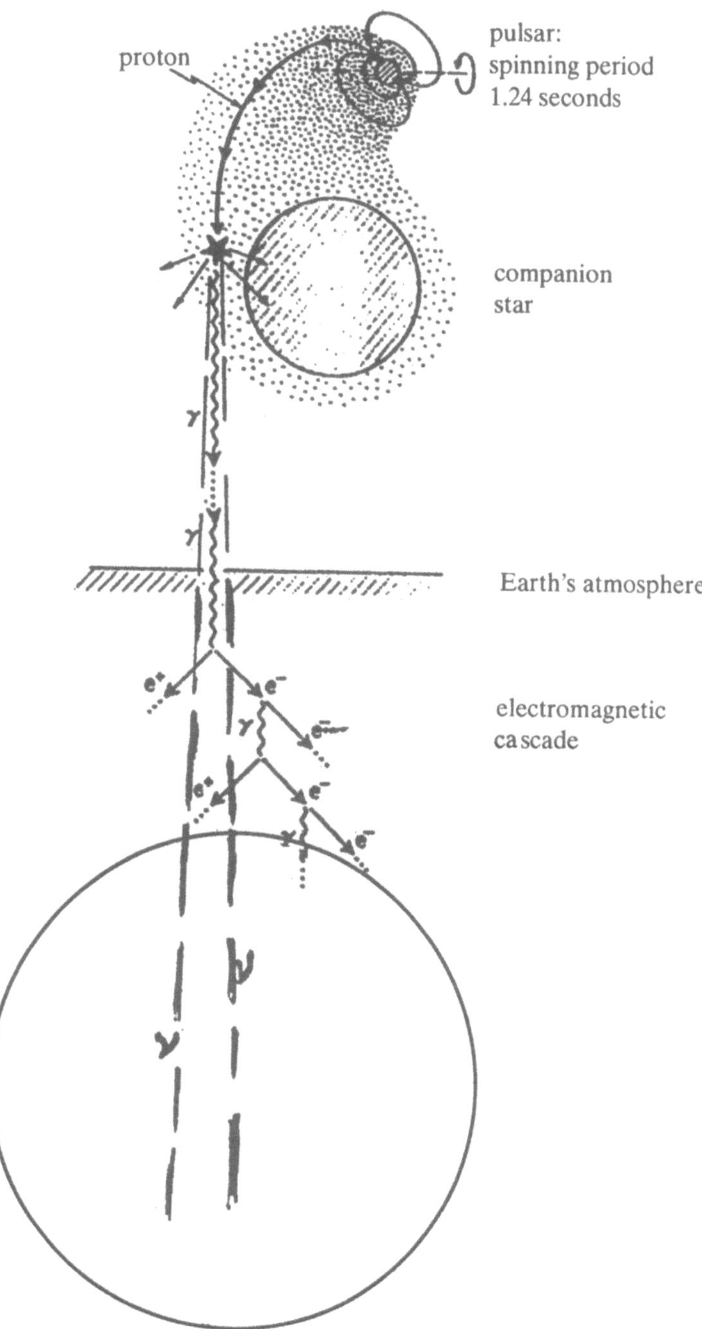

Fig. 5 Possible scenario for the production of gamma rays and neutrinos in binary x-ray sources. Beams accelerated by a pulsar interact with target material in an accretion disc formed by matter from the companion star falling into the collapsed neutron star.

MAJOR UNDERGROUND
INSTALLATIONS
OF PAST AND PRESENT

TABLE A

		depht (in m water equioralent)
AMANDA, South Pole	construction	1000
Baksan, Elbrus Mountains, Russia		850
BOREX, Gran Sasso, Italy	planned	3100
CWI, South Africa	shut-down	8200
DUMAND, Hawaï, USA	construction	4500
Frejus, France	shut-down	4400
GALLEX, Gran Sasso, Italy		3100
Homestake, Lead, South Dakota, USA		4200
ICARUS, Gran Sasso, Italy	planned	3100
IMB, Ohio, USA	shut-down	1570
Issik-Kul Lake, (Alma Ata), Kasakstan	construction	500
Kamiokande, Kamioka, Japan		2700
KGF, Kamataka, India		7600
Lake Baikal, Russia	construction	1300
LSD, Mont Blanc, Italy		5000
LVD, Gran Sasso, Italy	construction	3100
MACRO, Gran Sasso, Italy		3100
NESTOR, Pylos, Greece	planned	3800
NUSEX, Mont Blanc, Italy	shut-down	5000
HPW, Park City, USA	shut-down	1700
SNO, Sudbury, Canada	construction	6000
Soudan, Minnesota, USA		2100
Super Kamiokande, Kamioka, Japan	planned	2700
Utah, USA	shut-down	1500

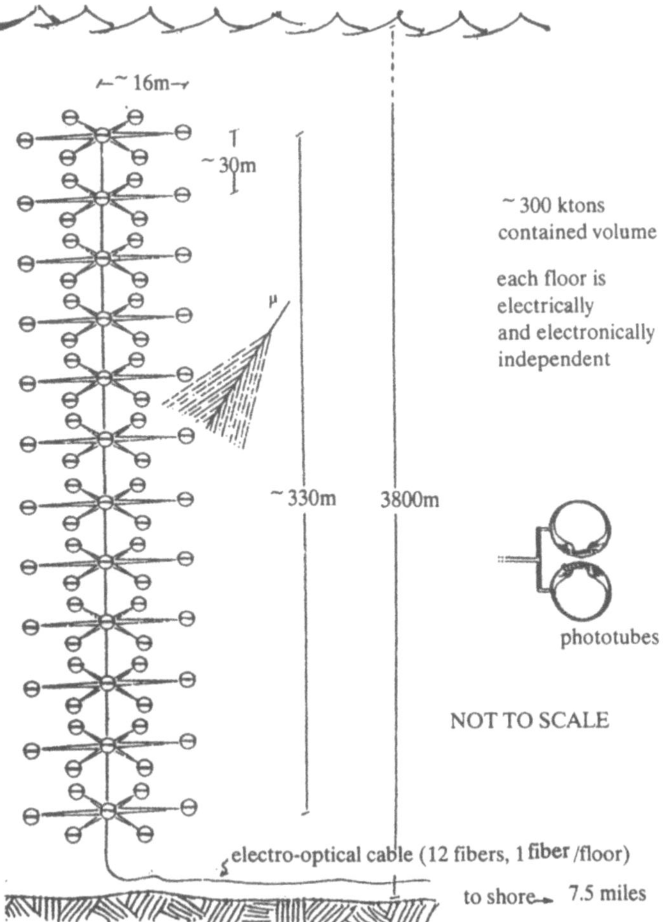

Fig.6 Nestor's tower Artist's view

CONCLUSIONS

Because of lack of space we couldn't review the High Energy "gamma" Astronomy by satellite or ground telescopes. The articles mentioned in the references give an indication of it. The majority of people involved in High Energy Physics and Astrophysics is in favor of the Standard Model and of the Big Bang. Nevertheless a minority is against them. We live everybody to have his opinion about.
The Author's opinion is that new theoretical and experimental ideas should be injected in the field to ensure it a bright future.

ACKNOWLEDGMENTS

The Author wants to express his gratitude to Miss. Sofia Magia for her assistance in the general editing of the paper.

REFERENCES

1. R.R.Wilson.The next generation of particle
   accelerators,ScientificAmerican,Vol 242,No1 (1980)
2. B.Richter.The future of electron physics.Beam Line,Vol20,No2 (1990).
3. Large Hadron Collider,CERN (1990)
4. CERN Courrier,Nov,(1992)
5. R.Ruth.The next linear collider,Beam Line,Vol20,No2 (1990)
6. K.Johnsen.The Eloisatron in:New Techniques for future
   AcceleratorsIII,G.Torellied.,Plenum Press,New York(1990)
7. Cosmic Rays 92,Proceedings of the 13th European Cosmic Ray Symposium,
   Nuclear Physics B, 33A,B,(1993)
8. B.Sadoulet and J.W.Cronin,Particle astrophysics,Physics Today,April,(1991)
9. C.F.McKee and William H.Press,Theoretical Astrophysics,Physics
   Today,April,(1991)
10 E.Anassontzis et all,Nestor: A Neutrino Particle Astrophysics underwater
   laboratory for the Mediterranean in Proc.of the 5th Workshop on Neutrino
   telescopes ,ed . Milla Baldo Ceolin ,Venice,(1993)

AN APPROACH TO FINITE-SIZE PARTICLES
WITH SPIN

Bronisław Średniawa

Institute of Physics, Jagiellonian University
Reymonta 4, 30-059 Cracow, Poland

1 Introduction

Quantum mechanics, quantum electrodynamics and quantum chromodynamics deal
with point particles. In quantum theory of fields the point particles are causes of
grave difficulties. Quantum field theory based on special theory of relativity should
satisfy three principles. It should :

I be Lorentz-invariant

II be gauge-invariant

III give finite results, which should agree with experimental data.

These three principles cannot be simultaneously satisfied. There were and there
are made attempts to improve quantum field theories. In most cases of improvement of
the theory two principles were not changed: Lorentz invariance and gauge-invariance.
There are two main ways to improve the theory in the third point:

1. the renormalization without changing the point-like character of the particles,

2. the passage to extended particles by introducing form factors,.

Key words: relativity, spin particles.

Frontiers of Fundamental Physics, Edited by M. Barone
and F. Selleri, Plenum Press, New York, 1994

The procedure 2 violates the principles I or II and therefore it was given up.

There were also made another attempts of introducing particles of finite sizes to the physics of elementary particles, between them:

1. theory of relativistic spin particles,

2. theory of strings and membranes.

Here we shall be concerned with the classical spin particles and their quantum analogues.

2 Classical equations of motion of dipole particles

2.1 Weyssenhoff's derivation

The equations of motion of dipole particles can be derived from different points of view. We begin with the most simple derivation (Weyssenhoff 1947c, 1958), resulting from the simple generalization of the mechanics of material points, by assuming that the energy-momentum vector G^μ [1] of the particle is not necessarily parallel to its four-velocity u^ν. Then the angular momentum is not conserved $(\cdot = \frac{d}{ds})$

$$\frac{d}{ds}(x_\mu G_\nu - x_\nu G_\mu) = x_\mu \dot{G}_\nu - x_\nu \dot{G}_\mu + u_\mu G_\nu - u_\nu G_\mu \ . \tag{1}$$

We shall restrict ourselves to the case of free particle. Then

$$\dot{G}_\mu = 0 \tag{2}$$

and the equation (1) reduces to

$$\frac{d}{ds}(x_\mu G_\nu - x_\nu G\mu) = u_\mu G_\nu - u_\nu G_\mu \ . \tag{3}$$

In order to save the law of the conservation of angular momentum we introduce the bivector $S_{\mu\nu}$, satisfying

$$\dot{S}_{\mu\nu} = u_\mu G_\nu - u_\nu G_\mu. \tag{4}$$

Then

$$\frac{d}{ds}(x_\mu G_\nu - x_\nu G\mu) + \dot{S}_{\mu\nu} = 0 \ . \tag{5}$$

Multiplying (4) by u^ν we get

$$G_\mu = mu_\mu + \frac{1}{c^2}S_{\mu\nu}\dot{u}^\nu. \tag{6}$$

where

$$m = \frac{1}{c^2}G_\mu u^\mu. \tag{7}$$

It can be seen that every bivector $S^{\mu\nu}$ may be written in the form

$$S^{\mu\nu} = s^{\mu\nu} + n^\mu u^\nu - n^\nu u^\mu, \tag{8}$$

---

[1] $\mu, \nu, ... = 0, 1, 2, 3, \ \ i, j, ... = 1, 2, 3.$
We assume the pseudoeuclidean metrics $+ - - -$. Then $u_\nu u^\nu = 1$.

where $s^{\mu\nu}$ is the bivector and

$$s^{\mu\nu}u_\nu = 0, \quad n^\mu u_\mu = 0. \tag{9}$$

We call $s^{\mu\nu}$ the spin, n^μ the dipole moment of the particle. The relation (9) gives in the proper system of the particle

$$s^{\mu\nu} = (\boldsymbol{s}, \boldsymbol{q}) \quad \text{and} \quad \boldsymbol{q} = \frac{1}{c}\boldsymbol{s} \times \boldsymbol{u}.$$

The equations (2), (4), (6), (7), (8), (9) form the system of classical equations of motion of the dipole particles. Two cases are important:

$s_{\mu\nu} \neq 0, n_\mu = 0$, spin particle (Frenkel 1926), (Mathisson 1936), (Weyssenhoff 1947 a,c),

$n_\mu \neq 0$, pole-dipole particle (Hönl 1939).

The equations of motion were also derived for dipole particles in external field of forces.

2.2 Mathisson's derivation

Mathisson (1936) considered the fundamental problem of general relativity (which was studied since 1919) of deriving the equations of motion of the particle in the gravitational field from the equations of this field as the equations of the world line of the singularity in this gravitational field. Mathisson's predecessors studied the case, when this singularity had (in a properly chosen coordinate frame) spherical symmetry and they got the equations of the geodesics. Mathisson was the first, who considered non-spherically singularities of gravitational field.

He introduced into his variational principle, based on the equations of gravitational field, the multipole moments of this singularity. Assuming the presence of only positive masses he obtained in the lowest approximation the equations of the geodesics. In the next approximation he got for positive masses the equations of motion of the spin particle, which for the gravitational field with metrics close to Minkowski's one, reduced to the equations (2) - (9). (See also (Średniawa 1983)).

2.3 Lubański's derivation

By expanding gravitational retarded potentials of gravitational multipoles (Lubański 1937) and using for the momentum-energy tensor $T^{\mu\nu}$ the relation

$$\nabla_\nu T^{\mu\nu} = 0 \ , \tag{10}$$

Lubański obtained (for positive masses) the equations of motion of the spin particle.

2.4 Weyssenhoff's and Raabe's derivation

Weyssenhoff and Raabe (Weyssenhoff and Raabe 1947 a) considered the incoherent fluid, for which they assumed the momentum-energy tensor of the form

$$T^{\mu\nu} = g^\mu u^\nu, \tag{11}$$

where g^μ was the density of momentum and assumed the existence of the intrinsic angular momentum density $s^{\mu\nu}$. They formulated the equations of motion of such a fluid and integrated them for its very small volume. They obtained again the equations of motion of the spin particle.

2.5 Hönl's and Papapetrou's derivation

Hönl and Papapetrou (1939) applied Lubański's method to a particle characterized
by the mass m, dipole moment n^λ, assuming $s^{\mu\nu} = 0$. They obtained the equations
of motion of the pole-dipole particles.

3 Solutions of the equations of motion of the spin particle and the pole-dipole particle

We assume that G^μ is a time-like vector.

3.1 Spin particle

The solution (Weyssenhoff and Raabe 1947 a) in the proper frame of reference of G^μ
(where $G^i = 0$) is in the uniform circular motion of radius

$$r = \frac{s_0 v}{m_0(c^2 - v^2)} \tag{12}$$

and frequency

$$\omega = \frac{m_0(c^2 - v^2)}{s_0} , \tag{13}$$

where
$s_0^2 = s_{\mu\nu}s^{\mu\nu} = const.$, m_0 is constant, $v^2 = (v^i)^2$

3.2 Pole-dipole particle

In order to obtain here the definite solution, supplementary condition must be assumed. Having put $n^\mu n_\mu = const.$, (Średniawa 1948) we get for the pole-dipole
particle in proper frame of G^μ also circular motion.

These solutions suggest that the whole circle should be regarded as "macroparti-
cle", in contradistinction to the "microparticle", which circulates along the circle.

4 Passage to the velocity of light for the spin particle and analogy to Dirac's electron

Up to now we considered dipole particles moving slower then light. Then the "microparticle" has still one arbitrary parameter, namely the velocity v, which can make
the radius of the circle arbitrary large. Therefore Weyssenhoff and Raabe (1947 b)
made the passage $v \to c$. Since then $ds \to 0$, and proper time cannot be used, they
introduced before the passage to the limit $v \to c$ the new parameter π, such that
$\frac{d\pi}{ds} > 0$. Having made the passage they choose the time t as π, and they got the
equations of motion, whose solution in the proper frame C of G^μ was the circle of the
definite radius

$$r_c = \frac{s_c}{cM_c} \tag{14}$$

and definite frequency

$$\omega_c = \frac{M_c c^2}{s_c}, \tag{15}$$

where

$$s_c = (s_c^i)^2, \; M_c = G^\mu u_\mu .$$

The spin particle moving with the velocity of light has following properties similar to those of Dirac's electron : (Weyssenhoff 1947 d):

1. in both theories the momentum G^μ is not parallel to the velocity u^μ,

2. kinematical velocity $v = c$ corresponds to the fact that the eigenvalues of the velocity operator in Dirac's theory are equal to $\pm c$,

3. in both theories angular momenta are not constants of motion, circular motion is the classical "picture" of the "Zitterbewegung",

4. if one puts $s_c = \frac{\hbar}{2}$ and M_c as the experimental mass of the electron, r_c is of the order of the Compton wave-length and ω_c of the order of "Zitterbewegung",

5. condition of stability of the circle in the electromagnetic field is the same as the condition of non forming electron pairs in this field in Dirac's theory.

These analogies suggest that the spin particle can in some sense be regarded as a classical relativistic model of Dirac's electron.

5 Canonical formalism with higher derivatives and attempts of quantization of spin particles

Up to now we considered classical, non quantized dipole particles. In order to try to quantize their theory by generalysing the standard way of quantization, the hamiltonian formalism with higher derivatives should be developed , since classical equations of motion are of the third order. This was done by Ostrogradski (1850) and for homogeneous equations of third order by Weyssenhoff (1951). They started from the lagrangian as a function of position, velocity and acceleration of the particle

$$L(\mathbf{r}, \dot{\mathbf{r}}, \ddot{\mathbf{r}}) \tag{16}$$

Then the canonical momenta to the variables

$$\mathbf{r}, \quad \mathbf{v} = \dot{\mathbf{r}}$$

are

$$p_i = \frac{\partial L}{\partial v_i} - \frac{d}{dt}\frac{\partial L}{\partial \dot{v}_i}, \quad s_i = \frac{\partial L}{\partial \dot{v}_i}, \tag{17}$$

and the hamiltonian is equal to

$$H = \mathbf{v} \cdot \mathbf{p} - \dot{\mathbf{v}} \cdot \mathbf{s} - L(\mathbf{r}, \mathbf{v}, \dot{\mathbf{v}}) \tag{18}$$

after elimination of $\mathbf{v}$.

Bopp (1946) and Weyssenhoff (1951) found the form of the lagrangian of the spin particle for $v = c$, $(' = \frac{d}{dt})$:

$$L = \sqrt{l_0}\,\sqrt[4]{\frac{w_\mu w^{\mu\prime}}{w_\mu w^\mu}} \; , \tag{19}$$

where $w^\mu = \frac{dx^\mu}{dt}$, l_0 is a constant of the dimension of length. From this lagrangian the corresponding hamiltonian could be obtained. Bopp (1948) introduced a complex spinor-like variable ξ on the place of w^μ and the wave function $\Psi(x,\xi,\xi^*)$. Then he extended the Jordan quantization rule to

$$\eta \to \frac{\hbar}{i}\frac{\partial}{\partial\xi^*} \quad , \quad \eta^\dagger \to \frac{\hbar}{i}\frac{\partial}{\partial\xi}$$

$$p_\mu \to \frac{\hbar}{i}\frac{\partial}{\partial x^\mu} \quad , \quad p_\mu^\dagger \to \frac{\hbar}{i}\frac{\partial}{\partial x^\mu} \tag{20}$$

and obtained a Dirac-like wave equation , which in simple cases could be reduced to Dirac equation.

Later it turned out (Infeld 1957), (Borelowski 1961), (Borelowski and Średniawa 1962) that from the postulates for L that

1. L should be a scalar not depending on x^μ ,

2. canonical p^μ and u^μ should not in general be parallel, a variety of models of free dipole particles could obtained.

<u>Remark</u>: One of the other ways of obtaining classical particles with spin leads through the consideration of the magnetic top(see (Barut et al. 1992)).

6   Conclusion

The general result of our considerations consists in the fact, that starting as well from general relativity as from special relativity and generalizing slightly the principles of mechanics by assuming that in general momentum and velocity are not parallel, one is lead to the flat structures of finite sizes. These classical structures, whose motion is described by differential equations of higher order, show before quantization some of the characteristic features of quantum particles. The standard methods of quantizations of the motions of this particles were, alas, not successful.

References

Barut, A.O., Božić, M., Marić, Z. (1992), Annals of Physics **214** 53.
Bopp, F. (1946), Zeits.f.Naturforschung **1** 196.
Bopp, F. (1948), ibid. **3a** 567
Borelowski, Z. (1961) Acta Physica Polonica **20** 619.
Borelowski, Z., Średniawa, B. (1962), ibid. **25** 608.
Frenkel, J. (1926), Zeits.f.Phys. **37** 243.
Hönl, H., Papapetrou, A.(1939), Annalen der Physik **112** 512.
Infeld, L. (1957), Bull.Ac.Pol.Sci., Math.-Astr.-Phys. **5** 979.
Lubański, J.K.(1937), Acta Physica Polonica **6** 356.
Mathisson, M. (1936), ibid. **6** 163.
Średniawa,B.(1948), Acta Physica Polonica **9** 99.
Średniawa,B.(1983), in Studies in the History of General Relativity, Birkhäuser, Boston 400-406.
Weyssenhoff, J., Raabe, A. (1947 a), Acta Physica Polonica **9** 7.
Weyssenhoff, J., Raabe, A. (1947 b), ibid. **9** 19.

Weyssenhoff, J.(1947 c), ibid. **9** 26.

Weyssenhoff, J.(1947 d), ibid. **9** 46.

Weyssenhoff, J.(1952), ibid. **11** 49.

Weyssenhoff, J.(1958), in Max Planck Festschrift **155** 168.

A NEW HIGH ENERGY SCALE?

Vladimir Kadyshevsky

Joint Institute for Nuclear Research
141980 Dubna, Moscow region, Russia

1

Our Conference is devoted to the frontiers of fundamental physics. What I intend to speak about is exactly in line with the topic of this Conference. I will dwell upon a radical enough attempt to modify the standard quantum field theory (QFT) in the region of superhigh energies and momenta or, which is the same, at supersmall space-time distances. A new theory, except $\hbar$ and c, contains one more universal constant – the "fundamental mass" M. An inverse quantity $\frac{\hbar}{Mc} = \ell$ is called the fundamental (minimal, elementary) length. To the standard QFT there corresponds a low-energy limit $E, \mid \mathbf{p} \mid, m \ll M$, or formally $M \to \infty$ $(\ell \to 0)$.

Thus, the fundamental mass M, like a frontier post, notifies approach to the territory that is under control of the new theory.

The ideas to be expounded have been working out for many years at Dubna in collaboration with A.Donkov, D.Fursaev, R.Ibadov, M.Mateev and M. Chizhov.

In fact, a great number of QFT models containing a new universal scale like M or ℓ were described in the literature. The hypothesis of the existence in Nature of a "minimal length", a so-called "atom of space", is much older than QFT: it originated here, on the earth of Hellas, in ancient times.

I want to give some reasonings in favour of that the QFT apparatus should include the parameter M. It is to be recalled that the modern QFT is a local Lagrangian theory forming a mathematical basis of the so-called **standard model** (SM). The leading actors in the SM are leptons, quarks, gluons, vector bosons $W^\pm$, Z^0, γ and the Higgs scalar H. In the initial Lagrangian of the model, all these particles are represented by **local fields** [1] , which implies their **elementary nature**, i.e., the absence of any structure. At the same time, they are specified by certain values of mass, spin, electric charge, colour, hypercharge, isotopic spin and others.

Intuitively, it is clear that elementary particles should carry rather small portions of different kind of "charges" and "spins". Theoretically, this implies that local fields

---

[1]Up to linear transformations of fields caused by the Higgs mechanism.

are related to the lowest representations of the relevant **compact** symmetry groups.

As for the particle mass m, this quantity is the Casimir operator of the **noncompact** Poincaré group, and in the representations of this group, which are used in QFT, it can take any values in the interval

$$0 \leq m < \infty. \tag{1}$$

Two particles that are thought of as elementary can have masses differing from each other by many orders. For instance, in the GUT models there appear vector bosons with mass $\sim 10^{15} - 10^{16}$ GeV whereas the electron mass amounts to $\sim 0.5 \; 10^{-3}$ GeV.

There arises a question: Up to what values of m the existing **conception of the local field** is applicable? Formally, QFT makes sense even in the case when an elementary act of interaction involves structureless objects whose masses are comparable with, say, car masses:

a)   b)

Suffice it, for instance, to choose an appropriate Higgs field. Such a far-going extrapolation of the **local** theory into the region of macroscopic masses looks like a pathology. It is doubtless that such monsters can serve as bricks of the Universe like, for instance, quarks. Nothing even remotely resembling this is observed experimentally. However, it is to be stressed: The modern QFT, in accordance with (1), does not forbid consideration of such exotic Feynman diagrams as a) and b). Maybe, it is a fundamental defect of the theory, its "the heel of Achilles"?

In 1985 M.A. Markov put forward a hypothesis[1] according to which the spectrum of masses of elementary particles should be cut off at "Planck's mass" $m_{planck} \approx 10^{19}$ GeV. Particles of the limiting mass $m = m_{planck}$, called by the author "maximons", are urged to play a special role in the world of elementary particles[2]. The conception of "maximon" underlies Markov's scenario of the early Universe[3].

However, I wish to note that with respect to QFT Markov's limitation $m \leq m_{planck}$ serves as an additional phenomenological condition. It does not affect the structure of the theory, and even a maximon is described by a standard field-theoretical method.

But we can go father and consider Markov's idea of the existence of an upper bound for the elementary particle mass as a **fundamental physical principle** that, like relativistic and quantum postulates, should form the very basis of QFT.

An attempt to realize this programme has been made by our Dubna group. We write down Markov's condition as

$$m \leq M, \tag{2}$$

where the limiting mass M, for the sake of generality, is not identified with m_{planck} but rather is considered as a new universal constant of the theory, "fundamental mass". As $M \to \infty$, the limitation (2) disappears and we come back to the spectrum (1).

The hypothesis (2) is equivalent to the statement that the Compton wave length of an elementary particle $\lambda_c = \frac{\hbar}{mc}$ cannot be smaller than the "fundamental length" ℓ.

According to Newton and Wigner[4], the parameter λ_c defines minimal linear dimensions of the space region in which a relativistic particle can be localized. Consequently, the fundamental length ℓ introduces into the theory a **universal limitation** on the accuracy of space localization of elementary particles.

Our further consideration will be based first on geometric arguments. By virtue of the equality

$$p_0^2 - p_1^2 - p_2^2 - p_3^2 = m^2 \tag{3}$$

the mass shell of a free particle is a two–sheeted hyperboloid imbedded into the four-dimensional momentum space. In the standard QFT, this space is Euclidean, i.e. homogeneous and infinite. Consequently, one can imbed in it hyperboloids (3) with an arbitrary large radius m . Sacrificing the pseudo-Euclidean geometry of p-space and postulating the momentum 4-space in QFT to be a de–Sitter one with the curvature radius M, one can incorporate the limitation (2) into a theory from the very beginning. Indeed, let us consider a 5-hyperboloid

$$p_0^2 - p_1^2 - p_2^2 - p_3^2 + p_5^2 = M^2, \tag{4}$$

whose surface is a realization of the de–Sitter four-space with the signature $(+ - - - +)$. For a free particle, by virtue of (3), $m^2 + p_5^2 = M^2$, i.e., (2) is fulfilled automatically. The main question is whether an adequate QFT can be worked out on a new arena (4)? Will it be possible to conserve, especially, a local nature of the theory, gauge invariance?

Our studies have shown that answers to these and similar questions turn out to be positive. In other words, the de– Sitter p-space (4) is an appropriate construction area for quantum field theory, a new formulation of QFT being physically deeper and more profound.

Here, I will consider, for simplicity, only a scalar model of the new theory. More realistic versions have been discussed in [5].

3

Let $\varphi(p_0, \mathbf{p})$ be a scalar field describing in the standard theory spinless particles of mass m. In a free case, we obviously have

$$(p_0^2 - \mathbf{p}^2 - m^2)\, \varphi\, (p_0, \mathbf{p}) = 0 \tag{5}$$

Upon passing from the Minkowski p-space to the de–Sitter p-space (4), it is convenient to use instead of $\varphi(p_0, \mathbf{p})$ the functional

$$\delta(p_0^2 - \mathbf{p}^2 + p_5^2 - M^2)\, \varphi\, (p_0, \mathbf{p}, p_5) \tag{6}$$

It is clear that defining in (6) one function $\varphi(p_0, \mathbf{p}, p_5)$ of five variables (p_μ, p_5) is equivalent to defining two independent functions $\varphi_1(p)$ and $\varphi_2(p)$ of the four–momentum p_μ

$$\varphi(p, p_5) = \begin{pmatrix} \varphi(p, |p_5|) \\ \varphi(p, -|p_5|) \end{pmatrix} \equiv \begin{pmatrix} \varphi_1(p) \\ \varphi_2(p) \end{pmatrix} \tag{7}$$

$$|p_5| = \sqrt{M^2 - p^2}$$

Appearance of a new discrete degree of freedom $\frac{p_5}{|p_5|}$ and hence doubling of the number of field variables is the most important specific feature of the approach developed. This fact should be taken into account even in deriving an equation of motion for a free field in the de–Sitter p-space.

Clearly, the standard Klein–Gordon equation is still valid

$$(p^2 - m^2)\varphi(p, p_5) = 0 \tag{8}$$

However, (8) has two obvious defects:

1. The condition (2) is not explicitly taken into account in it.

2. From (8) one cannot determine the dependence of the field on the new quantum number $\frac{p_5}{|p_5|}$ to distinguish between the components $\varphi_1(p)$ and $\varphi_2(p)$.

Noting that for the field $\varphi(p, p_5)$ (8) can be represented as

$$(M^2 \cos^2 \mu - p_5^2)\varphi(p, p_5) = 0 \tag{9}$$

we postulate the sought equation of motion in the form

$$2M(M \cos \mu - p_5)\varphi(p, p_5) = 0 \tag{10}$$

Eq. (10) is already free from the above drawbacks of (8) though the latter is still valid.

It follows from (10) and (7) that

$$2M(M \cos \mu - |p_5|)\varphi_1(p) = 0 \tag{11}$$

$$2M(M \cos \mu + |p_5|)\varphi_2(p) = 0 \tag{12}$$

which results in

$$\varphi_1(p) \simeq \delta(m^2 - p^2)\tilde{\varphi}_1(p) \tag{13}$$

$$\varphi_2(p) = 0 \tag{14}$$

Thus, the field $\varphi(p, p_5)$ given in the de–Sitter p-space (4) describes the same free scalar particles of mass m which in the Minkowski p-space were described by the field $\varphi(p)$ with the only difference that now $m \leq M$. The two–component structure (7) of the new field on mass shell does not manifest itself as for $\varphi_2(p)$ the "confinement" (14) takes place. However, this component will play an important role in treating field interactions, i.e., off the mass shell.

Now let us show that eq. (10) is connected with a certain Lagrangian formalism. First, noting that in (6) all the components of the 5-momentum can be treated on equal footing, we make the Fourier 5–transformation for this functional

$$\frac{2M}{(2\pi)^{\frac{3}{2}}} \int e^{ip_L x^L} \delta(p^2 + p_5^2 - M^2)\varphi(p, p_5)d^5p \equiv \varphi(x, x^5) \tag{15}$$

$$L = 0, 1, 2, 3, 5$$

Obviously, (15) satisfies the following differential equation in the configuration 5-space:

$$(\Box + \frac{\partial^2}{\partial x_5^2} + M^2)\varphi(x, x^5) = 0 \tag{16}$$

Integration over p_5 in (15) with (7) taken into account gives

$$\varphi(x, x^5) = \frac{M}{(2\pi)^{\frac{3}{2}}} \int_{p^2 \leq \tilde{M}} e^{-ipx} \frac{d^4p}{|p_5|}[e^{-i|p_5|x^5}\varphi_1(p) + e^{i|p_5|x^5}\varphi_2(p)] \tag{17}$$

which results in

$$i\frac{\partial \varphi(x, x^5)}{M \partial x^5} = \frac{1}{(2\pi)^{\frac{3}{2}}} \int_{p^2 \leq M^2} e^{-ipx}d^4p[e^{-i|p_5|x^5}\varphi_1(p) - e^{i|p_5|x^5}\varphi_2(p)] \tag{18}$$

380

Relations (17) — (18) are Fourier transformations converting the fields $\varphi_1(p)$ and $\varphi_2(p)$ into the configuration representation. Inverse transformations have the form

$$\varphi_1(p) = \frac{-i}{2M(2\pi)^{\frac{5}{2}}} \int d^4x\, e^{ipx} [\varphi(x, x^5) \overset{\leftrightarrow}{\frac{\partial}{\partial x^5}} e^{i|p_5|x^5}] \tag{19}$$

$$\varphi_2(p) = \frac{i}{2M(2\pi)^{\frac{5}{2}}} \int d^4x\, e^{ipx} [\varphi(x, x^5) \overset{\leftrightarrow}{\frac{\partial}{\partial x^5}} e^{-i|p_5|x^5}] \tag{20}$$

where the notation $f_1 \overset{\leftrightarrow}{\frac{\partial}{\partial x^5}} f_2 \equiv f_1 \frac{\partial f_2}{\partial x^5} - \frac{\partial f_1}{\partial x^5} f_2$ was used. The right-hand sides of relations (19) — (20) are independent of x^5 by virtue of eq.(16).

Using (17) and (18) one can easily establish that (11) and (12) are equivalent to the system of differential equations in the configuration 5–space

$$(\Box + M^2)\varphi(x, x^5) + iM \cos\mu \frac{\partial\varphi(x, x^5)}{\partial x^5} = 0 \tag{21}$$

$$i\frac{\partial\varphi(x, x^5)}{\partial x^5} = M \cos\mu\varphi(x, x^5) \tag{22}$$

Note that (16) is a consequence of this system.

As is seen from (22), the dependence of the free field $\varphi(x, x^5)$ on the fifth coordinate x^5 is stationary

$$\varphi(x, x^5) = e^{-iM \cos\mu x^5}\varphi(x, 0), \tag{23}$$

the "initial value" of $\varphi(x, 0)$, by virtue of (21), satisfying the standard Klein-Gordon equation

$$(\Box + m^2)\varphi(x, 0) = 0 \tag{24}$$

From (23) and (17) it immediately follows that $\varphi_2(p) = 0$ (cf (14)).

Now let us consider the functional

$$S = \int L_0(x, x^5) d^4x \equiv \frac{1}{2} \int [\frac{\partial\varphi(x, x^5)}{\partial x^\mu}\ \frac{\partial\varphi^+(x, x^5)}{\partial x_\mu} - m^2|\varphi(x, x^5)|^2 -$$

$$- |i\frac{\partial\varphi(x, x^5)}{\partial x^5} - M \cos\mu\varphi(x, x^5)|^2]d^4x \tag{25}$$

where the field $\varphi(x, x^5)$, as before, obeys (16). However, (16) being a second order differential equation with respect to x^5 does not impose any constraint on the variables $\varphi(x, x^5)$ and $\frac{\partial\varphi(x, x^5)}{\partial x^5}$ leaving them to be arbitrary initial Cauchy conditions [2] given at the fixed "time moment" x^5. Varying (25) with respect to $\varphi(x, x^5)$ and $\frac{\partial\varphi(x, x^5)}{\partial x^5}$ as independent arguments, one can easily be convinced that the condition of stationarity of this functional coincides with the system of equations (21) and (22). Consequently, S is the action for the field $\varphi(x, x^5)$ and $L_0(x, x^5)$ is the Lagrangian density. It should be emphasized that this density is **local in five dimensions**.

Finally, let us note two more important properties of the functional (25).

1. By virtue of eq.(16) this functional is explicitly independent of x^5, i.e.,

$$\frac{\delta S[\varphi(x, x^5), \frac{\partial\varphi}{\partial x^5}(x, x^5)]}{\partial x^5} = 0 \tag{26}$$

In other words, the action S is one of the integrals of motion of eq. (16).

---

[2] A correct formulation of the Cauchy problem for an equation of the ultrahyperbolic type, as (16), needs special comments that are omitted here.

2. In the Euclidean formulation of the theory, (25) turns into a positive definite functional $\int L_{0E}(x, x^5)\, d^4x_E \geq 0$.

If the field is not free and has, for instance, self-interaction, then to $L_0(x, x^5)$ one has to add one more term $L_{int}(x, x^5)$ that is naturally chosen as a local function of the initial data $\varphi(x, x^5)$ and $\frac{\partial\varphi(x,x^5)}{\partial x^5}$. This will allow one to conserve the additional condition (16) and, consequently, the whole formalism developed.

The behaviour of the model amplitudes at large energies and transfer momenta will depend essentially on the parameter M . Thus, this constant represents not only the "limiting" mass (see (2)) but also the new universal scale in the high energy region.

In conclusion, I would like to express my sincere gratitude to Prof. F.Selleri and Prof. A.Janussis for the invitation to this Conference and hospitality extended to me in Olympia.

REFERENCES

[1] M.A.Markov. Supplement of the Progress of Theoretical Physics, *in:* "Commemoration Issue for 30th Anniversary of Meson Theory by Dr. H.Yukawa", p. 85 (1965); M.A.Markov. *ZhETF* vol. 85, p.878 (1966) (in Russian).

[2] M.A.Markov. *Preprint INR* P-0208 (1981).

[3] M.A.Markov. *Preprint INR* P-0207 (1981); *Preprint INR* P-0286 (1983); M.A.Markov, V.F. Mukhanov. *Preprint INR* P-0331 (1984).

[4] T.D.Newton, E.P.Wigner. *Rev.Mod.Phys.*, vol. 21, p. 400 (1949).

[5] V.G.Kadyshevsky. *Nuclear Physics* B141, p. 477 (1978); V.G.Kadyshevsky, M.D.Mateev. *Phys.Lett.* 106B, p. 139 (1981); V.G.Kadyshevsky, M.D.Mateev. *Nuovo Cimento* 87A, No. 3, p. 324 (1985); V.G.Kadyshevsky. Quantum field theory and Markov's "maximon", in: "Quantum Theory of Gravitation" (Report on the III International Sem.), Moscow (1984); *Preprint JINR* P2-84-753 (1984); A.D.Donkov, M.D.Mateev, V.G.Kadyshevsky. Non–Euclidean momentum space and two-body problem, *Theor.Math.Phys.* vol. 50, No. 3, p. 36-369 (1982); R.M.Ibadov, V.G.Kadyshevsky. *Preprint JINR* P2-86-830 (1986) (in Russian); V.G.Kadyshevsky, D.V.Fursaev. On a generalization of gauge principle at high energies, *Teor.Mat.Fiz.* 83:197 (1990); V.G.Kadyshevsky, D.V.Fursaev, R.M.Ibadov. The rotation invariant gauge model with the compact momentum space, *in:* Proc. of 25th Int. Conf. on High Energy Physics, Singapore (1990); V.G.Kadyshevsky, D.V.Fursaev. Left-right components of bosonic field and electroweak theory, *JINR Rapid Communications*, No.6(57)-92, p.5 (1992).

ON THE SPACE-TIME STRUCTURE OF THE ELECTRON

Martín Rivas

Departamento de Física Teórica
Universidad del País Vasco, Euskal Herriko Unibertsitatea
Apdo. 644, 48080 Bilbao, Spain

INTRODUCTION

In previous works [1,2] we have found a lagrangian description of classical elementary spinning particles where the spin is produced by the *zitterbewegung* and rotational motion of the particle around its center of mass. The novelty with respect to other approaches is the definition of particle. The usual canonical formulation defines a classical particle as a system whose phase space is a homogeneous space of the Poincaré group. In our approach is the kinematical space of the system which is required to be a homogeneous space of the corresponding space-time kinematical group. This definition of particle leads for a general lagrangian to depend on time, position, velocity, acceleration, orientation and angular velocity of the particle. This dependence on second order derivatives of position makes neccesary to work in a generalized lagrangian formalism. One of the salient features for a general spinning particle is that the center of mass q does not match with the position r of the particle and is a function of the above observables.

Since position r and center of mass position q are different points one question arises. What does position vector r mean? It will represent the center of charge position, because when external fields are acting upon the particle, potentials and fields become functions of the r position. It thus seems in general that for charged particles the center of charge and center of mass are different points, so that the center of charge motion around the center of mass gives rise to the appearance of a normal magnetic moment.

This general discussion of a classical particle suggests that particles are systems of six degrees of freedom similarly as rigid bodies but with the condition that the center of charge is shifted from the center of mass.

In next section we discuss the classical system of a charged rigid body and its lagrangian description in terms of the center of charge position.

When quantizing generalized lagrangian systems, the wave function becomes a squared integrable function defined on the kinematical space of the system.[3] If we consider for quantization a system of six degrees of freedom such that the center of charge

Frontiers of Fundamental Physics, Edited by M. Barone
and F. Selleri, Plenum Press, New York, 1994

is spinning around the center of mass with the speed c, we obtain Dirac's equation. The analysis of Dirac's algebra shows that its structure is completely determined by the knowledge of the internal orientation, so that electron's internal structure is due to its spatial orientation.

THE CHARGED RIGID BODY

The kinematical description of a rigid body evolution is usually expressed in terms of its center of mass position q and the orientation of its principal axes of inertia centered at q, so that its free lagrangian can be written as a function $L_0(q, \dot{q}, \alpha, \omega)$, being α for instance the three Euler's angles and ω its angular velocity. Let us assume that the rigid body is charged and the charge distribution has no multipole moments so that it can be reduced to a point, the center of charge r. To describe the interaction with an external electromagnetic field we have to include in the lagrangian for instance the minimal coupling term $L_I = j^\mu(t, r) A_\mu(t, r)$, where the current depends on the charge motion and the potential fields are functions defined at point r where the center of charge is. Conditions of rigidity allow us to express r and the charge velocity in terms of q and orientation α and their time derivatives. In fact

$$r(t) = q(t) + R(\alpha(t))(r_0,$$

$$\frac{dr}{dt} = \frac{dq}{dt} + \omega \times R(\alpha)r_0,$$

where r_0 is the initial position of the center of charge. These relations must be used to replace the charge position r in the arguments of the functions entering in the coupling term to properly obtain from them Euler-Lagrange equations.

Alternatively we can describe the kinematics of a rigid body in terms of the evolution of any other point, not necessarily the center of mass, and the corresponding change of orientation will be related to the body frame centered at this new point. We can take the center of charge as the basic point so that the interaction term $L_I = j^\mu(t, r) A_\mu(t, r)$ will be the same as before with no restrictions on the r variables but the free part of the lagrangian when written in terms of r must give rise to a nonuniform motion for r and a uniform motion for the center of mass q. This suggests that the free lagrangian must depend on higher order derivatives of r.

In the mentioned references [1,2] we obtained for free particles a nonuniform motion for the position r that was interpreted as a zitterwebegung and a uniform motion for center of mass q. The lagrangians were dependent on the second order derivative of r.

QUANTIZATION

Quantization of generalized lagrangian systems is performed through Feynman's path integral method. Since the action function is a function of the initial and final point on the $X \times X$ manifold, being X the kinematical space of the system, this leads for Feynman's propagator to be in general a distribution on $X \times X$. When restricted to some arbitrary final point, leaving the initial point arbitrary, we obtain the wave function of the system as a squared integrable function on X.

In the case of relativistic systems we have considered [3] the classical system whose kinematical space is the nine-dimensional manifold spanned by the variables: time t, charge position r, charge velocity $u = dr/dt$ but $u = c$ and orientation α. This system represents a particle whose center of charge is describing a circle of radius $R_0 = S/mc$

around its center of mass at the speed of light. A general spinning particle wave function will be a function of ten variables $\Psi(t, \boldsymbol{r}, \boldsymbol{u}, \boldsymbol{\alpha})$ with domains $t \in \mathbb{R}$, $\boldsymbol{r} \in \mathbb{R}^3$, $\boldsymbol{u} \in \mathbb{R}^3$, but $u = c$ and $\boldsymbol{\alpha} = \alpha \boldsymbol{a}$ is the canonical parameterization of rotation group in terms of the rotated angle α and the unit vector $\boldsymbol{a}$ along rotation axis.

The wave function satisfies Dirac's equation and can be written as

$$\Psi = \sum_{i=1}^{4} \psi_i(t, \boldsymbol{r}) \Phi_i(\boldsymbol{\alpha}),$$

where the space-time parts $\psi_i(t, \boldsymbol{r})$ satisfy Klein-Gordon equation and the angular parts are the four eigenvectors of spin operators S^2, S_3 and Z_3, where the last two are the spin projections on third spatial axis and third body axis respectively. The four-dimensional Hilbert space spanned by functions $\Phi_i(\boldsymbol{\alpha})$ represents the internal structure space such that any internal observable like charge velocity and acceleration, spin and orientation become 4×4 matrices. This is what is called Dirac's algebra. The six spin projections S_i and Z_i, $i = 1, 2, 3$ and the nine components of the three unit vectors of body axis $\boldsymbol{e}_i$, $i = 1, 2, 3$, have the matrix representation in this basis:

$$\hat{\boldsymbol{S}} = \frac{\hbar}{2} \begin{pmatrix} \boldsymbol{\sigma} & 0 \\ 0 & \boldsymbol{\sigma} \end{pmatrix},$$

$$\hat{Z}_1 = \frac{\hbar}{2} \begin{pmatrix} 0 & \mathbb{I} \\ \mathbb{I} & 0 \end{pmatrix}, \quad \hat{Z}_2 = \frac{\hbar}{2} \begin{pmatrix} 0 & i\mathbb{I} \\ -i\mathbb{I} & 0 \end{pmatrix}, \quad \hat{Z}_3 = \frac{\hbar}{2} \begin{pmatrix} \mathbb{I} & 0 \\ 0 & -\mathbb{I} \end{pmatrix},$$

$$\hat{\boldsymbol{e}}_1 = \frac{1}{3} \begin{pmatrix} 0 & \boldsymbol{\sigma} \\ \boldsymbol{\sigma} & 0 \end{pmatrix}, \quad \hat{\boldsymbol{e}}_2 = \frac{1}{3} \begin{pmatrix} 0 & i\boldsymbol{\sigma} \\ -i\boldsymbol{\sigma} & 0 \end{pmatrix}, \quad \hat{\boldsymbol{e}}_3 = \frac{1}{3} \begin{pmatrix} \boldsymbol{\sigma} & 0 \\ 0 & -\boldsymbol{\sigma} \end{pmatrix},$$

where $\boldsymbol{\sigma}$ are the three Pauli matrices and $\mathbb{I}$ represents the 2×2 unit matrix.

These three spatial spin components S_i, the three body spin projections Z_j and the nine components of body frame $(\boldsymbol{e}_i)_j$, $i, j = 1, 2, 3$, together the 4×4 unit matrix $\mathbb{I}$ form a set of sixteen linearly independent hermitian matrices. They are a linear basis of Dirac's algebra, and satisfy the following commutation relations:

$$[S_i, S_j] = i\hbar\epsilon_{ijk}S_k, \qquad [Z_i, Z_j] = -i\hbar\epsilon_{ijk}Z_k, \qquad [S_i, Z_j] = 0,$$

$$[S_i, (\boldsymbol{e}_j)_k] = i\hbar\epsilon_{ikr}(\boldsymbol{e}_j)_r, \qquad [Z_i, (\boldsymbol{e}_j)_k] = -i\hbar\epsilon_{ijr}(\boldsymbol{e}_r)_k,$$

$$[(\boldsymbol{e}_i)_k, (\boldsymbol{e}_j)_l] = \frac{4i}{9\hbar}\left[\delta_{ij}\epsilon_{klr}S_r - \delta_{kl}\epsilon_{ijr}Z_r\right],$$

showing that $\hat{\boldsymbol{e}}_i$ operators transform like vectors under rotations but they are not commuting observables.

If we fix the couple of indices i, and j, then the set of four operators S^2, S_i, Z_j and $(\boldsymbol{e}_j)_i$ form a complete commuting set since the algebra of 4×4 matrices admits 4 diagonal and linearly independent matrices.

The basic observables satisfy the following anticommutation relations:

$$\{S_i, S_j\} = \{Z_i, Z_j\} = \frac{\hbar^2}{2}\delta_{ij}\mathbb{I},$$

$$\{S_i, Z_j\} = \frac{3\hbar^2}{2}(\boldsymbol{e}_j)_i,$$

$$\{S_i, (\boldsymbol{e}_j)_k\} = \frac{2}{3}\delta_{ik}Z_j, \qquad \{Z_i, (\boldsymbol{e}_j)_k\} = \frac{2}{3}\delta_{ij}S_k,$$

$$\{(e_i)_j, (e_k)_l\} = \frac{2}{9}\delta_{ik}\delta_{jl}\mathbb{I} + \frac{2}{3}\epsilon_{ikr}\epsilon_{jls}(e_r)_s.$$

If we define the dimensionless normalized matrices:

$$a_{ij} = 3(e_i)_j, \qquad s_i = \frac{2}{\hbar}S_i, \qquad z_i = \frac{2}{\hbar}Z_i,$$

together the 4×4 unit matrix $\mathbb{I}$, they form a set of sixteen matrices Γ_λ, $\lambda = 1, \ldots, 16$ that they are hermitian, unitary, linearly independent and of unit determinant.

The set of 64 unitary matrices of determinant $+1$, $\pm\Gamma_\lambda$, $\pm i\Gamma_\lambda$, $\lambda = 1, \ldots, 16$ form a finite subgroup of $SU(4)$. Its composition law can be obtained from:

$$a_{ij}\,a_{kl} = \delta_{ik}\delta_{jl}\mathbb{I} + i\delta_{ik}\epsilon_{jlr}\,s_r - i\delta_{jl}\epsilon_{ikr}\,z_r + \epsilon_{ikr}\epsilon_{jls}\,a_{rs}, \tag{1}$$

$$a_{ij}\,s_k = i\epsilon_{jkl}\,a_{il} + \delta_{jk}\,z_i, \tag{2}$$

$$a_{ij}\,z_k = -i\epsilon_{ikl}\,a_{lj} + \delta_{ik}\,s_j, \tag{3}$$

$$s_i\,a_{jk} = i\epsilon_{ikl}\,a_{jl} + \delta_{ik}z_j, \tag{4}$$

$$s_i\,s_j = i\epsilon_{ijk}\,s_k + \delta_{ij}\mathbb{I}, \tag{5}$$

$$s_i\,z_j = z_j\,s_i = a_{ji} \tag{6},$$

$$z_i\,a_{jk} = -i\epsilon_{ijl}\,a_{lk} + \delta_{ij}s_k \tag{7},$$

$$z_i\,z_j = -i\epsilon_{ijk}\,z_k + \delta_{ij}\mathbb{I} \tag{8},$$

and similarly we can use these expressions to derive the above commutation and anticommutation relations.

Dirac's algebra is generated by the four Dirac's gamma matrices γ^μ, $\mu = 0, 1, 2, 3$ that satisfy the anticommutation relations

$$\{\gamma^\mu, \gamma^\nu\} = 2\eta^{\mu\nu}\mathbb{I},$$

being $\eta^{\mu\nu}$ Minkowski's metric tensor.

Similarly it can be generated by the following four observables, for instance: S_1, S_2, Z_1 and Z_2. In fact by (5) and (8) we obtain S_3 and Z_3 respectively and by (6) the remaining elements.

Classically, the internal orientation of electron is characterized by the knowledge of the components of the body frame $(e_i)_j$, $i, j = 1, 2, 3$ that altogether constitute an orthogonal matrix. To completely characterize in a unique way this orthogonal matrix we need at least four of these components. In the quantum version, the knowledge of four $(e_i)_j$ matrices and by making use of $(1) - (8)$ allows us to recover the remaining elements of the complete Dirac's algebra. It is in this sense that *internal orientation* of electron completely characterizes its internal structure.

REFERENCES

1. M. Rivas, Elementary classical systems: I. Galilei free particles, *J. Phys. A* 18:1971 (1985).
2. M. Rivas, Classical relativistic spinning particles, *J. Math. Phys.* 30:318 (1989).
3. M. Rivas, Quantization of generalized spinning particles: New derivation of Dirac's equation, (to be published) *J. Math. Phys.* (1994).

PHYSICS WITHOUT PHYSICAL CONSTANTS

Edward Kapuścik

Department of Theoretical Physics,
H.Niewodniczański Institute of Nuclear Physics,
ul. Radzikowskiego 152, 31 342 Kraków, Poland
and
Institute of Physics and Informatics
Cracow Pedagogical University
ul. Podchorążych 2, Kraków, Poland

INTRODUCTION.

One of the most fundamental properties of both Newton's mechanics and Maxwell electrodynamics is the absence of any physical constants in their basic equations. All necessary constants appear only at the stage of applications of these theories to specific phenomena. This is one of the reasons of universality and generality of these theories since physical constants always reflect our ignorance in formulation of physical laws. Therefore primary equations of physics should not contain physical constants at all, including the fundamental ones.

Quantum mechanics and general relativity seem to be counterexamples of the above requirement since Schroedinger equation contains both the fundamental Planck constant and the mass of the particle and Einstein equation contains the gravitational constant. It is however possible to suspect that both these great theories may be in some sense secondary and their basic equations may be derived from more fundamental formulations in which all physical constants do not appear.

Universality and generality of any theory may be achieved only after adequate choice of its basic concepts. Having that in mind for each theory we introduce four collections of basic fields defined over space-time with different physical interpretation.

The first collection of basic fields denoted by $\psi_\alpha (x)$ (α stands for all indices necessary in the theory) describes all space-time properties of the considered physical systems including their space-time localization, their space-time symmetries and interactions present in the systems. The particular physical interpretation and the number of these fields depends on the particular kind of the theory.

The second collection of basic fields denoted by $\varphi_{\mu,\beta} (x)$ determines the evolution of the basic fields $\psi_\alpha (x)$ with respect to the coordinate labelled by the index μ. The possible constraints present in the considered systems may cause that the indices α and β will run over different sets.

Frontiers of Fundamental Physics, Edited by M. Barone
and F. Selleri, Plenum Press, New York, 1994

The first group of basic equations of any theory of matter and/or fields we shall write in the form

$$K_\beta^{\alpha\,\nu}_{\ \ \mu} \nabla_\nu \psi_\alpha (x) = \varphi_{\mu,\beta} (x) \tag{1}$$

where $K_\beta^{\alpha\,\nu}_{\ \ \mu}$, called below as a kinematical factor, is a dimensionless multiindexed quantity specific for each theory, ∇_ν denotes some kind of differentiation with respect to the ν-th coordinate and the summation over repeated indices is understood. The group of equations (1) defines the kinematics of any theory.

To formulate the dynamical laws we introduce a third collection of fields denoted by $\rho_\gamma (x)$ which should describe the influence of the external environment on the studied systems of matter and fields (here γ denotes a set of indices necessary to describe this influence). The fields $\rho_\gamma (x)$ define the balance equations for the collection of all dynamical fields $\pi_\gamma^\mu (x)$ which form the last fourth collection of basic fields and the second group of our basic equations acquires the form

$$\nabla_\mu \pi_\gamma^\mu (x) = \rho_\gamma (x). \tag{2}$$

We shall now prove that equations (1) and (2) may be treated as basic primary equations of any known physical theory provided we shall complete these equations by a suitable set of constitutive relations. First such proof was presented in [1] and in a more general form in [2]. A possible application was discussed in [3].

CLASSICAL MECHANICS OF MATERIAL POINTS

Newton ' s equations of classical mechanics follows from our basic equations (1) and (2) provided the basic fields $\psi_\alpha (x)$ depend solely on the time variable and are grouped into triplets of quantities which describe trajectories of each material point. The kinematical factor $K_\beta^{\alpha\,\nu}_{\ \ \mu}$ has to be chosen in the form

$$K_\beta^{\alpha\,\nu}_{\ \ \mu} = \delta_\beta^\alpha \, \delta_\mu^\nu, \tag{3}$$

and the derivative ∇_μ is simply the time derivative. The nonzero components of the fields $\varphi_{\mu,\beta}$ are then the components of the velocity, those of the fields $\pi_\gamma^\mu (x)$ which also depend only on the time variable are the components of the momentum and those of the fields $\rho_\gamma (x)$ are the components of the acting force.

WAVE EQUATIONS

All wave equations in flat space-time follow from (1) and (2) provided the kinematical factor is given by (3), the index γ coincides with the index α and the ∇_μ are the ordinary partial derivatives. In curved space-times the ∇_μ mean the corresponding covariant derivatives. Under these conditions equations (1) and (2) read

$$\partial_\mu \psi_\alpha (x) = \varphi_{\mu\,,\,\alpha} (x) \tag{4}$$

$$\partial_\mu \pi_\alpha^\mu = \rho_\alpha (x) \tag{5}.$$

The basic fields $\psi_\alpha (x)$ are now wave functions with specific properties (specified by the nature of the index α) under space-time transformations. The fields $\rho_\alpha (x)$

describe interactions of the fields and in the simplest case of free fields are simply proportional to the fields $\psi_\alpha (x)$. The type of the wave equation is determined by the dependence of $\pi_\alpha^\mu (x)$ on $\psi_\alpha (x)$ and $\varphi_{\mu,\alpha} (x)$. The following examples illustrate that.

a) Schroedinger equation for a scalar field

For scalar fields the index α is absent and we have only one wave function $\psi (x)$. The fields $\varphi_\mu (x)$ and $\pi^\mu (x)$ have four components which are related by the constitutive relations

$$\pi^0 (x) = i \, \hbar \, \psi (x) \tag{6}$$

$$\pi^j (x) = - \frac{\hbar^2}{2m} \, \varphi_j (x) \tag{7}$$

$$\rho (x) = - V (x) \, \psi (x) \tag{8}$$

where $V (x)$ is the usual non-relativistic potential. It easy to check that substituting these constitutive relations into (4) and (5) we get the usual Schroedinger equation.

b) Klein - Gordon equation

To obtain the Klein - Gordon wave equation with self - interaction we must assume the following constitutive relations

$$\pi^\mu (x) = g^{\mu \, \nu} \, \varphi_\nu (x) \tag{9}$$

where $g^{\mu \, \nu}$ is the Minkowski metric tensor and instead of (8) we put

$$\rho (x) = - \frac{m^2 \, c^2}{\hbar^2} \, \psi (x) + \mathcal{F} (\psi) \tag{10}$$

where $\mathcal{F}$ describes the selfinteraction of the field $\psi (x)$.

c) Dirac equation for a spinor field

The Dirac equation follows from (4) and (5) when the index α runs from 1 to 4 and

$$\pi_\alpha^\mu (x) = i \, (\gamma^\mu \, \psi (x))_\alpha \tag{11}$$

$$\rho_\alpha (x) = - \frac{m \, c}{\hbar} \, \psi_\alpha (x) \tag{12}$$

where γ^μ are the usual Dirac matrices. Adding to the right - hand side of (12) an interaction term we may obtain Dirac equation with arbitrary interaction.

Since every wave equation may be written in the form of the Dirac equation with an appropriate choice of the γ matrices, the present example shows in fact that an arbitrary wave equation may be obtained from our basic equations (4) and (5). Similarly, replacing the partial derivatives ∂_μ by the gauge covariant derivatives

$$D_\mu = \partial_\mu + i \, A_\mu (x) \tag{13}$$

with the gauge field A_μ (x) we may obtain all gauge covariant wave equations.

The approach to wave equations presented here has at least two big advantages over the standard approach. First, our approach shows that all quantum mechanical wave equations have the common root with classical mechanics and this connection is independent from the widely used canonical formalism. All these theories differs only by suitable choice of constitutive relations. Second, the basic equations (4) and (5) of any wave theory do not contain physical constants which however are always introduced by constitutive relations. These relations only for very simple systems operate with physical constants while for more complicated systems the constants are always replaced by some functions of space-time variables. In the cases of classical mechanics and classical electrodynamics such a replacement considerably extents the range of applicability of these theories. Using our approach to wave mechanics we may also replace all introduced constants by some phenomenological functions of space-time variables. In particular, we may replace the fundamental Planck constant $\hbar$ by a function $\hbar$ (x) what means that we may intensify or relax quantum effects in particular space regions and/or time intervals. Unfortunately, it is not known whether such quantum systems exists in Nature [4].

FIELD THEORIES

A common feature of all theories of matter is the universal form of the kinematical factor $K^\alpha_{\beta}{}^\nu_\mu$. It is no longer true for field theories for which this factor describes important symmetry properties. A common feature of field theories is the vanishing of the fields $\varphi_{\mu\,\alpha}$ (x). The general scheme for all field theories is therefore provided by the equations

$$K^\alpha_{\beta}{}^\nu_\mu \, \nabla_\nu \, \psi_\alpha \, (\, x \,) \; = \; 0 \tag{14}$$

$$\nabla_\mu \, \pi^\mu_\gamma \, (\, x \,) \; = \; \rho_\gamma \, (\, x \,). \tag{15}$$

We shall now specify the shape of the kinematical factor for electrodynamics and general relativity.

a) Maxwell electrodynamics.

In this case the basic fields ψ_α (x) coincide with the standard skew symmetric electrodynamical tensor of the second rank and we shall use for them the standard notation $F_{\mu\,\nu}$ (x). The kinematical factor has the form (α is now a pair of indices ($\lambda\,\epsilon$), β is a pair ($\omega\,\eta$))

$$K^{\lambda\;\epsilon\;\nu}_{\omega\;\eta\;\mu} \; = \; \delta^\lambda_\omega \, \delta^\epsilon_\eta \, \delta^\nu_\mu \; + \; \delta^\lambda_\eta \, \delta^\epsilon_\mu \, \delta^\nu_\omega \; + \; \delta^\lambda_\mu \, \delta^\epsilon_\omega \, \delta^\nu_\eta \tag{16}.$$

It is easy to check that with such kinematical factor equations (14) reduce to the first pair of Maxwell equations.

The second pair of Maxwell equations is obtained from equations (15) under the following identifications (the index γ is now called ν)

$$\rho_\nu \, (\, x \,) \; = \; j_\nu \, (\, x \,) \tag{17}$$

$$\pi^\mu_\nu \; = \; g^{\mu\;\lambda} \, H_{\lambda\,\nu} \, (\, x \,) \tag{18}$$

where j_ν (x) is the electromagnetic current and $H_{\lambda\,\nu}$ (x) is the electromagnetic tensor constructed from the electromagnetic fields in matter $\vec{D}$ (x) and $\vec{H}$ (x).

b) General relativity

In the case of general relativity we should identify the basic fields ψ_α (x) with the components of the curvature tensor $R^\alpha_{\mu\nu\lambda}(x)$. The kinematical factor is given by

$$K^{\beta\ \eta\ \lambda\ \epsilon\ \nu}_{\alpha\ \omega\ \kappa\ \xi\ \mu} = \delta^\beta_\alpha\,\delta^\eta_\omega\,(\,\delta^\lambda_\kappa\,\delta^\epsilon_\xi\,\delta^\nu_\mu\,+\,\delta^\lambda_\xi\,\delta^\epsilon_\mu\,\delta^\nu_\kappa\,+\,\delta^\lambda_\mu\,\delta^\epsilon_\kappa\,\delta^\nu_\xi\,) \tag{19}$$

With this factor equations (14) coincide with the well-known Bianchi identities.

It is clear that the famous Einstein equation cannot be obtained from equations (15). The only equation of general relativity which has the form of (15) is the conservation law

$$T^\nu_{\mu\,;\,\nu}\,(\,x\,)\,=\,f_\mu\,(\,x\,) \tag{20}$$

where $T_{\mu\nu}$ (x) is the energy - momentum tensor of the system and f_μ (x) is the density of the external force which acts on it. The Einstein equation is a non - differential relation between the Ricci tensor (constructed from the basic curvature tensor $R^\alpha_{\mu\,\nu\,\lambda}$ (x)) and the dynamical energy - momentum tensor and it introduces into the theory the gravitational constant. According to our classification of basic equations which should not contain physical constants and constitutive relations which introduce them into consideration Einstein equation has to be treated simply as a constitutive relation and not as a basic equation.

CONCLUSIONS

We have shown that basic equations of all physical theories can be derived from one universal and simple set of primary equations which do not contain any physical constant. All necessary constants appear through constitutive relations which define the physical situation to which the primary equations are applied. Our scheme allows the unification of all particular theories into one elegant and simple supertheory.

ACKNOWLEDGMENTS

The author expresses his gratitude to the Polish State Committee for Scientific Research for financial support provided under grant 2 0342 91 01. Many thanks to Profs. A. Janussis, R. M. Santilli and F. Selleri for valuable discussions.

REFERENCES

1. E. Kapuścik, Newtonian form of wave mechanics, *Acta Phys.Pol.* 17:923(1986)
2. E. Kapuścik, Physics without physical constants, *Proc. of the V Int. Symposium on Selected Problems of Statistical Mechanics*, A. A. Logunov et al. eds., World Scientific, Singapore (1991).
3. E. Kapuścik and A. Horzela. Another treatment of the relation between classical and quantum mechanics, *Ann. Fond. L. de Broglie* 18:155(1993).

THE RELATION BETWEEN INFORMATION, TIME AND SPACE INFERRED FROM UNIVERSAL PHENOMENA IN SOLID-STATE PHYSICS

Gerhard Dorda

Siemens AG
Corporate Research and Development
Munich, Germany

INTRODUCTION

In recent years, phenomena of a universal character have been observed in solid-state physics that, as in nuclear physics and astrophysics, permit the clarification of fundamental problems of physics. Examples are the quantum Hall effect[1], quantum transport[2,3], the Aharonov-Bohm effect[4] as well as 1/f noise[5]. All these phenomena are universal, i.e. material-independent.

It will be shown in this paper that these non-homogeneous experimental phenomena can be combined in the form of an energy equation. The essential point is that the characteristic states of this equation, known as the Electron Energy Paradigm (EEP), differ only in the coupling constant α having different powers[5,6]. To facilitate interpretation of the EEP, we will briefly treat the phenomena leading to formulating the EEP. The goal of the paper is the following comprehensive discussion of the significance of the particular energy states of this paradigm. Finally, a new way will be shown of gaining a better understanding of the wave-particle duality of the electron on the basis of the EEP.

Formulation of the EEP starts from the quantum Hall effect (QHE) and quantum transport in solids. The specific features of these phenomena, apart from universality, include their independence of the size of the sample studied[7], a totally unexpected phenomenon with regard to resistance properties. This experimentally observed fact will play an important part when classifying the properties of the electron.

THE FORMULATION OF THE ELECTRON ENERGY PARADIGM

The QHE, whose fundamental importance for physics lies in its application as a

Frontiers of Fundamental Physics, Edited by M. Barone
and F. Selleri, Plenum Press, New York, 1994

resistance standard, is given by the following equations:

$$N = i\,(e/h)\,B \tag{1}$$

or rather

$$D = i\,(e^2/h)\,B, \tag{1a}$$

paired with the current-voltage equation in the form

$$I = i\,(e^2/h)\,V \tag{2}$$

where N is the induced 2-dimensional electron density, D the displacement, B the magnetic flux density, I the electric current, V the voltage applied between source and drain, e the electric charge, h Planck's constant and i a quantum number.

The universal equation (2) that represents Ohm's law in quantum form could also be used to describe the quantum transport phenomena observed without a magnetic field both in disturbance-free materials[2] as well as in thin amorphous Si films[3]. The deeper meaning of this universally valid equation can be deciphered when we apply, to the square of the electric charge e^2, neither the definition of the SI system of units given by

$$e^2 = 2\,\varepsilon_0\,\alpha\,h\,c \tag{3}$$

nor that of the CGS system of units, given by

$$e^2 = \alpha\,h\,c\,/\,2\pi \tag{4}$$

but rather the new equation

$$e^2 = \alpha\,h\,. \tag{5}$$

Here ε_0 is the permittivity of vacuum, c the velocity of light and α the coupling constant. If we multiply (2) by the charge e and when using (5) and $i = 1$, we obtain Ohm's law in a new form, expressed by the reference energies of the current eI_0 and the voltage eV_0[6,8]

$$eI_0 = \alpha\,eV_0 \tag{6}$$

or

$$e^2 f_e = \alpha\,hf_e = \alpha\,e\,\phi\,f_e \tag{6a}$$

Here f_e is the reference frequency, given by the Rydberg energy Ry, $f_e = 2Ry/h$, and $\phi = h/e$ is the flux quantum, given in the unit system of (5) by $\phi = V_0/f_e$.

As already shown in another paper[6], definition (5) is in agreement with quantum electrodynamics and has priority over formulations (3) and (4). It leads to a new system of electromagnetic units, see also[8].

Formulation (5) is very important for our further discussion. It allows a new relationship to be set up between five reference energies of the electron via the coupling constant α. In this treatment we must include the relation between the reference energy of the atomic spectra eV_0 and the energy of the rest mass of the electron m_0 that has already been known for a long time without being understood. It is given by

$$eV_0 = 2Ry = \alpha^2 \, m_0 \, c^2 \tag{7}$$

where Ry is the Rydberg energy.

A further link in formulating the reference energy relation of the electron is provided by $1/f$ noise. As already shown elsewhere[5], the empirically deduced Hooge equation used worldwide describing $1/f$ noise and taking into account (5) leads to the energy relation

$$S_V = V^2 / Nf \tag{8}$$

where S_V is the noise level measured at constant voltage, V the applied voltage, f the frequency and N the number of coherent quantum states of a given transport state. In the system of electromagnetic units referred to (5), V^2/f is assigned the dimension of energy, but only in this system of units. Accordingly, S_V is a measure of quantized energy fluctuations of the externally applied electric field arising from extraneous disturbances. The reference energy $S_{0,V}$ is obtained from (8) for $N = 1$ and when using the reference values V_0 and f_e, which is permitted as the noise refers to electrons. We then get[5,6]

$$S_{0,V} = V_0^2 / f_e = \phi^2 f_e = \alpha^{-1} hf_e = \alpha^{-1} eV_0 \tag{9}$$

Another relation to eV_0 is obtained by determining the reference umklapp energy of the spin of the electron $E_{0,0}$. By using the Bohr magneton $\mu_B = eh/2m_0$ and the reference value of the magnetic flux density $B_0 = \phi/2\pi a_e^2$, where $2\pi a_e^2$ with $a_e = (2\pi f_e)^{-1} c = 7.25$ nm represents the reference area[8], we obtain

$$E_{0,0} = (eh/m_0) \, \phi/2\pi a_e^2 = \alpha^2 \, eV_0 \tag{10}$$

It should be noted that the theoretically derived reference area was verified experimentally on MOSFETs of ultra-small dimensions[8] and on the basis of experimental data from the Aharonov-Bohm effect[6].

If we now combine relations (6), (7), (9) and (10), we obtain the electron energy paradigm (EEP) relating five reference energies $E_{0,n}$, $n = 0,....4$, in the form

$$\boxed{E_{0,0} = \alpha \, E_{0,1} = \alpha^2 \, E_{0,2} = \alpha^3 \, E_{0,3} = \alpha^4 \, E_{0,4}} \tag{11}$$

where the reference energies are given by

$$E_{0,0} = 2\mu_B B_0, \qquad E_{0,1} = e \, I_0 = e^2 f_e, \qquad E_{0,2} = eV_0 = h \, f_e = e \, \phi \, f_e = 2Ry$$

$$E_{0,3} = V_0^2 / f_e = \phi^2 f_e, \qquad E_{0,4} = m_0 c^2 \tag{11a}$$

The relation between these reference energies, given by the coupling constant α of various powers, is shown in Fig.1.

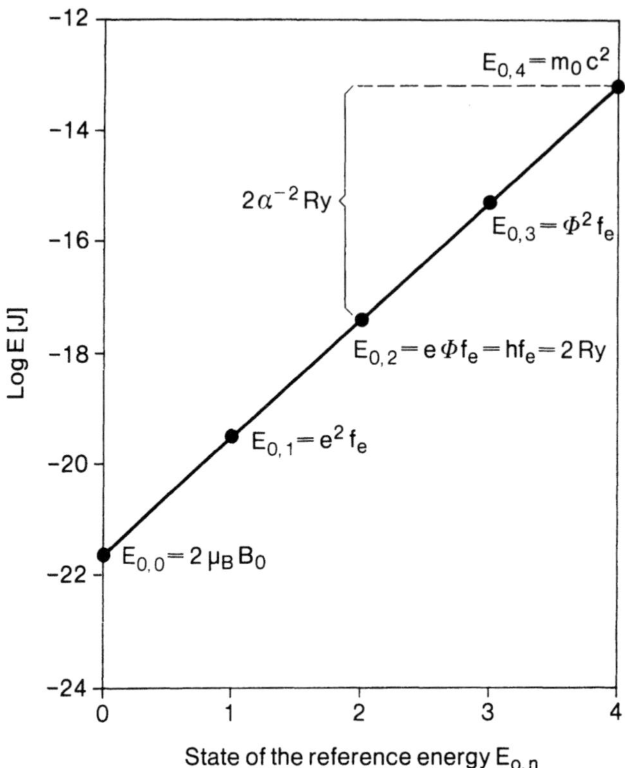

Figure 1. The electron energy paradigm.

Let us again expressly stress in this connection that the EEP represents an empirical relation, as it is obtained when taking into account (5) from the experimental data of the quantum Hall effect and the quantized transport phenomena, the 1/f noise, the spin and the atomic spectral lines. This EEP is an energy equation, and is thus invariant with respect to any system of electromagnetic units.

INTERPRETATION OF THE ELECTRON ENERGY PARADIGM
WITH RESPECT TO INFORMATION, TIME AND SPACE

The sequence of five reference energies, linked only by the coupling constant α, can be interpreted as a sequence of different forms of the electron, starting with $E_{0,0}$, that we try to interpret as "information", ending with the hitherto highest known state $E_{0,4}$, that we understand as "mass". It should be noted that the particular reference energy state (and its associated quantum states) also includes the specifications of the preceding reference energy states, but not yet those of the subsequent ones. This fact seems to be of great importance, as it opens a new approach to interpreting the wave-particle duality.

We will try to proceed along the sequence shown in Fig. 1 and interpret it on the basis of experimental data. The following considerations should be regarded as a suggestion and refer exclusively to the electron, although a generalization of these statements cannot be excluded.

As already shown, the $E_{n,0}$ energy (where the index n refers to the nth quantum state associated with the reference energy $E_{0,0}$) reflects the electron spin, that contains only the (+,-) or yes-no statement. It thus represents the digitalization of information. Because $E_{0,0}$ is the smallest reference energy value, information (which may be equated with the term "logos" used in philosophy and theology) can be regarded as the starting point of further development stages.

The next highest state $E_{n,1}$ corresponds to the familiar phenomenon of the electric current. The characteristic of this state becomes clear when we look at ballistic, i.e. disturbance-free boundary conditions of the kind shown in (2) for i = 1. The essential feature of this state is the inclusion of "frequency" (or "time"), i.e. of the temporal division between two events, or the number of events occurring during a particular time period. We imagine these events as being marked by the generation and/or annihilation of an electric charge. The current is thus a type of time-counting process. It should be noted that the term "position" or "space" plays no part with respect the $E_{n,1}$, as can also be seen in the formulation $e^2 f_e$.

Experience has shown that the coupling to space does not occur until the next highest state, i.e. in the state referred to the reference energy $E_{0,2}$. We know this state in the form of the light quantum or of the wave state of the electron. It covers both the terms information (spin) and time (frequency) as well as that of space in its simplest form as "length". The length, also declared as the wavelength, can be understood with respect to the states $E_{0,3}$ and $E_{0,4}$ as space in one-dimensional form. The combination of frequency and length (wavelength) then yields the term speed or light velocity that is characteristic of this state. It should, however, be noted that with respect to $E_{n,2}$, i.e. within the scope of this fundamental 1-dimensional space, we may indeed speak of lengths or length quanta and thus of speed, but this type of space does not include the terms position or localization. This interpretation agrees with the results obtained from the double-slit experiment.

The new property of space, known as "position", is not realized until a further state $E_{n,3}$ by the interaction of two electric flux quanta, as indicated by the process of the 1/f noise that represents the interaction of the externally applied voltage field with the fields of the disturbances present in the material[5]. We can thus assume in geometrical terms that in the $E_{n,3}$ state a coupling occurs between one 1-dimensional space and another 1-dimensional space independent of it, whereby the position or localization state results from their intersection. According to this idea, therefore, a 2-dimensional space results from the interaction between two flux fields.

In the $E_{0,4}$ state we have the highest state of reference energies hitherto known to us, that we see in the form of the "electron mass". Experience with all particles to date tells us that the concept of mass is closely linked to the phenomenon of (at least) 3-dimensional space, an idea we may also transfer to the electron mass. Seen from this viewpoint, the $E_{0,4}$ state reveals a 3-dimensional space that can be recognized by the observation of mass.

In summary, we interpret the EEP as a physical formulation of the assignment of five fundamental, independent observables such as information, time, length, location and mass to the electron energy. That means that on the basis of our analysis we can understand these five observables as properties of the electron energy. As equation (11) shows us, this assignment occurs solely via the coupling constant α of different powers. Seen from the standpoint of energy, information then represents the lowest state, whereas mass is the highest state. Thus we recognize mass as the form of the electron which can show a large number of different complex aspects.

THE WAVE-PARTICLE DUALITY

The double-slit experiment shows us that equi-energetic transitions are possible, for example from $\alpha^{-2} E_{0,2}$ to $E_{0,4}$ and vice-versa, that we know as the wave-particle duality. Owing to our analysis of the EEP, we came to the conclusion that the property of "position" cannot exist in the $E_{n,2}$ state, i.e. in the 1-dimensional space-time state, the wave state of the electron. This means that all experimental efforts to localize this wave, or a suspected part of it, are quite impossible in principle and must fail[9].

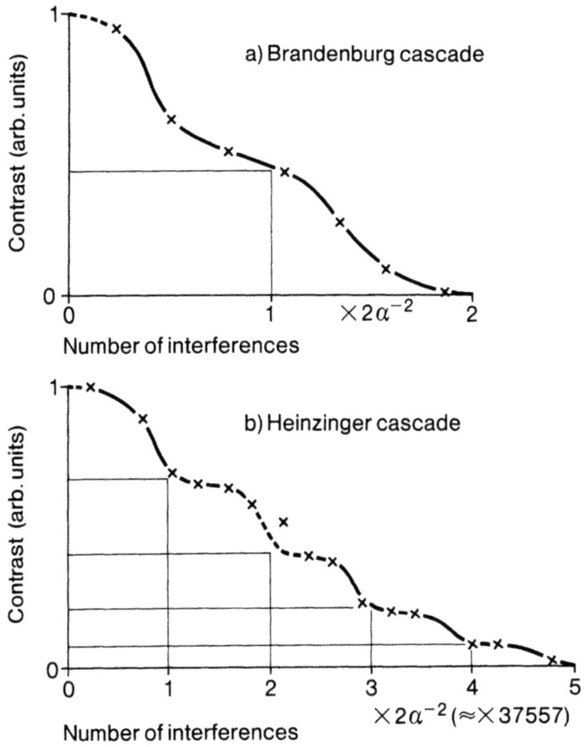

Figure 2. Contrast dependence of electron interference patterns (Redrawn after Schmid[10])

It is further clear that the transition between state $E_{n,2}$ and state $E_{0,4}$ must be related to the factor α^2, see Fig.1. Experimental evidence for these considerations has been obtained by analyzing the contrast of the interference patterns generated by an electron interferometer, showing quantum behaviour with respect to the factor $2i\,\alpha^{-2}$. As Fig. 2 shows[10], in the first experimental set-up $i = 1, 2$ and in the set-up with the greater density of electron current $i = 1, 2, 3, 4, 5$. We suppose that the quantum number i visualizes the clustered participation of i electrons in the interference process.

CONCLUSION

We showed in this paper that experimental results from the area of quantum transport in solid-state physics, of 1/f noise, the physics of atomic spectra and of electron spin can be combined in the form of an energy equation known as the Electron Energy Paradigm (EEP). When expressed in their energy form, these independent phenomena are distinguished only by the coupling constant α of different powers. The EEP, that has a purely empirical background, suggests that the seemingly independent terms such as information, time, length, position and mass (here referred to the electron) are merely various expressions of the electron energy represented by five specific energetic states. As shown, the impossibility of localizing an electron wave can be simply explained on the basis of the EEP.

REFERENCES

1. K.von Klitzing, G. Dorda, and M. Pepper, New method for high-accuracy determination of the fine-structure constant based on quantized Hall resistance, *Phys.Rev.Lett.45*, 494:497 (1980); R.E. Prange and S.M. Girvin, "The Quantum Hall Effect", Springer, Berlin (1987).
2. B.J. van Wees, H. van Houten, C.W.J. Beenakker, J.G. Williamson, L.P. Kouwenhoven, D. van der Marel, and C.T. Foxon, Quantized conductance of point contacts in a two-dimensional electron gas, *Phys.Rev.Lett.60*, 848:850 (1988).
3. J. Hajto, A.E. Owen, S.M. Gage, A.J. Snell, P.G. LeComber, and M.J.Rose, Quantized electron transport in amorphous-silicon memory structures, *Phys.Rev.Lett.66*, 1918:1921 (1991).
4. Y. Imry and R.A. Webb, Quantum interference and the Aharonov-Bohm effect, *Scientific American 260*, 36:42 (1989).
5. G. Dorda, The interpretation of 1/f noise from the point of view of the electron energy paradigm, in: "Noise in Physical Systems and 1/f Fluctuations," invited paper, *AIP Conf.Proc. 285*, P.H. Handel and A.L. Chung, ed., American Inst. of Physics, New York (1993), ps. 623:628.
6. G. Dorda, $e^2 = \alpha h$, the only physically justified formulation of electron charge and the resulting Electron Energy Paradigm, in: "Waves and Particles in Light and Matter," A. van der Merwe and A. Garuccio, ed., Plenum Publ. Corp., New York (1994).
7. F.F. Fang and P.J. Stiles, Quantized magnetoresistance in two-dimensional electron systems, *Phys.Rev.B27*, 6487:6488 (1983).
8. G. Dorda, Universal behaviour of the electronic transport, invited paper, Proc. 12th General Conf. of the Condensed Matter Division of the European Physical Society, Praha, *Physica Scripta T45*, 297:299 (1992).
9. J. Horgan, Quantum philosophy, *Scientific American 267*, 72:80 (1992).
10. H. Schmid, Ein Elektronen-Interferometer mit 300 µm weit getrennten kohärenten Teilbündeln zur Erzeugung hoher Gangunterschiede und Messung der Phasenschiebung durch das magnetische Vektorpotential bei metallisch abgeschirmtem Magnetfluß, PhD-Thesis, University of Tübingen, Tübingen, Germany (1985).

QUANTUM-LIKE BEHAVIOUR OF CHARGED PARTICLES IN A MAGNETIC FIELD AND OBSERVATION OF DISCRETE FORBIDDEN STATES IN THE CLASSICAL MECHANICAL DOMAIN

Ram K. Varma

Physical Research Laboratory
Ahmedabad 380 009
India

INTRODUCTION

The relationship between classical mechanics and quantum mechanics has been a subject of longstanding discussion ever since the advent of quantum mechanics. A related question has been the probabilistic nature of the quantum mechanical description, in particular, whether the probabilistic nature of QM description is intrinsic or is attributable to some unspecifiable "hidden parameters" belonging to the system.

The wave nature of matter that is characteristic of quantum behaviour is known to come into play for microscopic dimensions and small effective masses. The "classical limit" is generally taken to be $\hbar \to 0$. In this limit, the quantum description should pass over into the classical description. But the precise manner of transition is far from clear. In particular, it may be noted, in this connection, that classical mechanics is an initial value problem being governed by a second order ordinary differential equation in time for the position coordinates. They advance the initial values in time pertaining to positions and momenta. An important tenet of classical mechanics is that the whole continuum of initial values are allowed values. Consequently, all the possible states of motion constrained only by the equation of motion are allowed. In particular, all the continuum of energy states are allowed.

Quantum mechanics, on the other hand, presents a boundary value problem for a wave function, the space time evolution of which is governed by the Schrödinger wave equation (for nonrelativistic case). This is a second order partial differential equation with respect to space and first order in time. Because of its wave nature, the equation predicts discrete energy states by virtue of the boundary conditions or periodicity conditions.

It may appear heretical to ask whether there exists a system in the classical mechanical domain which exhibits a quantum mechanical like wave behaviour. There may not exist any *a priori* reason to suspect such a behaviour for the particular system

Frontiers of Fundamental Physics, Edited by M. Barone
and F. Selleri, Plenum Press, New York, 1994

described below. But motivated by certain intuitive considerations, the present author was led to construct first a heuristic derivation of a set of Schrödinger-like equations for the dynamical evolution of a certain specifically prepared ensemble of charged particles in a magnetic field (Varma, 1971), where the role of $\hbar$ is played by the gyroaction associated with the gyration of particles around the magnetic field. Later, the same set of equations were derived as amplitude representation of the classical Liouville equation for the system (Varma, 1985). To be sure, the derivation of a Schrödinger-like set of equations from the classical Liouville equation is not in accordance with the canonical conceptual understanding and framework of the physical world, but the derivation (Varma, 1985) does constitute a different mathematical representation of the Liouville equation which happens to be a set of Schrödinger-like equations. The physical and observational consequences of these set of wave equations do raise certain issues concerning the reconciliation between the quantum-like predicted consequences of the Schrödinger-like equations and those of the Liouville equation or the classical equation of motion. The discussion of these important aspects will be considered later.

However, one way (and, perhaps, the only one) to examine the physical significance of the set of Schrödinger-like equation is to carry out experiments to check the predictions of the equations.

CHARGED PARTICLE MOTION IN A MAGNETIC FIELD AND THE SCHRÖDINGER-LIKE EQUATIONS

If we inject an ensemble of charged particles in a magnetic field with a given common value of the gyroaction $\mu = \frac{1}{2}mv_\perp^2/\Omega$ ($v_\perp$ is the component of the velocity perpendicular to the magnetic field, and $\Omega = eB/mc$, is the gyrofrequency), a given common energy E at a given $X_\parallel$ (position along the magnetic field), then the motion of particles in the "adiabatic approximation" is governed by the "adiabatic equation of motion" (Northrop, 1963)

$$m\frac{dv_\parallel}{dt} = -\nabla_\parallel(\mu\Omega) \tag{1}$$

where, by adiabatic approximation, one means that the magnetic field B varies slowly enough in space, so that

$$\varepsilon \equiv \frac{v_\perp}{\Omega}\mid\frac{\nabla B}{B}\mid \ll 1$$

Under this approximation $\mu = \frac{1}{2}mv_\perp^2/\Omega$, has been known to be an "adiabatic invariant". The Schrödinger-like equations referred to above (Varma, 1971, 1985) are as follows:

$$\frac{i\mu}{n}\frac{\partial\Psi(n)}{\partial t} = -\frac{(\mu/n)^2}{2m}\frac{\partial^2\Psi(n)}{\partial X^2} + (\mu\Omega)\Psi(n), \quad n = 1,2,3\ldots\ldots \tag{2}$$

with the total probability density being given by

$$G(X,t) = \sum_n \Psi^*(n)\Psi(n) \tag{3}$$

where X is the coordinate along the magnetic field.

There are two obvious predictions of these one-dimensional set of equations, where the adiabatic potential $(\mu\Omega)$ (vide eqn.(1)) appears in the role of 'potential' in these equations and μ appears in the role of $\hbar$. One of them pertains to the existence of the multiplicity of residence times corresponding to the various equations $n = 1,2,3,\ldots$ for the adiabatically trapped particles in an adiabatic potential well described by an appropriate potential $\mu\Omega$. Such multiple residence times corresponding to $n = 1,2$ and

3 have indeed been determined experimentally with characteristics which are found to be completely in accordance with the eqn.(2) (Bora et.al., 1979, 1980, 1982). This establishes experimentally the existence of the modal behaviour as described by the modes $n = 1, 2, 3$ through the corresponding Schrödinger-like equations.

We, however, describe here another, perhaps, a more astonishing prediction of the equations. This pertains to the existence of one-dimensional interference effects predicted by the equation, which arises out of the amplitude character of the functions $\Psi(n)$. If we consider only the function $\Psi(1)$ for $n = 1$, corresponding to an electron beam from an electron gun propagating along a magnetic field, then it has been shown (Varma, 1989) that the probability density of finding the electrons after propagating a distance D along the field is given by

$$\overline{\Psi^*\Psi} = \sum_k \frac{\mid \hat{\Psi}(k) \mid^2}{k} + \sum_k 2R(K, \bar{k}) \sin\left[\frac{\bar{\Omega} D}{\bar{v}_\parallel} + \phi\right] \tag{4}$$

where $v_\parallel$ is the "parallel velocity"

$$v_\parallel = [2(E - \mu\Omega)/m]^{1/2}$$

We see that this expression has an oscillating term which oscillates with the magnetic field B, with the energy E, and the distance traversed D with other parameters remaining constant. The oscillating term is a consequence of the interference effects. The minima in the expression (5) of $\overline{\Psi^*\Psi}$ are interpreted as "forbidden states" of the charged particles in a magnetic field. The energies of the "forbidden states" are given by

$$E_j = \frac{1}{2}m \left(\frac{\Omega D}{2\pi}\right)^2 / (j + \phi_j/2\pi)^2 \tag{5}$$

where j denotes the "quantum number" labelling the forbidden energy states E_j, with ϕ_j being the "phase shift".

It may be remarked that the classical Lorentz equation of motion which governs the dynamics of charged particles in a magnetic field in macroscopic dimensions, does not predict any forbidden states, since classical mechanics admits the entire continuum of energy states as the allowed states. The prediction, therefore, of the existence of forbidden states as described by eqn. (5) is thus quite enigmatic as it is contrary to the behaviour expected à la the standard classical mechanical paradigm.

We describe below an experiment reported earlier (Varma and Punithavelu, 1993) and its results to check the prediction of the Schrödinger-like formalism as expressed by the relation (5).

THE EXPERIMENT AND ITS RESULTS

An electron beam of very low intensity ($< 0.1\mu A$) is injected from an electron gun along a magnetic field ($\sim$ 200-300 gauss) in an SS vacuum chamber of 27cm dia evacuated to $\sim 5.10^{-7}$ torr, and is received at a detector (Faraday Cup) at a distance ($L \sim 20$ cm) away from the gun. The injection is almost parallel to the field. The Faraday Cup Detector consists of a grounded collector plate of $\sim$ 25cm dia 1cm behind a grid which can be biased to any required potential. The electron current from the beam can be measured by the detector (current received by the plate) for various negative grid potentials from zero to $-\Phi_{max}$, where $\Phi_{max} > E/ \mid e \mid$, E being the electron beam energy. Since according to classical mechanics, the whole continuum of energy states are allowed states, the plate current is expected to exhibit a monotonically increasing response as the negative grid potential is swept from $-\Phi_{max}$ to zero.

The actual experimentally observed response is quite and astonishingly different from the expected one (à la the standard classical mechanical paradgm) and is shown in Fig. 1. The upper curve (a) is the plate current and quite clearly exhibits a series of sharply defined dips. The lower curve (b) is the grid current for the same set of parameters and also exhibits a series of dips which are found to be exactly correlated with the plate current dips. From the point of view of classical mechanics, these dips are quite unexpected and astonishing as they signify the existence of "forbidden states" which have no place in the formalism of classical mechanics.

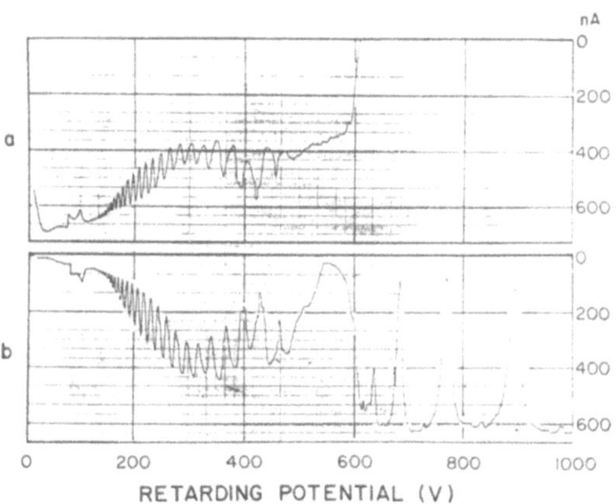

Fig. 1. Plate current (a) and grid current (b) as functions of the retarding potential. $\bar{B} = 170$ G, $L = 30$ cm and $E = 600$ eV.

To further check that these are indeed the forbidden states both the plate current and the anode current were measured simultaneously as the grid potential is swept (It should be explained that the anode which is grounded, is a part of the electron gun and which accelerates the electrons emanating from a negatively biased hot cathode placed about one cm away from it. Thus the electrons encounter the anode much before (~ 20 cm) on the way to grid-plate assembly). Fig. 2 shows both the simultaneously recorded plate and anode currents, the upper curve being the plate and the lower one being the anode current. The two curves are clearly anti-correlated, showing that electrons that did not (or could not) pass through the system in a certain energy state (forbidden state) have found their way to the anode, confirming in a sense the interpretation of the 'dips' in the plate current as forbidden states.

The experiment was repeated with different distances between the gun and the detector and magnetic fields. The plots so obtained are shown in Fig. 3. We see that the positions of the dips change both with the distance L and the magnetic field B. A little reflection shows that the dependence on the distance L is most enigmatic because it signifies a kind of wave-like nonlocality.

404

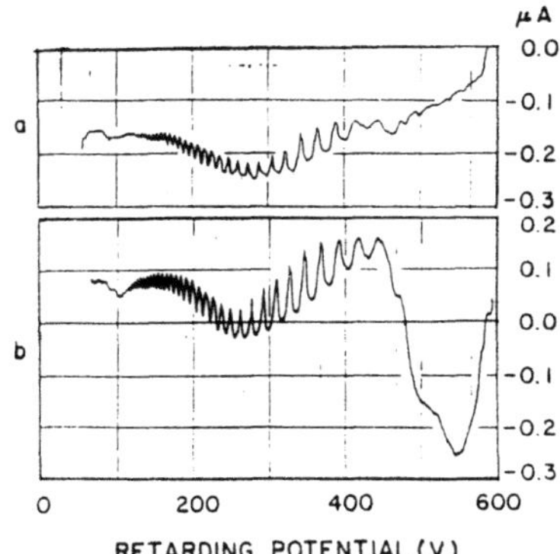

Fig. 2. Plate current (a) and anode current (b) as functions of the retarding potential. $\bar{B} = 177$ G, $L = 19$ cm and $E = 600$ eV.

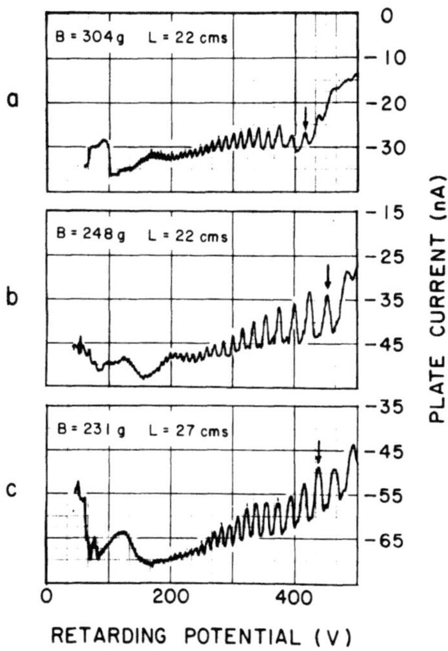

Fig. 3. Plate current as a function of the retarding potential for different values of $\bar{B}$ and L (as shown in the plots) and $E = 650$ eV.

One may now finally examine whether the dips so observed can be described by the relation (5). To do so, we read the positions of the dips (in energy) from the plots and using these values on the left of eqn. (5) with the magnetic field and distance D inserted on the right, we calculate the quantity $(j + \phi/2\pi)$ for every third dip counted from an (arbitrary) arrowed peak in the respective plots of Fig. 3. The results are shown in Table 1. (It should be pointed out that $D = 3L$ was used in the calculation for reasons to be explained elsewhere). The whole number in the value of $(j + \phi/2\pi)$ so obtained is identified with the "quantum number" characterising the dip and the fraction with $(\phi/2\pi)$. It is clearly seen that the j values do differ by 3 corresponding to the fact that every third dip was chosen. Since the different curves correspond to different B and L values, the dependence on B and, in particular, on the distance L, is well borne out by the experimental results.

Table 1

Peaks	Energy E_j (eV)			$j + \phi/2\pi$		
	plot 3a	plot 3b	plot 3c	plot 3a	plot 3b	plot 3c
N($\downarrow$)	417	453	438	41 + 0.15	31 + 0.32	36 + 0.56
N + 3	357	377	373	44 + 0.48	34 + 0.36	39 + 0.62
N + 6	313	317	323	47 + 0.45	37 + 0.47	42 + 0.57
N + 9	277	272	283	50 + 0.50	40 + 0.46	45 + 0.47
N + 12	247	237	250	53 + 0.48	43 + 0.35	48 + 0.41

In summary, it is important to highlight the following facts that we have demonstrated:

a. The discrete forbidden states of motion do exist in the domain of parameters where one would use classical equation of motion to determine the motion.

b. The energies of these "forbidden states" are well represented by the relation (5) which is obviously nonquantal as there is no Planck quantum h appearing in it.

c. The forbidden states E_j form a hydrogen-like sequence for which "quantum numbers" j and the phases can be identified as in Table 1.

d. The forbidden states E_j and the associated quantum numbers j depend on the distance between the gun and the detector. This is a manifestation of wave-like behaviour which is not known to be a characteristic of the standard initial value paradigm of classical mechanics.

CONCLUSIONS

Based on the above described experimental results, one may now conclude that the electrons moving in a magnetic field do appear to exhibit a wave-like behaviour in macroscopic dimension which is known to be a domain of operation of classical dynamics and which, therefore, admits a continuum of energy states. The existence of discrete forbidden states as described above appears to be in contradiction with the latter.

While the apparent paradox needs to be further understood, the above finding may throw some light on the classical-quantum relationship and the nature of quantum mechanics itself. Indeed the present author has obtained a generalized Schrödinger formalism based on these ideas which have been presented elsewhere (Varma 1985b, 1988).

REFERENCES

Northrop, T.G., 1963, "Adiabatic Motion of the Charged Particles" (New York : Interscience).

Varma, R.K., 1971, *Phys. Rev. Letts., 26*, 417.

Varma, R.K., 1985a, *Phys. Rev., A31*, 3951.

Bora et.al., 1979, *Phys. Lett., A75*, 60.

Bora et.al., 1980, *Plasma Phys., 22*, 653.

Bora et.al., 1982, *Phys. Fluids, 25*, 2284.

Varma, R.K., 1985b, *Had. Jour. Suppl., 1*, 437.

Varma, R.K., 1988, "Microphysical Reality and Quantum Formalism", A. Van der Merwe et.al., eds., Klewer Academic Publishers, 175-194.

Varma, R.K., 1989, Wave Mechanics of an Ensemble of Charged Particles in Inhomogeneous Magnetic Fields - A New Paradigm, in "A Variety of Plasmas", *Proceedings of the Int. Conf. on Plasma Phys., Delhi*, A. Sen and P.K. Kaw, eds., Ind. Acad. of Sciences, Bangalore, India, p. 235.

Varma, R.K. and Punithavelu, A.M., 1993, *Mod. Phys. Lett., A8*, 167.

UNIPOLAR INDUCTION AND WEBER'S ELECTRODYNAMICS

André K. T. Assis,[1,2] and Dario S. Thober[1]

[1]Instituto de Física 'Gleb Wataghin', Universidade Estadual
de Campinas, Unicamp, 13083-970, Campinas, São Paulo, Brasil
Internets: assis@ifi.unicamp.br, and thober@ifi.unicamp.br
[2]Also Collaborating Professor at the Department of Applied
Mathematics, IMECC, State University of Campinas,
13083-970 Campinas-SP, Brazil

INTRODUCTION

Unipolar induction is the generation of current on a conductor for the case in which the conductor and the magnet are in relative rotatory motion. A typical case of unipolar induction is shown in figure 1.

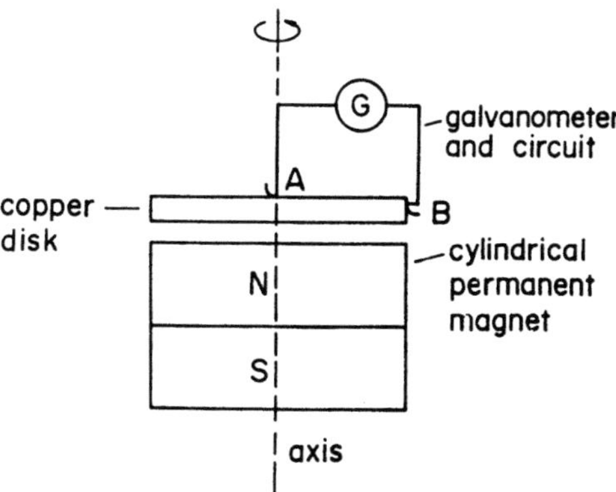

Figure 1. Apparatus used to investigate unipolar induction. The sliding contacts in A and B connect the galvanometer to the copper disk. Copper disk and magnet are free to rotate.

Since Faraday's experiments[1] of 1832 on electromagnetic induction on rotating systems there are intense debates concerning the location of the seat of the electromotive force (emf)[2].

In this work whenever we speak of "rotation" it should be understood "rotation relative to the earth or laboratory."

Let us see what happens in the laboratory. When we rotate only the disk an emf is produced on the galvanometer-disk circuit (the magnet is fixed in the laboratory), as wee can see on the galvanometer. When we rotate only the magnet (the disk is fixed in the laboratory) no current flows by the galvanometer. When we rotate both the disk and the magnet there is a current in the galvanometer.

These results have lead some scientists, like Kennard[3,4], to think of a special frame of reference at which the systems shows unexpected effects when under acceleration relative to it. A classical example is the disk-magnet system when we observe a polarization when both rotate together. Kennard makes no consideration about inductions on the galvanometer. This means that he does not consider the galvanometer as part of the seat of induction.

The galvanometer consideration has been made[5] but the seat of the emf (disk or galvanometer and circuit) is a matter of controversy. It should be observed that in all the experiments which have been made, the galvanometer is fixed in the laboratory.

WEBER'S ELECTRODYNAMICS

In the last few years there has been a renewed interest in Ampère's force between current elements[6,7] and in Weber's force between point charges[8,9,10]. As we know, Ampère's force between current elements can be derived from Weber's force between point charges. It has also been shown that Faraday's law of induction for closed circuits can be derived from Weber's force[11].

The renewed interest in the basic laws of electromagnetism prompted us to study unipolar induction.

Weber's force states that a charge q_j exerts a force on a charge q_i given by:

$$\vec{F}_{ji} = \frac{q_i q_j}{4\pi\epsilon_0} \frac{\vec{r}_{ij}}{r_{ij}^3} \left[1 - \frac{\dot{r}_{ij}^2}{2c^2} + \frac{r_{ij}\ddot{r}_{ij}}{c^2} \right], \tag{1}$$

where $\vec{r}_{12} \equiv \vec{r}_1 - \vec{r}_2$, $r_{ij} \equiv |\vec{r}_i - \vec{r}_j|$, $\hat{r}_{ij} \equiv \frac{\vec{r}_{ij}}{r_{ij}}$, $\dot{r}_{ij} \equiv \frac{dr_{ij}}{dt}$, $\ddot{r}_{ij} \equiv \frac{d\dot{r}_{ij}}{dt}$, $\epsilon_0 = 8.85 \times 10^{-12}$ F/m and c is the ratio of electromagnetic and electrostatic units of charge which was found experimentally to have the same value of the light velocity in vacuum.

UNIPOLAR INDUCTION BY WEBER'S LAW

We will analyse unipolar induction in a region of uniform magnetic field. This can be obtained rotating a uniformly charged spherical shell at a constant angular velocity.

Two shells of radius R and $R + dR$ made up of non-conducting material and uniformly charged with charges Q and $-Q$, respectively, are rotating with constant angular velocities $\vec{\omega}_M$ and $\vec{\omega}_M + \vec{\omega}_N$ ($\vec{\omega}_M$ and $\vec{\omega}_N$ at the same direction).

According to Weber's electrodynamics the force exerted by the first shell (radius R, charge Q, $\vec{\omega}_M$) on an internal charge q located at $\vec{r}$ ($r < R$), moving relative to the laboratory with velocity $\vec{v}$ and acceleration $\vec{a}$ is given by ([10]):

$$\vec{F}(r < R) = \frac{qQ}{12\pi\epsilon_0 c^2 R}\left[\vec{a} + \vec{\omega}_M \times (\vec{\omega}_M \times \vec{r}) + 2\vec{v} \times \vec{\omega}_M + \vec{r} \times \frac{d}{dt}\vec{\omega}_M\right]. \tag{2}$$

Since we are interested in the force upon a free charge of a spinning conductor we make $\vec{v} = \vec{\omega} \times \vec{r}$, where $\vec{\omega}$ is the angular velocity of the conductor.

The net force on the charge q is obtained by adding the contributions of the two shells. Considering that $dR << R$, that $d\vec{\omega}_M/dt = d\vec{\omega}_N/dt = 0$, and utilising (2) this yields ($\vec{\omega}_M = \omega_M \hat{z}$, $\vec{\omega}_N = \omega_N \hat{z}$, $\vec{\omega} = \omega \hat{z}$):

$$\vec{F}_1 = \frac{q_1 Q_M}{12\pi\epsilon_0 c^2 R}[\omega_N^2 + 2\omega_N(\omega_M - \omega)]\vec{\rho}. \tag{3}$$

In this expression $\vec{\rho}$ is the position vector to the axis of rotation, so that ρ is the distance between q and this axis.

Classically this situation of a double shell would give rise to a uniform magnetic field $\vec{B} = B\hat{z}$ inside the shells given by

$$B(r < R) = -\frac{\mu_0 Q \omega_N}{6\pi R}. \tag{4}$$

We may consider $\vec{\omega}_M$ as the rotation of the magnet itself as usually the positive charges are fixed in the lattice. So $\vec{\omega}_N$ may be considered as the drifting angular velocity of the electrons responsible for the current and for the magnetic field. In Faraday's experiments and in all other experiments on unipolar induction we had $\omega_N^2 << \omega_N(\omega_M - \omega)$, where ω represents the angular velocity of the copper disk. For this reason we can write (3) as (by (4)):

$$\vec{F}(r < R) = -qB[\omega_M - \omega]\vec{\rho}. \tag{5}$$

We can see that the force on the charge q is completely dependent of the relative motion between the magnet and q. Equation (5) is the basic expression for understanding unipolar induction with Weber's electrodynamics.

In equation (5) ω_M represents the rotation of the magnet relative to the laboratory, and B has been defined by equation (4). Moreover, ρ is the distance of the charge q (an electron) to the axis of rotation (z axis). When this electron belongs to a spinning disk we have $\omega = \omega_D$, where ω_D represents the rotation of the disk relative to the laboratory. In this case equation (5) reads

$$\vec{F}(r < R) = -qB[\omega_M - \omega_D]\vec{\rho}. \tag{6}$$

When this electron belongs to the circuit connected to the galvanometer (AGB in figure 1) we have $\omega = \omega_G$, where ω_G represents the rotation of this circuit relative to the laboratory. In this case we have

$$\vec{F}(r < R) = -qB[\omega_M - \omega_G]\vec{\rho}. \tag{7}$$

We can now analyse the situation of figure 1 by the device of figure 2, where the magnet has been replaced by the double shell of zero net charge.

There will be a polarization of the disk or of the open circuit connected to the galvanometer whenever $\omega_M - \omega_D \neq 0$ or $\omega_M - \omega_G \neq 0$, respectively. However when $\omega_D = \omega_G$ no current will flow in the closed circuit composed by the disk and the galvanometer. This is because even when $\omega_M - \omega_D \neq 0$ we will have in this case the same polarization of the disk and the open circuit, so that the net *emf* in the closed circuit is zero. A net *emf* only happens in the closed circuit when $\omega_D \neq \omega_G$.

From equations (6) and (7) we can construct the table 1 where "ω" and "0" represents the "presence" or "absence" of rotation relative to the laboratory, respectively. In table 1 in the column of the galvanometer the simbol (I) indicates that Weber's electrodynamics predicts a current through the galvanometer.

Let us calculate the *emf* in situation 2 of table 1, for the others the procedure is the same. From equation (6) we have, with $\omega_D \equiv \omega_0$

$$\vec{E} = \frac{\vec{F}}{q} = B\omega_0\vec{\rho}. \tag{8}$$

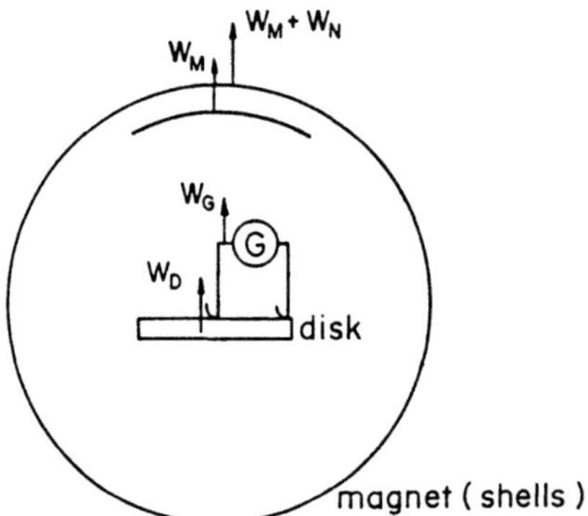

Figure 2. Two shells (Q and $-Q$) under rotation ($\vec{\omega}_M$ and $\vec{\omega}_M + \vec{\omega}_N$) generate a uniform magnetic field for $r < R$.

If the disk has the radius a the voltage between its center and the border will be

$$\Delta\phi = \int_0^a \vec{E} \cdot \vec{dr} = \frac{B\omega_0 a^2}{2}. \tag{9}$$

This will be the *emf* in the closed circuit. If there is a resistance R in the closed circuit composed of the galvanometer and disk, the current I flowing through the galvanometer will be given by

$$I = \frac{B\omega_0 a^2}{2R}, \tag{10}$$

When there is a current in table 1, this is its tipical predicted value.

Table 1. Predictions for the current in the galvanometer.

	ω_G	ω_D	ω_M	Galvanometer
1	0	0	0	0
2	0	ω	0	I
3	0	0	$-\omega$	0
4	$-\omega$	0	0	I
5	ω	ω	0	0
6	$-\omega$	0	$-\omega$	I
7	0	ω	ω	I
8	ω	ω	ω	0

The experiments which have been performed up to now, to our knowledge, had always the galvanometer at rest relative to the laboratory. These are situations 1, 2, 3 and 7 of table 1. The observed values of the currents agree with table 1 and equation (10).

With Weber's electrodynamics we can easily predict the situations 4, 5, 6 and 8 of table 1. If $\omega_G = \omega_0 \neq 0$ the predicted currents in these cases is given by (10), provided that $\omega_D = \omega_0$ or $\omega_M = \omega_0$ when they are also spinning relative to the laboratory.

We propose these experiments as a test of Weber's electrodynamics. A qualitative experiment of this kind might be easily performed if the galvanometer were replaced by a small lamp which is visible under a current of the order of equation (10).

ACKNOWLEDGMENTS

One of the authors (AKTA) wishes to thank FAPESP and CNPq for financial support in the past few years. He thanks also the Organizing Commitee of the International Conference Frontiers of Fundamental Physics for financial support.

REFERENCES

1. M. Faraday, "Experimental Researches in Electricity," Encyclopaedia Britannica, Great Books of the Werstern World, volume 45, Chicago (1952).

2. A. I. Miller, Unipolar induction: a case study of the interaction between science and technology, *Ann. of Science* 38:155 (1981).

3. E. H. Kennard, Unipolar Induction, *Phil. Mag.* 23:937 (1912).

4. E. H. Kennard, Unipolar induction: another experiment and its significance as evidence for the existence of the aether, *Phil. Mag.* 33:179 (1917).

5. J. P. Wesley, Weber electrodynamics, part II: unipolar induction, Z - antenna, *Found. Phys. Lett.* 3:471 (1990).

6. P. Graneau, "Ampere-Neumann Electrodynamics of Metals," Hadronic Press, Nonantum (1985).

7. P. G. Moyssides, Calculation of the sixfold integrals of Ampère force law in a closed circuit, *IEEE Trans. Mag.* 25:4307 (1989).

8. A. K. T. Assis, Deriving gravitation from electromagnetism, *Can. J. Phys.* 70:330 (1992).

9. T. E. Phipps Jr, Toward modernization of Weber's force law, *Phys. Essays* 3:414 (1990).

10. A. K. T. Assis, Centrifugal electrical force, *Comm. Theor. Phys.* 18:475 (1992).

11. J. C. Maxwell, "A Treatise on Eletricity and Magnetism," volume 2, chapter 23, Dover, New York (1954).

IMPACT OF MAXWELL'S EQUATION
OF DISPLACEMENT CURRENT ON ELECTROMAGNETIC LAWS
AND COMPARISON OF THE MAXWELLIAN WAVES
WITH OUR MODEL OF DIPOLIC PARTICLES

Lefteris A. Kaliambos

Institute of Larissa
Komninon 15
Larissa 41223 - Greece

ABSTRACT

Maxwell's reasons for introducing displacement current are considered and attention is drawn to a basic error in the formulation of Maxwell's equation of the displacement current between the plates of a capacitor as though it covers all the length of the circuit, which contains the capacitor.
It is emphasized that it is sufficient to apply the Biot-Savart law to all the real currents, and to ignore the fallacious idea of displacement current. Thus the troublesome hypothesis of self propagating fields in Maxwell's theory is compared with our theory of dipolic particles to interpret the dual view of the nature of light.

INTRODUCTION

A. French and J. Tessman[1] showed that the application of Maxwell's equation of displacement current for the internal discharge of a capacitor involves misconceptions. Particularly during the motion of ions in the ionized air in a capacitor the changing electric field of the discharged plates cannot produce a magnetic field.

Max Planck expressed this many years ago when he wrote[2] ".... the magnetic intensity of the field is calculated from the vector potential of the conduction currents without regard to the displacement currents...."

Furthermore, in our theory of dipolic particles[3] a mathematical analysis of Maxwell's postulation of displacement current showed also that changing electric fields cannot produce magnetic fields.

In this paper first we will prove why Maxwell's generalized form of Ampere's circuital theorem is incorrect, even in case we believe that a varying electric field in a capacitor can give rise to a magnetic field. In circuits containing capacitors it occurs because the distance between the capacitor plates is very small. Certainly the general equation $\oint B.dl=\mu_0 i$ can be applied, when the field is obtained from the contribution of all current elements, whereas for contributions to B of current elements in parts of circuits, it is sufficient to apply the Biot-Savart law. The application of the circuital theorem in such cases gives misleading or erroneous results.

Furthermore, we avoid here to use the differential equation (curl of B), because it is correct only for the motion of charge distributions having plane geometry[6]. Of course for quasi-steady conditions, that is, ignoring the fallacious idea of displacement current, the magnetic field can be calculated using the Biot-Savart law from a knowledge of the real currents alone.

Under this condition the hypothesis of self propagating fields in Maxwell's theory is incorrect.

Finally, comparison of the Maxwellian waves with our model of dipolic particles shows accurately why Maxwell's theory cannot explain the optical phenomena related to atomic physics.

INVALIDITY OF MAXWELL'S FORM OF DISPLACEMENT CURRENT

Suppose that the equation of current's continuity is to be satisfied in a circuit containing a capacitor. This assumption allows us to replace a parallel-plate capacitor having circular plates of radius R and separation 2b (Fig.1a) with a cylindrical conductor of the same radius R and length L=2b (Fig.1b). The cylinder and the capacitor are connected with long wires carrying the same current i. If the current density j through the cross section πR^2 of the cylinder is uniform, the equation of current's continuity allows us to write $i=j\pi R^2$.

For the magnetic field at P (Fig.1a and Fig.1b) in the two equivalent systems we should apply the Maxwell equation to find the two equal results $B=(\mu_0/2\pi a)(i+i_d)$ or $B=(\mu_0/2\pi a)(i+j\pi R^2)$. However in both cases the results are incorrect, because the displacement current i_d or the real current $j\pi R^2$ covers a short distance KK'=L=2b. For this quasi-steady condition B can be calculated using only the Biot-Savart law. For example the equation $B2\pi a=\mu_0(i+j\pi R^2)$ would be true, if the short cylinder were another very long conductor placed with the wire, as shown in Fig.1c. In this case of high symmetry Ampere's circuital theorem can be applied for the contribution of all current elements along the two infinitely long conductors[4].

Now let us calculate B at P.(Fig.1a). Using Biot-Savart's law[4] we integrate along the conductors Z'K and K'Z:

$$B_1=\frac{\mu_0 i}{2\pi a}\int_\theta^{\pi/2}\cos\theta\,d\theta=\frac{\mu_0 i}{2\pi a}(1-\sin\theta) \tag{1}$$

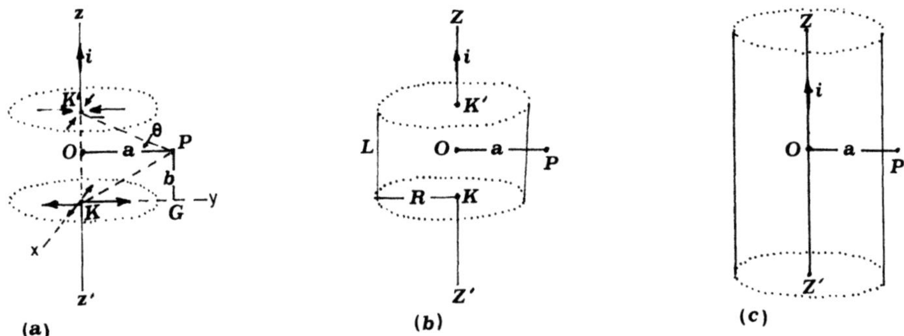

Fig. 1 (a) Magnetic field at P due to current in a long straight wire containing a capacitor. (b) Magnetic field at P due to current in a long straight wire containing a cylindrical conductor of radius R and length L. (c) A long cylindrical conductor of current $j\pi R^2$ is placed with a wire of current i.

Then if we accept Maxwell's displacement current i_d, the integration for the hypothetical current elements of i_d along K'K will give

$$B_2 = \frac{\mu \circ i_d}{2\pi a}\int_0^\theta \cos\theta d\theta = \frac{\mu \circ i_d}{2\pi a}\sin\theta \qquad (2)$$

Hence for the total field due to the two kinds of currents we write

$$B2\pi a = \mu \circ i\,(1-\sin\theta) + \mu \circ i_d\sin\theta \qquad (3)$$

or

$$B2\pi a = \oint B.\,dl = \mu \circ i\,(1-\sin\theta) + \mu \circ \epsilon \circ \frac{\partial E}{\partial t}\pi R^2\sin\theta \qquad (4)$$

The above detailed analysis, however, contradicts Maxwell's equation

$$B2\pi a = \oint B.\,dl = \mu \circ i + \mu \circ i_d \qquad (5)$$

It can be shown that in (5) the addition of the term $\mu_0 i_d$ would be justified, if the current i_d were an additional current flowing along a long wire.

To learn more about the theoretical errors in the formulation of (5) we must refer to the known equation for the modification of B due to the magnetic contributions of magnetic materials[4].

$$\oint B.\,dl = \mu \circ i + \mu \circ i_m \qquad (6)$$

This expression is mathematically similar to (5), because it contains the term $\mu_0 i_m$ in the same manner. But this situation cannot be compared with Maxwell's reasoning, because the amperian current i_m of the magnetic behaviour of matter is exactly an additional current covering all the length of the real current i in the conductors. The characteristic of i_m is that it modifies B. So if (5) were true, we should expect an analogous modification of B in case of a capacitor. At the absence of i, (5) must be written

$$B2\pi a = \oint B.\,dl = \mu \circ i_d \qquad (7)$$

But according to Ampere's law this form would be justified, if the plates of Fig.1a were separated by a long distance. In this case, however, the electric field at O, between the charged disks always would be zero.

MAGNETIC FIELD BETWEEN THE PLATES OF A CAPACITOR
DUE TO THE REAL MOTION OF CHARGES

At the position $P_{(ab)}$ in yz plane (Fig.1a and Fig.2a) the magnetic field due to a pair of charges in the charged disk K at the symmetrical points M and N moving along +x and -x will be zero. It happens because in the equal triangles PKN and PKM, PM=PN and $\varphi=\theta$. Thus we have

$$B = B_1 - B_2 = K_m\frac{q\upsilon}{(PM)^2}\sin\varphi - K_m\frac{q\upsilon}{(PN)^2}\sin\theta = 0 \qquad (8)$$

where $B_1 = B_2$ pointing in opposite directions. In general, any pair of charges moving oppositely at symmetrical points and in perpendicular direction to yz plane give at P zero field.

On the other hand, for the same point P the charges at the symmetrical points Q and W of the disk K moving along +y and -y (Fig.2b) give

$$B = B_1 - B_2 = K_m \frac{qv}{r_1^2} \sin\theta_1 - K_m \frac{qv}{r_2^2} \sin\theta_2 > 0 \qquad (9)$$

Since B_1 points out and B_2 into the paper, the net magnetic field $B = B_1 - B_2$ will point out of the paper (in x direction), because $r_1 < r_2$ and $\theta_1 > \theta_2$. The same net field B pointing outward will give also the moving charges at the symmetrical points Q' and W' of the disk K'. Now let us consider four moving charges in the disk K at the symmetrical points A, B, C, D (Fig.2c) at a distance r from K. Their radial velocity υ, which is directed outward from K, when the disk is charged, makes an angle φ with the axis x. The x components of υ will give zero field at P, because they are normal to the plane yz, while the field at P due to y components of υ can be calculated to give net field B pointing in x direction. The same net field pointing in x direction will give also the charges in the disk K' because they move toward the K' (Fig.1a). In general, the magnetic field at P produced by the radial motion of charges in the capacitor plates is opposite to that produced by the currents in the wire. According to Fig.1a the currents of Z'K and K'Z give at P magnetic field pointing in -x direction. Under this condition we may write

$$B < \frac{\mu_o i}{2\pi a} (1 - \sin\theta) \qquad (10)$$

That is, using the fundamental equation of Biot-Savart for the real motion of charges we find a field at P smaller than that given by (3), because in (3) Maxwell's hypothetical current i_d is parallel to i producing field in the same direction. This result, which can be confirmed by experiment, shows clearly the obvious incorrectness of Maxwell's displacement current.

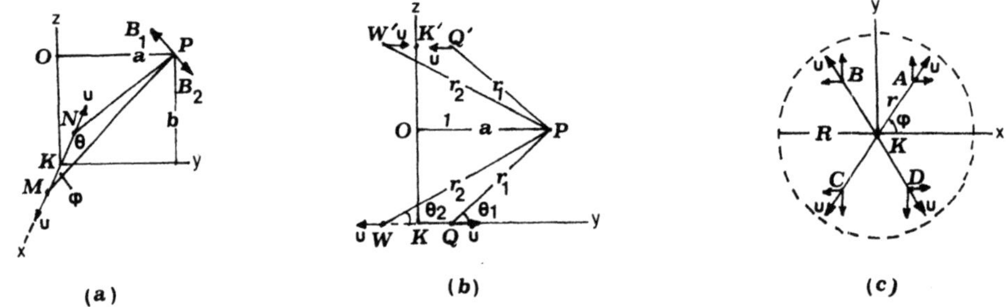

Fig.2 (a) The opposite motion of two equal charges at symmetrical points on the axes +x and -x gives zero magnetic field at any point on the plane yz. (b) The opposite motion of charges at the symmetrical points W, Q and W', Q' on the axes +y and -y gives at P (on the plane yKz) net magnetic field directed out of the paper. (c) Magnetic field due to the x and y components of the radial velocity υ of four charges at the symmetrical points A, B, C, D, in the disk K.

COMPARISON OF THE MAXWELLIAN WAVES
WITH OUR MODEL OF DIPOLIC PARTICLES

Consider a rotating dipole (Fig. 3a), which moves in free space at the speed of light with respect to the source of radiation[3] (reference frame xy). The charges +q and -q, separated by a small distance R, can produce at a point P on the plane xy of rotation an electric field E_y, given by Coulomb's law:

$$E_y = \frac{q}{4\pi\epsilon_0} \left(\frac{\sin\varphi_1}{r_1^2} + \frac{\sin\varphi_2}{r_2^2} \right) \tag{11}$$

Of course in the special case of E_y at points along the line between the charges +q and -q the above equation becomes more simple.

Purposely we ignore the field E_x, because for $\varphi=0$ the electromagnetic force (F_{em}) between the charges is equal to the electric force, according to the equation[3] (Fig. 3b)

$$F_{em} = K_e \frac{q^2}{R^2} \cos\varphi \tag{12}$$

which implies waves (concentration and rarefaction). That is, when $\varphi=0$ the charges of such dipoles are brought together and the resulting zero net charge is unable to cause electric fields.

Similarly at the same point P the magnetic field B_z (normal to the plane of rotation), caused by the moving charges, is given by Biot-Savart's law:

$$B_z = \frac{\mu_0 qc}{4\pi} \left(\frac{\sin\varphi_1}{r_1^2} + \frac{\sin\varphi_2}{r_2^2} \right) \tag{13}$$

Then, since $\mu_0\epsilon_0 = 1/c^2$ comparison of (13) with (11) gives

$$B_z = (\mu_0\epsilon_0)^{1/2} E_y = E_y/c \tag{14}$$

Here the rotation of dipoles makes E_y and B_z varying at the same time.

As known, (14) is also derived by using the law of induction and the fallacious idea of displacement current in Maxwell's hypothesis of "self propagating fields"[4].

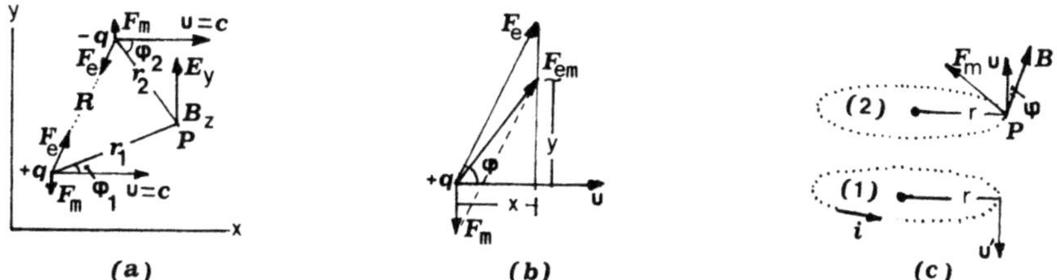

Fig. 3 (a) E_y and B_z at any point P in xy plane of rotation of a dipole moving at the speed of light. (b) the electromagnetic force F_{em} on +q is the vector sum of the electric and magnetic forces of interaction with -q of a dipole. (c) A current loop (1) and a wire loop (2) are coupled together by the effects of motional emf. We observe only magnetic forces due to the relative motion of one coil with respect to the other.

The varying fields E_y and B_z, produced by dipolic particles, are always in phase, and their connection with the moving dipoles is responsible for the propagation of localized energy in discrete small regions.

To explain the generation of such moving dipoles we should suppose that accelerated electrons at sources act on positive charges at rest with another kind of magnetic forces which appear to cause emf when currents are changed[6]. As a result the electrons at sources can carry away positive charges and move together as dipolic particles[3].

However, according to Maxwell's theory, the energy of radiation is due to strange travelling fields. As known the initial electromagnetic fields near the source are 90° out of phase[4], while at an arbitrarily chosen position farther away from the source they have an additional local source for propagating time-varying fields in phase. Moreover, according to this picture the energy is uniformly distributed over the whole wave front.

On the contrary, our theory of dipolic particles interprets also the quantum of energy $E=h\nu$, by assuming that the energy of rotation of dipoles is related with their constant angular momentum.

Let us now consider a moving dipole in a transparent medium (radiation in matter). In a material the electric forces F_e between the charges of a moving dipole, separated by a small distance R (Fig.3a), are decreased, because the rapidly varying E_y between the charges, rotated at the frequency of light, cause in matter some distortion of the atomic electronic cloud relative to the nucleus. So for a dielectric permittivity $\varepsilon=\varepsilon_0 K$, the electric force F_e on $+q$, caused by $-q$ (Fig.3b), according to Coulomb's law is

$$F_e = \frac{1}{4\pi\varepsilon_0 K}\frac{q^2}{R^2} \tag{15}$$

However the magnetic forces (F_m) on the moving charges usually cannot be affected by the material, because the most transparent materials are non magnetic[4]. So using Biot-Savart's law the magnetic force F_m on $+q$, caused by the other moving charge $-q$, will be

$$F_m = \frac{\mu_0}{4\pi}\frac{q^2}{R^2}\upsilon^2\sin\varphi \tag{16}$$

Comparison of (16) with (15) and using the relation $\mu_0\varepsilon_0=1/c^2$ we have

$$F_m = F_e\frac{\upsilon^2 K}{c^2}\sin\varphi \tag{17}$$

On the other hand, for the electromagnetic force F_{em} (vector sum of F_e and F_m) by the use of Fig.3b we write

$$F_{em}^2 = x^2 + y^2 \tag{17a}$$

Then writing $x^2=F_e^2\cos^2\varphi$ and $y^2=(F_e\sin\varphi-F_m)^2=F_e^2\sin^2\varphi+F_m^2-2F_eF_m\sin\varphi$ and using (17) we find

$$F_{em}=F_e\left(1+\frac{\upsilon^4 K^2}{c^4}\sin^2\varphi-2\frac{\upsilon^2 k}{c^2}\sin^2\varphi\right)^{1/2} \tag{18}$$

So for $c/\upsilon=K^{1/2}$, and substituting F_e in (15) the above equation leads to the following simple equation, which implies wave propagation:

$$F_{em}=\frac{1}{4\pi\varepsilon_0 K}\frac{q^2}{R^2}\cos\varphi \tag{19}$$

That is, the sinusoidal variation of F_{em} occurs, when the dipole moves in the material at the speed $\upsilon = c/K^{1/2}$. This also says, that the velocity $\upsilon = c/K^{1/2}$ of dipoles inside matter is the result of (19). Here the velocity υ is taken only with respect to the source because any moving observer (frame S') measures always the same magnetic forces[6]. Under this condition, if we wish to calculate the relation of E_y with B_z at P (Fig.3a) for radiation in matter, we must replace ε_0 by $\varepsilon_0 K$ in (14) to write

$$B_z = (\mu_0 \varepsilon_0 K)^{1/2} E_y \qquad (20)$$

In Maxwell's theory (20) is discussed[4] to explain electromagnetic waves in transparent materials at the velocity $\upsilon = c/K^{1/2}$, because it is believed, that the electronic polarization in transparent materials is due to the Maxwellian "self propagating fields", based on the equations of induction law and the fallacious idea of "displacement current".

However experiments showed, that the induced emf cannot exist without changing currents or moving charges. Furthermore, it can be shown, that in the case of the relative motion of charges changing magnetic fields cannot produce electric fields. Suppose (Fig.3c) the coil 2 (wire loop) were moved away from coil 1 (current loop) with a velocity υ reducing the magnetic flux through circuit 2. On a charge q at P appears a magnetic force F_m given by

$$F_m = q \upsilon B \sin\varphi \qquad (21)$$

where B is caused by the current i of coil 1. Multiplying F_m/q by $2\pi r$ we get

$$\frac{1}{q} 2\pi r F_m = \varepsilon = -\int \frac{\partial B}{\partial t} \cdot dS \qquad (22)$$

That is, a changing magnetic field produces magnetic force.

Furthermore, motion of coil 1 with coil 2 held at rest gives exactly the same magnetic force, because only the relative motion of one coil with respect to the other matters. But since in the last case the velocity of q at P with respect to the observer (frame S') is zero, the whole idea of special relativity says that the magnetic force becomes electric one. As known, it is described theoretically for a mental arrangement about the force on a charge q moving in the magnetic field of a long straight wire carrying a current i[1]. If we apply Einstein's idea of length contraction for the moving coil 1, we will be forced to accept the contraction of the cross-section area of the wire in the direction of the velocity υ'. As a result, the charge density of both positive ions and negative charges (conduction electrons) will increase to the same amount, because both positive and negative charges move with the same velocity υ' in the same direction. Thus, since the net charge on the wire of coil 1 remains zero, the force at P will be again magnetic. Even in case we expected to observe a net charge on the moving coil 1, of course we would measure a resulting electric force at P, which would be unable to produce emf, because its direction would be perpendicular to the wire of coil 2. So in any frame of reference we observe always magnetic forces.

This inconsistency of special relativity in electromagnetism can be shown also in one experiment about the magnetic forces inside a rotating charged cylinder. If an observer is rotating with the cylinder (frame S'), he will measure only magnetic forces inside. Electric forces cannot appear, because the electrostatic equations say, there, will be no electric fields inside[1].

But Einstein in his book "The Evolution of Physics" emphasizes that the perimeter of a rotating disk is reduced according to his idea of length contraction. Nevertheless it is of interest to ask that the special theory of relativity cannot be related with the basic laws of electromagnetism.

CONCLUSIONS

Although French and Tessman exposed the basis of the misconceptions regarding displacement current, however, in their concluding remarks they emphasized the importance of the displacement current in electromagnetic theory, because there was great difficulty for them to believe that Maxwell's theory could be invalid after the triumph of the quantum theory. Certainly under the influence of Maxwell's theory Rosser[5] tried to interpret the displacement current by attributing the postulated changing fields in intermediate courses to the vibration of charges at the source of radiation, whereas he ignored Maxwell's basic idea of self-propagating waves[4], which should be working according to the equation of displacement current and Faraday's induction.

Under this condition I must seek to make three main points. The first is to emphasize that in all cases of circuits containing capacitors, it is sufficient to apply the Biot -Savart law to all the real currents and to ignore what we call the generalized form of circuital theorem, because Ampere's law is unsuitable for calculating magnetic fields produced by currents covering only parts of circuits.

The second is that in combining the wrong postulation of displacement current with the fact that changing magnetic fields, cannot produce electric fields[6], one must be able to understand why the hypothesis of moving fields failed to give a sensible explanation in all the optical phenomena related to atomic physics.

The third is to look for a physically plausible explanation of electromagnetic radiation in one conceptual scheme between Maxwell's moving fields and Einstein's photons since nature works in only one way. I believe that a clear answer to the photon-wave dilemma[7] is given by me in the model of dipolic particles.

REFERENCES

1. A. P. French and J. R. Tessman Am.J.Phys.31,201 (1963)
2. Max Planck, translated by H.L.Brose, Introduction to theoretical Physics (MacMillan and Company Ltd, London 1932) Vol.III P.197.
3. L.A.Kaliambos Ind.J.Th. Phys. 35,257 (1987)
4. A.F.Kip Fundamentals of Electricity and Magnetism. Edited by inc 17 printing 1983 P.P. 36, 245, 251, 333, 451, 455, 461.
5. W.G.V. Rosser Am.J.Phys. 31,807 (1963).
6. L.A.Kaliambos Ind. J. Th. Phys. 38,253 (1990).
7. G.Holton Introduction to concepts and theories in physical science. Copyright Stephen G. Brush. First Addison-Wesley edition 1973, P.441.
8. R. P. Feynman, R. B. Leighton and M. L. Sands. Feynman lectures on Physics Vol II. Mainly Electromagnetism and Matter. Addison-Wesley Publishing Co. Sixth printing (1977). P 14-6.

DIRECT CALCULATION OF h AND OF THE COMPLETE SELF ENERGY OF THE ELECTRON FROM FLUID MODELS

William M. Honig

Curtin University, P.O.B. 361, S. Perth, 6151,
Western Australia.- (Direct Mail Address)
E-mail: RHONIGW@cc.Curtin.edu.au

INTRODUCTION

We present here two extremely simple derivations of h and of the complete self energy of the electron. This is based, however, on an extensive set of ideas consisting of a dual fluid plenum which retains relativistic invariance. These ideas have been presented extensively elsewhere and so for those not familiar with these ideas an extensive bibliography is appended to this paper. The most recent summary of these ideas will appear shortly for those who wish to see a capsule summary.[1]

Even so, a few words must be said about this fluidic approach relevant to h and to the fluid models for the electron and for electromagnetic dipole waves. The basic concept involves replacing the QM paradigm with a realistic dual fluid plenum consisting of continuous positive and negative fluids. These give a neutral vacuum space and give fluid models of the canonical particles and electromagnetic waves (mainly dipole waves which form the basis for all electromagnetic waves). Relativistic invariance is retained, at least on an empirical and realistic basis and this is explained extensively in the bibliography (see at least a-d in the bibliography).

The fluid model for the electron is a negative spinning droplet surrounded by equal amounts of positive fluid which characterise the fields and all other qualities of the electron. The prototype electromagnetic wave is the half wavelength dipole field distribution which is treated as a discontinuous independent entity which has been named the "Photex" and which is the seat for the explanation of all QM phenomena. Converting all these phenomena to fluid models makes possible the   realistic description for electromagnetic radiation from the fluidic electron model (Again see at least a-d in the bibliography).

Electromagnetic wave generation consists of the shedding of a toroidal vortex in either positive or negative fluid which then evolves very much as per the sketches first given by Hertz of dipole waves. This vortex shedding occurs upon the acceleration or deceleration of the fluidic electron models, principally in collisions. Thus if the fluidic electron is pictured like a tennis ball which rebounds back and forth between 2 walls, a "Photex" is emitted upon the electron contact with the wall and its momentary deceleration to zero velocity and then another "Photex" (of opposite charge and spin) upon its rebounding acceleration and velocity in the opposite direction. The meaning for h (or rather h/2) has been shown to permit an estimate of the energies of each of

Frontiers of Fundamental Physics, Edited by M. Barone
and F. Selleri, Plenum Press, New York, 1994

these "Photexi" of the order of a little less than 10^{-15} ev. Thus the decrement of energy for each collision of the moving electron is very small unless the number of such collisions is extremely large (like 10^{+15} collisions for a 1 ev decrement).

THE DERIVATION OF h

We start with remarks about the significance of deriving h. Dirac in discussing this has said:[2] "I think one is on safe ground if one makes the guess that in the physical picture we shall have at some future stage, e and c will be fundamental quantities and h will be derived. If h is a derived quantity instead of a fundamental one, our whole set of ideas about uncertainty will be altered .... [The] uncertainty relations(s) cannot play a fundamental role in a theory in which h itself is not a fundamental quantity. I think one can make a safe guess that uncertainty relations will not survive in the physics of the future."

We point out here how the very simple formula of Larmor can be used to get a very close estimate of h.

We start with the expression of h as energy per cycle per second and what we do is ask if we can get from Larmor's formula the energy emitted for one half cycle of radiation from an electron. Of course his formula giving the energy per unit time must be converted to energy per half cycle per second (sic!). Detailed discussions of these points are given in a-d, f, and g in the bibliography and in ref.1.

We start by noting the relation between $\underline{h}$ (h-bar) and h:

$$\underline{h} = h/2\pi \tag{1}$$

We then write the well-known relationship between h/2 and the other well known fundamental constants:

$$\alpha = e^2/4\pi\varepsilon_0\underline{h}\,c \tag{2}$$

Solving for $\underline{h}$:
$$\underline{h} = e^2/4\pi\varepsilon_0 c\,\alpha \tag{3}$$

or:
$$\underline{h} \cong e^2(137)/4\pi\varepsilon_0 c \tag{4}$$

or:
$$h \cong e^2(137)(2\pi)/4\pi\varepsilon_0 c \tag{5}$$

or
$$h/2 \cong e^2(137)(\pi)/4\pi\varepsilon_0 c \tag{6}$$

or
$$h/2 \cong [e^2/4\pi\varepsilon_0 c][430.4] \tag{7}$$

Eq. (6) will be compared with the forthcoming result of operating on the Larmor formula. The Larmor formula can be written as:[3]

$$S = [e^2\,\overline{\dot{v}^2}]/[6\pi\varepsilon_0 c^3] \tag{8}$$

with the units: energy/time and the v-dot term is the average of the square of the acceleration. We approximate the collision of the electron droplet with a wall (in the x direction) by means of a simple sinusoidal function where only a half cycle will be used for the first half of the motion; with fluids such a continuous function should be a good approximation for the motion. Thus the motion taken as:

$$X = A \sin \omega t$$
$$\dot{X} = v = \omega A \cos \omega t$$
$$a = \dot{v} = -\omega^2 A \sin \omega t$$

The average of $\sin^2 \omega t$ (from Eq.8) for an integral half cycle is 1/2, so that S is:

$$S = [e^2 \omega^4 A^2] [1/2] /[6 \pi \varepsilon_0 c^3] \tag{9}$$

The units of S are power or energy/second, but this can also be written as energy-cycle/second because the word cycle has been superfluous up to now and has been suppressed, i.e., energy/second is recovered when cycle is suppressed. The task now is try to convert S into a measure, say T which has the units energy per cycle per second or rather energy per half-cycle per second. It should be evident upon study that if S is multiplied by $1/2f^2$ it becomes T. This is because one f converts S to pure energy, the other f puts cycle per second in the denominator of S, and the 2 in the above factor finally converts S into energy per half cycle per second, thus:

$$T = [1/2f^2] [S] = [1/2f^2] [e^2 \omega^4 A^2] [1/2] /[6 \pi \varepsilon_0 c^3] \tag{10}$$

which becomes:

$$T = [e^2/4 \pi \varepsilon_0 c][519.5][A^2/2\lambda^2] \tag{11}$$

Now if the last factor is close to one, or $A = 2^{1/2} \lambda$ which appears to be a reasonable assumption, then Eq. (11) lies within 27% of Eq. (7). It should be mentioned that A.O. Barut[4] derived a related result from purely QM considerations in 1978.

The ideas presented here possess physically realistic and heuristic qualities which come from the fluid models. This appears to be the first indication that h is derivable from a fluidic electron model in a vortex shedding situation and relates well to the Dirac comment.

FINDING THE COMPLETE SELF ENERGY OF THE ELECTRON

Since only charged fluids are the means used for the construction of the various canonical particle models, a knowledge of the varying concentration of charge and velocity of the fluids of a model can be used to define a continuously variable charge and charge-velocity 4-vector function, J4. This also permits that an electrostatic energy and charge-motion energy 4-vector potential function, A4, also be defined or derivable. In the case of the fluidic electron model which is a rotating droplet of negative charge surrounded by a varying radial concentration of positive charge and charge velocity, estimates for the total energy necessary to assemble the fluidic electron should also give estimates for the mass of the electron.

This assumes that the $E = mc^2$ relation provides the connection between the assembly energy of the fluid model and its mass. Most of this is covered in a-d, f, and g of the bibliography. We give here the energy of various parts of the electron droplet model which tend to show that the complete energy for the assembly of the model gives the full self energy of the electron.

We start with the well known work on this subject. Sommerfeld[3] and many others have shown that an evaluation of the self energy of the classical electron falls short of the full self energy by amounts that are of the order of 25% of the self energy. This has been considered a pretty good estimate because of the ignorance that presently exists about the exact nature of the electron. The spinning droplet electron because of its nature permits a clearer idea of the nature of this extended model of the electron. One starts with the energy of the classical electron as the electrostatic energy outside a charged spherical sphere:

$$\text{Energy electron} = m_0 c^2 = e^2/[4 \pi \varepsilon_0 c \, a] \tag{12}$$

where a is the classical radius of the electron. The pictorial qualities of the droplet electron permit that an alternate way of assigning energies be made. This consists of allotting equal energies to the electrostatic (e.s.) and to the motional velocities of the fluids of the model. Thus instead of Eq.(12) above one may write:

$$\text{Energy}_{\text{electron}} = \text{Energy}_{\text{e.s.}} + \text{Energy}_{\text{motional}} \tag{13}$$

$$\text{and} \qquad \text{Energy}_{\text{e.s.}} = \text{Energy}_{\text{motional}} \tag{14}$$

The reason for doing this is that it has a desirable symmetry which is also aesthetic. One may rewrite Eq. (13) allotting equal energies to each component so that their sum will still be that shown in Eq.(12):

$$\text{Energy}_{\text{electron}} = \{e^2/[8\,\pi\,\varepsilon_0\,c\,a]\}_{\text{e.s.}} + \{e^2/[8\,\pi\,\varepsilon_0\,c\,a]\}_{\text{motional}} \tag{15}$$

where the total electrostatic energy of the 2-fluid droplet electron is the first term in Eq.(15) and has been explained in detail (Ref. 1 and a-d) and where the necessity for the '8' term is obvious. Bucherer[5] in 1905 derived the magnetic fields inside and outside a uniformly charged sphere spinning at the angular velocity $\underline{\omega}$. If one assumes that

$$\underline{\omega}\,a = c \tag{16}$$

this gives the droplet an equatorial velocity equal to the velocity of light and the evaluation of the magnetic field energy inside the droplet comes out to be 1/3 of the second term (the motional term) in Eq.(15).

Furthermore, Lorentz[6] has shown that the Poincare stress needs an energy which is also 1/3 of the second term in Eq.(15). In this case the stress is oppositely directed to that which Poincare consider since it prevents the positive external fluid of the droplet from falling into the droplet. Note that no force is required to keep the charge of the droplet from flying apart because of the nature of the dual fluid plenum.

Thus almost all of the energy of the droplet electron is accounted for: the electrostatic energy is easily shown to be equal to half of the self energy, this is the first of the terms in Eq.(15). Two thirds of the other half of the self energy is accounted for as explained above. This leaves an amount of 1/6 lf the self energy unaccounted for. It leaves out however the motional energy of the external fluids of the electron droplet model. It seems reasonable here to assume that this energy is also equal to the internal magnetic field energy of the droplet because equal and oppositely velocities for the internal and external fluids would appear to be necessary to the model's construction. All the above is discussed in some detail in item a of the bibliography.

FINAL REMARKS

We make some concluding remarks which may clarify our purposes a bit. It has been shown that in a global, abstract, and non-empirical sense QM is both inconsistent and contradictory. In a local and empirical sense, however, it is indeed fully consistent and empirically verified. Furthermore, Special Relativity has been shown to have a similar logical structure. It is well know that inconsistent theories can be used to literally prove anything. Thus, is not possible nor will it ever be possible to refute QM. It may however, be possible to come up with realistic type theories which provide predictions and which are fertile in new testable ideas and heuristic in that they will lead to further ideas and theories which are testable. QM, however, will always be superb, as it has been in the past, in postdictive explanations even for the predictions of new realistic theories such as the one advocated fully in the bibliography. Details of the logical considerations of QM and dual fluids are discussed in items g, h, and i of the bibliography and where in item i, nonlocal realistic fluid models are also discussed.

REFERENCES

1. Honig, W.M.,"Phenomenology of a Subquantum, Realistic, Relativistic Theory" To appear in the proceedings of the Sept '92 Trani conference entitled, <u>Waves and Particles in Light and Matter</u> from Kluwer Pubs. in '94, Edited by van der Merwe and Garuccio.
2. Dirac, P.A.M., <u>Sci. Am.</u>, May '63.
3. Sommerfeld, A., <u>Electrodynamics</u>, Vol.III of series, Academic '64.
4. Barut, A.O., <u>Z. Naturforsch.</u>, **33a**, 993-994 (1978).Another related paper appeared in <u>Old and New Questions in Physics, Cosmology, etc.</u>, Plenum '83, Edited by van der Merwe; Another unreferenced related paper also exists.
5. Bucherer, A.H., <u>Phys. Zeit.</u>, **6**, 225-227; 269-270; 833-834, all 1905. **7**, 256-257 (1906).
6. Lorentz, H.A., The Theory of Electrons, 213-214, Dover 1952.

BIBLIOGRAPHY

a. Honig, W.M.,"<u>The Quantum and Beyond</u>" published by Philosophical Library, Inc., 200 W. 57 St., N.Y., N.Y., 10019, USA, (1986). $30 USD. ISBN 8022-2517-9.
b.   "   , "An Electromagnetic World Picture, Part I: Massless Dual Charged Fluids for Modelling Vacuum Space, Fundamental Particles, and Electromagnetic Waves", <u>Physics Essays</u>, **4**, No. 4, 583-590 (1991).
c.   "   , "An Electromagnetic World Picture, Part II: Planck's Constant, and the Discrete Electromagnetic Wave Model - the Photex; a Physical Model for the QM Hidden Variable", <u>Physics Essays</u>, **5**, No. 2, 254-261, (1992).
d.   "   , "An Electromagnetic World Picture, Part III: Subjectivity of Space, Relative Metrics, and the Locality-Nonlocality Conundrum", <u>Physics Essays</u>, **5**, 514-525, (1992).
e.   "   , "Logical Organisation of Knowledge With Inconsistent and Undecidable Algorithms Using Imaginary and Transfinite Exponential Number Forms in a Non-Boolean Field: Part 1 - Basic Principles", <u>IEEE Trans. in Knowledge and Data Engineering</u>, **5**, No. 2, 190-203, April '93.
f.   "   , "Physical Models for Non-Local Particles, Hidden Variables, and All That", in <u>Problems in Quantum Physics</u>, pp. 120-147, World Scientific Publishing, Singapore & N.Y., (1988), the Proceedings of the '87 Gdansk Conference.
g.   "   , "On the Physical Meaning of Planck's Constant, h, From a Realistic Subquantum Theory", in <u>Problems in Quantum Physics</u>, pp. 575-581, World Scientific Publishing, Singapore & N.Y., (1988), the Proceedings of the '87 Gdansk Conference.
h.   "   , "The Locally Consistent and Globally Inconsistent Axioms in STR and QM: Using Exponential, Imaginary, and Transfinite Number Fields and the Forms $e^{i\varnothing}$", in <u>Nature, Cognition, and System, Volume 3</u>, Ed., M.E. Carvallo, Kluwer Publishing, London. To appear in '94.
i.   "   , "The Relativity of the Metric and of Geometry", in <u>Nature, Cognition, and System, Volume 3</u>, Ed., M.E. Carvallo, Kluwer Publishing, London. In '94.

k.   ” , "A Minimum Photon 'Rest Mass' - Using Planck's Constant and Discontinuous Electromagnetic Waves", <u>Foundations of Physics</u>, **4**, 367-380, Sept. '74.

m.   ” , "Godel Axiom Mappings in Special Relativity & Quantum-Electromagnetic Theory", <u>Foundations of Physics</u>, **6**, 37-57, '76.

n.   ” , "Transfinite Ordinals as Axiom Number Symbols of Quantum and Electromagnetic Wave Functions", <u>Int. Jour. Theor. Physics</u>, **15**, No. 2, 87-90, 1976.

p.   ” , "Photon Rest Frames and Null Geodesics", <u>Int. Jour. Theor. Physics</u>, **15**, No. 9, 673-676, 1976.

q.   ” , "Relativity of the Metric", <u>Foundations of Physics</u>, **7**, 549-572, '77.

r.   ” , "Quaternionic Electromagnetic Wave Equation and a Dual Charge-Filled Space", <u>Lettere al Nuovo Cimento</u>, **19**, 37-140, 28 May '77.

INTERBASIS "SPHERE–CYLINDER" EXPANSIONS FOR THE OSCILLATOR IN THE THREE–DIMENSIONAL SPACE OF CONSTANT POSITIVE CURVATURE

G.S.Pogosyan, A.N.Sissakian, and S.I.Vinitsky

Laboratory of Theoretical Physics
Joint Institute for Nuclear Research
141980 Dubna, Moscow region, Russia

INTRODUCTION

In recent years, systems with accidental degeneracy in spaces of constant curvature have been in the focus of attention of many researches due to their nontrivial symmetry.

These systems have first been considered by Schrödinger[1] who used the factorization method to solve the Schödinger equation and to find the energy spectrum for the harmonic potential being an analog of the Coulomb potential on the 3–sphere and showed that like in the case of flat space there occurs complete degeneracy of energy levels in orbital and azimuthal quantum numbers. Later, Infeld and Schild [2] have treated this problem for the three-dimensional space of constant negative curvature.

Essential advance in the theory of systems with accidental degeneracy in spaces with constant curvature made by Higgs[3], Leemon [4], and Kurochkin and Otchik [5] in 1979. They have shown that complete degeneracy of the spectrum of the Coulomb problem and oscillator on the three–dimensional sphere is caused by an additional integral of motion: Runge–Lenz's vector (for the Coulomb potential) and Fradkin's tensor (for the oscillator). However, in contrast with the flat space, commutation relations between the components of the Runge–Lenz operator and the Fradkin tensor on the sphere are of the nonlinear nature, which makes description of degeneracy of the energy spectrum by using classical algebras and groups impossible.

Quite recently, the authors from Ukraine[6,7] have proposed the quadratic Racah algebra QR(3) as an algebra of hidden symmetry for the Coulomb–Kepler problem and Higgs oscillator. This approach allowed them, without solving directly the Schrödinger equation, to show that the coefficients of interbasis expansions for "sphere–cylinder" transitions for the Higgs oscillator on the sphere and "sphere–parabola" transitions for the Coulomb interaction on a hyperboloid are expressed through Racah's polynomials introduced in papers by Wilson[8].

It is known from papers by Olevskii[9] that variables in the Laplace - Beltrami or

Frontiers of Fundamental Physics, Edited by M. Barone
and F. Selleri, Plenum Press, New York, 1994

Schrödinger equation on the three-dimensional sphere can be separated in six orthogonal coordinate systems. Therefore, it is interesting, for example, to find out in what coordinate system variables in the Schrödinger equation are separated for the oscillator. The answer to this question is rather unexpected: in all six coordinate systems. In the present paper, we restrict ourselves to two simple coordinate systems (which are independent of the dimensional parameter like, for example, the sphero-conical, two elliptic-cylindrical or ellipsoidal coordinate systems), spherical and cylindrical. The solution to the Schrödinger equation for the oscillator in the spherical coordinate system has been found in [3-5], whereas in the cylindrical one this investigation is the first. The Schrödinger equation in other coordinate systems on the three-dimensional sphere will be investigated, we hope, in our subsequent papers.

The present paper is aimed to solve the Schrödinger equation for the oscillator in the cylindrical coordinate system, and then using the cylindrical and spherical wave functions to calculate explicitly the coefficients of the "sphere–cylinder" expansion.

The paper is organized as follows. Section 2 is devoted to the solution of the Schrödinger equation for the oscillator in the spherical and cylindrical coordinate systems. Section 3 is the calculation of the coefficients of interbasis expansion over spherical and cylindrical bases of the oscillator.

SOLUTION OF THE SCHRÖDINGER EQUATION

The three–dimensional space of constant positive curvature can be realized on the three-dimensional sphere S_R^3 of radius R, $0 < R < \infty$ imbedded into the four–dimensional Euclidean space M_4, on the hypersurface

$$q_0^2 + q_i q_i = \mathbf{R}^2, \quad i = 1, 2, 3$$

The Cartesian coordinates q_i change in the range $q_i q_i \leq R^2$ and to each value of q_i there correspond two points on the sphere.

The coordinates of the space M_4 are related to the spherical coordinates $\Omega = \{\chi, \vartheta, \varphi\}$ describing motion on the three–dimensional sphere by the relations

$$q_1 = R \sin \chi \sin \vartheta \cos \varphi, \quad q_2 = R \sin \chi \sin \vartheta \sin \varphi, \quad q_3 = R \sin \chi \cos \vartheta, \quad q_0 = R \cos \chi,$$

$$0 \leq \chi \leq \pi, \ 0 \leq \vartheta \leq \pi, \ 0 \leq \varphi < 2\pi.$$

The second simple system of coordinates on the sphere is the cylindrical coordinate system that can be determined as follows:

$$q_1 = R \sin \alpha \cos \phi_1, \quad q_2 = R \sin \alpha \sin \phi_1, \quad q_3 = R \cos \alpha \sin \phi_2, \quad q_0 = R \cos \alpha \cos \phi_2,$$

$$0 \leq \alpha \leq \pi/2, \ 0 \leq \phi_1 < 2\pi, \ -\pi < \phi_2 < \pi.$$

The Schrödinger equation describing a particle motion in the potential field $V(\Omega, R)$ on S_R^3 at each value of R has the form

$$[H_0 + V(\Omega, R)] \Psi(\Omega, R) = E(R)\Psi(\Omega, R). \tag{1}$$

where the Hamiltonian of a free particle on the three–dimensional sphere H_0 has the form

$$H_0 = M_{\mu\nu} M_{\mu\nu} / 4\mu R^2$$

and

$$M_{\mu\nu} = -i\hbar(q_\mu \partial_\nu - q_\nu \partial_\mu), \quad \mu, \nu = 0, 1, 2, 3$$

430

are the generators of the symmetry group $O(4)$.

It is known[3] that the oscillator potential on the three–dimensional sphere is

$$V(\Omega, R) = \frac{\mu\omega^2 R^2}{2}\left[\frac{q_1^2 + q_2^2 + q_3^2}{q_0^2}\right] = \frac{\mu\omega^2 R^2}{2}tg^2\chi = \frac{\mu\omega^2 R^2}{2}\left\{\frac{1}{\cos^2\alpha\cos^2\phi_2} - 1\right\} \qquad (2)$$

In the form (2) the oscillator potential is symmetric between the two hemispheres and singular on the equator. This means that the motion in a field like that is possible only in the upper or lower hemisphere and has an identical nature. It is to be noted that due to the singularity on the equator, in the limit $\omega \to 0$ we do not get a free motion on the whole sphere but only in one of the hemispheres separated by the impenetrable barrier.

Now we proceed to solve the Schrödinger equation (1) in the spherical and cylindrical coordinate systems.

A) *Spherical coordinates*

Having chosen a wave function in the form

$$\Psi(\chi, \vartheta, \varphi; R) = Z(\chi; R)Y_{lm}(\vartheta, \varphi),$$

after separation of variables in the Schrödinger equation (1) in the spherical coordinate system we arrive at the following equation for the function $Z(\chi; R)$

$$\left\{\frac{1}{\sin^2\chi}\frac{\partial}{\partial\chi}\sin^2\chi\frac{\partial}{\partial\chi} + \frac{2\mu R^2}{\hbar^2}\left[E - \frac{\hbar^2}{2\mu R^2}\frac{l(l+1)}{\sin^2\chi} - \frac{\mu\omega^2 R^2}{2}tg^2\chi\right]\right\}Z(\chi; R) = 0 \qquad (3)$$

Then, upon introducing the notation

$$l(l+1) = k_1^2, \quad \frac{\mu\omega^2 R^4}{\hbar^2} = k_2^2, \quad \frac{2\mu R^2 E}{\hbar^2} + \frac{\mu^2\omega^2 R^4}{\hbar^2} + 1 = \mathcal{E} \qquad (4)$$

and changing the functions $Z(\chi; R)$ by

$$Z(\chi; R) = \frac{f(\chi; R)}{\sin\chi}$$

we arrive at the equation without a first derivative of the Pöschl–Teller type

$$\frac{d^2 f}{d\chi^2} + \left[\mathcal{E} - \frac{k_1^2}{\sin^2\chi} - \frac{k_2^2}{\cos^2\chi}\right]f = 0$$

Having introduced a new variable $t = \sin^2\chi$ and written down the sought equation in the form

$$f(t) = t^{1/2(1/2+\nu_1)}(1 - t)^{1/2(1/2+\nu_2)}w(t),$$

where $\nu_i = \frac{1}{2}\sqrt{1 + 4k_i^2} \geq 1/2$, we come to the hypergeometric equation

$$t(1 - t)\frac{d^2 w}{dt^2} + [c - (a + b + 1)t]\frac{dw}{dt} - abw = 0 \qquad (5)$$

with the coefficients

$$a = \frac{1}{2}(\nu_1 + \nu_2 + 1 - \sqrt{\mathcal{E}}), \; b = \frac{1}{2}(\nu_1 + \nu_2 + 1 + \sqrt{\mathcal{E}}), \; c = \nu_1 + 1$$

The general solution of equation (5) is well known[10] and can be written in the form of the hypergeometric function

$$w(t) = C_1 {}_2F_1\{a, b; c; t\} + C_2 t^{1-c} {}_2F_1\{a - c + 1, b - c + 1; 2 - c; t\}$$

where C_i are the constants.

The requirement for the wave function $Z(t)$ to be regular at $t = 0$ and $t = 1$ ($\chi = 0$ and $\pi/2$) imposes the limitation $C_2 = 0$, $a = -n_r$ or $b = -n_r$, i.e.

$$\mathcal{E} = (2n_r + \nu_1 + \nu_2 + 1)^2$$

which results in quantization of energy of the oscillator

$$E_N = \frac{\hbar^2}{2\mu}\left[\frac{(N+1)(N+3)}{R^2} + \frac{(2\nu-1)}{R^2}(N + 3/2)\right] \tag{6}$$

where $\nu \equiv \nu_2 = \sqrt{1 + \frac{4\mu^2\omega^2 R^4}{\hbar^2}}/2$, and the principal quantum number $N = 0, 1, \ldots$ is related with the radial and orbital quantum numbers by the relation $N = 2n_r + l$. It is seen from the last equation that quantum numbers N and l must have the same parity; therefore, the degree of degeneracy, as in the case of motion in the oscillator field, in the Euclidean space equals $(N+1)(N+2)/2$.

The solution of the equation (3) normalized in the interval $[0, \pi/2]$ can be expressed through the Jacobi polynomials in the form

$$Z_{Nl}(\chi; R) = \sqrt{\frac{2(N + \nu + \frac{3}{2})\Gamma(\frac{N-l}{2} + 1)\Gamma(\frac{N+l}{2} + \nu + \frac{3}{2})}{R^3\Gamma(\frac{N+l}{2} + \frac{3}{2})\Gamma(\frac{N-l}{2} + \nu + 1)}}$$
$$(\sin\chi)^l(\cos\chi)^{\nu+1/2}P_{\frac{N-l}{2}}^{l+1/2, \nu}(\cos 2\chi).$$

B) *Cylindrical coordinates*
Substituting into the Schrödinger equation the harmonic oscillator potential written in the cylindrical coordinate system and choosing the wave function in the form

$$\Psi(\phi_1, \alpha, \phi_2; R) = \Phi(\alpha)K(\phi_2)\frac{e^{im\phi_1}}{\sqrt{2\pi}}$$

after separation of variables we get two equations

$$\frac{1}{\sin\alpha\cos\alpha\,d\alpha}\frac{d}{d\alpha}\sin\alpha\cos\alpha\frac{d\Phi}{d\alpha} + \left[\left(\frac{2\mu R^2 E}{\hbar^2} + \frac{\mu^2\omega^2 R^4}{\hbar^2}\right) - \frac{m^2}{\sin^2\alpha} - \frac{I_1^2}{\cos^2\alpha}\right]\Phi = 0 \tag{7}$$

$$\frac{d^2 K}{d\phi_2^2} + \left[I_1 - \frac{\mu^2\omega^2 R^4}{\hbar^2}\frac{1}{\cos^2\phi_2}\right]K = 0 \tag{8}$$

Both these equations reduce to an equation of the Pöschl–Teller type. Indeed, having substituted

$$\Phi(\alpha) = \frac{w(\alpha)}{\sqrt{\sin\alpha\cos\alpha}}$$

into equation (7) and

$$\frac{1}{\cos^2\phi_2} = \frac{1}{4\sin^2\phi} + \frac{1}{4\cos^2\phi},$$

into equation (8) where $\phi = \phi_2/2 + \pi/4$ and $\phi \in [0, \pi/2]$ we get the following equations of the Pöschl–Teller type:

$$\frac{d^2 w}{d\alpha^2} + \left[\mathcal{E} - \frac{m^2 - 1/4}{\sin^2 \alpha} - \frac{I_1 - 1/4}{\cos^2 \alpha} \right] w = 0,$$

$$\frac{d^2 K}{d\phi^2} + \left[4I_1 - \frac{\mu^2 \omega^2 R^4}{\hbar^2} \left(\frac{1}{\sin^2 \phi} + \frac{1}{\cos^2 \phi} \right) \right] K = 0,$$

where $\mathcal{E}$ is determined by expression (4). Now, we can write down the spectrum of quantum constants

$$I_1 = (n_3 + \nu + 1/2)^2, \qquad \mathcal{E} = (2n + n_3 + |m| + \nu + 3/2)^2$$

where quantum numbers n_3 and n run the values $0, 1, 2 \ldots$. Assuming the principal quantum number N to be equal to $N = 2n + |m| + n_3$ we arrive at formula (6) for the oscillator energy.

For the functions $K_{n_3}^{\nu}(\phi_2; R)$ and $\Phi_{N'mn_3}^{\nu}(\alpha; R)$ normalized by the conditions

$$\int_{-\pi/2}^{\pi/2} K_{n_3}^{\nu}(\phi_2; R) K_{n_3'}^{\nu}(\phi_2; R) d\phi_2 = \delta_{n_3 n_3'}$$

$$R^3 \int_0^{\pi/2} \Phi_{Nmn_3}^{\nu}(\alpha; R) \Phi_{N'mn_3}^{\nu}(\alpha; R) \sin \alpha \cos \alpha \, d\alpha = \delta_{NN'}$$

can be expressed through the Jacobi polynomials in the form

$$\Phi_{N'mn_3}^{\nu}(\alpha; R) = \sqrt{\frac{2(N + \nu + 3/2)\Gamma(\frac{N-|m|-n_3}{2} + 1)\Gamma(\frac{N+|m|+n_3}{2} + \nu + 3/2)}{R^3 \Gamma(\frac{N+|m|-n_3}{2} + 1)\Gamma(\frac{N-|m|+n_3}{2} + \nu + 3/2)}}$$

$$(\sin \alpha)^{|m|}(\cos \alpha)^{n_3 + \nu + 1/2} P_{\frac{N-|m|-n_3}{2}}^{(|m|, n_3 + \nu + 1/2)}(\cos 2\alpha)$$

$$K_{n_3}^{\nu}(\phi_2; R) = \frac{\sqrt{(n_3 + \nu + 1/2)\Gamma(n_3 + 2\nu + 1)(n_3)!}}{2^{\nu}\Gamma(n_3 + \nu + 1)}(\cos \phi_2)^{\nu + 1/2} P_{n_3}^{(\nu, \nu)}(\sin \phi_2)$$

INTERBASIS EXPANSION

Let us write down a sought expansion of the spherical wave function over cylindrical in the form

$$\Psi_{Nlm}^{\nu}(\chi, \vartheta, \varphi; R) = \sum_{n_3 = 0, 1}^{N-m} W_{Nmn_3}^{l}(\nu; R) \Psi_{Nmn_3}^{\nu}(\phi_1, \alpha, \phi_2; R) \tag{9}$$

where the lower limit of summation n_3 depends on parity $(N - |m|)$.

To calculate the explicit form of the expansion coefficients $W_{Nmn_3}^{l}(\nu; R)$ suffice it to use the orthogonality of one of the functions in the cylindric wave function. Consequently, we can fix the second, at a point most convenient for us, the second variable not participating in integration. It is more convenient to use the orthogonality of the function $K_{n_3}^{\nu}(\phi_2; R)$. Now let us pass in the left-hand side of the expansion (i.e., in the spherical basis) from the spherical coordinates to the cylindrical ones according to the formulae

$$\cos \chi = \cos \alpha \cdot \cos \phi_2, \quad \sin \vartheta = \frac{\sin \alpha}{\sqrt{1 - \cos^2 \alpha \cos^2 \phi_2}}, \quad \phi = \phi_1$$

Taking the limit $\alpha \to 0$ and allowing for the relation

$$\cos \chi \to \cos \phi_2, \quad \sin \vartheta \to \frac{\sin \alpha}{\sin^2 \phi_2} \to 0$$

we immediately get that

$$Z^\nu_{Nl}(\chi; R) \;\to\; Z^\nu_{Nl}(\phi_2; R)$$

$$\Phi^\nu_{N|m|n_3}(\alpha; R) \;\to\; \sqrt{\frac{2(N + \nu + \frac{3}{2})(\frac{N+|m|-n_3}{2} + 1)!\,\Gamma(\frac{N+|m|+n_3}{2} + \nu + \frac{3}{2})}{R^3(\frac{N-|m|-n_3}{2} + 1)!\,\Gamma(\frac{N-|m|+n_3}{2} + \nu + \frac{3}{2})}} \frac{(\sin \alpha)^{|m|}}{|m|!}$$

$$Y_{lm}(\vartheta, \phi) \;\to\; \frac{(-1)^{\frac{m+|m|}{2}}}{2^{|m|}|m|!} \sqrt{\frac{2l + 1}{2} \frac{(l + |m|)!}{(l - |m|)!}} \frac{(\sin \alpha)^{|m|}}{(\cos \phi_2)^{|m|}}$$

Upon substituting the asymptotic formulae derived into the interbasis expansion (9), contracting by $(\sin \alpha)^{|m|}$ ¿from both sides in formula (9), and finally, using the orthogonality of the function $K^\nu_{n_3}(\phi_2; R)$ in the interval $-\pi/2 \le \phi_2 \le \pi/2$, we arrive at the following integral representation for the coefficients $W^l_{Nmn_3}(\nu; R)$:

$$W^l_{Nmn_3}(\nu; R) = \frac{(-1)^{\frac{m+|m|}{2}}}{2^{|m|+\nu}\Gamma(n_3 + \nu + 1)} \sqrt{\frac{2l + 1}{2R^3} \frac{(l + |m|)!(\frac{N-|m|-n_3}{2})!}{(l - |m|)!(\frac{N+|m|-n_3}{2})!} \frac{(n_3)!\,\Gamma(\frac{N-l}{2} + 1)}{\Gamma(\frac{N+l}{2} + 3/2)}}$$

$$\cdot \sqrt{\frac{\Gamma(\frac{N-|m|+n_3}{2} + \nu + \frac{3}{2})\Gamma(\frac{N+l}{2} + \nu + \frac{3}{2})\Gamma(n_3 + 2\nu + 1)}{\Gamma(\frac{N+|m|+n_3}{2} + \nu + \frac{3}{2})\Gamma(\frac{N-l}{2} + \nu + 1)(n_3 + \nu + 1/2)}} \; A^l_{N|m|n_3}(\nu; R) \qquad (10)$$

where

$$A^l_{N|m|n_3}(\nu; R) = \int_{-\pi/2}^{\pi/2} (\sin \phi_2)^{l-|m|}(\cos \phi_2)^{2\nu+1} P^{(l+1/2,\nu)}_{\frac{N-l}{2}}(\cos 2\phi_2) P^{(\nu,\nu)}_{n_3}(\sin \phi_2)\,d\phi_2 \qquad (11)$$

For a complete solution of the problem we have only to calculate the integral in formula (10). Consider separately the cases of even and odd quantum number n_3. Let us first divide the integration interval into $(-\pi/2, 0)$ and $(0, \pi/2)$; then, after the change in the first integral $\phi_2 \to -\phi_2$, we see that the integral is just doubled due to the parity $(l-|m|-n_3)$. Further, using the well-known transformation for the Jacobi polynomials[11]

$$P^{(\nu,\nu)}_{n_3}(x) = \begin{cases} \dfrac{\Gamma(n_3 + \nu + 1)(n_3/2)!}{\Gamma(n_3/2 + \nu + 1)(n_3)!} \, P^{(\nu,-1/2)}_{n_3/2}(2x^2 - 1) & \text{for } n_3 - \text{even} \\[4mm] \dfrac{\Gamma(n_3 + \nu + 1)\left(\frac{n_3-1}{2}\right)!}{\Gamma\left(\frac{n_3+1}{2} + \nu\right)(n_3)!} (\sin \gamma) \, P^{(\nu,1/2)}_{\frac{n_3-1}{2}}(2x^2 - 1) & \text{for } n_3 - \text{odd} \end{cases}$$

and making the substitution $x = \cos \phi_2$, we get the following integrals:

$$A^{l(+)}_{N|m|n_3}(\nu; R) = \frac{(-1)^{n_3/2}}{2^{\nu + \frac{l-|m|+1}{2}}} \cdot \frac{\Gamma(n_3 + \nu + 1)(n_3/2)!}{\Gamma(n_3/2 + \nu + 1)(n_3)!} \times$$

$$\times \int_{-1}^{1} (1 - x)^{\frac{l-|m|-1}{2}}(1 + x)^\nu P^{(l+1/2,\nu)}_{\frac{N-l}{2}}(x) P^{(-1/2,\nu)}_{n_3}(x)\,dx$$

$$A^{l(-)}_{N|m|n_3}(\nu; R) = \frac{(-1)^{\frac{n_3-1}{2}}}{2^{\nu + \frac{l-|m|}{2}+1}} \cdot \frac{\Gamma(n_3 + \nu + 1)\left(\frac{n_3-1}{2}\right)!}{\Gamma\left(\frac{n_3+1}{2} + \nu\right)(n_3)!} \times$$

$$\times \int_{-1}^{1} (1 - x)^{\frac{l-|m|}{2}}(1 + x)^\nu P^{(l+1/2,\nu)}_{\frac{N-l}{2}}(x) P^{(1/2,\nu)}_{\frac{n_3-1}{2}}(x)\,dx$$

The integrals we have put down are tabular; therefore, using the formula for integration of two Jacobi polynomials [12]

$$\int_{-1}^{1}(1-x)^{\tau}(1+x)^{\beta}P_n^{(\alpha,\beta)}(x)P_m^{(\rho,\beta)}(x)dx = \frac{2^{\beta+\tau+1}\Gamma(\alpha-\tau+n)\Gamma(\beta+n+l)}{(m)!(n)!\Gamma(\rho+1)\Gamma(\alpha-\tau)}$$

$$\cdot\frac{\Gamma(\rho+m+1)\Gamma(\tau+1)}{\Gamma(\beta+\tau+n+2)}\cdot {}_4F_3\left\{\begin{array}{c} -m,\ \rho+\beta+m+1,\ \tau+1,\ \tau-\alpha+1 \\ \rho+1,\ \beta+tau+n+2,\ \tau_\alpha-n+1 \end{array}\bigg|1\right\}$$

we can show that $A_{N|m|n_3}^{l(\pm)}(\nu;R)$ are expressed through the generalized hypergeometric function ${}_4F_3$ of the unit argument

$$A_{N|m|n_3}^{l(+)}(\nu;R) = \frac{(-1)^{\frac{n_3}{2}}}{\sqrt{\pi}}\frac{\Gamma(n_3+\nu+1)\Gamma(\frac{N+|m|}{2}+1)\Gamma(\frac{N-l}{2}+\nu+1)}{(n_3)!\Gamma(n_3/2+\nu+1)(\frac{N-l}{2})!\Gamma(\frac{l+|m|}{2}+1)}$$

$$\cdot\frac{\Gamma(\frac{l-|m|+1}{2})\Gamma(\frac{n_3+1}{2})}{\Gamma(\frac{N-|m|+3}{2}+\nu)}\cdot {}_4F_3\left\{\begin{array}{c} -\frac{n_3}{2},\ \frac{n_3+1}{2}+\nu,\ \frac{l-|m|+1}{2},\ -\frac{l+|m|}{2} \\ \frac{1}{2},\ \frac{N-|m|+3}{2}+\nu,\ -\frac{N+|m|}{2} \end{array}\bigg|1\right\} \qquad (12)$$

or analogously

$$A_{N|m|n_3}^{l(-)}(\nu;R) = \frac{(-1)^{\frac{n_3-1}{2}}}{\sqrt{\pi}}\frac{2\Gamma(n_3+\nu+1)\Gamma(\frac{N+|m|+1}{2})\Gamma(\frac{N-l}{2}+\nu+1)}{(n_3)!\Gamma(n_3/2+\nu+1/2)(\frac{N-l}{2})!\Gamma(\frac{l+|m|+1}{2})}$$

$$\cdot\frac{\Gamma(\frac{l-|m|}{2}+1)\Gamma(\frac{n_3}{2}+1)}{\Gamma(\frac{N-|m|}{2}+\nu+2)}\cdot {}_4F_3\left\{\begin{array}{c} -\frac{n_3-1}{2},\ \frac{n_3}{2}+\nu+1,\ \frac{l-|m|}{2}+1,\ -\frac{l+|m|-1}{2} \\ \frac{3}{2},\ \frac{N-|m|}{2}+\nu+2,\ -\frac{N+|m|-1}{2} \end{array}\bigg|1\right\} \qquad (13)$$

As is seen from the above relations (12) and (13), both the series are of the Saalschutz[12] type. Applying to the series (12) and (13) the known symmetry relations for the series ${}_4F_3(1)$[13]

$${}_4F_3\left\{\begin{array}{c} -n,\ b,\ c,\ d \\ e,\ f,\ g \end{array}\bigg|1\right\} = \frac{(f-b)_n(g-b)_n}{(f)_n(g)_n}{}_4F_3\left\{\begin{array}{c} -n,\ b,\ e-c,\ e-d \\ e,\ b-f-n+1,\ b-g-n+1 \end{array}\bigg|1\right\}$$

$$-n+b+c+d = 1+e+f+g,$$

after some obvious transformations we finally get a formula for the inerbasis expansion coefficients $W_{Nmn_3}^l(\nu;R)$

$$W_{Nmn_3}^l(\nu;R) = \frac{(-1)^{\frac{m+|m|}{2}}\sqrt{\pi}}{2^{l+\nu+1/2}}\frac{\sqrt{(2l+1)(n_3+\nu+1/2)(l+|m|)!(l-|m|)!}}{\Gamma(\frac{l-|m|-n_3}{2}+1)\Gamma(\frac{l+|m|-n_3}{2}+1)\Gamma(\nu+1)} \qquad (14)$$

$$\cdot\sqrt{\frac{\Gamma(\frac{N+l}{2}+\nu+\frac{3}{2})\Gamma(\frac{N-l}{2}+\nu+1)\Gamma(n_3+2\nu+1)(\frac{N-|m|-n_3}{2})!(\frac{N+|m|-n_3}{2})!}{\Gamma(\frac{N+l}{2}+\frac{3}{2})\Gamma(\frac{N-l}{2}+1)\Gamma(\frac{N-|m|+n_3}{2}+\nu+\frac{3}{2})\Gamma(\frac{N+|m|+n_3}{2}+\nu+\frac{3}{2})(n_3)!}} \qquad (15)$$

$$\cdot {}_4F_3\left\{\begin{array}{c} -\frac{n_3}{2},\ -\frac{n_3-1}{2},\ -\frac{N-l}{2},\ \frac{N+l}{2}+\nu+\frac{3}{2} \\ \nu+1,\ \frac{l+|m|-n_3}{2}+1,\ \frac{l-|m|-n_3}{2}+1 \end{array}\bigg|1\right\} \qquad (16)$$

Note that expression (14) is independent of parity of the quantum number n_3.

CONCLUSION

In conclusion, we have considered interbasis expansions over spherical and cylindrical wave functions for a quantum mechanical system with accidental degeneracy known as the Higgs oscillator. It is shown that the relevant expansion coefficients are expressed through hypergeometric functions ${}_4F_3$ of the unit argument. Further, we are planning to analyse solutions to the Schrödinger equation for the Higgs oscillator in all orthogonal coordinate systems admitting separation of variables, especially, interbasis expansions and integrals of motion arising in this case.

Acknowledgment

One of the authors (G.S.P.) is very grateful to Prof A.Janussis for his invitation to the Conference and kind hospitality extended to him in Olimpia.

References

[1] E.Schrödinger. *Proc.Irish.Acad. A.* 45:9 (1940).

[2] L.Infeld, A.Schild, A note on the Kepler problem in a space of constant negative curvature, *Phys.Rev.* 67:121 (1945).

[3] P.W.Higgs, Dynamical symmetries in a spherical geometry I, *J.Phys A.* 12:309 (1979).

[4] H.I.Leemon, Dynamical symmetries in a spherical geometry I, *J.Phys A.* 12:489 (1979).

[5] Yu.A.Kurochkin, V.S.Otchik, Analogue of the Runge–Lenz vector and energy spectrum in the Kepler problem on the three-dimensional sphere, *Vesti Akad. Nauk BSSR* 3:56 (1983).

[6] Ya.A.Granovsky, A.S.Zedanov, and I.M.Luzenko, Quadratic algebras and dynamics into the curvature space I. Oscillator, *Teor. Math. Phys.* 91:207 (1992).

[7] Ya.A.Granovsky, A.S.Zedanov, and I.M.Luzenko, Quadratic algebras and dynamics into the curvature space I. The Kepler problem, *Teor. Math. Phys.* 91:397 (1992).

[8] J.A.Wilson, Some hypergeometric orthogonal polynomials, *SIAM J. Math. Anal.* 11:690 (1980).

[9] M.N.Olevskii, Triorthogonal systems in space of constant curvature in which the equation $\Delta_2 u + \lambda u = 0$ allows the complete separation of variables, *Math. Sb.* 27:379 (1950) (in russian).

[10] S.Flügge, "Problems in Quantum Mechanics", V.1 Springer - Verlag, Berlin - Heidelberg - New York, (1971).

[11] G.Szegö, "Ortogonal Polynomials", Publisher by the American Mathematical, Society, New York, (1959).

[12] G.Bateman, A.Erdelyi. "Higher Transcedental Functions", MC Graw-Hill Book Company, INC. New York-Toronto-London, (1953)

[13] W.N.Bailey. "Generalized hypergeometric series", Cambridge Tracts, No 32, Cambridge, 1935.

PANCHARATNAM´S TOPOLOGICAL PHASE IN RELATION TO THE DYNAMICAL PHASE IN POLARIZATION OPTICS

Susanne Klein[1], Wolfgang Dultz[2], Heidrun Schmitzer[3]

[1] Universität Saarbrücken
[2] Forschungszentrum DBP-Telekom Darmstadt
[3] Universität Regensburg

INTRODUCTION

Interferometric comparison of waves is the perhaps most sensitive method in metrology: phase differences of only π lie between the constructive superposition and the destructive extinction of two waves, and because of the very good agreement of real waves with the trigonometric functions, very small fractions of this phase difference are measurable with the help of a computational fit to the experimental data. For optical telecommunication interferometry not only provides a very precise way to characterize optical fibers and other optical components, but also allows the construction of very economical optoelectronic switching devices and may become the design principle of the future all-photonic network.

The importance of interferometry in science and application was the reason why the concept of Berry´s phase[1] was taken up immediately and applied successfully to optical phenomena[2,3]. In this paper we want to deal with Pancharatnam´s phase[4], an example of Berry´s phase in optics discovered as early as in 1956 but neglected until the notice of Berry´s phase in the literature[5].

In 1956 S. Pancharatnam had investigated the question what the phase between light beams of general but different states of polarization could be. He found his interesting theorem by using the Poincaré sphere as a descriptive aid to show the transformation properties of elliptically birefringent, thin plates on the state of polarization of a coherent light beam. In the following we want to introduce the Poincaré sphere and discuss Pancharatnam´s theorem in the light of the new development initiated by Berry´s work. Our goal is, to develop a method for a better understanding of polarizing interferometers, and to name some of the problems involved.

THE POINCARÉ SPHERE

The general state of polarization of light is described by the two angles λ, ω and a $\pm$ sign depending on the handedness of the ellipse, (see Fig 1).

Frontiers of Fundamental Physics, Edited by M. Barone
and F. Selleri, Plenum Press, New York, 1994

The Poincaré sphere is a mapping of all the polarizations onto the surface of a sphere. Left handed elliptically polarized states for instance are mapped onto the northern hemisphere, linear polarized states onto the equator. Orthogonal states lie on mutually antipodic positions.

The change of the state of polarization of light in a transparent elliptically birefringent medium is described by a simple rotation on the Poincaré sphere:

In an elliptically birefringent plate (birefringent and optically active) two orthogonal eigenwaves S and $\bar{S}$ propagate independently with different phase velocities $c_S = \dfrac{c}{n_S}$; $c_S = \dfrac{c}{n_L}$. The orientation of the plate with respect to the reference system x, y, z can be described by the point S (the fast state of polarization) on the Poincaré sphere, (see

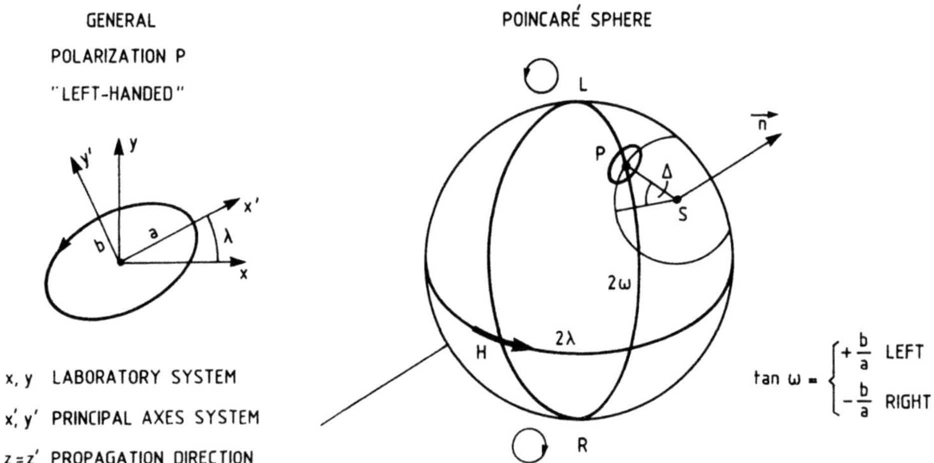

Fig. 1 Polarized light on the Poincaré sphere

Fig. 1). A general state of polarization is decomposed into a linear combination of these two eigenwaves in the crystal plate[4]:

$$P = \alpha S + \beta \bar{S}, \quad |\alpha|^2 = cos^2 \tfrac{1}{2}\widehat{PS}, \quad |\beta|^2 = sin^2 \tfrac{1}{2}\widehat{PS} \tag{1}$$

α and β are complex numbers. After propagating the distance $z \leq d$ the plate of thickness d has introduced a phase difference Δ between the two eigenwaves

$$P(z) = \alpha S exp\{-i2\pi z n_S / \lambda\} + \beta \bar{S} exp\{-i2\pi z n_L / \lambda\}$$

$$= exp\{-i\pi z(n_L + n_s)/\lambda\}\left(\alpha S\,exp\left(\frac{i\Delta}{2}\right) + \beta \bar{S}\,exp\left(-\frac{i\Delta}{2}\right)\right) \tag{2}$$

$$\Delta = 2\pi z(n_L - n_S)/\lambda$$

Equation (2) describes the state of polarization of the electrical displacement P(z) and the phase at every point z inside the plate. The corresponding position P(z) on the Poincaré sphere lies on the small circle on the sphere around the axis $S\bar{S}$ through P(0) at a distance angle Δ (counterclock wise) from P(0)[8,9]. We rewrite equation (2) within a coordinate system with the basis $\{S,\bar{S}\}$ in a spinor type form:

$$S = \begin{pmatrix} 1 \\ 0 \end{pmatrix},\ \bar{S} = \begin{pmatrix} 0 \\ 1 \end{pmatrix},\ P(0) = \begin{pmatrix} \alpha \\ \beta \end{pmatrix},\ \Delta = \Delta(z)$$

$$P(z) = exp\{-i\pi z(n_L + n_s)/\lambda\}\begin{pmatrix} exp\{i\Delta/2\} & 0 \\ 0 & exp\{-i\Delta/2\} \end{pmatrix}P(0) \tag{3}$$

$$= exp\{-i\pi z(n_L + n_s)/\lambda\}\,exp\left\{i\frac{\Delta}{2}\begin{pmatrix} 1 & 0 \\ 0 & -1 \end{pmatrix}\right\}P(0)$$

In close analogy to the transformation of a spin 1/2 particle under rotation we find for an arbitrary rotation axis $\vec{n}$[6,7]:

$$P(z) = \hat{R}(\vec{n},\Delta)P(0) = exp\{-i\pi z(n_L + n_s)/\lambda\}\,exp\left\{i\frac{\Delta}{2}\vec{n}\hat{\sigma}\right\}P(0) \tag{4}$$

$\hat{R}(\vec{n},\Delta)$ is the rotation operator, $\hat{\sigma}$ is the vector of the Pauli-matrices and $\vec{n}$ is the unitvector in the direction of the axis. Note: Equ. (3) is a special case of (4) with $\vec{n} = (0,0,1)$ and $\hat{\sigma}_z = \begin{pmatrix} 1 & 0 \\ 0 & -1 \end{pmatrix}$. In the following we want to evaluate the phase factor in equ. (4) in a more detailed manner.

DYNAMICAL AND TOPOLOGICAL PHASES

The situation in the system of the Poincaré sphere, is shown in Fig. 2.
The Poincaré sphere has the radius R=1, so that we only deal with angles. The position vector $\vec{P}$ is changed by an infinitesimal rotation $d\Delta$ around $\vec{n}$ into $\vec{P}+d\vec{P}$. We decompose this rotation into a rotation around $\vec{P}$ through the angle $d\Delta\cos\alpha$ and a rotation around $\vec{P}\times d\vec{P}$, perpendicular to $\vec{P}$ and $d\vec{P}$, through an angle $d\Delta\sin\alpha$.
The rotation around $\vec{P}$ does not shift $\vec{P}(0)$ on the Poincaré sphere but introduces a phase factor $exp\left(+i\frac{d\Delta}{2}\cos\alpha\right)$, since P(0) is an eigenvector of $\vec{P}\hat{\sigma}$ with eigenvalue 1.

$$\hat{R}(\vec{n}, d\Delta)P(0) = exp\left\{-i\pi dz(n_L + n_S)/\lambda\right\}exp\left\{i\frac{d\Delta}{2}cos\alpha\right\} \tag{5}$$

$$\times exp\left\{i\frac{d\Delta}{2}sin\alpha \cdot \frac{\vec{P}\times d\vec{P}}{\left|\vec{P}\times d\vec{P}\right|}\hat{\sigma}\right\}P(0)$$

$\hat{\sigma}$ are the Pauli-matrices and dz is the infinitesimal thickness of the crystal plate which introduces the phase difference $d\Delta$.

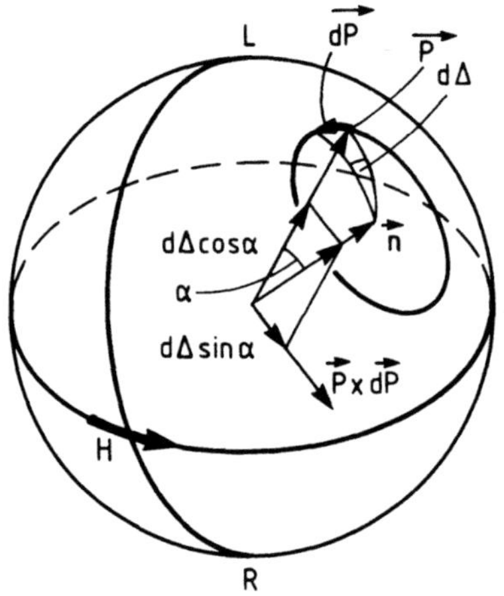

Fig. 2 Decomposition of an infinitesimal rotation

We can now iterate this step to lead $\vec{P}$ on a closed path on the Poincaré sphere.

$$\prod_j \hat{R}(\vec{n}_j, d\Delta_j)P(0) = \prod_j exp\left\{-i\pi d(n_L + n_S)/\lambda\right\} \times$$

$$\times \prod_j exp\left\{i\frac{d\Delta_j}{2}cos\alpha_j\right\} \tag{6}$$

$$\times \prod_j exp\left\{i\frac{d\Delta_j}{2}sin\alpha_j \frac{\vec{P}_j \times d\vec{P}_j}{\left|\vec{P}_j \times d\vec{P}_j\right|}\hat{\sigma}\right\}P(0)$$

The Pauli-matrices $\hat{\sigma}$ are associated with the Poincaré sphere. $\Sigma dz_j = d$ is the path length inside the crystal plate. Jordan[10] showed that the last product of equ. (6) equals the topological phase factor:

$$\prod_j exp\left\{ i\frac{d\Delta_j}{2} sin\alpha_j \frac{\vec{P}_j \times d\vec{P}_j}{\left|\vec{P}_j \times d\vec{P}_j\right|}\sigma \right\} P(0) = exp\{ i\Omega/2 \} \tag{7}$$

where Ω is the solid angle generated by the closed path on the Poincaré sphere. $\gamma = -i\frac{\Omega}{2}$ is called the topological (geometrical) phase or Pancharatnam´s phase. With (7) we find:

$$\prod_j \hat{R}(\bar{u}_j, d\Delta_j) P(0) = exp\left\{ -i\pi\frac{d(n_L + n_S)}{\lambda} \right\} exp\left\{ i\sum_j \frac{d\Delta_j}{2} cos(\alpha_j) \right\} exp\{ +i\gamma \} \tag{8}$$

In the case of the closed path being a circle (all $\alpha_j = \alpha$, $\sum_j d\Delta_j = 2\pi$, $\sum_j dz_j = d$), we find:

$$\prod_j \hat{R}(\bar{n}_j, d\Delta_j) P(0) = exp\{ -i\pi d(n_L + n_S)/\lambda \} \times \tag{9}$$

$$\times exp\{ i\pi cos\alpha \} exp\{ i\pi(1 - cos\alpha) \} P(0)$$

Equation (8) has three different phase factors: The first one is independent of the birefringence of the crystal plate and describes the phase which a light wave gains inside a plate of thickness $d = \sum_j dz_j$ and with an average refractive index $(n_L + n_S)/2$. The second factor describes a phase proportional to birefringence $(n_L - n_S)$ which was somewhat inaccurately called the dynamical phase γ_{din}[11]. In the limiting case of a light beam of polarization $S(\bar{S})$ falling onto a crystal plate with the eigenwaves $S, \bar{S}$, we can combine the second factor of equ. (8) with the first one and find

$$exp\{ -i\pi d(n_L + n_S)/\lambda \} exp\left\{ i\sum_j \frac{d\Delta_j}{2} cos\alpha_j \right\} = exp\{ -i2\pi dn_S(n_L)/\lambda \} \tag{10}$$

respectively, which is exactly the phase factor of a glass plate of thickness d and refractive index $n_S(n_L)$. Thus the combination of the first two phases in equation (8) should be called the dynamical phase.

The third phase factor in equation (8), which describes Pancharatnam´s phase, does only depend on the solid angle of the closed optical path on the Poincaré-sphere. In principle it does not depend on birefrigence, wave length of light or the manner in which the state of polarization is changed.

Pancharatnam´s phase can be introduced by birefrigent plates[12] or polarizers,[13] and its simple structure makes it likely that it will become a major help to a better understanding of the behaviour of polarized light.

This work is connected with the COST 241 action.

REFERENCES

1. M. V. Berry, Quantal phase factors accompanying adiabatic changes, Proc. R. Soc. Lond. <u>A 392</u> 45 (1984).
2. R. Y. Chiao and Y.S. Wu, Manifestations of Berry´s topological phase for the photon, Phys. Rev. Lett. <u>57</u> 933/(1986).
3. A. Tomita and R.Y. Chiao, Obervations of Berry´s topological phase by use of an optical fiber, Phys. Rev. Lett. <u>57</u> 937 (1986).
4. S. Pancharatnam, Generalized theory of interference, and its applications, Proc. Ind. Acad. Si <u>A 44</u> 247 (1956)
5. R. Ramaseshan and R. Nityananda, The interference of polarized light as an early example of Berry´s phase, Current Science India <u>55</u> 1225 (1986)
6. M. V. Berry, The adiabatic phase and Pancharatnam´s phase for polarized light, Mod. Opt. <u>34</u> 1401 (1987)
7. J. Byrne, Joung´s double beam interference experiment with spinor and vector waves, Nature <u>275</u> 188 (1978)
8. G. N Ramachandran and S. Ramaseshan, Crystal optics, *in: Handbuch der Physik Vol XXV/1, S. Flügge, ed, Springer, Berlin (1961)*
9. D. S. Kliger, J. W. Lewis, C. E. Randall, Polarized Light in Optics and Spectroscopy, Academic Press 1990
10. T. F. Jordan, Berry phases and unitery transformations, J. Math. Phys. <u>29</u> 2042 (1988)
11. M. Martinelli and P. Vavassori, A geometrie (Pancharatnam) phase approach to the polarization and phase control in the coherent optics circuits, Opt. Comm. <u>80</u> 166 (1990)
12. R. Bhandari, Observation of nonintegrable geometrie phase on the Poincaré sphere, Physics Letters <u>A</u> <u>133</u> 1 (1988)
13. H. Schmitzer, S. Klein and W. Dultz, Nonlinearity of Pancharatnam´s topological phase, Phys. Rev. Lett. <u>71</u> 1530 (1993)

ON THE CONNECTION BETWEEN

CLASSICAL AND QUANTUM MECHANICS

Andrzej Horzela

Department of Theoretical Physics
H.Niewodniczański Institute of Nuclear Physics
ul. Radzikowskiego 152, 31 342 Kraków, Poland

INTRODUCTION

In the hierarchy of physical theories classical Newton's mechanics may be considered as a "limit" of two more general theories: the relativistic classical mechanics and the non-relativistic quantum mechanics. In the first case, the limit is reached when in all relativistic formulae the velocity of light is going to infinity. The limiting procedure changes the shapes of formulae but does not change either the number or the physical interpretation of the corresponding quantities. In this sense we may say that we understand satisfactorily the limiting procedure both from the mathematical and physical point of view [1].

A similar conclusion is far to be true for the second case. All known ways of relating classical mechanics with "limits" of quantum mechanics suffer from the lack of precision [2]. In particular, it is not known, either from mathematical or physical points of view, what is the classical limit of the most fundamental object of quantum mechanics - the wave function - when the Planck constant is going to zero.

GENERAL FORMALISM

In the present contribution we are going to investigate the possibility of obtaining a well-defined classical limit of quantum mechanical wave function. We shall perform our investigation in a framework of presented in this volume scheme [3] which enables us to consider various fundamental equations of physics using a simple, mathematically consistent field theory. The scheme describes as particular cases standard classical field theories like electrodynamics and gravitation as well as it incorporates both the classical and the wave mechanics. Within our approach the primary physical notions in each theory are described in terms of four collections of basic fields. The first collection of fields $\psi_\alpha(x) = \psi_\alpha(x^0, x^1, x^2, x^3)(\alpha = 1, \ldots, N)$ contains all fields which in the case of mechanical theories describe the localization of physical objects.

Frontiers of Fundamental Physics, Edited by M. Barone
and F. Selleri, Plenum Press, New York, 1994

The second collection consists of fields $\phi_{\alpha\mu}(x)$ ($\mu = 0,1,2,3$) which determine the space-time evolution of fields $\psi_\alpha(x)$. Altogether, these two collections describe the kinematical aspects of each theory. The dynamical aspects are described by the third collection of fields $\pi_\alpha(x)$ which realize the dynamical quantities of the theory and by the fourth collection of fields $\rho_\alpha(x)$ which describe all external influences acting on the considered physical system.

By definition the basic fields satisfy the set of differential relations

$$\frac{\partial \psi_\alpha(x)}{\partial x^\mu} = \phi_{\mu\alpha}(x) \tag{1.a}$$

$$\frac{\partial \pi_\alpha^\mu(x)}{\partial x^\mu} = \rho_\alpha(x) \tag{1.b}$$

which are universal and do not contain any physical constants. In order to determine, in each case, the fields from these general relations we have to complete them adding constitutive relations which specify the model under consideration. These relations distinguish various physical theories and introduce all necessary physical constants into general scheme.

In particular, the equations of Newton's classical mechanics of a material point moving in one dimension are obtained from (1a-b) by taking a single field $\psi(t)(N = 1)$ depending on time variable only, and by adopting the following constitutive relation

$$\pi^\mu(t) = m \, \phi_\mu(t) \tag{2}$$

where m is the mass of the material point. In this case we may interpret the field $\psi(t)$ as the trajectory of the point and the nonzero component of the field $\phi_\mu(t)$ as its velocity. The non-vanishing component of the field $\pi^\mu(t)$ is then equal to the momentum of the point and the field ρ is the force acting on it.

The Schrödinger wave equation for the wave function $\psi(x)$ of a scalar particle is obtained from (1a-b) by taking the following constitutive relations

$$\pi^0(x) = i\hbar\psi(x) \tag{3.a}$$

$$\pi^i(x) = -\frac{\hbar^2}{2m} \, \phi_i(x) \tag{3.b}$$

$$\rho(x) = -V(x) \, \psi(x) \tag{3.c}$$

where $\hbar$ is the Planck universal constant, m-the mass of the particle and $V(x)$ is the usual potential acting on it.

In order to investigate the relation between the Newtonian and Schrödinger mechanics let us consider the constitutive relations combining the classical and quantum mechanical cases:

$$\pi^0(\vec{x},t) = m \, l_0^2 \, \phi_0(\vec{x},t) + i\hbar\psi(\vec{x},t) \tag{4.a}$$

$$\pi^i(\vec{x},t) = -\frac{\hbar^2}{2m} \, \phi_i(\vec{x},t) \tag{4.b}$$

$$\rho(\vec{x},t) = -V(\vec{x},t)\psi(\vec{x},t) + l_0^2 F(\psi(\vec{x},t),\phi_\mu(\vec{x},t),\vec{x},t) \tag{4.c}$$

where l_0 is an arbitrary parameter with the dimension of length, and F describes all external influences not taken into account by the quantum mechanical potential $V(\vec{x},t)$. Substituting (4) into (1) we obtain the following equation for the field $\psi(\vec{x},t)$:

$$ml_0^2 \, \frac{\partial^2 \psi}{\partial t^2} + i\hbar \, \frac{\partial \psi}{\partial t} - \frac{\hbar^2}{2m} \, \Delta\psi + V\psi = l_0^2 \, F \tag{5}$$

which is a generalization of the famous telegraphist's equation. It is now easy to see that in the limit $l_0 \to 0$ equation (5) goes to the Schrödinger equation

$$i\hbar \frac{\partial \psi}{\partial t} - \frac{\hbar^2}{2m} \Delta \psi + V\psi = 0 \tag{6}$$

while in the limit $\hbar \to 0, V \to 0$ it takes the form of the Newton's equation

$$m \frac{d^2 \psi}{dt^2} = F(\psi, \dot{\psi}, t) \tag{7}$$

Moreover if the limit $V \to 0$ is performed before the limit $\hbar \to 0$ the only quantity which defines the length scale of the problem is l_0. It means that for all solutions of (7) which slowly vary with x on distances comparable with l_0 the only significant dependence is their dependence on time and the function $\psi = \psi(t)$ may be identified with the classical trajectory of a particle moving under the force F. Therefore equation (5) provides a two parameter family of fields $\psi(\vec{x}, t; \hbar, l_0)$ which, as limiting points, contains both the quantum mechanical wave function $\psi_q(\vec{x}, t)$ and the classical trajectory $\psi_{cl}(t)$. We may therefore say that our approach is an alternative realization of the de Broglie idea of Double Solution [4]. In order to make the analogy between classical and quantum cases more exact, we must ensure that the force F in equation (7) and the potential V in equation (6) corresponds to the same physical interaction. We must therefore restrict ourselves to the field equation (5) in two dimensional space-time and put

$$F(\psi, \dot{\psi}, t)\Big|_{\psi(t)=z(t)} = -\frac{\partial V(x, t)}{\partial x}\Big|_{x = z(t)} \tag{8}$$

where the interpretation of the field $\psi(t)$ as the classical trajectory is explicitly taken into account. To maintain the possibility of taking in equation (5) the limit $V \to 0$ with $F \neq 0$ we must introduce into it one more dimensionless parameter, λ, which will multiply the potential V.

HARMONIC OSCILLATOR

As the simplest example of our approach let us now consider the harmonic oscillator for which all the requirements can easily be fulfilled. In this case we have the basic equation of the form

$$ml_0^2 \frac{\partial \psi^2}{\partial t^2} + i\hbar \frac{\partial \psi}{\partial t} - \frac{\hbar^2}{2m} \frac{\partial^2 \psi}{\partial x^2} + \frac{m\omega_0^2}{2} x^2 \psi + l_0^2 k\psi = 0 \tag{9}$$

where we have put

$$\omega_0^2 = \lambda \frac{k}{m} \tag{10}$$

Note that the harmonic oscillator case, due to the relation (8), is the only case in which the basic equation (5) is linear. Therefore the full discussion of classical limits of quantum mechanics may be performed only in the framework of nonlinear field equations which considerably complicates the problem.

In order to see some details of our limiting procedures, it is convenient to pass in equation (9) to dimensionless variables. The general form of such a change of variables is

$$t \to \tau = \omega_0 t \, A^\alpha \, B^\beta \tag{11.a}$$

$$x \to \xi \;=\; x\, l_0^{-1}\, A^\gamma\, B^\delta \tag{11.b}$$

where

$$A \;=\; \frac{\hbar}{m\omega_0 l_0^2} \tag{12.a}$$

and

$$B \;=\; \frac{k}{m\omega_0^2} \tag{12.b}$$

are two independent dimensionless constants constructed from our parameters. In the quantum limit the change of variables has to be independent of l_0 and therefore we must choose $\alpha \;=\; 0$ and $\gamma \;=\; -1/2$. Since in this case $\lambda \;=\; 1$, we have $B \;=\; 1$ and the powers β and δ are irrelevant. On the other hand, in the classical limit $\omega_0 \to 0$ and therefore $\beta \;=\; 1/2$. After the change of variables any solution of equation (9) depends on ξ, τ and in order to "loose" in it the x-dependence the remaining parameter δ must satisfy the inequality

$$\delta \le \frac{1}{4} \tag{13}$$

This illustrates our previous general remarks on passing from the general solution of equation (5) to a function which depends only on the time variable, as is required for solutions of the classical equations of motion. The traditional classical limit of quantum mechanics implemented by the limit $\hbar \to 0$ must therefore be performed after the limit $\omega_0 \to 0$ in order to keep meaning of the field $\psi(x,t)$. More generally, the switching off the quantum potential realized by the limit $\lambda \to 0$ has to be taken always before the limit $\hbar \to 0$. Therefore, in the classical limit all effects due to a "classical trace" of the quantum mechanical potentials disappear and in order to have a nontrivial classical limit of the quantum mechanical wave function we should in the usual approach start from the nonlinear Schrödinger equation without any potential term. Otherwise, in the usual approach it is meaningless to speak about the classical limit of quantum mechanical wave equations and wave functions.

Since we are treating our basic equation (9) as a field equation and we adopt the usual quantum mechanical interpretation only for the limiting wave function $\psi_q(x,t)$ we do not impose the usual square-integrability condition on the field $\psi(x,t;\hbar,l_0)$. Instead of that we shall consider only those solutions of (9) which are bounded in the whole space-time. The general such solution may be written as

$$\psi(x,t;\hbar,l_0) \;=\; \sum_n \left(c_n e^{i\omega_n^+ t} \;+\; d_n e^{i\omega_n^- t} \right) H_n\left(x\sqrt{\frac{m\omega_0}{\hbar}} \right) e^{-\frac{m\omega_0^2}{2\hbar} x^2} \tag{14}$$

where $H_n(\xi)$ are the usual Hermite polynomials and

$$\omega_n^\pm \;==\; \frac{1}{2ml_0^2}\left[\hbar \pm \sqrt{\hbar^2 + 4\hbar\omega_0 m l_0^2\left(n + \frac{1}{2}\right) + 4mkl_0^4} \right] \tag{15}$$

In the quantum limit $\omega_n^{(-)} \to \infty$ and hence $d_n = 0$, while

$$\omega_n^{(+)} \to \omega_0\left(n + \frac{1}{2}\right) \tag{16}$$

Thus, the solution (14) is going to the usual quantum mechanical oscillator wave function.

In the classical limit

$$\omega_n^{(\pm)} \to \pm \sqrt{\frac{k}{m}} \tag{17}$$

and taking the limit $\omega_0 \to 0$ before the limit $\hbar \to 0$ we get from (14) the classical solution

$$\psi_{cl}(t) = A \, \exp\left(i\sqrt{\frac{k}{m}} \, t \right) \, + \, B \, \exp\left(i\sqrt{\frac{k}{m}} \, t \right) \tag{18}$$

where

$$A = \sum_{n=0}^{\infty} c_n H_n(0) \qquad B = \sum_{n=0}^{\infty} d_n H_n(0) \tag{19}$$

The formula (15) for small but nonzero l_0^2 gives a simple formula for the deviations of frequencies from usual quantum mechanical frequencies

$$\Delta\omega_n \approx \frac{2l_0^2}{\hbar} \left[k - m\omega_0^2 \left(n + \frac{1}{2} \right) \right] \tag{20}$$

If such deviations really are observed, (20) will determine the experimental value of l_0.

Up to now in our discussion we have notoriously manipulated the dimensional parameters. In order to make such manipulations more precise let us consider the Fourier transform of the field $\psi(x, t; \hbar, l_0)$ in the case when $\omega_0 = 0$. Then the field equation (9) takes the form

$$\left[l_0^2 m \, (\omega^2 - \omega_{cl}^2) \, + \, \hbar \left(\omega - \frac{\hbar \kappa^2}{2m} \right) \right] \psi(\omega, \kappa; \hbar, l_0^2) = 0 \tag{21}$$

For a classical particle

$$\omega \approx \omega_{cl} = \sqrt{\frac{k}{m}} \tag{22}$$

and ω differs considerably from $\dfrac{\hbar \kappa^2}{2m}$. Therefore the only way to satisfy the field equation is to put $\hbar = 0$. On the contrary, in the quantum case the frequency ω significantly differs from ω_{cl} while according to the basic de Broglie idea

$$\omega \approx \frac{\hbar \kappa^2}{2m} \tag{23}$$

Hence the only way to satisfy the field equation is to put $l_0 = 0$. More exactly, we are close to the classical case when

$$m l_0^2 \left| \frac{\omega^2 - \omega_{cl}^2}{\omega - \dfrac{\hbar}{2m}\kappa^2} \right| \gg \hbar \tag{24}$$

while we approach the quantum case when

$$m l_0^2 \ll \left| \frac{\omega - \dfrac{\hbar}{2m}\kappa^2}{\omega^2 - \omega_{cl}^2} \right| \hbar \tag{25}$$

As we have seen, the parameter l_0 plays a crucial role in our approach and we need to give it a clear physical interpretation. To do this we would like to remind ,[5], that for any macroscopic theory it is necessary to define a finite length or resolution which is not explicitly present in the theory, but which determines the domain of applicability of the theory. Such a parameter should, however, appear explicitly in a theory which generalizes the given macroscopic theory. This is required in order to have the possibility to determine the domain of applicability of the coarser initial theory. Our theory generalizes both the classical and the wave mechanics, and should therefore contain lengths which determine the domains of applicability of both these particular theories. For the classical mechanics we should forget about all effects connected with the Compton length proportional to $\hbar$. Therefore in the classical limit $\hbar \to 0$. The theory remains classical independently of the distance at which we are looking on the physical phenomena. In the quantum case we will be able to see the details of order of the Compton length only when we are looking at them from sufficiently small distances. Therefore we connect the length l_0 with the minimal distance of observation of phenomena and it is now clear that in order to see all quantum effects we must go with l_0 to zero just as our formalism requires.

The considerations presented above open a new way for a sufficiently precise discussion of the interrelations between the classical and the wave mechanics. As we have already mentioned, more insight into this problem can be obtained only when the basic field equation (5) becomes nonlinear. This circumstance crucially complicates the problem however even if it cannot be solved it can at least be defined by our formalism.

ACKNOWLEDGMENTS

The author is very grateful to the Organizing Committee of the conference "Frontiers in Fundamental Physics" for the help obtained and to the Polish State Committee for Scientific Research providing under grant 2 0342 91 01 financial support of the research and the participation in the conference.

REFERENCES

1. J. M. Lévy-Leblond, Galilei group and Galilean invariance, *in*: Group theory and its Applications, v.2, E. M. Loebl, ed., Academic Press, N. Y. (1971).
2. L. V. de Broglie, " Les Incertitudes d' Heisenberg et Interprétation Probabiliste de la Mecanique Ondulatoire ", Bordas, Paris, (1982).
   V. P. Maslov, " Asymptotic Methods and the Perturbation Theory ", Nauka, Moscow, (1988), (in Russian).
3. E. Kapuścik; Physics without physical constants, this volume; see also Acta Phys. Polonica B17:923 (1986) and Physics without physical constants *in* Proc. of the 5th International Symposium on Selected Topics in Statistical Mechanics, A.A. Logunov et al. eds., World Scientific, (1990).
4. L. V. de Broglie; Nonlinear Wave Mechanics, a Causal Interpretation, Elsevier Publishing Company, (1960).
5. F. N. H. Robinson; Macroscopic Electromagnetism, Inter. Series of Monographs in Natural Philosophy V.57, Pergamon Press, (1973).

DISCRETE TIME REALIZATIONS OF QUANTUM MECHANICS

AND THEIR POSSIBLE EXPERIMENTAL TESTS

C. Wolf

Department of Physics
North Adams State College
North Adams, MA 01247

ABSTRACT

The possible existence of a microscopic uncertainty principle in time due to the uncertain tuning between a particle's sense of time (a particle being a microuniverse) and the frame of synchronous observers (the synchronous observers being described by an averaged out macroscopic time) suggests that the usual quantum formalism be modified to encompass this feature by allowing time derivatives to be replaced by discrete time differences. Such an alteration leads to a formalism that fits into a general mathematical structure described by Hadronic Mechanics. The consequences of such a generalization of Quantum Mechanics leads to high field modifications of spin-polarization precession frequencies, modifications of spin-resonance frequencies, modifications of spin-flip frequencies and spectral shifts in emission frequencies of systems that contain internal non-conventional characteristics such as dyon-dyon interactions, extra internal dimensions and extra compact dimensions. Such internal properties of elementary particles might provide excellent probes to compositeness which might provide more well defined evidence for sub-structure than the usual probes of anomalous magnetic moments, modifications of form factors and rare decays.

P.A.C.S. - 03.65 - Bz - Quantum theory, foundations, theory of measurement, miscellaneous theories.

INTRODUCTION

The underlying structure of Quantum Mechanics has undergone a metamorphosis and re-evaluation ever since the historical paper of Einstein, Padolsky and Rosen.[1] The competing formulations embodied in the statistical interpretation,[2] the Copenhagen interpretation,[3] the hidden variable approach[4] and the theory of "quantum potential"[5] have suggested that Quantum Mechanics may have either a complex or simple structure that to date has not been understood in all its detail. Probably one of the ultimate reasons for the problematic structure of Quantum Mechanics stems from the fact that their might not be a complete separation between the "arena" of space-time and the matter or particles that occupy it. Certainly if space-time emerges from some sort of combinatoric sequence of choices in the sense of "Wheeler"[6] or from the primitive structure of a "quantum net"[7] then fluctuations and deviations from the continuum might be hidden beneath an averaging process that experiment would find hard to probe at this point.[8] If at the most fundamental level each particle comprising the universe has characteristics and properties not

Frontiers of Fundamental Physics, Edited by M. Barone
and F. Selleri, Plenum Press, New York, 1994

initially correlated[9] with other particles then we might envision space and time in a Minkowski sense as being born of the correlations between individual particle properties. Actually Wootters has made a notable attempt at formulating this idea wherein he uses the correlation of spins of individual "fermions" and space and time then result from spin spin correlations.[10] Due to our ignorance of a "sub-quantum mechanics" we might seek to describe the fact that a particle "retains some of its individual characteristics separate from the rest of the universe" in the following manner; first the Schrondinger equation represents the first approximation to a theory with the hamiltonian and wave function representing a description of a particle with an averaged out sense of space and time representing a good approximation to the dynamical arena. However when we take into account the uncertain tuning between the particle's sense of time and the frame of synchronous observers (averaged out sense of space time) then the particle's wave function will respond a small finite time interval removed from the point of application of the hamiltonian to represent the uncertain tuning between the particle's frame and the frame of synchronous observers. Such an idea can be viewed as a fundamental uncertainty principle of the wave function's response in time because of the uncertain tuning between the particle's sense of time and the frame of synchronous observers. Thus we write

$$ H\psi(\vec{r},t) = i\hbar \frac{\left[\psi\left(t+\frac{\tau}{2}\right) - \psi\left(t-\frac{\tau}{2}\right) \right]}{\tau} \tag{1} $$

or

$$ H\psi(r,t) = i\hbar \int_{-\infty}^{\infty} \frac{\left(\psi\left(t+\frac{\tau}{2}\right) - \psi\left(t-\frac{\tau}{2}\right) \right)}{\tau} g(\tau)d\tau \tag{2} $$

(τ = discrete time interval)[11]

In Eq. (1) the response is centered at two points $t+\frac{\tau}{2}, t-\frac{\tau}{2}$, in Eq. (2) the response is at an infinity of points with weight function $g(\tau)$ weighing the response at $t+\frac{\tau}{2}, t-\frac{\tau}{2}$. Also in Eq. (1) and Eq. (2) we take the hamiltonian as that of the usual Schrodinger theory since we have lumped all of our uncertainty of knowledge into the wave-functions non-local response. A deeper understanding of the transition from a particle's sense of time to the "averaged out sense of time" might generate additional terms to be added to the hamiltonian which to date we have no clue to. The origin of Eq. (1) and Eq. (2) can be traced to Caldirola's papers[12, 13] and the interpretation in terms of a fundamental "uncertainty principle in time" was given by Recami (Ref. 9). When the discrete time difference on the right side of Eq. (1) is inverted to generate a non-local operator in (space, spin - -) on the left hand side, the right side looks like the time evolution embodied in the Schrodinger equation with a non-local left hand side.[14, 15] There has been an avalanche of interest in the mathematical structure of Eq. (1) and Eq. (2) with regard to its Lie-isotopic structure and axiom-preserving properties and Kadeisvilli[16] has discussed in connection between the properties of this discrete time difference Quantum Theory and that of normal quantum mechanics.[17] There has been a debate on whether or not discrete time difference quantum mechanics as expressed in Eq. (1) and Eq. (2) is consistent and it is my feeling that it is a matter of language. The criteria for a sensible quantum theory includes "unitarity" or a conservation principle for probability and a consistent operator representation of the operators in the theory. In what follows I demonstrate that a conserved probability density exists in such a theory and also derive an equation for the time evolution of the expectation value of an observables in such a theory. It turns out that if an operator representing an observable commutes with the hamiltonian then the expectation value of the observable will not change in time, this ensures the stationary nature of an eigenstate. In (Ref. 17), the difference between normal Quantum Mechanics and Hadronic Mechanics is illustrated with a modification of the commutations relations (q deformations) giving rise to a loss of Hermiticity of the hamiltonian, loss of probability conservation and loss of form invariance under transformation theory in Hadronic Mechanics. My contention is that the conventional commutation relations can be retained which will ensure the stationary nature of an eigenstate of the observable whose operator commutes with the hamiltonian. The explicit form for the time evolution of the expectation value takes on the form of the Heisenberg equation with the hamiltonian replaced by a non-local operator which is a function of H. In the discussion below I briefly discuss the existence of a conserved probability density in a discrete time difference theory along with the form that the time derivative of the expectation value of an observable takes on in such a theory. I then point out the various experimental probes to such a theory which include electron spin polarization precession,[18] gauge boson spin polarization precession,[19] and spectral shifts induced by a discrete time difference Quantum theory for the hydrogen atom and systems containing internal hidden quantum numbers.[20, 21, 22] Because of the non-linear relation between the transition frequency and the energy difference of two levels in such a theory quantum numbers that don't change in a transition still show up in the transition frequency, thus this theory can be used to probe the compositeness of elementary particles. I also discuss the possible origin of a discrete time interval τ using the central limit theorem for an ensemble of random variables[23] that label the individual preons comprising the elementary particle. Though present experiments can only set limits on a discrete time interval in quantum theory, the idea of a microscopic uncertainty principle in time emerging from the uncertain tuning between the particles sense of time (micro universe) and the time measured by a synchronous set of observers (averaged out time) would suggest the primordial beginnings for introducing a geometry which was preceeded by a statistical set of correlations between particles to generate an averaged-out notion of space and time. This view would be in accord with the incompleteness of a physical description in the spirit of Einstein Padolsky and Rosen[24] only at a level deeper than that probed by conventional experiments.

EXPERIMENTAL PROBES TO DISCRETE TIME DIFFERENCE THEORY

To illustrate the existence of a probability conservation law in a discrete time difference Quantum Mechanics we write down Eq. (1) for a free particle

$$-\frac{\hbar^2}{2m}\frac{\partial^2\psi}{\partial x^2} = i\hbar\frac{\left[\psi\left(t+\frac{\tau}{2}\right)-\psi\left(t-\frac{\tau}{2}\right)\right]}{\tau} \tag{3}$$

(τ = discrete time interval)

The right side of Eq. (3) can be written as

$$-\frac{\hbar^2}{2m}\frac{\partial^2\psi}{\partial x^2} = \frac{2i\hbar}{\tau}\left(\sinh\frac{\tau}{2}\frac{\partial}{\partial t}\right)\psi \tag{4}$$

If we invert the operator in Eq. (4) we have

$$\frac{2i\hbar}{\tau}\sinh^{-1}\left[-\frac{\tau i}{2\hbar}\left(-\frac{\hbar^2}{2m}\frac{\partial^2}{\partial x^2}\right)\right]\psi = i\hbar\frac{\partial\psi}{\partial t} \tag{5}$$

Firstly the operator

$$\frac{2i\hbar}{\tau}\sinh^{-1}\left[-\frac{\tau i}{2\hbar}\left(-\frac{\hbar^2}{2m}\frac{\partial^2}{\partial x^2}\right)\right]$$

is Hermitian, since H = H$^+$ and sinh^{-1}( ) contains only odd powers of iH so that whenever (i → -i), it cancels the substitution in front of O(H) (i → -i).

Secondly, following the usual derivation of the probability conservation law, we obtain from Eq. (5) and its complex conjugate upon multiplying by ψ^+ and ψ respectively and adding

$$\frac{2}{\tau}\psi^*\sinh^{-1}\left[-\frac{\tau i}{2\hbar}\left(-\frac{\hbar^2}{2m}\frac{\partial^2}{\partial x^2}\right)\right]\psi + \frac{2}{\tau}\psi\sinh^{-1}\left[\frac{\tau i}{2\hbar}\left(\frac{\hbar^2}{2m}\frac{\partial^2}{\partial x^2}\right)\right]\psi^* = \frac{\partial}{\partial t}(\psi^*\psi) \qquad (6)$$

the left hand side of Eq. (6) can be written as

$$-\frac{\partial S}{\partial x} = \frac{2}{\tau}\psi^*\sinh^{-1}\left[-\frac{\tau i}{2\hbar}\left(-\frac{\hbar^2}{2m}\frac{\partial^2}{\partial x^2}\right)\right]\psi + \frac{2}{\tau}\psi\sinh^{-1}[---]\psi^* \qquad (7)$$

(here Eq. (7) represents a partial differential equation for S) with the probability conservation law

$$\frac{\partial S}{\partial x} + \frac{\partial P}{\partial t} = 0 \qquad (P = \psi^* \psi) \qquad (8)$$

Certainly, the first term in the expansion of the $\sinh^{-1}[\ ]$ term in Eq. (7) gives the usual probability current. Since the right hand side of Eq. (4) involves an infinite set of derivatives in time the initial data must include the wave function and an infinity of time derivatives at t = 0 for all spatial points with restrictions on the derivatives given by Eq. (4). Such a construction would generate the complex addition of plane waves with an infinite number of phases. How the initial state of a quantum system is specified is one of the central problems associated with Eq. (1) and Eq. (2). IN three dimensions the construction similar to Eq. (6) will demonstrate the existence of a local conservation law of probability. If Eq. (5) is rewritten as

$$i\hbar\frac{\partial \psi}{\partial t} = \frac{2i\hbar}{\tau}\sinh^{-1}\left[-\frac{\tau i H}{2\hbar}\right]\psi = O(H)\psi \qquad (9)$$

we may write the following equation for the time evolution of the expectation value in such a theory, here X = operator with expectaion value $<X>$ (X independent of t).
Using Eq. (8) and $O^+ = O$ we have

$$\frac{d<X>}{dt} = \frac{i}{\hbar}\int \psi^*[OX-XO]\psi\, dX \qquad (10)$$

From the form of O(H) we see from Eq. (10) that if X commutes with H then X will commute with O(H) since O(H) represents an infinite expansion in H of odd powers of H.
Thus if XH - HX = 0, multiplying on the right by H

$$XH^2 - HXH = XH^2 - H^2X \qquad (11)$$

(after using Eq. 11) and so on

$$XH^3 - H^3X = 0$$

$$XH^4 - H^4X = 0$$

Thus at least for the simple case of one dimension we have a conserved probability density and a condition for the stationary nature of an eigenstate (XH - HX = 0).

The first application of the above theory is to electron spin polarization precession in a z component B field, we write

$$H = \frac{e}{m}(S_z B) \qquad (12)$$

for the hamiltonian and the wave function we write in a separated form as (Ref. 18)

$$\psi = \begin{pmatrix} a_1 \\ a_2 \end{pmatrix} T(t) \tag{13}$$

Looking for eigenstates, the discrete time difference Quantum Mechanical wave equation is (Ref. 18)

$$\frac{e}{m}(S_z B)\begin{pmatrix} a_1 \\ a_2 \end{pmatrix} T(t) = E\begin{pmatrix} a_1 \\ a_2 \end{pmatrix} T(t) = i\hbar \left[\frac{T\left(t+\frac{\tau}{2}\right) - T\left(t-\frac{\tau}{2}\right)}{\tau} \right]\begin{pmatrix} a_1 \\ a_2 \end{pmatrix}$$

The eigenvalues are

$$E_{\pm} = \pm \frac{e\hbar B}{2m}$$

with a mixed state with $<S_x>_{t=0} = \dfrac{\hbar}{2}$ represented by

$$\psi = \begin{pmatrix} \dfrac{1}{\sqrt{2}} e^{-\frac{2}{\tau} \sinh^{-1}\left(\frac{E_+ \tau}{2\hbar}\right)it} \\[2em] \dfrac{1}{\sqrt{2}} e^{-\frac{2}{\tau} \sin^{-1}\left(\frac{E_- \tau}{2\hbar}\right)it} \end{pmatrix}$$

If we evaluate $<S_y>$ at t we have

$$<S_y> = \psi^+ S_y \psi = \frac{\hbar}{2}\sin\left[\frac{4}{\tau}\sin^{-1}\left(\frac{E\tau}{2\hbar}\right)t \right]$$

where $E = E_+ = -E_-$. The precession frequency is

$$\omega = \frac{4}{\tau}\sin^{-1}\left(\frac{E\tau}{2\hbar}\right) = \frac{eB}{m} + \frac{e^3 B^3}{96 m^3}\tau^2 \tag{14}$$

Thus any anomalous dependence of ω on B^3 would signal the possible effects of discrete time difference Quantum theory. If Eq. (14) is applied to the charged lepton spectrum of e^-, u^-, τ^- and we think of the charged leptons as composed of n_i preons, (i labeled the generation that the lepton is in) then if τ emerges from an average of a sub-microscopic random variable the "central limit theorem for random variables would suggest that each preon is subject to a fundamental discrete time interval τ_0. Thus for n_i preons we would have $\tau_i = n_i \tau_0$ (Ref. 23). If τ_0 is the width of the random variable distribution then $\tau_i = \sqrt{n}\tau_0$. Then Eq. (14) would read

$$\omega_i \approx \frac{eB}{m_i} + \frac{e^3 B^3}{96 m_i^3} n_i^2 \tau_0^2 \quad or \quad \omega_i \approx \frac{eB}{m_i} + \frac{e^3 B^3}{96 m_i^3} n_i \tau_0^2 \tag{15}$$

The same idea has been applied to gauge boson spin polarization precession (Ref. 19) where a model of a gauge boson is given in terms of two spin 1/2 fermions[25] bound together by a spin-spin interaction of a Q.C.D. (or hypercolor) like interaction. The result of that analysis gives the following for the x spin polarization expectation value

$$\langle S_x \rangle = \frac{\hbar}{2}\cos\omega_1 t + \frac{\hbar}{2}\cos\omega_2 t \tag{16}$$

$$\omega_1 = a_1 - a_3, \quad \omega_2 = a_3 - a_2$$

$$a_1 = \frac{2}{\tau}\sin^{-1}\left(\frac{E_+ \tau}{2\hbar}\right) \;(S_z = 1), \qquad\qquad a_2 = \frac{2}{\tau}\sin^{-1}\left(\frac{E_- \tau}{2\hbar}\right) \;(S_z = -1),$$

$$a_3 = \frac{2}{\tau}\sin^{-1}\left(\frac{E_0 \tau}{2\hbar}\right) \;(S_z = 0),$$

$$E_+ = M_0 C^2 + \frac{n_1^2 h^2}{8mL^2} + \frac{n_2^2 h^2}{8mL^2} + \frac{e\hbar B}{2m} + \frac{g\hbar^2}{4}$$

here $M_0 C^2$ = rest mass parameter; m = heavy preon mass (m = $M_\omega/2$); M_ω = mass of gauge boson

$$E_- = M_0 C^2 + \frac{n_1^2 h^2}{8mL^2} + \frac{n_2^2 h^2}{8mL^2} - \frac{e\hbar B}{2m} + \frac{g\hbar^2}{4}$$

$$E_0 = M_0 C^2 + \frac{n_1^2 h^2}{8mL^2} + \frac{n_2^2 h^2}{8mL^2} + \frac{g\hbar^2}{4}$$

L = confinement scale of heavy preon
n_1, n_2 = spatial quantum numbers of heavy preons in square-well like potential.
g = spin-spin coupling constant; B = external magnetic field.

The experimental signature for discrete time effects would be two different sinusoidal terms in the x spin polarization with frequencies give by ω_1, ω_2. For small $\tau \to 0$ Eq. (16) becomes

$$\langle S_x \rangle \to \hbar\cos\left(\frac{eB}{M_\omega}\right)t \tag{17}$$

here m = $M_w/2$ is enforced to obtain the proper limit for normal Quantum Mechanics. This would also be a test for the composite structure of gauge bosons and depending on the model for the gauge boson structure different frequencies ω_1, ω_2 could be obtained.

The other experimental probes for a discrete time difference quantum theory are in spectral shifts of both conventional system (atoms, nuclei, etc.) and in spin flips of elementary particles in an external magnetic field when the internal quantum numbers show up because of the non-linear relation between the spin flip frequency and the energy difference between the final and initial state. When the usual formalism of time-dependent perturbation-theory is applied to Eq. (1) I have shown in a rather detailed calculation[26] that the transition frequencies for the resonance absorption and emission are

$$\omega_A = \frac{2}{\tau}\sin^{-1}\left(\frac{E_F T}{2\hbar}\right) - \frac{2}{\tau}\sin^{-1}\left(\frac{E_I T}{2\hbar}\right) \tag{18}$$

(absorption)

$$\omega_E = \frac{2}{\tau}\sin^{-1}\left(\frac{E_I \tau}{2\hbar}\right) - \frac{2}{T}\sin^{-1}\left(\frac{E_F \tau}{2\hbar}\right) \tag{19}$$

(emission)

If we apply Eq. (19) to the transitions in the hydrogen atom we have (Ref. 20)

$$\omega \approx \frac{E_I - E_F}{\hbar} + \frac{2}{\tau}\left[\frac{1}{3!}\left(\frac{E_I \tau}{2\hbar}\right)^3 - \frac{1}{3!}\left(\frac{E_F \omega}{2\hbar}\right)^3\right] \tag{20}$$

Since discrete time effects first enter in power of τ^2 we may use the unperturbed energy levels of E_F, E_I in the formula (Eq. 20).

Thus

$$\Delta\omega_{DT} \approx \frac{2}{\tau}\left[\frac{1}{3!}\left(\frac{E_I \tau}{2\hbar}\right)^3 - \frac{1}{3!}\left(\frac{E_F \tau}{2\hbar}\right)^3\right]$$

When this is compared to the Lamb shift for $n = 3 \rightarrow n = 2$

$$\Delta\omega_{DT} \approx \Delta\omega_{\mathcal{L}} \approx (1057.2)(10^6)(2\pi)$$

we obtain $\tau \approx 1.25 \times 10^{-18}$ sec. (here $\Delta\omega_{DT}$ = discrete time correction, $\Delta\omega_{\mathcal{L}}$ = Lamb shift) where the Lamb shift is of the order of magnitude of the hyperfine correction. Thus a rather large discrete time interval would cause spectral shifts comparable to that of the Lamb shift or hyperfine corrections in a hydrogen atom. Since the uncertainty in the parameters of nuclear theory is larger than that of atomic physics, atomic spectral shifts would provide a better window for discrete time effects. I have applied the same idea to spin-flip transitions of a particle in an external magnetic field when the particle itself contains internal structure, with preons carrying internal quantum numbers moving in the normal three space in extra dimensions (Ref. 22). The possible internal magnetic structure of an elementary particle has also been probed in this way with corrections due to discrete time quantum Mechanics being dependent on internal quantum numbers of the dyon-electric charge system that may or may not change in the spin-flip (Ref. 21). The use of this probe to internal structure as well as discrete time effects necessitates huge magnetic fields ($B \approx 10^{12}$ gauss) which can only be found in an astrophysical setting in the field of a pulsar. If Eq. (1) is applied to electron-spin resonance absorption[27] once again huge magnetic fields are required to generate observable effects. In Table I, I quote various experimental probes to discrete time difference Quantum Mechanics that may be used as upper limits to the discrete time interval when the experiment quoted in the relevant reference is performed:

CONCLUSION

From the discussion above, it is clear that the precise form that a discrete time difference quantum theory takes on is uncertain at this point. Nonetheless the existence of an "uncertainty principle at a sub-quantum level" arising because of the uncertain turning between the particle's

Table 1

Experiment[20, 28, 29]	Upper Limit Of Discrete Time Interval to be in Accord with Experiment at Left
Comparing Lamb shift in H with Discrete Time shifts in transition frequency [20]	1.25×10^{-18} sec
1% shift [28] in 1st Diffraction Minimum using Axion (10^{-10} GeV)	3×10^{-16} sec
1% shift [28] in 1st Diffraction Minimum using Electrons ($.5 \times 10^{-3}$ GeV)	4×10^{-23} sec
1% shift [28] in 1st Diffraction Minimum using Meson (.15 GeV)	2×10^{-25} sec
1% shift [28] in 1st Diffraction Minimum using Proton (.968 GeV)	3.3×10^{-26} sec
1% shift [28] in 1st Diffraction Minimum using Gauge Boson (80 GeV)	3.6×10^{-28} sec
$\Delta\omega/\omega = 10^{-4}$, discrete time correction to spin polarization precession frequency of electron in field of B = 10^{4} gauss (Ref. [29])	6.2×10^{-14} sec
$\Delta\omega/\omega = 10^{-4}$, discrete time correction to spin polarization precession frequency of electron in field of B = 10^{12} gauss (Ref. [29])	6.2×10^{-22} sec
Comparing corrections to ω induced by anomalous corrections to g of elecgron brought about by supersymmetry, compositeness, G.U.T. theory and technicolor with discrete time corrections to ω at 10^{4} gauss (Ref. [29])	1.3×10^{-15} sec
Comparing corrections to ω nduced by anomalous corrections to g of electron brought about by supersymmetry, composteness, G.U.T. theory and technicolor with discrete time corrections to ω at 10^{8} gauss (Ref. [29])	1.3×10^{-19} sec

sense of time (a microuniverse) and the statistically averaged sense of time (Frame of synchronous observers) remains an attractive and realistic possibility. Though this discussion borders on the field of pre-geometry it provides us with a gentle reminder that pre-geometric notions and foundational ideas for how space and time themselves were born might give rise to corrections in quantum based experiments that compete with purely quantum based corrections based on conventional Quantum Mechanics. If only a first step in trying to give a mathematical formulation to the above pre-geometric idea I feel that these investigations might be suggestive to future investigators who are inclined to think alike.

ACKNOWLEDGEMENT

I'd like to thank the Physics Departments at Williams College and Harvard University for the use of their facilities, and R. M. Santilli and Dave Park for their helpful comments and F. Selleri for asking me to contribute to this conference.

REFERENCES

1. A. Einstein, B. Padolsky and N. Rosen. *Phys. Rev.* 47:777 (1955).
2. L.E. Ballentine. *Rev. of Mod. Phys.* 42:358 (1970).
3. H.P. Stapp. *Am. J. of Phys.* 40:1098 (1972).
4. F.J. Belinfamte, "A Survey of Hidden Variable Theory," Pergamon, Oxford (1973).
5. C. Philippides, C. Dewdney and B. J. Hiley. *Il Nuovo Cimnento* 52:15 (1979).
6. J. Wheeler, *Chap. in Quantum Theory and Gravitation*, proc. of a symposium held at Loyola Univ. New Orleans, May 23 - 26, 1979, Academic, NY (1980).
7. D. Finkelstein. *Int. J. of Theoretical Phys.* 27:473 (1985).
8. L. Bombelli, J. Lee, D. Meyer and R. Sorkin. Preprint IASS NS-HEP-87/23 - Inst. of Adv. Study, Princeton, NJ (1987).
9. E. Recami, personal communication at Fourth Workshop on Hydronic Mechanics and Non-potential Interaction, 23-26 Aug. 1988, Skopje, Yugoslavia (Nova Science Pub., NY, 1990).
10. W. Wootters. *Int. J. of Theoretical Phys.* 23:701 (1984).
11. C. Wolf. *Hadronic J.* 15:321 (1992).
12. P. Caldirola. *Suppl Nuovo Cimento* 3:297 (1956).
13. P. Caldirola. *Lett. Nuovo Cimento* 16:151 (1976).
14. R. Mignami, H. C. Myung and R. M. Santilli. *Hadronic J.* 6:1973 (1983).
15. A. Jannussis, G. Brodimas, V. Papatheou, G. Karayannis, P. Panagopoulos and H. Ioannidou. *Hadronic J.* 6:1434 (1983).
16. J. V. Kadeisvili, *Agebras, Groups and Geometry* 9:283 and 9:319 (1992).
17. D. F. Lopez, Preprint Inst. of Basic Research (IBR-TH-622) July 20, 1993.
18. C. Wolf. *Il Nuovo Cimento Note Breve* 100B:431 (1987).
19. C. Wolf, to appear in *Annales de la Foundation Louis de Broglie* (1993).
20. C. Wolf. *Eur. J. of Phys.* 10:197 (1989).
21. C. Wolf. *Annales de la Foundation Louis de Broglie* 15:487 (1990).
22. C. Wolf. *Hadronic Journal* 14:321 (1991).
23. C. Wolf. *Hadronic J.* 13:22 (1990).
24. J. S. Bell. *Com. Atomic and Molecular Phys.* 9:121 (1980).
25. D. Grosser, P. Falkensteiner and F. Schoberl. *Phys. Lett.* B153:179 (1985).
26. C. Wolf. Unpublished work (1989).
27. C. Wolf. *Phys. Lett.* A123:208 (1987).
28. C. Wolf. Submitted to *Hadronic Journal* (1992).
29. C. Wolf. Proc. of Fourth Workshop on Hadronic Mechanics and Non-Potential Interactions, Aug. 22-26, 1988, Skojpe, Yugoslavia (Nova Sci. Pub., NY 1990), p. 298.

HERACLITUS' VISION - SCHRÖDINGER's VERSION

Pitter Gräff

Harkirchnerstr. 22
82335 Berg, Germany
Plenum Press
New York, NY 10013

A preliminary result: The existence of nature

In ancient times, when society started to define itself, there arose a need for law and order. Some of the earliest figures outstanding in this respect are still famous today: Hamurabi, Moses, Solon, Lycurgus. Furthermore, they also defined what a "law" is. One of its purposes is to act against corruption.

Is corruption natural? The cynical point of view is that it is. But the Greeks seem to be the first to have concluded instead: No! They found that there are laws which cannot be corrupted at any price. They were referring to the laws of nature. In a certain sense they discovered "nature" itself this way.

Concepts

If there are laws, nature can be thought. One just needs a starting point: a concept. Several have been offered, the most famous being perhaps that of Democritos: atoms in an empty space. Another one is due to Heraclitus (see below). But was there any way to find out the "true" one? Today we believe the answer is experiment.

Heraclitus

Unfortunately, we do not know very much about Heraclitus beyond "War is the father of all things" and "Everything is in flux". Otherwise he was considered as "obscure". - But this was also the fate of Nils Bohr's "correspondence principle" in Arnold Sommerfeld's view. And also Schrödinger (1954) - in a booklet on Greek philosophy - disregarded Heraclitus (perhaps influenced by his friend A. von Mörl,1948).

The idea of flow implies relativity of the motion of any part of a fluid. This is very much in accordance with Mach's ideas. Let us now consider different types of possible flows.

Frontiers of Fundamental Physics, Edited by M. Barone
and F. Selleri, Plenum Press, New York, 1994

Turbulent diffusion

Consider a fixed point in space $\mathbf{x}$, and velocity of a fluid, there as $\mathbf{u}(\mathbf{x},t)$. A "test particle" will be swept away according to

$$\frac{d\mathbf{x}}{dt} = \mathbf{u}(t, \mathbf{x}(t)). \tag{1}$$

This is the equation of motion of turbulent diffusion, Gräff (1987). We may call it Heraclitus' equation of motion. Compared with Newton's it is of first order in time. The discrepancy for the initial conditions in both concepts is repared by a Hamilton-Jacobi equation (see below).

If the velocity field is incompressible and free of vortices, it will be completely determined by its boundary conditions. A famous result at that level is the Jukowski profile of a wing.

Navier-Stokes

The acceleration $\mathbf{a}(\mathbf{x}, t)$ is again a vector field and hence can be split (Helmholtz) into a potential and a vortex part:

$$\mathbf{a} = \frac{d\mathbf{u}}{dt} = \frac{\partial \mathbf{u}}{\partial t} + (\mathbf{u}\nabla)\mathbf{u} = -\nabla P + \mathrm{rot}\ \mathbf{C},$$

where P is essentially the free energy. For the incompressible case (div $\mathbf{u} = 0$) P is proportional to the well-known pressure and is determined at any instant by

$$\Delta P = -\mathrm{div}((\mathbf{u}\nabla)\mathbf{u}).$$

The assumption $\mathbf{C} = 0$ yields Euler's equation, whereas $\mathbf{C} = -\nu\ \mathrm{rot}\ \mathbf{u}$ leads to the Navier-Stokes equation.

Maxwell's map

Let $\mathbf{G}(\mathbf{x}, t)$ be a set of 3 functions, not necessarily a vector, as the density, pressure and temperature of a gas: $\mathbf{G} = <n(\mathbf{x}, t), p(\mathbf{x}, t), T(\mathbf{x}, t)>$. With the definition

$$\mathbf{B} = -\mathrm{rot}\ \mathbf{G}, \qquad \mathbf{E} = \frac{\partial \mathbf{G}}{\partial t} + \nabla\phi,$$

where ϕ is still open we get

$$\begin{aligned}
\mathrm{div}\mathbf{B} &= 0, \\
\frac{\partial \mathbf{B}}{\partial t} &= -\mathrm{rot}\ \mathbf{E}
\end{aligned}$$

as a first set of Maxwellians. Defining furthermore the set

$$\rho = \mathrm{div}\ \mathbf{E}, \qquad \mathbf{j} = -\frac{\partial \mathbf{E}}{\partial t} + \gamma\ \mathrm{rot}\ \mathbf{B},$$

where $\gamma \sim c^2$ is a constant, we obtain a continuity equation

$$\frac{\partial \rho}{\partial t} + \mathrm{div}\ \mathbf{j} = 0$$

which tells that ρ need only be given initially. Let us call

$$\mathbf{G} \longrightarrow \mathbf{j}$$

the Maxwellian map, and $\mathbf{G}$ its generic functions.

The currents

So far the set of Maxwell equations is either a tautology or a pure definition. They do not provide laws of nature till we prescribe the currents $\mathbf{j}$. In the classical context, the current is assumed to be known explicitly $\mathbf{j}_0(t, \mathbf{x})$. More generally, we may assume $\mathbf{j} = \mathbf{f}(\mathbf{G})$ - see Barut (1980). Especially, $\mathbf{j} \sim \mathbf{G}$ yields vector bosons.

Instantanious solutions

Let

$$\mathbf{j} = -\frac{\partial \mathbf{E}}{\partial t}.$$

Then $\mathbf{B} = 0$ is allowed and hence rot $\mathbf{E} = 0$, which implies $\mathbf{E} = \nabla \Psi$ with

$$\nabla \Psi = \rho$$

This potential is caused by the instantaneous values of $\rho(\mathbf{x}, t)$ according to the ordinary Coulomb formula. No energy is transported with its temporary changes: $\mathbf{E} \times \mathbf{B} = 0$. This is reminiscent of the properties of the Bohm potential in quantum mechanics. Both fit together - till down to their paradoxes.

Dielectric and magnetic properties

They can be obtained by suitable feedbacks $\mathbf{j} = \mathbf{j}(\mathbf{G}, \mathbf{x})$.

A generalized Hamilton-Jacobi equation

A Helmholtz decomposition of $\mathbf{G} = \nabla S - \mathbf{A}$ with $\operatorname{div}\mathbf{A} = 0$ yields

$$\mathbf{B} = \operatorname{rot} \mathbf{A}, \quad \mathbf{E} = \nabla \left(\frac{\partial S}{\partial t} + \phi \right) - \frac{\partial \mathbf{A}}{\partial t}.$$

Call the bracket $-U(\mathbf{x}, t)$, then:

$$\frac{\partial S}{\partial t} + \phi + U = 0$$

with

$$\Delta U = -\rho$$

The freedom in ϕ and ρ allows a wide class of generic functions $\mathbf{G}$ of the Maxwellians - and also corresponding gauge-transformations.

Heraclitus-Hamilton-Jacobi equation

Assume our velocity field as proportional to a generic one:

$$\mathbf{u} = \frac{1}{m} \mathbf{G}$$

For the acceleration we thus obtain with Heraclitus equation (1)

$$\mathbf{a} = \frac{1}{m} \left\{ \frac{\partial \mathbf{G}}{\partial t} + \frac{1}{m}(\mathbf{G}\nabla)\mathbf{G} \right\}$$

From a wellknown relation of Clebsch one obtains

$$= \frac{1}{m}\left\{\frac{\partial \mathbf{G}}{\partial t} + \nabla\frac{G^2}{2m} + \mathbf{u} \times \mathbf{B}\right\}$$

$$= \frac{1}{m}\left\{\mathbf{E} + \nabla\left(\frac{G^2}{2m} - \phi\right) + \mathbf{u} \times \mathbf{B}\right\}$$

The special choice $\phi = (1/2m)G^2$ with equ. (1) yields the famous Lorentz formula

$$\frac{d^2\mathbf{x}}{dt^2} = \frac{1}{m}(\mathbf{E} + \frac{d\mathbf{x}}{dt} \times \mathbf{B})$$

and for the Hamilton-Jacobi-equation we get in particular

$$\frac{\partial S}{\partial t} + \frac{1}{2m}(\nabla S - \mathbf{A})^2 + U = 0$$

which shows the minimal coupling term. The above scheme applies for several types of fields: electromagnetic ones, inertial fields in rotating systems, gravitational fields (1. approximation to Einstein) and Bohm fields, see below, see Hughes(1992), Santilli(1978). For the Langrangian we find

$$L = \frac{dS}{dt} = -\frac{m}{2}u^2 - U + (\mathbf{u}\nabla)S$$
$$= \frac{m}{2}u^2 - U + \mathbf{u}\mathbf{A}$$

for the coupling to a Maxwell-type field. In particular $\mathbf{A} = 0$ and U arbitrary are allowed, which implies the Bohm-potential (h is Planck's constant $/2\pi$):

$$\mathbf{A}_B = 0, \quad U_B = -\frac{h^2}{2}\frac{\Delta\sqrt{n}}{\sqrt{n}}.$$

Schrödingers equation

Consider the continuity equation for our flow (= conservation of some density n):

$$\frac{\partial n}{\partial t} + \mathrm{div}(\mathbf{u} \cdot n) = 0$$

Introducing $\sigma = \ln n(\mathbf{x}, t)$ and assuming for simplicity $\mathbf{A} = 0$ and $m = 1$, we get

$$\frac{\partial \sigma}{\partial t} + \nabla S \nabla \sigma + \Delta S = 0 \tag{2}$$

and for the H-J equation

$$\frac{\partial S}{\partial t} + \frac{1}{2}(\nabla S)^2 + U = 0 \tag{3}$$

Take $(2)/2 + \alpha \cdot (3)$, where α is some dummy variable

$$\frac{\partial}{\partial t}\left(\alpha S + \frac{\sigma}{2}\right) + \frac{1}{2\alpha}\left(\nabla\left(\alpha S + \frac{\sigma}{2}\right)\right)^2 + \alpha U + \frac{1}{2}\Delta S - \frac{1}{8\alpha}(\nabla\sigma)^2 = 0$$

If the products of α make sense then a Ψ may be defined by

$$\Psi = e^{\alpha S + \sigma/2}$$

and we get:

$$\frac{1}{\alpha}\frac{\partial\Psi}{\partial t} + \frac{1}{2\alpha^2}\Delta\Psi + \left(U - \frac{1}{2\alpha^2}\frac{\Delta\sqrt{n}}{\sqrt{n}}\right)\Psi = 0.$$

Stability of the Bohr H-atom requires $\alpha = i/h$ and for the bracket in the last equation the Coulomb potential $\sim 1/r$. This is a point, where the experiment enters. Our H-J equation now has to be corrected by the Bohm-potential:

$$U = U_{Coulomb} + U_{Bohm}$$

Hence U is a superposition of two potentials with electric charge and Plancks constant as coupling parameters. The extension of the above results to arbitrary $\mathbf{A}$ and m is easy:

$$ih\frac{\partial \Psi}{\partial t} = \frac{1}{2m}\left(\frac{h}{i}\nabla - \mathbf{A}\right)^2 \Psi + U_{Coul}\Psi$$

is the Schödinger equation with minimal coupling.

Conclusions

Democritus concept seems more convenient from the Newtonian point of view. Heraclitus' concept requires for static solutions ($\mathbf{u} = 0$ or $U = const$) an imaginary value of α and hence fits more closely to Schrödinger's ideas without any "first quantization."

No invariance group (e.g. Lorentz) is determined before the currents are specified.

References

Barut, A. U. , 1980, "Electrodynamics and and classical theory of fields and particles" . Dover publications, New York.

Gräff, P. , 1987, Some asymptotic results in turbulent diffusion. *Z. Naturforsch.* **42 a**, 1079.

Hughes, R. J. , 1992, On Feynmans proof of the Maxwell equations. *Am. J. Phys.* **60**, 301.

von Mörl, A. , 1948, "Die große Weltordnung II" Paul Zsolnay Verlag, Berlin, Wien, Leipzig.

Santilli, R. M. , 1978, "Foundations of Theoretical Mechanics I." Springer Verlag, New York, Heidelberg, Berlin.

Schrödinger, E. , 1954, "Nature and the Greeks." Cambridge University Press, London.

IS IT POSSIBLE TO BELIEVE IN BOTH ORTHODOX QUANTUM THEORY AND HISTORY?

Euan J. Squires

Department of Mathematical Sciences
University of Durham
Durham City, DH1 3LE, England

THE PROBLEM OF HISTORY

In a recent conference at Oviedo in Spain, Roland Omnes[1] introduced us to an imaginary experimentalist who is describing an experiment he has performed (fig. 1): "A neutron, produced in a reactor, *passed through the slit S*, and proceeded to an interaction region..." What, asked Omnes, does the statement in italics mean? Rather, perhaps, we should ask what meaning can we give to it in order that it will be a correct statement? This is a question of history, which is very much a fashion in quantum theory these days[2,3,4,5], and which is appropriate in the wonderful surroundings in which we are privileged to be holding this conference. It is also an instructive thing to discuss in the context of trying to understand quantum theory. In such a context we often talk about the future, where we can ask what happens if we "do something" (e.g., measure), without being very clear about what it is that we do. In discussing the past, we do not have such a damaging[6] option. So, in this talk we shall ask:

Is the statement that the neutron went through S true?

Can we give it a meaning so as to make it true?

What are the implications of such a meaning for other ideas about history?

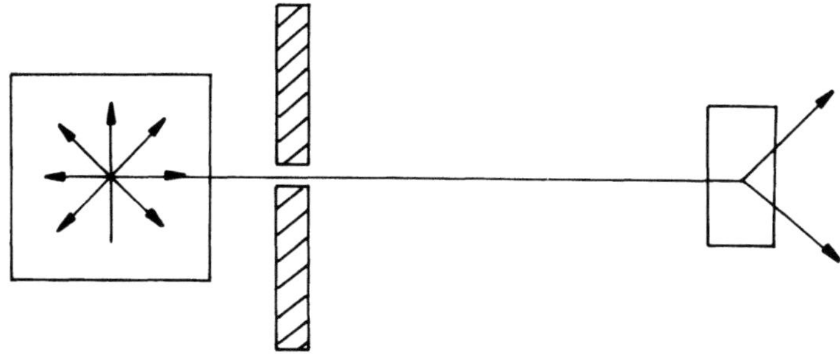

Figure 1. Showing how a neutron, produced in a reactor, passes through a slit, S, and is seen in an experiment, E.

First, let us suppose that the statement is, without any qualification, true, i.e., we suppose that it a simple fact of history that a neutron went through S. In this case then it seems reasonable to say that it was a fact when it happened. The claimed fact refers to the state of our world at a particular time, and whether or not it is a true fact surely cannot itself be something that depends on time. (Although politicians may try to "rewrite" history, I do not think even they would claim that in doing so they are actually changing it). Hence, at some time, say, t_1, the neutron passed through S.

The problem is now immediate. At a time t_2 after t_1, the experimentalist could have changed his mind and performed a different experiment, e.g., the experiment E' of fig.2. Then he would have observed interference, and he certainly would not have stated that the neutron went through the slit. (It does not matter greatly which text-book on quantum theory he used as a student; all would agree that interference means that the neutron did not go along one particular route).

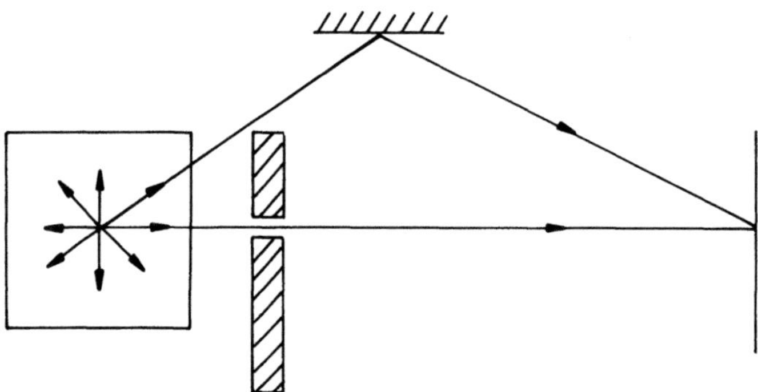

Figure 2. Showing how the same neutron as considered in fig.1 could have been used in a different experiment, E'.

Can we therefore conclude that the naive statement that the particle went through the slit is false? Not necessarily - there are at least two ways in which it might be saved. One is to use the deBroglie-Bohm pilot-wave model, in which neutrons (etc.) at all times move along trajectories. The different behaviour in the two experiments of figs.1 and 2 is accounted for by the presence of the non-local "quantum potential". Now this is a perfectly good solution to our problem (sometimes I believe it to be the only one that makes sense) but it is not the topic of this talk (recall the word "orthodox" in the title) so it will be ignored from hereon.

An alternative possibility for saving the naive statement is to use "superdeterminism": the world is totally determined and hence the experimentalist was not free to decide to change his experiment. The issue here goes beyond that of the conscious experience of free-will of course. It is necessary to say that the decision of which experiment, E or E', to do could not be decided by a "random" output of some other quantum experiment. Quite apart from the fact that using determinism in this way seems to destroy the whole idea of causality, and hence makes physics impossible, it is clear that it again implies the existence of some hidden information beyond orthodoxy, so is also ruled out by our title.

Thus, we have to conclude that the experimentalist was wrong to assert that the neutron went through the slit. Omnes reached a similar conclusion[1,4], but was prepared to allow the statement to be "reliable", where this is distinguished from being "true". Griffiths has discussed similar issues[2], and has suggested that the statement might be true within one "logic", but not another - somehow this allows him to say that it is true if we are talking about the experiment in fig. 1, but not about that in fig. 2.

For further discussion of these ideas I refer to the original references. (Later we shall see how, suitably interpreted, both can be regarded as correct). Here we shall adopt a different approach, following refs.[7,8] We shall accept that the naive statement is wrong. We need not be surprised about this; orthodox quantum theory, as John Bell reminded us[9], does not have particles with positions, so why should it have something passing through a slit? Instead of speculating what statement we can make about history, we shall *calculate* (that, after all, is what we are supposed to do as physicists). We shall first need to *define* history, and then we shall be able to see whether it is possible to give a suitable meaning to the experimentalist's statement so that it will be true. We do the calculation in the following steps, all of which will be explained below:

(i) Define an ontology.

(ii) Define the present.

(iii) Extrapolate backwards.

(iv) Add other branches.

Note that, since our aim here is to *define* history, we are not too concerned with whether the steps can actually be accomplished in practice.

THE ONTOLOGY

History is about *what was*. This is a *what is* type of question. As such it concerns ontology, so we must begin here. (Note again the contrast with the usual quantum discussions about the future, where we can avoid asking about *what is* and instead ask what happens if....). Now the ontology of orthodox quantum theory is clear: there are no hidden variables, so the only existing reality is the wavefunction. Sometimes it seems as though the orthodox wish to deny even this, but I am not aware of any alternative.

Having settled on this fact, there is no need to worry about which wavefunction; there is only one, namely, the wavefunction of the universe. All others that we use in our local calculations are some sort of approximation in which correlations are ignored. What is more, if we accept general relativity (in particular, general coordinate invariance), then this wavefunction satisfies the Wheeler-deWitt equation,

$$i\hbar\frac{\partial\Psi}{\partial t} = H\Psi = 0, \tag{1}$$

from which we see that the wavefunction of the universe is independent of time. Note that here we are again using orthodoxy, this time to forbid any non-Schrödinger evolution of the wavefunction, such as would for example cause it to collapse.

Thus, in orthodox quantum theory, it would seem that the complete description of the present, and of the past, and of the future, is a constant wavefunction Ψ. Clearly, though this may be true, it is not interesting. What is missing is the facts of the present world: me, my experiences, history books, etc. To introduce these things is the second step.

THE PRESENT

From the wavefunction of the universe, it is presumably possible to calculate the probability of, for example, my existing. The answer cannot be zero, though it could be very small. However, regardless of how small it is, it is only that part of the wavefunction in which I do exist that is of interest to me. Thus I am led to define "my world" not as Ψ, but rather as

$$\Psi_0^{ME} = P^{ME}\Psi, \tag{2}$$

where the suffix refers to the time, and $t = 0$ has been chosen as "now", and where P^{ME} is a suitable projection operator which in some way projects out the world that I know.

How do we define P^{ME}? It is tempting to suggest that we should put something like

$$P^{ME} = |\psi><\psi|, \tag{3}$$

where $|\psi>$ is the "state of my world". However, this does not mean anything; there is only one state that exists, and that is Ψ. "My world" does not have a state. Instead, I propose[7,8] a "minimal" assumption in the following way. I am aware of certain things, e.g., of my existence, of the desk in front of me, of the present time (position of hands of a clock in my room), etc. Now presumably my awareness is related to certain brain states, in the sense that without these brain states, then the awareness would not exist. (This is some sort of psycho-physical parallelism, and does not require any assertion that the brain states and awareness are identical- which they are not). Then I can define P^{ME} as the operator that projects onto that part of configuration space that contains these brain states. This is reasonably well-defined in that, given adequate knowledge of what brain states correspond to particular awareness, we could actually *calculate* it.

With this definition of P^{ME}, we can then form Ψ^{ME} as in eq. (2). A fascinating question immediately arises. We know that this state contains a brain that is aware of certain things. Does it also have a high probability of actually having those things? For example, I might be aware that X exists. Thus, P^{ME} will project onto a brain state that has the awareness of X. Does this imply that in Ψ^{ME}, X *actually* exists. In mathematical terms, is it true that

$$\frac{||P_X P_X^{ME}\Psi||^2}{||P_X^{ME}\Psi||^2} \approx 1, \tag{4}$$

where P_X^{ME} projects onto that part of configuration space containing a brain that is aware of X, and P_X projects onto that part of configuration space that contains X. This question is of course related to the old philosphical debate between realism and solipsism, i.e., to what extent is there an external reality corresponding to our observations, or are the experiences all that there is. However, it appears here in a different form. We are not denying a real world, but asking about its properties. And the question has an answer that, in principle, can be calculated. The left-hand side of eq.4 has a value that is a property of the real world. Of course, most of us naturally accept that our experiences of X are caused by the fact of X, so we can ask what is the property of the solution of the Wheeler deWitt equation that ensures that this belief is justified? This, however, is probably not a good way to put the question; rather we should perhaps say that a given solution Ψ determines the brain states for

which the belief would be true, and that (for evolutionary reasons) these are the ones
that correspond to awareness.

There are other questions we can ask about Ψ_0^{ME}. How much does it depend
upon my knowledge? Since I know about the existence of galaxies, it presumably has
a high probability of having galaxies, but would this still be true if instead of my
brain we used that of someone who was not aware of such things. We should expect
(or perhaps I should say here, "hope") that the answer is yes, because it is part of our
current understanding of cosmology that heavy nuclei (surely necessary for "brains"
of any sort) can only have been made in galaxies. However, these things depend on
the way things have changed with time, so this notion must now be introduced. It is
the topic of the next section.

BACKWARD EVOLUTION

Having defined "the present" to be the wavefunction Ψ_0^{ME}, we have an obvious
suggestion for "the past", namely,

$$\Psi_t^{ME} = \exp(-iHt)\Psi_0^{ME}, \tag{5}$$

where H is, presumably, the Wheeler-deWitt hamiltonian used in eq. 1. This wavefunction *is* my world at times t earlier than the present - it *defines* what I mean by
history (with, however, an important qualification to be discussed in the next section)

Notice that a genuine time dependence has reappeared because H will not in
general commute with P^{ME}. The basic philosphy here is that there is such a thing as
"external time" but that the wavefunction of the universe does not depend upon it.
This does not imply that everything is static because any "part" of Ψ, obtained by
projection as in eq. 2, will, in general, vary with time. Of course the actual parameter
that is used as time, and correspondingly, the form of the hamiltonian, will depend
on some initial foliation of the space-time manifold.

It is easy to see[10,11] that under suitable circumstances eq. 5, gives the expected
classical evolution as a good approximation. In other words it corresponds very closely
to what we actually *do* when deducing facts about history.

Before proceeding we note, however, that there are two alternative methods of
defining history which are worthy of consideration. In both of these we begin by
denying the existence of an external time. What I regard as the universe "now" is
the universe projected onto some particular state of an object which will play the role
of a "clock". The radius of the universe is sometimes given as an example here. Since
there is a real clock in the room in which I am now writing, I could perhaps rather
regard the positions of its hands (it is an old-fashioned clock) as suitable variables.
Then, instead of eq. 2, I would define my world at now by the equation:

$$\Psi_0^{ME} = P^{ME}P_{a_0}\Psi, \tag{6}$$

where a_0 is the present value of the clock variable(s), and P_{a_0} projects onto the world
containing the clock at this value. Then the world at an earlier "time", *defined* by
different clock variable(s) a_t, would be given by

$$\Psi_t^{ME} = P_{a_t}P^{ME}P_{a_0}\Psi. \tag{7}$$

Provided the solution of the Wheeler-deWitt equation has a suitable form (semi-
classical), the relation between Ψ_t^{ME} and Ψ_0^{ME} can be written in a manner similar to

eq. 5, so this alternative probably gives essentially the same physics (see Halliwell[12] for further details). There are interesting problems which arise if we need to use a relation between the parameter and the quantity we wish to regard as "time", which is not single valued. The obvious example here[13,14] would be the radius of the universe in a situation corresponding to a (classical) recollapsing cosmology. Then the projection operator P_a would give a wavefunction that is non-zero over two disjoint regions of configuration space, one corresponding to the expanding and one to the contracting phase. Some additional projector would therefore be required in order to give a unique specification of " my world".

A very different possibility (at least at first sight) is to suppose that projections like that in eq. 2 project onto (or close to) a "ridge" i.e., a region of configuration space where the wavefunction is large, and falls away rapidly in all directions except one. Then we could suppose that movement along the ridge corresponds to what we call movement in time. This might give the appearence of a classical world provided the ridge remains sufficiently sharply "peaked", so that for example classical objects have well-defined positions. Of course it is by no means obvious that the solution of the Wheeler-deWitt equation will have this form (even ridges on a two dimensional space, e.g., real mountains, eventually fade into plains).

We shall not use either of these two possible alternatives here, particularly as they seem to require rather special solutions to the Wheeler-deWitt equation, but instead adopt eq.5 as our definition of history. With such a definition we emphasise that a statement about "what happened" is a statement about a wavefunction. It is not, however, a statement about the wavefunction of the universe (which is rather dull) but about what I might call the wavefunction of "my universe", or, in the language of Many-Worlds quantum theory, the wavefunction of my branch of the universe.

Let us then return to the Omnes problem posed at the beginning. The experimentalist will be aware of the existence of his reactor and of the arrangements of slits, shielding, etc., so when he uses unitary evolution to extrapolate backwards he will find that most of the neutron wavefunction did actually pass through the slit S. In mathematical terms the wavefunction $\Psi_{t_1}^{EXP}$, defined analogously to eq.5, but with the experimentalist's awareness replacing mine, has a high probability of having the neutron in the interaction region of the experiment at time t_1. Hence,

$$||P_s \exp(iH(t_1 - t_2)\Psi_{t_1}^{EXP}|| \approx 1, \tag{8}$$

where P_s is the projection operator corresponding to the neutron in the slit, and $t_2 < t_1$ is the time when the neutron wavepacket reaches the slit. Thus the experimentalist's statement that the neutron passed through the slit is true (to a very good approximation with proper shielding), provided it refers only to the part of the wavefunction that contributes to his present world. What it means is that the part of the wavefunction that did not pass through the slit makes a negligible contribution to the world he has experienced, i.e., to his branch. Notice that this understanding of the meaning of the statement makes no use of any non-standard rules of logic, or of any reservations about what true statements are. Nevertheless, we could relate it to the ideas of Omnes, referred to above, if we regarded a "reliable" statement as one that is true about "my" world, in the sense defined above. Similarly we could interpret the different "logics" of Griffiths as being statements about different worlds. It by-passes the problem about what happens if the experimentalist did a different experiment, e.g., E' of fig.2, because then we would refer to a different part of the wavefunction and would expect to obtain a different answer. To put this another way, we can say that at the time when the neutron has its highest probability of being inside the slit, it has a probability distribution which is roughly peaked in some spherical shell around its point of production. The experiment E, however, only uses the part of this distribution that is inside the slit; another experiment would generally use a different part.

I believe this is the only satisfactory answer that can be given to the Omnes question within the confines of orthodox quantum theory. In going from a statement about a wavefunction, to a more classical-like statement about the path of a particle, we have gone from something that is exact to something that is approximate. However, it is only when the "error" is extremely small that the classical history would be used. This is indeed the situation in cases when we speak of classical histories.

So far our definition of history appears to be satisfactory. Unfortunately, as we shall see in the next section, it meets a serious problem.

THE NEED FOR OTHER BRANCHES

We continue to think about the neutron experiment, but we now extrapolate further backwards in time, in order to define the earlier history of our neutron. Eventually we reach the stage, inside the reactor, when it was emitted in some nuclear reaction. Now our neutron, which we recall passed through the slit, has a direction confined to a quite small solid angle. This means that it is made up of a set of carefully correlated angular momentum states. As we extrapolate backwards in time this requires similar almost exact correlations in the particles which participated in the collision giving rise to the nuclear reaction that produced the neutron. We have defined our history to contain a very special "initial condition", that causes a nuclear reaction to emit a neutron in a well-defined direction. But surely this is wrong! If we asked our experimentalist to say what is happening inside the reactor, he would tell us there are random collisions initiating nuclear reactions that emit neutrons in specific angular momentum states, i.e., spreading out over the whole solid angle.

It is easy to see why our definition gave the "wrong" answer: we failed to use all the neutron wave. To have obtained the "right" answer we would have needed to include the other branches, i.e., that part of the neutron wavefunction we threw away in defining "our world". This is the reason why it is often stated, e.g., ref [15], that unitary evolution according to the Schrödinger equation should not be used to go to earlier times. However, I am not aware of any satisfactory alternative way even to *define* history within orthodox quantum theory. The dilemma is now evident. How do we decide what to include and what to omit in deciding what is the history of our world? The part of the neutron wavefunction that did not pass through the slit is irrelevant to the experiment we decided to do, but nevertheless our prejudice tells us that the history we get by including it at the stage of the nuclear reaction is more reasonable than the one we get by excluding it. Should we then go to the other extreme and include all branches? Clearly not, because as we have seen that leads us to the wavefunction of the universe, which is a constant and gives no history at all. Of course, in simple cases the "answer" seems obvious. But why? Why is one history more "correct" than another? Is there some mathematical rule for telling us what to include to define our history? In the absence of such a rule, might it not be possible to select anything we like as a valid history?

As we saw in the example above, we were able to remove the need for an apparent "fine-tuning" of some initial condition, by including extra branches, so it is an obvious suggestion that many of the apparent "coincidences" that seem to be necessary for the existence of life, say, are only artifacts that arise from our failure to include all the branches. Indeed we might ask whether all of them can be removed in this way.

COSMOLOGY

Having given a definition of history, we are now in a position to use it to study the history of our universe - the subject of cosmology. In particular, it is of interest to see how some of the arguments of standard big-bang cosmology now appear. For example, it is usually argued that a big universe, containing many galaxies, is essential for the existence of creatures like ourselves because it is only in exploding galaxies that heavy nuclei can be formed (very few would be formed at the time when hydrogen was fusing to give helium). Thus we might expect that Ψ_0^{ME} has a high probability of having galaxies, even if I was not aware that galaxies existed. This would imply that

$$\frac{||P_{NG}\Psi_0^{ME}||}{||\Psi_0^{ME}||} << 1, \tag{9}$$

where P_{NG} projects onto a universe with no galaxies. Somehow we expect this fact to be a consequence of the solution of the Wheeler-deWitt equation - certainly, *in principle*, the left-hand side of this equation could be calculated if we knew the solution. In order to see if there are good reasons for expecting the result of this calculation to be a small number, we need to extrapolate backwards in time (since the classical argument relies on evolution since the time of the big bang). Let us suppose that the result of such an extrapolation does indeed take us back to a high temperature universe. Then the no-galaxy wavefunction defined by

$$\Psi_{NG}(t) = \exp(-iHt)P_{NG}\Psi_0^{ME} \tag{10}$$

would have large nuclei existing all the way back to the times when it becomes too hot for them to exist. There, they would be seen to have been created in highly special conditions (i.e., well away from statistical equilibrium, which the standard arguments tell us lead to a negligible number of large nuclei). It is precisely such conditions that are ruled out in the standard argument. The wavefunction with galaxies

$$\Psi_G(t) = \exp(-iHt)(1 - P_{NG})\Psi_0^{ME} \tag{11}$$

would, on the contrary, have few heavy nuclei at this stage, since they would have disappeared (we are running time backwards) in the centres of hot galaxies. Thus, for this wavefunction there would not be the need for such "special" conditions in the very high temperature universe. The "classical" argument looks very similar at this stage - we obtain it by the postulating that the initial conditions were close to thermal equilibrium; then the no-galaxy route could not work. In the quantum case both histories exist; the only way we can give one a preference over the other (as is required if we want to "understand" why we have galaxies) is to say that because $|\Psi_{NG}(t)|$ only differs from zero over a part of configuration space that is special, its integral over the whole of configuration space is "small". It is not at all clear, however, if such an argument is valid, or rather, what properties we are assuming for the solution of the Wheeler -deWitt equation when we use it. My wavefunction, eq.2, is surely already "special",so why should that in eq. 10 be so much more special that we can exclude it?

To give another illustration of the difficulties, we note that it seems reasonable to say the probability of the wavefunction containing (in the usual quantum sense) a

highly complex system is lower than the probability of its containing a less complex
system. Thus, for example, the probability that the wavefunction contains dinosaur
fossils should be much higher than the probability that it contains dinosaurs. This,
however, surely cannot be true: it would suggest that the existence of the fossils
cannot be used as evidence for the previous existence of dinosaurs (which would suit
some fundamentalists, but which most scientists would regard as surprising). Again
we can ask: what properties are we assuming for the wavefunction of the universe
when we use the traditional scientific arguments?

The issues raised in this section, and the previous one, show that the resolution of
classical problems related to the arrow of time and initial conditions of the universe,
must now be found in statements about the properties of solutions of the Wheeler
deWitt equation. We need to know what properties appear naturally, i.e., are true
for a wide class - defined in some way- of solutions, and what depend upon special
choices.

I am grateful to Julian Barbour, Bob Griffiths, Jonathan Halliwell, Claus Kiefer,
Roland Omnes and Dieter Zeh for discussions related to the topic of this talk.

REFERENCES

1. R.Omnes, to be published in the Proceedings of the Oviedo conference, Frontiers
of Fundamental Physics, ed. M.Ferrero, (1993).
2. R.B.Griffiths, J.Stat.Phys. 36, 219 (1984), and Phys. Rev., Letters, 70, 2201
(1993).
3. M.Gell-Mann and J.B.Hartle, in "Complexity, Entropy and the Physics of Information", ed. W.H.Zurek, Addison Wesley, New York, (1993).
4. R. Omnes, Rev. Mod. Phys., 64, 339 (1992).
5. A.Albrecht, Phys.Rev., D46, 5504 (1992).
6. J.S.Bell, in "62 Years of Uncertainty", ed. A.Miller, Plenum, New York (1989).
7. E.J.Squires, Found. of Phys. Letters, 5, 279 (1992).
8. E.J.Squires, in "E.Schrödinger, Philosophy and the Birth of Quantum Mechanics,
ed. M.Bitbol and O.Darrigol, Editions Frontieres, Gif-sur-Yvette (1993).
9. J.S.Bell, "Speakable and Unspeakable in Quantum Mechanics", Cambridge, p.187
(1987).
10. E.J.Squires, Phys. Letters, A155, 357 (1991).
11. P.D.B.Collins and E.J.Squires, Found. of Phys., 23, 913 (1993).
12. J.J.Halliwell, "Introductory lectures on quantum cosmology", in "Quantum Cosmology and Baby Universes", ed. T.Piran (1991).
13. H.D.Zeh, "Time (a-)symmetry in a recollapsing universe", in "Physical Origins of
Time Asymmetry", ed. J.J.Halliwell, J.Perez-Mercader and W.H.Zurek, CUP, Cambridge (1993).
14. C.Kiefer, "Quantum cosmology and the emergence of a classical world", in Proceedings of the FEST workshop, "Concepts of Space and Time", ed. E.Rudolph and
I.O.Stamatescu, Springer, Berlin (1993).
15. R.Penrose in "The Emperor's New Mind", O.U.P. Oxford, (1989), p.461.

A NEW LOGIC FOR QUANTUM MECHANICS ?

Eftichios Bitsakis

Department of Philosophy, University of Ioannina
Department of Physics, University of Athens

1.INTRODUCTION

The debate concerning the status of a logic adequate for Quantum Mechanics (Q.M.) continues, fifty years after the formulation of the so called "quantum logic" by Hans Reichenback[1]. The two dominant theses in this debate, are the following: 1) Quantum logic-a three valued logic- is the apparatus suitable for Q.M. 2) Classical, two valued logic is sufficient for Q.M. There is no quantum logic, but only a propositional calculus expressing the specific character of quantum mechanical phenomena[2]. However as I will try to show, a third answer to this question is not impossible.

Today, even specialists very often consider, Reichenbach, Weizsäcker, and other writers as the founders of the three-valued "quantum logic". However the first formulation of a three-valued logic is to be found in the pioneer work of the great polish logician J.Lukasiewicz. And if we go back in time, we will find the first intuitions concerning a non classical logic in the debate between the stoics and the epicurians, on the question of determinism and freedom.

In fact, Aristotelian two-valued logic (Aristotle's syllogistic), differs essentially from the stoic logic, which, according to Lukasiewicz, is the ancient prototype of modern propositional logic. However, both of them, by accepting the validity of the law of the exluded middle (LEM), constituted a solid foundation for the deterministic conception of the Universe. Epicurians, on the contrary, tried to transcend fatalism, by postulating a certain liberty to the atoms, and to the humans as well. (It is well known that the first hints concerning the possibility of a three-valued logic are to be found in Aristotles *De Interpretatione*, ch.9. , despite the fact that his name is associated only with two-valued logic).

It is not a matter of chance, if the contradiction between determinism and freedom constituted an essential motivation for the founder of the three-valued logic. As Lukasiewicz himself confesses: "I have always been most interested in the issue of determinism and indeterminism. I have associated it with the problem of many valued logics"[3] . In fact, Lukasiewicz was striving, since 1910, to construct a *non -Aristotelian logic**. Thus he focused his attention to the modal propositions and the categories of possibility and necessity connected with them. He tried, in particular, to refuse the law of bivalence, a fundamental law of the logic of identity, because for him bivalence is incompatible with freedom.

According to the Aristotelian tradition, there are four modes: possible, impossible, contingent and necessary. Studing possible but not necessary cases, Lukasiewicz arrived at considering propositions that were neither true nor false. These propositions, Lukasiewicz argues, must possess a third value, different from "O" or falsity and "1" or truth. He designated this value by "1/2". The new value represents "the possible" and joins "the true" and "the false" as a third value. The three-valued system of propositional logic, Lukasiewicz writes, owes its origin to this line of thought[4].

In that epoch (1918) the problem of indeterminism related to Q.M. had not been, evidently, posed. Yet Lukasiewicz, relating the validity of bivalence and of the law of the excluded middle to determinism, and the three valued logic to the categories of the contingent and the possible, anticipated the great debate concerning the validity of determinism in quantum physics. The three-valued calculus of Lukasiewicz opened the way for the introduction of the categories of possibility and potentiality in the interpretation of Q.M. and the elaboration of the so called quantum logic.

It is well known that the protagonists of "quantum logic" (of the propositional calculus related to Q.M.) rejected determinism together with the LEM. The realist school, on the contrary, strived to restore the LEM and with it a certain *dynamical* form of determination . As I will try to show, it is possible to transcend the antithesis between indeterminism and classical determinism, by introducing the concepts of possibility and potentiality in the interpretation of the formalism of Q.M. A new logic is then necessary transcending both: the formal two-valued logic and the three-valued propositional calculus.

2.CLASSICAL MECHANICS AND THE LIMITS OF THE LOGIC OF IDENTITY

Let us first consider the case of two-valued logic.

Classical Mechanics is founded, as is well known, on a number of ontological postulates:

- Matter is constituted of particles possessing mass as their unique attribute
- Particles exist in an infinite euclidean space, independently of matter. Likewise, time is considered absolute, independent of matter and motion.
- Particles interact by the mediation of forces transmitted with infinite valocity (the postulate of action-at-a-distance).
- Bodies conserve their state of motion, if no force acts upon them[5]. The galilean group of transformations is the formal expression of these postulates.

Classical physics accepts that energy is exchanged in a continuous way. Accordingly it is accepted that the disturbance of the system by the measuring apparatus can be made arbitrarely small (negligible). In conformity with this postulate it was considered possible to measure both, the position and the momentum of the particle and define its state as a point $P(p_i,q_i)$ in the phase space: Classical states are defined with a variance equal to zero. As a result of the preceding idealizations, classical propositions are considered as corresponding to Lebesgue-measurable sub-sets of the phase space. For every state there exists a class of propositions simultaneously true.

The compatibility of classical observables entails, on the other hand, the validity of the distributive law:

$$a \wedge (b \vee c) = (a \wedge b) \vee (a \wedge c) \quad \text{and} \quad a \vee (b \wedge c) = (a \vee b) \wedge (a \vee c)$$

The lattice of classical propositions is, therefore, a Boolean one. On the other hand, the Boolean structure is that of the formal, two-valued logic, and formal logic is the logic of *identity*. This is a consequence of the postulate that no new elements of reality are created as a result of the interaction of the system with the measuring apparatus.

The above idealizations are valid for *galilean particles*, that is to say for systems for which the observables of momentum, position and time are defined. The postulated conservation of the identity of such systems, entails the validity of the two-valued logic.

The (postulated) compatibility of the classical observables makes possible the measurement of $a \wedge b$. This is an essential difference with Q.M.: According to the inequalities of Heisenberg, it is impossible to measure two non-commuting observables. This is due to the fact that the measurement of one of them modifies the value of its conjugate observable. In the general case, therefore, the lattice of the quantum propositions is not a boolean one. The violation of the identity of the system by the measurement explains, on physical grounds, the fact that the distributive law is not valid in Q.M. and, also, that in its case, it is the syperposition principle that holds:

$$\Psi = \sum_i \lambda_i \Psi_i, \quad i = 1,2,\ldots,n$$

The validity of the superposition principle is a specific feature of Q.M. . However, superpositions of states exist in classical mechanics as well. But in its case they are superpositions of *actual states* and not superpositions of *potential* ones, *actualized* during the measurement, as it is the case of quantum superpositions[6]. (The Hilbert space corresponding to quantum superpositions is a space of potential states, a measure of the potentialities of the quantum ensemble under the given experimental conditions[7]).

Now, concerning the validity of the LEM: It is generally accepted that the inequalities of Heisenberg preclude the validity of the LEM in Q.M. I will contest this contention in part 4 of this paper, from more than one angle. However, there are also other phenomena, which seem to be incompatible as the validity of the LEM in Q.M. I will discuss them also in part 4, in the frame of the general question: the specific character of Q.M. makes necessary and possible a new logic, different from the formal, two-valued one? In particular: what is the status of the so called three-valued quantum logic?

3. THE ELABORATION OF THREE-VALUED LOGIC BY HANS REICHENBACH

The classical presuppositions for the validity of the LEM seem to be violated , in the general case, in Q.M. In fact: The inequalities of Heisenberg preclude the simultaneous measurement of non commuting observables. In that case, the LEM seems to fail. The famous two-slit experiment seems, in its turn, to violate the LEM. The postulated wave-particle duality constitutes a logical contradiction in the frame of formal logic. Finally, during quantized transitions, the LEM seems, once more, to fail. It is because the preceding features of Q.M. that eminent physicists and logicians arrived at the conclusion that two-valued logic is not valid in Q.M., and that a three-valued, non -classical logic, is neccessary in its realm.

It is impossible, even to outline here, the relevant debates and the resulting rich ideas and formalisms. I will, therefore, confine myself to a brief otuline of one of the first and more important attempts at elaborating "quantum logic": that of Hans Reichenbach. The presentation will be based on his two classical books: *The Philosophic Interpretation of Quantum Mechanics* (1944) and *The Direction of Time* (1956).

Observables, Reichenbach writes, correspond to macroscopic things. Statements about physical quantities of smaller scale, on the contrary, are derived by way of inferences based on macroscopic observables. Such phenomena may be considered as observable in a wider sense. Example: The coincidences between electrons, or electrons and photons. Elementary occurences of this kind are called *phenomena*[8]. Phenomena are not immediate objects of observation.

Phenomena are observable in the preceding wider sense. On the contrary, Reichenbach maintains, we shall consider as unobservable, all those occurences which happen between coincidences, such as the movement of an electron, or of a light ray from its source to a collision with matter. Reichenbach calls this class of occurences, *interphenomena*[9]. Inferences leading to them can be made only within the framework of quantum mechanics. It is evident that Reichenbach envisages here a limit of knowledge, in conformity with the completeness postulate of the Copenhagen Interpretation (CI).

These inferences, Reichenbach notes, presuppose a certain definition, or *extenstion rules* of language, which allow to extend the language of phenomena to that of interphenomena. Thus it would be possible to speak meaningfuly about occurences which cannot be observed, even when "observing" is meant in the wider sense. Being definitions, these rules are not true, nor false. However, we can ask what consequences arise from the use of a language constructed by these rules[10].

Let us try, for example, to assign, by any kind of rules, simultaneous values to non-commuting quantities, the distributions of which are governed by the Ψ function. Then, Reichnbach asserts, the resulting interphenomena are subject to *causal anomalies* [11]. Causal anomalies result whenever we use an exhaustive interpretation.

Let us define the above new concepts, by using an example. The term "particle" and the term "wave", both belong, according to Reichenbach, to interphenomena. They assert something about what happens between localized phenomena. Speaking of particles, we say that interphenomena are localized. Speaking of waves, we say that interphenomena, are spread over a wide area, although their final products, the phenomena, are localized. Both: the particle and the wave interpretation, are exhaustive [12].

Now, concerning the nature and the function of causal anomalies: Causal anomalies are, roughly speaking, "statements about quantum mechanical phenomena, which contradict classical physical laws for observable objects" (S. Haak). However it would be, in principle, possible to avoid causal anomalies. In fact, as Reichenbach maintains, "every exhaustive interpretation states too much, so far as it speaks of causal anomalies which have no bearing upon the world of phenomena". Causal anomalies have a ghostlike existence. Consequently they can always be banished from the part of the world in which we happen to be interested. But this is possible only if a three-valued logic is used. The three-valued language, Reichnbach, states, appears adequate to Q.M. "because the causal anomalies, as formulated in exhaustive interpretations, appear to be supperfluous complications" [13]. It is once more evident, that Reichenbach reduces our cognitive possibilities to quantum level, transforming any eventual deeper realities into an unknowable *thing in itself.*

In fact, the principle of anomaly, is related to Bohr's principle of complementarity, but it goes beyond that, in as much as it makes precise statements about unobservables in Q.M. . Consequently, in order to avoid causal anomalies Reichenbach confined his interpretation to the frame of the Copenhagen school:"In view of this result, one may decide not to speak of interphenomena, but to restrict the language of quantum mechanics to phenomena. Such a *restricted interpretation* can be carried out. Of this kind is the interpretaion developed by N. Bohr and W. Heisenberg, who regard statements about interphenomena as meaningless. An interpretation on terms of a three-valued logic which I have proposed, appears preferable, because it still permits the inlclusion of interphenomena, in some sense, in the language of physics" [14].

According to Reichenbach it would be possible to avoid causal anomalies by using a three-valued logic. In fact, he considered the statements about interphenomena not as meaningless, but as *indeterminate.* By this way, he succeeded in elaborating a "three-valued logic". From this point of view, Reichenbach does not accept the thesis of Bohr and Heisenberg. However, his general epistemological frame is that of the C.I. Thus, he considered the actual quantum mechanical description as *complete* and superfluous (or impossible?) any attempt to investigate the deeper causes of phenomena, or to assign any ontic status to entities or quantities of Q.M. with contradictory attributes.

As an example, let us consider the famous two-slit experiment. In order to avoid causal anomalies, Reichenbach rejected the LEM, that is to say, the possibility to describe this phenomenon in terms of space and time. In his own words: "The particle arrives on the screen. The wave is shallowed, so to speak, by the flash at C. This process of disappearance of the wave constitutes a *causal anomaly,* so far as it contradicts the laws established for observable occurences. We see that in this description the laws of interphenomena are different from the laws of phenomena; the given description, therefore, does not represent a normal system" [15]. The probability with which a particle passing through B, reaches A, depends on whether the slit is open. This is a causal anomaly [16].

In conformity with his empiristic epistemology, Reichenbach considered the actual formalism of Q.M. as complete and, consequently, the space-time description of the two-slit experiment, as impossible. More generally, the description D of a physical state given by the Ψ function, Reichenbach maintains, is the *most complete* description possible. He named this description, the *synoptic principle* of Q.M. If this principle is true, Heisenberg's indeterminacy is inescapable, because it is a mathematical consequence of this prnciple. But if the principle should not be true, the indeterminacy relations may be abandoned[17]. In that case determinism would be superimposed to Q.M. , but it would not be the determinism of classical physics. It would be a determinism in terms of action-at-a-distance and causal anomalies[18]. However, one could reasonably argue that the pre-relativistic action-at-a-distance is not an inevitability and that relativistic locality would be the frame of a new form of determination -the quantum statistical form of determinism. In that case, "causal anomalies" would be an inappropriate concept corresponding to processes taking eventually place in the sub-quantum level of organisatin of matter[19].

Reichenbach characterized as "causal anomaly", any eventual more complete description of quantum mechanical phenomena. Everything beyond the actual formalism was treated as anomaly, in conformity with the orthodox interpretation. Consequently, Reichenbach believed that there are no observables that could be included in the Ψ function. For him, as for the C.I., quantum-mechanical description is restricted only to statements about phenomena: the so called *restrictive interpretation*.

This was the general epistemological presuppositions for the elabotation of a three-valued logic. For Reichenbach, the three-valued logic is the logical apparatus suitable for Q.M., once the decision for the use of a restrictive interpretation is made[20]. In such a logic, the use of the "sharp" categories of *true* and *false* must be considered as idealizations. The truth value *indeterminate* represents a topologically different category[21].

By this way Reichenbach rejected the validity of the LEM for certain classes of quantum phenomena. This was inevitable, from the moment he accepted the CI and in particular the thesis that the actual quantum-mechanical description is complete.

Reichenbach considered the principle of the *tertium non datur* as one of the pillars of the traditional logic. Yet, once he introduced the truth value indeterminate, he inevitably rejected the LEM. One preliminary remark is needed here: The LEM is valid when the two, *logically incompatible* statements have a well defined truth value (0 or 1). Nevertheless, this condition is not satisfied in every possible case. And since logical axioms must be valid "in every possible world" (Leibniz), we conclude the LEM is not a fundamental law of logic. What kind of logic is, therefore, the three-valued one, constructed on the basis of the rejection of a law, non possessing the status of a law of logic? More than that: *What would be the status of the three-valued quantum logic, if it would be possible, at least in principle, to restore the validity of the LEM in the domain of quantum phenomena?*

In the following I will question the solidity of the foundations of the three-valued "quantum logic". I will try to show that it is in principle possible to restore, at least in some crucial cases, the validity of the LEM. Consequently I will support the thesis that the logic elaborated on the basis of its rejection is not a "Logic", but a propositional calculus valid in the epistemological context of the Copenhagen Interpretation.

4.IS A NEW LOGIC REALLY NEEDED FOR QUANTUM MECHANICS?

I will discuss now some quantum phenomena which seem to violate the LEM. I will try to show that in their cases the rejection of the LEM is not warranted.

4.1.The Inequalities of Heisenberg and the validity of the LEM

Let us discuss first, the case of the inequalities of Heisenberg. For quantum - mechanical propositions representing commuting observables, the two-valued logic is, evidently, valid and the corresponding sub-lattices are boolean. The problem of logic and, in particular, that of the validity of the LEM, is posited only in the case of observables represented by non commuting operators.

Let us recall the definition of the LEM: Of two contradictory propositions negating each-other, one is true and the other false. *Tertium non datur.*

We now pose the question: Do the Heisenberg's inequalities entail the rejection of the LEM in every case of non commuting observables and under all possible conditions? The LEM concerns contradictory propositions. However, we must make a distinction between two kinds of incompatibility: the *logical* and the *factual* one. The contradiction between waves and particles, for example, is a logical contradiction in the frame of formal logic. Two statements concerning the position and the momentum of a particle, on the contrary, are not logically incompatible: a particle can have a precise momentum and at the same time be in a precise position, even if the measurement of both is not practically possible. Thus, between the position and the momentum exists, eventually, a factual incompatibility. Consequently, it seems that in the case of the position and the momentum the validity of the inequalities of Heisenberg is irrelevant to the question of the LEM.

However, let us generalize the definition of the LEM, in order to include factually and not only logically incompatible propositions. In the case of factually incompatible propositions, one can be true, the other being indeterminate. However, this is not the point of view prevailing among the protagonists of the CI. In fact, it is possible to distinguish three points of view concerning the status of a pair of propositions related to non commuting observables: 1) If x is true, p_x is inexistent and *vice-versa*. (If a particle is in the Brookhaven accelerator with momentum p_x , then it is nonwhere, remarks ironically H.Mehlberg). This is the more radical thesis of the CI. 2) If x is true , p_x is meaningless, and *vice-versa*. This was often the thesis of Bohr and Heisenberg. 3) However, it would be more realistic to affirm that in that case, the conjugate observable is simply indeterminate or meaningless. This is the thesis of Reichenbach. The term "meaningless" of the Bohr-Heisenberg interpretation, is replaced by Reichnbach by the term "indeterminate"[22]. In fact, if x is meaningful, p_x is meaningless, according to the principle of complementarity (Bohr, 1927)[23]. In the frame of the CI, the state is indeterminate or meaningless. This contention, however does not in principle, preclude the possibility that p_x has a precise value, even if it is impossible to measure it.

Let us try now a more concrete discussion concerning the inequalities of Heisenberg. For the CI these inequalities constitute the *limit* of our knowledge for quantum phenomena. In that case, p∧p never holds. (On the contrary, p∨p is always valid). Suppose now that a *deus ex machina* achieves to measure both, the position and the momentum of a particle. In that case the boolean structure of our lattice would be restored (our non classical lattice would be embedded in a classical one) and the validity of two-valued logic would be extended to quantum mechanics as well. The obstacle has been removed by our demon.

However, it is not necessary to invoke the benevolent creature. Because, as is well known, the unquestionable validity of the inequalities of Heisenberg has been contested from many points of view. First of all: Heisenberg himself gave two mutually incompatible interpretations of his inequalities: The ontological one based on the questionable concept of the wave packed, and the operatonalist one, related to his famous microscope[24]. From the experimental point of view: The limit posed by the inequalities seems to be violatged in everyday practice. The knowledge, for example, of the position of the particle in a spectrograph is used for the computation of its momentum. In a similar way, the geometrical data of he path of a particle in the buble chamber, are used for the computation of its dynamical variables, etc. In these cases, the validity of the enlarged version of the LEM is not refuted .

I will propose now a thought experiment, which indicates that the problem of the interpretation of the inequalities of Heisenberg is more subtle and complicated that is in general thought.

Let us consider two EPR particles with zero total spin. We separate them and we measure, say, the $\vec{ox}$ component of the spin of the particle A. Using the value of the spin of this component, we can predict the value of the corresponding component of the particle B. At the same time we measure, with another apparatus, the component $\vec{oy}$ of the particle B. Using its value, we can predict the value of the spin of the corresponding component of the particle A. By this way we succeeded to measure the values of two components of the spin of our particles, contrary to the formalism of Q.M.

This possibility does not mean that the values of the three components of the spin are actual before the measurement. They are potential values, actualized *via* the interaction with the measuring apparatus[25].

Conclusions. 1) The inequalities of Heisenberg do not concern logically incompatible attributes of matter. 2) They are not necessarily valid in all possible cases. 3) Consequently, even if we enlarge the definition of the LEM in order to include factually incompatible attributes, it seems not legitimate to construct a logic on the basis of a law which is not valid in every possible world.

4.2. The Two-Slit experiment

We will discuss now a genuine case of logical incompatibility, related to the famous two-slit experiment.

Formal, two-valued logic claims that our particle passes either through the slit A, *or* through the slit B. *Tertium non datur.* However the probability density behind the screen, when both slits are open, does not obey the classical law:

$$\rho(x) = |\psi(x)|^2 = |\psi_A(x) + \psi_B(x)|^2 = \rho_A(x) + \rho_B(x) + \psi_A^*(x)\psi_B(x) + \psi_B^*(x)\psi_A(x) \neq$$

$$|\psi_{CA}(x)|^2 + |\psi_B(x)|^2$$

How to explain this fact? Let us accept that our system is a corpuscle. Then, it must pass either through A *or* through B. In that case, how the other slit affects the distribution of the particles? (How the particle "knows" that the other slit is open?). If, on the contrary, we suppose that our system is represented by a wave, then it is possible to explain the experimental fact. However, as it is possible to detect the particle in a suitable distance behind A *or* B, the whole argument seems to fail.

The enigma seems insoluble. The Orthodox School eludes the problem, by postulating the complementary, mutually exclusive character of the contradictory attributes or phenomena. Thus, if the particle passes through the slit A, any statement concerning B is meaningless, and *vice-versa.* According to a more agnostic argument, there is no process of passing of the particle through the slit at all. For Reichenbach, finally, if one of two complementary propositions is true, the other is indeterminate. The postulated indeterminacy is the departure point for a three-valued logic.

All these seemed sound in the time of the Como and Solvay Meetings (1927), in spite of the opposition of Einstein and de Broglie. However, what was impossible in 1927, became possible in our days: We put two counters in a fairly long distance behind the slits, and we localize the particle . We know now whether it has passed through the slit A or through the slit B. If A is true, B is false, and *vice-versa.* The validity of the law of the excluded middle has been restored, although the mystery of the influence of the slit B on the slit A and *vice-versa* remains intact.

Once more the rejection of the LEM is not experimentally justified. Consequently, the knowledge gap in the case of the two slit experiment does not legitimize the elaboration of a three-valued logic.

4.3. The wave-particle duality

For classical, two-valued logic a particle is not a wave and *vice-versa.* There is a logical incompatibility between particle and wave properties. However, "elementary" particles exhibit, as is well known, mutually exclusive properties: wave-like and particle-like. Therefore, the question: elementary particles are corpuscles, waves, or a certain kind of centaur (wave-particles) is a legitimate one.

According to the positivist dogma endorsed by the principle of complementarity, the above question has no meaning. It depends on our apparatus if the particle exhibits wave-like or particle-like properties . The two properties are mutually exclusive and our knowledge is restricted to these contradictory images of reality. Although accepting the general epistemological premises of the CI, Reichenbach-as already noted-postulated that if one of the complementary propositions is true (or false) the other is indeterminate. Thus Reichenbach rejected the LEM and justified the elaboration of its non classical logic.

The ontology of the CI was a positivist one and the empiricism of Reichenbach was not alien to the philosophy of the Vienna School. Nevertheless it is not unreasonable to postulate another ontology, which attributes a status of objectivity to the wave-particle duality. This was, for example, the case of the theory of double solution of L. de Broglie (1927). And it is well known that such an ontology is suggested today by neutron interferometry and by oher experimental evidence or theoretical models. Consequently, the renunciation of the LEM in the case of the wave-particle quality is not justified on physical grounds. Therefore, any attempt to use it in order to legitimize a three-valued logic, is founded on presuppositions non valid in every possible world.

4.4. Transitions and the three-valued logic

The last case to discuss here, concerns the quantum transitions. In that case also, we have to do with a genuine logical incomatibility. In fact, a particle cannot be in the state A and at the same time in the state B. Before the transition, the particle was in the state A. After the transition, it is in the state B. But what is the state *during* the time Δt of the transition between A and B?

If we accept the CI, our particle *constitutes*, during Δt, a unique and non-analyzable system with the measuring apparatus . Any question concerning its state during this time interval is meaningless. The Orthodox School elevates to the status of *unknowable* the process of transition from A to B. Other writers maintain that the state between A and B is indeterminate . In both cases the LEM seems to fail.

However, if we contest the dogma of completeness, it is legitimate to consider the "indeterminacy" as expressing a void of knowledge. It is not, therefore, a matter of chance, if the realists insist that it would be, in principle, possible to describe transitions as rapid, but not instantaneous processes. In such a case the LEM will be restored and the three-valued formalism will be devoid of any physical counterpart.

This is an old thesis of the realist School. Forty years ago, for example, E. Schrödinger considered the existing quantum formalism as a "mysterious, fit and jerk theory" and was expecting that transitions would be described as slow and actually describable processes. For Schrödinger, if one does not understand transitions but only stationary states, one understands nothing[26]. Today there are at our disposal models describing transitions as continuous and deterministic processes, having an "epaisseur temporelle". The last world is not yet spoken on this complicated question. Consequently, the realists have essential arguments in favour of the possibility to restore the LEM in the case of quantum transitions.

What would be then the state of the system during the time of transition? Formal logic cannot grasp such phenomena. Yet, it is reasonable to suppose that the time of transition does not correspond to *nothingness*. During Δt , A ceased to be A. It is not yet B. It tends to become B. The new reality is going to emerge from potentiality, as a result of successive transformations. Three-valued or two-valued logic are not adequate to describe the passage from the potential to the actual.

4.5. Conclusion

Complementarity considers contradictory attributes as mutually exclusive. In its frame, consequently, the LEM is not valid for certain categories of phenomena. A three-valued logic seems to correspond to the existence of indeterminate propositions. The formalism elaborated by Reichenbach and other logicians, presupposes the CI. However: 1)The LEM is not a fundamental law of logic, valid in "all possible worlds" 2)As I tried to show, it is in principle possible to restore the LEM in cases where its rejection seems inevitable. 3)Consequently, "quantum logic" is not a Logic. It is "a propositional calculus expressely suited to quantum mechanics"[27], in tis actual form and only for mutually incompatible observations.

5. BEYOND THE LOGIC OF IDENTITY

What is needed for Q.M. is not a classical logic, that is to say, a logic of *identity*, two or three-valued. Only a logic of *processes*, as anticipated by Schrödinger, de Broglie and others, would be adequate for the description of quantum phenomena . According to such a logic:
I. Physical phenomena are objective and irreversible processes realized in space and time. The categories of possibility and potentiality are necessary for a logic corresponding to processes and not only to stationary states.
II. Quantum transformations are not acausal, instantaneous jumps. They are deterministic processes, having an "epaisseur temporelle".

III. Phenomena define a direction of time. This conforms with Relativity and Thermodynamics.

IV. A logic suitable for such processes , must include in its "paradigm" the dialectics of the possible and the actual.

V. Such a logic must elaborate a generalisation of the categories of causality, determinism, chance and necessity, in order to include the new, specifically quantal modes of interaction and determination.

VI. The new logic must be open to the theories of hidden variables. Yet, such, theories do not necessarity presuppose a total embedding of the boolean lattice in a classical structure. Partial embeddings may correspond to such theories. More than that: Probabilistic hidden variable theories would be possible in the frame of such a logic, because the probabilistic form of law is not the negation of causality and determination.

VII. The logic of change recognizes the validity of classical logic, in every case in which the law of identity holds.

VIII. The new logic attributes to the category of contradiction an ontic and not simply an epistemic status. However, the logic of contradiction is not contradictory.

One can find the first core of such a logic in the philosophies of Heraclitus and Aristotle. This logic was further developed by Hegel and Marx, and the marxist tradition. *Question.* Such a logic, is a *Logic* in the formal sense of the term? Evidently not! *Conclusion.* We must transcend the classical paradigm. A new point of view and a new ontology are necessary for quantum mechanics and for the elaboration of a logic suitable for quantum phenomena.

Now, I want to conclude, by quoting an Utopian Philosopher: "Naturalism may describe these so called facts. They are as valuable and as superficial, as naturalism itself. Genuine realistic poetry deals with processes, isolating and manipulating the facts. The process requires a precise imagination to portray it and is directly connected with imagination [......]. It is because of the really possible, that the world is not made into a sophisticated book, but into a process dialectically mediated, therefore dialectically open". *E. Block, 1935.*

What is valid for poetry, may be valid for quantum mechanics as well.

References

1. H. Reichenbach, *The Philosophic Foundations of Quantum Mechanics,* Univ. of California Press, 1944.

2. See the pioneer works of J. Lukasiewicz (1988), E.L. Post (1921), A. Tarski (1930), P. Fevrier (1937). Also, the more recent works of H.Reichenbach, K. Von Weizsäcker, J. M. Jauch, C. Piron, S. Haak, H. Putnam, K. Hubner, P. Mittelstaedt, W. Stegmuller.

3. J. Lukasiewicz, *Selected works,* North Holland, 1970, p. 221.

4. J. Lukasiewicz, ibid, p. 165.

5. See: 1) G. Galilei, *The Two New Sciences,* Dover, N.Y. 2)I.Newton, *Principia* Univ. Of California Press, 1947. Id, *Opticks,* Dover, N.Y. 1952.

6. For the physical meaning of the superposition principle, see: E. Bitsakis, *Physics Essays,* 4, 124 (1991).

7. E. Bitsakis, in *Microphysical Reality and Quantum Formalism,* A. Van der Merwe *et al* (Eds), Kluwer Academic Publishers, Dordrecht, 1988, V.I., p. 47.

8. H. Reichenbach, *The Philosophic Foundations of Quantum Mechanics* op.cit. , p. 21. Id., *The Direction of Time,* Univ. Of California Press, 1956, p. 217.

9. H. Reichenbach, *The Philosophic Foundations of Quantum Mechanics,* op.cit., p. 21.

10.H. Reichenbach, *The Direction of Time*, op, cit., p. 217.

11. H.Reichenbach, Ibid, p. 218.

12. H.Reichenbach, ibid, p. 217.

13. H.Reichenbach, *The Philosophic Foundations of Quantum Mechanics*, op.cit., pp 40-43; For the definition of causal anomalies, S. Haack, *Philosophy of Logic*, Cambridge Univ. Press, 1988, p. 210.

14. H.Reichenbach, *The Direction of Time*, op.cit, p. 218.

15.H.Reichenbach, *The Philosophic Foundations of Quantum* Mechanics, op.cit. p. 26.

16. H.Reichenbach,, ibid, p. 28.

17. H.Reichenbach, *The Direction of Time*, op.cit, p. 214.

18. H.Reichenbach,, ibid, p. 220.

19.E. Bitsakis, *Found. of Physics*, *18*, 331 (1988). id., *Open Questions in Quantum Physics*, G. Tarozzi, A. Van der Merwe (Eds), Reidel, 1985. Id, *in Greek Studies in the Philosophy of Science*, P. Nicolacopoulos (Ed.) , Kluwer Acad. Publ., 1990.

20. H.Reichenbach, *The Philosophic Foundations of Quantum Mechanics*, op.cit. p.43.

21. H.Reichenbach, ibid, p. 147.

22. H.Reichenbach, ibid, p. 160. See also, P. Mettlstaedt, *Philosophical Problems of Modern Physics*, Reidel 1976, H. Putnam, in *A Portrait of Twenty Five Years* . K.Hübner, *Critique of Scientific Reason*, The Univ. Of Chicago Press, 1983.

23.See, N.Bohr, *Atomic Theory and the Description of Nature*, Cambridge Univ. Press, 1961. For a critical analysis of the principle of complementarity, see, E.Bitsakis, in *Bell's Theorem and the Foundations of Modern Physics*, A.van der Merwe et al (Eds), World Scientific, 1992.

24.See:W.Heisenberg, *The Physical Principles of Quantum Theory*, Dover, N.Y., 1949. For a critical analysis of the problem, see: E. Bitsakis, *Le Problème de Déterminisme en Physique*, Thèse d' Etat, Paris, 1976.

25.For the relation of actual and potential states, see, E. Bitsakis in:*Problems in Quantum Physics*, L. Kostro *et al* (Eds.) , World Scientific, 1988.Id, *Annalles de la Fondation Louis de Broglie*, *15*, 35 (1990).

26.E.Schrodinger, in :*Louis de Broglie, Physicien et Penseur*, Albin Michel, Paris, 1953.

27.K.Hübner, *Critique of Scientific Reason*, op.cit, p.

DANGEROUS EFFECTS OF THE INCOMPREHENSIBILITY IN MICROPHYSICS

Jenner Barretto Bastos Filho

Departamento de Fisica da
Universidade Federal de
Alagoas. Cep 57061
Maceio, Alagoas, Brasil

INTRODUCTION

Einstein said [1] "the most incomprehensible thing about the world is that it is comprehensible". On the other hand Bohr wrote [2] "...I advocated a point of view conveniently termed complementarity, suited to embrace the characteristic features of individuality of quantum phenomena, and at the same time to clarify the peculiar aspects of the observational problem in the field of experience. For this purpose, it is decisive to recognize that, *however far the phenomena transcend the scope of classical physical explanation, the account of all evidence must be expressed in classical terms"*.

The disagreement between Einstein and Bohr is not only centered on the important EPR (Einstein, Podolsky, Rosen) paradox. This discordance is strongly rooted in two different conceptions of Natural Philosophy. Einstein teaches us to believe that the world is understandable in agreement with the gnoseological optimism of all the high periods of the History of Knowledge: Old Greece, Italian Renaissance, the birth of Modern Science in the 17 th century, etc. . In spite of Niels Bohr being part of this rational tradition, there is something new and highly pessimist in his attitude: "Indispensable use of classical concepts ...even though classical physical theories do not suffice". [3]

Bohr said that the phenomena *transcend* the scope of classical physical explanation, but *must be* expressed in classical terms. We would like to pay attention to these terms: *must be*.

According to Bohr, classical categories (space, time, cause, wave, particle,...) *must be* used in order to make the phenomena inteligible; causal relations exclude space-time ones and vice versa; wave aspect excludes corpuscular one and vice versa; they are mutually exclusive concepts (see the skilful critical analysis of Selleri [4]).

In classical physics no dual object exists i. e. being, at the same time, wave and particle. Therefore, according to Bohr, any methodological choice consisting of considering quantum objects (protons, electrons, photons,...) as objectively dual i. e. wave *and* particle really existing at ontological level, *must be* rejected.

In short, Bohr's Complementarity Principle (C.P.) necessarily requires classical concepts. On the other hand, C.P.necessarily also requires mutual exclusion of these classical concepts. Einstein had never accepted this pessimist attitude consisting of considering science merely as an instrument. Concerning this situation, Stapp [5] has pointed out that "according to Bohr quantum theory must be interpreted, not as a description of nature itself, but merely as a tool for making prediction about observations appearing under conditions described by classical physics".

Popper [6] also considered this situation as a true brainwashing consisting of making incomprehensibility comprehensible.

Frontiers of Fundamental Physics, Edited by M. Barone
and F. Selleri, Plenum Press, New York, 1994

It is important to emphasize that Bohr's *must be* depends on only methodological choice. This choice theaches the people to believe that the microreality is not so understandable. As Feynman [7] claimed "it contains, only mystery".

We are firmly against the attitude consisting of claiming mysterious aspects of quantum reality. We claim that this attitude is dangerous for science, scientific education and social life.

THE TEACHING OF QUANTUM MECHANICS

With respect to the famous Bohm's 1952 papers, Bell [8] wrote: "...I have always felt since that people who have not grasped the ideas of those papers (and unfortunately they remain the majority) are handicapped in any discussion of the meaning of quantum mechanics".

Still, the quantum mechanics teaching has not significantly changed. The students learn what Bohr has said in the Como conference in 1927 as a last and definitive word on the subject. Apparently there is no opposition. Without presenting to the students both sides of the coin the quantum mechanics teaching remains an authoritarian discipline. In this context, the students continue to be handicapped in any conceptual discussion on quantum mechanics. Dyson [9] argues that the students arrive at the conclusion that there is nothing to be understood.

All the claims are presented to the students in the direction of the incomprehensibilities, impossibilities, miracles and mysteries. Gell Mann [10] has pointed out that "Bohr brainwashed a whole generation of physicists into believing that the problem had been solved fifty years ago". Born [11] also recognize that among the dissenters of the dominant interpretation there are some of the founders of quantum mechanics: Planck, Einstein, de Broglie and Schrödinger.

Why does the conflict between dissenters and adherents of the Copenhagen view not appear in conventional quantum mechanics courses ?

We can observe that the behavior of physicists and students oscillates between considering quantum mechanics as merely a tool or merely as a miracle. Both attitudes are dangerous for the development of the cognitive sciences. This is a disastrous and dangerous situation because the gap between microreality and social life becomes very large. In consequence of this, utilitarism, pragmatism, nihilism and mysticism are privileged instead of genuine cognitive philosophies.

One of the characteristics of this situation is the vagueness of the words and concepts. With respect to this situation, Bell [12] pointed out in 1989 that in spite of 62 years of quantum mechanics there are "some words which , however legitimate and necessary in application, have no place in a formulation with any pretention to physical precision: *system, apparatus, environment, microscopic, macroscopic, reversible, irreversible, observable, information, measurement.".* And the "worst of all is *measurement".*

In spite of quantum mechanics being a very successful discipline for making predictions, its comprehension definitively is not. As Gell Mann [13] pointed out, "quantum mechanics, that mysterious, confusing discipline which none of us really understands but which we know how to use".

It is a very peculiar situation: a good theoretical tool on the one side but on the other side a Tower of Babel - a dialogue between the deaf.

Fortunately, in the last years, the dissenters of the Copenhagen view have significantly increased. Several important results in favour of the realists and against the Copenhagen view have been obtained. The von Neumann's theorem forbidding the causal completion of quantum mechanics is proved to be not sufficiently general [14]. Some very simple examples are sufficient to demolish famous von Neumann's impossibility. Moreover, several phenomena considered by the non-realists as " absolutely impossible to explain..." can be understood in a realistic and rational manner: two slit experiment [15], neutron interferometry [16], Janossy and Dagenais/Mandel experiments [17] and many others.

The strong connection between Bohr's Complementarity and von Neumann's theorem had been made by Bohr himself [18]. However, as Selleri [19] pointed out, this fact is rarely emphasized. Indeed, Bohr's Complementarity and von Neumann's theorem are twin brothers; consequently, the refutation of one implies in the refutation of the other.

In virtue of the above reasons, the quantum mechanics teaching must be drastically changed. The implementation of the rational discussion in the spirit of Old Greece is necessary.

In the forthcoming section we argue that the Bohrian *must be* constitutes a methodological deviation responsible for the mysterious results of the spin-one particles. We make a comparison between two analogous "games": a classical and a quantum one involving spin-one particles. We argue that the radically different results between these "games" cannot in a rational and realistic way, be attributed to the mysteries of interference of amplitudes. We argue (against Feynmann) [20] that no mystery exists in these so called "strange" results. If one adopts the Copenhagen view the mysteries appear but if one adopts the view that spin-one particles are dual objects (i. e. particle + wave really existing) all the mysteries disappears.

CLASSICAL SITUATION

We consider an ensemble of classical indestructible mass-points obeying the conservation of the total number of particles. The role played by the S-machine (see fig.1, from 1-a to 1-e) consists of performing the separation of the incoming particles in three outgoing p-beams according to a given law, for example, x% in the first beam, y% in the second, and z% in the third, in such a way that

$$x\% + y\% + z\% = 100\% \text{ of the incoming particles} \tag{1}$$

In order to prepare N particles in a given initial beam, we put blocking masks in the other ones as fig. 1 shows. In the fig.1 we prepare N particles in the intermediate beam. As a second step N incoming particles will be separated by a T-machine in three other outgoing k-beams in such a way that

$$x'\% + y'\% + z'\% = 100\% \text{ of the N incoming particles} \tag{2}$$

being x'%, y'% and z'% respectively the fractions in each of the k-beams. As a third step incoming particles are separated by a second S-machine in three outgoing r-beams. We can put or remove blocking masks in order to select any possibility of the game and calculate its probability. The upper, intermediate and lower beams are denoted respectively by 1, 2, and 3. So,

$$p = 1, 2, 3 \; ; \; k = 1, 2, 3 \; ; \; r = 1, 2, 3 \tag{3}$$

The probability of a given particle initially prepared in a given p-beam (after the separation performed by the first S-machine) to be found in a given k-beam (after the separation performed by the T-machine) will be denoted by

$$(kT|pS) \tag{4}$$

In the case of the fig. 1-a, this probability is (1T|2S). The probability of a given particle initially prepared in a given p-beam (after the separation performed by the first S-machine) to be found in a given r-beam (after the separation performed by the second S-machine) will be

$$(rS|kT) (kT|pS) \tag{5}$$

where k is a beam in the intermediate step. In the case of fig 1-a this probability is (2S|1T)(1T|2S). The probability (5) corresponds to a given possibility of the game; the probability corresponding to all the possibilities of the game is

$$\sum (rS|kT) (kT|pS) = 1 \tag{6}$$

(All the k,r;p fixed)

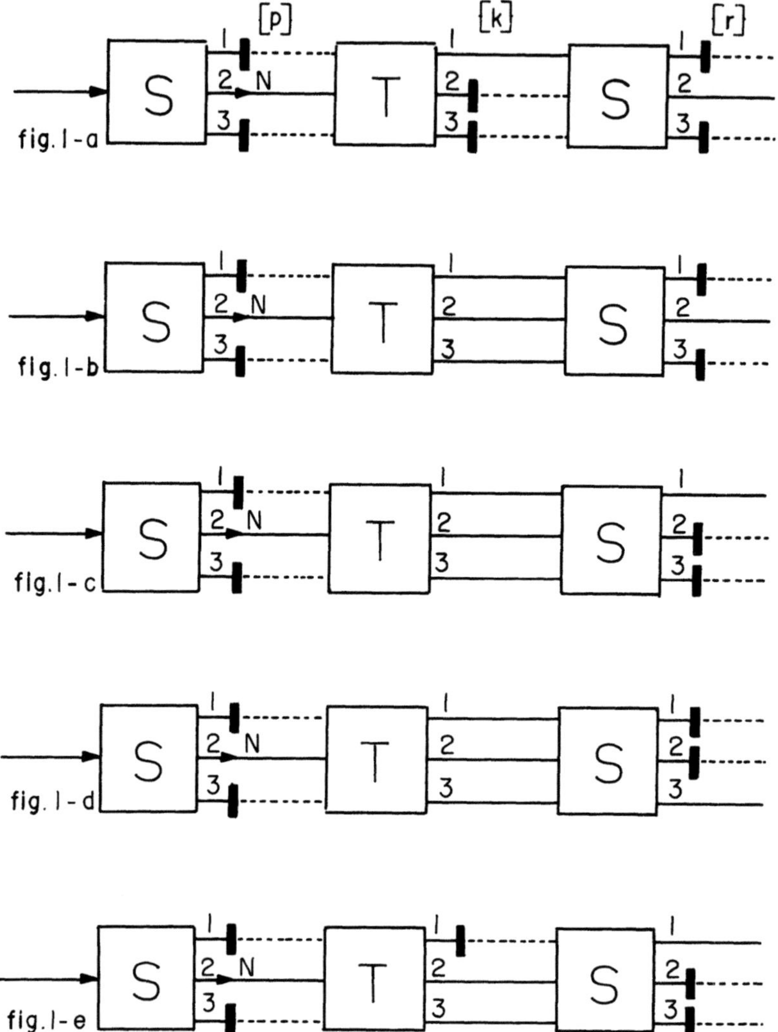

Figure 1. The figs. from (1-a) to (1-e) are valid for both classical and quantum games. Concerning the classical game, the role played by the S and T machines consists of separating the incoming beam in three outgoing ones according respectively to the S and T laws. Concerning the quantum game, S and T are inhomogeneous magnetic fields splitting the spin-one particles respectively in S and T basis. The vertical bars represent the blocking mask for both games.

The reason for which p in (6) is fixed is the following: all the probabilities are normalized according to the fact that N particles are prepared in a given initial p-beam. If p=2, the nine terms of (6) will be

$$
\begin{array}{lll}
(1S|1T)\ (1T|2S) \ ; & (1S|2T)\ (2T|2S) \ ; & (1S|3T)\ (3T|2S) \\
(2S|1T)\ (1T|2S) \ ; & (2S|2T)\ (2T|2S) \ ; & (2S|3T)\ (3T|2S) \\
(3S|1T)\ (1T|2S) \ ; & (3S|2T)\ (2T|2S) \ ; & (3S|3T)\ (3T|2S)
\end{array}
\tag{7}
$$

The probabilities relative to the situations showed from fig.1-a to fig.1-d are:

$$
\begin{aligned}
P_a &= (2S|1T)\ (1T|2S) & \text{(8-a)} \\
P_b &= (2S|3T)\ (3T|2S) + (2S|2T)\ (2T|2S) + (2S|1T)\ (1T|2S) & \text{(8-b)} \\
P_c &= (1S|3T)\ (3T|2S) + (1S|2T)\ (2T|2S) + (1S|1T)\ (1T|2S) & \text{(8-c)} \\
P_d &= (3S|3T)\ (3T|2S) + (3S|2T)\ (2T|2S) + (3S|1T)\ (1T|2S) & \text{(8-d)}
\end{aligned}
$$

$$P_b + P_c + P_d = 1 \tag{9}$$

It is important to note that, in general,

$$(mS|nT) \neq (nT|mS) \tag{10}$$

where n and m assume the values 1, 2, and 3. As a numerical example we suppose that the role played by the S-machine consists of separating the incoming particles in the following way: 27% in beam 1, 38% in beam 2 and 35% in beam 3. For the T-machine we suppose: 30% in beam 1, 19% in beam 2 and 51% in beam 3. For these numbers all the probabilities involved in the game are

$$
\begin{aligned}
&(1S|1T) = (1S|2T) = (1S|3T) = 0.27 \\
&(1T|1S) = (1T|2S) = (1T|3S) = 0.30 \\
&(2S|1T) = (2S|2T) = (2S|3T) = 0.38 \\
&(2T|1S) = (2T|2S) = (2T|3S) = 0.19 \\
&(3S|1T) = (3S|2T) = (3S|3T) = 0.35 \\
&(3T|1S) = (3T|2S) = (3T|3S) = 0.51 \\
&P_a = 0.114 \ ; \ P_b = 0.38 \ ; \ P_c = 0.27 \ ; \ P_d = 0.35
\end{aligned}
\tag{11}
$$

QUANTUM SITUATION

In our quantum game, S and T are inhomogeneous magnetic fields pointing in different directions. Our quantum particles are the spin-one particles with rest mass different from zero. According to quantum mechanics a particle of spin equal S_0 has $(2S_0+1)$ eigenstates of S_z. The role played the inhomogeneous magnetic field consists of splitting the incoming beam in $(2S_0+1)$ outgoing beams. For the spin-one particles with rest mass different from zero we have three outgoing beams. According to quantum mechanics, before the action of the inhomogeneous magnetic field a given particle of the incoming beam is described by a general quantum state

$$|X> = C_1 \, |1> + C_2 \, |2> + C_3 \, |3> \tag{12}$$

where $|1>$, $|2>$ and $|3>$ denote respectively the eigenstates of the quantum mechanics operator S_z. After this action there is a direct correspondence between particles in the eigenstates $|1>$, $|2>$ and $|3>$ and the beams respectively 1, 2, and 3; C1, C2, and C3 are respectively the amplitudes of probability $<1|X>$, $<2|X>$ and $<3|X>$, in general complex numbers. From the mathematical point of view, the passage from the splitting in the S-base to the splitting in the T-base constitutes a change of base in geometry.

By using blocking masks we can prepare the initial and final eigenstates and calculate all the probabilities relative to each possibility of the game.

For the situations exhibited from fig.1-a to fig.1-d, the corresponding amplitudes of probability are:

$$A_a = <2S|1T> <1T|2S> \tag{13-a}$$

$$A_b = <2S|3T> <3T|2S> + <2S|2T> <2T|2S> + <2S|1T> <1T|2S> \tag{13-b}$$

$$A_c = <1S|3T> <3T|2S> + <1S|2T> <2T|2S> + <1S|1T> <1T|2S> \tag{13-c}$$

$$A_d = <3S|3T> <3T|2S> + <3S|2T> <2T|2S> + <3S|1T> <1T|2S> \tag{13-d}$$

The formulas (8) and (13) are radically different. The classical quantities $(..|..)$ are real numbers in the range $[0,1]$ and obeying in general, the asymmetry expressed by (10). On the other hand, the quantum quantities $<..|..>$ are, in general, complex numbers obeying the quantum law

$$< mS|nT> = <nT|mS>* \tag{14}$$

where the * denotes complex conjugate; The quantities $|<..|..>|^2$ are defined in the range $[0,1]$.

In quantum mechanics the law holds that

$$\sum_{\text{(All the k)}} |kT\rangle \langle kT| = 1 \tag{15}$$

and, also the orthonormality of the eigenstates holds that

$$\langle rS|pS \rangle = \delta_{r,p} \tag{16}$$

where $\delta_{r,p} = 1$ if $r = p$ and $\delta_{r,p} = 0$ if $r \neq p$

In virtue of the above quantum laws, (13) becomes,

$$A_a = \langle 2S|1T\rangle \langle 1T|2S\rangle \tag{17-a}$$
$$A_b = 1 \tag{17-b}$$
$$A_c = 0 \tag{17-c}$$
$$A_d = 0 \tag{17-d}$$

DISCUSSION

According to quantum mechanics, the results (17) mean that in the case (17-a) *some* of the N particles prepared in the initial eigenstate $|2S\rangle$ get through the second inhomogeneous magnetic field $\underline{S}$. In the case (17-b) *all* the N particles prepared in $|2S\rangle$ get through the second inhomogeneous magnetic field, but *none* do so in the cases (17-c) and (17-d).

In order to discuss the above results we study the situation showed by the fig. 1-e . The situation (1-e) differs from situation (1-c) only by the circumstance in which we put a blocking mask in the beam k=1. The corresponding amplitude of probability for this case is

$$A_e = \langle 1S|3T\rangle \langle 3T|2S\rangle + \langle 1S|2T\rangle \langle 2T|2S\rangle \tag{18}$$

A_e is not equal to zero because (15) is valid only if *all* k-beams are free. *According to quantum mechanics, the amplitude of probability can increase from zero to a value different from zero, by putting blocking masks (see figs.(1-c) and (1-e)). This result is impossible from the classical point of view.*

Using the numerical example given in (11) the probabilities corresponding to the *classical* situation showed by figs.(1-c) and (1-e) are respectively,

$$(P_c)\text{class.} = 0.27 \quad ; \quad (P_e)\text{class.} = 0.189 \tag{19}$$

From the classical point of view, any obstruction of the beam *necessarily* implies a decreasing of the probability. Considering the probability a non negative quantity (defined in the range [0,1]) it is impossible to choose an example in which the result

$$(P_c)\text{class.} > (P_e)\text{class.} \tag{20}$$

is violated.

From the quantum point of view we have,

$$(P_a)\text{quant.} = |\langle 2S|1T\rangle|^2 |\langle 1T|2S\rangle|^2 = |\langle 2S|1T\rangle \langle 1T|2S\rangle|^2 \tag{21-a}$$
$$(P_b)\text{quant.} = 1 \tag{21-b}$$
$$(P_c)\text{quant.} = 0 \tag{21-c}$$
$$(P_d)\text{quant.} = 0 \tag{21-d}$$
$$(P_e)\text{quant.} > 0 \tag{21-e}$$

and consequently,

$$(P_c)quant. < (P_e)quant. \qquad (22)$$

The contradiction between (20) and (22) leads to the conclusion that quantum particles are not Newtonian mass-point particles. If we adopt the concept of particles in the Newtonian sense, no rational solution exists for the quantum result (22). However, this conclusion does not imply that causal and space-time descriptions are mutually incompatible.

We have, at least, two ways to follow; the first consists of accepting the mystery, as Feynmann wrote [20]: "This is the old, deep mystery of quantum mechanics - the interference of amplitudes"; the second one consists of searching for both a rational and a realistic solution for the problem. We think that the first is not a good one because mystery does not constitute a scope of science. Moreover, in accepting mysteries we adopt a position against the spirit of science. Science with mystery is not science in the Popperian sense because mysteries immunize the tests. Following the second way, a possible solution is to assume the *dual nature of quantum particles in the direction originally indicated by Einstein and de Broglie. So, we are able to explain the result (22) in a realistic way as developing causality in space-time.*

Denoting the results (13-b) and (21-b) by completely construtive interference (CCI) and the results (13-c), (13-d) and (21-c), (21-d) by completely destrutive interference (CDI) all the possible results can be summarized as follows:

1- CCI necessarily requires that the intermediate k-beams be *completely free.*

2- In the case of CDI there are two different possibilities:

   (i) That the intermediate k-beams be *completely free.*

   (ii) That the intermediate k-beams be *completely blocked.*

3- The results relative to final probability in the range (0,1) are obtained by removing or putting blocking masks in the intermediate k-beams in such a way that neither obstruction necessarily implies decreasing the number of particles nor the removal necessarily implies increasing this number. *It is important to emphasize that for this case neither the intermediate k-beams be completely free nor they be completely blocked.*

We argue that no difficulty exists in understanding the above results in a rational and realistic way.

I- With respect to the result 1 it is remarkable that no possibility exists in obtaining CCI when the beams are either totally or partially blocked (no surprise from a realistic point of view).

II- The result (2-i) shows simply a *CDI between quantum waves really existing and consequently none of the particles get through the second inhomogeneous field; the result (2-ii) shows that no real quantum wave exists in the region between the second and the third steps; this circumstance is important because it shows a causal property of these quantum objects.*

III- If we assume that quantum objects are Newtonian particles, no realistic explanation exists for the result 3. However, if we accept the dual nature of quantum objects, a simple realist solution is possible. *In fact, one blocked beam can imply an elimination of a CDI as well as one free beam can imply exactly its implementation.* .In this way, no mysteries are required to understand these quantum features in a rational and realistic way as developing causality in space-time.

Finally, we emphasize the following points:

(A)- The comparison between (20) and (22) shows that quantum objects are not particles in the Newtonian sense.

(B)- Rejecting the mystery, a possible acceptable solution consists of assuming the dual nature of the quantum objects.

(C)- The mathematical interference of amplitudes has a counterpart in the interference of real waves. In this way, these, at first sight "strange" results do not belong to the category of mysteries.

CONCLUDING REMARKS

A given methodological choice can lead to mystery. It is exactly the case of the mutual exclusion expressed in Bohr's Complementarity. Moreover, this mysterious character comes from the insistence in ascribing to quantum objects a simple nature. The adoption of another methodological choice as in considering quantum objects as dual ones (i.e. particle + wave really existing) is enough to eliminate all the mysteries. The dangerous situation is not properly centered on the Copenhagen solution. It is a legitimate tendency and a rational discussion on the subject must include all the tendencies. The danger comes from the monopoly of persuasion in favour of the incomprehensibility together with an almost total absence of confrontation between rival conceptions.This situation consists of a true brainwashing because mystery appears as a "necessary" part. Definitively, this situation is not in favour of a rational discussion.

Why the insistence in emphasizing mystery ?

Is it absolutely necessary ?

Surely, complexity of the reality does not imply in the absurdity of the world. Great thinkers like Einstein and de Broglie must be revisited.

ACKNOWLEDGEMENTS

I would like to thank Prof. Franco Selleri for his kind invitation to participate in the Olympia Conference.Thanks are due to Prof. Fernando Gama (Rector of the Universidade Federal de Alagoas) and to FAPEAL(Fundação de Amparo à Pesquisa de Alagoas) for the support. I would like also to thank Prof. Augusto Garuccio for his kind help in the preparation of the final form of this manuscript.

REFERENCES

[1] A. Einstein, cited from " Cosmology, .Physics and Philosophy", B.Gal-Or, Springer Verlag NY Inc (1981) p. 166

[2] N.Bohr, Discussion with Einstein on Epistemological Problems in Atomic Physics 1949, in: "Atomic Physics and Human Knowledge" Science Edition INC. NY (1961) p.39

[3] N.Bohr, Phys.Rev. 48:696 (1935) (p. 701)

[4] F.Selleri, Rivista d. Storia d. Scienza 2:545 (1985)

[5] H.P.Stapp, Light as foundation of being, in: "Quantum Implications (Essays in Honour of David Bohm)" Ed. by B.J.Hiley and F.D.Peat, Routledge /Kegan Paul, London and New York (1987)

[6] K.R.Popper, in "Le Grand Débat de la Théorie Quantique", Franco Selleri, Flammarion, Paris (1986), (Preface).

[7] R.P.Feynman,et.al., Lectures on Physics, Addison-Wesley, Reading, Massachusetts, p. 1-1

[8] J.S.Bell, Beables for quantum field theory, in: " Essays in Honour of David Bohm" cited in reference [5]

[9] F. Dyson, Scientific American, sept. 1958

[10]. M.Gell-Mann, in :" The Nature of Physical Universe" (Nobel Conference/1976), Wiley, NY 1979 (p.29)

[11] M.Born, (Nobel Conference/ 1954)

[12] J.S.Bell, Physics World, August 1990 p.33-40 (see p.34)

[13] M.Gell-Mann,"The Nature of Matter", Wolfson College Lectures, 1980 Claredon Press, Oxford, 1981

[14] F.Selleri, Le Grand Débat de la Théorie Quantique, Flammarion, Paris, 1986 ;(see also F.Selleri, ref. [4])

[15] F.Selleri and G. Tarozzi, Nuovo Cimento 43B:31 (1978).

[16] F.Selleri, Fisica senza dogma (La conoscenza scientifica tra sviluppo e regressione) , Edizioni Dedalo, Bari, Italy 1989 (see Ch. 4)

[17] F.Selleri,ref [16]

[18] Proceedings of the Conference "New Theories in Physics", Warsaw, 1938.

[19] F. Selleri, ref. [4]

[20] R.P.Feynman et. al., ref.[7] section 5-5 p.5-10

CLASSICAL INTERPRETATION OF QUANTUM MECHANICS

V.K. Ignatovich

Laboratory of Neutron Physics
Joint Institute for Nuclear Research
141980 Dubna, Moscow region, Russia

INTRODUCTION

Experiments with ultracold neutrons (UCN) in bottles reveal that the probability of UCN losses at single collision with the bottle walls is of two order of magnitude higher then it is expected on theoretical grounds. In attempts to explain this anomaly it was supposed that the neutron is described by a wave packet that contains plane wave components with high enough energy to propagate over the potential barrier. Then the losses can be ascribed to the probability of penetration through barrier in non tunneling fashion that is proportional to part of the energy spectrum which is shared by high energy components. To forbid the independent propagation of high and low energy components of the wave packet it is necessary to suppose that the packet is the neutron's immanent wave function which is not spreading with time.

The analysis shows that the singular de Broglie's wave function can be taken as a candidate for such a packet.

The immanent nonspreading wave packet has simultaneously precisely defined position and momentum. From such a description it comes out that the quantum dogma of impossibility to define simultaneously position and momentum of a particle is not necessary.

The acceptance of the particle's immanent wave function leads to conclusion that the wave function can be considered as a classical field. If such an interpretation is accepted then the second quantum dogma about impossibility to define trajectory in the presence of interference, is also shown to be not necessary.

It is possible to devise an experiment with the neutron interferometer, where the observation of interference does not forbid simultaneous determination of the neutron trajectory.

In the next paragraph the UCN anomaly is explained, after that the de Broglie's

Frontiers of Fundamental Physics, Edited by M. Barone
and F. Selleri, Plenum Press, New York, 1994

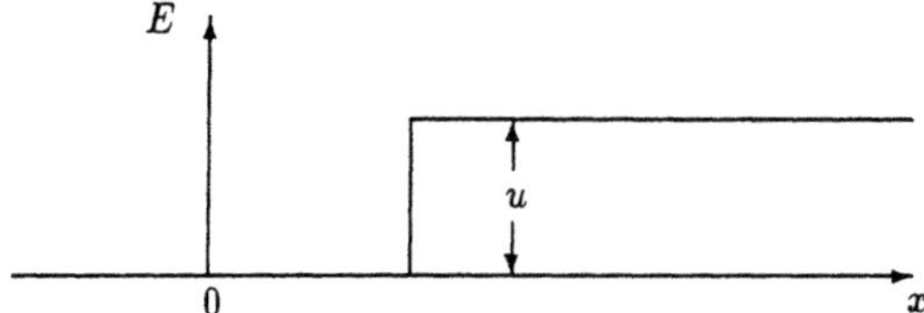

Figure 1. The optical potential u seen by the UCN at reflecting walls.

wave packets are presented and the problem of simultaneous definition of the position and momentum is discussed. In the subsequent section an experiment with observation of the interference and determination of the trajectory is discussed.

THE UCN STORAGE EXPERIMENT

A neutron is called ultracold (see the books[1-3]) if it can be stored in a bottle. Since the interaction of the neutron with matter is described by the potential $u \approx 10^{-7}$ eV the storage is possible if neutron's energy E is less than $E_{\lim} \equiv u$ (Figure 1). In that case neutrons are totally reflected from walls at any angle of incidence. The velocity of such neutrons is of $v \leq v_{\lim} \approx 5$ m/s and the wave length is nearly 1000 Å. These values are actually determined by the neutron — matter interaction $u = (\hbar^2/2m)4\pi N_0 b$, where m is the neutron mass, N_0 is the atomic density and b is the coherent neutron nucleus scattering amplitude. The interaction u varies from substance to substance, but the magnitudes $E_{\lim} \approx 10^{-7}$ eV and $v_{\lim} \approx 5$ m/s are the typical ones.

The neutrons of higher energies can be totally reflected from the walls only in a limited range of angles of incidence, and this range decreases with increasing energy.

Now we formulate the main UCN problem (anomaly). This problem is connected with the storage experiment, the scheme of which is shown on Figure 2. Here s_1 and s_2 denote two shutters that can be opened and closed. When s_1 is closed and s_2 is opened, neutrons fill the bottle. When the s_2 is closed the UCN are trapped in the bottle. After waiting some "exposition time," $t_{\exp}$, the shutter s_1 is opened and the number of UCN left in the bottle are counted by the detector.

The storage curve similar to the one found in[4], is plotted in Figure 3. It shows in logarithmic scale the number of neutrons in the bottle that survived after exposition time $t_{\exp}$. This curve can be represented by a function

$$N(t) = N_0 \exp(-t/\tau(t)),$$

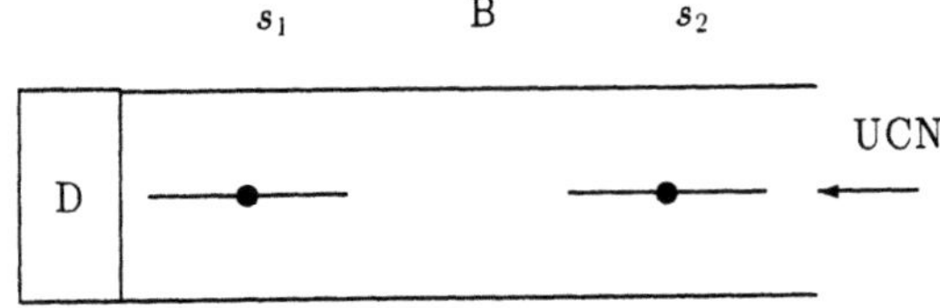

Figure 2. UCN storage experiment scheme: One opens the shutter s_2 (shutter s_1 is closed), fills the bottle, then closes it, waits some time $t_{\exp}$ called the exposition time, opens the shutter s_1 and counts the number of UCN left in the bottle.

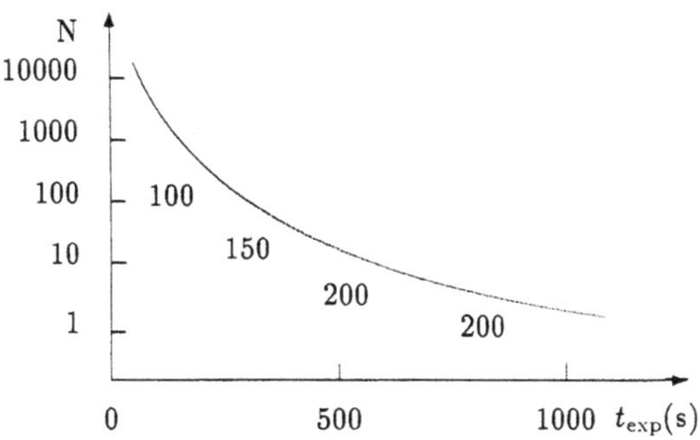

Figure 3. The storage curve: The different numbers show the storage time for different expositions.

where $\tau(t)$ is called "storage time." For monoline spectrum of stored neutrons τ is constant, but for the broad spectrum it depends on exposition time, because neutrons of different energies have different loss rate. The numbers on the picture denote the τ at different expositions.

The storage time is limited by the own life time of the neutron T_β with respect to β-decay and by losses in the walls:

$$1/\tau = 1/T_\beta + 1/T_l.$$

The loss time T_l is equal to t_f/μ, where t_f is a free-flight time between two collisions with the walls and μ is the probability of losses at a single collision with the wall. The last value for small velocity v of the neutron and for normal incidence (this case is chosen for simplicity) can be represented as

$$\mu = 2(v/v_{\lim})\eta, \tag{1}$$

where $v_{\lim}$ is the limiting velocity of UCN: $v_{\lim} \propto \sqrt{u}$ and η is the reduced loss coefficient (see any of cited textbooks on UCN) equal to the ratio: $\eta = b''/b'$ of imaginary and real parts of the amplitude b. The imaginary part b'' can be represented as $b'' = \sigma_l/2\lambda$, where λ is the neutron wave length and σ_l is the loss cross section.

Let us enumerate the factors responsible for losses (T_l).

1. Absorption cross section σ_a (own and impurities)

2. Inelastic scattering cross section $\sigma_{\text{inel}}^{\text{coh.inc}}$ (own, impurities and vacuum gas)

3. Geometry: dimension, surface roughnesses, gaps.

4. Gravity: change of trajectories, variation of spectrum along the height.

Taking into account all these factors, we can say that at present for the coldest, purest bottle (Beryllium walls covered with oxygen) with the smallest self-absorption, we observe[5] *an additional loss cross section of 1 barn ($\eta \approx 3 \times 10^{-5}$) of unknown nature, which is of two-three order of magnitude higher than the theoretical one.* And this is the main UCN problem or UCN anomaly. To explain the anomaly, it is necessary to go out of the enumerated factors.

DE BROGLIE'S WAVE PACKETS

To explain the anomaly it is natural to suppose that the neutron is described by a wave packet containing the high energy components. Then the part of the wave packet spectrum shared by these high energy components defines the probability of penetration through the potential barrier in a non tunneling fashion.

But for such an explanation it is necessary to suggest that the high and low energy components of the wave packet are not independent. That means that the wave packet should be an immanent property of the neutron and should not spread. Now the question arises whether it is possible to construct a nonspreading wave packet in the framework of the existing quantum mechanical theory?

The answer to this question is affirmative and the number of possible wave packets is infinite. But if we restrict ourselves only to spherically symmetrical case we come to two types of wave packets, describing a particle moving with velocity v and found by de Broglie[6].

One of them (non normalizable)

$$\Psi(\mathbf{r}, t) = \exp(i\mathbf{v}\mathbf{r} - i\omega t) j_0(s|\mathbf{r} - \mathbf{v}t|), \tag{2}$$

is the solution of the homogeneous Shroedinger equation

$$(i\partial/\partial t + \Delta/2)\Psi(\mathbf{r}, t) = 0,$$

and the other

$$\psi(s, \mathbf{v}, \mathbf{r}, t) = c \exp(i\mathbf{v}\mathbf{r} - i\omega t)\frac{\exp(-s|\mathbf{r} - \mathbf{v}t|)}{|\mathbf{r} - \mathbf{v}t|}, \tag{3}$$

is the solution of the inhomogeneous equation

$$(i\partial/\partial t + \Delta/2)\psi(\mathbf{r}, t) = -C(t)\delta(\mathbf{r} - \mathbf{r}(t)). \tag{4}$$

where s is the width of the wave packet that should give the penetration of the neutron over barrier and explain the UCN anomaly, v is the neutron velocity and c is the normalization constant defined by the equality:

$$\int d^3r |\psi(s, \mathbf{v}, \mathbf{r}, t)|^2 = 1. \tag{5}$$

To get the moving packet (2) it is useful to start from the spherically symmetrical stationary solution

$$\psi(r, t) = j_0(sr) \exp(-is^2t/2)$$

of the equation

$$(i\partial/\partial t + \Delta/2)\psi(r, t) = 0,$$

where

$$j_0(x) = \sin(x)/x$$

is the spherical Bessel function and then to make transformation to a moving reference frame of variables

$$x = x_0 + vt,$$

and of the wave function

$$\psi(\mathbf{r}_0, t) = \exp(-i\mathbf{v}\mathbf{r} + iv^2t/2)\Psi(\mathbf{r}, t).$$

In the result we get (2) with

$$\omega = (v^2 + s^2)/2,$$

where the first term describes the kinetic energy of the particle and the second is related to the energy that is required to hold wave packet of width s in non spreading state.

Position and Momentum

The nonspreading wave packet (refla) has precisely definite shape and velocity and with respect to these properties it legitimate to consider the possibility of simultaneous definition of position and momentum. The momentum of the wave packet is precisely defined because of precise definition of velocity v and of relation $p = mv$ with strictly defined mass m of the considered particle.

The position should be defined for the quantum object as a whole, that means for particle with its wave function. The last is extended in space, so we should define a position of the extended object.

To do that it is very useful to look at a classical analogy. Let us consider a classical object, say a ping pong ball. The position of it is a point which we in principle can identify with any point on the ball surface. But if the ball is ideally symmetrical it is natural to identify the position with the center of gravity, or with the center of the ball. The same can be done with the quantum system.

In the case of the wave function (2) the center of gravity coincides with the maximum of the amplitude. And it is seen that if this point is defined as the position of the particle, this position is precisely defined simultaneously with the momentum of the particle.

If we try to define position with the help of measurement, then even in classical physics we shall have uncertainty in the case of extended objects. If we are dealing with point object in classical physics, then we can define precisely its position with the help of measurements only if the particle is infinitely heavy, or if we prepare precisely defined test particles and deduce the previous position of the target particle from scattering measurements. But in this case the position point is again the matter of definition and such a definition is in the same way applicable to a quantum object with a wave function described by an infinitely narrow wave packet.

In conclusion we can say that the position and momentum of particle in quantum mechanics can be precisely defined simultaneously, so the quantum dogma of impossibility of such a definition is not necessary.

Now the question arises about the meaning of the Hisenberg uncertainty relations. It is possible to say that these relations reflect dimensions of the object or its field structure independently whether it is quantum or classical object. For instance these relations are applicable even to the classical electron with its electrostatic field. Indeed, if we define

$$< \Delta x^2 > = \int r^2 \phi^2 \, d^3 r,$$

where $\phi = e/r$ is Coulomb potential, we shall always get an infinity. And the same takes place in the case of the wave packet (2). In other words the simultaneous definition of position and momentum of the particle does not contradict to the uncertainty relations.

In the case of wave packet (3) position of the particle can be identified with the singularity point in analogy with classical physics where the position of classical electron coincides with the singularity point of the Coulomb potential.

The Choice of the Right Packet

Now we consider the packets (2) and (3) and chose the one that helps to explain

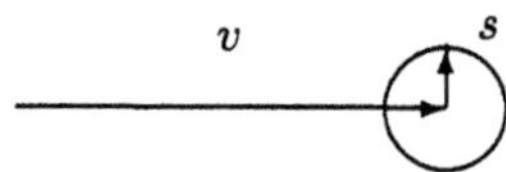

Figure 4. The spectrum of the nonsingular de Broglie's wave packet is represented by a sphere of radius s centered at the extremity of vector $\mathbf{v}$.

the UCN anomaly. We shall find that only the wave packet (3) is appropriate for such a purpose and it leads to some consequences.

To make a right choice it is necessary to consider Fourier spectra of these packets. The Fourier representation of the function (2) has the form

$$\exp(i\mathbf{v}\mathbf{r} - i\omega t)j_0(s|\mathbf{r} - \mathbf{v}t|) = \int \exp(i\mathbf{p}\mathbf{r} - ip^2 t/2)\delta((\mathbf{p} - \mathbf{v})^2 - s^2)\, d^3 p/2\pi s.$$

So its spectrum in the momentum space is like a sphere of radius s (Figure 4) centered at the end of the velocity $\mathbf{v}$ (in units $m/\hbar = 1$).

But, to describe small losses it is necessary to have small s, implying that the spectrum of such a packet is very narrow. Therefore the neutron with velocity v lower than $v_{\lim} - s$ (the majority of UCN have just such velocities) has no components with the energy higher than u and has no extra losses. It means that the parameter s is of no help in the case of the wave packet (2), and such a wave packet cannot explain the UCN anomaly.

The Fourier representation of the wave packet (3) is

$$\psi(s,\mathbf{v},\mathbf{r},t) = \frac{4\pi c}{(2\pi)^3} \int \frac{\exp[i\mathbf{p}\mathbf{r} - ip^2 t/2 + i((\mathbf{p} - \mathbf{v})^2 + s^2)t/2]}{(\mathbf{p} - \mathbf{v})^2 + s^2}\, d^3 p.$$

Its spectrum is centered around $p^2 = v^2$ and extends to infinity. A part of this spectrum is always higher than u. So, we can suppose that with the probability

$$\mu = \frac{|4\pi c|^2}{(2\pi)^3} \int_0^\infty \frac{1 - |R(p)|^2}{[(\mathbf{p} - \mathbf{v})^2 + s^2]^2}\, d^3 p \tag{6}$$

the neutron escapes through the wall. It is postulated that it is the neutron as a whole and not a part of it that escapes through the wall. This postulate looks very alike to the projection hypothesis in the theory of measurement and can be accepted if low and high energy parts of the wave packet can not propagate independently, which means that the wave packet is an immanent property of the particle and can not spread with time.

To estimate the parameter s entering the relations (3,6), we substitute $|R(p)| = 1$ for $p_\perp^2 < u$, where u is the wall potential, and $|R(p)| = 0$ for $p_\perp^2 > u$. Then, for small v and perpendicular incidence of the incoming wave, the expression (6) can be represented in the form

$$\mu = \frac{|4\pi c|^2}{(2\pi)^3} \int_0^\infty d^3 p\, \theta[(p_\perp + v)^2 > u]/(p^2 + s^2)^2 \approx 2s/\pi v_l, \tag{7}$$

where the θ function is equal to unity when the inequality in its argument is satisfied and to zero in the opposite case.

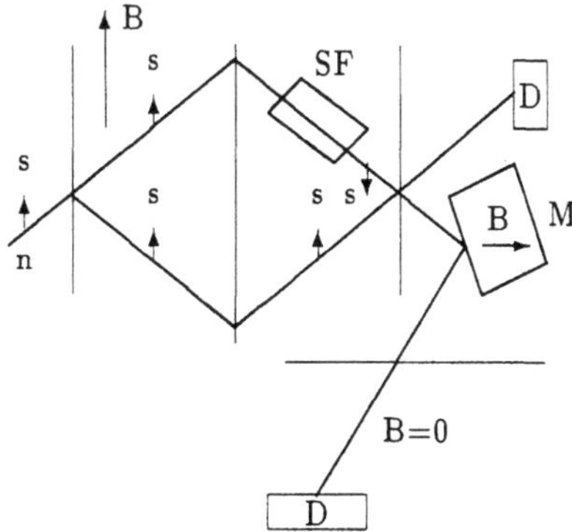

Figure 5. The experiment with 3-crystal interferometer for observation of interference and determination of trajectory.

Comparing with (1), we get

$$s = \pi \eta v \approx v \times 10^{-4}.$$

Thus the storage experiments give a possibility to estimate the parameter s of the
singular de Broglie solution.

CLASSICAL INTERPRETATION OF QM

The attempts to explain the UCN anomaly with the help of wave packets leads
to a singular wave function that is a solution of inhomogeneous Shrödinger equation
with a source. Thus it can be interpreted as a classical field created by the particle.
The motion of the particle should be determined by the interaction of this field with
surrounding objects in the same way as the motion of the electron is determined by
interaction of its electrostatic field with the environment.

In this way it is easy to understand the meaning of interference. The interference
can be observed even with classical particles. Indeed, let us consider a screen with two
holes and an electron, which moves through one of the holes, and we know which one
exactly. The motion of the electron depends on whether the other hole is open or not,
since it depends on the interaction of its field with the screen, and the last is different
for the two cases. So we see that the other hole interfere with the motion through the
first one, and this is the phenomena called "interference." In that respect the reasoning
about absence of interference in the case of propagation of classical electrons[7] are not
valid.

Of course, with the electrostatic field it is not possible to obtain an oscillating pattern, since there is no such a parameter as the wavelength, but the above example shows
that there is no principal difference between the wave function and an electromagnetic
field.

Trajectory in the Interference

If determination of trajectory in classical physics does not forbid interference the
same should take place in quantum physics too. Let us consider such an experiment
with the help of neutron interferometer. The scheme of the experiment is shown on
figure 5. It is a slight generalization of the experiment[8] with polarized neutrons (spin
$s \parallel B$), when in one of the paths the neutron's spin s is reversed by a radio frequency
spin flipper (SF). After reverse of the spin neutron's kinetic energy is not changed

(at the moment we neglect corrections due to Golub et all[9]), but its interaction with external magnetic field and the total energy is changed. So after recombination the neutron wave function can be represented in the form:

$$\Psi = e^{ikx - i\omega t}(\xi_{\downarrow} e^{i\omega_1 t} + \xi_{\uparrow} e^{-i\omega_1 t}),$$

where $\omega_1 = \mu B/\hbar$ and $\omega = \hbar^2 k^2/2m$ and $\xi_{\uparrow}$, $\xi_{\downarrow}$ denote spinors with spin parallel and antiparallel to the external field B. Thus after recombination neutron has the precessing polarization perpendicular to the external field. If at some distance from the third crystal there is placed a mirror (M) magnetized in a direction of arriving neutron, then there takes place the total reflection of recombined neutron from both paths. If the reflected neutron enters the free field region $(B = 0)$, then its total energy transforms into kinetic one, so the neutron's wave function becomes

$$\Psi = \xi_{\downarrow} \exp(ik_+ x - i\omega_+) + \xi_{\uparrow} \exp(ik_- x - i\omega_-),$$

where $\omega_{\pm} = \omega \pm \omega_1$ and $k_{\pm} = \sqrt{2m\omega_{\pm}/\hbar^2}$. It means that two halves of the neutron are separated in space and time and can observe be observed as two peaks in time of flight spectrum with opposite polarizations, that correspond to neutron coming over two different paths.

¿From the field point of view it can be understood because the neutron goes only over one of the paths but its field is extended and the motion of the neutron is determined by interaction of its field with the environment. Such an approach to quantum mechanics in some respect is very close to the one where particle is considered to be accompanied with an empty wave[10], but here the wave is considered to have a content and this content is the classical field.

Equations of Motion

Now, what does it means, that the wave function is a field, and how this field influences the motion of the particle? To describe all that it is necessary to introduce all the equations determining the field and coordinates of the particle. It seems to us that the field can be determined from boundary conditions and Eq. (8),

$$(i\partial/\partial t + \Delta/2 - u(\mathbf{r}))\psi(\mathbf{r},t) = -C(t)\delta(\mathbf{r} - \mathbf{r}(t)). \tag{8}$$

where $\mathbf{r}(t)$ are the coordinates of the particle. The dependence of these coordinates on time is determined by the equations

$$md^2\mathbf{r}/dt^2 = -\nabla V(\mathbf{r},t), \tag{9}$$

where the force (for the case of UCN it gives a reasonable result) can be put down tentatively as

$$\nabla V(\mathbf{r},t) \propto \mathbf{n}|\Psi(\mathbf{r},t)|^2 u(\mathbf{r}), \tag{10}$$

u being the usual potential, $\mathbf{n}$ being the vector normal to the interface. The solution of (8), entering (10), can contain also the part, satisfying the homogeneous Schrödinger equation.

Of course, one cannot solve these equations even in the simplest case (by the way, the analogous equations for the case of propagation of the classical electron have not been solved yet too), but it seems that, due to quantum mechanics it is not necessary.

Quantum mechanics gives us a very good route to avoid this, but we should pay for
that by dealing only with probabilities.

Acknowledgment

The author is very grateful to Prof F.Selleri, A.Garuccio and M.Barone for their
invitation to Olimpia conference, to Prof. A.A.Tyapkin for his interest and support
of this work, and to PhD student of Nuclear Engineering Department of Texas A&M
University Igor Carron for his invaluable contribution and to Soros Foundation for the
short term financial support.

References

[1] A.Steyerl, 1977, Very low energy neutrons, *in:* "Neutron Physics" (Springer Tracts
of Modern Physics, vol. 80), Part 2, p. 57, G.Höhler, ed. Springer-Verlag, Berlin,
N.Y.

[2] V.K.Ignatovich, 1990, "The Physics of Ultracold Neutrons". Clarendon Press, Oxford.

[3] R.Golub, D.Richardson, and S.K.Lamoreaux, 1991, "Ultra-Cold Neutrons". Adam
Hilger, Bristol.

[4] R.Golub et al., 1981, UCN for EDM, *Rutherford Appleton Laboratory Report* RL-
91-001.

[5] V.P.Alfimenkov et al., 1992, Anomalous interaction of UCN with the Surface of
Beryllium Bottles, *JETP Lett.* 55:92.

[6] L. de Broglie, 1960, "Non-Linear Wave Mechanics: A Causal Interpretation". Elsevier, Amsterdam.

[7] L.D.Landau and E.M.Lifshits, 1963, "Quantum Mechanics". Gosizdat Fizmatlit,
Moscow.

[8] G.Badurec, H.Rauch and J.Summhammer, 1988, Polarized Neutron Interferometry, *Physica B* 151:82.

[9] R.Golub et all., 1994, to be published in *Am. J. of Phys.*

[10] F.Selleri, 1990, "Quantum Paradoxes and Physical Reality". A.van der Merwe, ed.,
Kluwer Academic, Dordrecht.

RABI OSCILLATIONS DESCRIBED BY DE BROGLIAN PROBABILITIES

Mirjana Božić and Dušan Arsenović

Institute of Physics, Beograd, P.O. Box 57, Yugoslavia

I. INTRODUCTION

Superposition principle, as one of the basic principles in quantum mechanics, is essential for numerous quantum phenomena: interference, quantum beats, quantum interference, wave packets, fluorescence, Rabi oscillations, spin echo, superposition of spin states etc. In all those cases a wave function of a quanton is a superposition of two or more eigenstates. Characteristic features of those phenomena are determined by relative phases of different components in the above superposition. But, the standard interpretation of quantum mechanics avoids to attribute physical meaning to phases as well as to relative phases of wave functions. As a consequence, attempts to physically explain and understand those phenomena often encounter paradoxes, difficulties and inconsistencies discussed for example by Feynman[1] and Ballentine[2] in the case of interference, by Klein et al.[3] and Kaiser et al.[4] in connection with wave packets, by Wigner[5] in connection with spin state superposition, by Schrödinger[6] and Brewer and Schenzle[7] in relation to fluorescence, etc.

de Broglie[8], Schrödinger[6,9], Vigier[10], Selleri[11] and others have considered that the difficulties would disappear if one would assume that a wave function described a real wave in space and time. Moreover, for them the dependence of those phenomena on relative phases of components in the superposition was the proof of the reality of waves described by a wave function $\psi(\vec{r}, t)$.

In the paper "What is an elementary particle?" Schrödinger exposed his views by the following words[9]: "The waves, so we are told, must not be regarded as quite real waves. It is true that they produce interference patterns - which is the crucial test that in the case of light had removed all doubts as to the reality of the waves. However, we are now told that all waves, including light, ought rather to be looked upon as "probability waves". They are only a mathematical device for computing the probability of finding a particle in certain conditions, ...

Here I cannot refrain from mentioning an objection which is too obvious not to occur to the reader. Something that influences the physical behaviour of something else must not in any respects be called less real than the something it influences - whatever meaning we may give to the dangerous epithet 'real'."

In connections with the uncertainty relation, in the same paper Schrödinger wrote[9]: "Still, it does not necessarily follow that we must give up speaking and thinking in terms of what is really going on in the physical world."

Recently, the s.c. de Broglian probabilities were introduced in order to facilitate the speach about what is really going in quantum interference experiments[12], double slit experiment[13,14], two-particle interference experiments[15], during the motion of a wave packet through the interferometer[16]. In this paper we shall apply the concept of de Broglian probabilities in order to describe what is eventually going on in an atom immersed in a laser field.

Wave function for a two-level atom subjected to monochromatic electromagnetic radiation is given in Sec. II. In Section III are evaluated de Broglian probabilities for the same problem. For a particular two level model de Broglian probabilities are graphically presented. In Section IV we propose an answer, based on the use of de Broglian probabilities, to Schrödinger' s question "Are there quantum jumps?".

II. THE RABI MODEL

Let us consider a system described by a total Hamiltonian H made up of two parts:

$$H = H_0 + V(t) \tag{1}$$

a part H_0 describing an unperturbed atom whose eigenvectors and eigenvalues are assumed to be known, and a time varying part $V(t)$ describing the interaction of the atom with the laser field

$$V(t) = V_L \cos \omega_L t \tag{2}$$

One expands the solution of the time-dependent Schrödinger equation $\psi(\vec{r}, t)$ into the complete set of eigenfunctions ϕ_n of H_0

$$H_0 \phi_n(\vec{r}) = E_n \phi_n(\vec{r}) \tag{3}$$

so that

$$\psi(\vec{r}, t) = \sum_n c_n(t) \exp(-iE_n t/\hbar) \phi_n(\vec{r}) \tag{4}$$

The time-dependent probability amplitudes satisfy

$$\sum_n | c_n(t) |^2 = 1$$

since $< \psi \mid \psi > \, = 1$. Substitution of the series (4) into the time-dependent Schrödinger equation gives the exact equation of motion

$$\dot{c}_p(t) = -\frac{i}{\hbar} \sum_n V_{pn}(t) \exp(-iE_{np} t/\hbar) c_n(t) \tag{5}$$

where $E_{np} = E_n - E_p$ and $V_{pn} = < \phi_p \mid V(t) \mid \phi_n >$. For simplicity it is assumed that $V_{nn} = 0$ for all n.

In the Rabi model[17] one approximates the sum (4) by the sum which contains only the resonating pair of states, let us say, $\phi_1(\vec{r})$ and $\phi_2(\vec{r})$

$$\psi(\vec{r}, t) = c_1(t)\phi_1(\vec{r}) \exp(-iE_1 t/\hbar) + c_2(t)\phi_2 \exp(-iE_2 t/\hbar) \tag{6}$$

One applies also the rotating wave approximation which means that the oscillating perturbation (2) is replaced by a rotating perturbation

$$V(t) = V_L e^{i\omega_L t} \tag{7}$$

In this way, the equations (5) are approximated by the equations

$$\dot{c}_1 = -\frac{i}{2\hbar} V_0 \exp\{i(\omega_L - \omega_0)t\} c_2$$

$$\dot{c}_2 = -\frac{i}{2\hbar} V_0 \exp\{-i(\omega_L - \omega_0)t\} c_1 \tag{8}$$

where $\omega_0 = \frac{E_2 - E_1}{\hbar}$ and $V_0 = <\phi_2 \mid V_L \mid \phi_1>$. The Rabi solution of those equations for the initial conditions $c_1(0) = 1$, $c_2(0) = 0$ reads[17]:

$$c_2(t) = i\frac{V_0}{\Omega_R \hbar} e^{i\Delta t/2} \sin(\Omega_R t/2)$$

$$c_1(t) = e^{i\Delta t/2}\{\cos(\Omega_R t/2) - i(\Delta/\Omega_R)\sin(\Omega_R t/2)\} \tag{9}$$

where

$$\Delta = \omega_0 - \omega_L \tag{10}$$

is detuning and

$$\Omega_R = \sqrt{\Delta^2 + V_0^2/\hbar^2} \tag{11}$$

is the Rabi frequency. Probabilities $\mid c_2(t) \mid^2$ and $\mid c_1(t) \mid^2$ to find the atom in the states 2 and 1, respectively oscillate sinusoidally at the Rabi frequency Ω_R.

$$\mid c_2(t) \mid^2 = \frac{V_0^2/\hbar^2}{[(\omega_L - \omega_0)^2 + V_0^2/\hbar^2]} \sin^2 \frac{1}{2}[(\omega_L - \omega_0)^2 + V_0^2/\hbar^2]^{1/2} t$$

$$\mid c_1(t) \mid^2 = 1 - \frac{V_0^2/\hbar^2}{[(\omega_L - \omega_0)^2 + V_0^2/\hbar^2]} \sin^2 \frac{1}{2}[(\omega_L - \omega_0)^2 + V_0^2/\hbar^2]^{1/2} t \tag{12}$$

Oscillations of probabilities $\mid c_2(t) \mid^2$ and $\mid c_1(t) \mid^2$ - Rabi oscillations, are monitored by detecting the variation of fluorescence intensity[18] as a function of $\frac{V_0}{\hbar}t$.

It became customary to interpret the sinusoidal variation of probabilities as "flopping" of population to-and-fro between the states. This intuitive view, however is based on the notion that an atom at a specific time can only be found in one of two eigenstates[6,7,19] and ignores that the true wave function is a linear superposition of ϕ_1 and ϕ_2 for any t. Schrödinger was the first who pointed out to this inconsistency in the usual description of atomic processes in an electromagnetic field[6].

We are now going to propose the new interpretation (the compatible statistical interpretation) of atomic transitions, which is free of the above mentioned inconsistency.

III. DE BROGLIAN PROBABILITIES FOR THE RABI MODEL

Compatible statistical interpretation contains in addition to the standard quantum mechanical probabilities new kind of probabilities, called de Broglian probabilities (probability densities). In the case of the Rabi model, de Broglian probability density $\mathcal{P}_1(\vec{r}, t)$ is the probability density that electron energy is E_1 and that electron is situated around $\vec{r}$ at time t. Similarly, $\mathcal{P}_2(\vec{r}, t)$ is the probability density that electron energy is E_2 and that it is situated around $\vec{r}$ at time t.

The sum of those probability densities is equal to the probability density that electron is situated around $\vec{r}$ at time t.

$$| \psi(\vec{r}, t) |^2 = \mathcal{P}_1(\vec{r}, t) + \mathcal{P}_2(\vec{r}, t) =$$

$$= | c_1 |^2 | \phi_1 |^2 + | c_2 |^2 | \phi_2 |^2 + 2\mathrm{Re} \exp(i\omega_0 t) c_1 c_2^* \phi_1 \phi_2^* \tag{13}$$

the latter relation is the generalized Selleri-Tarozzi relation[20,13]. Similarly to the other cases studied previously, we shall assume that de Broglian probabilities satisfy also the following relation

$$\frac{\mathcal{P}_1(\vec{r}, t)}{\mathcal{P}_2(\vec{r}, t)} = \frac{| c_1 |^2 | \phi_1 |^2}{| c_2 |^2 | \phi_2 |^2} \tag{14}$$

By combining (13) and (14) one obtains

$$\mathcal{P}_1(\vec{r}, t) = | c_1 |^2 \cdot | \phi_1 |^2 \left[1 + \frac{2\mathrm{Re} \exp(i\omega_0 t) c_1 c_2^* \phi_1 \phi_2^*}{| c_1 |^2 | \phi_1 |^2 + | c_2 |^2 | \phi_2 |^2} \right]$$

$$\mathcal{P}_2(\vec{r}, t) = | c_2 |^2 \cdot | \phi_2 |^2 \left[1 + \frac{2\mathrm{Re} \exp(i\omega_0 t) c_1 c_2^* \phi_1 \phi_2^*}{| c_1 |^2 | \phi_1 |^2 + | c_2 |^2 | \phi_2 |^2} \right] \tag{15}$$

Those very simple expression are positive or zero. At Figs. 1 and 2 are presented graphically probability densities:

$$| \psi(r, t) |^2 r^2 = \frac{1}{4\pi} \int | \psi(\vec{r}, t) |^2 r^2 \sin\vartheta \, d\vartheta \, d\varphi$$

$$P_1(r, t)r^2 = \frac{1}{4\pi} \int \mathcal{P}_1(\vec{r}, t)r^2 \sin\vartheta \, d\vartheta \, d\varphi \tag{16}$$

$$P_2(r, t)r^2 = \frac{1}{4\pi} \int \mathcal{P}_2(\vec{r}, t)r^2 \sin\vartheta \, d\vartheta \, d\varphi$$

for the Rabi model in which $\phi_1(\vec{r})$ is 1s state and $\phi_2(\vec{r})$ is 3s state.

$$\phi_1(\vec{r}) = \phi_1(r) = \frac{2}{a_0^{3/2}} \exp\left(-\frac{r}{a_0}\right)$$

$$\phi_2(\vec{r}) = \phi_2(r) = \frac{2}{(3a_0)^{3/2}} \exp\left(-\frac{r}{3a_0}\right) \left[\frac{2}{3}\left(\frac{r}{3a_0}\right)^2 - 2\frac{r}{3a_0} + 1\right] \tag{17}$$

We have chosen those states because the graphical representation of the corresponding probability densities is transparent.

We see that $\mathcal{P}_i(\vec{r}, t)$ is a product of two terms. First term is $| c_i(t) |^2$ which is the probability to find eigenvalue E_i of electron energy at time t. Taking into account the definition of $\mathcal{P}_i(\vec{r}, t)$ we conclude that the second term, $| \phi_i |^2 [1 + 2\mathrm{Re} \exp(i\omega_0 t) c_1 c_2^* \phi_1 \phi_2^* / (| c_1 |^2 | \phi_1 |^2 + | c_2 |^2 | \phi_2 |^2)]$, represents the probability

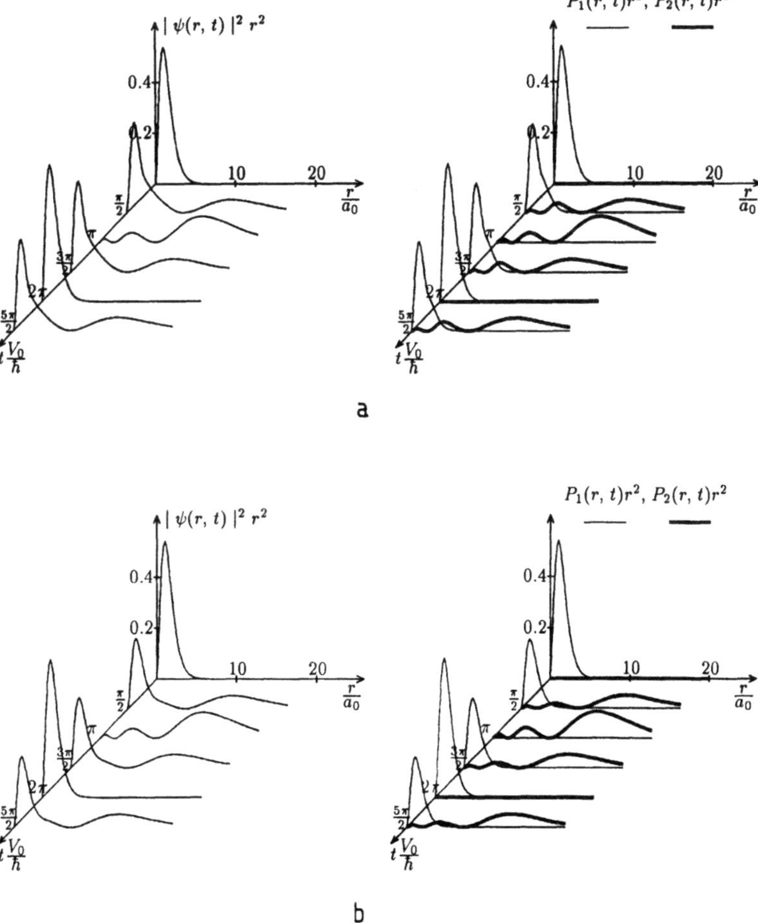

Figure 1. Probability densities $| \psi(r, t)^2 | r^2$, $P_1(r, t)r^2$, $P_2(r, t)r^2$ as functions of r/a_0 and $tV_0/\hbar$ for the Rabi model associated with 1s and 3s states in hydrogen. $E_1 = -13.6\,eV$, $E_2 = -(13.6/9)\,eV$, $a_0 = 0.52 \cdot 10^{-10}\,m$. Detuning Δ is equal to zero (resonance $\omega_0 = \omega_L$), a) $\omega_0 \hbar/V_0 = 1$; b) $\omega_0 \hbar/V_0 = 10$

density that electron with energy E_i is found around $\vec{r}$ at time t. Evidently that the latter probability density is different from the probability density $| \phi_i(\vec{r}) |^2$ in the eigenstate $\phi_i(\vec{r})$ of a free atom. This difference is due to the fact that $\psi(\vec{r}, t)$ is a linear superposition of $\phi_1(\vec{r})$ and $\phi_2(\vec{r})$. $\psi(\vec{r}, t)$ reduces to $\phi_1(\vec{r})$ once during Rabi's period. At resonance $\psi_2(\vec{r}, t)$ reduces also to $\phi_2(\vec{r})$ once during Rabi's period. This is seen clearly at Fig. 1.

IV. ARE THERE QUANTUM JUMPS?

With the aid of de Broglian probabilities it is possible to give new contribution to the theoretical and experimental search having the aim to answer to Schrödinger's question: are there quantum jumps?[6] Schrödinger asked this question because the idea that an atom in a field should be in one of stationary states between the transitions, (i.e. during "noninteracting periods when nothing happens"), seemed to him inconsistent with the evolution law. He insisted on the fact that any solution of the wave equation for an atom in a field is a superposition of stationary states. This remark has been always considered reasonable. Later, Schrödingers question was re-

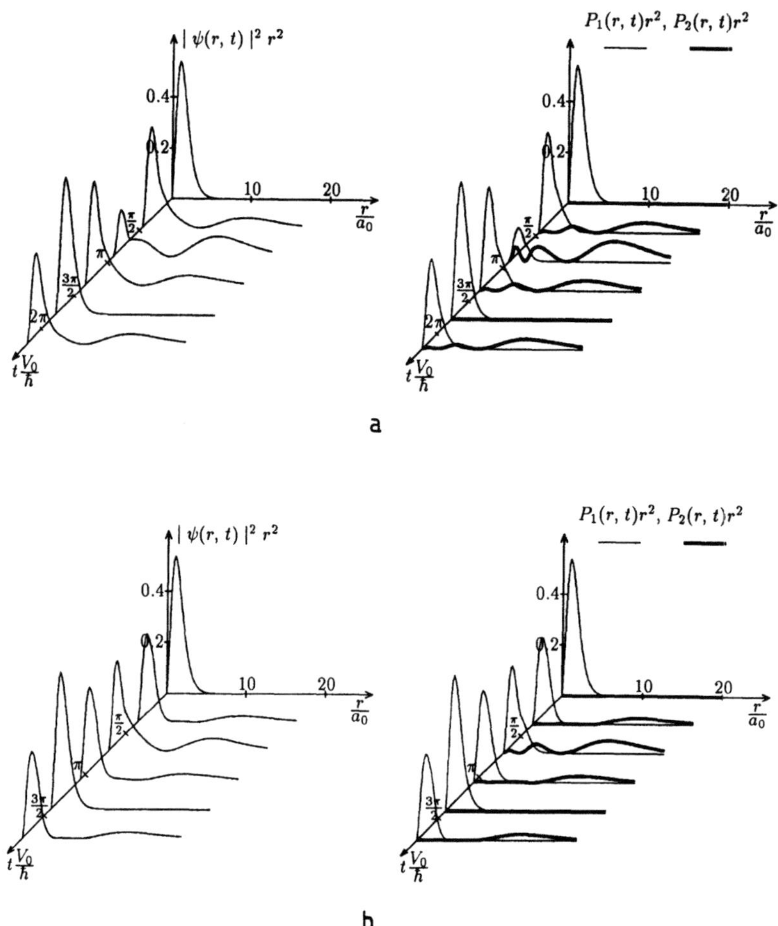

Figure 2. Probability densities $| \psi(r, t) |^2 r^2$, $P_1(r, t)r^2$, $P_2(r, t)r^2$ as functions of r/a_0 and $tV_0/\hbar$ for the Rabi model associated with 1s and 3s states in hydrogen. $E_1 = -13.6\,eV$, $E_2 = -(13.6/9)\,eV$, $a_0 = 0.52 \cdot 10^{-10}\,m$. Detuning Δ is different from zero. a) $\Delta\hbar/V_0 = 0.5$; $\omega_0\hbar/V_0 = 1$; b) $\Delta\hbar/V_0 = 1$, $\omega_0\hbar/V_0 = 10$

formulated and got the form[19]: Does a single atom, subjected to electromagnetic radiation, undergoes a continuous internal evolution as described by a Schrödinger equation, or discontinuities in time occur?

Recently measured time dependence of laser-excited fluorescence from a tree level atom has been considered to be a direct demonstration of the "instantaneous" internal transtions which happen in an atom interacting with light[19]. This implies that

the answer to Schrödinger's question should be: there are quantum jumps. What
about the inconsistency pointed out by Schrödinger? The concept of de Broglian
probabilites gives a possibility to amalgamate the concept of instantaneous transitions (quantum jumps) and continuous evolution of wave function $\psi(\vec{r}, t)$. The point
is that stationarity of energy eigenvalues does not imply stationarity of the electron
wave function and of the electron distribution in space. In our interpretation, an electron wave function is a time-dependent solution $\psi(\vec{r}, t)$ of the Schrödinger equation.
Electron distribution in space is determined by $\mathcal{P}_1(\vec{r}, t)$ if energy is E_1 or by $\mathcal{P}_2(\vec{r}, t)$
if electron energy is E_2. Neither $\mathcal{P}_1(\vec{r}, t)$ nor $\mathcal{P}_2(\vec{r}, t)$ are stationary.

So, we propose the following answer to Schrödinger's question: there are quantum
jumps which consist of "instantaneous" transitions between discrete energy levels and
of continuous changes of states.

REFERENCES

1. R.P. Feynman, R.B. Leighton and M. Sands, "The Feynman Lectures on Physics," Addison-
   Wesley, Reading (1964)
2. L.E. Ballentine, *Am. J. Phys.* 54:883 (1986)
3. A.G. Klein, G.I. Opat and W.A. Hamilton, *Phys. Rev. Lett.* 50:563 (1983)
4. H. Kaiser, R. Clothier, S.A. Werner, H. Rauch and H. Wolwitsch, *Phys. Rev.* A45:31 (1992)
5. E.P. Wigner, *Am. J. Phys.* 31:6 (1963)
6. E. Schrödinger, *The British Journal for the Philosophy of Science* 3:109 (1952)
7. R.G. Brewer and A. Schenzle, *Proc. 2nd Int. Symp. Foundations of Quantum Mechanics*,
   Tokyo, 257 (1986)
8. L. de Broglie, "Etude Critiques des Bases de l'Interpretation Actuelle de la Mecanique Ondulatoire," Gauthier-Villars, Paris (1963)
9. E. Schrödinger, *Endeavour* 9:109 (1950)
10. J.P. Vigier, *Physica* B151:386 (1988)
11. F. Selleri, "Quantum Paradoxes and Physical Reality," Kluwer, Dordrecht (1990)
12. M. Božić and Z. Marić, *Phys. Lett.* A158:33 (1991)
13. M. Božić, Z. Marić and J.P. Vigier, *Found. Phys.* 22:1325 (1992)
14. M. Božić, Compatible statistical interpretation of interference in double-slit interferometer,
    in: "Waves and Particles in Light and Matter," A. van der Merwe and A. Garuccio, eds.,
    Plenum, New York (1994)
15. M. Božić and Z. Marić, Probability and interference, to be published in *Courants, amers et
    écueils en microphysique*, G. Lochac, ed.
16. M. Božić and Z. Marić, to be published in *Found. Phys.*
17. P.L. Knight and P.W. Milonni, *Phys. Reports* 66:21 (1980)
18. L. Allen and J.H. Eberly, "Optical Resonance and Two-level Atoms," Dover, New York (1987)
19. Th. Sauter, R. Blatt, W. Neuhauser and P.E. Toschak, *Physica Scripta* T22:128 (1988)
20. F. Selleri and G. Tarozzi, *Il Nuovo Cimento* 43B:31 (1978)

A TEST OF THE COMPLEMENTARITY PRINCIPLE
IN SINGLE-PHOTON STATES OF LIGHT

Yutaka Mizobuchi and Yoshiyuki Ohtaké

Central Research Laboratory
Hamamatsu Photonics K.K.
Hamakita 434, Japan

INTRODUCTION

The complementarity principle, the way usually accepted was originally proposed
by Niels Bohr. First, he postulated that one could perceive the quantum world only
through classical measurements. But the classical measurements of a pair of physical
quantities (or "observables") such as position and momentum could only be made in
a mutually exclusive manner. Such a pair he called complementary. Thus he made
the complementarity principle. He even extended this principle to a pair of concepts
like the wave and particle pictures of light, which were confronted with each other in
classical physics. Namely, if one is to detect light as a wave, he must abandon the
possibility of detecting its particle nature, or vice versa. This sounds similar to the
uncertainty principle of Werner Heisenberg, which says that two mutually conjugate
observables cannot possibly be detected simultaneously beyond a certain level of inac-
curacy. The product of uncertainties of the conjugate observables cannot be smaller
than a certain universal constant called Planck's constant $\hbar$. However, Bohr's principle
is a wider and more vague concept. The wave and particle pictures of light are not such
a conjugate pair, yet Bohr tried to establish that they were not possibly detected in a
single experiment because of the complementarity principle. Was he right?

P. Ghose, D. Home and G. S. Agarwal[1] (GHA) suspected Bohr's conjecture of mutual exclusiveness between the wave and particle natures of light. They analyzed the
possibility of detecting the both pictures in a single experiment, and reached a conclusion that one could indeed observe both waves and particles by using the single-photon
states of light incident on a combination of two prisms similar to the one employed by
J. C. Bose[2] when he measured the refractive indexes of variable materials for microwave
radiations. In this paper we will report on an experiment proposed by GHA and performed at Hamamatsu Photonics, Japan. The essence of the experiment is described
in Ref. 3, and its interpretation is given in Ref. 4. The result of the experiment agreed
with the GHA analysis, and demonstrated the possibility of observing both the classical
wave and classical particle pictures of light in a single experiment. In the next section

we will discuss on Bohr's complementarity principle. Then in Section 3, we will briefly visit the roles of single-photon states of light, especially the difference between true single-photon states and low-intensity light. Section 4 will be devoted to describing the meaning of the double-prism experiment. Our experiment and its result will be described in Section 5. A brief discussion will follow in Section 6.

WAVE-PARTICLE DUALITY IN TERMS OF INTERFERENCE

There are three types of complementarity in Bohr's mind, as is pointed out in Ref. 4. The first two are complementarity between space-time coordination and causal description, and complementarity between position and momentum. Here we will concentrate ourselves in the third type of complementarity, namely, complementarity between wave and particle pictures, sometimes referred to as the wave-particle duality.

Traditionally the wave-particle duality of light has been argued in terms of interference experiments, e.g., Young's double-slit experiment and the Mach-Zehnder interferometer. The characteristics of these experiments are the following: There are two possible paths for the light beam, and we do not know the criterion of which path they take. To obtain the interference fringes, we should not detect which path the light beam actually takes. On the other hand, to detect the path of the light beam ("which-path" measurement), we must abandon the interference. Namely, the wave and particle pictures are mutually exclusive. But why do such things happen? Is the exclusiveness inevitable?

That the mutual exclusiveness of the wave and particle pictures is not inevitably the consequence of the complementarity principle but is inherent in the interference-type experiment, on which Bohr's argument was based, was first pointed out by Ghose et al. (GHA) in Ref. 1 and further discussed in Ref. 4. It is the choice of experimental methods—interference and which-path measurement—that causes the mutual exclusiveness, for one needs two paths and the other selects one out of the two. So they asked for the possibility of another method of detecting waves. A light wave has many properties; it is propagated in vacuum, diffracted by an edge, reflected off and refracted on a surface, and of course interferes with itself, to count a few. But from among all these properties, besides interference, tunneling is one of the rare cases which contrasts waves with particles with just one "path", because there is a certain spatial domain where a classical particle cannot enter but a wave can. Therefore GHA analyzed the possibility of the optical tunneling effect and found that a certain experiment could be done in which both the classical waves and particles were detected. This experiment will be discussed in detail in Section 4.

SINGLE-PHOTON STATES OF LIGHT

The particle nature of light is the most relevant when we make some photoelectric detection at an extremely low light intensity level. For example, if Young's double slit is exposed to such low intensity light,[5] we will see a sporadic pattern on the screen when the exposure time is short. This sporadic pattern will eventually grow into the fringe pattern due to interference, as time goes on. Each individual spot on the screen seems to present a single photon, and in fact it has been so explained to us in a classroom.

However, very important experiments by A. Aspect and collaborators[6,7] show that the single-photon states of light can be clearly distinguished from ordinary (or classical) light by measuring the mutual correlation functions. The idea is as follows. Photons are incident on a beam splitter. The reflected and transmitted portions are respectively detected with photoelectric devices, and the coincidence between them is counted. Call the number of photons which hit the beam splitter $N_1\,[\mathrm{s}^{-1}]$, the numbers of photons detected in the reflected and transmitted portions $N_r\,[\mathrm{s}^{-1}]$ and $N_t\,[\mathrm{s}^{-1}]$, respectively, and the number of coincidence counted $N_c\,[\mathrm{s}^{-1}]$. Then the second order correlation function is presented as

$$g^{(2)}(i_r, i_t) = \frac{\langle i_r i_t \rangle}{\langle i_r \rangle \langle i_t \rangle} = \frac{N_c N_1}{N_r N_t} \equiv \alpha$$

where $\langle i_r \rangle$ and $\langle i_t \rangle$ are the average intensities of the reflected and transmitted beams, respectively. From the Cauchy-Schwartz inequality,

$$\alpha \geq 1$$

where equality holds if the light source is in a coherent state. Namely, for the classical wave, there is a minimum rate of coincidence which corresponds to the coherent state. When the single-photon states of light is incident on the beam splitter, however, this α must vanish, because a photon cannot be divided into two. Aspect et al. experimentally demonstrated this quantum effect accompanying the single-photon states of light using photons from radiative cascade of the calcium atom.

DOUBLE-PRISM EXPERIMENT

First, we recapitulate J. C. Bose's double-prism experiment[2] with microwaves in modern words.[8] The microwave radiation striking one face of a cube of glass perpendicularly would be transmitted across the opposite face. If the cube is cut across the diagonal, two right-angled isosceles prisms (45° prisms) will be obtained. Waves directed at one of such 45° prisms are totally internally reflected from the hypotenuse face because the incident angle (45°) is greater than the critical angle (Figure 1 (a)). When the second prism is placed in contact with the first one, keeping the two hypotenuses parallel, the waves pass straight through again (Figure 1 (b)). If we now control the air gap between the prisms, maintaining the hypotenuses faces parallel, so that its thickness becomes 5 mm to 6 mm, about a half of the beam will be transmitted while the other half will be reflected (Figure 1 (c)). Here he wrote "transmitted", but actually it is the optical tunneling. The concept of tunneling, however, was first introduced in quantum mechanics. Therefore J. C. Bose, who did this experiment a few years before the beginning of old quantum theory, could not come across this concept.

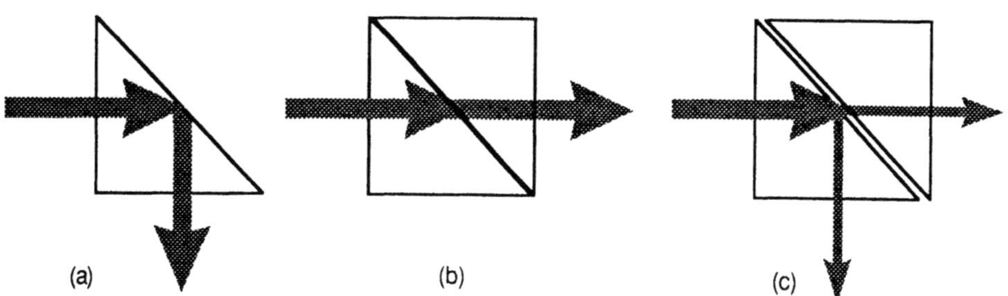

FIGURE 1. Penetration of internally reflected waves. (a) Totally reflected waves. (b) Transmitted waves. (c) Partially reflected and partially transmitted waves.

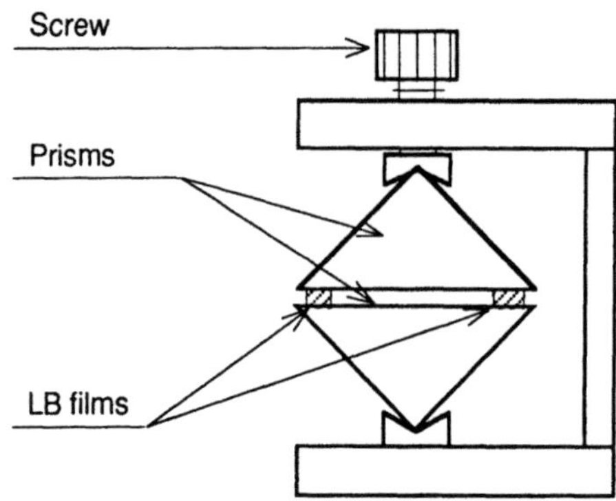

FIGURE 2. Gap controlling mechanism of the double prism.

These phenomena of the double prism can also be confirmed with ordinary light only if we are able to control the gap between two glass prisms within several tens of a nanometer—about one tenth of the wavelength of visible light. Clearly, this optical tunneling effect shows the classical wave nature of light.

In the GHA experiment, to make the double prism for visible light, we employed a Langmuir-Blodgett (LB) film for a spacer between the two hypotenuse faces of the prisms, so that the gap between the faces can be controlled within 10 nm. The gap controlling system is shown in Figure 2. When the thickness of the air gap reduces by tightening the screw, the rate of the transmitted portion increases while correspondingly the reflected portion decrease as is shown in Figure 3.

If light behaves like a classical wave, it should simultaneously be reflected and transmitted. Therefore at a sufficiently high intensity level, coincidence between reflection and tunneling can be observed. However, when the intensity is sufficiently low, the situation becomes a little complicated. At a low intensity level, the statistical property of each light source becomes important. If the light source is a classical one, there always remains a possibility of detecting coincidence between the signals from the reflected and tunneling beams, although the probability may be small. But, if the light source is in the single-photon states, what will happen? GHA argued on three cases:

(a) The "tunneling" phenomenon *occurs* and the two counters click in perfect *anti-coincidence*.

(b) The "tunneling" *occurs* and the two counters click in *coincidence*.

(c) The "tunneling" does *not* occur and *only* the counter for reflection clicks.

They showed that case (a) is the one favored by quantum optics and that in this case we can observe the wave and particle pictures simultaneously. Apparently, case (b) represents the classical wave picture (infinite divisibility), whereas case (c) the classical particle picture (no tunneling). The idea of case (a) is similar to the one proposed by Grangier, Roger and Aspect [6] and Aspect and Grangier [7] with a beam splitter. However, using the double prism makes the conditions different, because tunneling phenomenon manifestly presents the wave picture of light. And under this condition, we can detect the tunneling, but at the same time, we will see the inseparability of the photon which shows the particle nature, if case (a) is demonstrated. We will discuss on this experiment in next section.

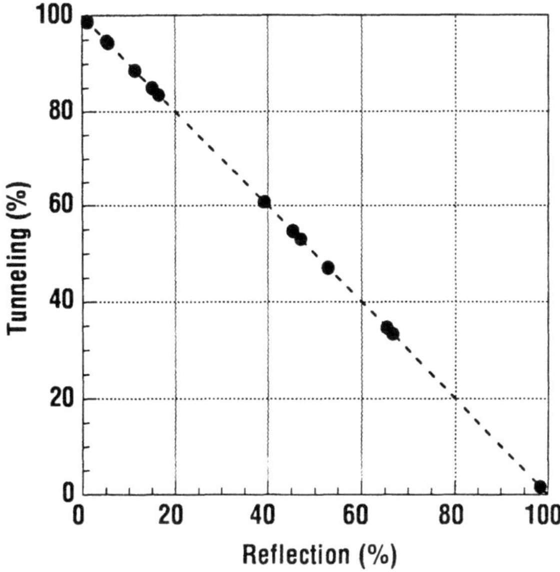

FIGURE 3. The relation between the ratios of reflection and tunneling for various gaps.

ANTICOINCIDENCE EXPERIMENT AND THE RESULTS

Now we describe the anticoincidence experiment proposed by GHA. To make this experiment we must prepare the single photon states of light. We adopted the parametric down conversion technique[9] to obtain photon pairs. Then one of the paired photons can serve as the single photon source. There is known to be another way of preparing the single photons. It is the atomic cascade method which was used in Refs. 6 and 7. Advantage with the down-converted photons over atomic cascades is in the fact that the former is more controllable and gives higher gain if a sufficiently strong laser is used.

The schematic diagram of our experimental arrangement is shown in Figure 4. We used the third harmonics of the pulsed Nd:YAG laser, a typical wavelength of which is 355 nm. The laser generates approximately 20 ps-long pulses which are cut off from

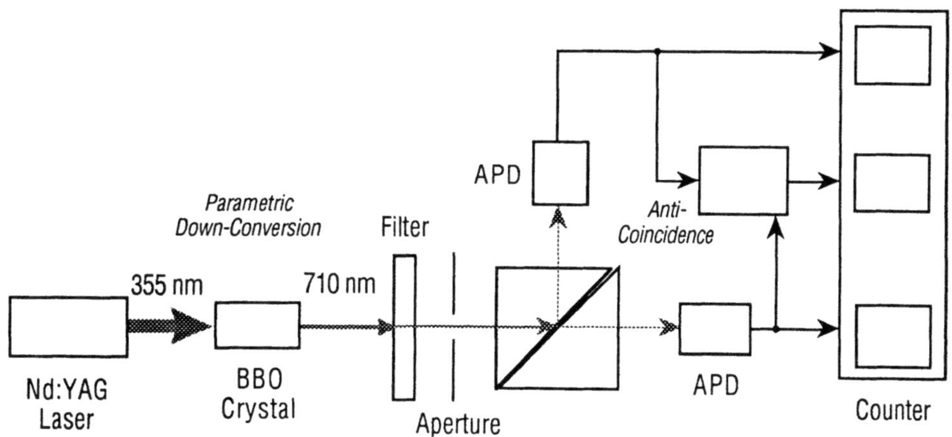

FIGURE 4. The experimental arrangement for anticoincidence experiment.

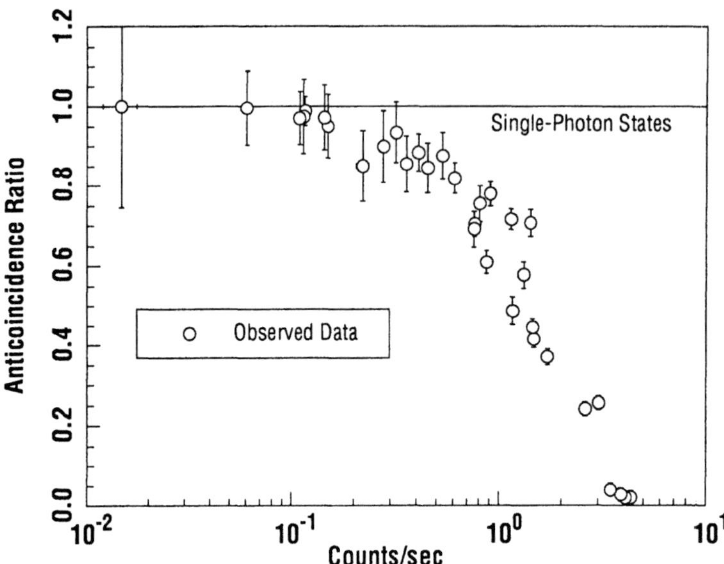

FIGURE 5. Measured anticoincidence ratio. The solid line represents perfect anticoincidence in the single photon states.

a train of pulses with the Pockels cell at 5 Hz repetition rate which was electronically controlled. After slightly focused with a lens ($f = 60$ cm), the beam was injected into a barium borate (β-BaB$_2$O$_4$, BBO) crystal, where the down-converted photon pairs of 710 nm were generated. One of the photons of a pair was selected with a pinhole and passed through a 640 nm cutoff filter to cut off the 355 nm pump beam. The intensity of the signal light was reduced with neutral density (ND) filters to the single-photon intensity level. Then the light was incident on the double prism which is discussed in Section 4. Each of the reflected or transmitted light was detected by an avalanche photo-diode (APD) single-photon detector (Hamamatsu: C4250), whose detection efficiency was measured to be 38%. The electric signals from the single-photon detectors (1.5 V pulse height and 90 ns rectangular pulses) were reshaped to 6 V, 600 ns rectangular pulses with a timing single-channel analyzer unit (EG&G ORTEC: 551 Timing SCA) and then led into the anticoincidence unit (EG&G ORTEC: 414A Fast Coincidence). The resolving time of anticoincidence was determined by the input rectangular pulse duration. The anticoincidence detection was made in the following way: Once a transmitted (reflected) photon signal was detected the clock started, and if no reflected (transmitted) photon signal was detected within 600 ns, a count was recorded. This is the output signal of the anticoincidence unit, which we call the anticoincidence signal. The anticoincidence signal and both the reflected and transmitted signals were accumulated in counter units (NAIG: E-541). The counter units were gated on for 20 μs, synchronized with each laser pulse to minimize the dark counts.

Figure 5 shows the result of the first run of the measurement. The horizontal axis is the detected signal counts per second at each detector. The vertical axis is the ratio of the anticoincidence counts to the number of signal counts from either the reflected or transmitted light which is pre-selected (here it is the transmitted or "tunneling" photon signal). This ratio (we call it the anticoincidence ratio) must be unity if there is perfect anticoincidence, while it must be zero if there is perfect coincidence. In our experimental result, the ratio increases around 1 count per second when we decrease

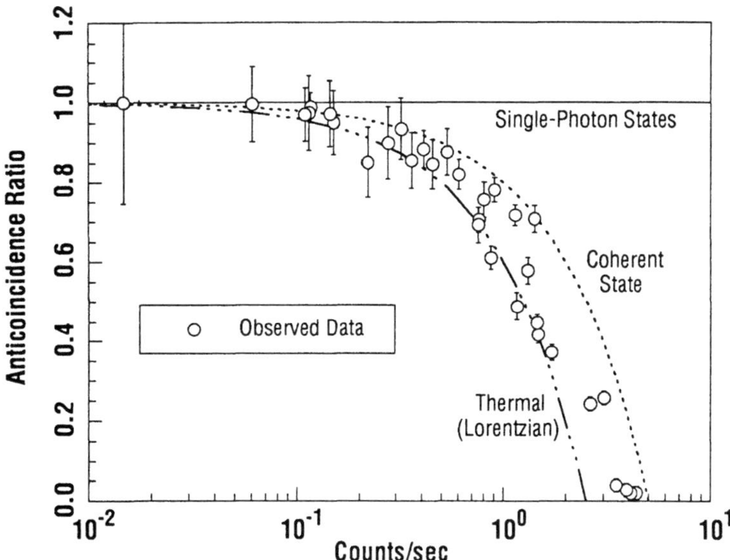

FIGURE 6. Anticoincidence ratio with the theoretical estimates for coherent states and thermal (chaotic) light source.

the number of incident photons, and eventually becomes unity. To see this in detail let us compare the observed trend with the theoretical estimate for a light in the coherent state. Since the parameter a in Section 3 becomes unity for the coherent state, the anticoincidence ratio is expressed as

$$1 - \frac{N_r}{N_1}$$

when the transmitted light is pre-selected. In our experiment $N_1 = 5$, because the repetition rate of the laser is 5 Hz. This curve is shown in a dotted line in Figure 6. Interestingly, the observed data are distributed below the "coherent-state" curve when the number of photons are large. This is because the down-converted photons are not coherent, although they are induced by a laser light. To see this, a theoretical estimate of the anticoincidence ratio is also plotted for the Lorentzian distribution, in which $\alpha = 2$, in Figure 6 (a dotted-dashed line). The observed data are mostly located above this "Lorentzian" curve, and approaches the "single-photon" line (a solid line) when the number of photons is sufficiently small. Unfortunately, our YAG laser does not have sufficient intensity stability, hence we could not obtain sufficient statistical accuracy. Nevertheless we can see that in the single-photon region the measurement showed the perfect anticoincidence, i.e., indivisibility of the photon, while at the same time the optical tunneling effect of classical light waves.

CONCLUDING REMARKS

We conclude that we were able to observe the complementary pair of pictures of light simultaneously in a single experiment; namely, light shows its indivisibility—manifestation of the property of classical particles, while still being able to tunnel across the gap between the prisms—manifestation of the property of classical waves. It is now clear that the mutual exclusiveness between waves and particles is not an

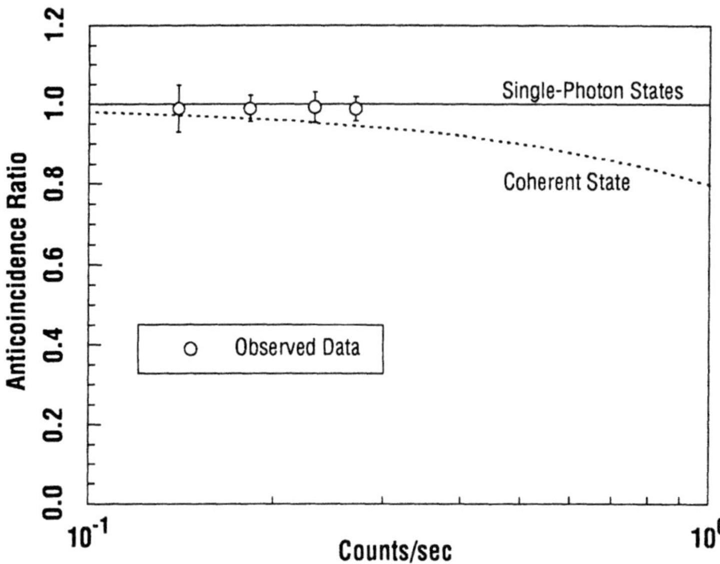

FIGURE 7. Anticoincidence ratio with an Ar-ion laser at 351nm as the pumping beam.

inevitable consequence of the complementarity principle. Rather, these two aspects must be united and the photon must be regarded as an ideal concept of this unity.[10] Namely, we must deal with the wave-particle unity instead of duality.

Finally, we would like to mention that we are now improving our experiment using an Ar-ion laser at 351nm instead of the Nd:YAG laser. We are trying to see the clear difference between the single-photon states and ordinary light with higher statistical accuracy in the intensity region where the coherent states would give, say, 90% of anticoincidence whereas the single-photon states give the 100% anticoincidence. A preliminary result is shown in Figure 7.

Acknowledgments: The research on which this paper is based has been supported in part by the Research Foundation for Opto-Science and Technology, Hamamatsu, Japan.

REFERENCES

1. P. Ghose, D. Home and G. S. Agarwal, *Phys. Lett. A* **153**, 403 (1991).
2. J.C. Bose, *Proc. Roy. Soc. London* **62**, 300 (1897).
3. Y. Mizobuchi and Y. Ohtaké, *Phys. Lett. A* **168**, 1 (1992).
4. P. Ghose, D. Home and G. S. Agarwal, *Phys. Lett. A* **168**, 95 (1992).
5. Y. Tsuchiya, E. Inuzuka, T. Kurono, and M. Hosoda, *Adv. Electron. Electron Phys. A* **64**, 21 (1985).
6. P. Grangier, G. Roger and A. Aspect, *Europhysics Lett.* **1**, 173 (1986).
7. A. Aspect and P. Grangier, *Hyperfine Interactions* **37**, 3 (1987).
8. R. P. Feynman, R. B. Leighton and M. Sands, in: *The Feynman Lectures on Physics*, Vol. II, Addison-Wesley, Reading, Mass. (1964), p.33-12. (Historical importance of J.C. Bose's double-prism experiment has been discussed in: S. Zhu, A. W. Yu, D. Hawley and R. Roy, *Am. J. Phys.* **54**, 601 (1986).)
9. D. C. Burnham and D. L. Weinberg, *Phys. Rev. Lett.* **25**, 84 (1970).
10. H. Yılmaz, a talk presented at this conference.

EXPERIMENTS WITH ENTANGLED TWO-PHOTON STATES

FROM TYPE-II PARAMETRIC DOWN CONVERSION:

EVIDENCE FOR WAVE-PARTICLE DUALITY

C.O. Alley,[1] T.E. Kiess,[1] A.V. Sergienko,[2] and Y.H. Shih[2]

[1]University of Maryland College Park, College Park, MD 20742
[2]University of Maryland Baltimore County, Baltimore, MD 21228

NOTE ON WAVE-PARTICLE UNITY

H. Yilmaz

Hamamatsu Photonics K.K. Hamamatsu City, 435 Japan
Electro-Optics Technology Center,
Tufts University, Medford, MA 02155, U.S.A.

INTRODUCTION

Although the abstract rules of quantum mechanics and quantum field theory seem to be able to describe the results of quantum optics experiments, it would be highly desirable to comprehend what is actually "going on" in terms of a physical picture of quantum phenomena. Actual experimental realizations in a simple manner of correlation of the Einstein-Podolsky-Rosen-Bohm type predicted for entangled two-photon states can focus attention on this "conceptual incompleteness" of quantum mechanics.

A very attractive physical picture has been conceived by Professor Hüseyin Yilmaz in which quanta are regarded as packets of phase waves. These ideas were presented by Prof. Yilmaz following the talk at the conference which is reported in this paper. A note by him on "Wave-Particle Unity" is at the end of this article.

Quantum mechanical entanglement[1,2] of a pair of light quanta generated from Type I spontaneous optical parametric down conversion was successfully demonstrated in an Einstein-Podolsky-Rosen-Bohm experiment[3,4] seven years ago.[5,6] Type I down conversion has drawn a great deal of attention recently, and has been used in different two-photon interferometry type experiments for demonstrating the quantum nature of light. The experimental study of Type II photon pairs was performed before Type I in our laboratory, however, the experimental results seemed to suggest that the orthogonally polarized signal and idler photon pair do not have the expected quantum entanglement. This phenomenon has troubled us and many other physicists with whom we have communicated in the past.[7] The entanglement of the Type II photon pair was finally demonstrated recently in our laboratory under two experimental conditions: (1) using a thin nonlinear crystal (2) detecting coincidences in narrow spectral bandwidth. In this paper, we wish first to report the experimental study of this crystal length and detection bandwidth dependent entanglement of Type II down conversion. A brief theoretical model is presented to explain this phenomenon. Then we report an experimental study of entangled two-photon states in Type II down conversion with linear, circular and elliptical polarizations.

Frontiers of Fundamental Physics, Edited by M. Barone
and F. Selleri, Plenum Press, New York, 1994

EXPERIMENTAL PROCEDURES

The experimental set up to study the effect of crystal length and detection bandwidth on the degree of entanglement is illustrated in fig. 1. A single mode CW Argon ion laser line of 351.1 nm was used to pump a BBO (β-BaB$_2$O$_4$) nonlinear crystal. The BBO was cut for a Type II phase matching condition to generate a pair of orthogonally polarized signal and idler photons collinearly and degenerately with 702.2 nm wavelength. Two BBO crystals with lengths of 5.65 mm and 0.5 mm, respectively, were used in the experiments. The 702.2 nm pairs were separated from the pumping beam by a quartz dispersion prism, then directed collinearly at a near normal incident angle to a polarization independent beam splitter which has 50% - 50% reflection and transmission coefficients. In each transmission and reflection output port of the beamsplitter a Glan Thompson linear polarization analyzer followed by a narrow bandwidth interference spectral filter were placed in front of a single photon detector. The photon detectors are dry ice cooled avalanche photodiodes operated in Geiger mode. The output pulses of the detectors were then sent to a coincidence circuit with a 3 nsec coincidence time window. The two detectors are separated by about 2 m, so that compared to the 3 nsec coincidence window, the detections are space-like separated events.

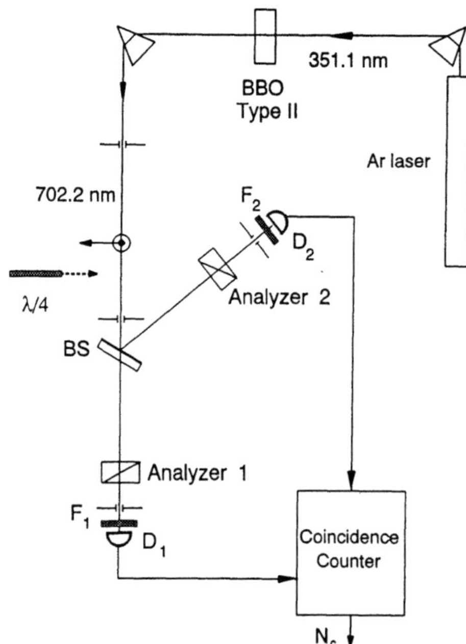

Figure 1. Schematic diagram of the experiment

The coincidence counting rates were studied as functions of angles θ_1 and θ_2, where θ_i is the angle between the axis of the ith polarization analyzer and the $\hat{x}$ direction, which is defined by the o-ray polarization plane of the BBO crystal. Keep in mind that a right-handed natural coordinate system with respect to the $\mathbf{k}_i$ vector as the positive $\hat{z}$ direction is employed for the discussions in this paper. The following form of coincidence rate as a function of θ_1 and θ_2 was observed in the experiments,

$$R_c = R_{c0}(\cos^2\theta_1 \sin^2\theta_2 + \sin^2\theta_1 \cos^2\theta_2 - \rho \sin\theta_1\cos\theta_2\sin\theta_2\cos\theta_1) \tag{1}$$

where ρ is a parameter which depends on the crystal length and the detection bandwidth, i.e., the bandwidth of the interference filters placed in front of the detector. If $\rho = 2$, eq. (1) reduces to,

$$R_c = R_{c0} \sin^2(\theta_1 - \theta_2) \tag{2}$$

which is the expected quantum correlation for the entangled two-photon EPR-Bohm state

$$| \Psi > = 1/\sqrt{2} \, (| X_1 > \otimes | Y_2 > - | Y_1 > \otimes | X_2 >) \tag{3}$$

$| \Psi >$ quantum mechanically indicates a two-photon polarization state which is a superposition of the quantum probability amplitudes: (1). <u>o-ray transmitted</u> $\otimes$ <u>e-ray reflected</u> and (2). <u>e-ray transmitted</u> $\otimes$ <u>o-ray reflected</u>, when the orthogonally polarized photon pair meets the beamsplitter. On the other hand, if $\rho = 0$ the interference cross term does not contribute, resulting in a classical interpretation of the experiment.

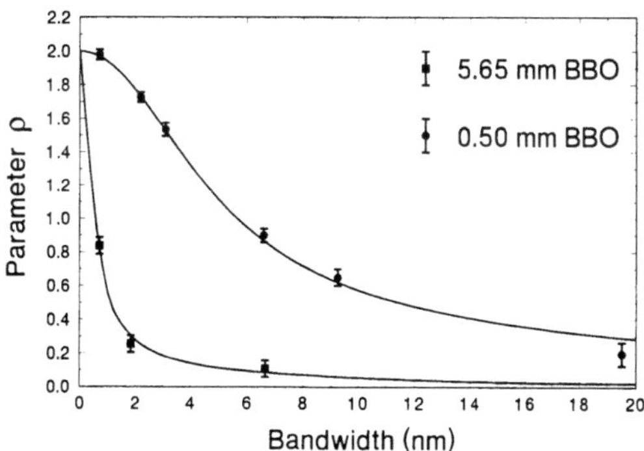

Figure 2. Crystal length and detection bandwidth dependent entanglement.

Fig. 2 reports the measured values of ρ for BBO crystals with lengths of 5.65 mm and 0.5 mm for different bandwidths of the filters. Note that for the 5.65 mm BBO crystal ρ was always substantially less than 2 for the filters that used in the measurements. For the 0.5 mm BBO, $\rho = 1.98$ was achieved with a 1 nm bandwidth spectrum filter.

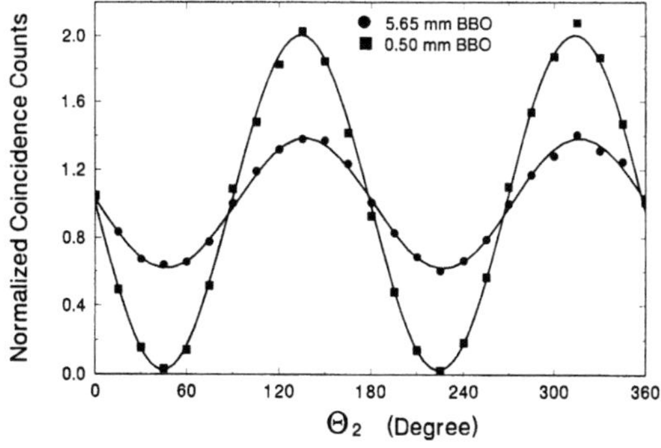

Figure 3. Coincidence measurements for linear polarization state when θ_1 was set equal to $45°$

The solid curves are the fits to a theoretical model which will be presented below. The values of ρ were obtained from the measurements of coincidence rate as functions of θ_1 and θ_2. Fig. 3 and fig. 4 show typical measurements which reflect the different coincidence behavior for 5.65 mm and 0.5 mm BBO crystals. In fig. 3, θ_1 was set to 45° and the coincidence rate was mapped out as a function of θ_2. In fig. 4, both θ_1 and θ_2 were changed, keeping the sum of θ_1 and θ_2 equal to 90°. In both fig. 3 and fig. 4 the filters were 1 nm bandwidth. By fitting many similar curves, $\rho = 0.72$ and $\rho = 1.98$ were determined for 5.65 mm and 0.5 mm crystals, respectively.

THEORETICAL ANALYSIS

One possible model to explain the above crystal length and detection bandwidth dependent entanglement in Type II down conversion is to consider that the generation of the photon pairs is highly localized in space and time inside the nonlinear crystal. The o-ray and e-ray of the signal and idler light quanta have different propagation velocities inside the nonlinear crystal. Therefore, the pairs generated close to the input surface of the crystal will suffer a larger

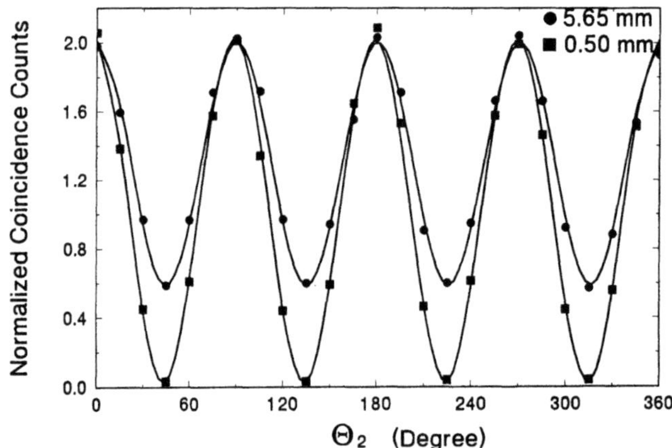

Figure 4. Coincidence measurements for linear polarization states when $\theta_1 + \theta_2 = 90°$ was preserved.

separation in time than these generated close to the output surface. The resulting non-overlap of the spatial part of the wavefunction for the pair reduces the degree of entanglement, i.e., producing a smaller value of the parameter ρ.

According to the standard theory of parametric down conversion, the two-photon state can be written as,[8,9]

$$| \Psi > = \int d\omega_p A(\omega_p) \int d\omega_1 d\omega_2 \delta(\omega_1 + \omega_2 - \omega_p) \delta(k_1 + k_2 - k_p) a_o^\dagger(\omega_1(k_1)) a_e^\dagger(\omega_2(k_2)) | 0 > \qquad (4)$$

where ω and k represents the frequency and the wave vector for signal (1), idler (2), and pump (p). The δ functions represent perfect phase matching of the down conversion. The subscript indices o and e for the creation operators indicate the ordinary and extraordinary rays of the down conversion. The defined $\hat{x}$ and $\hat{y}$ coordinate axes coincide with the o-ray and the e-ray polarization directions of the crystal. $A(\omega_p)$ is a spectral distribution function for the laser line, which is usually considered to be a Gaussian.

The fields at the detectors 1 and 2 are given by

$$E_1^{(+)}(t) = \alpha_t \int d\omega\, f(\omega)\, [\exp(-i\omega t_1^o)\, \hat{e}_1 \cdot \hat{e}_o\, a_o(\omega) + \exp(-i\omega t_1^e)\, \hat{e}_1 \cdot \hat{e}_e\, a_e(\omega)]$$

$$E_2^{(+)}(t) = \alpha_r \int d\omega\, f(\omega)\, [\exp(-i\omega t_2^o)\, \hat{e}_2 \cdot \hat{e}_o\, a_o(\omega) + \exp(-i\omega t_2^e)\, \hat{e}_2 \cdot \hat{e}_e\, a_e(\omega)] \tag{5}$$

where $\hat{e}_i$ is in the direction of the ith linear polarization analyzer axis, $a_o(\omega)$ and $a_e(\omega)$ are the destruction operators for the o-ray and the e-ray, α_t and α_r are the complex transmission and reflection coefficients of the beamsplitter, $f(\omega)$ is the spectral transmission function of the filters, $t_i^o = t - S_i^o/c$, $t_i^e = t - S_i^e/c$, $i = 1,2$, where $S_i^{o,e} = \int dr\, n^{o,e}(r)$, indicates the optical path for o-ray or e-ray of the ith beam, where $n^{o,e}$ is the refraction index. The use of pinholes, which limit the transverse width of the beams, allows a good one dimensional approximation. The choice of the directions of the beams will be used implicitly to replace the $\delta(k_1 + k_2 - k_p)$ function in the following calculations. The detectors will be treated as point detectors to be located at r_1 and r_2. An effective two-photon wavefunction which is realized by the coincidence measurement at the two detectors can be defined by

$$\Psi(t_1,t_2) = \langle\, 0\, |\, E_1^{(+)}(t_1)\, E_2^{(+)}(t_2)\, |\, \Psi\, \rangle \tag{6}$$

For Gaussian filters $f(\omega)$ with bandwidth σ and Gaussian spectral distribution of the pump field with bandwidth σ_p, it is straight forward to show from (4), (5), and (6),

$$\Psi(t_1,t_2) = \alpha_t\alpha_r\, [\hat{e}_1 \cdot \hat{e}_o\, \hat{e}_2 \cdot \hat{e}_e\, A(t_1^o,t_2^e) + \hat{e}_1 \cdot \hat{e}_e\, \hat{e}_2 \cdot \hat{e}_o\, A(t_1^e,t_2^o)] \tag{7}$$

where $A(t_1,t_2)$ is,[10]

$$A(t_1,t_2) = A_0\exp[-\sigma_p^2(t_1+t_2)^2/8]\exp[-\sigma^2(t_1+t_2)^2/4]\exp(-i\Omega_1 t_1)\exp(-i\Omega_2 t_2) \tag{8}$$

where Ω_i is the ith filter center frequency and related to the peak frequency of the pump, Ω_p, by $\Omega_1 + \Omega_2 = \Omega_p$. For the single frequency pump, $\exp[-\sigma^2(t_1+t_2)^2/8]$ can be taken equal to one for a good approximation, so that $|\Psi|^2$ can be written as,

$$|\Psi|^2 = \left| A_0\left\{\exp[-(\sigma/2)^2(\tau-\delta)^2]\cos\theta_1\sin\theta_2 - \exp[-(\sigma/2)^2(\tau+\delta)^2]\sin\theta_1\cos\theta_2\right\}\right|^2 \tag{9}$$

where $\tau \equiv T_1 - T_2$, T_i is the detection time of the ith detector, and $\delta \equiv (S_i^o - S_i^o)/c$. In (9) for simplicity we have assumed that the two detectors are equidistant from the beamsplitter. A phase shift of π due to reflection has been taken into account. The coincidence counting rate is given by

$$R_c = (1/T) \iint\limits_0^T dT_1 dT_2\, |\, \Psi(t_1,t_2)|^2\, S(T_1-T_2,\Delta T_c) \tag{10}$$

where T is the duration time of the measurement, $S(t,\Delta T_c)$ is a function that describes the coincidence circuit, ΔT_c is the time window of the coincidence circuit, for $T_1-T_2 > \Delta T$, $S(t,\Delta T_c) \cong 0$, and for $T_1-T_2 < \Delta T$, $S(t,\Delta T_c) \cong 1$. If the coincidence time window is large enough, the time integral can be taken from $-\infty$ to $+\infty$ and the coincidence counting rate becomes,

$$R_c = R_{c0}[\cos^2\theta_1\sin^2\theta_2 + \sin^2\theta_1\cos^2\theta_2 - 2\exp(-\sigma^2\delta^2/2)\sin\theta_1\cos\theta_1\sin\theta_2\cos\theta_2] \tag{11}$$

Eq. (11) clearly shows the dependence on σ, the bandwidth of the filters, and δ, the optical path difference between the o-ray and e-ray. This determines the degree of the entanglement. For a nonlinear crystal with thickness L, ρ can be estimated by averaging over δ,

$$\rho = [2/\delta(L)] \int_0^{\delta(L)} d\delta \, \exp(-\sigma^2 \delta^2/2) \tag{12}$$

The solid lines in fig. 2 are calculated from (12) for the 5.65 mm and 0.5 mm BBO crystals. The curves agree with the measured values of ρ within reasonable experimental error. One can achieved $\rho \cong 2$ with bandwidth filters less than 1 nm for a 0.5 mm BBO thin crystal.

TWO-PHOTON ENTANGLEMENT IN LINEAR, CIRCULAR AND ELLIPTICAL POLARIZATION

After the achievement of $\rho = 2$, measurements for two-photon polarization entangled states were made. The use of a quarter-wave plate and a beamsplitter can demonstrate the quantum mechanical entanglement of arbitrary elliptical polarization states easily in Type II down conversion. The experimental set up is the same as in fig. 1, except a quarter-wave plate is placed after the 0.5 mm BBO crystal. If the fast axis of the quarter-wave plate is oriented at angle Φ with respect to the $\hat{x}$ direction, the orthogonal linear polarization states $|\,X>$ and $|\,Y>$ are transformed to orthogonal elliptical polarization states:

$$|\,X > \Rightarrow \cos\Phi \,|\,X'> - i \sin\Phi \,|\,Y'> \tag{13}$$
$$|\,Y > \Rightarrow \sin\Phi \,|\,X'> + i \cos\Phi \,|\,Y'>$$

where $|\,X'>$ and $|\,Y'>$ are in the direction of the fast and slow axes of the quarter-wave plate, respectively. After the beamsplitter a two-photon entangled state with elliptical polarizations is produced,

$$|\,\Psi> = 1/\sqrt{2}\left[\begin{pmatrix}\cos\Phi \\ -i\sin\Phi\end{pmatrix}_1 \begin{pmatrix}\sin\Phi \\ -i\cos\Phi\end{pmatrix}_1 + \begin{pmatrix}\sin\Phi \\ i\cos\Phi\end{pmatrix}_1 \begin{pmatrix}\cos\Phi \\ i\sin\Phi\end{pmatrix}_2\right] \tag{14}$$

where state $|\,\Psi>$ is a superposition of the quantum probability amplitudes:

(1). $(\cos\Phi|X'> - i\sin\Phi|Y'>)$ transmitted $\otimes$ $(\sin\Phi|X'> + i\cos\Phi|Y'>)$ reflected,
(2). $(\sin\Phi|X'> + i\cos\Phi|Y'>)$ transmitted $\otimes$ $(\cos\Phi|X'> - i\sin\Phi|Y'>)$ reflected,

when the orthogonal elliptical polarized photon pair meets the beamsplitter.

The coincidence counting rate for linear polarization analyzers is then,

$$R_c = R_{co}[\,\sin^2(2\Phi)\cos^2(\theta'_1 + \theta'_2) + \cos^2(2\Phi)\sin^2(\theta'_1 - \theta'_2)] \tag{15}$$

where θ'_i is the angle between the axis of the ith polarization analyzer and the $|\,X'_i>$ direction. Care has to be taken to follow the rules of natural coordinate system, especially for the reflected beam. Note that the direction of $|\,X'_2>$ is opposite to that of $|\,X'_1>$.

If $\Phi = 0°$, state (14) becomes state (3) which is a two-photon linear polarization entangled state. Quantum correlations given by eq. (2) were observed. Bell's inequality violation of 22 standard deviation was demonstrated[11].

For $\Phi = 45°$. State (14) becomes the circular polarization EPR-Bohm state,

$$|\,\Psi> = 1/\sqrt{2}\,(|\,R_1>\otimes|\,R_2> + |\,L_1>\otimes|\,L_2>) \tag{16}$$

The expected quantum correlations

$$R_c = R_{c0}\cos^2(\theta_1 + \theta_2) = R_{c0}\cos^2(\theta'_1 + \theta'_2) \tag{17}$$

were measured experimentally. Fig. 5 reports the measured results.

524

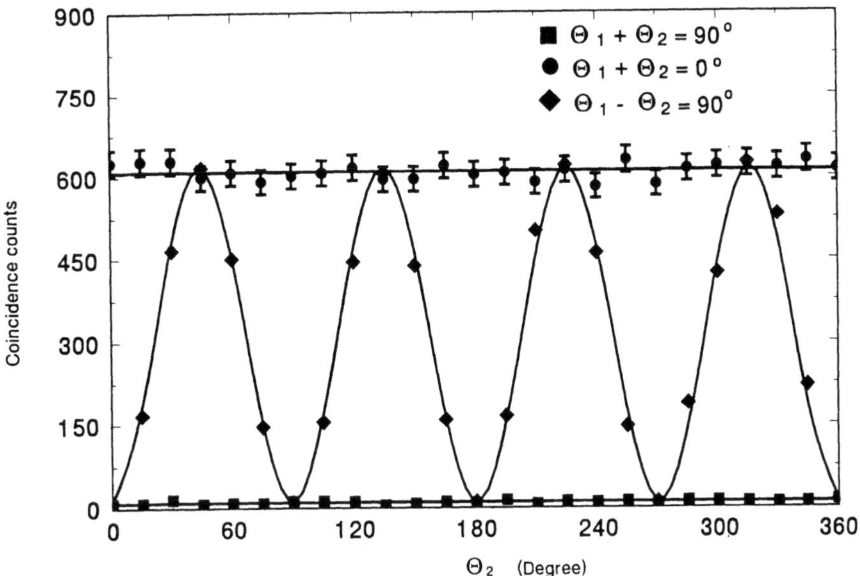

Figure 5. Coincidence measurements for circular polarization EPR-Bohm state.

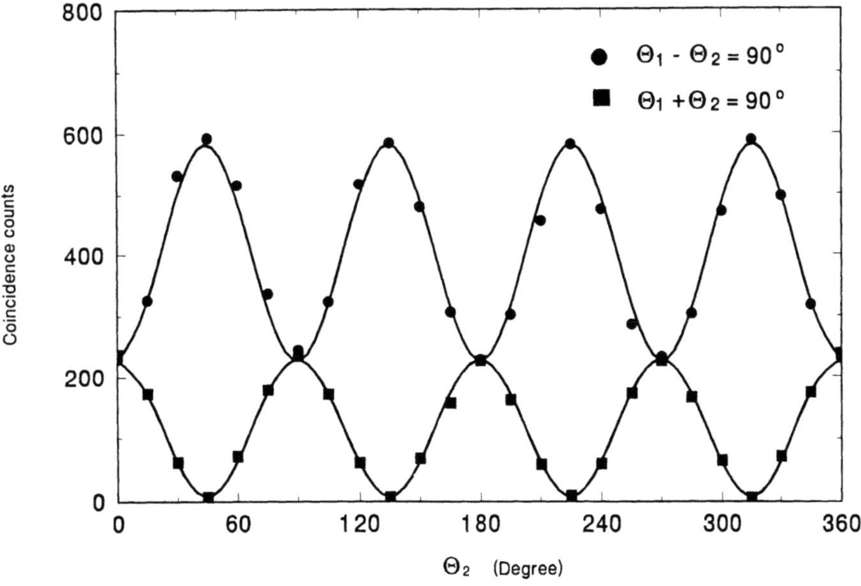

Figure 6. Coincidence measurements for elliptical polarization state with quarter waveplate oriented at 26.5°

When the quarter-wave plate was set to $\Phi = 26.5^\circ$ and 71.5°, fig. 6 and fig. 7 report four typical measurements which were taken under the conditions: $\theta_1' \pm \theta_2' = 90^\circ$. The solid lines in these figures are the fitting curves of (15), the measured coincidence counting rates agree with (15) within reasonable experimental errors. Note, here, we use θ' system to define the angles for the analyzers.

NOTE BY H. YILMAZ: WAVE-PARTICLE UNITY

Consider the set of five equations:

$$\lambda p = h \tag{1}$$

$$\frac{\partial \omega}{\partial k} = v_g = v = \frac{\partial E}{\partial p} \tag{2}$$

$$\frac{\omega}{k} = u_\varphi = u = \frac{E}{p} \tag{3}$$

$$uv = c^2 \tag{4}$$

$$p = mv \tag{5}$$

$$(+) \text{ Energy, } (+) \text{ Probability}$$

1) They contain both c and h. They seem to be compatible with Special Relativity and Quantum Mechanics. However, there seems to be a problem with $u=E/p$ in (3). Because:

    a) There are two velocities v, u for single particle
    b) The phase velocity u is in general greater than c

$$u = \frac{E}{p} = \frac{c^2}{v} \geq c$$

2) Yet the five equations above can be used to consistently derive

    a) Mass-Energy Relation: $E = pu = muv = mc^2$

    b) Energy-Frequency Relation: $E = pu = p\lambda \upsilon = h\upsilon$

    c) Increase of mass with speed.

    d) Uncertainty relations. $\Delta x \Delta(\frac{1}{\lambda}) \geq 1 \Rightarrow \Delta x \Delta p \geq h/2\pi$

    e) Relativistic Hamilton-Jacobi Equation.

    f) Lorentz transformations

    .

    .

    .

    (...) Conceivably all of the Relativity & Quantum Mechanics
        (Note that non-relativistic quantum mechanics is not compatible with (3) since there
        $u = v/2$ hence does not preserve the orthogonality of rays and wavefronts implied by
        $uv = c^2$.)

3) <u>Reconciliation</u>: Consider the pictorial representation below

as a quasistable wave configuration (essentially a finite energy solution of a (possibly nonlinear) wave equation) where the phase waves are defined only inside the envelope. The envelope moves with group velocity whereas the phase waves move with phase velocity. But the phase waves are defined only inside the envelope. Therefore the problem mentioned above can be resolved by assuming the pictorial representation to act as a unit with regards to measurement (interaction). This makes the introduced object non-local but achieves a certain unity which removes the contradiction between Special Relativity and Quantum Mechanics.

4) Mutual exclusivity is untenable. If the classical limit be imagined as $h \Rightarrow 0$ we can write

$$\lambda p = h \Rightarrow 0$$

$$p = \text{finite}, \ \lambda = 0 \ (\text{particle; not wave})$$

$$\lambda = \text{finite}, \ p = 0 \ (\text{wave; not particle})$$

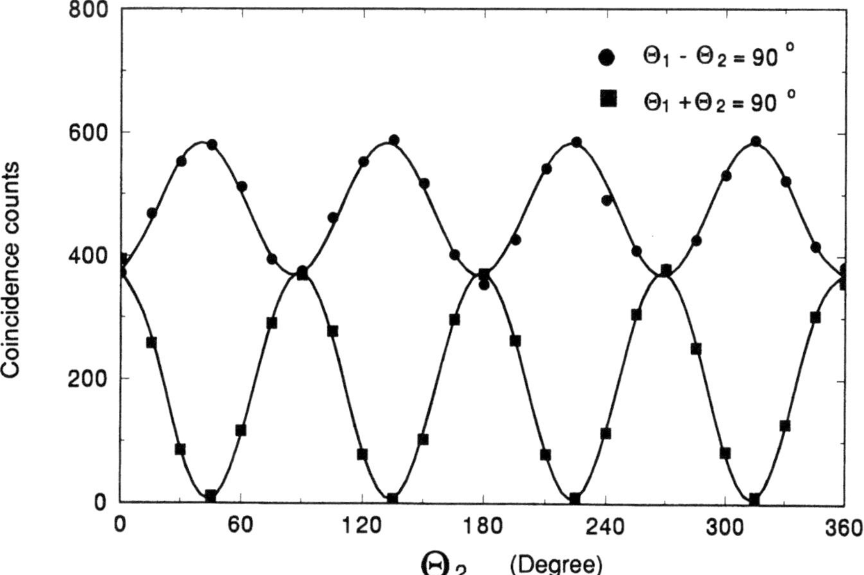

Figure 7. Coincidence measurement for elliptical polarization state with quarter waveplate oriented at 71.5°

Thus when $h \Rightarrow 0$ there are two distinct classical limits, not one. They are mutually exclusive as Bohr advocated. But Bohr wants to reconstruct Quantum Mechanics from such mutually exclusive limits by saying that they are also complementary. Complementarity belongs to $h \neq 0$, mutual exclusivity to $h = 0$. The letter is counter factual and can not be defended. We shall rather say:

The wave and particle aspects are mutually
compatible and complete each other into a higher order of
existence we may call Wave-Particle Unity.

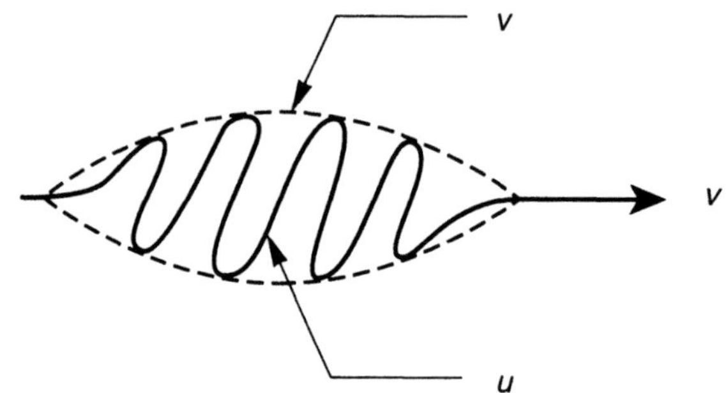

Figure 8. Schematic of wave packet.

CONCLUSIONS

The phenomena exhibited in these experiments and the analysis presented support the picture of the pair of photons being actually wave packets produced simultaneously at the same place in the crystal. This is consistent with the wave-particle unity idea. The influence on the degree of entanglement by crystal length and by filter bandwidth can be comprehended intuitively in terms of this picture.

Another experiment described at this conference by Dr. Yutaka Mizobuchi "A Test of the Complementarity Principle in Single Photon States of Light", also supports the wave-particle unity idea.[12] We also wish to suggest similarities with the ideas of Professor Asim Barut on an underlying "second quantum mechanics" of individual events presented at this conference.[13]

REFERENCES

1. E. Schrödinger, Naturwissenschaften, **23**, 807, 823, 844 (1935).
   A translation of these papers appears in J.A. Wheeler and W.H. Zurek, ed., *Quantum Theory and Measurement*, Princeton University Press, (1983).
2. M.A. Horne, A. Shimony, and A. Zeilinger, Phys. Rev. Lett. **62**, 2209 (1989).
3. A. Einstein, B. Podolsky and N. Rosen, Phys. Rev. **47**, 777 (1935).
4. D. Bohm, *Quantum Theory*, Prentice Hall, Englewood Cliffs, (1951).
5. C.O. Alley and Y.H. Shih, Foundations of Quantum Mechanics in the Light of New Technology, ed. M.Namiki et. al., p. 47 (1986).
   Y.H. Shih and C.O. Alley, Phys. Rev. Lett. **61**, 2921 (1988).
6. Z.Y. Ou and L. Mandel, Phys. Rev. Lett. **61**, 50 (1988).
7. Private communications with J.A. Wheeler, E.P. Wigner, A. Shimony, J.D. Franson and others.
8. W.H. Louisell, A. Yariv, and A.E. Siegman, Phys. Rev. **124**, 1646 (1961).
9. D.N. Klyshko, *Photons and Nonlinear Optics*, Gorden and Breach Science Publishers, N.Y., (1988).
10. Y.H. Shih and M.H. Rubin, Phys. Lett. A, **182**, 16 (1993).
11. T.E. Kiess, Y.H. Shih, A.V. Sergienko, and C.O. Alley, Phys. Rev. Lett., **71**, 3893 (1993).
12. Y. Mizobuchi, these proceedings.
13. A. Barut, these proceedings.

CORRELATION FUNCTIONS AND EINSTEIN LOCALITY

Augusto Garuccio,[1] and Liberato De Caro[2]

[1]Dipartimento di Fisica
Università di Bari
Via Amendola 173
I-70126 Bari, Italy
[2]Centro Nazionale Ricerca e Sviluppo Materiali
S.S. 7 Appia Km. 712
I-72100 Brindisi, Italy

INTRODUCTION

Among the problem concerning the foundation of quantum mechanics (QM), the disagreement between this theory and relativity is the main and most difficult one to solve. The argument was posed in 1935 by Einstein, Podolsky and Rosen in a famous paper[1] in which they proved the incompatibility among three hypotheses: a) quantum mechanics is correct; b) quantum mechanics is complete; c) local "elements of reality" exist associated with the atomic system that determine the results of performed measurements.

This paper opened a long and as yet unsettled debate about which one of the three hypotheses should be discarded. For example, Einstein suggested to abandon the completeness of QM, while Bohr[2] refused to suppose that "elements of reality" exist.

The most important step forward in this debate was taken in 1965 by Bell[3], who proved that the assumption (b) is unessential and that there is an incompatibility between the hypothesis that QM is correct and the hypothesis that local "elements of reality" exist. The paper of Bell opened the way to an experimental check of the validity of assumption (a) or (c), and in the last years many attempts have been carried out in order to solve this debate using the experimental results.

Unfortunately the inequality proved by Bell is based on two hypotheses hard to realise experimentally. The first one is that an experimental configuration yielding total correlation exists; the second one is that it is possible perform measurements on every physical system and that such measurements always yield well defined results.

The aim of research since 1964 was directed on one hand toward new experimental designs able to better approximate the ideal configuration, on the other hand toward the definition of new experimentally testable inequalities. The papers by Clauser, Horne, Shimony and Horne[4] , Clauser and Horne[5], and Garuccio and Rapisarda[6] represent a step toward that goal. In all these papers some "supplementary assumptions" have been introduced in order to compare the experimental data with Bell's inequality obtained in the ideal case.

In the present paper we will introduce a different approach[7] to the problem: starting from a general causal and local theory we will deduce a new Bell-type inequality for the case of real experiments. The mathematical formalism, which we will introduce in order to describe the most general experimental set-up, will include all the pervious supplementary assumptions as particular cases.

BELL'S INEQUALITY FOR THREE-VALUED OBSERVABLES

Schematically, an experiment on the EPR paradox based on polarization correlation measurements of optical photons consists of a source emitting polarizated photon pairs travelling in opposite directions and two measuring apparata, each composed of a polarizer followed by one or two photomultipiers.

Let A and B be two three-valued observables which can assume the values $\{\pm 1, 0\}$. For example, using a source of correlated photon pairs in a typical EPR experimental set-up, the values $A=\pm 1$ ($B=\pm 1$) can be associated with the transmission through the ordinary and extra-ordinary channels of polarizer 1 (2) with axis orientation a (b), respectively, and subsequent detection; whereas $A=0$ ($B=0$) can be associated with the case of no count in the detector 1 (2), which means either absorption in the polarizer or lack of detection. Let the result of measurements of A and B be dependent on the arguments a and b, respectively, which are assumed to be experimental parameters, and on the element of reality λ. Let λ be variable over the set L, with density $\rho(\lambda)$ such that

$$\int_\Lambda d\lambda \rho(\lambda) = 1. \tag{1}$$

The expectation value of a function A is, by definition

$$E[A] = \int_\Lambda d\lambda \rho(\lambda) A(a,\lambda) \tag{2}$$

and, anagously, the expectation value for the product of the obsevables A and B is:

$$E[A,B] = \int_\Lambda d\lambda \rho(\lambda) A(a,\lambda) B(b,\lambda) \tag{3}$$

Let us note that usually in the literature on this subject the expectation value (3) is referred to as correlation function. This is not generally correct, since the correlation function is expressed by formula (3) if and only if the expectation values (2) of the two observables A and B equal zero and the variances equal one, i.e. if we are dealing with a dichotomic case. Nevertheless, Bell's inequality has been always defined as a linear combination of the functions E[AB]. Therefore, in order to obtain results directly comparable with those of the dychotomic case, also for the three-valued case we will define the Bell's observable as a function of the expectation values E[AB]. Thus, let define the Bell's observable as follows:

$$E(a,a';b,b') = |E[A(a)B(b)]-E[A(a)B(b')]|+|E[A(a')B(b)]+E[A(a')B(b')]|. \tag{4}$$

From the properties of the integrals, from the definitions of $E(a,a';b,b')$ and E[AB], and from the property $A(a,\lambda) \in \{\pm 1, 0\} \ \forall \lambda \in \Lambda$, it immediately follows:

$$E(a,a';b,b') \leq \int_\Lambda d\lambda \rho(\lambda)[|A(a,\lambda)||B(b,\lambda)-B(b',\lambda)|+|A(a',\lambda)||B(b,\lambda)+B(b',\lambda)|] \leq$$

$$\leq \int_\Lambda d\lambda \rho(\lambda)[|B(b,\lambda)-B(b',\lambda)|+|B(b,\lambda)+B(b',\lambda)|]. \tag{5}$$

Now, let us define $\Lambda_0(b)$ and $\Lambda_0(b')$ the two subset of Λ in which $B(b,\lambda)=0$ and $B(b',\lambda)=0$, respectively. Thus, we can subdivide the set Λ into two subsets: the subset $\Lambda_0(b) \cap \Lambda_0(b')$ of l's such that both $B(b,\lambda)$ and $B(b',\lambda)$ are zero; and the subset $\Lambda' = \Lambda - \Lambda_0(b) \cap \Lambda_0(b')$ of l's such that either $B(b,\lambda)$ or $B(b',\lambda)$ or both are different from zero. As in the subset $\Lambda_0(b) \cap \Lambda_0(b')$ the function $|B(b,\lambda)-B(b',\lambda)|+|B(b,\lambda)+B(b',\lambda)|$ is always equal to zero, we can calculate the integral of Eq. (5) only over its complementary subset Λ'.

As we always have $B(b, \lambda) \in \{\pm 1, 0\}$ $\forall \lambda \in \Lambda'$, it is a simple matter to show that, in this subset, $|B(b, \lambda)-B(b', \lambda)|+|B(b, \lambda)+B(b', \lambda)| = 2$.

Thus, we obtain:

$$E(a,a';b,b') \leq 2 \int_{\Lambda'} d\lambda \rho(\lambda) \equiv 2\mu(\Lambda'), \tag{6}$$

where $\mu(\Lambda')$ denotes the measure of Λ'. From Eq.(1) and from the definition of Λ' it follows that we can put:

$$\mu(\Lambda')=1-\mu(\Lambda_0(b)\cap\Lambda_0(b'))\equiv 1-\mu_0(b,b'), \tag{7}$$

and, therefore, we have:

$$E(a,a';b,b') \leq 2(1-\mu_0(b,b')). \tag{8}$$

We wish to note that the local upper limit given by Eq. (8) is stronger than the analogous limit valid in the dychotomic case. Only if $\mu_0(b,b')=0$ the upper limit becomes 2. In the most general case $\mu_0(b,b')$ is different from zero and it will be necessary to evaluate its value in each case.

A local inequality for probabilities for three-valued observables

As the original Bell's inequality cannot be actually applied to a concrete experiment Clauser, Horne, Shimony and Holt (CHSH) in the 1969 introduced a deterministic approach in order to obtain a stronger inequality[4]. The CHSH inequality, given by the following equation, follows from the hypothesis of local realism and from the famous CHSH no-enhancement assumption:

$$-D(\infty,\infty) \leq CHSH(a,a';b,b')\equiv D(a;b)-D(a;b')+D(a';b)+D(a';b')-D(a',\infty)-D(\infty,b)\leq 0, \tag{9}$$

where $D(a,b)$ denotes the joint transmission and detection probability with the linear single-channel polarizers set to the angles a and b, and ∞ the physical situation with the correspondent polarizer removed.

Next we will present a generalisation of the CHSH approach to the three-valued case. First we will obtain a general local inequality with no further additional assumptions. Later, we will discuss how this inequality changes as a function of the additional hypotheses.

If we define $p(a_+,b_+)$, $p(a_+,b_-)$, $p(a_+,b_0)$ as the probabilities that the first photon of a pair will be detected in the ordinary channel, while the other photon is, respectively, detected in the ordinary channel, the extraordinary channel or absorbed, and if analogous definitions hold for $p(a_-;b_+)$, $p(a_-;b_-)$, $p(a_-;b_0)$, $p(a_0;b_+)$, $p(a_0;b_-)$ and $p(a_0;b_0)$, the following relations hold:

$$\Sigma_{ij}\, p(a_i;b_j)=1, \quad \Sigma_j\, p(a_i;b_j)=p_1(a_i), \quad \Sigma_i\, p(a_i;b_j)=p_2(b_j),$$

$$p(a_i;b_+)+p(a_i;b_-)=p(a_i;\infty), \quad p(a_+;b_j)+p(a_-;b_j)=p(\infty;b_j), \tag{10}$$

$$p(a_+;b_+)+p(a_+;b_-)+p(a_-;b_+)+p(a_-;b_-)=p(\infty;\infty),$$

where: the indexes $i, j \in \{+,-,0\}$; and the symbol ∞, differently from the one of Eq.(9), means the sum of detected counts in the ordinary and extra-ordinary channels of the correspondent polarizer.

It is important to stress that in Eqs. (10) we have assumed the rotational symmetry of the quantum state emitted by the source. This hypothesis is analogous to the one made by

CHSH ; moreover, the validity of the above assumption can be easily verified with an experimental test, in which the coincidences p(a+,∞), p(∞,b+) and so on, should be measured at different values of a and b.

Using Eqs. (10) it is a simple matter to show that the expectation value E(a;b), given by

$$E(a;b) \equiv p(a_+;b_+)\text{-}p(a_+;b_-)\text{-}p(a_-;b_+)+p(a_-;b_-),\tag{11}$$

and p(∞;∞) can be expressed, respectively, by the following equations:

$$E(a;b) = 4p(a_+; b_+)\text{-}2p(a_+; \infty)\text{-}2p(\infty; b_+)+p(\infty;\infty),\tag{12}$$

$$p(\infty;\infty) = 1\text{-}p(a_0; \infty)\text{-}p(\infty; b_0)\text{-}p(a_0; b_0) = 1\text{-}p_1(a_0)\text{-}p_2(b_0)+p(a_0; b_0).$$

Let us observe that in the first Eq. (12) we can simplify the notation neglecting the + symbols. Finally, inserting Eq. (12-a) in the definition of Bell's observable for the three-valued case, given by the Eq. (8), we obtain:

$$L(b,b') \leq CHSH(a,a';b,b') \leq U(b,b'),\tag{13}$$

where

$$L(b,b')=[\text{-}1\text{-}p(\infty,\infty)+\mu_0(b,b')]/2,$$

$$\tag{14}$$

$$U(b,b')=[1\text{-}p(\infty,\infty)\text{-}\mu_0(b,b')]/2.$$

The above inequality is quite different from the original CHSH one. This is due to the fact that the inequality (13) has not been obtained via the additional CHSH assumption.

In fact, if we put $\mu_0(b,b')=0$, as in the dichotomic Bell's inequality, the limits that bound CHSH(a,a';b,b') are too high in order to being experimentally violated, due to the small value that p(∞;∞) can actually reach. When we consider $\mu_0(b,b')\neq0$, as it is a measure of the number of photons which have been lost, we should expect that the smaller is p(∞;∞) the larger should be $\mu_0(b,b')$. Thus, in the lower and upper limits of the above inequality there are two terms that have opposite behaviour as a function of the quantum efficiency. In the next sections we will evaluate the lower and the upper limits of the inequality (13) for very important physical situations.

Evaluation of the boundary limits

In this section, we will evaluate the functions L(b,b') and U(b,b') in two important physical situations in which these functions assume the maximum and the minimum possible values.

The two physical situations considered will differ only in the additional hypotheses. Therefore, we will prove that lower and upper limits of the inequality (13-14) depend of the additional hypotheses. Thus, our approach is able to include any additional hypotheses in the deduction of inequalities from locality.

The first important physical case concerns the case in which $\Lambda_0(b)$ and $\Lambda_0(b')$ are completely random.

Let us denote with $\beta(b)$ and $\beta(b')$ the measure of the set $\Lambda_0(b)$ and $\Lambda_0(b')$, respectively. The function $\beta(b)$ is the probability for a photon to have its reality element λ belonging to the subset $\Lambda_0(b)$ of Λ. Moreover, the function $\mu_0(b,b')$ is the joint probability that λ will belong to $\Lambda_0(b)$ <u>and</u> to $\Lambda_0(b')$. Therefore, the conditional probability p(b'|b) that a λ, belonging to $\Lambda_0(b)$, will belong also to $\Lambda_0(b')$, is given by (if $\beta(b)\neq0$):

$$p(b'|b)=\mu_0(b,b')/\beta(b). \tag{15}$$

Thus, in the case of a $\Lambda_0(b)$ and a $\Lambda_0(b')$ completely randomly determined, the conditional probabilities are $p(b'|b)=\beta(b')$, $p(b|b')=\beta(b)$ and

$$\mu_0(b,b')=\beta(b)\beta(b'). \tag{16}$$

From now on we will refer to this physical situation with the phrase: "random selection of the subset of coincidences in the set of photon pairs"; or more briefly with: "random selection".

Moreover, from the assumed rotational symmetry property of the quantum state emitted by the source , we have that the sum of counts in the ordinary and in the extra-ordinary channels does not depend on b. Thus, if we denote with $\Lambda_+(b)$ and $\Lambda_-(b)$ the subsets of λ in which $B(b,\lambda)=1$ and $B(b,\lambda)=-1$, respectively, the above condition is equivalent to say that $\mu(\Lambda_+(b))+\mu(\Lambda_-(b))$ does not depend on b. Therefore, from the obvious condition $\Lambda_+(b)+\Lambda(b)+\Lambda_0(b)=\Lambda$ we can also conclude that, even if $\Lambda_0(b)$ depends on b, its measure does not depend on b. In other words, there are different reality elements which contribute to $\Lambda_0(b)$ and $\Lambda_0(b')$, but the probability β of no detection is always the same, no matter the values of b and b'. The above conclusion permits us to consider the functions L and U, defined in eqs. (13-14), as not depending on b and b', but only on β. Therefore, under the above symmetry conditions, we have:

$$\mu_0(b,b')=\beta^2. \tag{17}$$

In order to evaluate the functions U and L in the inequality (13) we have to calculate the values of $p(\infty;\infty)$ for the considered physical situation, or equivalently, given (12-b) the probabilities $p_1(a_0)$, $p_2(b_0)$ and $p(a_0,b_0)$. But the probability $p_1(a_0)$ of having no single count at the polarizer 1 is given by the measure of the subset $\Lambda_0(a)$, denoted with α. Analogously, $p_2(b_0)$ is given by β, and $p(a_0,b_0)$ is given by $\mu_0(a,b)=\mu(\Lambda_0(a)\cap\Lambda_0(b))=\alpha\beta$. Therefore:

$$p(\infty;\infty)=1-\alpha-\beta+\alpha\beta=(1-\alpha)(1-\beta). \tag{18}$$

Inserting Eq.(18) into Eqs.(14), for a "random selection" we have:

$$L(b,b')= L(\alpha,\beta)= -1 + (\alpha+\beta+\beta^2-\alpha\beta)/2,$$

$$U(b,b')=U(\alpha,\beta)=(\alpha+\beta -\beta^2-\alpha\beta)/2. \tag{19}$$

In Eqs. (19) we have the lower and the upper limit of the local inequality for three-valued observables as a function of two variables: α and β. At this point, in order to make a quantitative evaluation of the functions L and U, we make the simplifying hypothesis that $\alpha=\beta$ (symmetrical measuring apparatuses of the two observers). In this case

$$p(\infty,\infty)=(1-\beta)^2, \tag{20}$$

and the inequality (13-14) becomes:

$$-(1-\beta) = -\sqrt{p(\infty,\infty)} \leq CHSH(a,a';b,b') \leq (\beta-\beta^2) = \sqrt{p(\infty,\infty)}-p(\infty,\infty). \tag{21}$$

Thus, when we make the additional assumption that there is a "random selection of the subset of coincidences", not only the lower but also the upper limit for the observable CHSH(a,a';b,b') depends on $p(\infty,\infty)$.

The second case considered is when all the elements of reality λ (and, therefore, all the photon pairs) which give no count, are always the same, no matter the value of a, a' and b, b' for the first and second apparatus, respectively. This can be expressed by the following mathematical requirements:

$$\Lambda_0(a) \cap \Lambda_0(a') \equiv \Lambda_0(a) \equiv \Lambda_0(a') \quad \forall a,\, a',$$

$$\Lambda_0(b) \cap \Lambda_0(b') \equiv \Lambda_0(b) \equiv \Lambda_0(b') \quad \forall b,\, b', \tag{22}$$

$$\Lambda_0(a) \cap \Lambda_0(b) \equiv \min\{\Lambda_0(a), \Lambda_0(b)\} \quad \forall a,\, b,$$

where $\min\{\Lambda_0(a), \Lambda_0(b)\}$ denotes the smaller subset between $\Lambda_0(a)$ and $\Lambda_0(b)$. From now on, we will indicate this physical situation with the phrase: "polarizer-orientation independence of the selected subset of coincidences"; or more briefly with: "parameter-independent selection".

Then, denoting with $\beta(b)$ and $\beta(b')$ the measure of the set $\Lambda_0(b)$ and $\Lambda_0(b')$, respectively, from Eq. (22-b) we have: $\beta(b)=\beta(b')=\beta$. Thus, as in the previous sub-section, the probability for a photon to have the element of reality λ belonging to the subset $\Lambda_0(b)$ of Λ, does not depend on b. But, contrary to the previous case, if Eq. (22-b) holds, the conditional probability $p(b'|b)$ that a λ, belonging to $\Lambda_0(b)$, will belong also to $\Lambda_0(b')$, now is equal to one. Therefore, we have:

$$\mu_0(b,b')=p(b'|b)\beta(b)=p(b|b')\beta(b')=\beta \tag{23}$$

In order to evaluate the functions U and L in the inequality (13) we have to calculate the value of $p(\infty;\infty)$ for the considered physical situation. From Eqs. (12-b) and (23), it follows that, denoting with $\mu_0(a,b)$ the measure $\mu(\Lambda_0(a) \cap \Lambda_0(b))$, we have:

$$p(\infty,\infty)=1-\alpha-\beta+\mu_0(a,b). \tag{24}$$

Moreover, from the condition (22-c) it follows:

$$\mu_0(a,b)=\min\{\alpha,\beta\} \equiv m. \tag{25}$$

Inserting Eqs. (23), (24) and (25) into Eqs. (14), it leads to:

$$L(b,b')=L(\alpha,\beta)=-1 + (\alpha+2\beta-m)/2\ ,$$

$$\tag{26}$$

$$U(b,b')=U(\alpha,\beta)=(\alpha-m)/2.$$

Being $\Lambda_0(a) \cap \Lambda_0(b) \subseteq \Lambda_0(a)$, we always have $U(\alpha,\beta) \geq 0$. In particular, the sign of equality holds if

$$\Lambda_0(a) \cap \Lambda_0(b) \equiv \Lambda_0(a) \equiv \Lambda_0(b) \quad \forall a,\, \forall b, \tag{27}$$

because in this case we have surely $m=\alpha=\beta$. Let observe that the above condition can be satisfied if and only if the two analysing experimental apparatuses have polarizers, detectors and so on, with the same identical characteristics.

Thus, when we make the additional hypothesis of "parameter-independent selection" , and we are dealing with symmetrical experimental apparatuses of the two observers ($\alpha=\beta$), we have:

$$-(1-\beta) = -p(\infty,\infty) \leq \text{CHSH}(a,a';b,b') \leq 0. \qquad (28)$$

In other words, we re obtain an inequality very similar to the CHSH inequality, given by Eq. (9). When $\alpha \neq \beta$, we obtain its generalisation given by the Eqs. (13) and (26).

It is very important to stress that the boundary limits in the inequality (28) are much stronger than the corresponding ones in the inequality (21). But, these two inequalities have been obtained starting from the local inequality given by the Eqs. (13-14), and making different "ad-hoc" additional assumptions. This result simply shows the great influence of the additional hypotheses on the boundary limits of the CHSH-type inequalities. In particular, the value zero for the upper limit of the CHSH inequality is obtained when:

1) the joint detection probability $p(\infty;\infty)$ is equal to 1, i. e. in the dichotomic case;

2) we assume that Λ is the sum of two disjoint subsets, Λ' and Λ_0, such that for all $\lambda \in \Lambda'$ both photons of a correlated pair are detected, and for all $\lambda \in \Lambda_0$ both photons are not detected, no matter the value of a, a' and b, b'.

In fact, in this latter case, and only in this case, we can take into account only the photon pairs with $\lambda \in \Lambda'$. For this subset we can redefine a new density function

$$\rho'(\lambda)=\rho'(\lambda)/\mu(\Lambda'), \qquad (29)$$

and reduce our three-valued observables A and B to dichotomic ones.

QUANTUM MECHANICS VERSUS LOCAL REALISM

In order to make a quantitative comparison between the quantum mechanical predictions and the local realistic ones, we need to make some important considerations.

Let denote with η the quantum efficiency of the photomultipliers. Quantum Mechanics predicts for the joint and the single detection probability with no polarizers: $D(\infty;\infty)=\eta^2$ and $D_1(\infty)=D_2(\infty)=\eta$, respectively, if the detection apparatuses are symmetric. Therefore, Quantum Mechanics predicts that the conditional probability, that a photon of a pair will be detected from the detector 2 when the first has just been detected from detector 1, is given by:

$$p(\infty|\infty)\equiv D(\infty;\infty)/D_1(\infty)= \eta. \qquad (30)$$

Thus, Quantum Mechanics, when there are no polarizers, assumes a "casual selection of the subset of coincidences from all photon pairs"! This assumption is also considered valid when there are polarizers. In fact, in order to obtain the quantum mechanical predictions for the joint detection <u>and</u> transmission probabilities, it is sufficient to multiply the joint transmission probabilities by η^2. In other words, for a photon pair which emerges from the polarizers, Eq. (30) is still valid. Therefore, it should be correct to compare the quantum-mechanical results for the EPR paradox with the inequality (21), which has been obtained under the same additional physical hypotheses, concerning the independence of the two events of photon detection at the respective detectors.

On the other hand, we <u>cannot</u> compare the quantum-mechanical results with the inequality (28). In fact, this latter stands on the additional assumption of "parameter-independent selection" which is not compatible with the predictions of the Quantum Theory for single detection rates.

The quantum-mechanical prediction for the maximum value of observable CHSH(a,a';b,b') for a singlet state with positive parity as a function of $p(\infty;\infty)$ is:

$$\text{CHSH(a,a';b,b')}_{MQ} = \frac{\sqrt{2}-1}{2} p(\infty;\infty). \tag{31}$$

In the simplifying hypotheses of ideal one-channel polarizers and identical detectors h, the comparison between the above value and Eq. (21) shows that quantum-mechanical predictions (31) can violate the inequality (21) if and only if $\eta > 0.828$. Unfortunately this limit is too higher with respect to the actual values of the quantum efficiency of the photomultipliers, in order to do an experimental test of the inequality (21).

Let observe that all the experimental tests[8] made using atomic cascade photon sources and, therefore, photon pairs in singlet state have been compared with the CHSH inequality or, equivalently, with Eq. (28). The experimental results are almost all in agreement with the violation of the upper limit equal to zero of this inequality and with the QM predictions, therefore they have been considered as the experimental proof of the violation of locality and as the confirm of the validity of the quantum mechanical predictions concerning correlated atomic systems. But, in the previous section we have shown that the inequality (28) stands on an additional hypothesis which is in contradiction with Quantum Mechanics also for single detection predictions. Thus, the fact that Quantum Mechanics violates inequality (28) and the experimental results violate its upper limit, simply means that the additional hypothesis of "parameter-independent selection" is wrong, and nothing can be concluded about the validity of the hypothesis of locality because, in order to have a meaningful test of the locality, we need to have detectors with a quantum efficiencies grater than 0.828.

On other hand, the operation of normalization of the measured joint probabilities with respect the subset of coincidences, in order to raise the values of the observable CHSH(a,a';b,b'), is possible only if we assume a "polarizer-orientation independence of the selected subset of coincidences" , while if we assume a "casual selection of the subset of coincidences from all photon pairs" the subset of coincidence will not be always the same for each choice of the polarizer orientations (a,b), (a,b'), (a',b) and (a',b'). In fact, in the previous section we have stressed that, even if the probability β of no detection is always the same, the reality elements λ of Λ, which contribute to $\Lambda_0(b)$ and $\Lambda_0(b')$, could be different.

Therefore, if we assume, coherently with Quantum Mechanics, a "casual selection of the subset of coincidences from all photon pairs", we cannot normalise the joint probabilities $p(a,b)$ with respect to $p(\infty,\infty)$ in the framework of a meaningful local and realistic theory. In other words, we cannot redefine a new dychotomic observable.

A similar conclusion can be draw if we analyse the experiments on EPR paradox based on parametric down-conversion sources[7]

REFERENCES

1. A. Einstein, B. Podolsky and N. Rosen, *Phys. Rev.* 47:777 (1935).
2. N. Bohr, Phys. Rev. 48:696 (1935).
3. J. S. Bell, *Physics* 1:195 (1964).
4. J. F. Clauser, M.A. Horne, A. Shimony and R.A. Holt, Phys. Rev. Lett. 23: 880 (1969).
5. J. F. Clauser and M.A. Horne *Phys. Rev . D* .10:526 (1974)
6. A. Garuccio and V.A. Rapisarda Nuovo Cim. A 65:269 (1981).
7 L. De Caro and A. Garuccio Bell's inequalities for three-valued observables, preprint (1993).
8. S.J. Freedman and J.F.Clauser, *Phys. Rev. Lett.* 28: 938 (1972).
   E.S. Fry and R.C. Thompson, *Phys. Rev. Lett.* 37: 465 (1972).
   J.F. Clauser, *Phys. Rev. Lett.* 36, 1223 (1976).
   A. Aspect, P.Grangier and G. Roger, *Phys. Rev. Lett.* 47: 460 (1981);
   A. Aspect, J. Dalibard and G. Roger, *Phys. Rev. Lett.* 49: 1804 (1982).
   A.J. Duncan, W. Perrie, H.J. Beyer and H. Kleinpoppen, in: *Fundamental processes in atomic collision physics*; 555 (Plenum, New York, 1985).
   For a complete review on this subject see: D.Home and F. Selleri, *La Rivista del Nuovo Cimento*, 14: 9 (1991).

OPTICAL TESTS OF BELL'S INEQUALITIES CLOSING THE POOR CORRELATION LOOPHOLE

Susana F. Huelga[1], M. Ferrero[1] and E. Santos[2]

[1]Departamento de Física. Universidad de Oviedo. Spain.
[2]Departamento de Física Moderna. Universidad de Cantabria. Spain

Abstract. - The existence of local-hidden-variable (LHV) models for the optical tests of Bell's inequality which agrees with Quantum Mechanics (QM) for all measurable quantities, even in the domain of ideal apparatuses, proves that such tests are not suitable in order to discriminate between QM and the whole family of LHV theories. The existence of the models rests upon the poor angular correlation between the photon pairs emitted in an atomic decay. Here we propose a new experiment that would block this loophole, that is, LHV models for these experiments would not be possible, except relying on the low efficiency of presently available photodetectors.

1. INTRODUCTION

Bell's theorem states that LHV theories of QM are not possible [1]. In fact, predictions of LHV theories significantly differ from those of QM in certain situations and the question of discriminating between the whole family of LHV theories and QM can be brought into the experimental domain.

Following the previous work by Bell, Clauser and Horne [2] stated a new incompatibility theorem (isomorphic to Bell's original formulation) that gives an experimentally testable result in the context of a typical EPR arrangement.

The relevant quantities measurable in the experiment are the coincidence probabilities $P_{ij}(a,b)$ and the single probabilities $P_i(a)$ and $P_j(b)$ where i and j stand for 1 and 2 and a and b are taken to be angles specifying the orientation of the analyzers. Clauser and Horne showed that any LHV theory predicts that these quantities must be constrained by the inequality:

$$-1 \leq P_{12}(a,b) - P_{12}(a,b') + P_{12}(a',b) + P_{12}(a',b') - P_1(a') - P_2(b) \leq 0 \qquad (1.1)$$

comparing coincidence probabilities with single ones. For this reason we will refer to (1.1) as heterogeneous inequality of Clauser and Horne, or Bell's inequality, given that only Bell's locality condition is required for its derivation.

Frontiers of Fundamental Physics, Edited by M. Barone
and F. Selleri, Plenum Press, New York, 1994

By symmetry arguments we will assume that $P_i(a)$ and $P_j(b)$ are independent of a and b and that $P_{12}(a,b)=P_{12}(\phi)$, where $\phi=|b-a|$.

Then, choosing a configuration such that $|b-a|=|a'-b|=|b'-a'|$, (1.1) becomes:

$$-1 \leq P_{12}(\phi)-P_{12}(3\phi)-P_1-P_2 \leq 0 \tag{1.2}$$

The quantum mechanical predictions for the quantities involved in (1.2) are:

$$P_{12}(\phi)= \frac{1}{4}\, \eta^2\, \frac{\Omega}{4\pi}\, g\left(\varepsilon_+^2+\varepsilon_-^2 F\cos(2\phi)\right)$$

$$P_1=P_2= \frac{1}{2}\, \eta\, \frac{\Omega}{4\pi}\, \varepsilon_+ \tag{1.3}$$

where we have taken identical apparatuses in both sides for simplicity. Ω is the solid angle covered by the lens system, related to the half angle ϑ by:

$$\Omega=2\pi(1-\cos\vartheta) \tag{1.4}$$

η is the quantum efficiency of the photodetectors; $\varepsilon_\pm=\varepsilon_M \pm \varepsilon_m$, where $\varepsilon_{M,m}$ are the maximum and minimum transmitivity of the analyzers relative to an appropriate orthogonal basis; $F(\vartheta)$ is a factor that takes into account that the polarization correlation of a photon pair decreases when the angle between their wave vectors departs from π. In the range $0 \leq \vartheta \leq \frac{\pi}{6}$ the approximation:

$$F(\vartheta) \approx 1- \frac{2}{3}\left(1- \cos\vartheta\right)^2 \tag{1.5}$$

is valid. Finally, $g(\vartheta)$ is the conditional probability, or angular correlation factor, that if the $J=0 \rightarrow J=1$ emission enters apparatus 1, then the $J=1 \rightarrow J=0$ emission will enter apparatus 2, and it is given by:

$$g(\vartheta)= \frac{\Omega}{4\pi}\left(1+ \frac{1}{8} \cos^2\vartheta\, (1 + \cos\vartheta)^2\right) \tag{1.6}$$

Inserting quantum mechanical predictions (1.3) into Bell's inequality (1.2), it is easy to show that the condition for the violation of the upper bound is:

$$\eta g \varepsilon_+ \left(\sqrt{2}\left(\frac{\varepsilon_+}{\varepsilon_-}\right)^2 F(\vartheta) + 1\right) \geq 2 \tag{1.7}$$

Putting in (1.7) the expressions (1.5) and (1.6) given for $F(\vartheta)$ and $g(\vartheta)$, one finds that this inequality is never satisfied, for any value of ϑ, even if the apparatuses are ideal, that is, when $\varepsilon_+= \varepsilon_-=\eta =1$. Therefore, for cascade-photon experiments, the QM predictions are compatible with (1.2) even in the domain of ideal devices.

This was clearly pointed out already by Clauser and Horne in their 1974 paper, where they correctly identified the poor correlation of the photon pairs (consequence of the fact that an atomic cascade is a three body decay) as responsible for this compatibility, an important remark surprisingly and quickly forgotten in all the literature since 1974.

To get an incompatibility between the two formalisms, it is necessary to introduce an additional assumption relative to the behavior of the hidden variables, the so called no enhancement hypothesis:

$$0 \leq P_1(\lambda,a) \leq P_1(\lambda,\infty) \leq 1$$

$$0 \leq P_2(\lambda,b) \leq P_2(\lambda,\infty) \leq 1 \tag{1.8}$$

Then the same theorem used to get (1.1) gives:

$$P_{12}(\infty,\infty) \leq 3P_{12}(\phi)-P_{12}(3\phi)-P_{12}(a',\infty)-P_{12}(\infty,b) \leq 0 \tag{1.9}$$

where only coincidence probabilities appear. For this reason we will refer to (1.9) as the homogeneous inequality of Clauser and Horne.

Considering the QM predictions for joint probabilities when one or both polarizers are removed and inserting the corresponding expressions in (1.9), we get that the upper bound in (1.9) will be violated if

$$\varepsilon_+\left(\sqrt{2}(\varepsilon_-/\varepsilon_+)^2 F(\vartheta)+1\right) \geq 2 \tag{1.10}$$

is satisfied. This inequality is insensitive to both angular correlation factor g and the efficiency of the photodetectors η. Later experimental data violated the upper bound of the homogeneous inequality [3], opening the way to the claim that LHV theories have been empirically disproved.

After the discussion that followed the publication of these experimental results, only the existence of the loophole associated to the low efficiency of the photodetectors was widely recognized [4]. But, contrary to this received wisdom, a LHV model for the optical test of Bell's inequality in agreement with QM for all measurable quantities even with perfect polarizers and detectors was exhibited [5]. For the existence of these models is essential that the photon pairs have a poor angular correlation, emphasizing that, besides a good correlation in polarization (or other appropriate quantity), a high directional correlation is also required in any test of locality.

These theoretical results point out that besides the low efficiency loophole, another loophole, associated to the unsuitable angular correlation between the photon pairs, remains open: the poor angular correlation loophole. In the next sections, we will propose an atomic-cascade experiment in order to block this loophole.

2. ATOMIC-CASCADE-EXPERIMENT WITH DETECTION OF THE RECOIL ATOM
A. QUALITATIVE DESCRIPTION

In the domain of optics, a good angular correlation between the photon pairs could be obtained, even using a three body decay, if the recoil atom were detected. Our proposal is a modification of the previous photon polarization correlation tests in which the correlation is only measured on a restricted ensemble of photons produced in decays that leave the atom with a well defined (in direction) linear momentum. This technique leads to a proposal of a realizable scheme of event-ready-detectors. In such a scheme, which has been called for by J. S. Bell since 1971 [6], one knows, via some initiating event, when a pair has been produced. Qualitatively, the experiment would be as follows:

Consider an experimental arrangement analogous to that described, for example, by Aspect et al. (See Fig. 1). An atomic beam of calcium atoms coming from a thermal source is irradiated by two perpendicular lasers focused into the interaction region. Atoms are selectively pumped to the upper level by the non-linear absorption of the laser photons and then they spontaneously decay emitting two visible photons correlated in polarization. Photons go through their respective lens systems and polarizers until, eventually, they reach a detector, and something similar happen to the atoms. Given that the total momentum of the photon pairs is chosen in such a way that the photons travel in directions close to π, F will be close to 1. We select our restricted ensemble of emitted photon pairs recording only coincidence counts of an atom with one photon (single counts) or coincidence counts of an atom with two photons (double counts). The additional requirement that the recoil atom has to be detected allow us to increase the angular correlation of the photon pairs involved, gives a well operational definition for this ensemble and, provided that measuring devices have an efficiency above a certain threshold, quantum mechanical predictions for this ensemble would contradict Bell's inequality (1.2), making possible an experiment able to discriminate between Local Realism and QM at high efficiencies.

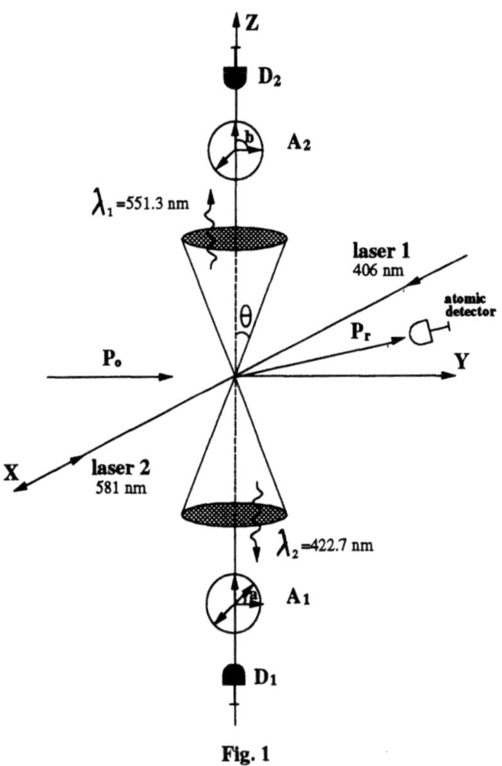

Fig. 1

B. NOTATION

In order to evaluate the QM predictions for the ensemble of interest, we assume that each photon has a well defined linear momentum. We shall denote by k_i (i=1,2) the corresponding wave vectors, whose direction is specified by the usual polar angles (θ_i, ϕ_i). The cones subtended by each detector aperture are taken to be equal, ϑ being the half angle. Note that with the chosen geometry, ϕ_i takes always values in the domain

$0 \leq \phi_i \leq 2\pi$.

Provided that the recoil atom is detected, we can differentiate three kinds of events that could yield single or double counts:

• We shall denote by R1 the events such that only the first emission enters the corresponding aperture. In terms of the polar angles that specify k_i we can describe this situation by the following bounds:

$$0 \leq \theta_1 \leq \vartheta \text{ and } 0 \leq \theta_2 \leq \pi\text{-}\vartheta \tag{2.1}$$

• R2 will refer to events such that only the second emission enters apparatus 2. Then:

$$\vartheta \leq \theta_1 \leq \pi \text{ and } \pi\text{-}\vartheta \leq \theta_2 \leq \pi \tag{2.2}$$

and finally
• B will refer to events such that:

$$0 \leq \theta_1 \leq \vartheta \text{ and } \pi\text{-}\vartheta \leq \theta_2 \leq \pi \tag{2.3}$$

that is, both emissions are collected for the corresponding lens system.

If there is no absorption in the corresponding polarizer, events Ri will contribute as single counts, while events B will produce:
i) Double counts if none of the photons is absorbed and
ii) Single counts, if one of the photons is not absorbed, independently of what happens to the other member of the pair.

If N is the number of atoms which have actually decayed in the source and N(B) and $N(R_i)$ are the number of events of the corresponding class, we can write the quantum mechanical predictions for measurable probabilities as follows:

$$P_1 = \frac{1}{2}\,\eta\,\varepsilon_+^1\left(\frac{N(B)}{N} + \frac{N(R_1)}{N}\right)\eta_a$$

$$P_2 = \frac{1}{2}\,\eta\,\varepsilon_+^2\left(\frac{N(B)}{N} + \frac{N(R_2)}{N}\right)\eta_a$$

$$P_{12}(\phi) = \frac{1}{4}\,\eta^2\,\frac{N(B)}{N}\left(\varepsilon_+^1\varepsilon_+^2 + \varepsilon_-^1\varepsilon_-^2\, F(\vartheta)\, \cos(2\phi)\right)\eta_a \tag{2.4}$$

Substitution of these predictions in (1.2) gives that QM predictions will violate Bell's inequality if the following inequality:

$$\eta\left(\varepsilon_+^1\varepsilon_+^2 + \sqrt{2}\, F\,\varepsilon_-^1\varepsilon_-^2\right) \geq \varepsilon_+^1 + \varepsilon_+^2 + \frac{\varepsilon_+^1 N(R_1) + \varepsilon_+^2 N(R_2)}{N(B)} \tag{2.5}$$

is satisfied.
Taking for simplicity $\varepsilon_\pm^1 = \varepsilon_\pm^2 = \varepsilon_\pm$ in (2.5) we get:

$$\eta\varepsilon_+\left(1 + \sqrt{2}\, F\left(\frac{\varepsilon_-}{\varepsilon_+}\right)^2\right) - 2 \geq t \tag{2.6}$$

as condition for violation, where

$$t = \frac{N(R_1) + N(R_2)}{N(B)} \tag{2.7}$$

Defining the function

$$k(\vartheta) = \varepsilon_+\left(1 + \sqrt{2}\, F\left(\frac{\varepsilon_-}{\varepsilon_+}\right)^2\right) \tag{2.8}$$

we can write the condition (2.6) for incompatibility between the two formalisms in the compact form:

$$\eta \geq \frac{t+2}{k(\vartheta)} \tag{2.9}$$

Therefore, the minimum η required for getting violation of Bell's inequality will be obtained with an arrangement in which the atomic detector is placed in such a way that the ratio t is minimized. We shall denote this location by OR (optimal region). We have found that the situation of the atomic detector that minimizes t, and hence η, is the one that makes the number of atoms corresponding to events with only one emission entering the apertures the lowest possible (events R_i). Therefore, let us to emphasize that placing the atomic detector in the OR does not ensure that N(B),the number of the atoms actually detected corresponding to emissions that have been both collected, is maximized.

3. PRELIMINARY RESULTS

The crucial point of finding the so called OR was solved in the following steps: In a first approximation, we considered an atomic beam with neither longitudinal nor transversal dispersion in the initial velocity. This corresponds for a typical thermal source to an incoming linear momentum $P_o=4.648\ 10^{-18}$ (cgs) along the Y direction. For the cascade 0-1-0 of calcium, $P_1=1.201\ 10^{-22}$ and $P_2=1.566\ 10^{-22}$ in the same units. In order to simplify the notation, we will write all the momenta normalized to 10^{-22} and without explicit units. This gives us: $P_0=46480$, $P_1=1.201$ and $P_2=1.566$. The half angle ϑ was taken to be 15°.

For an atomic decay with emission along $\pm Z$ the conservation law for total linear momentum gives:

$$P_{rx}=P_{ry}=0 \text{ and } P_{rz}=0.365 \tag{3.1}$$

for the components of the recoil atom, this effect being "singular" in the sense that only emissions in opposite directions along the Z axis will produce a recoil with coordinates (3.1). Considering this recoil as reference and assuming that the first [second] photon has a wave vector given by (2.1) [the second given by (2.2)], with variables θ and ϕ uniformly distributed in the corresponding intervals, a simulation was generated in order to calculate the spherical coordinates of the second [first] emission and the recoil component along Y. This method allowed us to select the interval:

$$P_{rx} \in [-0.415,0.415] \text{ and } P_{rz} \in [0.343,0.361] \tag{3.2}$$

in which t is minimized, as suitable for simulating a true statistics, when both emissions are randomly generated assuming uniform distribution over the sphere, and the momentum of the recoil atom is calculated.

A simulation with $N=5\ 10^6$ events gives: $R_1+R_2=12.6356$ and $B=232.803$. This corresponds to $t=5.43$ % for recoil components verifying (3.2). The angular correlation has been taken into account giving different weights to different emissions, according to the value of the angle (k_1,k_2).

The solid angle subtended by the atomic detector is:

$$\Omega=\frac{\Delta P_{rx}\,\Delta P_{rz}}{P_0^2} = 6.92\ 10^{-12} \text{ sr.} \tag{3.3}$$

which corresponds to an area of 6.92 μm^2 for a radial distance $R=1$ m from the interaction region.

As a second step, certain longitudinal dispersion was considered. For that, we assumed that the incident momentum is normally distributed with parameters (μ,σ) equal to (700, 28). No transversal dispersion was considered in this first approach. In the simulation, photons were randomly generated assuming uniform distribution in the corresponding hemisphere (it can be checked that if both photons are emitted in the same hemisphere, the recoil is not collected by the selected solid angle) and the half angle was increased to 25^0 in order to improve the ratio t.

Applying the conservation law for the total linear momentum, we evaluated if the recoil momentum belongs to the selected Ω, that is, if the atom recoils through the OR, giving bounds to the angles Prx/Pry and Prz/Pry. Whenever a recoil through OR is obtained, the corresponding emissions are evaluated in order to determine if one or two photons had parameters θ_i and ϕ_i corresponding to be collected by the lens system [Eqs (2.1) and (2.2) or (2.3) respectively]. This method gives the following result for a simulation with $N_{total} = 5\ 10^6$ events:

B	R1	R2
5215.63	15.9144	713.488

A slight modification of the limits (3.2) relatives to the z-recoil component, but maintaining the same value for the solid angle, allow us to obtain a region with a symmetrical behavior for the events Ri

Replacing (3.2) by $P_{rx} \in [-0.415, 0.415]$ and $P_{rz} \in [0.319, 0.337]$ as specification for the OR, successive simulations with $N_{total} = 5\ 10^6$ each gives us:

B	R1	R2	t
4645.87	201.310	170.201	0.07997
4634.00	220.826	123.885	0.07439
4550.50	195.036	137.367	0.07048
4558.41	212.557	175.088	0.08504
4748.97	243.735	154.280	0.08381
4456.55	216.223	154.907	0.08327
4546.22	189.077	142.895	0.07022

Substitution now in (2.8) of the typical values for $\varepsilon_+ = 0.99$, $\varepsilon_- = 0.94$ and $\sqrt{2}F = 1.4059$, which corresponds to the selected half angle of $25°$, gives for the function $k(\vartheta)$ the value 2.2448. Finally, putting the value of t just obtained in each case in the inequality (2.9), we get the quantum efficiency of the photodetectors required for getting violation of Bell's inequality. This allow us to state that the quantum efficiency of the photodetectors has to satisfy: $\eta \geq 0.926 \pm 0.002$ for quantum mechanical predictions to violate the Bell's inequality.

A detailed study concerning the specific experimental conditions required for the scheme proposed here to be feasible is in progress.

4. CONCLUSIONS

After the optical tests of Bell's inequality two loopholes remained open, one connected with the low efficiency of optical photodetectors and the second due to the unsuitable angular correlation between the photon pairs issuing from an atomic decay. The experiment proposed here provides a possibility for blocking the former, in the sense that, for this experiment, no LHV model in agreement with QM could be exhibited for efficiencies above the obtained threshold and, in particular, the agreement even with perfect apparatus showed for the previous atomic cascade experiments is no longer valid. Only a loophole due to the static character of the experiment would remain open, which could be blocked using an scheme analogous to the employed by Aspect et al. in their third experiment.

ACKNOWLEDGMENT

We acknowledge financial support of DGICYT Project No. PB-92-0507 (Spain) and Universidad de Oviedo Project No. DF-93-214-42 (Spain).

REFERENCES

[1]   J. S. Bell, *Physics* (N. Y.) **1**, 195 (1964).
[2]   J. F. Clauser and M. A. Horne, *Phys. Rev. D* **10**, 938 (1972).
[3]   S. J. Freedman and J. F. Clauser, *Phys. Rev. Lett.* **28**,460 (1981).
       A. Aspect, P. Grangier and G. Roger , *Phys. Rev. Lett.* **47**, 460 (1981).
       A. Aspect, P. Grangier and G. Roger , *Phys. Rev. Lett.* **49**, 91 (1982).
       A. Aspect, J. Dalibard and G. Roger , *Phys. Rev. Lett.* **49**, 1804 (1981).
[4]   E. Santos, T. W. Marshall and F. Selleri, *Phys. Lett.* **98** A, 5 (1983).
       M. Ferrero and E. Santos, *Phys. Lett.* **116** A, 356 (1986).
       M. Ferrero, T. W. Marshall and E. Santos, *Am. J. Phys.* **58**, 683 (1990).
[5]   E. Santos, *Phys. Rev. Lett.* **66**, 1388 (1991); **68**, 894 (1992), Phys. Rev. A **46**, 3646 (1992).
[6]   J. S. Bell, *Speakable and Unspeakable in Quantum Mechanics*, Cambrigde University Press, 1988. pp. 29 and 105.
       J. F. Clauser and A. Shimony, *Rep. Prog. Phys.* **41**, 1981 (1978).

ATOMIC CASCADE EXPERIMENTS WITH TWO–CHANNEL POLARIZERS
AND QUANTUM MECHANICAL NONLOCALITY

M. Ardehali*

Box 6572
Stanford, Ca 94305

INTRODUCTION

In 1935, Einstein, Podolsky, Rosen (EPR)[1] used their famous criteria of realism and locality to conclude that the wave function does not provide a complete description of physical reality. Their argument (adapted to Bohm's[2] gedanken experiment for a pair of photons in the singlet state) is based on the following three premises:

1 – Quantum mechanical (QM) perfect correlations: If the polarizations of photons 1 and 2 are measured along the same axis, then the outcomes are perfectly (anti) correlated.

2 – EPR's criterion of realism: "If without in any way disturbing a system, we can predict with certainty (i.e., with probability equal to unity) the value of a physical quantity, then there exists an element of physical reality corresponding to this physical quantity."

3 – Einstein's locality:[3] "But on one supposition, we should, in my opinion absolutely hold fast: The real factual situation of the system S_2 is independent of what is done with the system S_1 which is spatially separated from the former."

EPR use these premises to conclude that quantum mechanics is not a complete theory. Their argument can be summarized as follows:

QM perfect correlations + Realism + Locality $\rightarrow$ *QM is not a complete theory.*

EPR conclude their paper by noting that: "While we have shown that the wave function does not provide a complete description of physical reality, we left open the question of whether or not such a description is possible. We believe, however, that such a theory is possible." Bell's[4] genius was to prove that such a theory *is not possible*. Bell proved his theorem by slightly modifying EPR's first premise from QM (perfect) correlations along parallel axes to QM correlations along arbitrary axes. Bell's argument is based on the following three premises:

1 – QM correlations along arbitrary axes,

2 – EPR's criterion of realism,

3 – Einstein's locality.

Bell then shows that the conjunction of these three premises leads to a contradiction, i.e.,

QM correlations along arbitrary axes + Realism + Locality → Contradiction.

Bell's theorem of 1965 is of paramount importance for understanding the conceptual foundation of quantum mechanics because it rigorously formulates EPR's premises and shows that these premises are quantitatively incompatible with quantum theory.

Bell's argument, however, can not be experimentally tested because it relies on the existence of a pair of detectors with 100% efficiency. In an actual (laboratory) experiment, the less than perfect efficiency of real detectors weakens the observed correlations from the strong ideal form on which Bell's argument is based. Faced with this problem, Clauser and Horne (CH)[5] derived a correlation inequality for systems which do not achieve perfect correlations but which do achieve a necessary minimum correlation. Unfortunately their inequality is based on a supplementary assumption that is not a consequence of local realism. In this letter, we show that a different supplementary assumption, weaker and more general than the no enhancement assumption of CH, is sufficient to make the existing atomic cascade experiments applicable as a test of local hidden–variable theories.

BELL'S INEQUALITY

We start by considering the Bohm's version of EPR gedanken experiment in which an unstable source emits pairs of photons in a singlet state. Figure 1 shows the experimental set up: S is a source of EPR photons, and P_1 and P_2 are two channel polarizers monitored by detectors $D_1^+ (D_2^+)$ and $D_1^- (D_2^-)$ put on the ordinary (extraordinary) beams. This experimental set up corresponds very closely with the ideal gedanken experiment considered by Bell to derive his inequality.

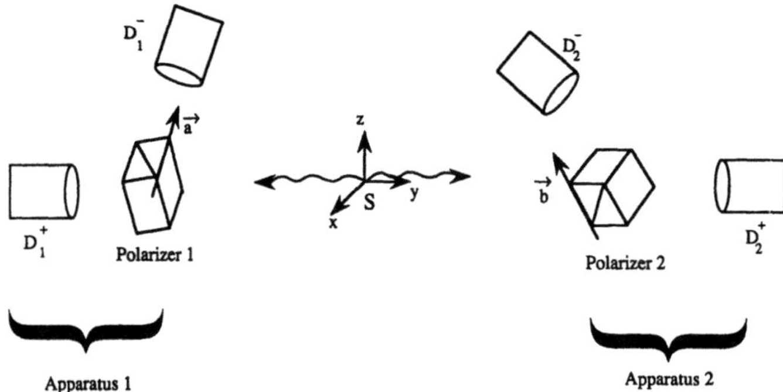

Figure. 1 Outline of an experiment to test the violation of the strong Bell's inequality with two–channel polarizers.

During a period of time, the unstable source emits, say, N pairs of photons. Assuming that N is sufficiently large, and indicating with $\vec{a}$ ($\vec{b}$) the orientation of the first (second) polarizer, with $N^{\pm\pm}(\vec{a}, \vec{b})$ the number of simultaneous counts from the detectors $D_1^{\pm}$ and $D_2^{\pm}$, and with $N^{++}(\infty, \infty)$ the number of simultaneous counts from the detectors D_1^{+} and D_2^{+} in the absence of polarizers P_1 and P_2, one defines the ensemble probabilities $p^{\pm\pm}(\vec{a}, \vec{b})$ and $p^{++}(\infty, \infty)$ as

$$p^{\pm\pm}(\vec{a}, \vec{b}) = \frac{N^{\pm\pm}(\vec{a}, \vec{b})}{N},$$

$$(1)$$

$$p^{++}(\infty, \infty) = \frac{N^{++}(\infty, \infty)}{N}.$$

We consider a particular pair of photons and specify its state with a parameter λ. Following Bell, we do not impose any restriction on the complexity of state λ: "It is a matter of indifference whether λ denotes a single variable or a set, or even a set of functions, and whether the variables are discrete are continuous".[4] As state λ evolves, it may or may not trigger counts from the detectors 1 and 2. Let $p^{\pm\pm}(\lambda, \vec{a}, \vec{b})$ be the joint probability that the state of emission is λ and two simultaneous counts are triggered from the detectors $D_1^{\pm}$ and $D_2^{\pm}$. The ensemble probabilities given by Eq. (1) are obtained by summing or integrating $p^{\pm\pm}(\lambda, \vec{a}, \vec{b})$ over the state of emission λ, i.e.,

$$p^{\pm\pm}(\vec{a}, \vec{b}) = \int d\lambda \, p^{\pm\pm}(\lambda, \vec{a}, \vec{b}).$$

$$(2)$$

According to Bayes' theorem, $p^{\pm\pm}(\lambda, \vec{a}, \vec{b})$ is defined as

$$p^{\pm\pm}(\lambda, \vec{a}, \vec{b}) = p(\lambda) \, p^{\pm}(\vec{a} \mid \lambda) \, p^{\pm}(\vec{b} \mid \lambda, \vec{a}).$$

$$(3)$$

Substituting (3) in (2), we obtain,

$$p^{\pm\pm}(\vec{a}, \vec{b}) = \int d\lambda \, p(\lambda) \, p^{\pm}(\vec{a} \mid \lambda) \, p^{\pm}(\vec{b} \mid \lambda, \vec{a}).$$

$$(4)$$

Equation (4) can be stated in physical term. The ensemble probability for triggering simultaneous counts from the detectors $D_1^{\pm}$ and $D_2^{\pm}$ [that is $p^{\pm\pm}(\vec{a}, \vec{b})$] is equal to the sum, or integral, of the probability that the state of emission is λ [that is $p(\lambda)$], times the probability that if the state of emission is λ, then a count is triggered from the first detector $D_1^\pm$ [that is $p^\pm(\vec{a} \mid \lambda)$], times the probability that if the state of emission is λ, and if the first polarizer is oriented along axis $\vec{a}$, then a count is triggered from the second detector $D_2^\pm$ [that is $p^\pm(\vec{b} \mid \lambda, \vec{a})$].

Note that the formulas (2), (3), and (4) are quite general and follow from the standard rules of probability theory. No assumption has yet been made that is not satisfied by quantum mechanics.

Hereafter, we focus our attention only on those theories which satisfy Einstein's criterion of locality (see premise 3). Einstein's locality can be translated into the following mathematical relation:

$$p^\pm(\vec{b} \mid \lambda, \vec{a}) = p^\pm(\vec{b} \mid \lambda), \tag{5}$$

where $p^\pm(\vec{b} \mid \lambda)$ is the probability that if the state of emission is λ, then a count is triggered from the second detector. Equation (5) is the hallmark of Einstein's locality and should be fulfilled by every local hidden variable theory; it states that the probability of triggering a count from the second detector is independent of the orientation of the first polarizer. Substituting (5) in (4), we obtain

$$p^{\pm\pm}(\vec{a}, \vec{b}) = \int d\lambda \, p(\lambda) p^\pm(\vec{a} \mid \lambda) p^\pm(\vec{b} \mid \lambda). \tag{6}$$

Equation (6) accounts for the correlations in terms of information jointly available to the photons when they left their common source. The representation (6) is the most general form of locality that accounts for the correlations by attributing them to some unspecified parameters λ, subject only to the requirement that a count at the first (second) detector for a given λ not depend on the choice of orientation of the second (first) polarizer. Obviously Eq. (6) should be fulfilled by all local realistic theories.

Before proceeding any further, it is worth to describe the difference between Eq. (5) and the CH's criterion of locality.[5] CH write the assumption of locality [see their Eq. (2')] as

$$p^{++}(\lambda, \vec{a}, \vec{b}) = p^+(\lambda, \vec{a}) p^+(\lambda, \vec{b}). \tag{7}$$

Apparently by $p^{++}(\lambda, \vec{a}, \vec{b})$, CH mean the conditional probability that if the state of emission is λ, then simultaneous counts are triggered from the detectors D_1^+ and D_2^+. However, what they call $p^{++}(\lambda, \vec{a}, \vec{b})$, in probability theory is usually written as $p^{++}(\vec{a}, \vec{b} \mid \lambda)$ (note that in probability theory, $p(x, y, z)$ is the joint probability of x, y, and x, whereas $p(x, y \mid z)$ is the conditional probability that if z, then x and y). Similarly, by $p^+(\lambda, \vec{a})$ [$p^+(\lambda, \vec{b})$], CH mean the conditional probability that if the state of emission is λ, then a count is triggered from the detector D_1^+ [D_2^+]. Again, what they call $p^+(\lambda, \vec{a})$

$[p^+(\lambda, \vec{b})]$, in probability theory is usually written as $p^+(\vec{a} \mid \lambda)$ $[p^+(\vec{b} \mid \lambda)]$. Thus according to the standard notations of probability theory, CH's criterion of locality can be written as

$$p^{++}(\vec{a}, \vec{b} \mid \lambda) = p^+(\vec{a} \mid \lambda)\, p^+(\vec{b} \mid \lambda). \tag{8}$$

Now it is easy to show that Eq. (8) reduces to Eq. (5). According to Bayes' theorem, $p^{++}(\vec{a}, \vec{b} \mid \lambda)$ is defined as

$$p^{++}(\vec{a}, \vec{b} \mid \lambda) = p^+(\vec{a} \mid \lambda)\, p^+(\vec{b} \mid \lambda, \vec{a}). \tag{9}$$

Substituting (9) in (8), we obtain

$$p^+(\vec{b} \mid \lambda, \vec{a}) = p^+(\vec{b} \mid \lambda), \tag{10}$$

which, for the ordinary beam, is the same as Eq. (5).

Having described the difference between Eq. (5) and the CH's criterion of locality, we now proceed to show that the representation (6) leads to validity of an inequality that is sometimes grossly violated by the quantum mechanical probabilities. This inequality can be deduced by considering the following algebraic theorem.

Theorem: If $x_1^+, x_1^-, x_2^+, x_2^-, y_1^+, y_1^-, y_2^+, y_2^-$, u, and v are ten real numbers such that

$$\left| x_1^+ - x_1^- \right| \le u, \qquad\qquad \left| x_2^+ - x_2^- \right| \le u,$$

$$\left| y_1^+ - y_1^- \right| \le v, \qquad\qquad \left| y_2^+ - y_2^- \right| \le v, \tag{11}$$

then the function Z, defined as

$$Z = x_1^+ y_1^+ + x_1^- y_1^- - x_1^+ y_1^- - x_1^- y_1^+ + x_2^+ y_1^+ + x_2^- y_1^- - x_2^+ y_1^- - x_2^- y_1^+ +$$
$$x_1^+ y_2^+ + x_1^- y_2^- - x_1^+ y_2^- - x_1^- y_2^+ - x_2^+ y_2^+ - x_2^- y_2^- + x_2^+ y_2^- + x_2^- y_2^+, \tag{12}$$

is bounded by

$$Z \le 2uv. \tag{13}$$

Proof: The function Z can be written as

$$Z = \left[\left(x_1^+ - x_1^- \right) + \left(x_2^+ - x_2^- \right) \right]\left(y_1^+ - y_1^- \right) + \left[\left(x_1^+ - x_1^- \right) - \left(x_2^+ - x_2^- \right) \right]\left(y_2^+ - y_2^- \right) \tag{14}$$

Let

$$x_1^+ - x_1^- = q, \qquad\qquad x_2^+ - x_2^- = r,$$

$$y_1^+ - y_1^- = s, \qquad\qquad y_2^+ - y_2^- = t. \tag{15}$$

Obviously

$$(q+r)s + (q-r)t \le (|q+r| + |q-r|)v. \tag{16}$$

But note that

$$|q+r| + |q-r| \le 2u. \tag{17}$$

Substituting (17) in (16), we obtain

$$(q+r)s + (q-r)t \le 2uv. \tag{18}$$

and this proves the theorem.

Now let $\vec{a}$ and $\vec{a'}$ be two arbitrary orientations of the first polarizer, and let

$$x_1^+ = p^+(\vec{a}\,|\,\lambda), \quad x_1^- = p^-(\vec{a}\,|\,\lambda), \quad x_2^+ = p^+(\vec{a'}\,|\,\lambda), \quad x_2^- = p^-(\vec{a'}\,|\,\lambda). \tag{19}$$

Similarly let $\vec{b}$ and $\vec{b'}$ be two arbitrary orientations of the second polarizer, and let

$$y_1^+ = p^+(\vec{b}\,|\,\lambda), \quad y_1^- = p^-(\vec{b}\,|\,\lambda), \quad y_2^+ = p^+(\vec{b'}\,|\,\lambda), \quad y_2^- = p^-(\vec{b'}\,|\,\lambda). \tag{20}$$

Obviously for each value of λ, we have

$$\left| p^+(\vec{c}\,|\,\lambda) - p^-(\vec{c}\,|\,\lambda) \right| \le 1. \qquad\qquad (\vec{c} = \vec{a},\ \vec{a'},\ \vec{b},\ \vec{b'}). \tag{21}$$

Inequalities (21) and (13) give

$$\begin{aligned}
Z = {}& p^+(\vec{a}\,|\,\lambda)p^+(\vec{b}\,|\,\lambda) + p^-(\vec{a}\,|\,\lambda)p^-(\vec{b}\,|\,\lambda) - p^+(\vec{a}\,|\,\lambda)p^-(\vec{b}\,|\,\lambda) - p^-(\vec{a}\,|\,\lambda)p^+(\vec{b}\,|\,\lambda) + \\
& p^+(\vec{a}\,|\,\lambda)p^+(\vec{b'}\,|\,\lambda) + p^-(\vec{a}\,|\,\lambda)p^-(\vec{b'}\,|\,\lambda) - p^+(\vec{a}\,|\,\lambda)p^-(\vec{b'}\,|\,\lambda) - p^-(\vec{a}\,|\,\lambda)p^+(\vec{b'}\,|\,\lambda) + \\
& p^+(\vec{a'}\,|\,\lambda)p^+(\vec{b}\,|\,\lambda) + p^-(\vec{a'}\,|\,\lambda)p^-(\vec{b}\,|\,\lambda) - p^+(\vec{a'}\,|\,\lambda)p^-(\vec{b}\,|\,\lambda) - p^-(\vec{a'}\,|\,\lambda)p^+(\vec{b}\,|\,\lambda) - \\
& p^+(\vec{a'}\,|\,\lambda)p^+(\vec{b'}\,|\,\lambda) - p^-(\vec{a'}\,|\,\lambda)p^-(\vec{b'}\,|\,\lambda) + p^+(\vec{a'}\,|\,\lambda)p^-(\vec{b'}\,|\,\lambda) + \\
& p^-(\vec{a'}\,|\,\lambda)p^+(\vec{b'}\,|\,\lambda) \le 2.
\end{aligned} \tag{22}$$

Multiplying both sides by $p(\lambda)$ and integrating over λ gives

$$\int d\lambda\, p(\lambda) Z \leq 2. \tag{23}$$

Finally using the representation (6), we obtain

$$
\begin{aligned}
& p^{++}(\vec{a}, \vec{b}) + p^{--}(\vec{a}, \vec{b}) - p^{+-}(\vec{a}, \vec{b}) - p^{-+}(\vec{a}, \vec{b}) + p^{++}(\vec{a}', \vec{b}) + p^{--}(\vec{a}', \vec{b}) - \\
& p^{+-}(\vec{a}', \vec{b}) - p^{-+}(\vec{a}', \vec{b}) + p^{++}(\vec{a}, \vec{b}') + p^{--}(\vec{a}, \vec{b}') - p^{+-}(\vec{a}, \vec{b}') - \\
& p^{-+}(\vec{a}, \vec{b}') - p^{++}(\vec{a}', \vec{b}') - p^{--}(\vec{a}', \vec{b}') + p^{+-}(\vec{a}', \vec{b}') + p^{-+}(\vec{a}', \vec{b}') \leq 2. \tag{24}
\end{aligned}
$$

The validity of inequality (24) is a necessary constraint on the statistical predictions of all local realistic theories. In the following, we shall show that inequality (19) is always fulfilled in the atomic cascade experiments. In the 0–1–0 atomic cascade, an unstable atom, emits pairs of photons in a cascade passing through a state of $J = 1$ toward a state of $J = 0$. The ensemble probabilities, as defined by Eq. (1), are given by

$$
p^{++}(\vec{a}, \vec{b}) = \eta_1^{+}\, \eta_2^{+} \left(\frac{\Omega}{8\pi}\right)^2 g(\theta, \varphi) \left[T_1^{+} T_2^{+} + T_1^{-} T_2^{-} F(\theta, \varphi) \cos 2(a-b) \right],
$$

$$
p^{--}(\vec{a}, \vec{b}) = \eta_1^{-}\, \eta_2^{-} \left(\frac{\Omega}{8\pi}\right)^2 g(\theta, \varphi) \left[R_1^{+} R_2^{+} + R_1^{-} R_2^{-} F(\theta, \varphi) \cos 2(a-b) \right],
$$

$$
p^{+-}(\vec{a}, \vec{b}) = \eta_1^{+}\, \eta_2^{-} \left(\frac{\Omega}{8\pi}\right)^2 g(\theta, \varphi) \left[T_1^{+} R_2^{+} - T_1^{-} R_2^{-} F(\theta, \varphi) \cos 2(a-b) \right], \tag{25}
$$

$$
p^{-+}(\vec{a}, \vec{b}) = \eta_1^{-}\, \eta_2^{+} \left(\frac{\Omega}{8\pi}\right)^2 g(\theta, \varphi) \left[R_1^{+} T_2^{+} - R_1^{-} T_2^{-} F(\theta, \varphi) \cos 2(a-b) \right],
$$

where

$$
\begin{aligned}
T_i^{+} &= T_i^{\parallel} + T_i^{\perp}, & T_i^{-} &= T_i^{\parallel} - T_i^{\perp}, & (i = 1, 2), \\
R_i^{+} &= R_i^{\parallel} + R_i^{\perp}, & R_i^{-} &= R_i^{\parallel} - R_i^{\perp}, & (i = 1, 2).
\end{aligned}
$$

$T_i^{\parallel}\,(T_i^{\perp})$ represents the transmittance along the transmitted path for incoming light polarized parallel (perpendicular) to the transmitted–channel polarization plane. Similarly $R_i^{\parallel}(R_i^{\perp})$ represents the transmittance along the reflected path for incoming light polarized parallel (perpendicular) to the transmitted–channel polarization plane; $\eta_i^{\pm}$ are the quantum efficiencies of the detectors $D_i^{\pm}$ $(i = 1, 2)$; Ω is the solid angle covered by the apertures of the lens systems; $g(\theta, \varphi)$ is the angular correlation function; $F(\theta, \varphi)$ is the so–called depolarization factor, which is due to noncolinearity of the photons, and $(a - b)$ is the angle between polarizers $\vec{a}$ and $\vec{b}$. Substituting (25) in (24), one finds that inequality (24) is always fulfilled in the case of atomic cascade experiments.

BELL'S INEQUALITY WITH A SUPPLEMENTARY ASSUMPTION

In order to obtain a new inequality violated by quantum mechanical probabilities for atomic cascade experiments with low efficiency detectors, we make the following supplementary assumption [for a supplementary assumption that is stated for an ensemble of photons, see Ref. (6)]: *For every photon in state λ, the absolute value of the difference of the detection probabilities in the ordinary and extraordinary beams emerging from a two–channel polarizer is less than or equal to the detection probability of the beam by* $D+$ *with the polarizer removed.* Calling $p^+(\infty \mid \lambda)$ $[q^+(\infty \mid \lambda)]$ the conditional probability that if a photon is in the state λ, then a count is triggered from the detector D_1^+ $[D_2^+]$ in the absence of polarizer P_1 $[P_2]$, we can translate our supplementary assumption into the following relations:

$$\left| p^+\left(\vec{d} \mid \lambda\right) - p^-\left(\vec{d} \mid \lambda\right) \right| \leq p^+\left(\infty \mid \lambda\right), \qquad \left(\vec{d} = \vec{a},\, \vec{a'}\right),$$

$$\left| p^+\left(\vec{e} \mid \lambda\right) - p^-\left(\vec{e} \mid \lambda\right) \right| \leq q^+\left(\infty \mid \lambda\right), \qquad \left(\vec{e} = \vec{b},\, \vec{b'}\right). \tag{26}$$

Now the same argument that was applied to (21) and led to inequality (24), when applied to (26) leads to the following inequality

$$p^{++}\left(\vec{a}, \vec{b}\right) + p^{--}\left(\vec{a}, \vec{b}\right) - p^{+-}\left(\vec{a}, \vec{b}\right) - p^{-+}\left(\vec{a}, \vec{b}\right) + p^{++}\left(\vec{a'}, \vec{b}\right) + p^{--}\left(\vec{a'}, \vec{b}\right) -$$
$$p^{+-}\left(\vec{a'}, \vec{b}\right) - p^{-+}\left(\vec{a'}, \vec{b}\right) + p^{++}\left(\vec{a}, \vec{b'}\right) + p^{--}\left(\vec{a}, \vec{b'}\right) - p^{+-}\left(\vec{a}, \vec{b'}\right) - p^{-+}\left(\vec{a}, \vec{b'}\right) -$$
$$p^{++}\left(\vec{a'}, \vec{b'}\right) - p^{--}\left(\vec{a'}, \vec{b'}\right) + p^{+-}\left(\vec{a'}, \vec{b'}\right) + p^{-+}\left(\vec{a'}, \vec{b'}\right) \leq 2\, p^{++}\left(\infty, \infty\right), \tag{27}$$

where $p^{++}(\infty, \infty)$ is the ensemble probability for double detection of photons by detectors D_1^+ and D_2^+ in the absence of polarizers (see Eq. 1), and is given by

$$p^{++}\left(\infty, \infty\right) = \eta_1^+ \eta_2^+ \left(\frac{\Omega}{4\pi}\right)^2 g(\theta, \varphi). \tag{28}$$

If we accept the supplementary assumption (26), then the validity of inequality (27) is a necessary constraints for the observed correlations to be consistent with the requirement of local realism. Quantum mechanics, however, violates this inequality for certain choices of axes. In particular, the magnitude of violation is maximized if we choose the following set of orientations $\left(\vec{a}, \vec{b}\right) = \left(\vec{a}, \vec{b'}\right) = \left(\vec{a'}, \vec{b}\right) = 22.5°$, and $\left(\vec{a'}, \vec{b'}\right) = 67.5°$. In this case, for perfect detectors and polarizers, we obtain $2\sqrt{2} \leq 2$ which is certainly impossible. The violation of inequality (27) therefore refutes all those local hidden variables theories which agree with the supplementary assumption (26).

It is important to emphasize that inequality (24) was derived only on the basis of Einstein's locality, whereas inequality (27) was derived on the basis of Einstein's locality and a supplementary assumption. Hence, it is useful to refer to inequality (24) as weak Bell's inequality and refer to inequality (27) as strong Bell's inequality.[7,8]

Finally we wish to describe the difference between the supplementary assumption of this paper and the no enhancement assumption of CH.[5] The CH supplementary assumption is

$$p^+\left(\vec{a}\mid\lambda\right)\leq p^+\left(\infty\mid\lambda\right). \tag{29}$$

In contrast the supplementary assumption of this paper is

$$\left|p^+\left(\vec{a}\mid\lambda\right)-p^-\left(\vec{a}\mid\lambda\right)\right|\leq p^+\left(\infty\mid\lambda\right). \tag{30}$$

Since $p^-\left(\vec{a}\mid\lambda\right)\geq 0$, the supplementary assumption (30) is weaker and more general than the CH no enhancement assumption. Thus an experiment based on (30) [i.e, inequality (27)] refutes a larger family of hidden variable theories than an experiment based on inequality (29).

CONCLUSION

We have demonstrated that a supplementary assumption, weaker and more general than that of CH, is sufficient to make the atomic cascade experiments applicable as a test of local realistic theories. In the past a large number of experiments have been carried out to test the violation of the strong Bell's inequality with one channel polarizers.[9] These experiments have overwhelmingly refuted all those local hidden variable theories which agree with the CH no enhancement assumption. However, from a general point of view, one can maintain that none of these experiments have refuted hidden variable theories which agree with the supplementary assumption (30).

Endnote

* Present address: Microelectronics research Laboratories, NEC Corporation, Sagamihara, Kanagawa 229, Japan.

REFERENCES

1. A. Einstein, B. Podolsky, and N. Rosen, Phys. Rev. 47: 777 (1935).
2. D. Bohm. "Quantum Theory," Prentice–Hall, Englewood Cliffs, N. J. (1951) p. 614–619.
3. A. Einstein, in: "Albert Einstein, Philosopher Scientist," P. A. Schilpp, ed., Library of Living Philosophers, Evanston, Illinois (1949) p. 85.
4. J. S. Bell, Physics 1: 195 (1965).
5. J. F. Clauser and M. A. Horne, Phys. Rev. D 10: 526 (1974).
6. M. Ardehali, Phys. Lett. A 176: 285 (1993); M. Ardehali, ibid. 180: 461 (1993).

7. F. Selleri, in: "Quantum Mechanics versus Local Realism. The Einstein–Podolsky–Rosen Paradox," F. Selleri, ed., Plenum, New York (1988).

8. F. Selleri, in: "Problems in Quantum Physics, Gdansk '87.," L. Kostro, A. Posiewnik, J. Pykacz, and M. Zukowski, eds., World Scientific, Singapore (1988) p. 256.

9. J. Freedman and J. F. Clauser, Phys. Rev. Lett. 28: 938 (1972); J. F. Clauser, ibid. 36: 1223 (1976); E. S. Fry and R. C. Thompson, ibid. 37: 465 (1976); A. Aspect, J. Dalibard, and G. Roger, ibid. 49: 1804 (1982); W. Perrie, A. J. Duncan, H. J. Beyer, and H. Kleinpoppen, ibid. 54: 1790 (1985); Z. Y. Ou and L. Mandel, ibid. 61: 50 (1988); Y. H. Shih and C. O . Alley, ibid. 61: 2921 (1988).

NEW TESTS ON LOCALITY AND EMPTY WAVES

R. Risco-Delgado

Departamento de Fisica Aplicada,
Universidad de Sevilla, 41012 Sevilla, Spain

INTRODUCTION

Though the problems of realism and locality in physical theories is at least as old as modern quantum mechanics, the interest for that subject decreased considerably after its great achivements. However, in the last decades, a very important research has been developed in this line, and we can safety say that neither the idea of reality, nor the locality, have been eliminated in physics, although this fact is still unkwon for the mayority of the scientific community. von Neumann's theorem on the theoretical side, and Aspect's experiment on the experimental one, have created this misunderstanding. The construction of explicit models have clarified highly the scope of these results[1,2]. But the construction of a model is not only important for refutal purpouses, of course. A model tries to reprent how Nature works, being of fundamental importance when one wants to deepen it and to really understand it.

In this work I present a set of recent experiment performed to test locality[3]. They are of very different nature. Some were done by Mandel et al. with non linear cristals. The others were done by Stirling group with a set-up similar to that of Aspect. I shall also display a set of models that try to adscribe physical endowment to their results. It is not necessary to say that these models have only a preliminary and exploratory character, however they fit quite well the experimental results. We are firmly convinced that a deep undestanding of Nature implies the constraction of such models of reality.

MANDEL'S EXPERIMENTS

From 1988 Mandel and coworkers have been publishing a set of experiments in which they measured the visibility of the interference pattern produce by the two photons emerging from a parametric down process[3]. We shall deal here with two of them, the most significative ones.

Figure 1 shows the arrangement of the first one. A laser pump the cristal and two photons are produced. Then they hit the same screen and the corresponding visibility is measured. Classical electromagnetism predicts a visibility below 50%. The visibility that they obtained was above this value, and they impute this fact to the local features of our world. We shall see how this high visibility can be obtained by a perfectly local

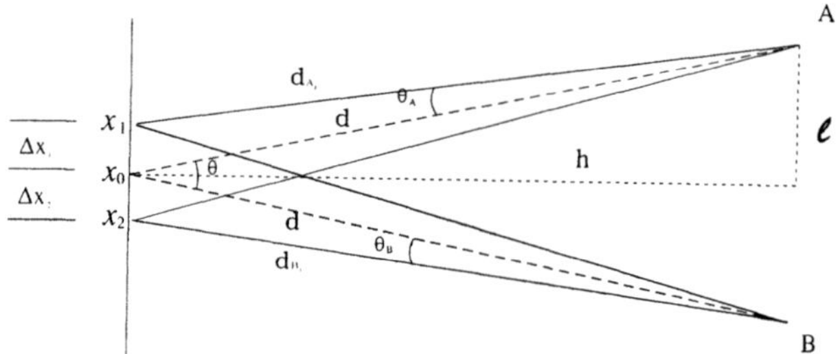

Figure 1. Interference on a screen in Mandel's experiment.

theory. The theory that we propose is the Einstein-de Broglie theory of empty waves. In this theory, a photon consists of a particle enveloped by a surrounding wave packet. The density of the present photons is proportional to the intensity of the wave

$$P(\vec{x}, t) = \mid \vec{\phi}(\vec{x}, t) \mid^2, \tag{1}$$

where $\vec{\phi}(\vec{x}, t)$ is the wave already properly normalizaed between 0 and 1. Classical electromagnetism deals with two fields, the electrical and magnetical. Both are subject to a conservation law of the type $|\vec{E}|^2 + |\vec{B}|^2 = const$. In our scheme we shall say that besides the field $\vec{\phi}$ there is another field $\vec{\epsilon}$ such that

$$|\vec{\phi}|^2 + |\vec{\epsilon}|^2 = 1 \tag{2}$$

For the moment we have moved very closely to the Eistein-de Broglie theory. Now we introduce a slight but essential new feature in this scheme. We say that there exists a variable detection probability. This means that each photon has a proper probability of been detected, contrary to what is commonly accepted. The variable detection probability approach was the first proposal to explain Aspect's experiments in local term, known as the low efficiency loophole. For us it was natural to think that also in the Mandel's experiments, even whether it is very different, a phenomenon of this kind was taking place. And effectively we shall see how this hipothesis can reproduce the high visibility with a very simple model.

In the parametric down conversion process photons are produced in couples. Let us suposse that there are two kinds of photon, those that have a detection probability proportional to $|\vec{\phi}|^2$, and those in which the detection probability is proportinal to $|\vec{\epsilon}|^2$,

$$D_\phi = n|\vec{\phi}|^2 \quad and \quad D_\epsilon = n|\vec{\epsilon}|^2 \tag{3}$$

The two photons of each couple are highly correlated in several senses, so it is natural to think that both two have the same type of detection probability.

Let us suppose that at a certain time a couple of "ϕ -type photons" are produced in the non-linear cristal. The probability of the photon coming from A been detected by detector 1 is D_{ϕ_1}. The probability of the photon coming from been detected by detector 2 is D_{ϕ_2}. In the same way these probabilities are D_{ϵ_1} and D_{ϵ_2} in the case that both photons are of the "ϵ" type. This means that the average probability of a joint detection is

$$P_{12} = \frac{D_{\phi_1} D_{\phi_2} + D_{\epsilon_1} D_{\epsilon_2}}{2} \tag{4}$$

However this joint detection probability is conditional to the case that photons were there. The probability of the photon coming from A reaches the point x_1 is $|\phi_1|^2$. The probability that the photon coming from B reaches the point x_2 is $|\phi_2|^2$. So,

556

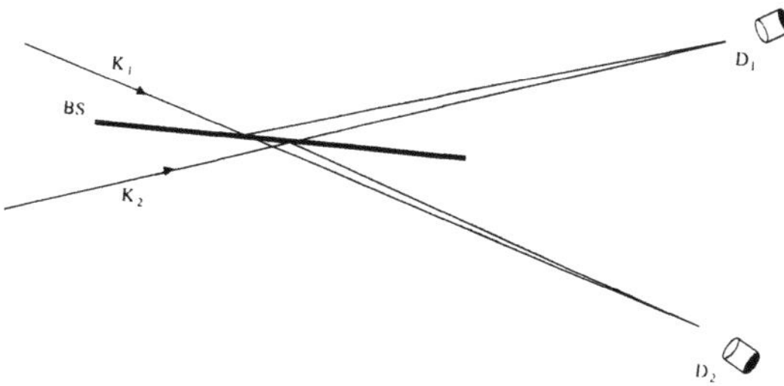

Figure 2. Interference experiment with beam-splitter.

the probability of photons arriving in the detectors and being detected is simply the product of probability:

$$D_{12} = \frac{D_{\phi_1} D_{\phi_2} + D_{\epsilon_1} D_{\epsilon_2}}{2} |\phi_1|^2 |\phi_2|^2 \tag{5}$$

But this is not the final expression. In a parametric down conversion process the idler and signal photons' phases are shifted by a random quantity Δ. This means that we must average over all values of Δ ($0 \leq \Delta \leq 2\pi$). Therefore we write:

$$\langle D_{12} \rangle_\Delta = \langle \frac{D_{\phi_1} D_{\phi_2} + D_{\epsilon_1} D_{\epsilon_2}}{2} |\phi_1|^2 |\phi_2|^2 \rangle_\Delta \tag{6}$$

If Ψ_0 and $\Psi_0 e^{i\Delta}$ are the photonic fields at points A and B at points x_1 and x_2 of the screen one has

$$\Psi(x_1) = \Psi_0 e^{i\vec{k}_a \vec{r}_{a_1}} + \Psi_0 e^{i\vec{k}_b \vec{r}_{b_1} + \Delta} \quad and \quad \Psi(x_2) = \Psi_0 e^{i\vec{k}_a \vec{r}_{a_2}} + \Psi_0 e^{i\vec{k}_b \vec{r}_{b_2} + \Delta} \tag{7}$$

where $\vec{k}_a$ and $\vec{k}_b$ are the wave-number vectors of the waves arising from the points A and B of Fig. 1, respectively and $\vec{r}_{a_i}$, $\vec{r}_{b_i}$ are the vector distances from A and B to the point x_i on the screen ($i=1,2$). Here the waves reaching the screen have been approximated with plane waves. Therefore:

$$|\vec{\phi}(x_1)|^2 = \frac{1}{2}[1 + \cos(\theta_1 + \Delta)] \tag{8}$$

$$|\vec{\phi}(x_2)|^2 = \frac{1}{2}[1 + \cos(\theta_2 + \Delta)] \tag{9}$$

with $\theta_i = |\vec{k}_b \vec{r}_{b_i} - \vec{k}_a \vec{r}_{a_i}|; (i = 1, 2)$

In (9,10) we have already properly normalized the $\vec{\phi}$ functions, in such a way that they span the whole range of their permissible values.

Substituting the values of $|\phi_1|^2$ and $|\phi_2|^2$ into the expression for $\langle D_{12} \rangle_\Delta$, taking into account (1) and (2), and averaging over Δ one gets

$$\langle D_{12} \rangle_\Delta = K[9 + 8\cos(\theta_1 - \theta_2) + 2\cos^2(\theta_1 - \theta_2)] \tag{10}$$

where K is a new constant. The visibility of the interference figure predicted by the last expression is 73%, in good agreement with the experimental results[4,5].

In a more recent experiment[4,5] they inserted a beam-splitter in the trayectories of the light rays (see Fig. 2). The conclusion was similar to that of the former experiment. The presence of the beam-splitter introduce new interference terms; however we shall see that the behaviour of our model also predicts a high visibility, without appeling, of course, to non local effects.

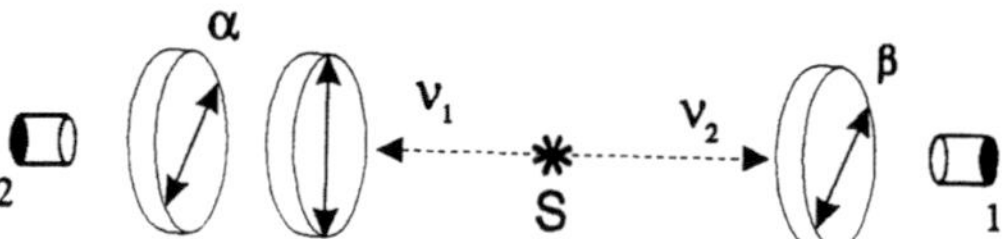

Figure 3. Set up with three polarizers in Stirling experiment.

Sources are again the signal and idler photons coming from a parametric down conversion process. The wave packets associated with these photons have again a ramdom relative phase shift Δ. As before, we suppose a high correlation within each pair of photons, and so, sometimes the detection probability is $D_{\phi_1} D_{\phi_2}$ and sometimes $D_{\epsilon_1} D_{\epsilon_2}$, with the same rate. If $|\vec{\phi}_1|^2$ and $|\vec{\phi}_2|^2$ are the probabilities of photons arriving to the detectors, then we have

$$\langle D_{12}\rangle_\Delta = \langle \frac{D_{\phi_1} D_{\phi_2} + D_{\epsilon_1} D_{\epsilon_2}}{2}|\vec{\phi}_1|^2|\vec{\phi}_2|^2\rangle_\Delta \qquad (11)$$

that expressed as a function of $|\vec{\phi}_1|^2$ and $|\vec{\phi}_2|^2$ is:

$$\langle D_{12}\rangle_\Delta = K\langle|\vec{\phi}_1|^2|\vec{\phi}_2|^2 - |\vec{\phi}_1|^2|\vec{\phi}_2|^4 - |\vec{\phi}_1|^4|\vec{\phi}_2|^2 + 2|\vec{\phi}_1|^4|\vec{\phi}_2|^4\rangle_\Delta \qquad (12)$$

The squared moduli of $\vec{\phi}_1$ and $\vec{\phi}_2$ are

$$|\vec{\phi}_1|^2 = \frac{1}{2}(1 + \frac{i}{2}V_1 V_2^* e^{i\alpha} - \frac{i}{2}V_1^* V_2 e^{-i\alpha}) \qquad (13)$$

$$|\vec{\phi}_2|^2 = \frac{1}{2}(1 - \frac{i}{2}V_1 V_2^* e^{i\beta} + \frac{i}{2}V_1^* V_2 e^{-i\beta}) \qquad (14)$$

(after a proper normalization, like in the former case)
respectively, in case of beams of equal intensities in points 1 and 2. Here $\alpha = (\vec{k}_1 - \vec{k}_2)\vec{r}_a$, $\beta = (\vec{k}_1 - \vec{k}_2)\vec{r}_b$. V_1 and V_2 are the complex amplitudes of the waves emerging from the non-linear cristal.

If $\gamma = \alpha - \beta$, then

$$\langle D_{12}\rangle_\Delta = K[9 - 8\cos\gamma + 2\cos^2\gamma] \qquad (15)$$

As before, a visibility of 73% is obtained.

STIRLING EXPERIMENTS

Here we tackle with a very different kind of experiment. In essence the experimental set-up is that of Aspect, but some optical devices can be introduced between the source and the detectors[3]. In particular we consider the case where an additional polarizer is inserted.

We propose to explain this kind of experiment with the same empty wave model[6]. As before, a photon consist in a particle surroundind by a wave packet. Again the important point is to say that the dection probability depends on the intensity of the field. However we should explain the action of the polarizer.

We describe a polarizer in the following terms. An ideal polarizer changes the polarization plane of the photon, that is, the direction of $\vec{\phi}(\vec{x},t)$ in such a way that the outcoming photon vibrates in the direction of the polarizer axis. On the other hand, it reduces by a factor $\cos^2\theta$ the intensity of these oscillations, where θ is the angle between the polarizer axis and the polarization of the incoming photon. For the moment all we have done is to recover Malus' law. We now introduce the important point of the variable detection probability; a photon, after traversing a polarizer is forever marked in a double sense:

A) The detection probability is increased to

$$D(\vec{x},t) = \frac{4}{3}\mid \vec{\phi}(\vec{x},t)\mid^2 \tag{16}$$

B) Following polarizers only reduce by $\cos\theta$ the intensity of the oscillation. This feature is a requirement to reproduce the single-photon physics, considered by us as a permanent boundary condition.

The action of a non ideal polarizer is as follows. As usual, a non ideal polarizer is described by two perpendicular axes, the principal and the secondary. In our model photons come out vibrating either in the direction of the principal axis, with an ϵ_M probability, or in the direction of the secondary axis, with an ϵ_m probability. In both cases the corresponding reduction of the intensity of the vibration also takes place of course. It can be shown that with this picture the well known single-photon physics is reproduced. This study has been done elsewhere. The experiment of Aspect were tested, in good agreement with the reported data. Also another experiment of Ou and Mandel was tested. These authors obtained a visibility of 75% and claimed that a local realistic explanation of such a high visibility was impossible; but, far from being so it can be seen through our model that you can set a visibility reading of 84%.

Let us analize this new experiment with three polarizers. The ratio $\frac{R(\beta,\alpha)}{R(\beta,\infty)}$ was measured, where $R(\beta,\alpha)$ is the detection rate when the polarizers 1 and 3 are placed at β and α degrees respectively to the polarizer 2. $R(\beta,\infty)$ is the corresponding rate when polarizer 1 is removed. The quantum mechanical prediction is

$$\frac{R(\beta,\alpha)}{R(\beta,\infty)}\mid_{Q.M.} = \frac{1}{2}[\epsilon_+^3 + \epsilon_-^3 \frac{\epsilon_M^2 P - \epsilon_m^2 Q}{\epsilon_M^2 P + \epsilon_m^2 Q}\cos 2\alpha - \Delta(\beta,\alpha)] \tag{17}$$

where

$$P = [\epsilon_+^1 + \epsilon_-^1 \cos 2\beta] \tag{18}$$

$$Q = [\epsilon_+^1 - \epsilon_-^1 \cos 2\beta] \tag{19}$$

$$\Delta(\beta,\alpha) = \frac{(\epsilon_M^2 \epsilon_m^2)^{1/2}\epsilon_-^1 \epsilon_-^3}{\epsilon_M^2 P + \epsilon_m^2 Q}\sin 2\beta \sin 2\alpha \tag{20}$$

The predictions of our model follows directly from our previous hipotheses:

$$\frac{R(\beta,\alpha)}{R(\beta,\infty)}\mid_{L.R.} = \frac{\langle\omega_{12}\rangle_3}{\langle\omega_{12}\rangle_2} \tag{21}$$

where

$$\langle\omega_{12}\rangle_2 = [1 + \epsilon_+^1\epsilon_+^2 + \frac{8}{9}\epsilon_-^1\epsilon_-^2\cos 2(a_1 - a_2) + \frac{1}{18}\epsilon_+^1\epsilon_+^2\cos 4(a_1 - a_2)]\eta_1\eta_2 \tag{22}$$

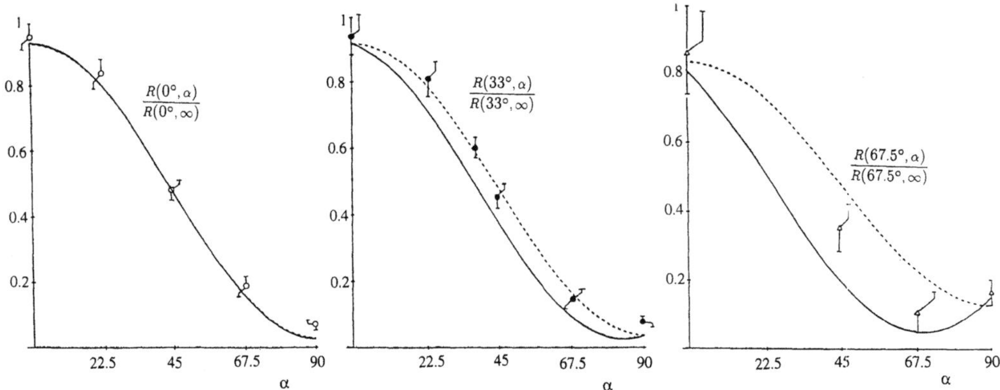

Figure 4. Experimental points for $\beta=0°$, $\beta=33°$ and $\beta=67.5°$. The dashed lines represent the predictions of our local model and the solid ones represent quatum mechanics.

and

$$\langle\omega_{12}\rangle_3 = \frac{\eta_1\eta_2}{8}[(\epsilon_+^1\epsilon_+^2\epsilon_+^3 + \epsilon_+^1\epsilon_-^2\epsilon_-^3\cos 2\alpha)+\frac{8}{9}(\epsilon_-^1\epsilon_-^2\epsilon_+^3 + \epsilon_-^1\epsilon_+^2\epsilon_-^3\cos 2\alpha)\cos 2\beta$$
$$+\frac{1}{18}(\epsilon_+^1\epsilon_+^2\epsilon_+^3 + \epsilon_+^1\epsilon_-^2\epsilon_-^3\cos 2\alpha)\cos 4\beta] \quad (23)$$

In the figures we represent the reported experimental points, the quantum mechanical predictions (solid line) and the predictions of our local model (dashed line). We see that, contrary to what the authors of this experimental work claimed in the abstract of their paper, the "conclusive evidence against the local realistic model which are capable of explaining all existing two-polarizers-type experiments", this model is not only capable but also tries to adscribe physical reality to the concepts that it involves.

REFERENCES

1. "Quantum Mechanics Versus Local Realism: The Einstein- Podolsky-Rosen paradox" F. Selleri ed., Plenum, New York (1988).
2. E. Santos, Critical analysis of the empirical tests of local hidden-variable theories, *Phys. Rev. A.* 46:3646 (1992).
3. "Bell's Theorem and the Foundations of Modern Physics" A. van der Merwe, F. Selleri and G. Tarozzi eds., Plenum, New York (1992).
4. F. Selleri, Two-photon interference and the question of empty waves *in*: "Wave-Perticle Duality", Plenum, New York (1992).
5. R. Risco-Delgado and F. Selleri, Variable detection probability in two-photon interference, *Found. Phys. Lett.* 6:225 (1993). R. Risco-Delgado, The variable detection approach: A wave particke model, *Found. Phys. Lett.* 6:399 (1993).
6. R. Risco-Delgado, A local realistic analysis of the polarization correlation of radiation from a two-photon deuterium source, *Hadr. Jour. Suppl.* 8:301 (1993).

WAVE-PARTICLE DUALITY

Marius Borneas

Department of Physics
Technical University
Timisoara, Romania

In this paper we advance a "lax" interpretation scheme (LIS) of quantum theory, based on a realistic point of view. One can admit the statement that observation represents the chief knowledge, yet description of something happening between the observations, which does not contradict them, also has meaning.

In order to develop our LIS let us begin from direct experimental data. Consider an actual example: place a sensible screen in front of a β–active source. One can observe spots on the screen. The source can be constructed in such a manner that in the detector, on the screen, the spots appear successively. Thus we can say that a microsystem (MS) appears as particle in some cases, namely when it interacts with some detecting device, where it ends its course. On the other hand, if between the source and the detector there is a proper device (e.g. a thin metallic sheet), the spots on the screen concentrate on some regions, forming diffraction patterns. The fact that the diffraction patterns appear even when the MS-s fall on the screen one by one, allows the conclusion that the individual MS presents in some circumstances a wave character. Thus MS has a dual character, being in general wave and/or particle. However, in the view of LIS, that-what-comes toward the detector, the free MS is only a wave, being subdued to specific wave phenomena; but in the detector there are particles. Moreover the free MS is not a probability wave. We state that the MS is an objective complicated wave of associated fields - a combined fields wave (CFW). MS formed by more waves has been advocated by several authors (e.g. refs. 1-5).

One can obtain more associated fields in a natural way if one considers a basic field, say W, which behaves according to equations with higher derivatives. By decomposition (methods of decomposition e.g. in refs. 6-8) one can write usual wave equations, of first and second order, for fields formed by self-interaction of the field W. The field W is not an ordinary field directly observable; it is a universal nonlocal field, generating the observable fields, the constituents of CFW, as well as other fields interacting with the given MS. We

Frontiers of Fundamental Physics, Edited by M. Barone
and F. Selleri, Plenum Press, New York, 1994

strongly emphasize the nonclassical statement that MS is a part of a universal system, a correlation existing, as shown, between MS and the field W.

Let us watch MS in its course. It propagates in space as a wave. Reaching the detector it interacts with an external field linked to the detector; the interaction is governed by the universal field W through a field in the MS sector, generated by it, say F, with singularities, in the sense of de Broglie. These singularities result from the nonlinear interactions. The combination of the CFW, the field of the detector, and the field F, leads to the building up of the particles.

As the field W is universal, the F singularities, i.e. the places where the particles occur, their state, depend on the state of the whole universe. One can say that nature is nonseparable, in the sense that one event may be influenced by another event from which is separated by spacelike interval. Locally the apparition of particles is random. But a basic difference vis-à-vis the conventional quantum theory is that instead of the probability wave ψ there is the real CFW, one of the components being proportional to ψ (just as in de Broglie's views), and the act of observation is more complex.

The particle can be conceived as a region of very high intensity of the field, as de Broglie states, following a kind of reduction of the CFW by the interaction, during a time short but different from zero. Of course the observations show only the final result (see also refs. 9-11). Thus our knowledge of the position of the particle is really a collapse.

In order to explain why the reduction of the CFW takes place in discrete quantities, one can imagine the phenomenon of particle generation as follows. As soon as CFW enters in interaction with the field of the detector, a system of stationary waves is formed from one of the CFW waves and the field of the detector, i.e. a kind of system with discrete levels. Particles occur as another component of CFW accumulates sufficient energy to surpass these levels. For a small energy only one level is surpassed, only one particle appears. It is meaningless to speak of the order in time of the formation of the wave-detector system and the building up of particles, because, as Mugur-Schächter[12] says, "there is a reflexive, double-way causality, carrying influences with any velocity, superluminal or infraluminal, both from the object-state to the measuring devices, and from measuring devices to the object-state". Clearly nonclassic view.

Concerning the energy, the conservation holds because the measuring of the energy is only possible at the source and at the detector, where it is the same. One must not forget that MS is a part of a larger system comprising the field W.

The established fact that in interaction with an external field CFW generates particles, suggests to admit that the emission of MS-s from the source is also an emission of particles. However, the free MS being only wave, it results that immediately after the emission, departing from their source, the particles disintegrate. The fact that the particles are the result of combinations of several fields makes understandable the existence of the process of desintegration. The idea of appearance and disappearance of particles has been already advanced in some works (e.g. refs. 13-15).

The detailed explanation, by the LIS scheme, of some actual experiments is presented elsewhere.

It must be underlined that the results within the frame of LIS are possible only because the universal field W obeys an equation with higher derivatives, allowing the decomposition in two, or three, or more usual wave equations.

One can speculate that if the field W behaves in some conditions as obeying equations with derivatives of order tending towards infinity, it could produce phenomena beyond the usual ones - (paraphenomena?).

As a general observation remember that de Broglie[16] said that the establishment of an orthodoxy was always fatal for the progress of science.

Bearing in mind de Broglie's ideas, we venture to the following continuation of LIS. Admit that usually the field W is weak enough and slowly variable. Then the fields composing the free CFW are linear and the phenomena occur as presented so far. Nevertheless in some conditions the field W can be very strong and rapidly variable, such that nonlinear terms appear in the equations of the free CFW. This nonlinearity creates a self-interaction field of the MS, say S. It is stated by several authors (e.g. refs. 17, 18) that nonlinearities can produce self-fields. The field S produces some changes in the behaviour of MS. It does not affect the wave character of CFW, yet it does affect the building up of particles at the detector. Remember that the formation of the particles with a weak and slowly variable W field was governed by the field F emerging from the nonlocal W. Now, with a strong and rapidly variable W, the field S born in these conditions "swallows" the field F in a local field, applied to the sector of the given MS, determining the state of the MS, for example its position at the detector. One can no more speak of a natural probability. The MS in these cases has a determined state; we call it saphion. The saphions could be detected in the future in some experiments.

The proposed LIS is trying to advance a new philosophico-physical sight on the microsystems, enveloping the conventional results and the nonconventional views.. The attempt is in agreement with the experimental data known up to day, but opens also the possibility for new discoveries.

I express my gratitude to Professor Franco Selleri who offered me the opportunity to present this communication.

REFERENCES

1. L. de Broglie, "Une tentative d'interprétation causale et non-linéaire de la mécanique ondulatoire: la théorie de la double solution", Gauthier-Villars, Paris (1956).

2. L. Kostro, A wave model of the elementary particle, a three waves hypothesis (1978, *unpublished*); A three waves model of the elementary particle, *Phys. Lett.* **107A**:429 (1985).

3. R. Horodecki, De Broglie wave and its dual wave, *Phys. Lett.* **87A**:95 (1981); The extended wave-particle duality, *Phys. Lett.* **96A**:175 (1983).

4. R.C. Jennison, A class of relativistic rigid proper docks, *J. Phys.* **A19**:2249 (1986);   The non-particulate nature of matter and the universe, in "Problems in quantum physics. Gdansk '87", World Scientific, Singapore-New Jersey-Hong Kong (1988).

5. F. Selleri,   On the direct observability of quantum waves, *Found.Phys.* **12**: 1087 (1982);   On the possibility of a rationalistic approach to microphysics, in "Foundations of Mathematics & Physics" Perugia, Italy (1990).

6. H. Stumpf, On some physical properties of a unified lepton-hadron field model with composite bosons, *Z. Naturforsch.* **37a**:1295 (1982).

7. M. Borneas, Perspectives in the integrated-space-field theory, Analele Univ. Timisoara (ser.fiz.) **20**:85 (1982).

8.   D. Grosser, On the decomposition of spinor fields which satisfy a nonlinear higher order equation, Z. *Naturforsch.* **38a**:1293 (1983).

9.   J.W.G. Wignal, De Broglie waves and the nature of mass, *Found. Phys.* **15**:207 (1985).

10.   D. Bedford and D. Wang, Toward an objective interpretation of quantum mechanics, *Nuovo Cimento* **26B**:313 (1975).

11.   N.V.Pope and A.D.Osborne, Instantaneous relativistic action-at-a-distance, in "Physical Interpretations of Relativity Theory", London (1990).

12.   M. Mugur-Schächter, What is the problem in the locality problem, *Found. Phys.* **18**:461 (1988).

13.   H.Aspden, The finite lifetime of the electron, *Spec. Sci. Technol.* **7**:3 (1983).

14.   M. Borneas, Some observations on microparticles, *Sem.Mat.Fiz.* *IPTVT*, Nov. 87 (1985).

15.   L.G. Sapogin, On unitary quantum mechanics, *Nuovo Cimento*, **53A**:251 (1971).

16.   L. de Broglie, "Certitudes et incertitudes de la science" Albin Michel, Paris (1966).

17.   D. Censor, Nonlinear wave-mechanics and particulate self-focusing, *Found. Phys.* **10**:555 (1980).

18.   P. Mathieu and T.F. Morris, Existence conditions for spinor solitons, *Phys. Rev.* **D30**:1835 (1984); Charged spinor solitons, *Can. J. Phys.* **64**:232 (1986).

QUANTUM CORRELATIONS FROM A LOGICAL POINT OF VIEW

N. A. Tambakis (*)

Interdisciplinary Research Group
Department of Physics
University of Athens

ABSTRACT

Quantum mechanical correlations have been thought of as a paradox or, alternatively, as a sign of *incompleteness* of standard quantum theory. Certainly, from a logical point of view, paradox and incompleteness are not equivalent. The concept of "paradox" is thus analyzed and the possible ways out discussed. The so-called *EPR-paradox* is recast in an elementary logical form of a "debate argument", i.e., where two seemingly sound arguments contradict each other. A way out of it, that of *adding new dimensions*, is discussed and a very general mathematical formulation is offered.

1. ON PARADOXES AND WAYS OUT OF THEM

The word "paradox" is the occidental version of the Greek word $\pi\alpha\rho\alpha\delta o\xi ov$, a composite of $\pi\alpha\rho\alpha$ and $\delta o\xi\alpha v$, i.e., "against $\delta o\xi\alpha v$", where $\delta o\xi\alpha$ meant the common opinion. The attribute "common" warns that what is a "paradox" to the ignorant may well have a rational explanation in the intellectual frame of a wise man. In our day, however, by talking about paradoxes in Physics we mean facts or concepts which do not fit well in the conceptual frame of contemporary physicists.

This incongruity may take either a soft or a hard form. We confront a soft paradox when we don't see some apparent *extrapolation* from the already known to the unknown imposed on us by the paradox, but it is quite possible that the sought-for extrapolation may hit us some time later. This is the way "normal" science works [1]. A hard paradox, in contrast, presents us with a persisting *contradiction* with respect to well-established knowledge. Defeating this contradiction may in some rare moments lead to a 'revolution' in science, i.e., to a new conceptual frame [2]. Contradiction may also take another form, the one between two arguments which *both seem sound*, i.e., both are apparently based on well-established knowledge but nevertheless contradict each other. (Then each party is inclined to say "My argument sounds right, so yours must be wrong!") It is this last kind of paradox, call it a *"Debate Paradox"*, which will mainly concern us here from a logical point of view. I will shortly present an example of such a paradox, but I would like first to briefly describe the possible ways out of a paradox in general, whether it be soft or hard.

We have already referred to two; first, the normal *extrapolation* of already existing knowledge which is achieved by finding out unknown elements of the existing conceptual frame. Second, the revolutionary *replacement* of the old conceptual frame, or part of it.

A third way involves making an *extension* of the existing conceptual frame itself. (It should be a *conservative* extension, i.e., one which keeps invariant the properties of the old structure.) Equivalently, it is said that *new dimensions* (geometrical or abstract) are added to the old structure, or that the *incompleteness* of the old structure is eliminated (or more precisely, reduced [3]). The third way will occupy us in some detail in the last part of this paper.

Finally, there is, or seems to be, a fourth way out of a paradox, which is, however, a paradox itself! This is the suggestion that in order to overcome a paradox we have to change our way of thinking. For example, it is suggested that the paradoxes of Quantum Theory need a "Quantum Logic,", i.e., brand new rules of reference. Such a "new logic" is *itself paradox*, by the very definition of a paradox. It seems to me, then, that trying to kill paradox by using another one is simply a tautological procedure, and in fact trivial. Furthermore, I think this approach is *inconsistent*. This may be seen from a model-theoretical viewpoint. In fact, in order for a theory to be consistent it is necessary that there exist a model of it *within a theory* already known to be consistent. It is however, clear that the new rules of inference cannot find a model in standard logic, for the simple reason that they refute it.

To this argument it may, perhaps, be opposed that things are the other way round, i.e., that Quantum Logic *is* consistent whereas standard logic is inconsistent.

I think, however, that at this point there is also a useful lesson: for some years after its discovery, Lobachevskian Geometry was thought of rather as a curiosity than as a useful, geometry, since its consistency was questionable. It was not until Beltrami's pseudosphere proved to be a model of Lobachevskian Geometry *within* Euclidean Geometry that Hyperbolic Geometry assumed a secure and useful status; it was the *consistency* of Euclidean Geometry that assured that [4]. And of course, Euclidean Geometry's consistency was unquestionable, since up to then the whole edifice of physicomathematical sciences was based upon this consistency. In a similar way, standard logic is certainly consistent on the basis of its tremendous success during the centuries of human evolution and the development of science and mathematics.

2. EPR - PARADOX IN A FORCED LOGICAL FORM

Let us now return to an example of a *Debate Paradox*, the sort of paradox that I referred to earlier, concerning the debate between two arguments which both seem sound. There is a quite famous example where the notions both of paradox and of completeness play crucial roles. Needless to say I refer to the so-called EPR-paradox, although it was never called that by its creators [5]. I will present it in quite schematic form, totally stripped of its subtleties and ramifications, but which makes its logical structure transparent [6]. The arguments, therefore, of the two sides, acceptably simplified run as follows:

Local Realism Argument

  E_1 (E for Einstein): Reality is not paradoxical
  E_2: Quantum Mechanics implies paradoxical situations (i.e., non-local correlations)
 $\therefore$ E_3: Hence, Quantum Mechanics is defective (i.e., does not correctly predict some aspects of reality).

Quantum Theoretical Argument

  B_1 (B for Bohr): $B_1 = E_1$
  B_2: Quantum mechanics correctly predicts reality
 $\therefore$ B_3: Hence, Quantum Mechanics is non-paradoxical.

Now, the head -to-head debate between the two camps becomes clear, $E_3 = \neg\, B_2$ and $B_3 = \neg\, E_2$, i.e., the conclusion of each part refutes the second premise of the other.

Let us now briefly comment on the above formulation of EPR:

(i) The first premise is common, $E_1 = B_1$. Of course, in their real-life arguments, Einstein and Bohr were at odds about the notion of Reality, but this does not concern us here: for whatever they believed about the nature of Reality, it seems to me that they *both* agreed that it should not be paradoxical. (By the way, I think they also agreed on the old rule that "real" is what is measured; what they disagreed about was the concept of "measurement" [7].)

(ii) Einstein's second premise, E_2, was *"selbst-verständlich"* for him (and hence, he didn't himself use the word "paradox"). He wrote: "...an external influence on A has no immediate influence on B. This is the *proximity principle* which is only consequently applied in field theory. A complete repeal of this principle would make impossible the idea of the existence of (quasi-) closed systems and thus of the establishment of empirically provable laws in the current sense..."[8].

(iii) Bohr's second premise, B_2, reflects his standard argument about the "practical completeness" of Quantum Mechanics. (He also supported it by arguing from his own concept of Reality, but this does not concern us here[9].)

(iv) It seems to me that the conclusion B_3 is not necessarily watertight: a theory may predict well, but that does not always mean that this theory is part of Reality, since its ontology may well be an *artifact of its successful mathematics* [3]. But if the theory is not a part of Reality, it cannot have all the attributes of Reality and, hence, the way from B_1 to B_3 is blocked.

Disregarding, however, our last objection (iv), the issue here is that *two arguments, which both seem sound, contradict each other.* According, then, to our views about what a "paradox" is, we must see the EPR debate as being one itself. The question then arises: which one of the described ways out seems to be most adequate for the EPR-paradox? We have already rejected for several reasons the solution calling for a "new logic"; what about the other three?

The softest, the "normal" extrapolation of existing knowledge, seems inappropriate to the EPR-paradox, being a subject of hot debate for more than half a century. What then about the second way out, the "revolutionary" one, which suggests the replacement of the old conceptual frame? The answer is far from simple, but I think there is here also some general reason for avoiding this choice: it is the danger that even the partial demolishing of the old conceptual structure is likely to leave without support some aspects of experience which previously fit in nicely. Our network of physical theories has reached such a successful intricacy that it is difficult even to think of the replacement of a tiny part of it without risking fatal damage elsewhere.

We then have to look for a theory which leaves intact all, or most of, the relations already existing in the body of knowledge, i.e. a *conservative extension* of the existing conceptual frame. We are, therefore, led to the third way out of a paradox, that of *adding dimensions to the old structures*, but before dealing with it we should ask ourselves how can we be sure that "adding new dimensions" does not lead us back to the situation of "new ways of thought", which we have already rejected?

This is certainly a thorny question which has to be answered each time for the theory under discussion. For example, I think that a physical theory with ten space--time dimensions does not necessarily impose a new way of thought since we may accept that at least in principle, the six extra dimensions are somehow 'curled and degenerated in the microscopic domain. (Of course for such a theory other important epistemological issues remain, such as its being irrefutable, which need not concern us here.)

In contrast, a physical theory which admits action-at-a-distance as a new "dimension" implies a new and "magical" way of thought.

I will now make some further remarks on the way out of a paradox by adding dimensions.

3. CORRELATIONS IN THE LIGHT OF NEW DIMENSIONS

Although there are always pitfalls [10] in the way of thinking "as if I were...", let me give an example of thinking as if I were a 2-dimensional intelligent creature. Let us also suppose that the 2-dimensional manifold on which this intelligent agent lives has, in

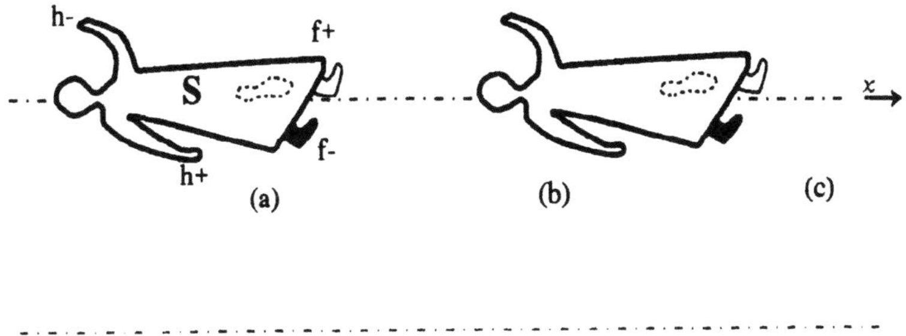

FIG. 1

addition, a third degenerated dimension that allows him to observe minute deformations along the third dimension; we may, e.g., imagine the 2-dimensional manifold as a very thin sandy surface.

What is observed on the 2-dimensional surface is: (fig. 1)

First, a sequence a-b-c... of twin deformations (f., f.) of the sand; the form f. points to the left along the direction x of the sequence, whereas the f. points to the right.

Secondly, the deformations f., f. are always followed by a darkness pattern S which is almost symmetric with respect to the x-axis except for the two protruding parts h. and h., whose orientations seem to be in close relationship with those of f. and f.. Hard as the 2-dimensional agent tries to explain appearances, he fails unless a 3-dimensional Big Brother puts things right for him: everything falls into place if we imagine a solemn humanoid walking sternly on the 2-dimensional sandscape, which is illuminated by a sun in the 3-dimensional firmament. (fig. 2)

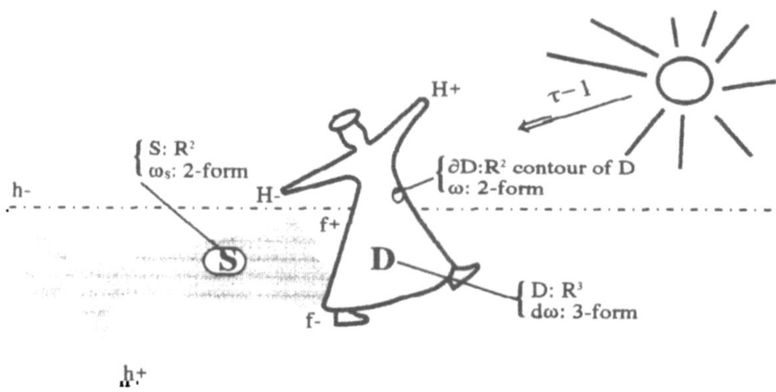

FIG. 2

Whereas   the 2-dimensional agent is ready to talk about the "correlations" between h. and f. (and also between h. and f.) Big Brother simply sees two footprints, f. and f., followed by the shadows h. and h., of the two hands, H. and H., helping with the walking rhythm.

The moral is that a *new dimension* was enough to explain a paradoxical situation. Needless to say, this new dimension is not necessarily a spatial one; it can also be a *new category* in the space of lower dimensionality.

Leaving our example and thinking about the enigmatic particle correlations in modern Quantum Theory(if they are actually experimentally confirmed beyond any doubt [11]), some explanation may exist in the light of "new dimensions". For instance, the existence of some objective background of higher space-time dimensionality may be responsible for a quantum correlation[12]. Also, instead of new space--time dimensions, some new property may be added to the singlet state pair in the usual 3-dimensional space. One may speculate that the twin particles of the pair are equipped with some "generic" mechanism, in the style of the uniovular twins.

Let us now recast the above way out of a paradox (by adding dimensions) in a quite general mathematical form. What we need is some general mathematical method in the form of a differential or an integral equation-- in order to relate some characteristic of the lower dimensionality space with another of the higher dimensionality space.

What comes to mind is the powerful mathematical theorem which establishes a connection between a given space and another of one less dimension, namely the contour of the first space. This is the generalized Stokes theorem [13]:

$$\int_{\partial D} \omega = \int_{D} d\omega$$

(1)

It states that the integral of a differential form ω over the boundary ∂D of the domain D is equal to the integral of its exterior derivative $d\omega$ over the domain D. This theorem encompasses the usual Gauss and Stokes theorems. Also, for ω a o-form, from (1) we get the "fundamental theorem of calculus."

Let us then see how our previous illustration with the walking figure (Fig. 2) appears in the light of the Stokes theorem., eq. (1).

To begin with, let the correlation function between h- and f+ be expressed as a differential 2-form ω in the flat 2-space S. Let τ^{-1} be the projection which transforms the contour ∂D of the moving volume D to the flat surface S:

$$\partial D \xrightarrow{\tau^{-1}} S$$

Then, inversely, $\partial D = \tau(S)$, i.e., the flat 2-dimensional surface S is transformed by τ to the 2-dimensional manifold ∂D; this projective transformation τ also induces a transformation τ^* on the differential form ω_s, expressed in terms of the coordinate differentials of the flat surface S, to a differential form $\omega = \tau^*(\omega_s)$, expressed in terms of the coordinate differentials of the 2-dimensional manifold ∂D.

So finally we may write the Stokes formula (1) which relates information about the 2-dimensional flat world to information about the 3-dimensional real world. Perhaps, in this way, some of the magic surrounding the poor 2-dimensional creature has been removed. In general, the use of the Stokes theorem seems to be a possible mathematical treatment of adding new dimensions as a way to the resolution of a hard paradox.

NOTES AND REFERENCES

[1]   In the Kuhnian sense; T. S. Kuhn, *The Structure of Scientific Revolution*, (University of Chicago Press, 1962).

[2]   Ref. [1].

[3]   Since every physical theory is incomplete: N. A. Tambakis, "On the Empirical Law of Epistemology; Physics as an Artifact of Mathematics", invited talk presented at the International Symposium on Fundamental Problems in Quantum Physics, Oviedo Spain, 29 August- 3 September, 1993. To appear in the Proceedings edited by Kluwer Academic Press.

[4]   See, for example, M. Kline, *Mathematics, the Loss of Certainty* (Oxford U. Press, New York, 1980) Ch.. IV.

[5]   A. Einstein, B. Podolsky and N. Rosen, Phys. Rev. *47*, 777 (1935).

[6]   In a similar form it is presented in a letter by D. T. Gillespie, to Am. J. Phys., *57*, 12 (1989), p. 1065.

[7]   H.-H.v. Borzeszkowski and R. Wahsner, Found. Phys., *18*,6, (1988)

[8]   A. Einstein, "Physik und Realität", in Journ. Franklin Inst. Philadelphia, *221*, 313 (1936).

[9]   See H. Stapp, Am. J. Phys. *53*, 4 (1985)

[10]   See the critiques following another well-known "if I were" thinking: T. Nagel, "What Is It Like to Be a Bat?", in *The Mind's I*, D. Hofstadter-D. Bennett, (Penguin, 1982)

[11]   M. Ferrero et. al., "Bell's Theorem: Local Realism Versus Quantum Mechanics", Am. J. Phys., *58*, 7 (1990)

[12]   This role may be attributed to the "vacuum". See B. J. Hiley's "Vacuum or Holomovement" in the *Philosophy of Vacuum*, S. Saunders and H. R. Brown, (eds.) (Oxford U. Press, New York 1991).

[13]   See for instance, H. Flanders, *Differential Forms*, (Academic, New York, (1963).

*Mailing Address: N.A. Tambakis, P.O. Box 65032, Psychiko, Athens 154 10, Greece

LOCAL REALISM AND THE CRUCIAL EXPERIMENT

Yoav Ben-Dov

Cohn Institute for the History and Philosophy of Science
Tel-Aviv University
69 987 Tel-Aviv, Israel

It is well known that the history of quantum mechanics is riddled with conceptual debates. The most famous debate is that which took place between Einstein and Bohr from 1927 onwards, and in which the two protagonists can be taken as representing two possible answers to what is perhaps the most central question concerning quantum mechanics. Both sides in this debate agree that the usual quantum formalism, as developed by Heisenberg, Schroedinger and Dirac, is inconsistent with what was believed, from the 17th century onwards, to be the central prerequisites of a sound physical theory: determinism, locality (which means that things don't act at places where they are not present), clear separation between object and subject (what we shall call here "realism" - the view that physical reality can be defined "as it is" without any reference to an observer), and so on. They differ, however, in their reaction to this situation. Should we look for an alternative theory, which would satisfy these prerequisites while still accounting for the observed quantum phenomena (Einstein's position), or should we rather accept quantum mechanics as it is, and try to adapt our epistemological prerequisites to the new formalism (Bohr's position)?

From a historical point of view, one may see some point in Bohr's claim. Einstein's prerequisites are modelled after the classical physical theories, for example Newtonian mechanics. But the acceptance of Newtonian mechanics was a historical event, and the prerequisites associated with it were introduced along with the same event. In fact, Newton's mechanics was at that time inconsistent with the previously accepted prerequisites of Aristotelian science, for example with the demand that a theory describe the purpose of things' being as they are, and not only their quantitative behavior. Thus, epistemological prerequisites change with the current scientific theories (so that they are actually "post-requisites"). Why should not the same hold for quantum mechanics?

Supporters of Einstein's view may, however, turn this argument around, and claim that Aristotelian science was much less successful than Newtonian science exactly because it tried to satisfy the wrong prerequisites. From the 17th century onwards, modern science arrived at immense achievements by following the classical prerequisites. Doesn't their definitive abandonment involve the risk of undermining the basis of the whole modern scientific enterprise? In other words, isn't a truly non-classical science simply a contradiction in terms?

Of course, Einstein's position must be abandoned if one can prove that no theory which satisfies the classical prerequisites can possibly account for the observed phenomena. The question is, do we have such a proof? As for the prerequisite of determinism (or "classical causality"), we know today that Von Newmann's alleged "proof" to this effect is invalid[1], and with the de Broglie-Bohm "pilot wave" theory[2] we even have a counter-example of a fully deterministic theory which perfectly reproduces all the quantum phenomena. However, many physicists and philosophers today believe that the violations of Bell's inequality[3], such as are manifested in Aspect's[4] and other experiments, are inconsistent with the conjunction of realism and locality, so that at least one of these two classical prerequisites should be abandoned.

Presented thus, Aspect's experiments appear as a "crucial experiment" in the original sense of Francis Bacon: in its progress, science arrives at a crossroad, where two mutually exclusive possibilities are open. An experiment is devised to decide between them, so that one can be discarded and the other accepted as true. Here, the alternatives are "Nature can be described by a local realistic theory" and "Nature cannot be described by a local realistic theory", and Aspect's experiment decides in favor of the second. It is in this sense that John Bell was called, in one obituary, "the man who proved Einstein wrong": any attempt to retain the complete list of classical prerequisites, and in particular, any attempt to devise a successful local realistic theory, is doomed to fail.

However, we should perhaps be more careful. At the beginning of this century, the great physicist and historian of science, Pierre Duhem, tried to introduce the following distinction between physics and metaphysics[5]. In his view, physics concerns the classification and prediction of experimental results, while metaphysics concerns the nature of reality itself. Following Duhem's distinction, we can claim that a physical theory can never be regarded as absolutely definitive, because our belief in it is based only on a finite series of experiments, namely those which have been performed so far. It is always possible that a different theory will still explain the same results, and also account for the results of future experiments for which the present theory proves inadequate. In this sense, a physical theory can be "correct" at some moment and "incorrect" later. In contrast, a metaphysical claim is supposed to be absolute, that is, "true" or "false" for all time, irrespectively of any experiment which we are able to conduct at a given moment.

According to this distinction, Aspect's experiments belong to the domain of physics, that is, they should be accounted for by physical theory. On the other hand, the statement "no local realistic theory is possible" is a metaphysical claim, as it asserts something about reality itself. The question is, whether a physical experiment can decide in a definite manner a metaphysical question - that is, whether a "crucial experiment" whose verdict is absolute is at all possible. Indeed, there are some who, apparently accepting a distinction not very different from Duhem's, refer to Bell's inequality and Aspect's experiments as "experimental metaphysics".

Duhem, however, regarded "experimental metaphysics" as an impossible expression. In his view, Metaphysical statements and experimental results exist in two different worlds. To bring them together, much elaboration and interpretation is required, and this involves many assumptions which can always be eventually revised. even if we work out our metaphysical positions into specific hypotheses cast in the mathematical language of physics, still "an *experimentum crucis* (to choose between them, Y.B.) is impossible in physics". The reason is twofold.

First, a single hypothesis by itself is not testable by experiment. At most, the experiment can test a complete theory, which always involves many additional hypotheses, some of them (perhaps the most important, and probably the most problematic) implicit. Suppose for example that the experiment has clearly condemned a certain theory. Still, we do not know which of the many hypotheses entering the theory is the one at fault, and we can never be sure that it is the one that we originally wanted to test. Second, whenever we formulate a choice between hypotheses as a true dilemma - either p is true, or q - it is possible to cast a doubt on the nature of the dilemma itself: perhaps an additional hypothesis has implicitly entered our considerations, or our imagination is too weak; how can we be absolutely sure that these are really the only two possibilities - that some intermediate (or otherwise different) possibility cannot be found, or that the very terms in which the question was posed are for ever beyond revision?

To illustrate his point, Duhem gives a historical example, which the years have rendered even more striking. At the beginning of the 19th century, two rival hypotheses concerning the nature of light existed in physics: the corpuscular theory, which is usually ascribed to Newton, and the ondular theory, which originated with Descartes and Huygens, and was revived by Young and Fresnel. According to the first theory, light is a stream of tiny particles; according to the second, it is a disturbance propagating in a medium, the luminiferous Ether. These two theories could account, each in its own terms, for many experimental phenomena, including the refraction of light as it passes from one transparent medium to another. But there was an important point of difference between them: the corpuscular theory assumed for its explanation of refraction (ironically, first formulated by Descartes) that the speed of light in a transparent medium (such as glass or water) is greater than the speed of light in empty space. The ondular theory, on the other hand, assumed exactly the opposite. On the basis of this difference, François Arago proposed a "crucial experiment" which would compare the two speeds, thus deciding once and for all whether light is a particle or a wave. Only some decades later, experimental technique was refined enough so that Léon Foucault could devise an actual experimental set-up along these lines. The experiment was carried out in 1850, with clear results: light travels more slowly in water that in air (which, for this purpose, is approximately equivalent to empty space). Therefore, as the physicists then concluded, light is a wave, and the corpuscular hypothesis can be discarded, not to be bothered with any more.

But for Duhem, this conclusion is very doubtful. First, he points out that Foucault's experiment did not decide between two isolated hypotheses, but between two complete theoretical systems. True, the corpuscular theory as held by early 19th-century Newtonians is condemned. But it is not inconceivable that a future theory might be built upon the corpuscular hypothesis, with the aid of some new auxiliary hypotheses which would be different from those entering the Newtonian system. In such a theory, the refraction of light might be explained in a different manner, so that it would be possible to account for the results of Foucault's experiment while still maintaining that light is a particle. Second, it is not at all certain that the current concepts "wave" and "particle" are the only possible ones; perhaps a new concept might be formulated, which would go beyond this dichotomy, possibly by combining some aspects of both concepts.

When Duhem formulated his arguments, this was only an abstract possibility. However, shortly afterwards, Einstein has re-introduced into physics the idea of a corpuscular nature of light, starting with his 1905 paper on light quanta. This paper has paved the road to present-day quantum mechanics. But as we know, the debate on the nature of quantum objects (including light quanta) and on the wave- particle duality still goes on. Thus, we can say that more than a century after Foucault's "crucial" experiment was supposed to decide it once and for all, the question of the nature of light is still open. Moreover, we do not even expect any more a simple answer such as "only wave" or "only particle" in the classical sense, a situation which provides a striking example of the acuteness of Duhem's vision. We should also note that nowhere in the present debate on the nature of quantum objects is Foucault's result even mentioned as evidence. Of course, no one doubts the result itself, that is, we are still convinced that light travels more slowly in water than in air, and that any future physical theory should account for this fact. But the experimental result itself has lost all relevance to the question of the nature of light. In Duhem's terms, today we interpret Foucault's set-up and his results as an experiment in physics, not in metaphysics.

Perhaps similar considerations can also be applied to Aspect's experiments. Between the metaphysical statement "a local realistic theory is impossible" and the actual experimental set-up there is a huge gap, which can only be bridged with the aid of many auxiliary hypotheses. Any one of these could be wrong. Proceeding from the experimental side, we can, for example, point out that there are "experimental loopholes"[6] in Aspect's experiments, which, if investigated further, might turn out to be responsible for the result. We can also suspect the existence of some "selection effect" which influences the detection probabilities, so that Aspect's experiments do not actually test Bell's inequality[7]. We can accept the possibility that some additional implicit assumption has entered into the mathematical derivation of Bell's inequality, or doubt that the mathematical criteria used in this derivation are accurate and complete translations of the metaphysical concepts "realism" and "locality". And surely, there are many more possibilities, which we cannot see from within the network of present-day physical concepts.

All this, of course, is not meant to undermine the value of Aspect's experiments. These are surely beautiful physical experiments. But those who interpret them as "experimental metaphysics" seem to ignore the force of Duhem's arguments. It is quite possible that exactly like Foucault's experiments, physicists in the next centuries would interpret Aspect's experiments as physically valid, but metaphysically irrelevant. Thus, notwithstanding Aspect's experiments, it cannot be maintained that a local realistic theory which will account for the observed phenomena is definitely ruled out.

REFERENCES

1. J.S. Bell, Rev. Mod. Phys. 38:447 (1966).
2. D. Bohm, Phys. Rev. 85:166 and 180 (1952).
3. J.S. Bell, Found. of Phys. 12:989 (1982).
4. A. Aspect, J. Dalibard and G. Roger, Phys. Rev. Lett. 49:1804 (1982).
5. P. Duhem, "La théorie physique, son objet - sa structure", Vrin, Paris (1989).
    English translation: "The Aim and Structure of Physical Theory", Atheneum,
    NY (1974).
6. A. Shimony, An exposition of Bell's theorem, in: "62 Years of Uncertainty", A.I. Miller,
    ed., Plenum Press, NY (1990).
7. F. Selleri, Einstein-de Broglie waves and two-photon detection, in: "Bell's Theorem and
    the Foundations of Modern Physics", A. van der Merwe, F. Selleri and G. Tarozzi,
    eds., Plenum Press, NY (1992).

THE SPACE OF LOCAL HIDDEN VARIABLES CANNOT BE A METRIC ONE AND WHAT NEXT?

Milan Vinduška

ISC-Engineering
Badeniho 3
160 00 Prague 6
Czech Republic

INTRODUCTION

Metric content of the Bell inequalities for the wide range of phenomena can be interpreted as a consequence of the isotropy of the hidden variable space defined by measuring devices.

The destroyed isotropy of such a space is described by the relative measure of probability (= RM)[1] ,which permits one to restore the quantum mechanical results in the language of local hidden variables.

The results of the recent investigations of the problems considered are shortly recapitulated in Sec. 2-3. The simple relativistic model is introduced in Sec. 4, where the problem of locality is considered together with the classical and nonclassical features of RM.

METRIC BELL' S INEQUALITIES AS A COSEQUENCE OF THE ISOTROPY OF THE HIDDEN VARIABLE SPACE

In the Bell' s scheme the correlations are described as[2]

$$P(a,b) = \int A(a,\lambda)B(b,\lambda)\rho(\lambda)d\lambda, \tag{1}$$

where $A(a,\lambda)$ and $B(b,\lambda)$ denote the experimental outcomes for apparatuses with pointer positions a and b , $\rho(\lambda)$ is a normalized probability density of the hidden variables λ.

Moreover, the following conditions are postulated

$$A(a,\lambda) \neq f(b) , B(b,\lambda) \neq f(a) , \tag{2}$$
$$\text{and}$$

Frontiers of Fundamental Physics, Edited by M. Barone
and F. Selleri, Plenum Press, New York, 1994

$$\rho(\lambda) \neq f(a,b) , \tag{3}$$

which express the isotropy of the hidden variable space.

In the literature both conditions are erroneously identified with "locality". While it is true for (2), the condition (3) rather expresses the isotropy of λ in respect to the space elements defined by measuring devices.

The actual content of the Bell' s inequalities for a wide range of phenomena (like the measurements of the projections of linear polarizations of photons, - of the spin projections in static or precessing singlet systems[3] , - of flavour characteristics of the oscillating K and B mesons[3] , - of phases of interfering particles) can be suitably expressed in terms of affine geometry.

The "pointer positions" of the measuring devices a, b, ...,.. can be taken as a vectors and/or points in certain space. Then there can be defined two functions D(a,b) and H(a,b) , the first one being a certain conditional probability and the second one a mean conditional information entropy

$$D(a,b) = \frac{1}{2.P(a,a)}\{P(a,a) - P(a,b)\},$$

$$X(a,b) \ \langle \tag{4}$$

$$H(a,b) = -\tfrac{1}{2}[P(a,b) + 1]\log\left\{\tfrac{1}{2}[P(a,b) + 1]\right\} -\tfrac{1}{2}[1 - P(a,b)]\log\left\{\tfrac{1}{2}[1 - P(a,b)]\right\},$$

which both have the properties of the metric distance.

The first three conditions of the metricity, i.e.,

$$X(a,b) = X(b,a) , \tag{5}$$

$$X(a,b) \geq 0 , \tag{6}$$

$$X(a,b) = 0 , \text{ for } a = b , \tag{7}$$

follow from the symmetry properties of the correlation function of P(a,b) (and they hold for the quantum correlations, too) , while the Bell' s scheme (1) and (2) supplements them with the triangular inequality

$$X(a,b) + X(b,c) - X(a,c) \geq 0. \tag{8}$$

It can be shown, that the extremal values of D(a,b) and H(a,b) correspond to the spherical and Riemannian metrics on the spherical surfaces and the necessary and sufficient conditions for D(a,b) to be exact metric distance in such geometries can be given (for that it is sufficient if $A(a,\lambda)$ and $B(b,\lambda)$ behave as the signs of properly defined scalar products of a, b and λ).

It is well-known, that inequalities for the quantum mechanical correlations have a nonmetric form and, therefore, that the Bell's scheme of hidden variables must be modified. In the following section we shall introduce such a modification for the case when the symmetry of the hidden variable space is different from that of our macroscopic world.

RELATIVE MEASURE OF PROBABILITY

For the further discussion, we need some generalization of the Bell' s scheme. In terms of affine geometry it can be done by identifying hidden variables with vectors in certain space. The values of $A(a,\lambda)$ and $B(b,\lambda)$ can be evaluated as signs of properly defined scalar products and the distribution $\rho(\lambda)d\lambda$ can be described as a distribution of ends of λ on some surface

$$\rho(\lambda) = (\mathbf{n}.\lambda)dS , \tag{9}$$

here $\mathbf{n}$ is a normal to surface S, chosen in such a way that $(\mathbf{n}.\lambda) \geq 0$, The normalization condition takes the form

$$\int (\mathbf{n}.\lambda)dS = 1. \tag{10}$$

We shall indicate now that sufficient condition for the correlation function $D(a,b)$ to be a metric distance is isotropy of hidden variable space. This property has any space with constant curvature.

The invariant description of the considered correlations is guaranteed by the using the proper scalar product for evaluating $A(a,\lambda)$ and $B(b,\lambda)$ and surface S must be taken as a spherical surface in such a space. According to the Theorem presented in Appendix, $D(a,b)$ is then an exact metric distance.

As an illustration the case of correlations of the linear photon polarizations in singlet systems can be shown.

Here

$$A(a,\lambda) = A(a,\lambda_1) = \mathrm{sign}(\lambda_1 . \lambda_1^a), \tag{11}$$

and

$$B(b,\lambda) = B(b,\lambda_2) = \mathrm{sign}(\lambda_2 . \lambda_2^b). \tag{12}$$

The vector λ_i^n is constructed in such a way that $\mathbf{n}$ is a.bissectrisse of the angle λ_i, λ_i^n. Hidden vectors λ_1 and λ_2 are linked as $\lambda_1 \uparrow\uparrow \lambda_2$ or $\lambda_1 \perp \lambda_2$ for states with even and odd parity respectively and S is a circle lying in the polarization plane, perpendicular to the photon momenta.

It can be shown, that the isotropy of the hidden varisabvle space is also necessary condition for Bell' s inequality to be hold.

Let us suppose that symmetry of the hidden variable space (e.g. for the description of the linear photon polarization) corresponds to the Minkowskian $(1,1)$ plane geometry. For treating such a situation the invariants under hyperbolic rotations must be used and, consequently, the "surface S" must be identified with the four branches of hyperbolas.

Explicitly

$$A(a,\lambda_1) = \mathrm{sign}(\|\lambda_1^a\|_h^2), \tag{13}$$

$$B(b,\lambda_2) = \mathrm{sign}(\|\lambda_2^b\|_h^2), \tag{14}$$

where $\|\lambda_i^r\|_h^2$ denotes the square of the hyperbolic norm of

$$\|\lambda_i^r\|_h^2 = (\lambda_i)_x^2 - (\lambda_i)_y^2 , \tag{15}$$

when axis x is identified with $\mathbf{r}$. The normalized probability density takes the form

$$\rho_a(\lambda) = c(\mathbf{n}.\lambda)_e = \|\lambda^a\|_h^2 . \tag{16}$$

The subscript "e" in the scalar product stresses the using of scalar product as it is defined in Euclidean geometry. $\|\lambda^a\|_e^2$ is the square of the ordinary Euclidean norm, and $\|\lambda^a\|_h^2$ is the square of hyperbolic norm for $\mathbf{a}\!\uparrow\!\uparrow\mathbf{x}$.

Using the basic relation (1), we obtain

$$P(a,b) = +/- \|a^b\|_h^2 = +/- \cos(2\varphi_{ab}) , \tag{17}$$

in accordance with the quantum mechanical results.

In the example considered above the isotropy of the hidden variable space is destroyed by the using the Minkowskian geometry itself, because we must in a certain way orientate our coordinate system in respect to the real experimental situation. In the example considered we must identify x or y axis of the subquantum space with orientations of measuring devices $\mathbf{a}$ or $\mathbf{b}$. Consequently, we use for the correlation function

$$P(a,b) = P_\mathbf{n}(a,b) = \int A(a,\lambda)B(b,\lambda)\rho_\mathbf{n}(\lambda)d\lambda , \ \mathbf{n} = \mathbf{a} \text{ or } \mathbf{b,} \text{ instead of (1).} \tag{18}$$

The last expression limits the possibility to describe different experiments with the use of common probability measure (it is not connected with the question how many experimental settings are really needed).

The difference between classical statistics and the quantum statistics , which uses the concept of RM , is presented in Fig.1. It can be shown, that this concept invalidates also other no-go theorems for hidden variables as von Neumann theorem[6], GHZ correlations[7] and Feynman's theorem[8].

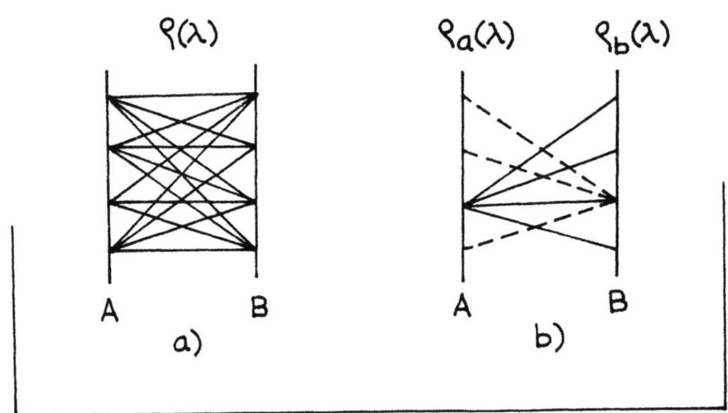

Fig.1

The statistical scheme for four different settings of apparatuses A and B: a) classical statistics; b) quantum statistics with RM.

THE PROBLEM OF LOCALITY

We have not touched the problem of locality of the proposed picture up to now, although we have used explicitly the true condition of locality (2) in our model.

There have arised some objections concerning the locality properties of RM before and also during the time, when it was elaborated. They can be summarized in the following manner:

(i) RM is not local because the change of the distribution of λ_1 of the first particle leads to the change of the distribution of λ_2 of the second one due to the link between them and it is equivalent to the nonlocal influence[9,10];

(ii) or "the source must peer at the future for creating the correct ρ_n"[11]

(iii) or there is no free will and specific orientation of the measuring devices is determined by the source itself[9].

We shall show that all these arguments are not valid after introducing a simple relativistic model which is certainly local.

Let us consider following model (See Fig.2).

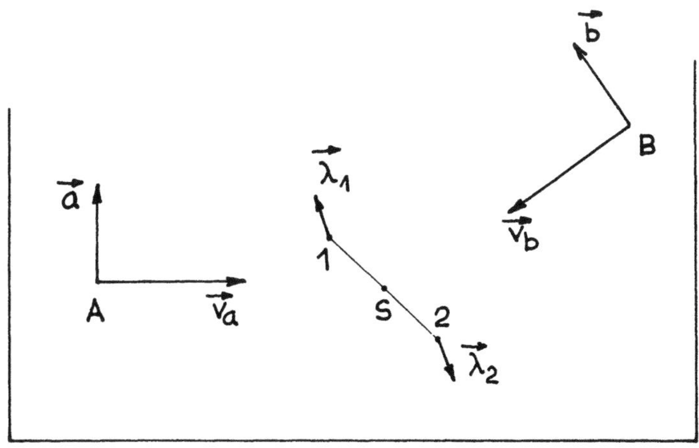

Fig.2

The relativictic plane model for correlations.

The source S creates two particles "1" and "2" which are bearing hidden vectors λ_1 and λ_2 , $\lambda_1 \uparrow\downarrow \lambda_2$. For the whole ensemble of such particles the pairs of opposite vectors are equally distributed on the circle, lying in the plane. For the simplicity sake let us suppose that the particles are at rest in relation to the frame of reference related to the source when the measurements are realised.

The first experimenter obeying the device A is moving with velocity v_a in direction to the source and his apparatus is oriented perpendicularly to his direction of motion. He is measuring the value $A(a,\lambda) = A(a,\lambda_1) = \text{sign}(\,a.\lambda_1\,)$,where λ_1 is hidden vector of the first particle.

Similarly the second experimenter with device B moving with velocity v_b in another direction measures value of λ_2 of the second particle onto b

$$B(b,\lambda) = B(b,\lambda_2) = \text{sign}(b.\lambda_2).$$

The measurements of both experimenters are realised in space-like intervals and it is supposed that they do not disturb each other.

Let us remark that this plane model remind us of a spin singlet system because

$$P(a,a) = -1 \text{ and } P(a,-a) = +1 .$$

What distribution of λ_1 and λ_2 will be found by A and B when the measuring procedure will be repeated many times? Taking into account the Lorentz transformation it is not difficult to realise that the first experimenter will find

$$\rho_a(\lambda_1) = \sqrt{1-\beta^2} \cdot (1 - \beta^2\cos^2(\lambda_1 - \tfrac{\pi}{2}))^{-1} , \quad \beta = \tfrac{v_a}{c} , \quad 0 \le \lambda_1 \le \pi , \tag{19}$$

i.e. the distribution of λ_1 picked around the direction of **a** as it is demonstrated in Fig.3.

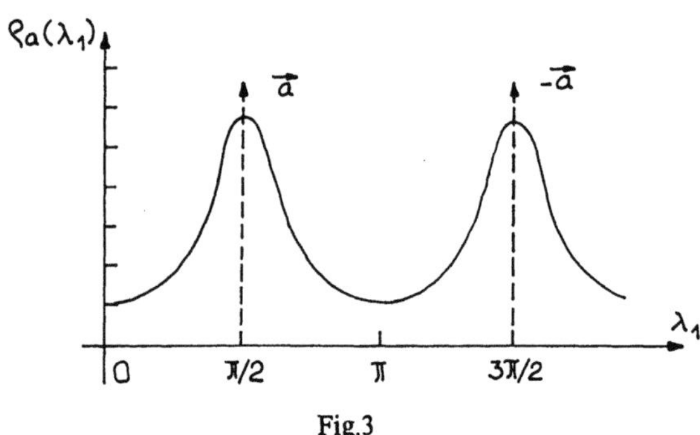

Fig.3

The distribution of $\rho_a(\lambda_1)$ for $\beta = 0,9$ .The distribution of $\rho_a(\lambda_2)$ has the same form; only the shift $\lambda_2 = \lambda_1 + \tfrac{\pi}{2}$ must be taken into account. The distribution of $\rho_b(\lambda_2)$ can be received by interchanging the notations $a \Leftrightarrow b$ and $\lambda_1 \Leftrightarrow \lambda_2$.

Due to the property of the Lorentz transformation (the straight line on which λ_1 and λ_2 lie remains the straight line in any frame of reference) the distribution of $\rho_a(\lambda_2)$ as it seen by the first experimenter is also picked around the direction of **a** , although there is no interaction between device A and particle "2". Because the apparatuses are equivalent, it is clear, that the same reasoning relates to the second experimenter with device B. Here both distributions $\rho_b(\lambda_1)$ and $\rho_b(\lambda_2)$ are picked around **b**. There really exists great similarity between model introduced here and that which explains the quantum correlations in the language of RM(Cf. citation[5]).

Inspecting objections (i) - (iii) formulated above, we conclude that they fail.

What can we say about the correlation function and Bell's inequalities in our case? Because each real event is invariant in respect to the different reference frame, it is clear, that $D(a,b)$ is a metric of spherical geometry on the circle and, therefore, the Bell's inequalities are satisfied if only the angle (or "distance") between **a** and **b** is measured in the frame reference of the source. It is not so in the case when this angle is measured in relation with one apparatus to the other because here the hidden destroyed isotropy of the whole picture appears.

We can conclude this section by comparison of the characteristics of the model presented with that connected with RM. The common features can be seen in the symmetry of both frames of reference, related with the corresponding devices and understanding each real event as an invariant.

But we do not think that the quantum properties are connected with some hidden velocities, etc.,. In the model with RM the principle of covariant description is highlighted to the more fundamental level, the invariance of the description in the different reference frames is here postulated and causes that the transformations of the RM forms only a cyclic group,etc., but this is another story[5].

APPENDIX

Theorem.

Let $D(0) = 0$, $D(\kappa) = 1$ and $A(a,\lambda)$, $B(b,\lambda)$ and $\rho(\lambda)$ guarantee the rotational invariance of $D(a,b)$, i.e. $D(a,b) = D(\varphi_{ab})$.

Then $D(a,b)$ is the exact metric distance of the spherical or Riemannian geometry if and only if the sequence

$$A(a_1,\lambda), A(a_2,\lambda), ...,A(a_n,\lambda)$$

for an ordered set of space elements

$$a_1,a_2,.....,a_n \; ; \varphi_{a_1 a_n} \leq \kappa,$$

change sign no more than once for each λ.

For one-dimensional space of elements the ordering simply means placing of a_i in correspondence with the increasing index, for a spherical surface,e.g., it means that a_i are placed similarly on the main circle. The symbol κ denotes one half of the period.

REFERENCES

1. A.A.Tyapkin and M.Vinduška: *Found.Phys.*21,185 (1991).
2. J.S.Bell: *Physics* 1,195 (1964).
3. M.Vinduška: Bell's scheme of LHV and time-dependent quantum phenomena,*Proc. Trani Conf.* (1992), (in press).
4. M.Vinduška: *Phys.Lett.A* 174,9 (1993)
5. M.Vinduška: Metric Bell inequalities,relative measure of probability and the geometry of hidden variable space, *Proc.Third Winter School on Measure Theory*, Liptovský Ján (1993),
   Tatra Mountains Math.Publ., (in press).
6. J.von Neumann: *The Mathematical Foundations of Quantum Mechanics*,Princeton, N.J., (1955).
7. D.M.Greenberger, M.A.Horne and A.Zeilinger,in: *Bell's Theorem, Quantum Theory, and Conception of Universe*, Dordrecht (1989).
8. R.P.Feynman: *Int.J.Theor.Phys.* 21,467 (1982).
9. J.S.Bell: (private communication)
10. J.A.Smorodinsky: (private communication).
11. The anonymous referee of citation[1].

HOW THE QUANTUM OF ACTION CAN LIMIT NON-LOCALITY

Constantin Antonopoulos

National Technical University of Athens
Division of Humanities, Social Sciences and Law
15773 Athens-Greece

i. What I am about to propose is relatively easy to state. I will argue that Bohr's philosophy is perfectly compatible with the case made by EPR. It is, however, just as relatively easy to misunderstand. What I am contending is that it is Bohr' s *philosophy* which is so compatible. Not that his Reply to *EPR* is. In fact, and despite a considerable bulk of literature floating about, I consider Bohr's Reply to EPR utterly unsuccessful in itself and thoroughly incompatible with the EPR case. It was so designed to be.

If however, I still claim that Bohr's quantum philosophy is compatible with EPR, whereas his Reply is not, I thereby imply that these two are incompatible with one another. They have to be, if one of them is compatible with what the other isn't. Bohr's Reply to EPR is in effect a giant step outside his standard quantum philosophy, what I will be referring to as his "Official Doctrine". For that matter, this Reply is essentially a giant step outside quantum mechanics itself. My reasons for saying this are as follows.

Prior to entering issues of locality, EPR chiefly and primarily raise an issue of *completeness.* The completeness of the complementary account of quantum mechanics, to be exact. And this account rests on the demand of a mutual exclusion between a pair of conjugate classical concepts, such as p with q and E with t, although the latter pair may not be called conjugate in strict orthodoxy. Now mutual exclusion is a quite familiar relation in Logic; it is the relation of *incompatibility.* Consequently, complementarity should itself be subject to the logical rules which govern incompatibility.

And incompatibility is of *two* kinds, exclusive of one another. The statements A‹B and B‹A are incompatible by *definition,* and there is not one single case conceivable of their simultaneous truth. So they are incompatible in all possible cases. This is logical incompatibility. But not all cases of incompatibility are like that. Far from it. Consider the infamous command of the bandit, "your money *or* your lives!'. "Having one's money" and "having one's life" are perfectly compatible states in ordinary situations, though they can be driven to mutual exclusion in view of specific circumstances, as the one just cited. And this can happen to any two states, if the appropriate conditions are posited. But such incompatibility will be *withdrawn,* when these specific conditions are withdrawn. So, whenever incompatible in this latter sense, they will still be incompatible only *sometimes.* This is factual incompatibility. And as no two concepts, or states, can be both, incompatible in all possible cases *and* also incompatible only sometimes, it follows that no two concepts, or states, can be both, logically as well as factually incompatible at the same time.

But Bohr's Official Doctrine, already constructed prior to the EPR case, implies that the conjugate concepts are *factually* incompatible and his Reply to EPR strongly implies that they are *logically* incompatible instead. Whence Bohr's Reply to EPR is nothing like the Official Doctrine.

Frontiers of Fundamental Physics, Edited by M. Barone
and F. Selleri, Plenum Press, New York, 1994

ii. Bohr appears to be taking hold of several arguments in his Reply, many of which belong to quantum theory. But the heart of the point is focused on his excessive emphasis on "*moveable* diaphragms", suitable for determining the momentum, as *opposed* to "rigid supports",[1] suitable for determining the position. Of such opposition we have almost grotesquely mechanical illustrations in later works.[2] Mechanical, yes. *Quantum* mechanical, no. As Horatio would no doubt say, "we need no (quantum) ghost come from the grave to tell us that", viz., that moveable diaphragms exclude rigid supports and vice versa. This much we know independently of *any* quantum assumption. Consider the following, excellent summary of Bohr's argument by Karl Popper:

> Bohr, in his reply, operates with the idea that measurement of a position can be achieved only by '*some instrument rigidly fixed to the support, which defines the space frame of reference*' while measurements of the momentum would be doneby a '*moveable diaphragm* whose momentum is measured before as well as after the passing of the particle*'. Bohr operates with the argument that in choosing one of these two systems of reference 'we cut ourselves off from any possibility' of using the other, in connection with the same physical system under investigation.[3]

Now this is one of Popper's clearest of moment's, when commenting on Bohr, however preciously few they may be. And the *absence of any assumption peculiar to quantum mechanics* is particularly striking in his summary. There is in fact no argument offered, except the self evident truth that rigid and moveable diaphragms are impossible to obtain simultaneously in any experimental arrangement, for the simple reason that nothing can be in motion and at rest at the same time. Which is not physics but *logic*. The strangeness of this approach has not gone unnoticed. C.A.Hooker has also emphasized that "conceptually, the *logic* of the situation is clear: precise position descriptions preclude velocities and precise velocity descriptions preclude positions"[4] a point which, he points out, "was already recognized by *Zeno* two millennia ago". Motion and rest are two *logically* incompatible situations, if analyzed closely enough. Is then Bohr trying to confront the EPR case with a modernized version of Zeno's paradoxes? However unbelievable it might seem, this is exactly what he does, a fact which has not escaped Hooker's attention. Hooker explains that, normally, "it is actual quantal connections between atomic systems"[5] that is to say "*the finite quantum of interaction* which is responsible for the incompatibility of application of the various classical concepts" which would provide the *factual* basis for complementarity. But those are absent, as Bohr himself acknowledges, since the two EPR objects are no longer in any physical contact at all. To conclude that "Bohr's general analysis and examples do not explain how the simultaneous applicability of classical concepts is restricted *because* of the quantum of action", a fact which, according to Hooker, "would seem to vitiate his reply". [5, 222].

I would certainly say it would. This is not the defense of completeness which his Doctrine requires, or quantum mechanics itself for that matter. There are, indeed, some feeble remarks here and there trying to *relate* his Zenonian analysis of the EPR case with his Official Doctrine, and several commentators have turned to those rather, but once what Bohr does is seen as what it really is, all these are futile. Once, that is, it is established that position and momentum determinations are recognized as *logically* incompatible demands, all other arguments automatically cease. Logical incompatibility is utterly self-sufficient and intolerant to supplementation. For supplementation would only imply that the concepts involved would not be incompatible without *additional* assumptions, and that would make their incompatibility *factual*; not logical, as Bohr strives to establish. For the very same reason, all other interpretations of Bohr's Reply, whatever their physical or philosophical basis, are equally out of place in the face of his Zenonian strategy.

But since it is the completeness of *quantum* theory, which is requisite here, and since no logical *tautology* will suffice for its establishment, Bohr's Reply to EPR fails. Plus conflicting irreconcilably with his Official Doctrine. To this latter I now turn.

iii. The EPR argument may be presented in the form of a reduction ad absurdum, leading to a particular dilemma:

(i) Either admit that any two systems, *A* and *B* are separable from one another at any time after their interaction, in which case whatever happens to the source-system *A* will not gave the slightest effect on the object-system *B*, (without in anyway disturbing the system), 6 whereupon the physical states represented by complementary quantities in the formalism, determined exclusively via the *source* system *A*, would *both* be there for the object-system *B* and the complementary account of quantum mechanics *incomplete* ; or, (ii) the complementary account of QM is complete, but then the two systems should no longer be considered *separable* at any time after their initial interaction.

Option (ii) is the absurd horn of the dilemma, or was to EPR themselves, when they first constructed their argument. Today, it seems to be the alternative to opt for, but that is something they hadn't anticipated, expecting of every sound man to opt for the former. To me, a defender of Bohr, not of Einstein, alternative (ii) is still as absurd as he would ever consider it, and would never adopt it of the sake of Complementarity. I just don't think it *threatens* Complementarity in the slightest. Not if complementarity expresses *factual* incompatibility between its sets of concepts. It does threaten it, if complementarity expresses *logical* incompatibility between them, for then such concepts would be incompatible *independently* of whether there is an interaction in process, and then the EPR dilemma would apply to at it any time after the interaction. But it would not affect it in the slightest if complementarity were *limited to transitions*. And this, I claim, is exactly what the case demands.

Energy values are discrete. That is to say, between any such two, consecutive values no *intermediate* value can be ascribed to an object without breach of the quantum postulate. Now assume such an object in transition from any such discrete state to the text, say from E_1 to E_2, whose difference E_1-E_2, or E, is therefore that of a single quantum. Given also that this transition cannot be completed in zero time, since then the system would have to be in two discrete states at the same time, there must be a temporal period t>0, over which the said system will complete its transition from the former state to the latter. Since, however, E is the *minimal* transition that the system can undergo, and since t is the time required for this transition, it follows that the *optimal* description of the process, or action, in question in terms or components E and t cannot be better than the *ultimate* energy unit E changing *over the interval t*, viz. no better than the product Et, which in accordance with our initial specifications has the dimensions of the quantum of action. Therefore we get $\Delta E \Delta t \geq h$.

Similarly, if we substitute momenta for energies and spatial intervals for temporal ones. Then the transition from p_1 to p_2, occurring over the distance q, and itself subject to the minimal conditions placed on energy, cannot be better analyzed that the *whole* value p_1-p_2, or p, changing *over the distance q*. But given that p is the *minimal* transition that the system can undergo, and q the distance required for this transition to be completed, it follows that the *optimal* description of the process, or action, in question in terms of components p and q cannot be any better than the product pq, viz. no better than the quantum of action. Therefore once more we get $\Delta p \Delta q \geq h$. Times and energies or positions and momenta are impossible to co-ordinate below the level of the quantum, in other words, during quantized transitions. This is Bohr's complementarity, merely expressed above in quantitative terms.

Is this what Bohr is actually saying? For when one hears of Bohr and complementarity, one invariably thinks of waves and particles superimposed on each other and therefrom on a single atomic object. What else is there besides duality in the structure of Complementarity? Here are some samples of what else there is:

> Complementarity is a term suited to embrace the features of *individuality* of quantum phenomena. [2,39]

> The *indivisibility* of the quantum of action... forces us to adopt a mode of description designated as complementary. 7

> The essence of quantum theory may be expressed in the quantum postulate which attributes to any atomic process an essential *discontinuity*, or rather *individuality*...This postulate implies a renunciation as regards the causal space-time *coordination* of atomic processes. [7,53].

> The causal-mechanical coordination of experience can be accomplished only in cases where the action involved is large compared with the quantum and where, therefore, *a subdivision of the phenomena is possible*. [2,7].

No mention of duality here; only mention of indivisibility and complentarity. Bohr goes so far, even, as to contend that "the characteristic new feature in quantum physics *is merely the restricted divisibility of the phenomena*".8 Thus complementarity is the consequence of processes, that is to say, transitions, which cannot be *analyzed* any further. When it comes to the *completeness* of this description, any suggestion to the contrary must needs make the following issue perfectly clear. What else is there, which the complementary account has failed to take into consideration? *Additional* physical states, not accounted for by the two uncertainties? But to seek for those is to presuppose that there can be *other* dynamical states, *besides* the initial and the final, contained in the process, and omitted in the account. But the two states were taken as *successive* and *discrete.* And to call them discrete means that no other state in-between them is admissible, in consistency with Planck's hypothesis. Consequently, any one demanding an analysis of the transition more thorough than the previous account permits must no less than *subdivide the quantum.* For myself, I can see other way out. But insofar as the quantum stays indivisible and insusceptible of further analysis, to expect a *finer* description is to misunderstand not Bohr's Doctrine, really, but the quantum of action itself.

Now to our other question; namely, what sort of incompatibility is the one expressed by the two uncertainty relations, as here derived. Is it perhaps *logical?* Far from it. It has not rested on *definitions.* It did not presuppose that times and energies or positions and momenta are *logically* incompatible. It only assumed that the transition in question was indivisible. Assume otherwise, for example, that in-between the two discrete energy states there are as many intermediate states as you care to mention; indeed, as many as there are temporal instants in which the interval t can be analyzed, as was the classical assumption. However you tried then, you would never obtain energies at the exclusion of times or times at the exclusion of energies. Therefore, complementarity as here conceived, is derivable on *condition* that (a) the transition is indivisible, and (b) that such a transition is indeed in process. And cannot be obtained otherwise. But any two concepts incompatible on condition that X, are simply compatible otherwise, viz. when X is removed. And this attests to their merely *factual* incompatibility.

Indeed, in the preceding analysis it was *nowhere* supposed, or even implied, that energies and times or positions and momenta did not *co-exist* before the transition, or once the transition is over. Quite the contrary. In fact their co-existence was implicitly *presupposed.* We have admitted among our premisses that *before* the transition, the system had energy E_1. But "before" is itself a temporal notion; it specifies a certain time, during which the system had exactly E_1. Then take *any* chosen, *split* instant of that period, *during* which the system had exactly E, e.g. a t'. Would we not, in speaking thus, imply the possibility of a *simultaneous* determination of both, energies and times? I submit we would. Nor does it matter in the least, in the context of the present approach, that this determination is merely hypothetical. For if times and energies were *logically* incompatible, they would be so even hypothetically, and to say then that they may have co-existed, even hypothetically, would constitute a logical contradiction. But not only does it not constitute a contradiction in this approach; in actual fact the attempt to deny their co-existence before or after the transition would reduce the present argument to utter incoherence.

The classical concepts are therefore only *factually* incompatible in the context of complementarity and therefore are complementary *only during transitions.* Nor is this conclusion in conflict with the principles of quantum mechanics. After all, it is *action* which is quantized and when no action is in process, no quantum principle should be involved. If the complementarity of the classical concepts is an exclusive consequence of energy, and thereby of action quantization, then no complementarity should exist and none should be expected, *when no action is in process.* In view of this realization, we may now turn to the EPR paradox.

Once the quantum conditions responsible for the complementarity of the classical concepts are *removed,* namely , when no transition is deemed to be in process any more, all obstacles hitherto preventing the co-existence, and I stress the word co-existence, not the word *co-measurability,* of the classical concepts, are completely removed. In this way, EPR's offered dilemma, crucially depending on the *cessation* of all interaction, and therefore of all transition, could not even be posited. And consequently, their absurd alternative (ii) could never arise. Once no transition is in process, what otherwise-would-have-been complementary states would now be there for either particle in *any* case. And then the strategy of driving the argument to an absurd alternative would be entirely redundant, if not indeed impossible. For the EPR argument, in its present form, proceeds on the assumption that the concepts in question are *still* complementary, even after the interaction, to then point out that in this case they can be (still) complementary only at the price of an utter paradox; namely, non-locality. Yet once it is conclusively shown that complementarity is limited to transitions, the *otherwise* complementary concepts would no longer be so, and then there would be no justification whatsoever in one's arguing as if they still were. In other words, no justification whatsoever for struggling to deduce absurd consequences from a premise which does not even obtain.

EPR would then have no reason to proceed with their either-or dilemma. For that matter, once no transition is in process, there is no reason to deny, none so far as the interests of complementarity go, that very possibly both interacting systems return to *classically describable states*, locality being one of them. For it is this, essentially, which the preceding approach implies. A result consonant with both, the EPR demand for locality and the epistemological prospects raised in topic 1 of the present Conference.

On the other hand, this result is wholly incompatible with the so-called wave-particle duality. The latter singularly entails the *logical* incompatibility of the classical conjugate concepts and this alternative I have given my reasons for rejecting. Nor is it even true that Bohr's doctrine requires it. This is another epistemological prospect raised by topic 1 of the present Conference.

Why Bohr failed to provide a reply along the previous lines, choosing to argue that complementary concepts were incompatible *all* the time instead, is something for which I have no ready answers. My best guess is that in this case he *confused* between logical and factual incompatibility, as so many others have done before him, and was unduly alarmed. And led to the ever so suspect endeavour of attempting to deduce complementarity even in *absence* of interactions altogether, in the only way left open, namely by treating the concepts involved as if logically incompatible.

iv. However the EPR dilemma did raise the issue of locality, and one cannot undo what has been done. So an answer is still due. On this matter much has been invested in Bohr's doctrine of *wholeness*, the alleged source of non-local approaches to the theory. Bohrian wholeness may encourage non-locality. But hardly in the sense demanded by non-local approaches.

Bohr did think that during measurements atomic object and measuring device, or any two quantally interacting objects, become inseparably linked. But there is nowehere in his texts a correlative affirmation, or even a hint, that this inseparability can go on *indefinitely*. Once I unpack this doctrine in they way I see fit, this will become more than evident.

Make the following assumption, together with the usual principles of quantum mechanics. Namely, that the principle of *energy conservation* is valid in all circumstances. Now let us see how their combination is supposed to work.

Consider any measuring device suitable for measuring the energy of a single photon, subsequently to be found equal with E. Now this unit of energy possessed by the photon is *indivisible*. And no fractions of it can be obtained. It will therefore be transferred to the device as a *whole*. (Observe how naturally talk of wholes already enters the description). But it cannot be transferred to the device in *zero* time, for then the photon would possess and not possess this energy at one and the same time. So, despite its indivisibility, this energy unit will take a certain time $t>0$ for its transfer. But then, what will the photon's and, indeed, what will the energy state of the device be, just *after* the transfer has begun and so *before* it is yet completed? According to Feyerabend and Hooker, Bohr's (implicit) proposal is that we cannot but conclude that QM-wise the energy of both interacting systems is completely *indeterminate*.[9]

Now combine this with the other premise, that energy is always conserved. Then the energy of the *composite* system, [photon+device] is still determinate and in fact equal to E. But how can any composite system have a *definite* energy , when none of its separate components has? It can, if and only if its separate components are no *longer separable*. Clearly the minimal energy unit in question cannot be *distributed* among the two interacting systems. Thus neither one in *itself* considered, and in *isolation* from the other and their interaction, can be said to possess it. They posses it only if taken *together*, as an undivided, unanalyzable *whole*. Thus, if energy is to remain constant and determinate throughout the interaction, the parts of the composite system *cannot be coherently separable.*

This, in a nutshell, is Bohr's doctrine of wholeness. It is strongly reminiscent of older, metaphysical, conceptions according to which there are wholes, which are *greater* than the mere sum of their parts. And this is just one such case, since the state possessed by the two systems, when taken in unison, cannot be *reduced* to the states independently possessed by each separate component. "A whole which is greater than the sum of its parts". [10] These are the very words employed by Dr. Basil Hiley, one of Prof. Bohm's distinguished associates, when he refers to the same phenomenon. But it is the *same* phenomenon?

Dr. Hiley and Prof. Bohm actually use these words in connection with their *own* approach, the Quantum Potential Approach. Let us have a quick preview of its properties. "The quantum potential" they say "does not produce a *vanishing* interaction between the two particles, as their distance reaches infinity. In other words, *distant* systems will still be in a strong *interconnection.* This " they claim "runs contrary to the requirement of classical physics, that when two particles are sufficiently apart, they will be have *independently*". [11] As a consequence of this potential, we are faced with "a new notion of physical reality, the unbroken wholeness of the *totality of the universe,* which denies the classical ideal of *analyzability* of the world into separate and *independently* existent parts". [11,sect.1]. And it is to *Bohr's* doctrine that they trace the origin of these ideas. But the case of quantum inseparability earlier expounded does not speak of *distant* systems at all. Whatever assertions it made, certainly obtained *during* the quantized interaction. But do they still obtain, once it is over? Nothing of the sort became apparent in the structure of the previous argument; and in point of fact it is a prospect *precluded* by it.

Bear in mind that quantum wholeness, as I deduced it, required indeterminateness, for only then can one proceed to raise the issue, how can two indefinite energy states yield a definite one. But indefiniteness of energy should be expected only *during* transitions, as already emphasized earlier on. Nor can the discrete states *themselves* be indefinite. Should we expect this to extend *beyond* the limits of the interaction and therefore even *after* the transfer is completed? Take the similar case of two interacting systems, A and B, possessing nominal values 2 and 1 respectively, and whose difference is that of a single quantum, which values are redistributed after the interaction, so that now A has 1 and B has 2. Can the inseparability argument be rerun? Quite clearly not. Once the transfer is over, both systems return to definite states, and then the energy value possessed by the composite system, 3, is no *longer* more than the sum of its parts. It is, quite trivially, *equal* to their sum. Hence, no quantum wholeness should be expected, after the interaction. To suppose that the two interacting systems are still inseparable *long* after the interaction, perhaps even *years* later, can never be derived as a consequence of the previous account of wholeness.

But this is not all. It would seem to me that, quite generally, it cannot be derived from any normal *quantum* assumption either, as I at least would use the word. Apparently, wholeness of the sort demanded here should at all times be able to indicate its dependence on the quantum of action. But physical action is a *product,* defineable as either Et or pq. Take the former definition. In order that a physical action of magnitude Et be defined, *both* of its constituent parameters must be so defined. So the duration t, of the process, must be defined. But t itself cannot be defined, unless of a *finite duration!* That is to say, cannot be defined, unless the action in question is sooner or later deemed to have been conclusively *over.* And if wholeness is an *exclusive* consequence of quantized action, then wholeness itself will thereby have been over.

It is no different if we consider the alternative definition, pq. Again the dimensions of such action cannot be determined, unless either parameter is determined. But the distance q *cannot* be determined , unless itself of *finite dimensions!* That is to say, cannot be defined unless the action in question is deemed to have been conclusively over. And once *all* action is deemed to have been conclusively over, *quantal* action as well is deemed to have been over. The joint conclusion of these remarks is that, if wholeness is an exclusive consequence of quantized action, wholeness itself cannot possibly extend *beyond* the temporal and spatial boundaries of such action, that is to say, wholeness thus understood must be *confined the finite temporal interval* t and *within the finite spatial region* q. Beyond it, no wholeness can exist. And the wholeness which does exist, viz. the one confined to interactions, is in perfect consistency with the EPR demand for locality, itself founded on the assumption that the two particles are no longer interacting.

Thus Einstein would seem to be vindicated, but not at *Bohr's* expense. Much rather at the Quantum *Potential's* expense. Bohr need not retract a single word of his unique Doctrine. He need only limit it to transitions. Then the evident absurdity of unlimited non-locality would not arise. And no *incompleteness* either. Of course, this approach has an additional consequence, one Bohr has never explicitly defended; viz. it must make room for the existence of classical states *together* with quantum ones. Which would defy a *unified* description of the macro and the micro, as was Einstein's, and perhaps every physicist's dream. But I am not at all sure that such unified description is desirable. Let alone warranted. If such a drive stems from the assumption that, since large scale objects are constructed of atoms, what applies to the latter should equally apply to the former, then this is a suspect assumption if not indeed an outright irrational one. No one should expect of *indivisible* entities to behave just like subdivisible ones do, or vice versa. Had that been so, then the *unit*, 1, in the field of natural numbers, -zero not included - should *itself* be representable as the sum of any other two naturals, since all other naturals certainly are. But it isn't. Because it is itself indivisible, whereas no other natural is.

To put it simply, a bucket placed beneath a dripping faucet will eventually fill up with water. And once this process is completed, the water in the bucket will turn up a perfectly *continuous* structure. But the drops themselves of which it is constituted, were not. They were in fact, discrete. I don't see why this cannot be the case with quantum and classical mechanics. I don't even see why this cannot be the case with wholeness and complementarity.

References

1. N. Bohr, "Can Quantum Mechanical Description of Physical Reality be Considered Complete?", *Physical Review,* **48** 1935, pp. 697-700.
2. N. Bohr, *Atomic Physics and Human Knowledge,* J. Wiley & Sons, New York 1958, p.48.
3. K.R. Popper, *The Logic of Scientific Discovery,* Hutchinson and Co, London 1975, p. 446.
4. C.A. Hooker, "Metaphysics and Modern Physics", *Contemporary Research in the Foundations and Philosophy of Quantum Theory,* Reidel, Dordrecht 1973, p.188.
5. C. A. Hooker, "The Nature of Quantum Mechanical Reality", *Paradigms and Paradoxes,* University of Pittsburgh Press 1972, p.195.
6. A. Einstein, B. Podolsky, N. Rosen, "Can Quantum Mechanical Description of Quantum Mechanical Reality Be Considered Complete?", *Physical Review,* **47** 1935, p.777.
7. N. Bohr, *Atomic Theory and the Description of Nature,* Cambridge University Press 1934, p.10.
8. N. Bohr, *Essays 1958-1962 on Atomic Physics and Human Knowledge,* Richard Clay and CO, Suffolk 1963, p.92.
9. For a more detailed analysis of this point, and of its sources, see C.A. Hooker, "Energy and the Interpretation of Quantum Mechanics", *Australasian Journal of Philosophy,* **49** 1971.
10. B.J. Hiley, "Causality and Measurement in the Quantum Potential Approach to Quantum Phenomena", *Determinism in Physics,* Gutenberg, Athens 1985, p.203.
11. D. Bohm, B. Hiley, "On the Intuitive Understanding of Non-Locality as Implied by Quantum Theory", *Fundamenta Scientiae,* **36** 1976, sect.2. As I am quoting directly from the manuscript, I cite number of sections.

"... we want more than just a formula. First we have an observation, then we have numbers that we measure, then we have a law which summarizes all the numbers. But the real *glory* of science is that *we can find a way of thinking* such that the law is *evident.*"
"The Feynman lectures on physics",
Addison-Wesley, MA, 1966, p.26-3.

THE GHOSTLY SOLUTION OF THE QUANTUM PARADOXES AND ITS EXPERIMENTAL VERIFICATION*

Raoul Nakhmanson

Frankfurt am Main, Germany†

This conference is entitled "Frontiers of fundamental physics". What does this mean? Is it the frontiers of today's physical knowledge, or is it the frontiers of physics itself as a science?

In my paper I shall try to show that today it is the same: the frontiers of contemporary physical knowledge coincide with the conceptual frontiers of physics as a science regarding the behaviour of so-called inanimate matter and even cross over to invade into the kingdom of ghost. Such a point of view permits a very natural interpretation of quantum phenomena, and suggests essentially new experiments in which information plays the principal rôle.

The microworld has surprised the "classical" physicists with the following paradoxes:[1,2]

1) Before quantum mechanics (QM) was created: quantization of mass, charge, energy, angular momentum; the identity of particles of the same type; wave-particle duality.

2) In QM: statistical predictions, Heisenberg's uncertainty principle, Pauli's exclusion principle.

3) In standard (Copenhagen) interpretation of QM: rejection of the classical realism, a ban on speaking about non-measured parameters, trajectories, etc.; Bohr's complementarity principle, collapse of the wave function.

The Copenhagen interpretation is only a translation of the mathematical formalism of QM to the ordinary language but not an interpretation in a common sense, because it does not explain how, why, and in which frames this formalism works. Feynman told his students that the quantum world was not like anything that we know; and although everybody knows QM, many people use it, some of them develop it, but nobody understands it.

In discussions about QM the "Gedankenexperimente" play an important rôle. We will discuss three of them which were really performed:

1) Delayed-choice experiment.[3] In one arm of an interferometer a Pockels cell is placed which closes the path of photons at the short moment when they can pass the cell. In accordance with old local-realistic concept each photon flies only in one arm of the interferometer. If it is the arm with the cell the photon will be absorbed and nothing will be registered. If it is another arm, the short work of the cell placed far away does not act on the photon and the same interference as without the cell must be registered. But no interference was found in accordance with QM.

---

* Shortened version of a report which was read on September 30, 1993 in Olympia, Greece.
† Present address: Waldschmidtstrasse 131, 60314 Frankfurt, Germany.

2) Aharonov-Bohm effect.[4] In accordance with QM the frequency of wave-function oscillation depends on the energy. If the particle has different energies in different arms of the interferometer, it leads to an additional phase shift and changes the interference pattern. The experiments were performed with an electron interferometer and a magnetic vector potential and justified the predictions of QM. It is of interest that in the experiments the electrons did not cross the magnetic field. From the old classical point of view it looks like non-local action at a distance.

3) Einstein-Podolsky-Rosen (EPR) experiment. It was suggested in[5] and modernized by Bohm.[6] Here two particles emitted simultaneously have common non-factorisable wave function and are measured after parting by a large distance. There is some correlation between the results measured. Bell has shown[7] that any local realistic theory (i.e. theory with hidden parameters and restricted velocity of interaction) estimates the uppermost limit of such correlation, and this limit is smaller than predicted by QM. The experiments being performed[2] are in accordance with QM, and today's dominant opinion is that local realism has been disproved and one must refuse either reality lying beyond the measurements (like Copenhagen) or locality. Later I will show that this conclusion as well as Bell's theorem itself do not have the generality being ascribed to them.

The EPR-scheme raises a question about separability. "Common sense" prompts that after some time and distance the "magic" correlation between particles must disappear, i.e. the factorisation of the wave function must take place. But how? The analogous question is connected with measuring procedure itself: If interaction between particle and apparatus allows several output results, the QM forecasted end state is a superposition of these results. But in practice the result of each measurement is a pure state, and the result of the series is a statistical mixture. It seems as if QM does not describe the whole measurement process.[8]

There are some explanations of the EPR paradox. From the *Copenhagen* point of view it is so as it is. Speaking about some hidden parameters of particles, e.g. directions of spins, before the measurement, has no sense, and Bell's theorem and experiments justify this.

Non-local theories with hidden parameters.[9] Here an instantaneous action at a distance is provided by instantaneous collapse of the wave function in all space. The critics emphasize that these theories only rewrite the Schrödinger's equation in a more complex form, giving the same results and nothing new.

Action of future on the past.[10] If such action is possible, the future conditions of measurement can act on the hidden parameters of particles at the moment of their departure to tune them for correct correlation. Up to now there is no complete theory ready to defy critique. But common sense prompts that such a world can not be stable.

Fatalism. This possibility was noted particularly by Bell.[11] In the spirit of Laplace it is possible to think that everything is pre-determined, particularly our choice of position of analyser. Here we are confronted with the old problem of "free will". If free will exists man (and not only he) can control the choice of alternatives taking into account physical and social conditions. The following chain of syllogisms supports the existence of free will:

> $\rightarrow$ Useful changes are selected and consolidated by evolution.
> \+ During evolution the volume of the human brain increases.
> = The volume of brain is a useful quantity.
> \+ Intelligence depends on the brain volume; as a rule,
>   the greater the volume, the higher the intelligence.
> = Intelligence is useful.
> \+ Intelligence can develop itself only if it can choose among several
>   alternatives; only in such situations can intelligence be useful.
> = *Free choice, i.e. free will exists.*

One can reply that the increase of the brain volume as well as evolution itself are included in the fatalistic scenario. But if one considers the existence of free will *ad hoc* as an axiom, then, in accordance with these syllogisms, *free will gives intelligence a chance to evolve.*

The roots of free will do not lie in the macroworld which is ruled by deterministic laws. They lie in the microworld, and quantum uncertainty points to it. Human intelligence is not the only product of free will. It is possible that earler, the free will created some intelligence at the level of its roots, i.e. in microworld. Because the time (measured not in seconds but in events) flowed there much faster, this intelligence had a longer evolution period. Perhaps the golden age of it is over, and now we have to do it only with a "relict" intelligence (so called by Cochran[12]). The additional pointers on intelligent matter are the Einstein's formula $E = mc^2$, the informational character of the wave function ψ, the principle of the least action, and quantum-mechanical stochastics.[13]

The development of quantum physics was a step across the boundary between matter and ghost drawn by Descartes. Physicists felt it and spoke about the free will of electrons and ghost (spirit, consciousness, intelligence) in matter. Similar meanings were expressed by Charles Galton Darwin, Eddington, Heisenberg, Schrödinger, Pauli, Jordan, Margenau, Wigner, Charon, Cochran, and others. Feynman said that it looks as if a computer is in each point of space. Cambrige University Press has published a book touching this theme[11] containing interviews with Bell, Bohm, Wheeler, Peierls, Aspect, and others.

Some interesting analogies between microworld and people have been noticed. Niels Bohr saw the manifestation of his complementarity principle in human thinking. Margenau wrote about Pauli's exclusion principle:[14]

> "Prior to that time, all theories had affected the individual nature of so-called 'parts'; the new principle regulated their social behaviour... The particles, though initially assumed to be free, are seen to avoid each other... In a crude manner of speaking, each particle wants to be alone; each runs away then it 'smells' the other, and its sense of smell is keener the more nearly its velocity equals to the other's."

This was said about Fermi-particles. Such behaviour is typical for scientists: each of them tries to find his own theme. Sometimes people's behaviour is like a Bose-particle. Phenomena such as fashion in dress or music, and applause or coughing in concert halls, are examples of Bose-condensation. The same man can manifest himself as a Bose- or Fermi-person. For particles this was only possibe in "big bang" time. Are we now at the same stage of evolution?

The next example concerns the EPR-experiment. Let us suppose there are twins, Ralf and Rolf, both of whom live in Frankfurt and work for Lufthansa as pilots. They fly all over the world but mainly to England and Greece. For Lufthansa (not for their families!) they are indistinguishable "particles". The twins always try to dress alike, they believe that this brings them happiness. Because they are often in different countries, they agree on an order of sartorial priority: cold before warmth and rain before dry spell.

Figure 1. Einstein-Podolsky-Rosen (EPR) experiment and the apparent non-local interaction.

God, who is observing the twins, sees as a rule the striking correlation: the twins dress alike! For example, Ralf and Rolf arrive in England and Greece, respectively. If it is cold in England, not only Ralf but also Rolf wears the overcoat in spite of the warm weather; if it is raining in Greece, not only Rolf but also Ralf hides beneath an umbrella, regardless of whether it is raining or not (Fig.1); etc.. "What is the matter?" - thinks God, - "I estimate experimental conditions, namely, weather in England and Greece and the twin's financial status, telephoning is too expensive for them. It seems there is a non-local interaction between the twins. I am sure it is a new escapade of the devil!"

God's conclusion was only half true. In his heavenly chariot he fell behind the technical progress of the 20th century. He was right suspecting the devil. But up to now the devil does not realize non-local interaction. Instead, he has invented television, power computer for meteorology, and communication satellite. Because of it the twins watch TV at every evening for a good tomorrow world weather *forecast* .

Although our behaviour occurs in real space-time, the strategy of it is not there. It is in our consciousness, which controls our behaviour, taking into account physical and social laws and circumstances. To develop a strategy we use our knowledge only about the past, and propagate it on the future. The thoroughness of the forecast depends on the information taken into account and the power of the intelligence.

But let us come back to physics. Unfortunately the idea about intelligent matter is not developed up to now. They, who spoke about ghost in matter, did not go beyond such a statement and did not suggest any hypotheses and schemes which could be tested experimentally. From another side physicists using QM do not see the necessity of such an idea and follow the principle thought as old as Aristoteles but named after William of Ockham "Ockham's Razor":

"entities should not be multiplied beyond necessity".

Niels Bohr said, that there are trivial and deep statements. To be asked "What is a deep statement?" he answered: "It is such a statement, that an opposite statement is also a deep one." If one accepts the Ockham's principle as a deep statement then, according to Bohr,

"entities should not be canceled beyond necessity"

must also be accepted as a deep statement. Besides, the practical necessity is not the only or main criterion of theory.

A consistent development of the idea of intelligent matter naturally interprets quantum paradoxes as well as QM itself within the limits of local realism, and suggests essentially new experiments with microparticles and atoms in which information plays the principal rôle.

In the new conception the wave function ψ is a strategy-function. It reflects an optimal behaviour of particles. It is not in the real 3-dimensional space. It is in imaginary configuration space, which, in its turn, is in the imagination (consciousness) of the particle. When the particle receives new information (it takes place by any interaction with micro- or macro-objects), it can change its strategy. Thus occurs the collapse of the wave function. It occurs not in the real (infinite) space, but in the consciousness of particles. The consequent time is determined by the rapidity of this consciousness. Therefore, compared with space-time conditions of experiment, collapse is local and instantaneous. Von Neumann and Wigner suggested that human consciousness has influence on the collapse of ψ - function. It is not so: in the human consciousness only the human knowledge about the ψ - function collapses. The laws of both collapses lie beyond physics.

The wave-particle duality is a mind-body one. In the space there exists only the particle; the wave exists in its consciousness, as well as the reflection of the whole world. If there are many particles, their distribution in accordance with the ψ - function looks as a real wave in real space.

Particles are artifical things. Division into different sorts or species with internal identities is typical for mass products. It simplifies production, usage, repairs, and replacement of such objects. Technics, plants and animals illustrate it very well. In the last two cases the production is ruled at the genetic level. For example, people have a very narrow statistical distribution of sizes and masses; the world records in sport differ from the middle results not more than twice. The identity of particles of one sort in QM is analogous to the identity of vehicles of one sort with respect to traffic rules. The individual differences lies beyond QM.

Because of free will the behaviour of particles is not strictly determined. In situations allowing alternative outputs the theory gives only a distribution of priorities. Taking this into account the particle makes its choice. The optimal tactics of proportional proving of all possibilities by an ensemble of disconnected particles is randomization of this choice. To do it the particles must have the generators of random signals.

If some theory and random generator (RG) are used to choose the alternative, it looks like a complete algorithm. Well, but where is the free will now? Is it only to change the RG?

The answer is, that purpose and means create a dependence. Really free is he who has no purpose and no desire including the desire of freedom. Therefore there is a danger that in the "Konsumgesellschaft" we transform ourselves into some kind of automata. Perhaps the microworld did not avoid it. But the new turn of development can be connected with a change of purpose or new information. Besides, Gödel's theorem prompts, that the space of correct statements can be manifold. In such a case to reach a new fold one must make a "quantum jump". "Do not sin against logic, one reaches nothing new", - said Einstein.

Pauli's principle and Bose- and Fermi-particles were discussed above. These types of social behaviour are optimal for searching (fermions) and power action (bosons). In the last case some macroscopical effects can be observed (in superconductors, superfluids, lasers).

With respect to Heisenberg's uncertainty principle: In the new conception it reflects not the reality but QM as a theory of measurement. In reality the particle has definite coordinates, impulse, trajectory, etc.. But during an interaction with the measurement apparatus it has a possibility to choose the next state. It solves this problem using its intelligence (reflected in the ψ - function), random generator, and freedom (e.g. reflected in the choice of RG). Neither QM nor any other theory predicts a particular result: it would be a refusal of free will.

In spite of this, the dream of Einstein and other realists, to know the values of all parameters included in a theory can become true. Particles remember what happened and tell it to others. To do this, they must have synchronised clocks, measure rules, and reference points for space and time. In this sense it is possible to speak about absolute coordinates and time, like Greenwich's ones. If we can communicate with particles[13,15], they can say everything about their parameters and forecast their and our future.

The new concept includes the previous realistic ones: *empty waves*[2] and *parallel worlds*[16] exist, but not in the real world: as virtual possibilities they exist in the consciousness of a particle. Not the real[10] but a forecasted *future acts on the past* . The above mentioned danger of total algorithmisation looks like a stochastic *fatalism* .

The new explanation of delayed-choice and EPR experiments, and suggestions how to have "non-QM-results", were done in[13]. The essence is, that particles are well informed about the world and its development. The Aharonov-Bohm effect has the same explanation. Besides, this effect emphasizes a priority of potential against field (in classical physics they enjoy equal rights). From the new point of view it is natural, because potential just contributes to the action function whose minimum as a function of trajectory is wanted. It should be observed that the idea of *forecasting* the conditions on this trajectory is also included in the least action principle. The change from integral form to a differential one does not solve the problem: the notion of derivative is connected with two points, and if we are in one of them, we know only the past conditions in the second point and must extend this into the present.

The proof of Bell's theorem is based on the next assertion: if a particle *1* is measured in the point *A* having a condition (e.g. angle of analyser) α , and P_a is a probability of result *a* , then a condition β existing in a distant point *B* , there is a measured particle *2* , has no influence on the P_a , and vice versa. Here Bell and others saw the indispensable requirement of local realism. Mathematically it can be written as

$$P_{ab}(\lambda_{1i},\lambda_{2i},\alpha,\beta) = P_a(\lambda_{1i},\alpha)\times P_b(\lambda_{2i},\beta) \quad , \qquad \text{(Bell)} \tag{1}$$

where P_{ab} is the probability of the join result ab , and λ_{1i} and λ_{2i} are hidden parameters of particles *1* and *2* in an arbitrary local-realistic theory. Under the influence of Bell's theorem and the following experiments some "realists" reject locality. In this case an instantaneous action at a distance is possible, and one can write

$$P_{ab}(\lambda_{1i},\lambda_{2i},\alpha,\beta) = P_a(\lambda_{1i},\alpha,\beta)\times P_b(\lambda_{2i},\beta,\alpha) \quad . \quad \text{(non-locality)} \tag{2}$$

In principle such a relation permits a description of any correlation between a and b , particularly predicted by QM and observed in experiments. But in the frame of local realism the condition (1) is not indispensable. Instead, one can write

$$P_{ab}(\lambda_{1i},\lambda_{2i},\alpha,\beta) = P_a(\lambda_{1i},\alpha,\beta')\times P_b(\lambda_{2i},\beta,\alpha') \quad , \qquad \text{(forecast)} \tag{3}$$

where α' and β' are the conditions of measurements in points A and B , respectively, as they can be forecast by particles at the moment of their parting. If the forecast is good enough, i.e. $\alpha' \approx \alpha$ and $\beta' \approx \beta$, then (3) practically coincides with (2) and has all its advantages plus locality.

On the issue of separability: The EPR-particles have a common strategy. It can continue as long as they can forecast the future. But particles can also have so intensive interactions (e.g. with detectors) that initial strategy is not important anymore. In both cases the consciousness of the particle has an ability to cut off and forget the old partnership.

QM is "microsociology". Like its humane sister, it makes only probabilistic forecasts. The transition to classical physics is the transition from sociology of persons to sociology of crowds: the level of freedom decreases and behaviour becomes deterministic. Feynman's statement "quantum world is not like anything that we know" is right only if we do not take into account living beings. If a baby, having more experience with his parents than with "inanimate" matter, could make experiments, the behavior of microparticles would appear to it to be very natural.

REFERENCES

1. M.Jammer. *The Conceptual Development of Quantum Mechanics*, McGraw-Hill, New York (1966).
2. F.Selleri. *Quantum Paradoxes and Physical Reality,* Kluwer, Dordrecht (1990); *Wave-Particle Duality,* F.Selleri, ed., Plenum Press, New York (1992).
3. J.Baldzuhn, E.Mohler, and W.Martienssen. *Z. Phys.-Cond. Matt.* **77B**, 347 (1989) and Refs. cited there.
4. Y.Aharonov, D.Bohm. *Phys.Rev.* **115**, 485 (1959).
5. A.Einstein, B.Podolsky, and N.Rosen. *Phys.Rev.* **47**, 777 (1935).
6. D.Bohm. *Quantum Theory*, Prentice Hall, New York (1951).
7. J.S.Bell. *Physics* 1, 195 (1965).
8. E.Wigner, in *The Scientist Speculates*, I.J.Good, ed., London (1962); G.Ludwig, in *Werner Heisenberg und die Physik unserer Zeit*, Vieweg, Braunschweig (1961).
9. L. de Broglie. *J.Phys.Radium* **8**, 225 (1928); D.Bohm. *Phys.Rev.* **85**, 166 (1952).
10. O.Costa de Beauregard. *Nuovo Cimento* **B42**, 41 (1977).
11. *The Ghost in the Atom.* P.C.W.Davies and J.R.Brown, eds., Cambridge University Press (1986).
12. A.A.Cochran. *Found. Phys.* 1, 235 (1971).
13. R.Nakhmanson, to be published. in *Waves and Particles in Light and Matter,* A.van der Merwe and A.Garuccio, eds., Plenum Press, New York
14. H.Margenau. *The Nature of Physical Reality.* McGray-Hill, New York (1936).
15. R.Nakhmanson. *Preprint 38-79*, Institute of Semiconductor Physics, Novosibirsk (1980); see also Ref.13 and A.Berezin and R.Nakhmanson. *J. of Physics Essays* **3**, 331 (1990).
16. H.Everett III. *Rev.Mod.Phys. 29*, 454 (1957); see also B.S. DeWitt. *Phys.Today* **9**, 30 (1970).

INDEX

Aberration, 209
Absolute motion, 217, 225
Action at a distance, 210, 252, 479, 592
Action of future on past, 592
Additional assumptions, 529, 531, 532, 534-536, 539, 546, 552, 553, 573, 584
Adiabatic invariant, 402
Age-redshift connection, 10
Aharonov-Bohm effect, 393, 395, 592, 595
Anthropic principle, 73-81
Apeiron, 209
Arp's theory, 51
Asymptotic freedom, 52
Atomic clock, 125, 127
Aufbau principle, 343

Background radiation, 99, 206, 207, 252
Ballistic theory, 218
Barrier penetration, **342, 345**
Bell's inequality, 524, 529, 530, 532, 537-540, 543, 546, 552, 553, 572, 574-577, 580
Bell's theorem, 592, 595, 596
Benioff zones, 293, 336
Berry's phase, 437
Bianchi identity, 391
Bifurcation, 277
Big bang, 1, 27, 54, 63, 75, 76, 80, 101-103, 108, 252, 269, 271, 360, 364, 368, 472, 593
Binary pulsar, 120, 139, 140, 146

Biogeographic anomaly, 264
Biot-Savart law, 415, 416, 418-420, 422
Black holes, 27, 28, 32, 33, 63, 215, 216, 254, 324
BLLAC object, 28
Blueshift, 29, 33-35, 41, 113
Brans-Dicke theory, 126

Caldirola's chronon, 353, 450
Causal anomalies, 478, 479
Causality, 177, 204, 207, 211, 234, 466, 483, 562
CHSH inequality, 531, 536
Classical limit, 443
Clock retardation, 176, 204
Cluster:
    Coma, 14
    Fornax, 9, 15, 17, 105, 106
    UMA, 17
    Virgo, 3-6, 8, 16-19, 25, 26, 37
Cold fusion, 348
Collapse of wave-f., 237, 591
Complementarity, 163, 169, 478, 481, 482, 485, 486, 491, 511, 512, 518, 527, 528, 583-587, 589, 591
Completeness (of qm), 477, 486, 529, 565, 567, 573, 583, 585, 589
Complex speed, 352
Complex time, 351-353, 356
COMPTEL, 6, 7
Compton wavelength, 214, 448
Connecting material, 8
Consciousness, 594

Galaxies:
    chain of, 37, 39
    clustering of, 39
    compact, 1, 10
    companion, 2
    dominant, 2
    dwarf, 63
    ejecting, 2
    Iwanowska's, 38
    radio, 3, 28, 96
    Seyfert, 28
    spiral, 15, 24, 25
Galilei covariance, 174
Gamma rays, 1, 2, 6-10
Genotopies, 44, 347, 348, 351, 355
GHA experiment, 511-515
GHZ correlations, 578
Global Navigation, 126, 128
Global Positioning, 126-129
GLONASS, 129
Gravimetric anomalies, 261, 292
Graviton, 83, 85
Gödel's theorem, 595
Green's function, 341
Group velocity, 527
Gyroaction, 402

Hafele-Keating experiment, 122, 182
Hall effect, 393, 396
Heisenberg inequalities, 476-480
Helmholtz,s derivative, 165
Hertz electrodynamics, 164-169
Hertzsprung-Russell diagram, 253
Hidden variables, 401, 449, 467, 483,
    537, 546, 553, 575-577, 592
Higgs particles, **344**
Higgs oscillator, 435, 436
HI profiles, 14
Homogeneity of space-time, 181,
    182, 211
Hoek experiment, 226-229
Hooge equation, 395
Hubble law, 68, 69, 75, 112
Hubble Space Telescope, 5, 32
Hydrides, 315-319
Hydrogen cloud, 5

Idealism, 233
Inflation theory, 364
Island arcs, 335-337

Isodoppler law, 41, 51-55
Isoeuclidean geometry, 41, 52, 53
Isominkowskian geometry, 41, 46-
    49, 53, 347-351, 355
Isoriemannian geometry, 41
Isoredshift, 42, 50
Isospecial relativity, 49-55
Isospaces, 45, 46
Isotopic lifting, 156-159
Isotopies, 43-47, 51, 56, 347, 348,
    355
Isotrigonometry, 48
Ives-Stillwell experiment, 176, 226,
    227

Kadeisvili's classes, 44, 50
Kaluza-Klein ansatz, 158
Kelvin's gyrostats, 109
K-meson regeneration, 154

Lamb shift, 455, 456
Large Hadron Collider, 362
Larmor formula, 424
Laser ranging, 126, 129, 284, 323
Lie-admissible theory, 347, 349, 351-
    355
Lie-isotopies, 45, 46, 49, 153-159, 350
Lienard-Wiechert potential, 205
Local realism, 535, 540, 560, 566,
    571, 572, 574, 592
Lorentz:
    contraction, 87, 175, 192, 204,
        219-221
    gauge, 119, 223, 342
    non-invariance, 153-155, 158, 159
    relativity, 217

M49, 4-6
M87, 3-6, 15
Mach's principle, 111
Magnetic anomaly, 261, 263, 265,
    267
Magnetic moment, 52, 61
    anomalous, 42
Magnetic monopole, 364
Mandel experiments, 555, 556
Markarian 1320, 29
Mass-energy relation, 526
Materialism, 233
Meme, 241-244